18 VIII A
						4,0 2 He
13 III A	14 IV A	15 V A	16 VI A	17 VII A		
10,8 5 B	12,0 6 C	14,0 7 N	16,0 8 O	19,0 9 F		20,2 10 Ne
27,0 13 Al	28,1 14 Si	31,0 15 P	32,1 16 S	35,5 17 Cl		39,9 18 Ar

10	11 I B	12 II B							
58,7 28 Ni	63,5 29 Cu	65,4 30 Zn	69,7 31 Ga	72,6 32 Ge	74,9 33 As	79,0 34 Se	79,9 35 Br	83,8 36 Kr
106,4 46 Pd	107,9 47 Ag	112,4 48 Cd	114,8 49 In	118,7 50 Sn	121,8 51 Sb	127,6 52 Te	126,9 53 I	131,3 54 Xe
195,1 78 Pt	197,0 79 Au	200,6 80 Hg	204,4 81 Tl	207,2 82 Pb	209,0 83 Bi	103 a 209 84 Po	8,3 h 210 85 At	222 86 Rn

150,4 62 Sm	152,0 63 Eu	157,3 64 Gd	158,9 65 Tb	162,5 66 Dy	164,9 67 Ho	167,3 68 Er	168,9 69 Tm	173,0 70 Yb	175,0 71 Lu
8·10⁷a 244 94 Pu	7400a 243 95 Am	16·10⁷a 247 96 Cm	1400a 247 97 Bk	800a 251 98 Cf	276 d 254 99 Es	106 d 257 100 Fm	55 d 258 101 Md	58 min 259 102 No	3 min 260 103 Lr

CHEMISTRY AND LIFE

An Introduction to

GENERAL, ORGANIC, AND BIOLOGICAL CHEMISTRY

CHEMISTRY AND LIFE

An Introduction to
GENERAL, ORGANIC, AND BIOLOGICAL CHEMISTRY

John W. Hill
University of Wisconsin–River Falls

Dorothy M. Feigl
Saint Mary's College

Stuart J. Baum
State University of New York, College at Plattsburgh

Macmillan Publishing Company
New York

Maxwell Macmillan Canada
Toronto

Editor: Paul F. Corey
Production Supervisor: Elisabeth H. Belfer
Production Manager: Sandra Moore
Text Designer: Marsha Cohen
Cover Designer: Jane Edelstein
Cover photographs and frontispiece: Index Stock Photography, Inc.
Inside Front Cover: By P. Menzel. © Ernst Klett Schulbuchverlag GmbH, Federal
    Republic of Germany, 1984. Available in poster and wall sizes from Science Import,
    Quebec, Canada G1T 2R8 (418-527-1414).
Photo Researcher: Diane Austin
Illustrations: Precision Graphics
This book was set in Times Roman and Helvetica by York Graphic Services, Inc.,
printed and bound by Von Hoffmann Press. The cover was printed by Lehigh Press,
Inc.

Macmillan Publishing Company
866 Third Avenue, New York, New York 10022

Macmillan Publishing Company is part of the Maxwell Communication Group of Companies.

Maxwell Macmillan Canada, Inc.
1200 Eglinton Avenue East
Suite 200
Don Mills, Ontario M3C 3N1

Library of Congress Cataloging-in-Publication Data

Hill, John William,
     Chemistry and life : an introduction to general, organic, and
biological chemistry. — 4th ed. / John W. Hill, Dorothy M. Feigl,
Stuart J. Baum.
       p.   cm.
     Includes index.
     ISBN 0-02-354541-0
     1. Chemistry. 2. Biochemistry.   I. Feigl, Dorothy M. II. Baum,
Stuart J. III. Title.
QD31.2.H56 1993
540—dc20   92-11254
                                                                          CIP

Printing: 1 2 3 4 5 6 7 8   Year: 3 4 5 6 7 8 9 0 1 2

PREFACE

Our world has been transformed by science and technology. The impact of science on the quality of human life is profound. Yet, to beginning students, the scientific disciplines that daily influence their lives often seem mysterious and incomprehensible. Those of us who enjoy the study of science, however, find it a fascinating and rewarding experience precisely because it *can* provide reasonable explanations for seemingly mysterious phenomena.

Chemistry and Life has been written in that spirit. Apparently obscure phenomena are explained in an informal, readable style. We assume that the student has little or no chemistry background and clearly explain each new concept as it is introduced. Chemical principles and biological applications are carefully integrated throughout the text, with liberal use of drawings, diagrams, and photographs.

Our selection of topics and choice of examples make the text especially appropriate for students in health and life sciences, but it is also suitable for anyone seeking to become a better-informed citizen of our technological society. The text provides ample material for a full year's course. The ten Special Topics cover optional material for added flexibility. They may be omitted or assigned as outside reading without loss of continuity. We have also included many short essays that focus on interesting applications of topics presented in the text. We have consciously increased the sophistication of chemical understanding as the student progresses through the chapters.

Changes in the Fourth Edition

In response to the suggestions of users and reviewers of the third edition and our own experience, we have made the following modifications in the fourth edition.

- We have gone to a four-color layout to heighten student interest and to emphasize important features. It is now a much more attractive and informative text.
- There are a greater number of marginal notes to supplement chapter material.
- Key words are boldfaced and are also listed in the Glossary (Appendix IV).
- Much of the material from former Special Topic B (Energy Sublevels and Orbitals) has been incorporated into Chapter 2 (Atoms). The sections dealing with the historical development of the atomic theory have been condensed. The concept of atomic weight is now introduced in this chapter.
- Some of the sections from former Special Topic C (Nuclear Power) have been incorporated into Chapter 3, which is now entitled "Nuclear Processes." This chapter contains material on radioactivity and nuclear medicine. Some instructors prefer that their students acquire this knowledge early in the course, whereas others prefer to teach it just prior to the organic chemistry chapters. The latter group can elect to cover this chapter after Chapter 12.
- Former Chapter 8 (Oxidation and Reduction) is now Chapter 6. We felt that the material in this chapter readily follows the chapter on Chemical Reactions.
- Former Chapter 13 (Inorganic Chemistry) has been rewritten and is now Special Topic C, following Chapter 12 (Electrolytes).
- Chapter 13 (Hydrocarbons) now contains sections on halogenated hydrocarbons and on polymerization. Special Topic D (Petroleum) has been added at the end of this chapter.

- The chapter on polymers has been deleted. Stereoisomerism, a Special Topic in the third edition, is now Chapter 18 and includes enantiomers, geometric isomers, and biochemical significance of stereochemistry. Special Topic G (Sight and Odor) follows this chapter.
- Chapter 23 (Nucleic Acids) has been expanded and now contains a more detailed description of DNA replication, protein synthesis, mutations and genetic diseases, and genetic engineering. Special Topic J (Viruses and Cancer) follows this chapter.
- Former Chapters 25 (Energy and Life) and 27 (Carbohydrate Metabolism) along with some sections from former Special Topic L (Digestion) are now combined within two new carbohydrate metabolism chapters. Chapter 24 contains sections on ATP, carbohydrate digestion, blood-sugar levels, and an extensive treatment of anaerobic glycolysis. Chapter 25 continues with aerobic metabolism, the respiratory chain, and oxidative phosphorylation.
- Chapter 28 (Body Fluids) is now the final chapter. It has been expanded to include sections on osmotic pressure, blood pressure, clotting of blood, and blood buffers.
- Several of the special topics have been updated and expanded. Special Topics E and F (on drugs) now contain sections on synthetic narcotics, endorphins and enkephalins, marijuana, Halcion, and designer drugs. Special Topic H (Hormones) contains sections on anabolic steroids, oral contraceptives, estrogen replacement therapy, and prostaglandins.
- The number of exercises at the ends of the chapters has been substantially increased, and there has been an expansion of the worked-out examples within the chapters (especially in the first portion of the text). Many examples are now accompanied by practice exercises for immediate reinforcement. Answers to Practice Exercises and selected Exercises follow the appendixes.
- The health-related topics from the third edition have been retained. We have expanded our discussion of many of these and have added numerous other relevant materials. For example, nuclear medicine, antiseptics, relative humidity, anhydrides, fuel cells, minerals and health, greenhouse effect, ozone depletion, smog, acid rain, octane ratings, anesthetics, phosphate esters, pheromones, chemistry of vision and odor, fiber in the diet, soaps and detergents, eutrophication, cell membranes, atherosclerosis, cholesterol and cardiovascular disease, electrophoresis, nerve poisons, chemotherapy, medical uses of enzymes, DNA fingerprinting, AIDS, carcinogens, diabetes, alcoholism, muscle action, exercise and diets, the chemistry of sports, jaundice, and sickle cell anemia.

Supplements

- As with the third edition, the Study Guide contains a key word list, chapter summary, study hints, additional worked-out examples and practice exercises, and a self-test for each chapter. New to this edition is the incorporation of a Solutions Manual within the Study Guide. It includes the answers to all the end-of-chapter exercises for each chapter.
- A laboratory manual, also published by Macmillan Publishing Company, complements the text very well. *Chemistry and Life in the Laboratory: Experiments in General, Organic, and Biological Chemistry,* 3rd ed., by Heasley, Christensen, and Heasley, also uses examples from students' lives to communicate the science of chemistry. The manual covers the same general topics as the textbook, instructions are clear and thorough, and the experiments are well written and imaginative. All the experiments work! There are 33 experiments to choose from, and all have been thoroughly tested over many years with students in the laboratory. Special care has been taken in the third edition to reduce the cost of chemicals required in the experiments and to make the experiments even safer. An *Instructor's Manual* is available from the publisher.

- The *Instructor's Manual* to accompany the text is available from the publisher for adopters of the text. It offers a selection of multiple-choice questions for each chapter. The Test Bank is also available in computer form on a Macintosh or IBM disk.
- A set of full-color transparencies of illustrations and tables from the text is available from the publisher.

Acknowledgments

In preparing this edition we have had the benefit of a number of thoughtful reviews. We acknowledge with gratitude the contributions from

Michael A. Bobrik
Northern Virginia Community College–
  Alexandria
Lorraine C. Brewer
University of Arkansas
Lawrence K. Duffy
University of Alaska–Fairbanks
Martin Gouterman
University of Washington
A. Donald Glover
Bradley University
Marvin L. Hackert
The University of Texas
Josephine Williams Phillips Van Kook
University of South Alabama
C. V. Krishnan
State University of New York–Stony Brook
David Lippman
Southwest Texas State University
Maria O. Longas
Purdue University–Calumet

Robert E. Maruca
Alderson-Broaddus College
W. Robert Midden
Bowling Green State University
Frazier Nyasulu
University of Washington
Stephen Ott
College of Eastern Utah
James R. Paulson
University of Wisconsin–Oshkosh
Mary Ellen Schaff
University of Wisconsin–Milwaukee
William M. Scovell
Bowling Green State University
Mahesh B. Sharma
Columbus College
Laverne Weidler
Black Hawk College
Donald H. Williams
Hope College

We are grateful to Noella Watts for typing the manuscript and ancillary materials and to Cheryl Jones for proofreading the galleys and helping in the preparation of the instructor's test bank. As usual, we received substantial support and encouragement from the editorial staff at Macmillan. Chemistry editor Paul Corey provided guidance and wise counsel, and copy-editor Jo Satloff gave immeasurable assistance. Once again, as in our earlier texts, production supervisor Elisabeth Belfer has converted a bundle of words, tables, and figures into a book we are proud of.

No book—or other educational device—can replace a good teacher; thus we have designed this book as an aid to the classroom teachers. The only valid test of this or any text is in a classroom. We would greatly appreciate receiving comments and suggestions based on your experience with this book.

J. W. H.
D. M. F.
S. J. B.

CONTENTS

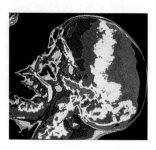

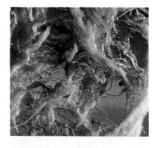

TO THE STUDENT

What is chemistry?

Chemistry is such a broad, all-encompassing area of study that people almost despair in trying to define it. Indeed, some have taken a cop-out approach by defining chemistry as "what chemists do." But that won't do; it's much too narrow a view.

Chemistry is what we all do. We bathe, clean, and cook. We put chemicals on our faces, hands, and hair. Collectively, we use more than 10,000 consumer chemical products in our homes. Professionals in the health and life sciences use thousands of additional chemicals as drugs, antiseptics, or reagents for diagnostic tests.

Your body itself is a remarkable chemical factory. You eat and breathe, taking in raw materials for the factory. You convert these supplies into an unbelievable array of products, some incredibly complex. This chemical factory—your body—also generates its own energy. It detects its own malfunctions and can regenerate and repair some of its component parts. It senses changes in its environment and adapts to these changes. With the aid of a neighboring facility, this fabulous factory can create other factories much like itself.

Everything you do involves chemistry. You read this sentence; light energy is converted to chemical energy. You think; protein molecules are synthesized and stored in your brain. All of us are chemists.

Chemistry affects society as well as individuals. Chemistry is the language—and the principal tool—of the biological sciences, the health sciences, and the agricultural and earth sciences.

Chemistry has illuminated all of the natural world, from the tiny atomic nucleus to the immense cosmos. We believe that a knowledge of chemistry can help you. We have written this book in the firm belief that beginning chemistry can be related immediately to problems and opportunities in the life and health sciences. And we believe that this can make the study of chemistry interesting and exciting, especially to nonchemists.

For example, an "ion" is more than a chemical abstraction. Mercury ions in the wrong place can kill you, but calcium ions in the right place can keep you from bleeding to death. "$P \times V = c$" is an equation, but it is also the basis for the respiratory therapy that has saved untold lives in hospitals. "Hydrogen bonding" is a chemical phenomenon, but it also accounts for the fact that a dog has puppies while a cat has kittens and a human has human offspring. Hundreds of similar fundamental and interesting applications of chemistry to life can be cited.

A knowledge of chemistry has already had a profound effect on the quality of life. Its impact on the future will be even more dramatic. At present we can control diabetes, cure some forms of cancer, and prevent some forms of mental retardation because of our understanding of the chemistry of the body. We can't *cure* diabetes or cure *all* forms of cancer or *all* mental retardation, because our knowledge is still limited. So learn as much as you can. Your work will be enhanced and your life enriched by your greater understanding.

Be prepared. Something good might happen to you—and to others because of you.

Chapter 1

MATTER AND MEASUREMENT

The appeal of colorful substances can stimulate an interest in investigating their properties.

This book is called *Chemistry and Life*. What does chemistry have to do with life? What is chemistry? For that matter, what is life?

The last question is more than rhetorical. Progress in science, technology, and medicine has blurred the distinction between life and death. Is someone whose heart has stopped beating dead? Is someone whose vital functions are being maintained by machine alive? We won't even attempt in this book to supply a definitive answer to the question "What is life?" We'll simply note its critical significance for our society.

How about the first question, "What does chemistry have to do with life?" A chemist would say, "Just about everything." The human body, for example, is the most extraordinarily complicated, most elegantly designed, and most efficiently operated chemical laboratory there is. Our attempts to answer that first question will fill most of this text.

That leaves the middle question, "What is chemistry?" And that is the subject of this first chapter. We shall see how science in general and chemistry in particular have developed from earlier human endeavors. Out study will include a consideration of the methods of science and the manner of its progress. Finally, we shall also develop some basic concepts necessary to our study of chemistry and its relationship to life.

1

1.1
SCIENCE AND THE HUMAN CONDITION

We are taught in elementary school that people have three basic needs: food, cloth-ing, and shelter. Certainly those three things—if adequate in quantity and quality—are enough to keep us alive. Most of us, however, would also agree to adding two more requirements for the good life: reasonable health and some chance for happiness.

In early human societies, nearly all human efforts were directed toward the hunting and gathering of food, the making of clothing, and the provision of shelter. Our early ancestors had no knowledge of the biological and chemical basis of illness, and they could do little about their health except pray and make sacrifices to their gods. With the coming of civilization, some people gained enough leisure to turn their thoughts to the human condition and to the natural world around them. Over the centuries what we now call science grew out of their speculations. As this scientific study of the material universe progressed, the responsibility for adding to the growing body of knowledge was divided among various disciplines. Among these disciplines was chemistry.

Modern chemistry's roots are firmly planted in alchemy, a kind of mystical chemis-try that flourished in Europe during the Middle Ages (Figure 1.1). And modern chemists

Figure 1.1

The Alchemist, a painting done by the Dutch artist Cornelis Bega around 1660, depicts a laboratory of the seventeenth century.

Figure 1.2
Theophrastus Bombastus von Hohenheim, better known as Paracelsus.

have inherited from the alchemists an abiding interest in those aspects of their study that relate to human health and to the quality of life.

Consider, for example, the fact that alchemists not only searched for a "philosopher's stone" that would turn cheaper metals into gold but also sought an "elixir" that would confer immortality on those exposed to it. Alchemists never achieved their primary goals, but they did discover many new chemical substances. As early as the ninth century, a Persian alchemist, Al-Razi, described the use of plaster of Paris for casts to set broken bones. And alchemists perfected techniques such as distillation and extraction that are still useful in our time.

It was a Swiss physician, Theophrastus Bombastus von Hohenheim (1493–1541), who urged alchemists to turn away from their attempts to make gold and to seek instead medicines with which to treat disease (Figure 1.2). Possessed of a monstrous ego, von Hohenheim (who preferred the self-chosen name Paracelsus) alienated many of his contemporaries. His followers, however, were numerous enough to ally forever the science of chemistry with the art of medicine.

By the seventeenth century, a changed attitude, characterized by a reliance on experimentation, had been adopted by astronomers, physicists, physiologists, and philosophers. It was this change in orientation that signaled the emergence of chemistry from alchemy. The English philosopher Francis Bacon (1561–1626) had visions of these new scientific methods endowing human life with new inventions and wealth (Figure 1.3).

By the middle of the twentieth century, it appeared that science and its application in technology had made the dreams of Bacon and von Hohenheim a reality. Many diseases had been virtually eliminated. The coordination of chemistry and medicine had also made the difference between operations in which four strong men were employed to hold the patient down and ones in which the patient was painlessly anesthetized. Fertilizers, pesticides, and scientific breeding had made food more abundant. Nutritionists were applying their science to designing diets that would produce healthier, stronger people. New materials were being developed to improve our clothing and shelter. Industry was offering an almost endless variety of products at relatively low cost to the average consumer.

Indeed, it seemed that, despite its sometimes less than honorable intentions, science could do no wrong. For example, during World War I, when the German armies' supply of ammonia (which they needed to make nitrate explosives) was cut off, the Haber process

Figure 1.3
Francis Bacon, English philosopher and Lord Chancellor to King James I.

Figure 1.4
Fritz Haber (1868–1934), the German chemist who invented a process for manufacturing ammonia.

provided them with an alternative supply (Figure 1.4). Fritz Haber's work probably lengthened the war, but it is far more significant for its influence on modern agriculture. Ammonia and nitrates are the stuff of which fertilizers are made, and fertilizers are essential to modern high-yield farming. In fact, most of the ammonia made by the Haber process today goes into fertilizer.

Much of the technology of the early twentieth century grew out of scientific discoveries. New technological developments were used in turn by scientists as tools to make new discoveries. These developments in science and technology became hallmarks of the modern world. Hardly anyone questioned that science in the early part of the twentieth century had significantly improved the human condition.

1.2
PROBLEMS IN PARADISE

If during the first half of the twentieth century science was viewed as humankind's savior, during the latter half it is sometimes viewed as quite the opposite. Those anesthetics that made surgery painless for the patient have caused female anesthesiologists, surgeons, and surgical nurses to suffer a high percentage of miscarriages compared with other health personnel. Fertilizer runoff from farms has polluted streams, and insecticide residues have had a devastating effect on wildlife. Some of those industrial workers making modern products for our use have died from diseases caused by the chemicals they worked with, and chemical waste dumps seem to threaten the health of us all.

One solution to these problems would be simply to throw out science. But do we really wish to return to surgery without anesthetics? Most of us don't. We need scientists, for it is they who will search for safer anesthetics, for approaches to increased agricultural production compatible with the natural environment, and for analytical techniques that will ensure healthful working conditions for industrial personnel.

The simple fact is that chemistry and its products, both good and bad, are so intimately involved in determining the quality of life that to ignore the subject is to court disaster. It will take an educated, informed society to ensure that science is used for the human good.

1.3
THE WAY SCIENCE WORKS—SOMETIMES

Textbooks often define science as a "body of knowledge," and it is frequently taught as a finished work rather than an ever-changing approach to learning. Science is organized into concepts. For example, even though we will often speak of atoms as if they were readily observed, the atomic concept is merely a convenience that successfully describes many observable facts in a metaphorical way. It is not the "body of facts" that characterizes science but the *organization* given to those facts. To be useful, concepts must have predictive value. If the atomic theory is to be useful, it should enable a scientist to predict how matter will behave.

The most distinguishing characteristic of science is its use of processes or methods. The making of observations and the cataloging of facts are bare, though necessary, beginnings to these intellectual processes. Scientists must be able to make careful measurements, but they must also be able to grasp the central theme of these observations. They must recognize the variables and be able to note the effect of changing one variable at a time. Scientists must be able to sort out the useful aspects of information and ignore irrelevancies. Perhaps basic to these intellectual processes is the ability to formulate testable hypotheses. Even an educated guess is of little value to scientists unless an experiment can be devised to test the guess. In fact, if a hypothesis cannot be tested, the question is generally considered to lie outside the realm of science.

Science is not totally different from other disciplines. For example, creativity is central to both science and the humanities. Science does not involve cold logic to the exclusion of other human characteristics. Albert Einstein recognized that there was no *logical* path to some of the laws that he formulated. Even he relied on intuition based on experience and understanding.

It is important that you realize there is no single "scientific method" that, when followed, produces guaranteed results. Scientists observe, gather facts, and make hypotheses, but somewhere along the way they test their hunches and their organization of facts by experimenting. Scientists, like other human beings, use intuition and may generalize from a limited number of facts. Sometimes they are wrong. One of the strengths of science lies in the fact that results of experiments are published in scientific journals. These results are read—and often checked—by other scientists in all parts of the world. To become an accepted part of the "body of knowledge," the results must be reproducible. Scientists also extend each other's work, sometimes to the point that we see a "bandwagon" effect. One breakthrough sometimes results in the unleashing of vast quantities of new data and leads to the development of new concepts. For example, early in the nineteenth century it was thought that certain chemical substances, called *organic* compounds, could be produced only by living tissue, such as someone's liver or the leaf of a plant. These substances were in contrast to other materials, labeled *inorganic,* which could be prepared by a chemist in a laboratory. In 1828, a German chemist named Friedrich Wöhler (1800–1882) succeeded in making an organic compound from an inorganic one in the laboratory. The belief that such a compound could not be prepared in this manner was so strong that Wöhler did the same thing over and over again to assure himself that he had really done the "impossible." When he finally published his work, other chemists quickly repeated it and then proceeded to make hundreds of thousands of organic compounds. That bandwagon is still rolling today, with chemists making hormones, vitamins, and even genes in the laboratory.

Thus, contrary to an often-expressed popular notion, scientific knowledge is not absolute. Science is cumulative, and the "body of knowledge" is dynamic and constantly changing. Old concepts or even old "facts" are discarded as new tools, new questions, and new techniques reveal new data or generate new concepts. To truly understand what science is, one has to observe what the entire worldwide community of scientists has done over a period of several years rather than look over the shoulder of a single scientist for a few days.

1.4
WHAT IS CHEMISTRY? SOME FUNDAMENTAL CONCEPTS

Chemistry is often defined as the study of matter and the changes that it undergoes. Changes in matter are accompanied by changes in energy. Since the entire universe is made up of nothing more than matter and energy, the field of chemistry extends from atoms to stars, from rocks to living organisms. Matter and energy are such fundamental concepts that definitions are difficult. **Matter** is the stuff that makes up all material things.

Figure 1.5

Astronaut John W. Young leaps from the lunar surface, where gravity pulls at him with only one-sixth the force on Earth.

Matter occupies space and has mass. Wood, sand, water, and people have mass and occupy space. So does air, although one usually needs a flat tire or a 50 km/hr wind to emphasize the point.

Mass is a measure of the quantity of matter in an object. **Weight,** on the other hand, measures a force. On Earth, it measures the force of attraction—gravity—between Earth and the object in question. For most of its history, the human race was restricted to the surface of the Earth. The terms *mass* and *weight* can be used interchangeably in such restricted circumstances. When manned exploration of space began, however, everyone with a television set could see that mass and weight are not the same thing. The mass of an object does not vary with location. An astronaut has the same mass on the moon as on Earth. The matter that makes up the astronaut does not change. John Young's weight on the moon, however, was one-sixth his weight on Earth because the moon's pull is only one-sixth as strong as the Earth's (Figure 1.5). Weight varies with gravity; mass does not. In this text, we investigate processes that occur on Earth, and therefore mass and weight are used interchangeably.

EXAMPLE 1.1

On Mars, gravity is one-third that on Earth.
a. What would be the mass on Mars of a person who has a mass of 60 kilograms (kg) on Earth?
b. What would be the weight on Mars of a person who weighs 120 kg on Earth?

SOLUTION
a. The person's mass would be the same (60 kg) as on Earth; the quantity of matter has not changed.
b. The person would weigh only 40 kg; the force of attraction between planet and person is only one-third that on Earth.

Matter is characterized by its *properties*. This means that we can distinguish water from gasoline by the characteristic properties of the two liquids. They differ in odor, flammability, and many other properties. **Chemical properties** describe how one substance reacts with other substances. To demonstrate a chemical property, a substance must undergo a change in composition. Sulfur burns (combines with oxygen) to form an acrid gas called sulfur dioxide. Sulfur also combines with carbon to form a liquid called carbon disulfide and with iron to form a solid called iron sulfide. Sulfur dioxide, carbon disulfide, and iron sulfide have different properties than sulfur. Each also has a different composition. Substances that undergo a change in chemical properties are said to undergo a **chemical change.**

Physical properties are those that can be observed and specified without reference to any other substance. Characteristics such as color, hardness, density, melting point, and boiling point are physical properties (Figure 1.6). A physical property of sulfur, for example, is that it is a brittle yellow solid under usual conditions. Another is that sulfur is denser than water; that is, sulfur sinks in water. Sulfur melts into a liquid form at 115 °C. When sulfur melts it has undergone a physical change; it has changed from a solid to a liquid. However, its *composition* has not changed; it has *not* changed chemically. A **physical change** occurs without a change in composition, and it does not alter the chemical nature of the substance. Although liquid sulfur looks different from solid sulfur and displays different physical properties, its chemistry has not changed.

A burning candle demonstrates both physical and chemical changes. After the candle is lit, the solid wax near the burning wick melts. This melting is a physical change because the composition of the wax is the same in both the solid and the liquid form. Some of the melted wax is drawn into the burning wick where a chemical change occurs. The wax combines with oxygen in the air to form carbon dioxide gas and water vapor. As the candle burns and the wax undergoes this chemical change, the candle becomes smaller and smaller. The apparent disappearance is not necessarily a sign of chemical change. When water evaporates, it seems to disappear. However, it merely changes from a liquid to a gas; it has undergone a physical change. The composition of the water does not change during evaporation. It is not always easy to distinguish a physical change from a chemical change. The critical question is: Has the composition of the substance been changed? In a chemical change it has; the substance has combined with another substance or has been decomposed to simpler substances. In a physical change no new combinations are formed.

An *acrid* gas is one with a sharp, irritating odor and taste. You can smell sulfur dioxide when you strike a match. Sulfur dioxide is an important air pollutant (Section C.7).

Modern chemists can transform matter in ways that would astound the alchemists. They can change ordinary salt into lye and laundry bleach. They can change sand into transistors and computer chips, and they can convert crude oil into plastics, pesticides, drugs, detergents, and a host of other products. Perhaps the ultimate example of this "modern alchemy" was achieved by James E. Butler of the Naval Research Laboratory in Washington, D.C. Butler made diamonds from sewer gas. Many of the products of modern chemistry are more valuable than gold.

Figure 1.6
A comparison of the physical properties of two elements. Copper, obtained as pellets, can be hammered into a thin foil or drawn into a wire. When hammered, lumps of sulfur crumble into a fine powder.

EXAMPLE 1.2

Which of the following describes a chemical change, and which describes a physical change?

a. Your hair is trimmed by a barber.
b. A female mosquito uses your blood as food from which she produces her eggs.
c. Water boils to form steam.
d. Bacteria convert milk to yogurt.
e. Water is broken down into hydrogen gas and oxygen gas.

SOLUTION

a. Physical change: the hair is not changed by clipping.
b. Chemical change: eggs differ in composition from blood.
c. Physical change: liquid water and water vapor (steam) have the same composition; water merely changes from a liquid to a gas.
d. Chemical change: yogurt and milk differ in composition.
e. Chemical change: new substances, hydrogen and oxygen, are formed.

There are three familiar **states of matter:** solid, liquid, and gas. **Solid** objects ordinarily maintain their shape and volume regardless of their location. A **liquid** occupies a definite volume, but assumes the shape of the occupied portion of its container. If you have a 12-oz soft drink, you have 12 oz whether the soft drink is in a can, a bottle, or, through some slight mishap, on the floor—which demonstrates another property of liquids. Unlike solids, liquids flow readily. **Gases** maintain neither shape nor volume. They expand to fill completely whatever container one puts them in. Gases can be easily compressed. For example, enough air for many minutes of breathing can be compressed into a steel tank for underwater diving. We shall take up the topic of the states of matter in more detail in later chapters.

Matter can be classified in yet other ways. For example, we can subdivide matter into pure substances and mixtures (Figure 1.7). A **pure substance** has a definite, or fixed, composition. Water is a pure substance; it always contains 11% hydrogen and 89% oxygen by weight. Similarly, pure gold is pure gold, that is, 100% gold. The composition of a **mixture** may vary. A milk shake is a mixture. The proportions of milk, ice cream, and flavorings change depending on who is preparing the shake. Mixed nuts are a mixture; the ratio of peanuts to pecans depends on how much you are willing to pay per pound. A *homogeneous mixture* (solution) has an even distribution of all mixed parts. So, in sugar water a drop of the solution has the same composition whether taken from the top, the middle, or the bottom of the beaker. On the other hand, a *heterogeneous mixture* varies from one part to another as in oil-and-vinegar salad dressing or dirt in water.

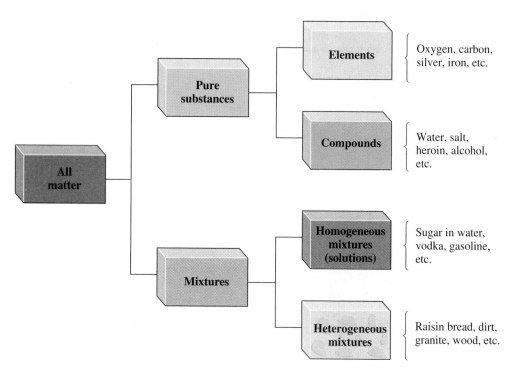

Figure 1.7
A scheme for classifying matter.

1.5
ELEMENTS AND COMPOUNDS

Pure substances may be either elements or compounds. **Elements** are those fundamental substances from which all material things are constructed. **Compounds** are pure substances that are made up of two or more elements chemically combined in constant or fixed proportions. Our ideas about elements have changed during historical times. Fire was once considered an element, but it is now regarded as nonmaterial, that is, as a form of energy. Water was once thought to be an element, but we now know it to be a compound composed of two elements, hydrogen and oxygen. We presently regard as elements a few more than 100 pure substances that cannot be broken down by chemical means into simpler substances. Sulfur, oxygen, carbon, and iron are elements. Sulfur dioxide, carbon disulfide, and iron sulfide are compounds.

Because elements are so fundamental to our study of chemistry, we find it useful to refer to them in a shorthand form. Each element can be represented by a **chemical symbol** made up of one or two letters derived from the name of the element (or, sometimes, from the Latin name of the element). The first letter of the symbol is always capitalized; the second is always lowercase. (It makes a difference. For example, Hf if the symbol for hafnium, an element, but HF is the formula for hydrogen fluoride, a compound. Similarly, Co is cobalt, an element; CO is carbon monoxide, a compound.) Symbols are the alphabet of chemistry. Symbols for the more important elements studied in this book are given in Table 1.1.

A chemical symbol in a formula stands for one atom of the element. If more than one atom is to be indicated in a formula, a subscripted number is used after the symbol. For example, the formula Cl_2 represents two atoms of chlorine, and the formula CCl_4 stands

The modern definition of an element is based on *atomic number*. The definition is given and explained in Chapter 2.

At present 109 elements are known. We deal with only about one-third of them in any detail in this book.

Table 1.1
Names, Symbols, and
Physical Characteristics of
Some Common Elements

Name (Latin name)	Symbol	Selected Properties
Aluminum	Al	Light, silvery metal
Argon	Ar	Colorless gas
Arsenic	As	Grayish white solid
Barium	Ba	Silvery white metal
Beryllium	Be	Steel gray, hard, light solid
Boron	B	Black or brown powder; several crystal forms
Bromine	Br	Reddish brown liquid (Br_2)
Calcium	Ca	Silvery white metal
Carbon	C	Soft black solid (graphite) or hard, brilliant crystal (diamond)
Chlorine	Cl	Greenish yellow gas (Cl_2)
Copper (Cuprum)	Cu	Light reddish brown metal
Fluorine	F	Pale yellow gas (F_2)
Gold (Aurum)	Au	Yellow, malleable metal
Helium	He	Colorless gas
Hydrogen	H	Colorless gas (H_2)
Iodine	I	Lustrous black solid (I_2)
Iron (Ferrum)	Fe	Silvery white, ductile, malleable metal
Lead (Plumbum)	Pb	Bluish white, soft, heavy metal
Lithium	Li	Silvery white, soft, light metal
Magnesium	Mg	Silvery white, ductile, light metal
Mercury (Hydrargyrum)	Hg	Silvery white, liquid, heavy metal
Neon	Ne	Colorless gas
Nickel	Ni	Silvery white, ductile, malleable metal
Nitrogen	N	Colorless gas (N_2)
Oxygen	O	Colorless Gas (O_2)
Phosphorus	P	Yellowish white waxy solid or red powder (P_4)
Plutonium	Pu	Silvery white, radioactive metal
Potassium (Kalium)	K	Silvery white, soft metal
Silicon	Si	Lustrous gray solid
Silver (Argentum)	Ag	Silvery white metal
Sodium (Natrium)	Na	Silvery white, soft metal
Sulfur	S	Yellow solid (S_8)
Tin (Stannum)	Sn	Silvery white, soft metal
Uranium	U	Silvery, radioactive metal
Zinc	Zn	Bluish white metal

for one atom of carbon and four atoms of chlorine. Table 1.1 gives actual formulas (Br_2, P_4, S_8, and so on) for some of the elements as they occur in the elemental form. You need not concern yourself with such formulas now; they are included in the table for your future reference.

Figure 1.8
A boulder on a cliff represents potential energy. A falling boulder represents kinetic energy. The falling boulder can do work (smash a house, for example). The boulder on the cliff has a potential for doing the same thing.

1.6
ENERGY AND ENERGY CONVERSION

Energy is the capacity for doing work. **Work** must be done to make something happen that wouldn't happen by itself. Getting out of bed, building a house, or mining coal all require energy. Eating requires energy; that forkful of spaghetti would never make it to your mouth by itself. Energy is the basis for change in the material world. When something moves or breaks or cools or shines or grows or decays, energy is involved.

Energy exists in several forms. Energy due to position or arrangement is called **potential energy** (sometimes referred to as stored energy). A boulder on a cliff has potential energy because of gravitational attraction (Figure 1.8). Simply because it is at the top of the cliff and not at the bottom, the boulder has potential energy, which it can use to do work. The boulder can fall from the cliff and destroy a house at the base (thereby doing work), whereas a boulder sitting at the base of the cliff cannot.

Kinetic energy (KE) is energy an object has because of its motion. The boulder at the base of the cliff could destroy the house if it were rolling along the ground at a good clip. Its ability to do work in this case would depend on its motion. If you were really determined to destroy the house using kinetic energy, your best bet would be to use a boulder rather than a stone. Given the choice, it would also be better to get the boulder moving at a brisk rather than a leisurely pace. That is just another way of saying that kinetic energy depends on mass and velocity. The bigger an object is and the faster it is moving, the more kinetic energy it has and the more work it can do. Mathematically, the kinetic energy is one-half the mass times the square of the velocity.

$$KE = \tfrac{1}{2}mv^2$$

Energy is often classified in other ways. These classifications are based on some characteristic of the energy being considered, for example, its source. It is easier to discuss some of these types of energy by indicating their significance in the overall pattern of energy flow on Earth.

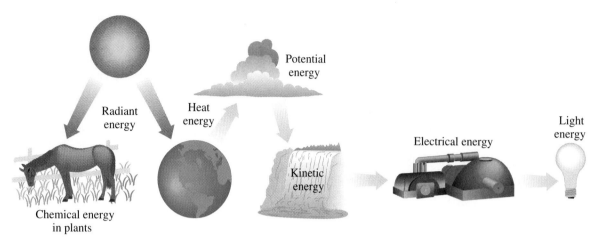

Figure 1.9
Energy can be changed
from one form to another.

The source of nearly all our energy is the sun. **Solar energy** radiates through space. A small portion of this **radiant energy** reaches Earth, where part of it is converted to **heat energy.** This heat causes water to evaporate and then rise to form clouds. The water in the clouds has potential energy. As the water falls through the air and then flows in rivers, the potential energy is converted to kinetic energy.

The kinetic energy of the flowing stream can be used to turn a turbine, which converts a part of the kinetic energy to **electrical energy.** The electricity thus produced can be transported by wires to homes and factories, where it can be converted to **light energy** or to **heat** or to **mechanical energy** (Figure 1.9).

Some of the solar energy striking the Earth is absorbed by green plants, which use a complicated chemical process called photosynthesis to convert solar energy into **chemical energy.** The chemical energy stored by plants—now and in ages past—is used by animals and humans for food and fuel. Nearly all of the vast quantities of energy being used in our modern civilization came originally from the sun by way of green plants. Plants of the current age are harvested by foresters and farmers. Those of ancient ages are reaped as fossil fuels—coal, oil, and gas.

Another form of energy is **nuclear energy.** This type of energy is not derived from the sun. Rather, it was stored in the Earth's crust when the solar system was formed some 4 or 5 billion years ago. We recover it for use when we mine uranium and build nuclear reactors or atomic bombs. (The latter is seen by many as a misuse of nuclear energy.)

EXAMPLE 1.3

What kind of energy does each of the following possess?

a.   a thrown softball
b.   a softball resting at the edge of a table
c.   gasoline

SOLUTION
a.   A thrown softball has kinetic energy; it is moving through the air.
b.   A softball at the edge of a table has potential energy by virtue of its position; it could release energy by falling.
c.   Gasoline contains chemical energy (a form of potential energy); the energy is released when the gasoline burns.

EXAMPLE 1.4

What energy changes occur during each of the following events?

a.   Fuel oil is burned.
b.   A softball falls off the edge of a table.
c.   Sunlight falls on your back.

SOLUTION
a.   Chemical energy is converted to heat energy.
b.   Potential energy is converted to kinetic energy.
c.   Radiant energy is converted to heat energy.

1.7
ELECTRIC FORCES

To deal with energy transformations, chemistry often borrows fundamental concepts from its neighboring discipline, physics. One such concept is force. A **force** is a push or a pull that sets an object in motion, or stops a moving object, or holds an object in place. **Gravity** is a force. Objects—including us—are held to the surface of the Earth by gravity, the attraction of Earth's mass for our mass. The weight of an object is the force of gravity that exists between it and the Earth.

Electric forces are extremely important in chemistry. Some particles of matter are neutral, and some bear an electric charge, either positive (+) or negative (−). No one can really tell exactly what an electric charge is. We simply accept the fact that a particle with a "charge" can exert a force, that is, can push or pull another particle that also has a "charge" on it. The particles do not have to be touching to attract or repel one another. For this reason, we say that charged particles have *force fields* around them. Even at a distance they attract and repel one another, although these forces get weaker as the particles get farther apart. Particles with **like** charges (both positive or both negative) repel one another. Those with **unlike** charges (one positive and one negative) attract one another (Figure 1.10).

This phenomenon of charged substances is not unfamiliar to you. Anyone who has pulled clothes from an automatic dryer on a cold winter day has probably seen what commercials like to call "static cling"—pieces of clothing sticking to one another. The "cling" is due to the attraction of *unlike* charges. If, on the other hand, you brush your

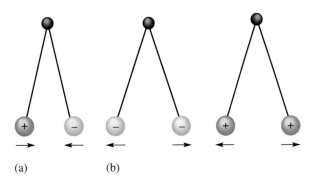

(a)   (b)

Figure 1.10
(a) Particles with unlike charges attract one another. (b) Those with like charges repel one another.

hair vigorously on a cold day, your hair may become "unmanageable" (another term used often in advertisements). Instead of lying flat against your head, the hair sticks out in all directions, each strand seemingly trying to get away from all the other strands. That is exactly what is happening; the strands have *like* charges on them and repel one another.

1.8
MEASUREMENT: THE MODERN METRIC SYSTEM

Accurate measurement of such quantities as mass, volume, time, and temperature is essential to the compilation of dependable scientific "facts." Such facts may be used by a chemist interested in basic research, but similar information is of critical importance in every science-related field. Certainly we are all aware that measurements of both temperature and blood pressure are routinely made in medicine. It is also true that modern medical diagnosis depends on a whole battery of other measurements, including careful chemical analyses of blood and urine.

The system of measurement used by most scientists is an updated metric plan called the **International System of Measurements,** or **SI** (from the French *Système International*). Indeed, SI has been adopted worldwide for everyday use. Even the United States is committed to change to SI, but conversion is voluntary and has proceeded rather slowly so far. To date, the United States, Liberia, and Burma are the only countries that still use the old English system (pounds and inches).

The beauty of SI is that it is based on the decimal system. This makes conversion from one unit to another rather simple. The SI has only a few basic units. For example, the basic unit of length is the **meter** (m), a distance only slightly greater than a yard. The SI unit of mass is the **kilogram** (kg), a mass that on Earth has a weight slightly greater than 2 pounds. All other units for length, mass, and volume can be derived from these basic units. For example, area can be measured in square meters (m^2) and volume in cubic meters (m^3).

A disadvantage of the basic SI units is that they are often of awkward magnitude. We seldom work with kilogram quantities in the laboratory. A cubic meter of liquid would fill a very large test tube. More convenient units can be derived by the use of (or, in the case of the kilogram, the deletion of) prefixes (Table 1.2). For example, in the laboratory we may work with grams (g) or milligrams (mg) of material. The prefix **milli-** means 1/1000 or 0.001. Thus, one milligram equals 1/1000 gram or 0.001 gram. The relationship between milligrams and grams is given by

$$1 \text{ mg} = 0.001 \text{ g}$$

or

$$1000 \text{ mg} = 1 \text{ g}$$

You should learn the more common prefixes (those in color in Table 1.2) right away.

For volume we may choose cubic centimeters (cm^3) or cubic decimeters (dm^3) as more convenient units. The cubic decimeter has a special name, the **liter** (L). From that is derived the milliliter (mL), a unit that is the same as the cubic centimeter. (The symbol cm^3 is sometimes written as cc.)

$$1 \text{ dm}^3 = 1 \text{ L}$$
$$1 \text{ L} = 1000 \text{ mL}$$
$$1 \text{ mL} = 1 \text{ cm}^3 \text{ (or 1 cc)}$$

Table 1.2
Approved Numerical Prefixes

Exponential Expression	Decimal Equivalent	Prefix	Phonic	Symbol
10^{12}	1,000,000,000,000	Tera-	ter′ a	T
10^9	1,000,000,000	Giga-	ji′ ga	G
10^6	1,000,000	Mega-	meg′ a	M
10^3	1,000	Kilo-	keel′ o	k
10^2	100	Hecto-	hek′ to	h
10	10	Deka-	deck′ a	da
10^{-1}	0.1	Deci-	dess′ i	d
10^{-2}	0.01	Centi-	sen′ ti	c
10^{-3}	0.001	Milli-	mil′ i	m
10^{-6}	0.000 001	Micro-	my′ kro	μ
10^{-9}	0.000 000 001	Nano-	nan′ o	n
10^{-12}	0.000 000 000 001	Pico-	peek′ o	p
10^{-15}	0.000 000 000 000 001	Femto-	fem′ to	f
10^{-18}	0.000 000 000 000 000 001	Atto-	at′ to	a

Most commonly used units printed in color.

Conversions within the SI system are much easier than those using the more familiar units of pounds, feet, and pints. This is best shown by examples.

EXAMPLE 1.5

Convert 0.742 kg to grams.

SOLUTION

$$0.742 \ \cancel{kg} \times \frac{1000 \ g}{1 \ \cancel{kg}} = 742 \ g$$

Note: If you are not familiar with the use of conversion factors or would like to review the mathematics of conversions, see Special Topic A immediately following this chapter.

EXAMPLE 1.6

Convert 0.742 lb to ounces.

SOLUTION

$$0.742 \ \cancel{lb} \times \frac{16 \ oz}{1 \ \cancel{lb}} = 11.9 \ oz$$

EXAMPLE 1.7

Convert 1247 mm to meters.

SOLUTION

$$1247 \ \cancel{mm} \times \frac{1 \text{ m}}{1000 \ \cancel{mm}} = 1.247 \text{ m}$$

EXAMPLE 1.8

Convert 1247 in. to yards.

SOLUTION

$$1247 \ \cancel{in.} \times \frac{1 \text{ yd}}{36 \ \cancel{in.}} = 34.64 \text{ yd}$$

In conversions involving pounds, ounces, inches, and yards, you multiply and divide by numbers such as 16 or 36. In metric conversions, you multiply by 10 or 100 or 1000, and so on; you need only shift the decimal point.

In the United States, it is sometimes necessary to convert a measurement from one systems' units to the other's. If you need to learn how to do this, such conversions are discussed in Special Topic A following this chapter. It is useful to have some idea of the relative sizes of comparable units. Some comparisons are shown in Figure 1.11. Other comparisons easily remembered are that the U.S. ten-cent coin, the dime, is about 1 mm thick, and the U.S. five-cent piece, the nickel, has a mass of about 5 g.

$$1.0 \text{ m} = 39 \text{ in. or } 1.1 \text{ yd}$$
$$1.0 \text{ L} = 1.1 \text{ qt}$$
$$1.0 \text{ kg} = 2.2 \text{ lb}$$

1.9
MEASURING ENERGY: TEMPERATURE AND HEAT

The SI unit for temperature is the **kelvin** (K), but for much of their work scientists still use the **degree Celsius** (°C). On the Celsius temperature scale, the freezing point of water is 0 °C and its boiling point is 100 °C. The scale between these two reference points is divided into 100 equal divisions, each a Celsius degree. (The Celsius scale used to be called the Centigrade scale because of this 100-degree interval.)

The Kelvin scale is called an **absolute scale** because its zero point is the coldest temperature possible, absolute zero. (This fact was determined by theoretical considerations and has been confirmed by experiment.) The zero point on the Kelvin scale, 0 K, is

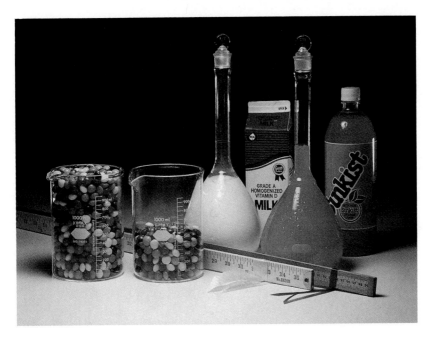

Figure 1.11
A comparison of metric and customary units of measure. The beaker at the left
contains 1 kilogram (kg) of candy, and the one at the right contains 1 pound (lb)
(1 kg = 2.2046 lb). The yellow ribbon is 1 inch (in.) wide and is tied around a
stick 1 yard (yd) long. The green ribbon is 1 centimeter (cm) wide and is tied
around a stick 1 meter (m) long. (1 m = 1.0936 yd and 1 cm = 0.3937 in.) The
flasks each hold 1 liter (L) when filled to the mark. The flask on the right and
the bottle behind it each contain 1 L of orange juice. The flask on the left and
the carton behind it each contain 1 quart (qt) of milk. (1 L = 1.057 qt.)

equal to -273 °C. A kelvin is the same size as a degree Celsius, so the freezing point of
water on the Kelvin scale is 273 K. Note that the Kelvin scale has no negative temperatures and we don't use a degree sign with the K.

To convert from degrees Celsius to Kelvin, you merely add 273 to the Celsius
temperature.

$$K = °C + 273$$

EXAMPLE 1.9

What is the boiling point of water on the Kelvin scale? The boiling point of water
is 100 °C.

SOLUTION

$$100 \ °C + 273 = 373 \ K$$

Figure 1.12

A comparison of the Fahrenheit and Celsius temperature scales.

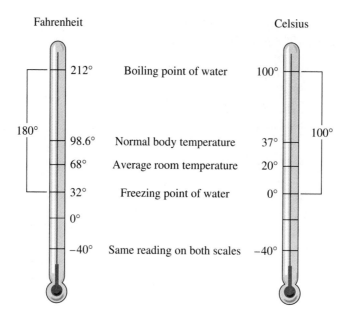

In the United States, weather reports and cooking recipes still use the **Fahrenheit scale,** with units of **degrees Fahrenheit** (°F). This scale defines the freezing temperature of water as 32 °F and the boiling point of water as 212 °F. Figure 1.12 compares the Fahrenheit and Celsius scales. Some additional temperature equivalents are found in Appendix I.

The difference between the freezing point of water and the boiling point of water is 100° on the Celsius scale and 180° on the Fahrenheit scale. It follows that a Celsius degree is 1.80 times as large as a Fahrenheit degree. But remember that the Fahrenheit freezing point of water is 32 degrees above zero. Thus, Fahrenheit temperatures (°F) and Celsius temperatures (°C) can be converted to each other by using the following equations.

$$°F = 1.80(°C) + 32$$
$$°C = (°F - 32)0.555$$

Some books use the following equations to interconvert Fahrenheit and Celsius.

$$°F = \frac{9}{5}(°C) + 32$$
$$°C = (°F - 32)\frac{5}{9}$$

EXAMPLE 1.10

Before the concern for energy conservation, many people kept their homes at a temperature of 72 °F. What is that on the Celsius scale?

SOLUTION

$$°C = (°F - 32)0.555$$
$$= (72 - 32)0.555$$
$$= 22$$

EXAMPLE 1.11

Convert the temperature $-20\ ^\circ C$ to the Fahrenheit scale.

SOLUTION

$$°F = 1.80(°C) + 32$$
$$= 1.80(-20) + 32$$
$$= -4.0$$

Scientists often need to measure amounts of heat energy. You should not confuse heat with temperature. *Heat* is a measure of quantity, that is, of how much energy a sample contains. *Temperature* is a measure of intensity, that is, of how energetic each particle of the sample is. A glass of water at 70 °C contains less heat than a bathtub of water at 60 °C. The particles of water in the glass are more energetic, on the average, than those in the tub, but there is far more water in the tub, and its total heat content is greater.

The SI unit of heat is the **joule** (J), but we will use the more familiar **calorie** (cal).

$$1\ cal = 4.184\ J$$
$$1000\ cal = 1\ kcal = 4184\ J$$

A **calorie** is the amount of heat required to raise the temperature of 1 g of water 1 °C. There is a more precise definition, but this version will do for our purposes.

Some substances have greater heat capacities than others. The **specific heat** of a substance is the amount of heat required to raise the temperature of 1 g of the substance by 1 °C. The definition of the calorie indicates that the specific heat of water is 1 cal/(g · °C). Water is a poor conductor of heat. Metals are good heat conductors. A sample of water must absorb much more heat to raise its temperature than must a metal sample of similar mass. The metal sample may become red hot after absorbing a quantity of heat that will make the water sample only lukewarm.

EXAMPLE 1.12

How many calories would it take to raise the temperature of 250 g of water from 24 °C to 100 °C?

SOLUTION
The specific heat of water is

$$\frac{1\ cal}{(1\ g)(1\ ^\circ C)}$$

The temperature change is 100 °C − 24 °C or 76 °C, and the amount of water is 250 g.

$$250\ g \times 76\ ^\circ C \times \frac{1\ cal}{(1\ g)(1\ ^\circ C)} = 19,000\ cal$$

EXAMPLE 1.13

If a hot-water bag contains 1 kg of water at 65 °C, how much heat will it have supplied to someone's aching muscles by the time it has cooled to 20 °C?

SOLUTION

$$1 \text{ kg} = 1000 \text{ g}$$

$$1000 \text{ g} \times (65 - 20)°C \times \frac{1 \text{ cal}}{(1 \text{ g})(1 \text{ °C})} = 1000 \text{ g} \times 45 °C \times \frac{1 \text{ cal}}{(1 \text{ g})(1 \text{ °C})}$$

$$= 45,000 \text{ cal or } 45 \text{ kcal}$$

Calculations such as the above can also be made using the equation

Heat absorbed or released = mass × specific heat × ΔT

where ΔT is the change in temperature. The setup and mathematics are the same.

For measuring the energy content of foods, the *large* **Calorie** (note the capital *C*), or **kilocalorie** (kcal), is sometimes used. A dieter might be aware that a banana split contains 1500 Calories. If the same dieter realized that this means 1,500,000 calories, giving up the banana split might be easier (Table 1.3).

Table 1.3
The Caloric Content of Some Desserts

Desserts	Portion	kcal
Cake		
Angel food cake	$\frac{1}{12}$ of tube cake	135
Devil's food cake, with		
chocolate icing	$\frac{1}{16}$ of two-layer cake	235
Coffee cake	$\frac{1}{6}$ cake, $2\frac{1}{2}$ oz.	230
Gingerbread	$\frac{1}{9}$ of 8-in. cake	175
Fruit cake, dark	$\frac{1}{2}$ oz.	55
Pound, loaf	$\frac{1}{7}$ of a loaf	195
Brownies, with nuts	$1\frac{3}{4}$ in. × $1\frac{3}{4}$ in.	95
Cookies		
Chocolate chip, commercial	1	50
Oatmeal, with raisins	1	60
Vanilla wafers	1	18
Gingersnap, 2 in. diameter	1	22

1.10
DENSITY

An important property of matter, particularly in scientific work, is *density*. When one speaks of lead as "heavy" or aluminum as "light," one is referring to the density of these metals. **Density** is defined as the amount of mass per unit of volume.

$$d = \frac{m}{V}$$

Rearrangement of this equation gives

$$m = d \times V \quad \text{and} \quad V = \frac{m}{d}$$

These equations are useful for calculations. Densities are usually reported in grams per milliliter (g/mL) or grams per cubic centimeter (g/cm^3).

EXAMPLE 1.14

What is the density of iron if 156 g of iron occupies a volume of 20 cm^3?

SOLUTION

$$\frac{156 \text{ g}}{20 \text{ cm}^3} = 7.8 \text{ g/cm}^3$$

EXAMPLE 1.15

What is the mass of a liter of gasoline if its density is 0.66 g/mL?

SOLUTION

$$\frac{0.66 \text{ g}}{1 \text{ mL}} \times 1 \text{ L} \times \frac{1000 \text{ mL}}{1 \text{ L}} = 660 \text{ g}$$

EXAMPLE 1.16

What volume is occupied by 461 g of mercury? The density of mercury is 13.6 g/mL.

SOLUTION

$$461 \text{ g} \times \frac{1 \text{ mL}}{13.6 \text{ g}} = 33.9 \text{ mL}$$

The density of water is 1 g/mL, or 1 g/cm^3 (remember that 1 mL = 1 cm^3). This nice round number for the density of water is not an accident. The metric system was originally set up in such a way as to ensure that this was the case.

A term related to density is *specific gravity*. **Specific gravity** is the ratio of the mass of any substance to the mass of an equal volume of water. The specific gravity of water itself, therefore, is 1. Mercury (the silvery liquid in thermometers) has a specific gravity of 13.6. That means it has a density 13.6 times as great as water's. The specific gravity of alcohol is 0.8; thus, alcohol is less dense than water. The specific gravity of human fat is 0.903. If mixed with water, the fat would float on top. The higher the proportion of body fat in a person, the more buoyant that person is in water. Because it is the ratio of two values, specific gravity is a number without units, whereas density is reported in units of mass per volume. And because the density of water is 1 g/mL, the specific gravity of a substance is numerically the same as its density.

$$\text{Specific gravity of a substance} = \frac{\text{density of the substance}}{\text{density of water}}$$

EXAMPLE 1.17

The density of chloroform is 1.5 g/mL. What is its specific gravity?

SOLUTION

$$\frac{1.5 \ \text{g/mL}}{1.0 \ \text{g/mL}} = 1.5$$

Specific gravity of a liquid is frequently measured by a device called a **hydrometer.** The hydrometer is placed in the liquid. How far it dips down into the liquid is determined by the density of the liquid (Figure 1.13). The stem of the hydrometer is calibrated in such a way that the specific gravity can be read directly at the surface of the liquid. Hydrometers can be used to measure the strength of the "battery acid" in your car, the percentage of antifreeze in the radiator, sugar content in maple syrup, dissolved solids in urine, alcohol content of wine, and many other properties of solutions that are related to specific gravity.

For those who are interested and those whose work might require it, measurement is discussed further in Appendix I, and a variety of conversion tables are collected there for convenient reference.

Figure 1.13
A hydrometer. The one shown here measures specific gravities over a range of 0.700 to 0.770.

EXERCISES

1. Define each of the following terms.
   a. chemistry   b. matter
   c. mass   d. weight
   e. energy   f. heat
   g. temperature h. calorie
   i. density   j. specific gravity
2. What is a hypothesis? What kind of hypothesis is necessary in science?
3. Which of the following are examples of matter?
   a. iron   b. air
   c. love   d. the human body
   e. gasoline   f. an idea
4. Explain the difference between mass and weight.

5. Two samples are weighed under identical conditions in a laboratory. Sample A weighs 1 lb, and sample B weighs 2 lb. Does B have twice the mass of A?
6. What has changed when a person completes a successful diet: the person's weight or the person's mass?
7. Sample A, which is on the moon, has exactly the same mass as sample B, which is on Earth. Do the two samples weigh the same?
8. Distinguish between chemical and physical properties.
9. Which of the following describes a physical change, and which describes a chemical change?
   a. Sheep are sheared, and the wool is spun into yarn.
   b. Silkworms feed on mulberry leaves and produce silk.

10. Which describes a physical change, and which describes a chemical change?
    a. Because a lawn is watered and fertilized, it grows more thickly.
    b. An overgrown lawn is manicured by mowing it with a lawn mower.

11. Which describes a physical change, and which describes a chemical change?
    a. Ice cubes form when a tray filled with water is placed in a freezer.
    b. Milk, which has been left outside a refrigerator for many hours, turns sour.

12. How do gases, liquids, and solids differ in their properties?

13. Which is the most compressible form of water: steam, ice, or liquid?

14. What is the difference between a pure substance and a mixture?

15. All samples of the sugar glucose consist of 8 parts (by weight) oxygen, 6 parts carbon, and 1 part hydrogen. Is glucose a pure substance or a mixture?

16. Identify each of the following as a pure substance or a mixture.
    a. carbon dioxide   b. oxygen
    c. smog   d. gasoline
    e. mercury   f. soup
    g. sodium chloride (salt) h. a carrot
    i. 24-karat gold   j. champagne

17. What is an element?

18. What is a chemical compound?

19. Without consulting Table 1.1, give names for the elements with the following symbols.
    a. He   b. N   c. F
    d. Mg   e. Si   f. S
    g. K   h. Fe   i. Cu
    j. Br   k. Sn   l. Ba
    m. Au   n. P   o. Pu

20. Without consulting Table 1.1, give symbols for the following elements.
    a. hydrogen   b. carbon   c. oxygen
    d. sodium   e. aluminum   f. phosphorus
    g. chlorine   h. calcium   i. cobalt
    j. zinc   k. silver   l. iodine
    m. mercury   n. lead   o. uranium

21. Which of the symbols or formulas represent elements, and which represent compounds?
    a. H   b. He   c. HF
    d. C   e. CO   f. Ca
    g. CO_2 h. Cl_2 i. NaBr

22. What is the difference between kinetic energy and potential energy?

23. Which has the greater kinetic energy: a sprinter or a long-distance runner (assume the two weigh the same)?

24. Which has the greater kinetic energy: a cannonball or a bullet, both fired at the same speed?

25. Which has the greater kinetic energy: a bicyclist traveling at 15 mi/hr or an automobile traveling at 40 mi/hr?

26. Which has the greater kinetic energy: a 110-kg football tackle moving slowly across the field or an 80-kg halfback racing quickly down the field?

27. Which has the greater potential energy: a diver on the 1-m board or the same diver on the 10-m platform?

28. Which has the greater potential energy: an elevator stopped at the twentieth floor or one stopped at the twelfth floor?

29. Which has the greater potential energy: a roller coaster as it starts to climb the first hill or as it reaches the top?

30. Describe what happens to two particles with like charges when they are brought close together. What happens to particles with unlike charges when they are brought close together?

31. What are the basic units of length, volume, and mass in the SI system? What derived units are more often employed in the laboratory?

32. How many millimeters and how many centimeters are there in 1 m?

33. For each of the following, indicate which is the larger unit.
    a. mm or cm   b. kg or g   c. dL or μL

34. For each of the following, indicate which is the larger unit.
    a. L or cm^3 b. cm^3 or mL

35. For each of the following, indicate which is the larger unit.
    a. in. or m   b. lb or g   c. L or gal

36. For each of the following, indicate which is the larger unit.
    a. °C or °F   b. cal or Cal

37. How many meters are there in each of the following?
    a. 50 km   b. 25 cm

38. How many millimeters are there in each of the following?
    a. 1.5 m   b. 16 cm

39. How many deciliters are there in each of the following?
    a. 1.0 L   b. 20 mL

40. How many liters are there in each of the following?
    a. 2056 mL   b. 47 kL

41. Make the following conversions.
    a. 15,000 mg to g   b. 0.086 g to mg

42. Make the following conversions.
    a. 0.149 L to mL   b. 47 mL to L

43. Make the following conversions.
    a. 1.5 L to mL   b. 18 mL to L

44. How many milliliters are there in 1 cm^3? In 15 cm^3?

45. Order the temperatures from coldest to hottest: 0 K, 0 °C, 0 °F.

46. Convert the following to degrees Fahrenheit.
    a. 37 °C   b. −100 °C   c. 273 °C

47. Which is hotter, 100 °C or 100 °F?

48. Convert the following to degrees Celsius.
    a. 98 °F   b. 5 °F   c. −5 °F

49. How many calories are there in each of the following?
    a. 2.75 kcal   b. 0.74 Cal

50. How many calories are required to raise the temperature of 50 g of water from 20 °C to 50 °C?

51. How many calories are required to raise the temperature of 13 g of water from 15 °C to 95 °C?

52. How much heat would be released by 2.0 kg of water cooling from 90 °C to 20 °C?

53. How much water can be heated from 20 °C to 50 °C by 800 cal of energy?

54. One gram of fat provides 9 kcal of energy. A nutrition article says that margarine, a fat, provides 100 kcal per tablespoon (T). How many grams of margarine are there in 1.0 T of margarine?

55. What is the density of a salt solution if 50 cm³ of the solution has a mass of 57 g?

56. What is the density of a liquid that has a mass of 60.2 g and a volume of 25.0 mL?

57. A 1.0-L container of carbon tetrachloride has a mass of 1.6 kg. What is the density of carbon tetrachloride?

58. A piece of tin has a volume of 16.4 cm³. The density of tin is 5.75 g/cm³. What is the mass of the tin?

59. What is the mass of 50.0 mL of mercury? The density of mercury is 13.6 g/mL.

60. If the density of a normal urine sample is 1.02 g/mL, what is its specific gravity?

61. What is the density of a urine sample if 150 mL has a mass of 157 g?

62. A 59.0-g piece of lead has a volume of 5.20 cm³. Calculate the density of lead.

63. The density of ethyl alcohol at 20 °C is 0.789 g/mL. What is the mass of a 50.0-mL sample?

64. What volume is occupied by 253 g of bromoform? The density of bromoform is 2.90 g/mL.

65. In Exercise 54, you calculated the mass of a tablespoon (T) of margarine. If 1.0 T has a volume of 15 mL, what is the density of the margarine?

66. A tall, layered drink called a pousse-café is made of four different liqueurs. The liqueurs (20.0 mL of each) are carefully poured into a single container so that they do not mix. The names, colors, and masses (per 20.0 mL) of the liqueurs follow.

Liqueur	Color	Mass (g)
Cassis (black currant)	dark red	23.2
Crème de Menthe	green	21.8
Irish Mist	amber	20.4
Triple Sec	white	21.2

Calculate the density of each liqueur and make a diagram of the drink, indicating the color of each layer.

67. The density of an antifreeze solution is 1.1044 g/mL. What is its specific gravity?

Unit Conversions

Sometimes we need to convert a measurement from one kind of unit to another. A powerful method for doing these unit conversions is the **factor-label method.** (The approach is also often called **dimensional analysis,** although the units employed are not always dimensions.) Whatever we call it, the method employs the units, such as L, cm/ft, mi/hr, or g/cm^3, as aids in setting up and solving problems. The general approach is to multiply the known quantity (and its units!) by one or more conversion factors so that the answer is obtained in the desired units.

$$\begin{array}{c}\text{Known quantity} \\ \text{and unit(s)}\end{array} \times \begin{array}{c}\text{conversion} \\ \text{factor(s)}\end{array} = \begin{array}{c}\text{answer} \\ \text{(in desired units)}\end{array}$$

The method is best learned by practice. We urge you to learn it now to save yourself a lot of time and wasted effort later.

A.1
CONVERSIONS WITHIN A SYSTEM

Quantities can be expressed in a variety of units. For example, you can buy beverages by the 12-oz can or by the pint, quart, gallon, or liter. If you wish to compare prices, you must be able to convert from one unit to another. Such a conversion changes the numbers and units, but it does not change the quantity. Your actual weight, for example, remains unchanged whether it is expressed in pounds, ounces, or tons.

You know that multiplying a number by 1 doesn't change its value. Multiplying by a fraction equal to 1 also leaves the value unchanged. A fraction is equal to 1 when the numerator is equal to the denominator. For example, we know that

$$1 \text{ ft} = 12 \text{ in.}$$

Therefore,

$$\frac{1 \text{ ft}}{12 \text{ in.}} = 1$$

Similarly,

$$\frac{12 \text{ in.}}{1 \text{ ft}} = 1$$

If we wish to convert from inches to feet, we can do so by choosing one of the fractions as a **conversion factor.** Which one do we choose? The one that gives us an answer with the right units! Let's illustrate by an example.

EXAMPLE A.1

My bed is 72 in. long. What is its length in feet? You know the answer, of course, but let's see how the answer is obtained by using unit conversions.

SOLUTION

We need to multiply 72 in. by one of the fractions that relate inches to feet. Which one? The known quantity and unit is 72 in.

$$72 \text{ in.} \times \text{conversion factor} = ? \text{ ft}$$

For the conversion factor, choose the fraction that, when inserted into the equation, cancels the unit *inch* and leaves the unit *feet*.

$$72 \text{ in.} \times \frac{1 \text{ ft}}{12 \text{ in.}} = 6.0 \text{ ft}$$

Now let's try the other conversion factor.

$$72 \text{ in.} \times \frac{12 \text{ in.}}{1 \text{ ft}} = 860 \text{ in.}^2/\text{ft}$$

Absurd! How can a bed be 860 in.^{2}/ft? You should have no difficulty choosing between the two possible answers.

Please note that conversion factors are not usually given in a problem. Rather, they must be obtained from tables such as those in Appendix I. For the following examples, please refer to those tables for conversion factors that you don't know already.

It is possible (and frequently necessary) to manipulate units in the denominator as well as in the numerator of a problem. Just remember to use conversion factors in such a way that the unwanted units cancel.

EXAMPLE A.2

At a track meet a runner completes the mile in 4.0 min. How fast is this in miles per hour?

SOLUTION

The time for 1 mi is 4.0 min, or 1 mi per 4.0 min, or

$$\frac{1 \text{ mi}}{4.0 \text{ min}}$$

This is the known quantity and unit. The unit desired is miles per hour.

$$\frac{1 \text{ mi}}{4.0 \text{ min}} \times \text{conversion factor} = ? \frac{\text{mi}}{\text{hr}}$$

To change minutes to hours, we need the equivalence

$$1 \text{ hr} = 60 \text{ min}$$

Now arrange this equivalence as the fractions

$$\frac{1 \text{ hr}}{60 \text{ min}} \quad \text{and} \quad \frac{60 \text{ min}}{1 \text{ hr}}$$

Choose for your conversion factor the fraction that cancels the unit *minute* and leaves the unit *hour* in the denominator.

$$\frac{1 \text{ mi}}{4.0 \text{ min}} \times \frac{60 \text{ min}}{1 \text{ hr}} = \frac{15 \text{ mi}}{1 \text{ hr}} \text{(or 15 mi/hr)}$$

Problems may involve the use of several conversion factors, as in the following example.

EXAMPLE A.3

If your heart beats at a rate of 72 times per minute and your lifetime will be 70 years, how many times will your heart beat during your lifetime?

SOLUTION

Two equivalences are given in the problem.

$$72 \text{ beats} = 1 \text{ min}$$
$$1 \text{ lifetime} = 70 \text{ years}$$

Three others that you will need you can recall from memory.

$$1 \text{ year} = 365 \text{ days}$$
$$1 \text{ day} = 24 \text{ hr}$$
$$1 \text{ hr} = 60 \text{ min}$$

Start now with the factor 72 beats/1 min (the known quantity and unit) and apply conversion factors as needed to get an answer in beats/lifetime (the desired unit).

$$\frac{72 \text{ beats}}{1 \text{ min}} \times \frac{60 \text{ min}}{1 \text{ hr}} \times \frac{24 \text{ hr}}{1 \text{ day}} \times \frac{365 \text{ days}}{1 \text{ year}} \times \frac{70 \text{ years}}{1 \text{ lifetime}}$$
$$= 2,600,000,000 \text{ beats/lifetime}$$

Most problems that you will have to deal with will be much simpler than Example A.3.

Now let's do some conversions within the metric system. They are much easier.

EXAMPLE A.4

How many milliliters are there in a 2.0-L bottle of soda?

SOLUTION

From memory or from the tables in Appendix I you find that

$$1 \text{ L} = 1000 \text{ mL}$$

$$2.0 \text{ L} \times \frac{1000 \text{ mL}}{1 \text{ L}} = 2000 \text{ mL}$$

Notice that we picked the conversion factor that allowed us to cancel liters and obtain an answer in the desired unit, milliliters.

EXAMPLE A.5

In the United States, the usual soft drink can holds 360 mL. How many such cans could be filled from one 2.0-L bottle?

SOLUTION

The problem tells us that

$$1 \text{ can} = 360 \text{ mL}$$

Using that equivalence, we can calculate the answer.

$$2.0 \text{ L} \times \frac{1000 \text{ mL}}{1 \text{ L}} \times \frac{1 \text{ can}}{360 \text{ mL}} = 5.6 \text{ cans}$$

EXAMPLE A.6

How many 325-mg aspirin tablets can be made from 875 g of aspirin?

SOLUTION

The problem tells us that

$$1 \text{ tablet} = 325 \text{ mg}$$

From memory or the tables, we have

$$1 \text{ g} = 1000 \text{ mg}$$

$$875 \text{ g} \times \frac{1000 \text{ mg}}{1 \text{ g}} \times \frac{1 \text{ tablet}}{325 \text{ mg}} = 2690 \text{ tablets}$$

A.2
CONVERSIONS BETWEEN SYSTEMS

To convert from one system of measurement to another, you need a conversion table such as the one in Appendix I. Let's plunge right in and work some examples.

EXAMPLE A.7

You know that your weight is 140 lb, but the job application form asks for your weight in kilograms. What is it?

SOLUTION

From the table we find

$$1.00 \text{ lb} = 0.454 \text{ kg}$$

So your weight is

$$140 \text{ lb} \times \frac{0.454 \text{ kg}}{1.00 \text{ lb}} = 63.6 \text{ kg}$$

EXAMPLE A.8

A recipe calls for 750 mL of milk, but your measuring cup is calibrated in fluid ounces. How many ounces of milk will you need?

SOLUTION

$$750 \text{ mL} \times \frac{1.00 \text{ fl oz}}{29.6 \text{ mL}} = 25.3 \text{ fl oz}$$

EXAMPLE A.9

A doctor puts you on a diet of 5500 kilojoules (kJ) per day. How many kilocalories (food Calories) is that?

SOLUTION

$$5500 \text{ kJ} \times \frac{1000 \text{ J}}{1 \text{ kJ}} = \frac{1.000 \text{ kcal}}{4184 \text{ J}} = 1315 \text{ kcal}$$

EXAMPLE A.10

A sprinter runs the 100-m dash in 11 s. What is her speed in kilometers per hour?

SOLUTION

The given speed is 100 m per 11 s, giving us the fraction

$$\frac{100 \text{ m}}{11 \text{ s}}$$

The conversion factors that we need are found in the tables or recalled from memory.

$$\frac{100 \text{ m}}{11 \text{ s}} \times \frac{1 \text{ km}}{1000 \text{ m}} \times \frac{60 \text{ s}}{1 \text{ min}} \times \frac{60 \text{ min}}{1 \text{ hr}} = 33 \text{ km/hr}$$

Note that the first conversion factor changes meters to kilometers. It takes two factors to change seconds to hours. Note also that we could have first applied the factors that convert seconds to hours and then converted meters to kilometers. The answer would be the same.

A.3
DENSITY PROBLEMS

Almost any type of problem can be solved using dimensional analysis. Let's try some involving density.

EXAMPLE A.11

A bottle filled to the 25-mL mark contains 75 g of bromine. What is the density of bromine?

SOLUTION

If you know that density is ordinarily reported in grams per milliliter (Chapter 1), then you know how to set up the problem. Arrange the data so the answer will come out in grams per milliliter. The density of bromine is

$$\frac{75 \text{ g}}{25 \text{ mL}} = 3.0 \text{ g/mL}$$

EXAMPLE A.12

The density of alcohol is 0.79 g/mL. What is the mass (weight) of 150 mL of alcohol?

SOLUTION

We know that mass is measured in grams, so we solve the problem this way.

$$150 \, \text{mL} \times \frac{0.79 \, \text{g}}{1.0 \, \text{mL}} = 120 \, \text{g}$$

EXAMPLE A.13

What volume is occupied by 451 g of mercury? The density of mercury is 13.6 g/mL.

SOLUTION

This time we want an answer in units of volume, so we set it up this way.

$$451 \, \text{g} \times \frac{1.00 \, \text{mL}}{13.6 \, \text{g}} = 33.2 \, \text{mL}$$

You can always be reasonably confident of your answer if it has the proper units.

We will use unit conversions to solve other problems as we proceed through subsequent chapters. For now, though, you should improve your proficiency by working through the following exercises.

EXERCISES

1. How many meters are there in each of the following?
   a. 3.4 km   b. 570 cm
   c. 4.76 in.   d. 100 yd
2. How many liters are there in each of the following?
   a. 3300 mL   b. 51 dL
   c. 51 qt   d. 81 gal
3. Make the following metric conversions.
   a. 1.29 kg to g   b. 1575 mg to g
   c. 0.421 g to mg   d. 255 cm to m
   e. 1.83 cm to mm   f. 4.22 L to mL
4. Make the following conversions. (You may refer to the conversion tables in Appendix I when necessary.)
   a. 16.4 in. to cm b. 4.17 qt to L
   c. 1.61 kg to lb   d. 9.34 g to oz
   e. 2.05 fl oz to mL f. 105 lb to g
5. Football player William "The Refrigerator" Perry weighs 310 lb. What is his weight (mass) in kilograms?
6. Basketball player Manute Bol is 7.5 ft tall. What is his height in meters? In centimeters?
7. How many pounds are there in 144 kg?
8. How many centimeters are there in 18.4 in.?
9. State your weight in pounds. Calculate your weight (mass) in kilograms.
10. State your height in feet. Calculate your height in centimeters.
11. An aspirin tablet has a mass of 325 mg. What is its mass in grams? In ounces?
12. How would you describe a young man who is 160 cm tall and has a mass of 90 kg?
13. Convert 12 km/min to miles per minute.
14. Convert 5.8 lb/gal to kilograms per liter.
15. Convert 45 mi/hr to kilometers per hour.

16. The speed limit in the Saskatchewan countryside is 88 km/hr. What is the speed in miles per hour?
17. The speed of light is 186,000 mi/s. What is this speed in meters per second?
18. The density of gasoline is 5.9 lb/gal. What is its density in grams per milliliter?
19. Milk costs $2.09 per gal and 59¢ per L. Which is cheaper?
20. One gram of fat provides 9 kcal of energy. How many kilocalories are furnished by 1 lb of fat?
21. According to the *Guinness Book of Records,* the heaviest recorded man in Great Britain weighed 53 stone, 8 lb (1 stone = 14 lb). What was his weight in pounds?
22. When a service station in Toronto converted to the metric system, the price of gasoline changed from 95.9¢ per imperial gal to 21.1¢ per L. Did the price increase, decrease, or remain unchanged? (There are 4.545 L in 1.000 imperial gal.)
23. A slug expends about 0.10 mL of mucus per 1.0 m crawled. (It is necessary for the slug to generate mucus continually in order to move.) Calculate the slug's progress in miles per gallon.
24. The food Calorie is actually 1000 cal (1 kcal). In SI, we measure our food energy intake in kilojoules (1.00 kcal = 4.18 kJ). How many kilojoules of energy are there in a can of soda that has 165 Cal (165 kcal)?

Note: Use unit conversions (rather than the formulas given in Chapter 1) to solve Exercises 25 and 26.

25. The density of iron is 7.86 g/cm³. What is the mass of a piece of iron that has a volume of 80.5 cm³?
26. What volume is occupied by 55.5 g of chloroform? The density of chloroform is 1.48 g/mL.

Chapter 2
ATOMS

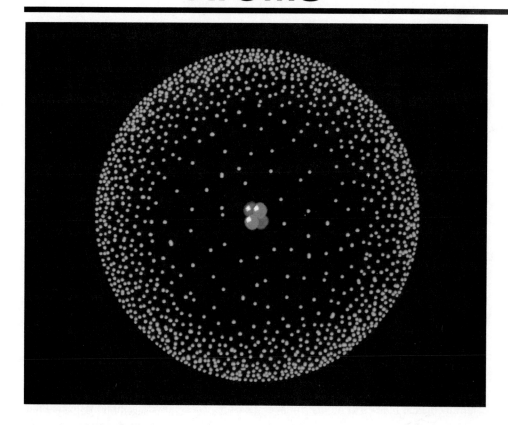

A computer-generated graphic of the helium atom. Two protons and two neutrons are contained within the nucleus, and these are surrounded by a "cloud" of electrons.

An **atom** is defined as the smallest characteristic particle of an element. This concept has been around for thousands of years. For most of that time, however, it was not a popular idea. The development of chemistry as a science was largely a matter of accepting the existence of atoms and then working to define the properties of the atom more and more precisely.

As originally proposed by both ancient Hindu and Greek scholars, the concept of the atom was as philosophical as it was practical. Both groups spoke of earth, water, fire, and air as elements composed of eternal and unchanging atoms in perpetual motion. For the Hindus, it was Brahma, the soul of the universe, who set the atoms in motion. Some Greek philosophers believed that atoms were in motion because they were falling toward the center of the universe. The atomistic view of matter fell into disfavor when the greatest of the Greek philosophers, Aristotle, elected to side with those who believed that matter was continuous, that is, infinitely divisible. According to Aristotle and his followers, there was no particle of matter so small that it could not be subdivided into still smaller pieces. It is a tribute to Aristotle that his view prevailed for 2000 years.

A gradual accumulation of data that could not be explained within Aristotle's philosophy finally forced a reevaluation of his views. The reassessment and ultimate rejection of

Atom—the smallest chemically obtainable portion of an element

Aristotle's concept of the nature of matter was a major victory for experimental science. From that point onward, scientific theories were accepted or rejected on the basis of their ability to explain experimental data and not because of their beauty or elegance or even their appeal to common sense.

2.1
DALTON'S ATOMIC THEORY

The man whose scientific theory did explain all the available experimental data was John Dalton (1766–1844), an English schoolteacher (Figure 2.1). Dalton accepted the atomistic view of matter and presented the modern atomic theory. The theory is called "modern" because it was based on the best evidence available in Dalton's time (the early nineteenth century) rather than on ideas formulated in Aristotle's time (fourth century B.C.). The most important points of Dalton's atomic theory are

1. All elements are made up of small, indestructible, and indivisible particles called atoms.
2. All atoms of a given element are identical, but the atoms of one element differ from the atoms of any other element.
3. Atoms of different elements can combine to form compounds.
4. A chemical reaction involves a change not in the atoms themselves, but in the way atoms are combined to form compounds.

What did Dalton mean when he said that all atoms of a given element are the same? Were the atoms of an element all the same color or the same shape? Rather—and this was Dalton's most important insight—he proposed that the mass of an atom identifies it uniquely. Atoms are extremely minute. In the nineteenth century, it was impossible to determine actual masses of atoms. Indirect measurements, however, could indicate their

Figure 2.1
John Dalton.

relative masses. Dalton set up a table of relative masses for the elements based on hydrogen, the lightest element. He believed that an oxygen atom had a mass six times the mass of a hydrogen atom.

Dalton was wrong. He was wrong about the mass of oxygen atoms. He was wrong about all atoms of the same element being identical. He was even wrong about atoms being indivisible and indestructible. The most amazing thing about Dalton is that he was wrong about so many things, yet he is still regarded, and rightly so, as one of the giants of modern chemistry. Other scientists took up his ideas and modified them, and a new era in chemistry began. Science is not simply an accumulation of correct bits of information; it is the gradual development of a model for our universe, a model that enables us to understand the workings of nature. Ideas are important if they help us understand. Some ideas have to be corrected or modified as our understanding increases, but they are nonetheless valuable. Without them it might not be possible to advance to a higher level of understanding. This situation is not unique to chemistry. Accepted medical treatment of 50 (to say nothing of 100) years ago is likely to be unacceptable today. Yet such now outdated treatment did once keep people alive and helped medical researchers develop better, safer, more effective procedures.

These relative masses are now expressed in terms of **atomic mass units (amu)**. This unit of mass was invented just for atoms.

2.2
THE NUCLEAR ATOM

Although Dalton may have erred in some details, he started us along the right path. If one begins with the premise that atoms exist, then it is possible to devise experiments to find out more about these atoms. In the century following the work of Dalton, many such experiments were conducted. Data continued to accumulate, and many of the results could not be explained by Dalton's atomic theory. An international array of scientists (see Figure 2.2) were discovering that the atom was not indivisible. It could still be regarded as the smallest characteristic unit of an element, but it certainly was not the smallest particle of matter.

(a)   (b)   (c)   (d)   (e)

Figure 2.2

(a) Sir William Crookes, a British scientist who invented the cathode-ray tube. (b) Antoine Henri Becquerel, a French physicist whose experiments showed that certain atoms emit subatomic particles, some of which are positive and some negative. (c) Marie Curie, the Polish scientist who named the phenomenon of radioactivity. (d) Ernest Rutherford, a British scientist who found that atoms are made up mostly of empty space. (e) James Chadwick, a student of Rutherford, who discovered the neutron.

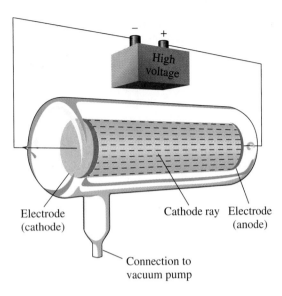

Electrode   Cathode ray   Electrode
(cathode)   (anode)

Connection to
vacuum pump

Figure 2.3
A simple discharge tube.

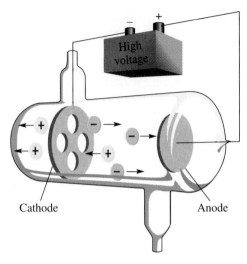

Cathode   Anode

Figure 2.4
When a gas is in the discharge tube, positive
particles are formed.

Electron—a subatomic particle with a negative charge and
essentially no mass

Proton—a subatomic particle
with a positive charge; 1837
times as massive as an electron

Certain atoms naturally decay
and release subatomic particles.

The studies of a German physicist, Eugen Goldstein, and an English physicist, William Crookes, demonstrated that atoms could be torn apart. In an apparatus called a
discharge tube (Figure 2.3), electric energy was used to kick small pieces from atoms of
the metal making up one electrode. The tiny *subatomic* particles were shown to be negatively charged and could be detected as a beam of particles (called a **cathode ray**) that
crossed from one electrode in the apparatus to the other. No matter what element the
electrode was made of, identical negatively charged subatomic particles were kicked out
to form the cathode ray. These particles were named **electrons.**

When a small amount of gas was admitted into the tube, the cathode-ray particles
struck the atoms of the gas, knocking more electrons loose from these atoms (Figure 2.4).
Those atoms that lost negatively charged electrons were left with a positive charge. Although all the negatively charged particles were identical, the positively charged particles
differed depending on what gas was in the tube. For example, helium produced more
massive positively charged particles than did hydrogen. The lightest positive particles
found were those from hydrogen and were given a special name, **protons.** The charges on
the proton and the electron are the same size (although opposite in sign), but the proton
was found to be 1837 times as massive as the electron. Thus, the smallest positive particle
isolated was many times heavier than the negatively charged electron.

A French physicist, Antoine Henri Becquerel, discovered that some atoms fall apart
all by themselves. Atoms of the element uranium, for example, emitted ''rays,'' some of
which proved to be subatomic particles that were a fraction of the size of the original
atoms (Figure 2.5). Some of the emitted particles were negatively charged and were
shown to be identical to the electrons formed in a cathode-ray tube. Uranium also emitted
positively charged particles much smaller than the uranium atom itself but much more
massive than the electrons. A Polish-born chemist, Marie Sklodowska Curie, named these
phenomena **radioactivity** (see Chapter 3).

A British scientist, Ernest Rutherford, used radioactive atoms as ''atomic guns,''
with the tiny emitted particles playing the role of bullets (Figure 2.6). By shooting such
bullets at other atoms, he discovered that the target atoms were mostly empty space. Most

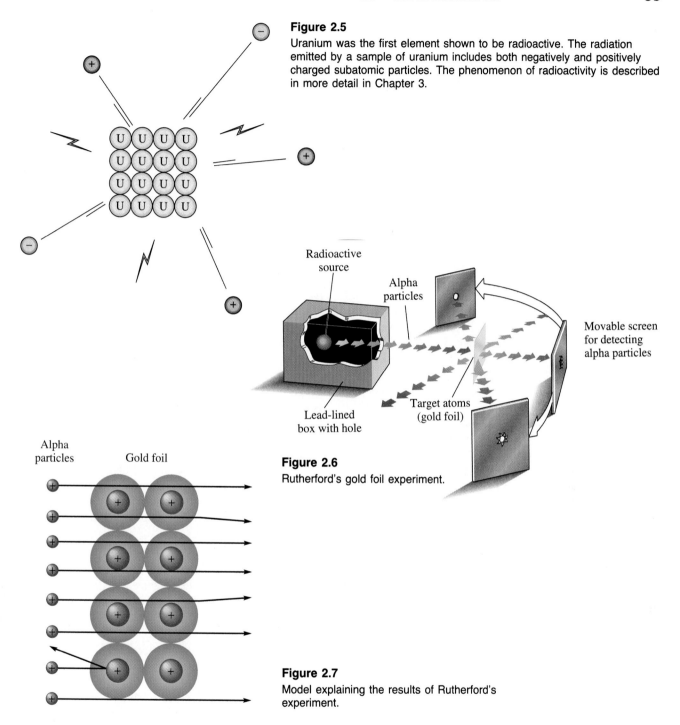

Figure 2.5
Uranium was the first element shown to be radioactive. The radiation emitted by a sample of uranium includes both negatively and positively charged subatomic particles. The phenomenon of radioactivity is described in more detail in Chapter 3.

Radioactive source

Alpha particles

Movable screen for detecting alpha particles

Lead-lined box with hole

Target atoms (gold foil)

Figure 2.6
Rutherford's gold foil experiment.

Alpha particles   Gold foil

Figure 2.7
Model explaining the results of Rutherford's experiment.

of the bullets passed right through the target atoms. Some of the bullets were deflected, however, suggesting that there was a concentrated bit of "solid" material in the atom (Figure 2.7).

A student of Rutherford, James Chadwick, discovered a subatomic particle unlike those previously found. This one, called a **neutron,** was neutral, being neither positively nor negatively charged.

Radioactivity—the emission by atoms of subatomic particles

Table 2.1
Subatomic Particles

Particle	Symbol[a]	Approximate Mass (amu)	Charge	Location in Atom
Proton	p^+	1	$1+$	Nucleus
Neutron	n	1	0	Nucleus
Electron	e^-	0	$1-$	Outside nucleus

[a]The superscripts are often omitted. We will keep them to remind us of the charge of each particle.

Neutrons are similar in mass to protons but have no charge.

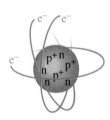

Figure 2.8
The lithium atom.

Atomic number = number of protons

An element's isotopes have the same number of protons but different numbers of neutrons.

Early in this century a new model of the atom was proposed to account for all these new data. According to this model there are three types of subatomic particles: protons, neutrons, and electrons. A proton and a neutron have virtually the same mass, 1.007276 amu and 1.008665 amu, respectively. This is equivalent to saying that two different people weigh 100.7 lb and 100.9 lb; the difference is so small that it can be ignored. Thus, for many purposes, we assume that the masses of the proton and the neutron are 1 amu each. The proton has a charge equal in magnitude but opposite in sign to that of an electron. This charge on a proton is written as $1+$. The electron has a charge of $1-$ and a mass of 0.000549 amu. The electrons in an atom contribute so little to its total mass that their mass is usually disregarded—electrons are treated as if they have zero mass.[1] These properties of the subatomic particles are summarized in Table 2.1.

When combined to form an atom, the more massive particles (protons and neutrons) are crowded into the **nucleus,** a tiny volume of space at the very center of the atom (Figure 2.8) The electrons are scattered throughout the remaining volume of the atom. To visualize how ''empty'' an atom is, picture a balloon that stands 10 stories high. If the balloon were an atom, then its nucleus would be the size of a BB. The rest of the space in the balloon-atom is the domain of the electrons, whose mass is smaller than that of the nucleus by a factor of thousands.

The number of protons in the nucleus of an atom of any element is equal to the **atomic number** of that element. This number determines the kind of atom, that is, the identity of the element. Dalton said that the mass of an atom determines the element. We now say it is not the mass but the number of protons that determines the element. For example, an atom with 26 protons (one whose atomic number is 26) is an atom of iron (Fe). An atom with 92 protons is an atom of uranium (U).

The number of neutrons in the nuclei of atoms of a given element may vary. For example, most hydrogen (H) atoms have a nucleus consisting of a single proton and no neutrons (and therefore a mass of 1 amu). About 1 hydrogen atom in 5000, however, does have a neutron as well as a proton in the nucleus. This heavier hydrogen is called **deuterium** and has a mass of 2 amu. Both kinds are hydrogen atoms (any atom with atomic number 1, that is, with one proton, is a hydrogen atom). Atoms that have this sort of relationship—the same number of protons but different numbers of neutrons—are called **isotopes** (Figure 2.9). A third, rare isotope of hydrogen is **tritium,** which has two neutrons and one proton in the nucleus (and thus a mass of 3 amu). Most, but not all, elements exist in nature in isotopic forms. This fact also requires a major modification of Dalton's original theory. He said that all atoms of the same element have the same mass. We now know that most elements have several isotopes, that is, atoms with different numbers of neutrons and therefore different masses.

The number of electrons in an atom equals the number of protons. Therefore, positively and negatively charged particles balance in an atom. The atom as a whole is neutral.

[1]To emphasize how small subatomic particles are, we list here their masses in grams: proton = 1.673 × 10^{-24} g (0.0000000000000000000001673 g); neutron = 1.675 × 10^{-24} g; electron = 9.107 × 10^{-28} g. (Appendix C reviews exponential notation.)

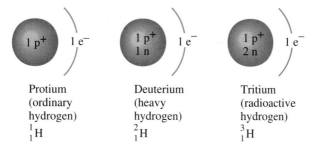

Figure 2.9
The isotopes of hydrogen.

Protium
(ordinary
hydrogen)
$_1^1 H$

Deuterium
(heavy
hydrogen)
$_1^2 H$

Tritium
(radioactive
hydrogen)
$_1^3 H$

2.3
NUCLEAR ARITHMETIC

To represent the different isotopes of the elements, symbols with subscripted and superscripted numbers are used. In the generalized symbol

$$_Z^A X$$

the subscript Z represents the **atomic number,** that is, the number of protons. The superscript A represents the **mass number,** that is, the number of protons plus the number of neutrons. From the symbol

neutrons + protons
$$_{17}^{35} Cl$$
protons

we know that the number of protons in a chlorine atom is 17. The number of neutrons is $35 - 17 = 18$.

Atomic number, Z = number of protons
Mass number, A = number of protons + number of neutrons

EXAMPLE 2.1

Write the symbol for an isotope with a mass number of 58 and an atomic number of 27 and identify the element.

SOLUTION
In the general symbol $_Z^A X$, A is the mass number, which is 58 in this case.

$$_Z^{58} X$$

Z is the atomic number, or 27.

$$_{27}^{58} X$$

From the periodic table we can determine that the element with atomic number 27 is cobalt.

$$_{27}^{58} Co$$

Practice Exercise
Write the symbol for an isotope with a mass number of 90 and an atomic number of 42.

EXAMPLE 2.2

Indicate the number of protons, neutrons, and electrons in a neutral atom of the isotope $^{235}_{92}U$.

SOLUTION

The atomic number gives the number of protons and electrons in a neutral atom of the isotope.

$$\text{Atomic number} = \text{protons} = \text{electrons} = 92$$

The number of neutrons is obtained by subtracting the atomic number from the mass number.

Mass number	235
Atomic number	− 92
Number of neutrons	143

Practice Exercise

Indicate the number of protons, neutrons, and electrons in a neutral atom of the isotope $^{90}_{38}Sr$.

EXAMPLE 2.3

Which of the following (a) are isotopes of the same element; (b) have identical mass numbers; (c) have the same number of neutrons? We are using the letter X as the symbol for all so that the symbol will not identify the elements.

$$^{16}_{8}X \qquad ^{16}_{7}X \qquad ^{14}_{7}X \qquad ^{14}_{6}X \qquad ^{12}_{6}X$$

SOLUTION

a. $^{16}_{7}X$ and $^{14}_{7}X$ are isotopes of the element nitrogen (N). $^{14}_{6}X$ and $^{12}_{6}X$ are isotopes of the element carbon (C).

b. $^{16}_{8}X$ and $^{16}_{7}X$ have the same mass number. The first is an isotope of oxygen, and the second is an isotope of nitrogen. $^{14}_{7}X$ and $^{14}_{6}X$ also have the same mass number. The first is an isotope of nitrogen, and the second is an isotope of carbon.

c. $^{16}_{8}X$ (16 − 8 = 8 neutrons) and $^{14}_{6}X$ (14 − 6 = 8 neutrons) have the same number of neutrons.

Practice Exercise

Which of the following are isotopes of the same element?

$$^{90}_{37}X \qquad ^{90}_{38}X \qquad ^{88}_{37}X \qquad ^{88}_{36}X \qquad ^{93}_{38}X$$

A specific isotope can be designated by its nuclear symbol (e.g., $^{12}_{6}C$) or by the element name followed by the mass number (e.g., carbon-12).

The atoms of the carbon-12 isotope are defined as having a mass of 12 amu. The relative masses of all other atoms are determined by comparing them to this standard. An atom of the isotope $^{1}_{1}H$ has one-twelfth the mass of an atom of $^{12}_{6}C$; thus, atoms of $^{1}_{1}H$ have an atomic mass of 1 amu. Atoms of deuterium ($^{2}_{1}H$) have one-sixth the mass of atoms of $^{12}_{6}C$; the atomic mass of deuterium atoms is 2 amu.

The atomic weight of an element, as listed in the periodic table, is related to the atomic mass in the following way. The **atomic weight** is the average of the masses of the atoms in a representative sample of an element. Each atom in the sample has a particular atomic mass. For example, in a sample of the element bromine, about half the atoms have a mass number of 79 amu and half have a mass number of 81 amu. The atomic weight of bromine is about 80 amu, the average of 79 amu and 81 amu. (The atomic weight of bromine is listed in the periodic table as 79.9 amu, indicating that there is slightly more of the bromine-79 isotope than of the bromine-81.) Another example: About three-fourths of the atoms of the element chlorine have a mass number of 35 amu, and about one-fourth have a mass number of 37 amu. The atomic weight of chlorine is 35.5 amu; the average mass in this case is much closer to that of the chlorine-35 isotope because there is more of the isotope in the sample (Figure 2.10).

An atom of carbon-12 is defined to have a mass of exactly 12 amu. The masses of the other atoms are compared to that of carbon-12.

Atomic masses (in amu) of atoms in sample

```
 81
 79
 79
 81
 79
 79
 81
 81
———
640 amu = total mass of sample
```

$$\text{Average mass} = \frac{640 \text{ amu}}{8 \text{ atoms}} = 80 \text{ amu} = \text{atomic weight}$$

Sample
of bromine

Figure 2.10

Atomic weights of elements are averages of isotopic masses.

Atomic masses (in amu) of atoms in sample

```
 35
 37
 35
 35
 35
 37
 35
 35
———
284 amu = total mass of sample
```

$$\text{Average mass} = \frac{284 \text{ amu}}{8 \text{ atoms}} = 35.5 \text{ amu} = \text{atomic weight}$$

Sample
of chlorine

Figure 2.11

The colors of a fireworks display result from changes in electron energy levels in atoms.

2.4
THE BOHR MODEL OF THE ATOM

Have you ever watched colored fireworks (Figure 2.11) or thrown chemicals into a fireplace to color the flames? If you have, then you have seen a phenomenon that puzzled scientists and triggered another modification of our concept of the atom. Light is pure energy. Light of different color is light of different energy. For example, blue light packs more energy than red light. If you throw compounds containing lithium into a fire, you see the flames colored red. With sodium compounds (like ordinary table salt), the light is yellow-orange; with strontium, red; with potassium, lavender (Figure 2.12). Why? That is what early-twentieth-century scientists asked themselves. A Danish

Figure 2.12

Certain chemical elements can be identified by the characteristic colors that their compounds impart to flames. Five examples are shown.

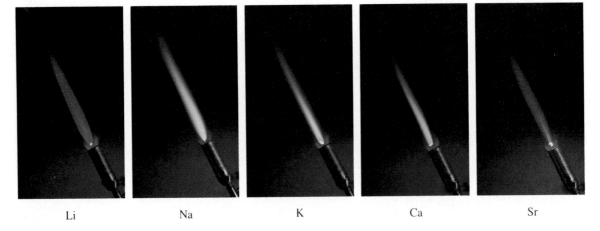

Li   Na   K   Ca   Sr

Figure 2.13
Niels Bohr.

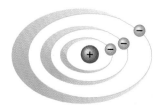

Figure 2.14
Bohr visualized the atom as
planetary electrons circling
a nuclear sun.

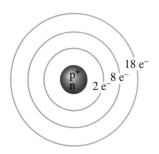

Figure 2.15
A Bohr diagram for the
atoms of an element pic-
tures specified numbers of
electrons in distinct energy
levels.

physicist, Niels Bohr (Figure 2.13), provided the answer. He said that electrons cannot be
located just anywhere outside the nucleus. They must move around only in well-defined
paths or orbits (Figure 2.14). An electron in one of these orbits has a characteristic amount
of energy (a certain amount of kinetic energy because of its motion around the orbit and a
certain amount of potential energy because it is located a certain distance from the nucleus). If an electron changes from one orbit to another, its energy also changes. If it
moves from a higher energy orbit to a lower energy orbit, it loses energy, and that lost
energy appears as light. The color of the light depends on the difference in energy between
the two orbits. Atoms in which all electrons are in their lowest possible **energy levels** are
said to be in the *ground state*. If one or more electrons occupy higher energy levels than in
the ground state, the atom is in an *excited state*.

Bohr also said that different orbits within an atom can handle only a certain number
of electrons (Figure 2.15). We shall simply state Bohr's findings in this regard. The
maximum number of electrons that can be in a given energy level is given by the formula
$2n^2$, where n is equal to the energy level being considered. For the first energy level
($n = 1$), the maximum population is $2(1^2)$, or 2. For the second energy level ($n = 2$), the
maximum number of electrons is $2(2^2)$, or 8. For the third level, the maximum is $2(3^2)$,
or 18.

Energy levels—regions outside the atomic nucleus in
which electrons have different
energies

EXAMPLE 2.4

What is the maximum number of electrons in the fifth energy level?

SOLUTION
For the fifth level, $n = 5$, so we have

$$2 \times 5^2 = 2 \times 25 = 50$$

Practice Exercise
What is the maximum number of electrons in the fourth energy level?

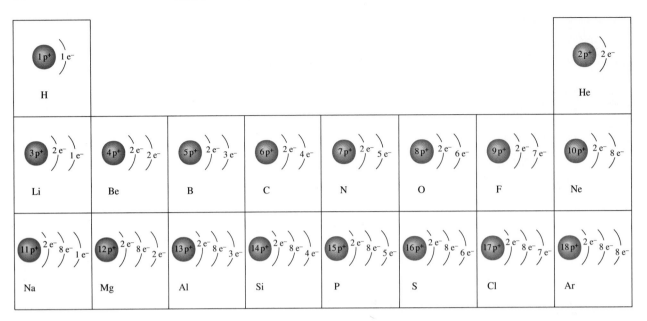

Figure 2.16
Bohr diagrams of the first
18 elements.

Imagine building up atoms by adding one electron to the proper energy level as protons are added to the nucleus. Electrons will go to the lowest energy level available. For hydrogen (H), with a nucleus of only one proton, the one electron goes into the first energy level. For helium (He), with a nucleus of two protons (and two neutrons), two electrons go into the first energy level. According to Bohr, the maximum population of the first energy level is two electrons; thus, that level is filled in the helium atom. With lithium (Li), which has three electrons, the extra electron goes into the second energy level. This process of adding electrons is continued until the second energy level is filled with eight electrons, as in the neon (Ne) atom (which has a total of 10 electrons, two in the first level and eight in the second). This buildup is diagrammed in Figure 2.16. One could continue to build atoms (in one's imagination) in this manner and, with a few modifications, build up the entire collection of known elements.

EXAMPLE 2.5

Draw a Bohr diagram for fluorine (F).

SOLUTION
Fluorine is element number 9; it has nine protons and nine electrons. Two of the electrons go into the first energy level, the remaining seven into the second.

$$\text{(F)} \quad 2\,e^- \quad 7\,e^-$$

EXAMPLE 2.6

Draw a Bohr diagram for sodium (Na).

SOLUTION

Sodium is element number 11. It has 11 protons and 11 electrons. Two of the electrons can go into the first energy level. Of the nine remaining electrons, eight go into the second energy level. That leaves one electron for the third energy level.

$$\left(Na\right) \quad 2\,e^- \quad 8\,e^- \quad 1\,e^-$$

2.5
ELECTRON ARRANGEMENTS

The Bohr model of the atom has been replaced for many purposes by more sophisticated models that explain more data in greater detail than does the simpler planetary model of Bohr. This presents us with something of a quandary. We can more accurately interpret the nature of matter only by using a model that is more difficult to understand. Fortunately, however, we need not understand the laborious mathematics that generate these models to make use of some of the results.

Quantum mechanics is a highly mathematical discipline used in the late 1920s by the Austrian physicist Erwin Schrödinger to develop elaborate equations that describe the properties of electrons in atoms. The results provide a measure of the probability of finding an electron in a given volume of space. The definite planetary orbits of the Bohr model are replaced by shaped volumes of space in which electrons move. The term **orbital** (replacing Bohr's *orbits*) was used for this new description of the location of electrons.

Orbital—the defined volume of an atom where an electron exists

Suppose you had a camera that could photograph electrons (there is no such thing, but we are just supposing) and you left the shutter open while an electron zipped about the nucleus. When you developed the picture, you would have a record of where the electron had been. Doing the same thing with an electric fan that was turned on would give you a picture in which the blades of the fan look like a disk of material. The blades move so rapidly that their photographic image is blurred. Similarly, electrons in the first energy level of an atom would appear in our imaginary photograph as a fuzzy ball (Figure 2.17).

Figure 2.17
Charge-cloud representations of atomic orbitals.

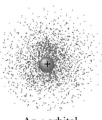

An *s* orbital

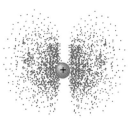

A *p* orbital

Figure 2.18

Electron orbitals. In these drawings the nucleus of the atom is located at the intersection of the axes. The eight electrons that would be placed in the second energy level of Bohr's model are distributed among these four orbitals in the current model of the atom, two electrons per orbital.

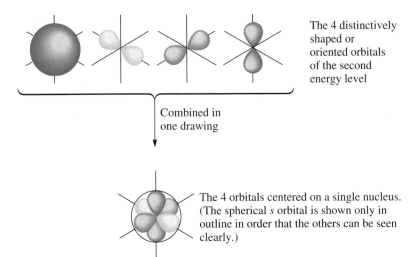

The 4 distinctively shaped or oriented orbitals of the second energy level

Combined in one drawing

The 4 orbitals centered on a single nucleus. (The spherical *s* orbital is shown only in outline in order that the others can be seen clearly.)

An orbital can hold no more than two electrons.

The fuzzy ball (frequently referred to as a *charge cloud* or *electron cloud*) is the rough equivalent of an orbital.

Like Bohr, Schrödinger concluded that only two electrons could occupy the first energy level in an atom. In the quantum mechanical atom, this level is referred to as the 1*s* orbital (which is spherical, that is, shaped like a ball). Also like Bohr, Schrödinger stated that the second energy level of an atom could contain a maximum of eight electrons. However, Schrödinger concluded that these eight electrons must be located in four different orbitals. Each individual orbital could contain a maximum of two electrons. One of the orbitals of the second energy level is spherical in shape and is referred to as the 2*s* orbital. The remaining three orbitals of the second level have identical dumbbell shapes (Figure 2.17). They differ in their orientation in space (i.e., the direction in which they point; see Figure 2.18). As a group these orbitals are called the 2*p* orbitals, and to distinguish them from one another they are individually referred to as the $2p_x$, $2p_y$, and $2p_z$ orbitals. Electrons in these three orbitals all lie at the same energy level and possess slightly more energy than the electrons in the 2*s* orbital. Thus, the quantum mechanical atom distinguishes between main energy levels (1, 2, 3, and so on) and sublevels (2*s* and 2*p*, for example). See Tables 2.2 and 2.3.

Table 2.2

Energy Levels and Orbitals

Energy Level	Orbital(s)
1	*s*
2	*s, p*
3	*s, p, d*

Table 2.3

Electrons in Orbitals

Orbital Type	Orientation	Number of Orbitals	Electrons per Orbital	Total Number of Electrons
s	Spherical	1	2	2
p	Perpendicular	3	2	6
d	Perpendicular	5	2	10

The third energy level of Bohr's model, and of Schrödinger's, could hold 18 electrons. In the quantum mechanical atom, however, the third main energy level is divided into three sublevels totaling nine orbitals: one 3*s* orbital, three 3*p* orbitals, and five 3*d* orbitals. Each higher main energy level adds another sublevel. The orbitals in each new sublevel have their own special shapes.

Just as one could build up atoms for all the elements by fitting electrons into the orbits of the Bohr model of the atom, so it is possible to construct a table of the elements

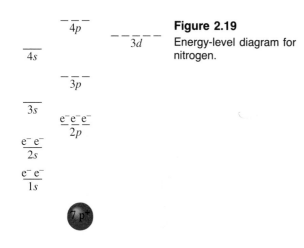

Figure 2.19
Energy-level diagram for nitrogen.

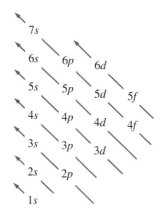

Figure 2.20
An order-of-filling chart for determining the electron configuration of atoms. The first energy level has only s orbitals, the second has s and p orbitals, the third has s, p, and d orbitals, and so on.

by placing electrons into the orbitals of the quantum mechanical model. The energy-level diagram for nitrogen (which has seven electrons) is shown in Figure 2.19. The quantum mechanical description of the nitrogen atom focuses on a more detailed description of its **electron configuration** (arrangement): $1s^2 2s^2 2p^3$. The superscripts indicate the total number of electrons contained in a particular energy sublevel. For example, in nitrogen there are a total of three electrons in the $2p$ orbitals. Table 2.4 lists the electron configurations of the first 20 elements. Notice that the orbitals fill in order of increasing energy: first the $1s$, then the $2s$, $2p$, $3s$, and so on. The d orbitals introduce a minor complication because the $3d$ orbitals turn out to be at a higher energy level than the $4s$ orbitals. Figure 2.20 illustrates the order in which orbitals are filled.

Electron   configuration—the specific ordering of electrons in an atom

Name	Atomic Number	Electron Structure
Hydrogen	1	$1s^1$
Helium	2	$1s^2$
Lithium	3	$1s^2 2s^1$
Beryllium	4	$1s^2 2s^2$
Boron	5	$1s^2 2s^2 2p^1$
Carbon	6	$1s^2 2s^2 2p^2$
Nitrogen	7	$1s^2 2s^2 2p^3$
Oxygen	8	$1s^2 2s^2 2p^4$
Fluorine	9	$1s^2 2s^2 2p^5$
Neon	10	$1s^2 2s^2 2p^6$
Sodium	11	$1s^2 2s^2 2p^6 3s^1$
Magnesium	12	$1s^2 2s^2 2p^6 3s^2$
Aluminum	13	$1s^2 2s^2 2p^6 3s^2 3p^1$
Silicon	14	$1s^2 2s^2 2p^6 3s^2 3p^2$
Phosphorus	15	$1s^2 2s^2 2p^6 3s^2 3p^3$
Sulfur	16	$1s^2 2s^2 2p^6 3s^2 3p^4$
Chlorine	17	$1s^2 2s^2 2p^6 3s^2 3p^5$
Argon	18	$1s^2 2s^2 2p^6 3s^2 3p^6$
Potassium	19	$1s^2 2s^2 2p^6 3s^2 3p^6 4s^1$
Calcium	20	$1s^2 2s^2 2p^6 3s^2 3p^6 4s^2$

Table 2.4
Electron Structures for Atoms of the First 20 Elements

To check whether you understand the fundamentals of orbitals and electron configurations, work through the following examples.

EXAMPLE 2.7

Write the electron configuration for carbon, atomic number 6.

SOLUTION

The atomic number indicates that there are six electrons in the carbon atom. The first energy level can accommodate only two of these electrons in its $1s$ orbital.

$$1s^2. . .$$

The remaining four electrons must occupy orbitals in the second main energy level. The lowest energy orbital of the second level is the $2s$ orbital, which can accommodate two electrons.

$$1s^2 2s^2. . .$$

The remaining two electrons must go to the $2p$ orbitals, which can handle up to six electrons if necessary.

$$1s^2 2s^2 2p^2$$

Practice Exercise

Write out the electron configuration for fluorine (F).

EXAMPLE 2.8

What is the electron configuration for phosphorus, atomic number 15?

SOLUTION

We have 15 electrons to distribute. Two go into the first main energy level ($1s^2$), eight go into the second ($2s^2 2p^6$), and five go into the third ($3s^2 3p^3$). In summary, the configuration of phosphorus is

$$1s^2 2s^2 2p^6 3s^2 3p^3$$

Practice Exercise

Write out the electron configuration for chlorine (Cl).

When an atom gains or loses electrons, it becomes charged and is referred to as an **ion.** Thus, if an atom of lithium were to lose an electron, its electron configuration would be $1s^2 2s^0$. It would then be Li^+. In the same way, if an atom of fluorine were to gain an electron, it would become F^- with the electron configuration $1s^2 2s^2 2p^6$. More on ions in Chapter 4.

Quantum mechanics offers to scientists detailed descriptions of the electron structure of atoms. However, we shall make most use of the picture it paints of electrons as clouds of negative charge. Sometimes the shape of the cloud (that is, the orbital) is presented simply in outline. (Figure 2.18 uses shaded drawings to present the combined orbitals of the second energy level.) Later we shall see how the shape and orientation of the electron cloud can determine the shape of what we call molecules.

2.6
THE PERIODIC TABLE

As our picture of the atom becomes more detailed, we find ourselves in a dilemma. With more than 100 elements to deal with, how can we keep all this information straight? One way is by using the **periodic table of the elements.** (See the inside front cover of this book.) The periodic table neatly tabulates information about atoms. It records how many protons and electrons the atoms of a particular element contain. It permits us to calculate the number of neutrons in the most common isotope for most elements. It even stores information about how electrons are arranged in the atoms of each element. The most extraordinary thing about the periodic table is that it was largely developed before anyone knew there were protons or neutrons or electrons in atoms.

Not long after Dalton presented his model for the atom (an indivisible particle whose mass determined its identity), chemists began preparing listings of elements arranged according to their atomic weights. While working out such tables of elements, these scientists observed patterns among the elements. For example, it became clear that elements that occurred at specific intervals shared a similarity in certain properties. Among the approximately 60 elements known at that time, the second and ninth showed similar properties, as did the third and tenth, the fourth and eleventh, the fifth and twelfth, and so on.

In 1869, Dmitri Ivanovich Mendeleev (Figure 2.21), a Russian chemist, published his periodic table of the elements. Mendeleev prepared his table by taking into account both the atomic weights and the periodicity of certain properties of the elements. The elements were arranged primarily in order of increasing atomic weight. In a few cases, Mendeleev placed a slightly heavier element before a lighter one. He did this only when it was necessary in order to keep elements with similar chemical properties in the same row. For example, he placed tellurium (atomic weight = 128) ahead of iodine (atomic weight = 127) because tellurium resembled sulfur and selenium in its properties, whereas iodine was similar to chlorine and bromine.

Mendeleev left a number of gaps in his table (Table 2.5). Instead of looking upon those blank spaces as defects, he boldly predicted the existence of elements as yet undiscovered. Furthermore, he even predicted the properties of some of these missing elements. In succeeding years, many of the gaps were filled in by the discovery of new

The periodic table of the elements provides a wealth of information.

Figure 2.21
Dmitri Mendeleev, a Russian chemist who devised the periodic table of the elements.

Table 2.5
Mendeleev's Periodic Table

Reihen	Gruppe I. — R²O	Gruppe II. — RO	Gruppe III. — R²O³	Gruppe IV. RH⁴ RO²	Gruppe V. RH³ R²O⁵	Gruppe VI. RH² RO³	Gruppe VII. RH R²O⁷	Gruppe VIII. — RO⁴
1	H = 1							
2	Li = 7	Be = 9,4	B = 11	C = 12	N = 14	O = 16	F = 19	
3	Na = 23	Mg = 24	Al = 27,3	Si = 28	P = 31	S = 32	Cl = 35,5	
4	K = 39	Ca = 40	— = 44	Ti = 48	V = 51	Cr = 52	Mn = 55	Fe = 56, Co = 59, Ni = 59, Cu = 63.
5	(Cu = 63)	Zn = 65	— = 68	— = 72	As = 75	Se = 78	Br = 80	
6	Rb = 85	Sr = 87	?Yt = 88	Zr = 90	Nb = 94	Mo = 96	— = 100	Ru = 104, Rh = 104, Pd = 106, Ag = 108
7	(Ag = 108)	Cd = 112	In = 113	Sn = 118	Sb = 122	Te = 125	J = 127	
8	Cs = 133	Ba = 137	?Di = 138	?Ce = 140	—	—	—	— — — —
9	(—)	—	—					
10	—	—	?Er = 178	?La = 180	Ta = 182	W = 184	—	Os = 195, Ir = 197, Pt = 198, Au = 199
11	(Au = 199)	Hg = 200	Tl = 204	Pb = 207	Bi = 208	—	—	
12	—	—	—	Th = 231	—	U = 240	—	— — — —

elements, and their properties were often quite close to those Mendeleev had predicted. The predictive value of this great innovation led to the wide acceptance of Mendeleev's table.

It is now known that properties of an element depend mainly on the number of electrons in the outermost energy level of the atoms of the element (Chapter 4). Sodium atoms have one electron in their outermost energy level (the third). Lithium atoms have a single electron in their outermost level (the second). The chemical properties of sodium and lithium are similar. The atoms of helium and neon have filled outer electron energy levels, and both elements are inert, that is, they do not undergo chemical reactions readily. Apparently, not only are similar chemical properties shared by elements whose atoms have similar electron **configurations** (arrangements) but also certain configurations appear to be more stable (less reactive) than others. This is a point we shall explore in Chapter 4.

In Mendeleev's table, the elements were arranged by atomic weights for the most part, and this arrangement revealed the periodicity of chemical properties. Because the number of electrons determines the element's chemical properties, that number should (and now does) determine the order of the periodic table. In the modern periodic table, the elements are arranged according to atomic number. Remember, this number indicates both how many protons and how many electrons there are in a neutral atom of the element. The modern table, arranged in order of increasing atomic number, and Mendeleev's table, arranged in order of increasing atomic weight, parallel one another because an increase in atomic number is generally accompanied by an increase in atomic weight. In only a few cases (noted by Mendeleev) do the weights fall out of order. Atomic weights do not increase in precisely the same order as atomic numbers because *both protons and neutrons* contribute to the mass of an atom. It is possible for an atom of lower atomic number to have more neutrons than one with a higher atomic number. Thus, it is possible for an atom with a lower atomic number to have a greater mass than an atom with a higher atomic number. Thus the atomic mass of Ar (no. 18) is more than that of K (no. 19), and Te (no. 52) has a mass greater than that of I (no. 53); see the periodic table.

Columns in the periodic table are called groups or families.

The modern periodic table (Figure 2.22) has vertical columns called **groups** or **families.**[2] Each group includes elements with the same number of electrons in their outermost energy levels and, therefore, with similar chemical properties. The horizontal rows of the table are called **periods.** Each new period indicates the opening of the next main electron energy level. For example, sodium starts row three, and the outermost electron in sodium is the first electron to be placed in the third energy level. Because each row begins a new energy level, we can predict that the size of atoms increases from top to bottom. And since electrons are easier to remove when farther from the nucleus, we can also predict that the larger the atom the lower its *ionization energy,* the energy needed to remove an electron.

Rows in the periodic table are called periods.

[2] Each group is identified by a number. The numbering system in the United States is similar to that of Mendeleev. That is, each group is designated by a Roman numeral and a letter, either A or B. (These are shown directly above the top of each group in the periodic table.) In Europe a different A and B group designation has been used. In 1986 the International Union of Pure and Applied Chemistry (IUPAC, see Section 13.5) attempted to reconcile the differences by adopting an alternative system. In the IUPAC system, the groups are numbered left to right from 1 to 18. (These numbers are shown above the Roman numerals in Figure 2.22.) Many chemists are opposed to the change, and the issue has not yet been settled. Therefore, we will continue to use the Roman numerals with the A and B designations.

Figure 2.22 also illustrates another area of lack of international agreement. Elements 104–109 are synthetic. When 104 and 105 were synthesized, two groups claimed priority, and different names were proposed. Thus, element 104 is sometimes called rutherfordium (Ru) and sometimes khurchatovium (Ku); element 105 is often called hahnium (Ha). The IUPAC therefore devised a system of provisional names (unnilquadium, unnilpentium, etc.) whose three-letter symbols are shown in Figure 2.22.

EXAMPLE 2.9

Write out the sublevel notation for the electrons in the highest main energy level for strontium (Sr) and arsenic (As).

SOLUTION

Strontium is in Group IIA and the fifth period of the periodic chart. Its outer electron configuration is $5s^2$. (Each period of the periodic table corresponds to the filling, or at least partial filling, of a main energy level.) Arsenic is in Group VA and the fourth period. Its five outer (valence) electrons have the configuration denoted by $4s^24p^3$.

Practice Exercise

Write out the sublevel notation for the electrons in the highest main energy level for rubidium (Rb) and selenium (Se).

Certain groupings of elements in the periodic table are known by special names. The heavy stepped diagonal line on the table divides the elements into two major classes. Those to the left of the line are called **metals,** and those to the right, **nonmetals.** Group IA elements are known as **alkali metals;** Group IIA are **alkaline earth metals;** Group VIIA, **halogens.** Group VIIIA at the extreme right of the table contains the **noble gases.** All the Group B elements are called **transition metals.** The periodic table inside the front cover depicts the physical state of each element.

Figure 2.22
The modern periodic table.

1 Group IA		2 Group IIA	3 Group IIIB	4 Group IVB	5 Group VB	6 Group VIB	7 Group VIIB	8 Group	9 Group VIIIB	10 Group	11 Group IB	12 Group IIB	13 Group IIIA	14 Group IVA	15 Group VA	16 Group VIA	17 Group VIIA	18 Group VIIIA
1 H 1.00794																		2 He 4.002602
3 Li 6.941		4 Be 9.01218											5 B 10.811	6 C 12.011	7 N 14.0067	8 O 15.9994	9 F 18.998403	10 Ne 20.179
11 Na 22.98977		12 Mg 24.305											13 Al 26.98154	14 Si 28.0855	15 P 30.97376	16 S 32.066	17 Cl 35.453	18 Ar 39.948
19 K 39.0983		20 Ca 40.078	21 Sc 44.9559	22 Ti 47.88	23 V 50.9415	24 Cr 51.9961	25 Mn 54.9380	26 Fe 55.847	27 Co 58.9332	28 Ni 58.6934	29 Cu 63.546	30 Zn 65.38	31 Ga 69.723	32 Ge 72.59	33 As 74.9216	34 Se 78.96	25 Br 79.904	36 Kr 83.80
37 Rb 85.4678		38 Sr 87.62	39 Y 88.9059	40 Zr 91.22	41 Nb 92.9064	42 Mo 95.94	43 Tc (98)	44 Ru 101.07	45 Rh 102.9055	46 Pd 106.42	47 Ag 107.8682	48 Cd 112.41	49 In 114.818	50 Sn 118.710	51 Sb 121.757	52 Te 127.60	53 I 126.9045	54 Xe 131.29
55 Cs 132.9054		56 Ba 137.33	57 La 138.9055	72 Hf 178.49	73 Ta 180.9479	74 W 183.84	75 Re 186.207	76 Os 190.23	77 Ir 192.22	78 Pt 195.08	79 Au 196.9665	80 Hg 200.59	81 Tl 204.383	82 Pb 207.2	83 Bi 208.9804	84 Po (209)	85 At (210)	86 Rn (222)
87 Fr (223)		88 Ra 226.0254	89 Ac 227.0278	104 Unq (261)	105 Unp (262)	106 Unh (263)	107 Uns (262)	108 Uno (265)	109 Une (266)									

Metals ◄── ──► Nonmetals

	58 Ce 140.12	59 Pr 140.9077	60 Nd 144.24	61 Pm (145)	62 Sm 150.36	63 Eu 151.96	64 Gd 157.25	65 Tb 158.9254	66 Dy 162.50	67 Ho 164.9304	68 Er 167.26	69 Tm 168.9342	70 Yb 173.04	71 Lu 174.967
	90 Th 232.0381	91 Pa 231.0359	92 U 238.0289	93 Np 237.0482	94 Pu (244)	95 Am (243)	96 Cm (247)	97 Bk (247)	98 Cf (251)	99 Es (252)	100 Fm (257)	101 Md (258)	102 No (259)	103 Lr (260)

2.7
WHICH MODEL TO USE

For our purposes, we will use whatever model is more helpful for understanding a particular concept. In trying to evaluate the behavior of gases, chemists often use Dalton's hard, indivisible (billard ball) model. In discussing how atoms combine to form molecules, we will use the Bohr model. To explain the shape of molecules (very important in explaining the action of drugs, for instance), we will use an extension of the Bohr theory called the **valence shell electron pair repulsion** (VSEPR) theory (Section 4.14), for the most part. Occasionally we will employ the orbital atom theory. Don't let this inconsistent use disturb you. Remember that a model or a theory is used to explain phenomena. When a better model or theory is invented, the old one is sometimes discarded or pushed into a secondary role, but occasionally it can be used to clarify some point that a newer, more complicated model only obscures.

EXERCISES

1. What is the distinction between the atomistic view and the continuous view of matter?
2. If foods were described as atomistic or continuous, which designation would you use for each of the following?
   a. peas   b. mashed potatoes
   c. milk   d. hot dogs
   e. hard-boiled eggs f. scrambled eggs
3. Outline the main points of Dalton's atomic theory.
4. An atom of calcium has a mass of 40 amu, and an atom of cobalt has a mass of 59 amu. Are these findings in agreement with Dalton's atomic theory?
5. An atom of calcium has a mass of 40 amu, and an atom of potassium has a mass of 40 amu. Are these findings in agreement with Dalton's atomic theory?
6. An atom of calcium has a mass of 40 amu, and another atom of calcium has a mass of 44 amu. Are these findings in agreement with Dalton's atomic theory?
7. How did the discovery of radioactivity contradict Dalton's atomic theory?
8. Give the distinguishing characteristics of the proton, the neutron, and the electron.
9. Should a proton and an electron attract or repel one another?
10. Should a neutron and a proton attract or repel one another?
11. Which subatomic particles are found in the nuclei of atoms?
12. What are the extranuclear (outside the nucleus) subatomic particles?
13. Compare Dalton's model of the atom with the nuclear model of the atom.
14. If the nucleus of an atom contains 10 protons,
    a. how many electrons are there in the neutral atoms?
    b. how many neutrons are there in the nucleus of the atom?

15. What are isotopes?
16. The table below describes four atoms.

	Atom A	Atom B	Atom C	Atom D
Number of protons	10	11	11	10
Number of neutrons	11	10	11	10
Number of electrons	10	11	11	10

Are atoms A and B isotopes? Are atoms A and C isotopes? A and D? B and C?

17. What are the masses of the atoms in Exercise 16?
18. Use the table inside the front cover to determine the number of protons in an atom of the following elements.
    a. helium (He)   b. sodium (Na)
    c. chlorine (Cl)   d. oxygen (O)
    e. magnesium (Mg)   f. sulfur (S)
19. How many electrons are there in the neutral atoms of the elements listed in Exercise 18?
20. Write the symbols for protium, deuterium, and tritium (which are hydrogen-1, hydrogen-2, and hydrogen-3, respectively).
21. Write the symbol for an isotope with a mass number of 8 and an atomic number of 5.
22. Write the symbol for an isotope with $Z = 35$ and $A = 83$.
23. Write the symbol for an isotope with 53 protons and 72 neutrons.
24. Write the symbols for the following isotopes. You may refer to the periodic table.
    a. gallium-69   b. molybdenum-98
    c. molybdenum-99   d. technetium-98

25. Indicate the number of protons and the number of neutrons in atoms of the following isotopes.
a. $^{62}_{30}Zn$ **b.** $^{241}_{94}Pu$
c. $^{99}_{43}Tc$ **d.** $^{81}_{36}Kr$

26. Which of the following sets represent isotopes?
a. $^{70}_{34}X$, $^{70}_{33}X$ **b.** $^{57}_{28}X$, $^{66}_{28}X$
c. $^{186}_{74}X$, $^{186}_{74}X$ **d.** $^{8}_{2}X$, $^{6}_{4}X$
e. $^{22}_{11}X$, $^{44}_{22}X$

27. The two principal isotopes of lithium are lithium-6 and lithium-7. The atomic weight of lithium is 6.9 amu. Which is the predominant isotope of lithium?

28. Out of every five atoms of boron, one has a mass of 10 amu and four have a mass of 11 amu. What is the atomic weight of boron? Use the periodic table only to check your answer.

29. Fill in the blanks for the six elements listed.

	Atomic No.	No. of Protons	No. of Electrons	No. of Neutrons	Mass No.
Pb	___	___	___	126	208
Sr	___	38	___	50	___
N	7	___	___	___	14
Cr	___	___	24	___	52
Ag	47	___	___	60	___
As	___	___	33	42	___

30. How did Bohr define the nuclear model of the atom?

31. What subatomic particles travel in the "orbits" of the Bohr model of the atom?

32. According to Bohr, what is the maximum number of electrons in the fourth energy level ($n = 4$)?

33. If the third energy level of electrons contains two electrons, what is the total number of electrons in the atom?

34. Define the terms
a. ground state   **b.** excited state

35. When light is emitted by an atom, what change has occurred within the atom?

36. Which atom absorbed more energy: one in which an electron moved from the second energy level to the third energy level, or an otherwise identical atom in which an electron moved from the first energy level to the third?

37. Draw Bohr diagrams for the elements listed in Exercise 18.

38. The following Bohr diagram is supposed to represent a neutral atom of an element. The diagram is incorrectly drawn. Identify the error.

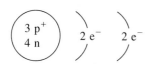

39. In the quantum mechanical notation $2p^6$,
a. how many electrons are described?
b. what is the general shape of the orbitals described?
c. how many orbitals are included?

40. Use quantum mechanical notation to describe the electron configuration of the atom represented in the following Bohr diagram.

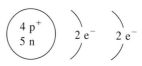

41. Give the electron configurations (using quantum mechanical notation) for the elements in Exercise 18.

42. What elements have the following electron configurations?
a. $1s^2 2s^2$ **b.** $1s^2 2s^2 2p^3$
c. $1s^2 2s^2 2p^6 3s^2 3p^1$

43. Give the atomic numbers of the elements described in Exercise 42.

44. None of the following electron configurations is reasonable. In each case explain why.
a. $1s^2 2s^2 3s^2$ **b.** $1s^2 2s^2 2p^2 3s^1$
c. $1s^2 2s^2 2p^6 2d^5$

45. What is the difference in the electron configurations of oxygen (O) and fluorine (F)?

46. What is the difference in the electron configurations of fluorine (F) and sulfur (S)?

47. If three electrons were added to the outermost energy level of a phosphorus atom, the new electron configuration would resemble that of what element?

48. If two electrons were removed from the outermost energy level of a magnesium atom, the new electron configuration would resemble that of what element?

49. Elements are defined on a theoretical basis as being composed of atoms that share the same atomic number. On the basis of this theory, would you think it possible for someone to discover a new element that would fit between magnesium (atomic number 12) and aluminum (atomic number 13)?

50. Referring only to the periodic table, indicate what similarity in electron structure is shared by fluorine (F) and chlorine (Cl). What is the difference in their electron structures?

51. Where are the metals and nonmetals located in the periodic table?

52. Identify the following elements as metals or nonmetals. You may refer to the periodic table. The numbers in parentheses are the atomic numbers of the elements.
a. sulfur (16)   **b.** chromium (24)
c. iodine (53)

53. Indicate the group number or numbers of the following families.
a. alkali metals   **b.** transition metals
c. halogens   **d.** alkaline earth metals

54. Identify the period of each of the following elements. You may refer to the periodic table. The numbers in parentheses are the atomic numbers of the elements.
a. chlorine(17)   **b.** osmium(76)
c. hydrogen(1)

55. Which of the following elements are halogens?
   a. Ag   **b.** At   **c.** As

56. Which of the following elements are alkali metals?
   a. K   **b.** Y   **c.** W

57. Which of the following elements are noble gases?
   a. Fe   **b.** Ne   **c.** Ge
   d. He   **e.** Xe

58. Which of the following elements are transition metals?
   a. Ti(22)   **b.** Tc(43)   **c.** Te(52)

59. Which of the following elements are alkaline earth metals?
   a. Bi   **b.** Ba   **c.** Be   **d.** Br

60. How many electrons are in the outermost energy level of the halogens?

61. How many electrons are in the outermost energy level of Group IIA elements?

62. In what period of elements are electrons first introduced into the fourth energy level?

63. What element is found at the intersection of
   a. the fourth row and Group VA?
   b. the third row and the halogens?
   c. the period representing the second energy level and the alkaline earth metal group?

64. Which is larger?
   a. K or Rb   **b.** Ne or Na
   c. Zn or Zr

65. Which will ionize more easily?
   a. Li or Cs   **b.** N or Sb
   c. Ar or Ac

Chapter 3
NUCLEAR PROCESSES

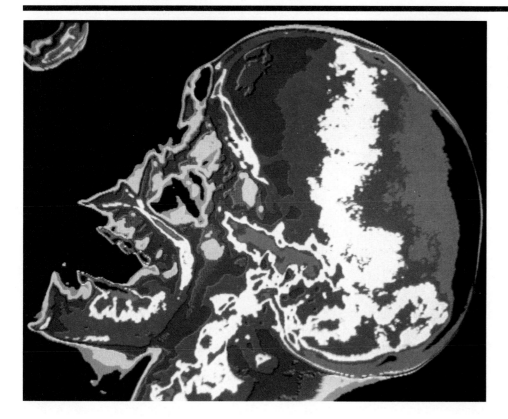

Knowledge of the properties of atomic nuclei has made it possible to obtain images that show the distribution pattern of various body functions.

In our consideration of chemical bonding we shall emphasize the importance of the outermost electrons in atoms. Through most of this text, we shall focus on chemical reactions, and in chemical reactions only these outer regions of atoms are changed. For the moment, however, let us turn our attention to that tiny speck of matter called the nucleus. The volume of a whole atom is about 10,000 times that of its nucleus, yet it is the nucleus that holds the power that has become the symbol of our age.

Nuclear power confronts us with a great paradox. Although nuclear power unleashed in wrath can destroy cities and perhaps civilizations, controlled nuclear power can provide the energy necessary to run our cities and maintain our civilization. Yet even the peaceful uses of nuclear power are not without potential danger. As citizens of the nuclear age, we have difficult decisions to make. Nuclear bombs may kill, but nuclear medicine saves lives. Our knowledge of nuclear science gives us both power and responsibility. How we exercise that responsibility will determine how future generations remember us—even whether there will be future generations to remember us.

3.1
DISCOVERY OF RADIOACTIVITY

In Chapter 2, we discussed the existence of isotopes. Isotopes usually are not considered in ordinary *chemical* reactions. Such reactions involve the outer electrons of an atom, and differences in the number of neutrons buried deep in the heart of the atom would not be expected to have a major effect. In **nuclear reactions,** reactions that do involve the heart of the atom, however, isotopes are of utmost importance, as we shall see.

The discovery of the first nuclear reactions was triggered by the discovery of X-rays, which are not nuclear phenomena. In 1895, a German scientist, Wilhelm Konrad Roentgen, found that he could produce a form of radiation that passed right through solid materials. This mysterious radiation was like visible light in that it was a form of pure energy. Also, like visible light, it resulted from electrons moving from one energy level to another. It was unlike visible light in that you could not see it. For want of a better name, Roentgen called the invisible, penetrating radiation **X-rays.** X-rays are high-energy radiation. They contain more energy than even the most energetic visible light. It is their high energy that gives X-rays such penetrating power.

The medical community immediately recognized the significance of the penetrating X-rays. The X-ray picture shown in Figure 3.1 was taken in February 1896, within two months of the publication of Roentgen's discovery. The round black dots are gunshot pellets. This picture was used to establish their position for removal by surgery.

This phenomenon fascinated other scientists, also. One, a Frenchman named Antoine Henri Becquerel, had been studying a different phenomenon called fluorescence. Fluorescent compounds, after exposure to strong sunlight, continue to glow even when taken into a dark room. Becquerel wondered if there were any substances that would give off X-rays when they fluoresced. In trying to answer this question, Becquerel tested a large number of materials—among them, uranium compounds. Becquerel found that uranium did emit invisible, penetrating rays. These rays, however, proved to have nothing to do with fluorescence or with X-rays. Becquerel had discovered a totally new phenomenon.

At this point, Marie and Pierre Curie and Ernest Rutherford entered the picture. We first mentioned these scientists in our discussion of models for the atom. Marie Curie (Figure 3.2) named the phenomenon discovered by Becquerel **radioactivity.** The Curies went on to discover a number of new radioactive elements, including radium.

X-rays—high-energy radiation

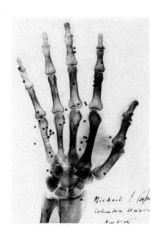

Figure 3.1
An early example of the use of X-rays in medicine. Professor Michael Pupin of Columbia University made this X-ray photograph in 1896 to aid in the surgical removal of gunshot pellets (dark spots) from the hand of a patient.

Figure 3.2
Marie Sklodowska Curie in her laboratory and Marie and Pierre Curie on a French postage stamp.

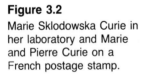

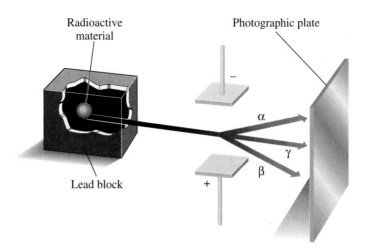

Radioactive material

Photographic plate

Lead block

Figure 3.3
Behavior of radioactive rays in an electric field.

3.2
TYPES OF RADIOACTIVITY

Scientists soon realized that the radiation emanating from uranium and other radioactive elements was of three types. When this radiation was passed through a strong magnetic field, one portion was deflected in one direction, another portion was deflected in the opposite direction, and a third was not deflected at all. Rutherford named these portions **alpha** (α) **rays, beta** (β) **rays,** and **gamma** (γ) **rays,** respectively (Figure 3.3).

Only the gamma rays are radiation in the same sense that visible light and X-rays are, that is, pure radiant energy. Gamma rays have the highest energy and are the most penetrating form of lightlike radiation yet discovered.

Both alpha and beta rays were found to be streams of tiny particles. An **alpha particle** has a mass four times that of a hydrogen atom and a charge twice that of a proton. These properties make alpha particles identical to helium nuclei ($^{4}_{2}$He). Beta rays were shown to be streams of electrons. Thus, a **beta particle** ($_{-1}^{0}$e) has a mass only 1/1837 that of a proton and has a negative charge. As a form of radiant energy, gamma rays have no mass and no charge.

We can describe in detail the forms of radioactivity, but we must still answer the question What is it? What is happening to produce radioactivity? The answer is that some nuclei are unstable as they occur in nature. Radium atoms with a mass number of 226, for example, break down spontaneously, giving off alpha particles (Figure 3.4). Since alpha particles are identical to helium nuclei, this process can be summarized by the equation

Alpha (α) particle
$^{4}_{2}$He

Beta (β) particle
$_{-1}^{0}$e

$$^{226}_{88}\text{Ra} \longrightarrow \overset{226}{\underset{88}{^{4}_{2}\text{He} + {}^{222}_{86}\text{Rn}}} \quad \text{or} \quad {}^{226}_{88}\text{Ra} \longrightarrow \alpha + {}^{222}_{86}\text{Rn}$$

The new element, with two fewer protons, is identified by its atomic number (86) as radon (Rn). Note that the mass number of the starting material must equal the total of the mass numbers of the products. The same is true for atomic numbers. The symbol $^{4}_{2}$He for the alpha particle is preferred to the symbol α because it allows us to check the balance of mass and atomic numbers more readily.

Figure 3.4

Nuclear emission of (a) an alpha particle and (b) a beta particle.

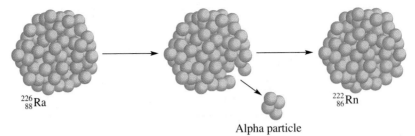

$^{226}_{88}Ra$ Alpha particle $^{222}_{86}Rn$

(a) Nuclear changes accompanying alpha decay.

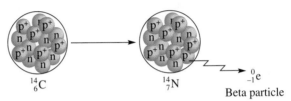

$^{14}_{6}C$ $^{14}_{7}N$ $^{0}_{-1}e$ Beta particle

(b) Nuclear changes accompanying beta decay.

Tritium nuclei are also unstable. Tritium is one of the heavy isotopes of hydrogen first mentioned in Section 2.2. Like all hydrogen nuclei, the tritium nucleus contains one proton. Unlike the most common isotope of hydrogen, however, the tritium nucleus contains two neutrons, and its mass is therefore 3 amu ($^{3}_{1}H$). Tritium decomposes by what is termed **beta decay.** Since a beta particle is identical to an electron, this process can be written

$$^{3}_{1}H \longrightarrow {}^{0}_{-1}e + {}^{3}_{2}He \quad\text{or}\quad {}^{3}_{1}H \longrightarrow \beta + {}^{3}_{2}He$$

The product isotope is identified by its atomic number as helium (He). How can the original nucleus, which contains only a proton and two neutrons, emit an electron?

We can envision one of the neutrons in the original nucleus splitting into a proton and an electron.

$$^{1}_{0}n \longrightarrow {}^{1}_{1}p + {}^{0}_{-1}e$$

The new proton is retained by the nucleus (therefore, the atomic number of the product increases by one), and the almost massless electron or beta particle is kicked out (the product nucleus has the same mass as the original). A second example of beta decay is pictured in Figure 3.4.

The third type of radioactivity is **gamma decay.** No particle is emitted; gamma rays are pure radiant energy. Gamma emission involves no change in atomic number or mass. The process is analogous to the emission of light from an atom. Visible light is emitted when an electron changes from a higher to a lower energy level. In gamma emission, a nucleus in a higher energy state drops to a lower energy state. Thus,

$$^{153}_{64}Gd \longrightarrow {}^{153}_{64}Gd + \gamma$$

This type of decay is particularly useful when **radioisotopes** (radioactive isotopes) are used for diagnostic purposes in medicine. We discuss such uses in Section 3.9.

The emission of an alpha or beta particle is often accompanied by the emission of a gamma ray.

Table 3.1
Radioactive Decay and
Nuclear Change

Type of Radiation	Symbol	Mass Number of Particle	Charge on Particle	Change in Mass Number of Emitting Nucleus	Change in Atomic Number of Emitting Nucleus
Alpha	α	4	2+	Decreases by 4	Decreases by 2
Beta	β	0	1−	No change	Increases by 1
Gamma	γ	0	0	No change	No change

The major types of radioactive decay and the resulting nuclear changes are summarized in Table 3.1.

EXAMPLE 3.1

Plutonium-239 emits an alpha particle when it decays. What new element is formed?

SOLUTION

$$^{239}_{94}\text{Pu} \longrightarrow \, ^4_2\text{He} + ?$$

Mass and charge are conserved. The new element must have a mass of 239 − 4 = 235 and a charge of 94 − 2 = 92. The nuclear charge (atomic number) of 92 identifies the element as uranium (U).

$$^{239}_{94}\text{Pu} \longrightarrow \, ^4_2\text{He} + \, ^{235}_{92}\text{U}$$

Practice Exercise
Fermium-250 undergoes alpha decay. What new isotope is formed?

EXAMPLE 3.2

Protactinium-234 undergoes beta decay. What new element is formed?

SOLUTION

$$^{234}_{91}\text{Pa} \longrightarrow \, ^{\;\;0}_{-1}\text{e} + ?$$

The new element still has a mass number of 234. It must have a nuclear charge of 92 in order for the charge to be the same on each side of the equation. The nuclear charge identifies the new atom as another isotope of uranium (U).

$$^{234}_{91}\text{Pa} \longrightarrow \, ^{\;\;0}_{-1}\text{e} + \, ^{234}_{92}\text{U}$$

Practice Exercise
Selenium-85 undergoes beta decay. What new isotope is formed?

3.3
PENETRATING POWER

Radioactive materials are dangerous because the radiation emitted by decaying nuclei can damage living tissue (see Section 3.8). The ability of the radiation to inflict damage depends, in part, on its penetrating power.

All other things being equal, the more massive the particle, the less its penetrating power. Of alpha, beta, and gamma rays, alpha rays are the least penetrating. These are streams of helium nuclei, each particle with a mass number of 4. Beta rays are more penetrating than alpha rays. The electrons that make up the stream of beta particles are assigned a mass number of 0. These particles are not really massless, but they are very much lighter than alpha particles. Gamma rays are high-energy radiation and, like light, truly have no mass. This is the most penetrating form of nuclear radiation.

That the biggest particles make the least headway may seem contrary to common sense. Consider that penetrating power reflects the ability of the radiation to make its way through a sample of matter. It is as if you were trying to roll some rocks through a field of boulders. The alpha particle acts as if it were a boulder itself. Because of its size, it cannot get very far before it bumps into and is stopped by other boulders. The beta particle acts like a small stone. It can sneak between boulders and perhaps ricochet off one or another until it has made its way farther into the field (Figure 3.5). The gamma ray can be compared to an insect that can get through the smallest openings: although the insect may brush against some of the boulders, it can, in general, make its way through most of the field without being stopped.

We said that penetrating power is determined by the mass of the particle, all other things being equal. But all other things are not always equal. The faster a particle moves or the more energetic the radiation is, the greater its penetrating power.

If a radioactive source is located *outside* of the body, an alpha emitter of low penetrating power is least dangerous, and a gamma emitter, most. The alpha particles are stopped by the outer layer of skin, whereas the gamma rays can penetrate to damage vital internal organs. On the other hand, if a radioactive source is *within* the body, then the nonpenetrating alpha particles can do great damage. All such particles are trapped within the body, which must then absorb all the energy released by the particle. Alpha rays inflict all their damage in a very small area because they do not travel far. Beta rays distribute the damage over a somewhat larger area because they travel farther. Tissue may recover from limited damage spread over a large area; it is less likely to survive very concentrated damage.

Failure to distinguish between damage done by a radiation source external to the body and one that is internal frequently leads to quite different assessments of potential

Alpha rays—least penetrating

Gamma rays—most penetrating

Figure 3.5

Shooting radioactive particles through matter is like rolling rocks through a field of boulders—the larger ones are more easily stopped.

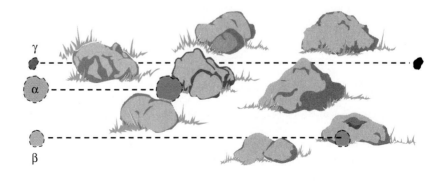

Figure 3.6
Small amounts of radioactive material were released during an emergency shutdown of the nuclear generator at Three Mile Island in Pennsylvania in 1979.

danger. In 1979, radioactive material was released during an emergency shutdown of a nuclear generator at Three Mile Island in Pennsylvania (Figure 3.6). Two different groups disputed the danger associated with this release. One group argued that the people in neighboring communities were exposed to little radiation, not much more than the normal **background radiation** (the radiation that comes from natural sources like radioisotopes in the Earth's crust). On the opposite side of the argument, another group maintained that the released radioactive isotopes would ultimately enter the body through food or inhaled air. The radioisotopes would then concentrate within the body, these people argued, and remain to do long-term damage to various organs. Only the future can tell us which group is correct in its assessment.

3.4
RADIATION MEASUREMENT

Several units of measurement are associated with the phenomenon of radioactivity. The *rate* at which nuclear disintegrations occur in a particular sample is measured in **curies** (Ci), named in honor of Marie Curie. A sample whose activity is hundreds or thousands of curies might be used as a source of externally applied radiation for the treatment of cancer. A sample of 10 mCi can be taken internally for diagnostic purposes by an adult, whereas a sample administered internally to a child might be measured in microcuries.

The *effect of radiation on matter* can be measured in a number of ways. The **roentgen** (R) is a measure of the ability of a source of X-rays or gamma rays to ionize an air sample. The **rad** (*rad*iation *a*bsorbed *d*ose) measures the amount of energy absorbed by any form of matter from any ionizing radiation. Alpha, beta, and gamma rays, as well as X-rays, are forms of **ionizing radiation,** that is, they cause the formation of ions from neutral particles. In the body, ionizing radiation most often interacts with water molecules. The reactive particles formed from water attack other molecules essential to proper cell function. Ionizing radiation damages living tissue by this route.

Techniques for measuring radiation range from the simple to the sophisticated. Individuals who work with radioactive materials wear **film badges** on their pockets or at their

Figure 3.7

Film badges are worn by people working around radioactive materials.

Figure 3.8

The Geiger counter. Schematic diagram at right.

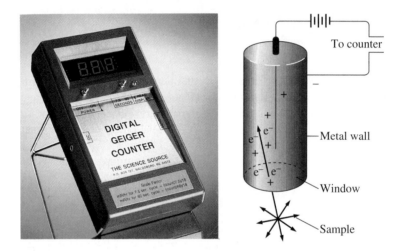

waist or as rings (Figure 3.7). The film in these badges reacts to radiation from radioactive isotopes or X-ray sources just as photographic film reacts to light. A certain dose of radiation will cause the film in the badge to become exposed, alerting the wearer to the potential danger. Sophisticated electronic devices, like the **Geiger counter** (Figure 3.8), measure the ionizing effects of radiation and translate these into an observable signal (a meter reading or a clicking sound or a light flash). Other detectors provide a permanent visual record of the intensity of the radiation. Such detectors play an important role in medical diagnosis (Section 3.9).

3.5
HALF-LIFE

Radioactivity results when nuclei decay. We cannot predict when a particular nucleus will decay, but we can accurately predict the **rate** of decay of large numbers of radioactive atoms. (Life insurance companies cannot tell exactly when you will die, but their business depends on being able to predict how many of their clients will die in a particular period of time.)

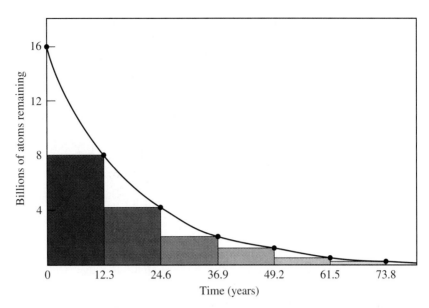

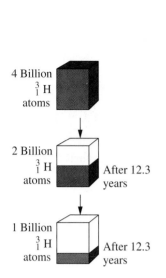

Figure 3.9
Life span of $^3_1$H.

Figure 3.10
The radioactive decay of $^3_1$H.

A radioactive isotope can be characterized by a quantity called its **half-life.** The half-life of a radioactive isotope is that period of time in which one-half of the radioactive atoms present undergo decay. Suppose, for example, that you had 4 billion atoms of the radioactive hydrogen isotope, $^3_1$H. The half-life of this isotope is 12.3 years. This means that in 12.3 years, one-half, or 2 billion, of the original 4 billion atoms would have undergone radioactive decay (Figure 3.9). In another 12.3 years, half of the remaining 2 billion atoms would have decayed. That is, after two half-lives, 1 billion atoms or one-fourth of the original atoms would remain unchanged. Two half-lives do not make a whole! The concept of half-life is shown graphically in Figure 3.10.

It is impossible to say exactly when *all* the $^3_1$H will have decayed. For practical purposes one may assume that nearly all the radioactivity is gone after about 10 half-lives. For the tritium sample considered here, 10 half-lives would be 123 years, at which time only a fraction of a percentage of the original atoms would still be present.

Half-life $(t_{1/2})$—the time in which one-half of the radioactive atoms present undergo decay. The half-life of an element can be very long (millions of years) or extremely short (tiny fractions of a second). The half-life of uranium-238 is 4500 million years; that of boron-9 is 8×10^{-19} s.

EXAMPLE 3.3

Phosphorus-30 has a half-life of 3 min. How much of a 16.0-mg sample would be left after 15 min?

SOLUTION
There are five half-lives in 15 minutes. Therefore the 16-mg sample would decrease by half, five consecutive times: 16/2 = 8; 8/2 = 4; 4/2 = 2; 2/2 = 1; 1/2 = 0.5 mg is left at the end of 15 min.

Practice Exercise
You obtain a 20.0-mg sample of mercury-190, half-life 20 min. How much of the mercury-190 sample remains after 2 hr?

The half-life of a carbon isotope can be used to date old artifacts. Such dating is based on the fact that there is a constant amount of carbon-14 in the ecosphere. Even though some is always lost as it decays, more is regenerated through other reactions so that an equilibrium amount of carbon-14 is maintained. As long as the plant or animal is alive and therefore a part of the ecosphere, its carbon-14 content will be the same as outside its system. Upon death, however the exchange with the ecosphere ceases, and the carbon-14 slowly diminishes, making it possible to measure the organism's age.

EXAMPLE 3.4

A bone sample of the extinct American saber-toothed tiger was found to have 4 counts per minute (cpm) compared with 16 cpm for fresh bone. Estimate the age of the remains. The carbon-14 half-life is 5730 years.

SOLUTION
One half-life ago, the sample would have had 8 cpm, and another half-life back it would have had 16 cpm. The age is then two half-lives: $2 \times 5730 = 10,740$ years.

Practice Exercise
How old is a piece of cloth that has a carbon-14 activity one-sixteenth that of new cloth fibers? The half-life of carbon-14 is 5730 years.

By the same technique, the Shroud of Turin (a religious artifact believed by many to have been the burial cloth of Jesus Christ) was recently determined to be no older than the Middle Ages. Laboratories in England, Switzerland, and Arizona were each given three unmarked pieces of cloth without being told which, if any, was a sample of the shroud. Their independent conclusions were that it dates from about A.D. 1200.

Much of the concern over nuclear power centers on the long half-lives of some isotopes that can be released in a nuclear accident or that are simply left as by-products of the normal operation of a nuclear reactor. In the first instance, the fear is that large areas could be rendered uninhabitable for thousands of years if an explosion released long-lived radioisotopes into the atmosphere. Even with no accidents, normal operation of a reactor produces nuclear wastes that must be safely stored for thousands of years.

3.6
ARTIFICIAL TRANSMUTATION
AND INDUCED RADIOACTIVITY

Artificial transmutation:
Artificial—not occurring in nature
Transmutation—one element changes into another

The forms of radioactivity encountered thus far occur in nature. The helium we use to fill balloons has accumulated over billions of years from the alpha decay of radioactive elements in the Earth's crust (Figure 3.11). It is possible to bring about nuclear reactions not encountered in nature. Such reactions are referred to as **artificial transmutations** because in the process one element is changed into another. These reactions are brought about by bombardment of stable nuclei with alpha particles, neutrons, or other subatomic particles. These particles, given sufficient energy, penetrate the target nucleus and trigger a nuclear reaction.

Figure 3.11
Bart Simpson and the other balloons in the parade are filled with helium, a product of alpha decay of radioactive elements.

Figure 3.12
Frédéric and Irène Joliot-Curie discovered artificially induced radioactivity in 1934. (Frédéric changed his name to Joliot-Curie when he married Irène.)

Ernest Rutherford studied the bombardment of the nuclei of a variety of light elements with alpha particles. One such experiment, in which he bombarded nitrogen nuclei, resulted in the production of protons.

$$^{14}_{7}\text{N} + ^{4}_{2}\text{He} \longrightarrow ^{17}_{8}\text{O} + ^{1}_{1}\text{H}$$

(Recall that the hydrogen nucleus is a proton; hence the alternative symbol for the proton is $^{1}_{1}\text{H}$.) Notice that the sum of the mass numbers on the left of the equation equals the sum of the mass numbers on the right of the equation. The atomic numbers are also balanced.

Irène Curie (daughter of Marie and Pierre) and her husband, Frédéric Joliot (Figure 3.12), studied the bombardment of aluminum nuclei with alpha particles. The reaction yielded neutrons and an isotope of phosphorus.

$$^{27}_{13}\text{Al} + ^{4}_{2}\text{He} \longrightarrow ^{30}_{15}\text{P} + ^{1}_{0}\text{n}$$

Much to the surprise of the Joliot-Curies, the target continued to emit particles after the bombardment was halted. The phosphorus isotope was radioactive. It emitted particles equal in mass but opposite in charge to the electron.

$$^{30}_{15}\text{P} \longrightarrow ^{0}_{+1}\text{e} + ^{30}_{14}\text{Si}$$

The $^{0}_{+1}\text{e}$ particle is called a **positron.** Once again the question arises: if the nucleus contains only protons and neutrons, where does this particle come from?

Previously we accounted for a beta particle (an electron) popping out of a nucleus by saying that a neutron changed into a proton and an electron. Perhaps a similar happening can account for the appearance of a positron. Imagine a proton in the nucleus changing into a neutron and a positron. (Remember that a proton is the same as a hydrogen nucleus.)

$$^{1}_{1}\text{H} \longrightarrow ^{0}_{+1}\text{e} + ^{1}_{0}\text{n}$$

Figure 3.13

Nuclear emission of a positron, $_{+1}^{0}e$.

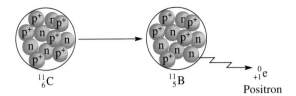

$^{11}_{6}C$ $^{11}_{5}B$ $_{+1}^{0}e$
Positron

The equation balances nicely. When the positron is emitted, the nucleus suddenly has one less proton and one more neutron than before. Therefore, the mass of the product nucleus is the same, but its atomic number is one less than that of the original nucleus. Figure 3.13 presents another example of nuclear decay—positron emission.

EXAMPLE 3.5

When potassium-39 is bombarded with neutrons, chlorine-36 is produced. What other particle is emitted?

SOLUTION

$$^{39}_{19}K + ^{1}_{0}n \longrightarrow ^{36}_{17}Cl + ?$$

To balance the equation, a particle with a mass number of 4 and an atomic number of 2 is required—that is, an alpha particle.

$$^{39}_{19}K + ^{1}_{0}n \longrightarrow ^{36}_{17}Cl + ^{4}_{2}He$$

Practice Exercise

Technetium-97 is produced by bombarding molybdenum-96 with a deuteron (hydrogen-2 nucleus). What other particle is emitted?

EXAMPLE 3.6

Carbon-10, a radioactive isotope, emits a positron when it decays. Write an equation for this process.

SOLUTION

$$^{10}_{6}C \longrightarrow ^{0}_{+1}e + ?$$

To balance the equation, a particle with a mass number of 10 and an atomic number of 5 (boron) is required.

$$^{10}_{6}C \longrightarrow ^{0}_{+1}e + ^{10}_{5}B$$

Practice Exercise

Gold-188 decays by positron emission. Write a balanced nuclear equation for the process.

Nuclear medicine depends on the availability of a broad range of radioisotopes, and many of these are artificially produced. Later in this chapter we will look into some aspects of nuclear medicine.

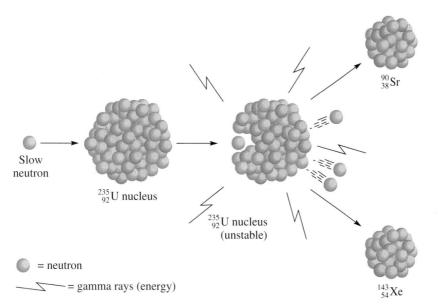

Figure 3.14
The splitting of a uranium atom. The neutrons produced in this fission can split other uranium atoms, thus sustaining a chain reaction. The splitting of one uranium-235 atom yields 8.9×10^{-18} kWh of energy. Fission of 1 mol of uranium-235 (6.02×10^{23} atoms) produces 5,300,000 kWh of energy.

Slow neutron

$_{92}^{235}$U nucleus

$_{92}^{235}$U nucleus (unstable)

$_{38}^{90}$Sr

$_{54}^{143}$Xe

= neutron

= gamma rays (energy)

3.7
FISSION AND FUSION

In the nuclear reaction called **nuclear fission,** a large unstable nucleus is bombarded with relatively slow moving neutrons. In the resulting nuclear reaction, the large nucleus does not simply emit a small particle; it breaks apart, leaving two medium-sized nuclei and releasing more neutrons. A typical fission reaction is

$$_{92}^{235}\text{U} + _{0}^{1}\text{n} \longrightarrow _{38}^{90}\text{Sr} + _{54}^{143}\text{Xe} + 3\,_{0}^{1}\text{n}$$

In nuclear fission the nucleus breaks apart to form two nuclei.

Because it was bombardment with neutrons that triggered the reaction in the first place, the product neutrons can go on to cause more of the large nuclei to fission (break apart). Neutrons produced from this second wave of reactions will trigger more reactions, and so on. Thus, nuclear fission is a chain reaction (Figure 3.14). Each reaction in the chain releases energy. If the chain of reactions is carried out in a controlled manner, then the energy released can be used to generate electric power, such as is done in a nuclear reactor (Figure 3.15). The reaction can also be carried out in such a way that all the energy is released in one gigantic explosion. What one then has is a bomb—a nuclear bomb (Figure 3.16). It is not possible for a nuclear reactor to give a nuclear-bomblike explosion because of the special conditions required for an effective bomb. The products of fission, however, whether from a bomb or a reactor, are radioactive. When the bomb explodes, these products are thrown into the atmosphere and eventually reach the ground as dangerous "fallout." In a reactor, these products must be periodically removed and transferred to safe storage facilities.

A second important nuclear reaction is called **fusion.** In this case, two smaller nuclei are combined or fused into a larger nucleus. Again, this process is accompanied by the release of vast amounts of energy. A typical fusion reaction is

$$_{1}^{2}\text{H} + _{1}^{3}\text{H} \longrightarrow _{2}^{4}\text{He} + _{0}^{1}\text{n}$$

In nuclear fusion several nuclei combine to form one nucleus.

Figure 3.15
Diagram of a nuclear power plant.

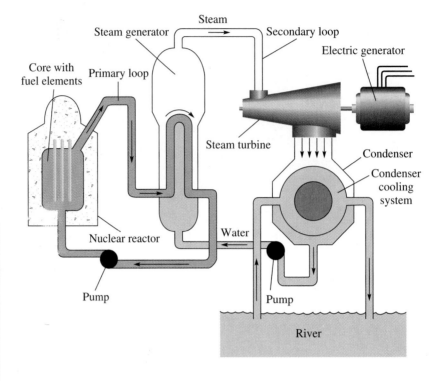

Figure 3.16
The now-familiar mushroom cloud that follows a nuclear explosion.

Our sun is powered by fusion reactions occurring in its core. Fusion reactions also generate the awesome destructive energy of a thermonuclear or hydrogen bomb. Scientists are currently attempting to bring fusion reactions under sufficient control to permit them to be used to generate power for peaceful uses. The problem is that to get two positively charged nuclei together, one has to overcome their repulsion for one another (like charges repel). One does this by slamming them into one another at high speed—and high speed means high temperature, in this case millions of degrees. The need to produce and contain these high temperatures is slowing the development of controlled fusion. (Note that this problem does not exist in fission reactions, which are triggered by neutrons, lasers, etc.; these neutral particles are not repulsed by the fissionable nucleus.)

3.8
RADIATION AND LIVING THINGS

High doses of radiation (on the order of 2000 rads) cause gross destruction of tissue. A person exposed to about 2000 rads passes into shock and dies in a few hours. With an exposure of 500 rads, the person usually survives long enough to exhibit the several phases of acute radiation syndrome.

1. A short latent period (a few hours) when no effects are observed.
2. A period of nausea and vomiting (which usually ends in 24 hr) and a drop in the white blood cell count.
3. A period of few symptoms. A low-grade fever may persist.
4. The last period. Loss of hair begins rather abruptly anywhere from about the seventeenth through the twenty-first day. Increased discomfort sets in, with loss of appetite

Figure 3.17
Radiation burns suffered from the Chernobyl accident.

and diarrhea. The body temperature rises, and the patient complains of pain in the throat and gums. Emaciation sets in, the general condition deteriorates, and the person dies.

With smaller doses, recovery can take place (Figure 3.17), but such recovery may not be complete. Malignant growths may appear even years later, and genetic effects may appear in succeeding generations.

Some cells are more susceptible to radiation than others. Those that are being constantly and rapidly replaced are affected most. These include the intestinal mucosa, germ cells, embryonic cells, blood cells, and the organs responsible for producing blood cells, such as the bone marrow. Damage to reproductive cells will show up as abnormalities in the descendants of affected persons.

The **rem** (*r*oentgen *e*quivalent, *m*an) is a measure of the relative biological damage produced by a particular dose of radiation. For our purposes, the roentgen and the rad are about equivalent because 1 R generates about 1 rad of energy when absorbed in muscle tissue. It is estimated that a whole-body exposure of about 500 rads would kill most of us. The International Commission on Radiological Protection recommends that adults whose occupations expose them to ionizing radiation limit their exposure to 5 rem in any one year.

People working with radioactive materials can do several things to protect themselves (Table 3.2). The simplest is to move away from the source, for intensity of radiation decreases with distance from the source. Workers can also be protected by shielding.

Primitive organisms such as yeasts, bacteria, and viruses have much greater ability to withstand radiation than do mammals.

1. Distance: The more distant the source, the greater the safety.
2. Sample size: The smaller the radiating sample, the greater the safety.
3. Radiation: The less penetrating the radiation, the greater the safety. Thus, for external sources safety increases in the order $\gamma < \beta < \alpha$.
4. Half-life: The longer the half-life, the greater the safety.
5. Time: Generally, the shorter the time of exposure, the greater the safety.
6. Frequency: The fewer the exposures, the greater the safety.

Table 3.2
Protection from Nuclear Radiation

Figure 3.18

The relative penetrating power of alpha, beta, and gamma radiation. Alpha particles are stopped by a sheet of paper. Beta particles will not penetrate a sheet of aluminum foil. It takes several centimeters of lead to block gamma rays.

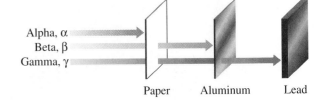

A sheet of paper will stop most alpha particles. A block of wood or a thin sheet of aluminum will stop beta particles. But it takes several meters of concrete or several centimeters of lead to stop gamma rays (Figure 3.18).

3.9
NUCLEAR MEDICINE

Nuclear medicine involves two distinct uses of radioisotopes: therapeutic and diagnostic. In radiation therapy, an attempt is made to treat or cure disease with radiation. Cancer is not one disease but many. Some forms are particularly susceptible to radiation therapy. Radiation is carefully aimed at the cancerous tissue, and exposure of normal cells is minimized (Figure 3.19). If the cancer cells are killed by the destructive effects of the radiation, the malignancy is halted. But persons undergoing radiation therapy often get quite sick from the treatment. Nausea and vomiting are the usual symptoms of radiation sickness. (Remember that the intestinal mucosa is particularly susceptible to

Figure 3.19

A cobalt-60 unit for radiation therapy.

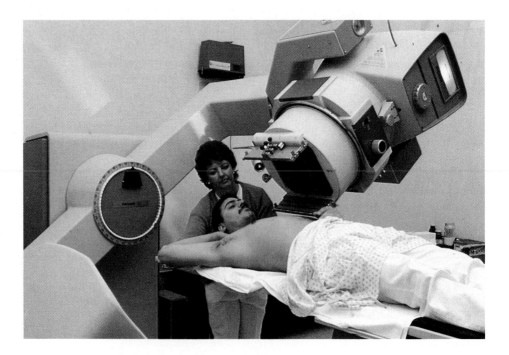

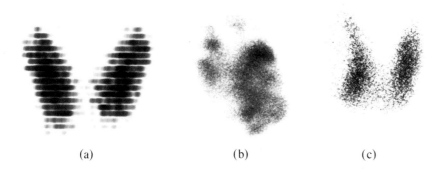

(a)   (b)   (c)

Figure 3.20
A linear photoscanner produced these pictures of (a) a normal thyroid, (b) a multinodal goiter, and (c) a thyroid adenoma.

radiation.) Thus, the aim of radiation therapy is to destroy the cancerous cells before too much damage is done to healthy tissue.

Radioisotopes are also used for diagnostic purposes, to help provide information about the type or extent of illness. Radioactive iodine-131 ($^{131}_{53}I$) is used to determine the size, shape, and activity of the thyroid gland as well as to treat cancer located in this gland and to control a hyperactive thyroid. First, the patient drinks a solution of potassium iodide, KI, incorporating iodine-131. The body concentrates iodide in the thyroid. Large doses are used for treatment of thyroid cancer; the radiation from the isotope concentrates in the thyroid cancer cells even if the cancer has spread to other parts of the body. For diagnostic purposes, however, only a small amount is needed. A radiation detector records the uptake of the isotope by body tissue and translates this information into a permanent visual record. The "picture" that results is called a **photoscan** (Figure 3.20).

One can imagine an ideal isotope for diagnostic scanning. The isotope should be a gamma emitter because gamma rays have high penetrating power. For diagnosis, radiation must escape from the body so the scanner can detect it. It would be best if neither alpha nor beta particles were emitted. These are not needed for detection and would simply do unnecessary damage to the body. The gamma rays should have just the right amount of energy: they should be neither so weak that we would have to wait around forever to get enough of a response from our detector nor so strong that they would impart unnecessary energy to the body in making their way out. The half-life also should be just right: neither so short that the activity is gone before we can measure it nor so long that significant activity remains long after the scan has been obtained.

Happily, an isotope much like this ideal one is available. It is the synthetic element technetium-99m ($^{99m}_{43}Tc$). The *m* stands for **metastable,** which means that this isotope will give up some energy to become a more stable version of the same isotope (same atomic number, same mass number). The energy it gives up is the gamma ray we need for detecting the isotope while scanning.

$$^{99m}_{43}Tc \longrightarrow \, ^{99}_{43}Tc + \gamma$$

The energy of the gamma ray is just about right, and so is the half-life, about 6 hr. Much effort has gone into making technetium-99m available in a variety of preparations. In some forms it is useful for kidney scanning; in other forms it will be concentrated by the liver and spleen. It can be used for brain scans or for lung scans. Thus, this isotope can replace a number of less-than-ideal radioisotopes. Table 3.3 lists some radioisotopes in common use in medicine.

Thallium-201 is used to study blood flow in the coronary arteries. Cardiologists frequently inject the radioisotope immediately after administering a stress test.

Table 3.3
Some Radioisotopes and
Their Application in
Medicine

Isotope	Name	Use
^{74}As	Arsenic-74	Location of brain tumors
^{51}Cr	Chromium-51	Determination of volume of red blood cells and total blood volume
^{58}Co	Cobalt-58	Determination of uptake of vitamin B_{12}
^{60}Co	Cobalt-60	Radiation treatment of cancer
^{131}I	Iodine-131	Detection of thyroid malfunction; measurement of liver activity and fat metabolism; treatment of thyroid cancer
^{59}Fe	Iron-59	Measurement of rate of formation and lifetime of red blood cells
^{32}P	Phosphorus-32	Detection of skin cancer or cancer of tissue exposed by surgery
^{226}Ra	Radium-226	Radiation therapy for cancer
^{24}Na	Sodium-24	Detection of constrictions and obstructions in the circulatory system
^{99m}Tc	Technetium-99m	Imaging of brain, thyroid, liver, kidney, bone, lung, and cardiovascular system
^{3}H	Tritium	Determination of total body water

3.10
MEDICAL IMAGING

Medical imaging provides a means of looking at internal organs without resorting to surgery. The history of medical imaging dates back to the discovery of X-rays at the turn of the century. Penetrating X-rays were used almost immediately to visualize skeletal structure. Radiation from an external X-ray source passes through the body (except where it is absorbed by more dense structures, such as bone) and exposes film, thus providing a picture that distinguishes the more dense structures from less dense tissue. Softer tissue can be visualized by introducing material that absorbs X-rays into the area to be studied. For example, compounds of barium are used to view portions of the digestive tract (Figure 3.21).

Figure 3.21
Barium sulfate ($BaSO_4$) is insoluble in water and opaque to X-rays. When swallowed, this salt can be used to outline the stomach for X-ray photographs.

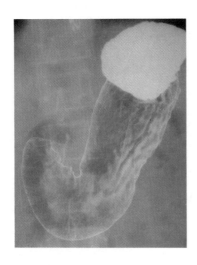

Radioisotopes can also be used to visualize internal organs. In this technique, the source of the radiation is inside the body, and the radiation (usually gamma rays) is detected as it emerges from the body.

Both X-ray technology and nuclear imaging have been coupled with computer technology to provide versatile and powerful imaging techniques. In *computed tomography* (referred to as CT or, sometimes, CAT scanning), many X-ray readings are obtained, processed by a computer, and then displayed. The resulting pictures present cross-sectional slices of a portion of the body. A series of these pictures gives a three-dimensional view of organs such as the brain.

Positron emission tomography (PET) can be used to measure dynamic processes occurring in the body, such as blood flow or the rate at which oxygen or glucose is being metabolized. PET scans are used to pinpoint the area of the brain damage that triggers severe epileptic seizures and to detect tiny blockages in blood vessels and hard-to-spot tumors. Compounds incorporating positron-emitting isotopes, such as carbon-11, are inhaled or injected prior to the scan. Before the emitted positron can travel very far in the body, it encounters an electron (in any ordinary matter there are numerous electrons), and two gamma rays are produced.

$$^{11}_{6}C \longrightarrow {}^{11}_{5}B + {}^{0}_{+1}e$$
$$^{0}_{+1}e + {}^{0}_{-1}e \longrightarrow 2\ \gamma$$

The gamma rays exit from the body in exactly opposite directions and are recorded by detectors positioned on opposite sides of the patient. If the recorders are set so that two simultaneous gamma rays must be "seen," gamma rays resulting from natural background radiation are ignored. A computer is then used to calculate the point within the body at which the annihilation of the positron and electron occurs, and an image of that area is produced.

Both X-rays and nuclear radiation are ionizing radiations, so there is always some tissue damage involved. Modern techniques keep this damage to a minimum. Other imaging techniques use nonionizing radiation.

Perhaps you are familiar with *ultrasonography*. In this technique, high-frequency sound waves are bounced off tissue, and the echo of the sound wave is recorded. Once again, a computer is used to process the data and produce an image of the tissue. (The technique is related to sonar detection of submarines.) Because no ionizing radiation is involved, ultrasonography is used extensively in obstetrics for following fetal development (Figure 3.22). A 1984 conference on the use of ultrasonography (sponsored by the

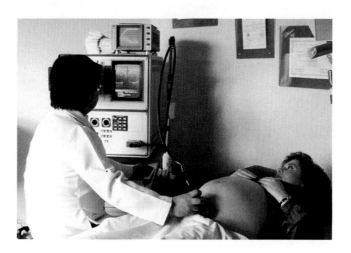

Figure 3.22
An ultrasonogram test on an eight months pregnant woman.

National Institutes of Health) concluded, however, that even this technique should not be used casually. The conference recommended that an ultrasonogram be obtained only if there is a good medical reason for doing so (for example, to evaluate fetal growth in mothers who have diabetes or hypertension).

Another nonionizing technique is *magnetic resonance imaging* (MRI). This technique depends on the fact that some nuclei behave as if they were little magnets; placed in a strong magnetic field (e.g., between the poles of a much more powerful magnet), the nuclei line up in a certain manner. When supplied with the right amount of energy, the nuclei absorb the energy and flip over in the magnetic field. The energy required to flip the nuclei is provided by nonionizing radiation. The absorption of the energy by the nuclei can be detected by appropriate equipment, and an image of the tissue in which the nuclei reside is produced. (Again, a computer is used to process the image.) The technique has demonstrated its potential for providing not only images of organs but also information about the metabolic activity in particular tissues. MRI is particularly adept at detecting very small tumors, blockages in blood vessels, or damage to vertebral discs.

The cost of most of these computer-based technologies is very high, ranging into the millions of dollars for a single installation (Figure 3.23). The great advantage of these

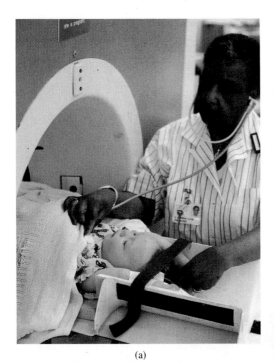

(a)

(b)

(c)

Figure 3.23
The mighty machines of modern medicine.
(a) CT scanner; (b) MRI scanner; (c) PET scanner.

techniques is that they provide information that could otherwise be obtained only by subjecting the individual to the risks of surgery. In some instances, the information provided by these techniques is not available by any other route.

3.11
OTHER APPLICATIONS

In the last several decades nuclear research has emphasized peaceful rather than military applications. Aside from the uses already mentioned, many other applications have been found.

Elements that do not appear in nature can be prepared through transmutation. Examples of synthetic elements include elements 43, 61, and 85 and the transuranium elements. Chemical reaction mechanisms (the manner in which atoms interact to make a new product) can often be determined only with radioactive tracers. In photosynthesis, for instance, it has been found that carbon dioxide does indeed play a major role. The calcium uptake in metabolism has been found to be 90% in the young but only 40% in older animals and humans. Polycythemia vera (formation of too many red blood cells) has been studied with the aid of iron-59. It was found that ten times as much iron as the body can use is assimilated by patients with this disease. Through similar studies it has also been learned that the uptake of trace elements by trees is quite pronounced in the winter. Thus, zinc moves up the trunk of a tree at the rate of about 2 ft/day. The effectiveness of industrial lubricants can be measured by monitoring the concentration of metal residues in the oils. By tagging the metal with radioactive isotopes, concentrations as low as 10^{-19} g/L can be detected.

Radioisotopes are also used as sources for the irradiation of foodstuffs as a method of preservation (Figure 3.24). The radiation destroys microorganisms that cause food spoilage. Irradiated food shows little change in taste or appearance. Some people are concerned about possible harmful effects of chemical substances produced by the radiation, but there is no good evidence of harm to laboratory animals fed irradiated food, nor are there any known adverse effects in humans in countries where irradiation has been used for several years. There is no residual radiation in the food after the sterilization process.

Figure 3.24
Gamma radiation delays the decay of mushrooms. Those on the right were irradiated; those on the left were not.

3.12
THE NUCLEAR AGE REVISITED

We live in an age in which fantastic forces have been unleashed. The threat of nuclear war was a constant specter from World War II until the collapse of the Soviet Union in 1991. Nuclear bombs have been used to destroy cities—and men, women, and children. Science and scientists have been greatly involved in it all. Would the world be a better place had we not discovered the secrets of the atomic nucleus? Consider this: More lives have been saved through nuclear medicine than have been destroyed by nuclear bombs. Also, nuclear power, with all its attendant problems, still may be one of our best hopes for a plentiful energy supply until well into the twenty-first century. Imagine yourself to be Dalton. Could this gentle Quaker, whose formal education ended when he was 11 years old, have anticipated the chain of scientific developments that would follow from his original speculations on the nature of matter?

EXERCISES

1. Define or identify each of the following.
   a. isotope
   b. alpha particle
   c. beta particle
   d. gamma ray
   e. half-life
   f. positron
   g. curie (Ci)
   h. roentgen (R)
   i. rad
   j. rem
   k. film badge
   l. Geiger counter
   m. background radiation
   n. ionizing radiation
   o. fission
   p. fusion
   q. radioisotope
   r. artificial transmutation

2. How are X-rays and gamma rays similar? How are they different?

3. Draw the nuclear symbols for the following subatomic particles.
   a. alpha particle
   b. beta particle
   c. neutron
   d. positron

4. When a nucleus emits a beta particle, what changes occur in the mass number and atomic number of the nucleus?

5. Lead-209 undergoes beta decay. Write a balanced equation for this reaction.

6. When a nucleus emits an alpha particle, what changes occur in the mass number and atomic number of the nucleus?

7. Thorium-225 undergoes alpha decay. Write a balanced equation for this reaction.

8. When a nucleus emits a gamma ray, what changes occur in the mass number and the atomic number of the nucleus?

9. Write an equation representing the emission of a gamma ray by gold-186.

10. Radioactive radon gas in homes could be a lung cancer risk. One set of reactions involves two successive alpha decays from radon-222. What are the products formed?

11. Selenium-82 undergoes a rare reaction in which two beta particles are emitted. What is the product?

12. C. E. Bemis and colleagues at Oak Ridge National Laboratory confirmed the synthesis of element 104, the half-life of which was only 4.5 s. Only 3000 atoms of the element were created in the tests. How many atoms were left after 4.5 s? After 9.0 s?

13. Krypton-81m is used for lung ventilation studies. Its half-life is 13 s. How long will it take the activity of this isotope to reach one-fourth of its original value?

14. When a nucleus emits a positron, what changes occur in the mass number and the atomic number of the nucleus?

15. Write a balanced equation for the emission of a positron by sulfur-31.

16. When a nucleus emits a neutron, what changes occur in the mass number and the atomic number of the nucleus?

17. Write a balanced equation for the emission of a neutron by bromine-87.

18. When a nucleus emits a proton, what changes occur in the mass number and the atomic number of the nucleus?

19. Write a balanced equation for the emission of a proton by magnesium-21.

20. Complete the following equations.
   a. $^{179}_{79}\text{Au} \rightarrow \ ^{175}_{77}\text{Ir} + \ ?$
   b. $^{23}_{10}\text{Ne} \rightarrow \ ^{23}_{11}\text{Na} + \ ?$

21. Complete the following equations.
   a. $^{10}_{5}\text{B} + \ ^{1}_{0}\text{n} \rightarrow \ ^{4}_{2}\text{He} + \ ?$
   b. $^{12}_{6}\text{C} + \ ^{2}_{1}\text{H} \rightarrow \ ^{13}_{6}\text{C} + \ ?$
   c. $^{121}_{51}\text{Sb} + \ ? \rightarrow \ ^{121}_{52}\text{Te} + \ ^{1}_{0}\text{n}$
   d. $^{154}_{62}\text{Sm} + \ ^{1}_{0}\text{n} \rightarrow 2 \ ^{1}_{0}\text{n} + \ ?$

22. When magnesium-24 is bombarded with a neutron, a proton is ejected. What new element is formed? (*Hint:* Write a balanced nuclear equation.)

23. A radioactive isotope decays to give an alpha particle and bismuth-211. What was the original element?

24. Compare nuclear fission and nuclear fusion.

25. Which subatomic particles are responsible for carrying on the chain of reactions that are characteristic of nuclear fission?

26. Why are such high temperatures required for nuclear fusion reactions?

27. What dangers are there in using nuclear fission to generate power?

28. Abandoned salt mines are often cited as good places to store nuclear waste. What are the pros and cons of such disposal?

29. List two ways in which workers can protect themselves from the radioactive materials with which they work.

30. A pair of gloves would be sufficient to shield the hands from which type of radiation, the heavy alpha particles or the massless gamma rays?

31. Heavy lead shielding is necessary as protection from which type of radiation, alpha, beta, or gamma?

32. Plutonium is especially hazardous when inhaled or ingested because it emits alpha particles. Why would alpha particles cause more damage to tissue than beta particles?

33. Explain how radioisotopes can be used for therapeutic purposes.

34. Which radioisotope has been used extensively for treatment of overactive or cancerous thyroid glands?

35. Describe the use of a radioisotope as a diagnostic tool in medicine.

36. What are some of the characteristics that make technetium-99m such a useful radioisotope for diagnostic purposes?

37. The activity of a radiation source is 500 Ci. The activity of another source is 10 mCi. To be used properly, one source is taken into the body and the other remains outside the body during treatment. Which is likely to be the internal source, and which the external?

38. Patient A takes internally an iodine-131 sample with an activity of 150 mCi. Patient B takes internally a dose of 15 μCi. In which patient is the iodine-131 being used to treat a malignancy, and in which patient is the isotope used for imaging the thyroid gland? Both are adults.

39. About 2 mCi of thallium-201 is given by intravenous administration for imaging the heart. It is estimated that the total body radiation dose in humans is about 0.07 rad per mCi of thallium-201. How does the radiation dose from this procedure compare to the lethal dose for humans?

40. What form of radiation is detected in CT scans?

41. What form of radiation is detected in PET scans?

42. What is the advantage of using nonionizing radiation for medical imaging?

43. Name two imaging techniques that do not use ionizing radiation.

44. Discuss the impact of nuclear science on the following.
    a. war and peace   **b.** medicine
    c. our energy needs

45. Cesium is one of the components of nuclear fallout. Why is it a particularly dangerous threat to the environment?

46. Nuclear wastes typically need to be stored for 20 half-lives to be safe. This translates into hundreds of years of storage time. (For instance, cesium-137 and strontium-90 have half-lives of about 30 years.) Would the shooting of such wastes into outer space be a responsible solution?

47. The following coded telephone conversation took place at 3:25 P.M. on December 2, 1942, between Dr. Compton from the University of Chicago and Dr. Conant at Harvard University.

    Dr. Compton: "The Italian navigator has landed in the new world."

    Dr. Conant: "How were the natives?"

    Dr. Compton: "Very friendly."

    Can you unravel the code?

Chapter 4
CHEMICAL BONDS

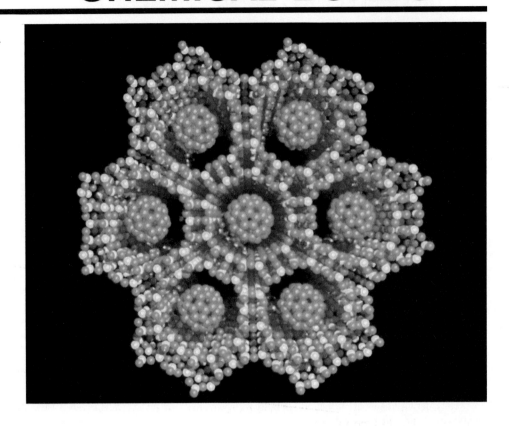

A computer graphics representation of buckminsterfullerene, a novel form of carbon (green spheres), fitting inside the channels of a zeolite polymer. Chemical bonds hold atoms in place in molecules, giving structure to matter.

There are slightly more than 100 chemical elements. There are millions of chemical compounds. To form these compounds, atoms of different elements must be held together in specific combinations. **Chemical bonds** are the forces that maintain these arrangements. Chemical bonding also plays a major role in determining the state of matter. At room temperature, water is a liquid, carbon dioxide is a gas, and table salt is a solid—because of differences in chemical bonding. Bonds even determine the physical shape and flexibility of molecules, that is, whether they are spherical or flat, rigid or wobbly.

As scientists developed an understanding of the nature of chemical bonding, they gained the ability to manipulate the structure of compounds. Dynamite, birth control pills, synthetic fibers, and a thousand other products were fashioned in chemical laboratories and have dramatically changed the way we live. We are now entering an era that promises (some would say threatens) even greater change.

The DNA molecule—the chemical basis of heredity—carries its genetic message in its bonds. Two DNA molecules in turn are joined to each other by a special kind of bonding (see Section 23.3). Whether an organism is fish, fowl, hippopotamus, or human is determined by the arrangement of the bonds in DNA. Scientists already have the ability to rearrange these bonds, and this ability has given them limited control over the structure of living matter. Genetic engineering now enables scientists to custom-tailor certain organisms.

4.1
THE ART OF DEDUCTION: STABLE ELECTRON CONFIGURATIONS

In our discussion of the atom and its structure (Chapter 2), we followed the historical development of some of the more important atomic concepts. Some of the nuclear concepts (Chapter 3) were approached in the same manner. We could continue to look at chemistry in this manner, but that would require several volumes of print—if we got very far—and perhaps more of your time than you care to spend. We won't abandon the historical approach entirely, but we will emphasize that other aspect of scientific enterprise: deduction.

The art of deduction works something like this.

1. *Fact*   The noble gases, such as helium, neon, and argon, are inert (i.e., they undergo few, if any, chemical reactions).
2. *Theory*   The inertness of the noble gases is due to their electron structures.
3. *Deduction*   If other elements could alter their electron structures to become more like the noble gases, they would become less reactive.

To illustrate, let's look at an atom of the element sodium (Na). It has 11 electrons, two in the first energy level, eight in the second, and one in the third. If the atom could give up an electron, it would have the same electron configuration as an atom of the noble gas neon (Ne).

Recall that neon has the structure

If a chlorine atom (Cl) could gain an electron, it would have the same electron configuration as argon (Ar).

The electron configuration of the argon atom is

Chlorine does not *become* argon. The chloride ion and argon atom have the same electron configuration, but the chloride ion has 17 protons in its nucleus and a net charge of 1−. The argon atom has 18 protons in its nucleus and no net charge; it is electrically neutral. Nor does sodium become neon by giving up an electron. The sodium ion has the same electron configuration as neon, but the nuclei differ in the number of protons.

Positively charged ions are called **cations** and negatively charged ions are called **anions** (see Section 12.1).

The sodium atom, having lost an electron, becomes positively charged. It has 11 protons (11+) and only 10 electrons (10−). It is written Na^+ and is called a *sodium ion*. The chlorine atom, having gained an electron, becomes negatively charged. It has 17 protons (17+) and 18 electrons (18−). It is written Cl^- and is called a *chloride ion*. Note that a positive charge, as in Na^+, indicates that one electron has been lost per atom. Similarly, a negative charge, as in Cl^-, indicates that one electron has been gained per atom. **Ions** are charged units, that is, units in which the number of electrons is *not* equal to the number of protons.

EXAMPLE 4.1

What is the charge on an aluminum atom that has lost three electrons?

SOLUTION
The neutral aluminum atom has 13 electrons and 13 protons (its atomic number is 13). The ion would have 13 protons (13+) and 10 electrons (10−). The net charge on the aluminum ion would be 3+. The symbol is Al^{3+}.

Practice Exercise
What is the charge on a sulfur atom that has gained two electrons?

4.2
ELECTRON-DOT FORMULAS

In forming ions, the cores of sodium atoms and chlorine atoms do not change. It is convenient, therefore, to let the symbol represent the *core* of the atom (nucleus plus inner electrons). The **valence electrons** (the electrons in the outermost energy level) are then represented by dots. The equations of the preceding section can then be written as follows.

$$Na\cdot \longrightarrow Na^+ + 1\,e^-$$

and

$$:\overset{\cdot\cdot}{\underset{\cdot\cdot}{Cl}}\cdot + 1\,e^- \longrightarrow :\overset{\cdot\cdot}{\underset{\cdot\cdot}{Cl}}:^-$$

Such forms, in which the symbol of the element represents the core and dots stand for valence electrons, are called **electron-dot formulas.** They are also called Lewis dot structures (or Lewis structures) in honor of G. N. Lewis (1875–1946), an American chemist who first proposed this method.

It is especially easy to write electron-dot formulas for most of the main group elements. The number of valence electrons for most of these elements is equal to the group number. Table 4.1 gives the electron-dot formulas for selected elements. Notice that there is a pattern to the way in which the dots are drawn. For elements with four or fewer outer electrons, the electrons are isolated from one another. With the appearance of the fifth electron, a pairing up begins. This is a useful convention—as we shall see shortly.

It doesn't matter whether you draw lithium in any of the following ways. They are all correct.

$$\overset{\cdot}{Li} \qquad Li\cdot \qquad Li \qquad \cdot Li$$

IA	IIA	IIIA	IVA	VA	VIA	VIIA	Noble Gases
H·							He:
Li·	·Be·	·B·	·C·	:N·	:O·	:F·	:Ne:
Na·	·Mg·	·Al·	·Si·	:P·	:S·	:Cl·	:Ar:
K·	·Ca·				:Se·	:Br·	:Kr:
Rb·	·Sr·				:Te·	:I·	:Xe:
Cs·	·Ba·						

Table 4.1
Electron-Dot Formulas for Selected Main Group Elements

And magnesium can be drawn

$$\overset{\bullet}{Mg}\cdot \qquad \underset{\bullet}{Mg} \qquad \cdot \overset{\bullet}{Mg} \qquad \cdot Mg\cdot \qquad \overset{\bullet}{Mg}$$

As long as you show two outer electrons, you're right. As we get into a discussion of compounds, you will see that sticking to one choice of structures will sometimes simplify writing formulas for compounds.

Symbolism is a convenient, shorthand way of conveying a lot of information in compact form. It is the chemist's most efficient and economical form of communication. Learning this symbolism is much like learning a foreign language. Once you master a certain basic "vocabulary," the rest is easier.

EXAMPLE 4.2

Without referring to Table 4.1, give electron-dot formulas for calcium, oxygen, and phosphorus.

SOLUTION
Calcium is in Group IIA, oxygen is in Group VIA, and phosphorus is in Group VA. The electron-dot formulas are therefore

$$\cdot Ca\cdot \qquad :\overset{\bullet}{\underset{\bullet}{O}}: \qquad :\overset{\bullet}{P}\cdot$$

Practice Exercise
Without referring to Table 4.1, give electron-dot formulas for each of the following elements.

a. Ar b. Sr c. F
d. N e. K f. S

Figure 4.1
Sodium, a soft silvery metal, reacts with chlorine, a greenish yellow gas, to form sodium chloride (ordinary table salt), a white crystalline solid.

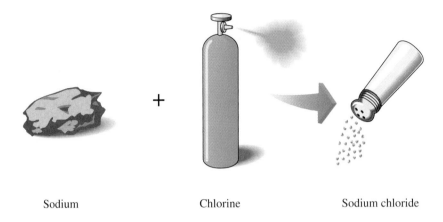

Sodium   Chlorine   Sodium chloride

4.3
SODIUM REACTS WITH CHLORINE: THE FACTS

Sodium is a highly reactive metal. It is soft enough to be cut with a knife. When freshly cut, it is bright and silvery, but it dulls rapidly because it reacts with oxygen in the air. In fact, it reacts so readily in air that it is usually stored under oil or kerosene. Sodium reacts violently with water also, producing so much heat that it melts. A small piece will form a spherical bead after melting and race around on the surface of the water as it reacts.

Chlorine is a greenish yellow gas. It is familiar as a disinfectant for swimming pools and city water supplies. (The actual substance added may be a compound that reacts with water to form chlorine.) Who hasn't been swimming in a pool that had ''so much chlorine in it that you could taste it''? Chlorine is quite irritating to the respiratory tract. In fact, chlorine was used as a poison gas in World War I.

When a piece of sodium is dropped into a flask containing chlorine gas, a violent reaction ensues. A white solid that is quite unreactive is formed. It is a familiar compound—sodium chloride (table salt) (Figure 4.1).

4.4
THE SODIUM–CHLORINE REACTION: THEORY

Actually, chlorine gas is composed of Cl_2 molecules. Each atom of the molecule takes an electron from a sodium atom. Two sodium ions and two chloride ions are formed.

$$Cl_2 + 2\,Na \longrightarrow 2\,Cl^- + 2\,Na^+$$

The sodium–chlorine reaction illustrates a basic tendency of nature. More reactive (more energetic) substances tend to become less reactive (less energetic) substances, releasing energy in the process. A sodium atom becomes less reactive by *losing* an electron. A chlorine atom becomes less reactive by *gaining* an electron. What happens when sodium atoms come into contact with chlorine atoms? Chlorine extracts an electron from a sodium atom. Using electron-dot formulas, the equation for this reaction is written

$$Na\cdot \; + \; :\overset{..}{\underset{..}{Cl}}\cdot \; \longrightarrow \; Na^+ \; + \; :\overset{..}{\underset{..}{Cl}}:^-$$

Figure 4.2
The arrangement of ions in a sodium chloride crystal.

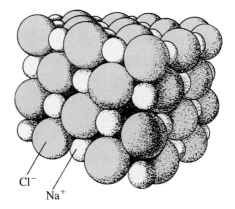

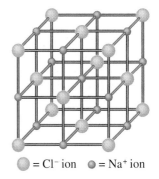

$\bigcirc = Cl^-$ ion $\bullet = Na^+$ ion

Figure 4.3
Ball-and-stick model of a sodium chloride crystal.

The two ions formed from sodium and chlorine atoms have opposite charges. They are strongly attracted to one another. Remember, however, that in even a tiny grain of salt, there are billions and billions of each kind of ion. These ions arrange themselves in an orderly fashion (Figure 4.2). The arrangements are repeated many times in all directions—front and back, left and right, top and bottom—to make a crystal of sodium chloride. Each sodium ion attracts (and is attracted by) six chloride ions (the ones to its front and back, its top and bottom, and its two sides). Each chloride ion attracts (and is attracted by) six sodium ions. The forces holding the crystal together (the attractive forces between positive and negative ions) are called **ionic bonds.**

Scientists sometimes use different models to represent the same system. The model employed in Figure 4.2 is a space-filling model showing the relative sizes of the sodium and chloride ions. Sometimes a ball-and-stick model is employed to better show the geometry of the crystal (Figure 4.3). From this model it is easy to see the cubic arrangement of the ions. In the crystal as a whole, for each sodium ion there is one chloride ion; thus, the ratio of ions is $1:1$, and the simplest formula for the compound is NaCl. The symbols Na and Cl, written together, stand for the compound sodium chloride. The formula is also used to represent one sodium ion and one chloride ion.

A **crystal** is a solid substance with a regular arrangement of its constituent particles. The solid as a whole has a well-defined, regular shape.

4.5
IONIC BONDS: SOME GENERAL CONSIDERATIONS

As we might expect, potassium, a metal in the same family as sodium, also reacts with chlorine. The reaction yields potassium chloride (KCl).

$$K\cdot \ + \ \cdot\ddot{\underset{\cdot\cdot}{Cl}}: \ \longrightarrow \ K^+ \ + \ :\ddot{\underset{\cdot\cdot}{Cl}}:^-$$

And potassium reacts with bromine, a reddish brown liquid in the same family as chlorine, to form a stable white crystalline solid called potassium bromide (KBr).

$$K\cdot \ + \ \cdot\ddot{\underset{\cdot\cdot}{Br}}: \ \longrightarrow \ K^+ \ + \ :\ddot{\underset{\cdot\cdot}{Br}}:^-$$

Sodium also reacts with bromine to form sodium bromide.

EXAMPLE 4.3

Use electron-dot formulas to show the transfer of electrons from sodium atoms to bromine atoms to form ions with noble gas configurations.

SOLUTION

Sodium has one valence electron, and bromine has seven. Transfer of the single electron from sodium to bromine leaves each with a noble gas configuration.

$$\text{Na}\cdot \ + \ \cdot\ddot{\text{Br}}\!: \ \longrightarrow \ \text{Na}^+ \ + \ :\!\ddot{\text{Br}}\!:^-$$

Practice Exercise

Use electron-dot formulas to show the transfer of electrons from lithium atoms to fluorine atoms to form ions with noble gas configurations.

Magnesium, a Group IIA metal, is harder and less reactive than sodium. Magnesium reacts with oxygen, a Group VIA element (a colorless gas), to form another stable white crystalline solid called magnesium oxide (MgO).

$$\cdot\text{Mg}\cdot \ + \ \cdot\ddot{\text{O}}\!: \ \longrightarrow \ \text{Mg}^{2+} \ + \ :\!\ddot{\text{O}}\!:^{2-}$$

Magnesium must give up two electrons and oxygen must gain two electrons for each to have the same configuration as the noble gas neon.

An atom such as oxygen, which needs two electrons to complete a noble gas configuration, may react with potassium atoms, each of which has only one electron to give. In this case, two atoms of potassium are needed for each oxygen atom. The product is potassium oxide (K_2O).

$$
\begin{array}{ccccc}
\text{K}\cdot & & & \text{K}^+ & \\
 & + \ \cdot\ddot{\text{O}}\!: & \longrightarrow & & + \ :\!\ddot{\text{O}}\!:^{2-} \\
\text{K}\cdot & & & \text{K}^+ &
\end{array}
$$

By this process, each potassium atom achieves the argon configuration. Oxygen again assumes the neon configuration.

EXAMPLE 4.4

Use electron-dot formulas to show the transfer of electrons from magnesium atoms to nitrogen atoms to form ions with noble gas configurations.

SOLUTION

$$
\begin{array}{ccccc}
\cdot\text{Mg}\cdot & & & \text{Mg}^{2+} & \\
 & \cdot\ddot{\text{N}}\cdot & & & :\!\ddot{\text{N}}\!:^{3-} \\
\cdot\text{Mg}\cdot & + & \longrightarrow & \text{Mg}^{2+} & + \\
 & \cdot\ddot{\text{N}}\cdot & & & :\!\ddot{\text{N}}\!:^{3-} \\
\cdot\text{Mg}\cdot & & & \text{Mg}^{2+} &
\end{array}
$$

Each of three magnesium atoms gives up two electrons (a total of six), and each of the two nitrogen atoms acquires three (a total of six). Notice that the total positive and total negative charges on the products are equal (6+ and 6−). Magnesium reacts with nitrogen to give magnesium nitride (Mg_3N_2).

Practice Exercise
Use electron-dot formulas to show the transfer of electrons from aluminum atoms to oxygen atoms to form ions with noble gas configurations.

Generally speaking, elements on the left side of the periodic chart (especially those on the far left) react with elements on the far right (excluding the noble gases) to form stable crystalline solids. The theory is that the elements on the left (the *metals*) tend to give up electrons to the elements of the right (*nonmetals*). The crystalline solids are held together by the attraction of oppositely charged ions (i.e., ionic bonds).

4.6
NAMES OF SIMPLE IONS AND IONIC COMPOUNDS

Names of simple monatomic (single-atom) positive ions are derived from those of the parent elements by addition of the word *ion*. A sodium atom (Na), upon losing an electron, becomes a *sodium ion* (Na^+). A magnesium atom (Mg), upon losing two electrons, becomes a *magnesium ion* (Mg^{2+}). Names of simple monatomic negative ions are derived from those of the parent elements by changing the usual ending to *-ide* and adding the word ion. A chlor*ine* atom (Cl), upon gaining an electron, becomes a chlor*ide* ion (Cl^-). A sulf*ur* atom (S), upon gaining two electrons, becomes a sulf*ide* ion (S^{2-}).

Names and symbols for several important monatomic ions are given in Table 4.2.

Group	Element	Name	Symbol for Ion
IA	Hydrogen	Hydrogen ion[a]	H^+
	Lithium	Lithium ion	Li^+
	Sodium	Sodium ion	Na^+
	Potassium	Potassium ion	K^+
IIA	Magnesium	Magnesium ion	Mg^{2+}
	Calcium	Calcium ion	Ca^{2+}
IIIA	Aluminum	Aluminum ion	Al^{3+}
VA	Nitrogen	Nitride ion	N^{3-}
VIA	Oxygen	Oxide ion	O^{2-}
	Sulfur	Sulfide ion	S^{2-}
VIIA	Fluorine	Fluoride ion	F^-
	Chlorine	Chloride ion	Cl^-
	Bromine	Bromide ion	Br^-
	Iodine	Iodide ion	I^-
IB	Copper	Copper(I) ion (cuprous ion)	Cu^+
		Copper(II) ion (cupric ion)	Cu^{2+}
	Silver	Silver ion	Ag^+
IIB	Zinc	Zinc ion	Zn^{2+}
VIIIB	Iron	Iron(II) ion (ferrous ion)	Fe^{2+}
		Iron(III) ion (ferric ion)	Fe^{3+}

Table 4.2
Symbols and Names for Some Monatomic Ions

[a]Does not exist independently in aqueous solution.

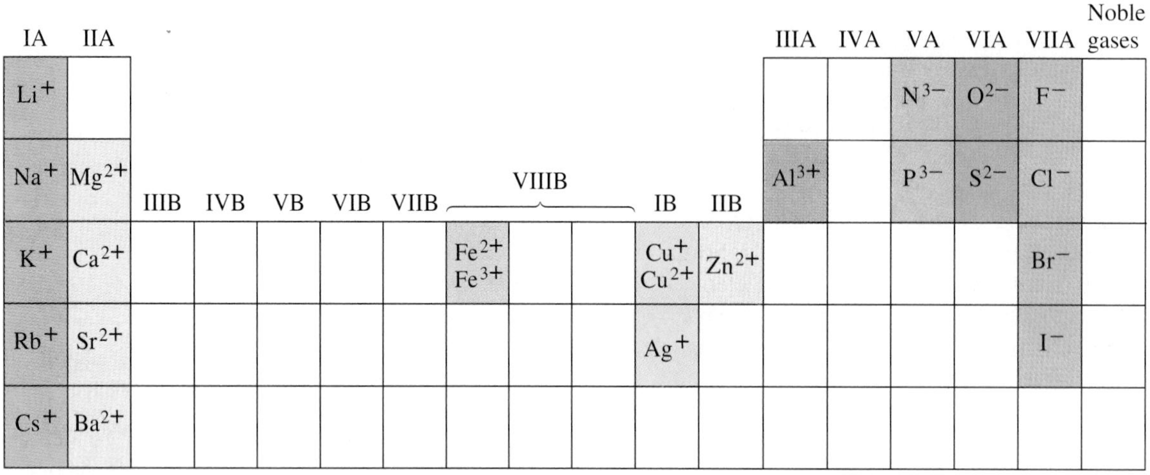

IA	IIA	IIIB	IVB	VB	VIB	VIIB		VIIIB		IB	IIB	IIIA	IVA	VA	VIA	VIIA	Noble gases
Li^+														N^{3-}	O^{2-}	F^-	
Na^+	Mg^{2+}											Al^{3+}		P^{3-}	S^{2-}	Cl^-	
K^+	Ca^{2+}						Fe^{2+} Fe^{3+}			Cu^+ Cu^{2+}	Zn^{2+}					Br^-	
Rb^+	Sr^{2+}									Ag^+						I^-	
Cs^+	Ba^{2+}																

Figure 4.4
The periodic relationships of some monatomic ions.

Since the older terminology (-ous, -ic) is still in use, you should become familiar with both methods of nomenclature.

Note that the charge on an ion of a Group IA element is $1+$ (usually written simply as $+$). The charge on an ion of a Group IIA element is $2+$, and that on an ion of a Group IIIA element is $3+$. You can calculate the charge on the negative ions in the table by subtracting 8 from the group number. For example, the charge on the oxide ion (oxygen is in Group VIA) is $6 - 8 = -2$. The charge on a nitride ion (nitrogen is in Group VA) is $5 - 8 = -3$. The periodic relationship of these monatomic ions is shown in Figure 4.4.

There is no simple way to determine the most likely charge on ions formed from elements in B subgroups. Indeed, you may have noticed that these elements can form ions with different charges. In such cases, chemists use Roman numerals with the names to indicate the charge. Thus, **iron(II) ion** means Fe^{2+}, and **iron(III) ion** means Fe^{3+}. An older terminology called Fe^{2+} **ferrous ion** and Fe^{3+} **ferric ion.** See similar names for the two copper ions in Table 4.2.

Compounds such as sodium chloride, potassium bromide, magnesium oxide, and potassium oxide are called **ionic compounds.** The constituent units of these compounds are charged particles—ions. Yet the compound as a whole is electrically neutral. We can use this principle of electrical neutrality to determine the combining ratio of ions. Potassium ions (K^+) would combine with bromide ions (Br^-) in a ratio of one to one. The formula KBr expresses this ratio and represents the compound potassium bromide. The combining ratio ($1:1$) and the ionic charges are understood.

Let's try another example. One calcium ion (Ca^{2+}) combines with *two* chloride ions (Cl^-). This ratio is expressed in the formula $CaCl_2$. In this formula, the ionic charges are understood. As with a coefficient of 1 in algebra, a subscript of 1 is understood, so, in the formula $CaCl_2$, the subscript 1 for calcium ion is understood, but the 2 for chloride is written explicitly (not Ca_1Cl_2 but $CaCl_2$). Thus, $CaCl_2$ not only gives us the combining ratio ($1:2$) but also stands for the ionic compound calcium chloride. It is a shorthand way of writing (1 Ca^{2+}) and (2 Cl^-).

You can use the charges on the ions in Table 4.2 to determine formulas for compounds of these elements. You can use the periodic table to predict the charge on ions formed from subgroup A elements, with Group IVA being the dividing line between positive and negative ions.

EXAMPLE 4.5

What is the formula for sodium sulfide?

SOLUTION

First, write the symbols for the ions (positive ion first). Sodium is in Group IA; therefore, the charge on the sodium ion is 1+. Sulfur is in Group VIA, and the charge on the sulfide ion is 2- (6 - 8 = -2). The symbols are Na^+ and S^{2-}. The smallest number into which both charges can be evenly divided, that is, the **least common multiple** (LCM), is 2. The least common multiple simply indicates the smallest number of electrons that can be evenly exchanged between the two elements. The subscript for each symbol can be determined by division of its charge (without the plus or minus) into the least common multiple. This step determines how many atoms of each element are needed to supply or accept the smallest common number of electrons. For Na^+,

$$\frac{2 \text{ (LCM)}}{1 \text{ (charge)}} = 2$$

For S^{2-},

$$\frac{2 \text{ (LCM)}}{2 \text{ (charge)}} = 1$$

Thus, we have the formula Na_2S_1 (2 Na^+ and 1 S^{2-}), or Na_2S.

EXAMPLE 4.6

What is the formula for aluminum oxide?

SOLUTION

The symbols are Al^{3+} and O^{2-} (Al is in Group IIIA, and O is in Group VIA). The LCM is 6. For Al^{3+},

$$\frac{6}{3} = 2$$

For O^{2-},

$$\frac{6}{2} = 3$$

The formula is therefore Al_2O_3 (2 Al^{3+} and 3 O^{2-}).

Practice Exercise

What is the formula for calcium nitride?

Another way to look at this is the cross-multiplication or cross-over method: the charge number on one ion becomes the subscript on the other. Thus, in Example 4.6, the charge on the aluminum ion is 3, which becomes the subscript for oxygen in aluminum oxide; and the charge on oxygen is 2, which becomes the subscript for the aluminum ion.

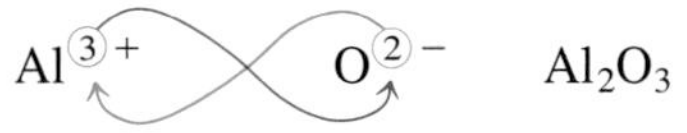

$$Al^{3+} \bowtie O^{2-} \quad Al_2O_3$$

Thus, two aluminum ions have six positive charges, and three oxygen ions have six negative charges. The resulting compound, Al_2O_3, is therefore neutral—as all compounds are.

Naming these binary ionic compounds is simple. First write the name of the positive ion and then the name of the negative ion. (The word *ion* is not used. It is understood in each case.)

EXAMPLE 4.7

What is the name for the compound Na_2S?

SOLUTION
Find the constituent ions in Table 4.2. They are sodium ion (Na^+) and sulfide ion (S^{2-}). The compound is sodium sulfide.

EXAMPLE 4.8

What is the name for the compound FeS?

SOLUTION
There are two kinds of iron ions. Since sulfur exists as the S^{2-} ion and one iron ion is combined with it, the iron ion in this compound must be Fe^{2+}. The name of FeS is iron(II) sulfide.

EXAMPLE 4.9

What is the name of the compound $FeCl_3$?

SOLUTION
Since the charge on the chloride ion is $1-$ and three of these ions are combined with one iron ion, the iron ion must be Fe^{3+}. The name of $FeCl_3$ is iron(III) chloride.

Practice Exercise
What is the name for the compound $CuBr_2$?

Ionic compounds generally exist as crystalline solids. However, many of them are soluble in water. Ionic compounds are found dissolved in all natural waters—including the water in the cells of our bodies, where they are involved in such critical functions as the transmission of nerve impulses.

We cannot emphasize too strongly the difference between ions and the atoms from which they are made. They are as different as a whole peach (an atom) and a peach pit (an ion). The names and symbols may look a lot alike, but the substances themselves are quite different. Unfortunately, the situation is confused because people talk about needing "iron" to perk up "tired blood" and "calcium" for healthy teeth and bones. What they really mean is iron(II) *ions* (Fe^{2+}) and calcium *ions* (Ca^{2+}). You wouldn't think of eating iron nails to get "iron." Nor would you eat highly reactive calcium metal. Although persons who are not chemists do not always make a careful distinction, we try to use precise terminology here.

Some manufacturers enrich their cereals with very fine specks of iron filings. These iron filings dissolve in the acidic environment of the stomach, producing iron ions.

4.7
COVALENT BONDS: SHARED ELECTRON PAIRS

One might expect a hydrogen atom, with its one electron, to acquire another electron and assume the helium configuration. Indeed, hydrogen atoms do just that in the presence of atoms of a reactive metal such as lithium—that is, a metal that finds it easy to give up an electron.

$$\text{Li} \cdot \ + \ \text{H} \cdot \ \longrightarrow \ \text{Li}^+ \ + \ \text{H} :^-$$

But what if there are no other kinds of atoms around? What if there are only hydrogen atoms? One atom can't gain an electron from another because hydrogen atoms all have an equal attraction for electrons. They can compromise, however, by *sharing a pair* of electrons.

$$\text{H} \cdot \ + \ \cdot \text{H} \ \longrightarrow \ \text{H} : \text{H}$$

By sharing electrons, the two hydrogen atoms form a hydrogen molecule. The bond formed by a shared pair of electrons is called a **covalent bond.**

> A shared single pair of electrons is called a **single bond.**

H : H

⌐ covalent bond (shared pair of electrons)

Consider next the case of chlorine. A chlorine atom will pick up an extra electron from anything willing to give one up. But, again, what if the only thing around is another chlorine atom? Chlorine atoms too can attain a more stable arrangement by sharing a pair of electrons.

$$: \overset{\cdot \cdot}{\underset{\cdot \cdot}{\text{Cl}}} \cdot \ + \ \cdot \overset{\cdot \cdot}{\underset{\cdot \cdot}{\text{Cl}}} : \ \longrightarrow \ : \overset{\cdot \cdot}{\underset{\cdot \cdot}{\text{Cl}}} : \overset{\cdot \cdot}{\underset{\cdot \cdot}{\text{Cl}}} :$$

Each chlorine atom in the chlorine molecule has eight electrons around it, an arrangement like that of the noble gas argon. This stable *octet* of electrons is the arrangement characteristic of all the noble gases except helium. Most of the covalently bonded atoms that we shall consider, except hydrogen, follow the **octet rule:** they seek an arrangement that will surround them with eight electrons. The shared pair of electrons in the chlorine molecule is another example of a covalent bond; they are called a **bonding pair.** The other electrons that stay on one atom and are not shared are called **nonbonding pairs.**

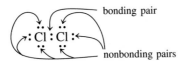

For simplicity, the hydrogen molecule is often represented as H_2 and the chlorine molecule as Cl_2. In each case, the covalent bond between the atoms is understood. Sometimes the covalent bond is indicated by a dash, H—H and Cl—Cl. Nonbonding pairs of electrons often are not shown.

4.8
MULTIPLE COVALENT BONDS

Atoms can share more than one pair of electrons. Consider, for example, the oxygen atom. Its electron-dot symbol is

$$:\overset{\cdot}{\underset{\cdot}{O}}\cdot$$

Oxygen is in Group VI, and therefore it has six valence electrons. It could share a pair of electrons with another oxygen atom to form an oxygen molecule.

$$:\overset{\cdot}{\underset{\cdot\cdot}{O}}\cdot \; + \; \cdot\overset{\cdot}{\underset{\cdot\cdot}{O}}: \; \longrightarrow \; :\overset{\cdot}{\underset{\cdot\cdot}{O}}:\overset{\cdot}{\underset{\cdot\cdot}{O}}: \qquad (\textit{incorrect structure})$$

Each atom in this arrangement has only seven electrons around it, and seven is not eight. To solve this dilemma, each oxygen could share an additional pair of electrons, for a total of two bonding pairs.

$$:\underset{\cdot\cdot}{O}::\underset{\cdot\cdot}{O}: \qquad (\text{or} \quad O{=}O)$$

Note that each atom has an octet of electrons about it as a result of this sharing.[1] We say that the atoms are joined by a **double bond,** a covalent linkage in which the two atoms share two pairs of electrons.

Atoms also can share three pairs of electrons. Nitrogen is in Group V, and it has five valence electrons. It could share a pair of electrons with another nitrogen atom and would then look like this.

$$:\overset{\cdot}{N}\cdot \; + \; \cdot\overset{\cdot}{N}: \; \longrightarrow \; :\overset{\cdot}{N}:\overset{\cdot}{N}: \qquad (\textit{incorrect structure})$$

To satisfy the octet rule, each nitrogen atom shares three pairs of electrons with the other.

$$:N:::N: \qquad (\text{or} \quad :N{\vdots}{\vdots}N: \quad \text{or} \quad N{\equiv}N)$$

The atoms are joined by a **triple bond,** a covalent linkage in which two atoms share three pairs of electrons. Each nitrogen also has a nonbonding (unshared) pair of electrons. Note that we could have drawn the unshared pair of electrons above or below the atomic symbol. Such a drawing would represent the same molecule.

> Covalent bonds often are represented as dashes. We can therefore show the three kinds of bonds as follows.
>
> Cl—Cl   O=O   N≡N

4.9
UNEQUAL SHARING:
POLAR COVALENT BONDS

So far, we have seen that atoms combine in two different ways. Some that are quite different in electron structure (from opposite ends of the periodic table) react by the complete transfer of an electron from one atom to another to form an ionic bond. Atoms that are identical combine by sharing one or more pairs of electrons to form covalent bonds. Now let's consider bond formation between atoms that are different, but not different enough to form ionic bonds.

---

[1] Unfortunately, although this electron-dot formula for the oxygen molecule satisfies the octet rule, it does not adequately describe the properties of oxygen. Experiments have shown that O_2 has two unpaired electrons. There is no satisfactory electron-dot formula for O_2.

Hydrogen and chlorine react to form a colorless gas called hydrogen chloride. We use the individual atoms in order to focus on the sharing of electrons to form a covalent bond.

$$H\cdot \ + \ \cdot \ddot{\underset{..}{Cl}}: \ \longrightarrow \ H:\ddot{\underset{..}{Cl}}: \qquad (or \ H\!-\!Cl)$$

Both hydrogen and chlorine need an electron to achieve a noble gas configuration, so they share a pair and form a covalent bond.

Both hydrogen and chlorine consist of diatomic molecules; the reaction is more accurately represented by the scheme

$$H:H + \ :\ddot{\underset{..}{Cl}}:\ddot{\underset{..}{Cl}}: \ \longrightarrow \ 2\,H:\ddot{\underset{..}{Cl}}:$$

EXAMPLE 4.10

Use electron-dot formulas to show the formation of a covalent bond between (a) two fluorine atoms; (b) a fluorine atom and a hydrogen atom.

SOLUTION

a. $\ :\ddot{\underset{..}{F}}\cdot \ + \ \cdot\ddot{\underset{..}{F}}: \ \longrightarrow \ :\ddot{\underset{..}{F}}\!:\!\ddot{\underset{..}{F}}:$ ⟵ bonding pair

b. $\ H\cdot \ + \ \cdot\ddot{\underset{..}{F}}: \ \longrightarrow \ H\!:\!\ddot{\underset{..}{F}}:$ ⟵ bonding pair

Practice Exercise
Use electron-dot formulas to show the formation of a covalent bond between (a) two bromine atoms; (b) a hydrogen atom and a bromine atom; (c) an iodine atom and a chlorine atom.

One might reasonably ask why the hydrogen molecule and the chlorine molecule react at all. Have we not just explained that they themselves were formed to provide a more stable arrangement of electrons? Yes, indeed, we did say that. But there is stable, and there is more stable. The chlorine molecule represents a more stable arrangement than two separate chlorine atoms. Nevertheless, given the opportunity, a chlorine atom would rather form a bond with a hydrogen atom than with another chlorine atom. Read on.

For the sake of convenience and simplicity, the reaction of hydrogen (molecule) and chlorine (molecule) to form hydrogen chloride often is reduced to

$$H_2 \ + \ Cl_2 \ \longrightarrow \ 2\,HCl$$

Molecules of hydrogen chloride consist of one atom of hydrogen and one atom of chlorine. These unlike atoms share a pair of electrons. Sharing does not mean sharing equally, though. Chlorine atoms have a greater attraction for a shared pair of electrons than do hydrogen atoms; chlorine is said to be more *electronegative* than hydrogen. The shared electrons are held more tightly by the chlorine atom, and this results in the chlorine end of the molecule being more negative than the hydrogen end (see Section 4.13).

When the electrons in a covalent bond are not equally shared, the bond is said to be **polar.** Thus, the bonding in hydrogen chloride is described as **polar covalent,** whereas the bonding in the hydrogen molecule or in the chlorine molecule is **nonpolar covalent.** The polar covalent bond is not an ionic bond. In an ionic bond, one atom completely loses an electron. In a polar covalent bond, the atom at the positive end of the bond (hydrogen in HCl) still has some share in the bonding pair of electrons (Figure 4.5). To distinguish this arrangement from that in an ionic bond, the following notation is used.

$$\overset{\delta+ \ \ \delta-}{H\!-\!Cl}$$

The line between the two atoms represents the covalent bond, a pair of shared electrons.

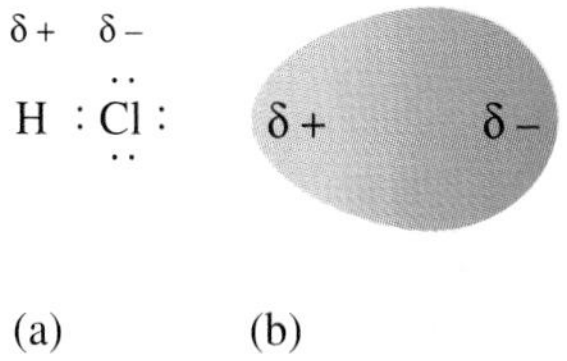

$$\overset{\delta+ \ \ \delta-}{H}:\ddot{\underset{..}{Cl}}:$$

(a)   (b)

Figure 4.5

Representation of the polar hydrogen chloride molecule. (a) The electron-dot formula, with the shared electron pair shown nearer the chlorine atom. The symbols $\delta+$ and $\delta-$ indicate partial positive and partial negative charges, respectively. (b) A diagram depicting the unequal distribution of electron density in the hydrogen chloride molecule.

The $\delta+$ and $\delta-$ (read "delta plus" and "delta minus") signify which end is partially positive and which is partially negative (the word *partially* is used to distinguish this charge from the full charge on an ion). This unequal sharing of electrons has a marked effect on the properties of a compound.

The gas hydrogen chloride dissolves readily in water. The aqueous solution formed is called hydrochloric acid (sometimes muriatic acid). This acid is used for, among other things, cleaning toilet bowls and removing excess mortar from new brick buildings. Hydrochloric acid is also the well-known "stomach acid." Acids are defined and further discussed in Chapter 10.

4.10
POLYATOMIC MOLECULES: WATER, AMMONIA, AND METHANE

To obtain an octet of electrons, an oxygen atom must share electrons with two hydrogen atoms, a nitrogen atom must share electrons with three hydrogen atoms, and a carbon atom must share electrons with four hydrogen atoms. In general, many nonmetals often form a number of covalent bonds equal to eight minus the group number. Oxygen, which is in Group VIA, forms $8 - 6 = 2$ covalent bonds in most compounds. Nitrogen, in Group VA, forms $8 - 5 = 3$ covalent bonds in most of its compounds. Carbon, in Group IVA, forms $8 - 4 = 4$ covalent bonds in most carbon compounds, including the great host of organic compounds. These simple rules will enable you to write formulas for a large number of molecules.

Water is one of the most familiar chemical substances. The early electrolysis experiments (Section 12.4) and ample evidence since that time indicate that the molecular formula for water is H_2O. In order to be surrounded by an octet, oxygen shares two pairs of electrons. But a hydrogen atom tends to share only one pair of electrons. An oxygen atom must therefore bond with two hydrogen atoms.

$$\cdot \ddot{O}: \; + \; 2\,H\cdot \; \longrightarrow \; H:\ddot{O}:\\ \qquad\qquad\qquad\qquad\quad H$$

This arrangement completes the octet in the outer energy level of oxygen, giving it the neon electron configuration. It also completes the outer energy level of the hydrogen atoms, each of which now has the helium electron configuration.

We have chosen to represent the water molecule with the hydrogen atoms arranged at an angle rather than on a straight line with the oxygen atom. This arrangement is necessary to explain the *polar* nature of the water molecule (Section 4.17).

An atom of the element nitrogen has five electrons in its outer energy level. It can assume the neon configuration by sharing three pairs of electrons with *three* hydrogen atoms. The result is the compound ammonia.

$$\cdot \ddot{N}\cdot \; + \; 3\,H\cdot \; \longrightarrow \; H:\ddot{N}:H\\ \qquad\qquad\qquad\qquad\quad H$$

In ammonia, the bond arrangement is that of a tripod, with a hydrogen atom at the end of each "leg" and the nitrogen atom with its unshared pair of electrons sitting at the top. This shape is necessary to explain the polar nature of the ammonia molecule (Section 4.17).

Ammonia (NH_3) is a gas at room temperature. Vast quantities of it are compressed into tanks and used as fertilizer. Ammonia is quite soluble in water. It forms a basic solution (Chapter 10). Such aqueous preparations are familiar household cleansing solutions.

An atom of carbon has four electrons in its outer energy level. It can assume the neon configuration by sharing pairs of electrons with four hydrogen atoms, forming the compound methane.

$$\cdot \overset{\displaystyle .}{\underset{\displaystyle .}{C}} \cdot \ + \ 4\,H\cdot \ \longrightarrow \ H \overset{\displaystyle H}{\underset{\displaystyle H}{\overset{..}{\underset{..}{:C:}}}} H$$

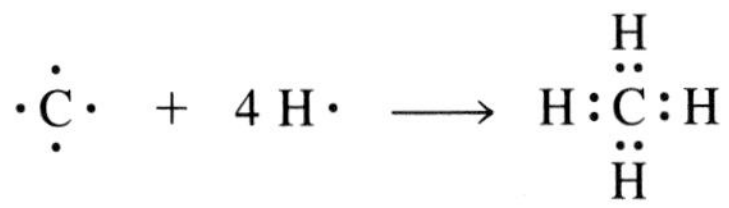

The methane molecule has a tetrahedral shape (Section 4.17).

Methane (CH_4) is the simplest of the hydrocarbons, a group of organic compounds discussed in detail in Chapter 13. It is the principal component of natural gas, used as a fuel.

4.11
NAMES FOR COVALENT COMPOUNDS

Many molecular compounds have common and widely used names. Examples are water (H_2O), methane (CH_4), and ammonia (NH_3). For other compounds, the prefixes *mono-, di-, tri,* and so on, are used to indicate the number of atoms of each element in the molecule. A list of these prefixes for up to ten atoms is given in Table 4.3.

Simply use the prefixes to indicate the number of each kind of atom. For example, the compound N_2O_4 is called *dinitrogen tetroxide*. (The *a* often is dropped from tetra- and other prefixes when it precedes another vowel.) We often leave off the mono- prefix (NO_2 is nitrogen dioxide), but do include it to distinguish between two compounds of the same pair of elements (CO is carbon monoxide; CO_2 is carbon dioxide).

Table 4.3
Prefixes That Indicate the Number of Atoms of an Element in a Compound

Prefix	Number of Atoms
Mono-	1
Di-	2
Tri-	3
Tetra-	4
Penta-	5
Hexa-	6
Hepta-	7
Octa-	8
Nona-	9
Deca-	10

EXAMPLE 4.11

What is the name for SCl_2? For SF_6?

SOLUTION
With one sulfur atom and two chlorine atoms SCl_2 is sulfur dichloride. With one sulfur atom and six fluorine atoms, SF_6 is sulfur hexafluoride.

EXAMPLE 4.12

Give the formula for carbon tetrachloride.

SOLUTION
The name indicates one carbon atom and four chlorine atoms. The formula is CCl_4.

Practice Exercise
Give the formula for dinitrogen pentoxide.

EXAMPLE 4.13

Give the formula for tetraphosphorus hexoxide.

SOLUTION
The name indicates four phosphorus atoms and six oxygen atoms. The formula is P_4O_6.

Practice Exercise
Give the formula for tetraphosphorus triselenide. (The symbol for selenium is Se.)

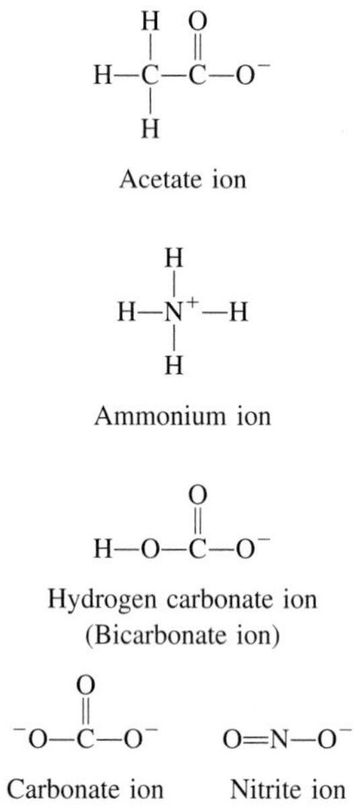

Acetate ion

Ammonium ion

Hydrogen carbonate ion
(Bicarbonate ion)

Carbonate ion   Nitrite ion

Figure 4.6
Polyatomic ions have both covalent bonds (dashes) and ionic charges (+ or −).

4.12
POLYATOMIC IONS

Many compounds contain both ionic and covalent bonds. Sodium hydroxide, commonly known as lye, consists of sodium ions (Na^+) and hydroxide ions (OH^-). The hydroxide ion contains an oxygen atom covalently bonded to a hydrogen atom, plus an ''extra'' electron. Whereas the sodium atom becomes a cation by giving up an electron, the hydroxide group becomes an anion by gaining an electron.

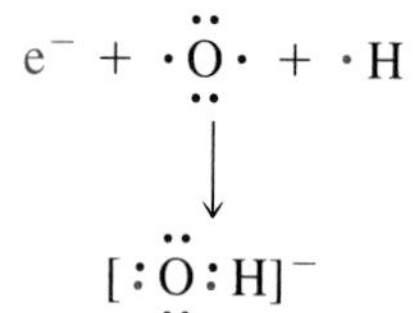

$$e^- + \cdot \overset{\cdot\cdot}{\underset{\cdot\cdot}{O}} \cdot + \cdot H$$

$$\downarrow$$

$$[\overset{\cdot\cdot}{\underset{\cdot\cdot}{:O}} : H]^-$$

Hydroxide ion

The formula for sodium hydroxide is NaOH; for each sodium ion there is one hydroxide ion.

There are many groups of atoms that (like the hydroxide ion) remain together through most chemical reactions. **Polyatomic ions** are charged particles containing two or more covalently bonded atoms (Figure 4.6). A list of common polyatomic ions is given in Table 4.4. You can use these ions, in combination with the monatomic ions in Table 4.2, to determine formulas for compounds that contain polyatomic ions.

EXAMPLE 4.14

What is the formula for sodium sulfate?

SOLUTION
First, write the formulas for the ions.

$$Na^+ \qquad SO_4^{2-}$$

Crossing over,

we get

$$Na_2^+(SO_4)_1^{2-}$$

Table 4.4
Some Common
Polyatomic Ions

Charge	Name	Formula
1+	Ammonium ion	NH_4^+
	Hydronium ion	H_3O^+
1−	Hydrogen carbonate (bicarbonate) ion	HCO_3^-
	Hydrogen sulfate (bisulfate) ion	HSO_4^-
	Acetate ion	$CH_3CO_2^-$ (or $C_2H_3O_2^-$)
	Nitrite ion	NO_2^-
	Nitrate ion	NO_3^-
	Cyanide ion	CN^-
	Hydroxide ion	OH^-
	Dihydrogen phosphate ion	$H_2PO_4^-$
	Permanganate ion	MnO_4^-
2−	Carbonate ion	CO_3^{2-}
	Sulfate ion	SO_4^{2-}
	Monohydrogen phosphate ion	HPO_4^{2-}
	Oxalate ion	$C_2O_4^{2-}$
	Dichromate ion	$Cr_2O_7^{2-}$
3−	Phosphate ion	PO_4^{3-}

Then, dropping the charges, we have

$$Na_2(SO_4)_1 \quad \text{or simply} \quad Na_2SO_4$$

Practice Exercise
What is the formula for potassium phosphate?

EXAMPLE 4.15

What is the formula for ammonium sulfide?

SOLUTION
The ions are

$$NH_4^+ \qquad S^{2-}$$

Crossing over, we get

$$NH_4^+ \qquad S^{2-}$$

giving us

$$(NH_4)_2^+ S_1^{2-}$$

Dropping the charges gives

$$(NH_4)_2S$$

The parentheses with a subscript 2 indicate that the entire ammonium unit is taken twice; there are two nitrogen atoms and eight ($4 \times 2 = 8$) hydrogen atoms.

Practice Exercise
What is the formula for calcium acetate?

EXAMPLE 4.16

What is the name for the compound NaCN?

SOLUTION
The ions are

$$Na^+ \qquad CN^-$$

The name is sodium cyanide.

Practice Exercise
What is the name for the compound $CaCO_3$?

EXAMPLE 4.17

What is the name for KH_2PO_4?

SOLUTION
The ions are

$$K^+ \qquad H_2PO_4^-$$

The name is potassium dihydrogen phosphate.

Practice Exercise
What is the name for $K_2Cr_2O_7$?

4.13
ELECTRONEGATIVITY

Before we look at the structure of any more molecules, we should consider in more detail the concept of *electronegativity*. When we speak of the **electronegativity** of an atom, we describe its tendency to attract shared electrons to itself. The atoms to the right in the periodic table are, in general, more electronegative than those to the left. The ones on the right are precisely those atoms that, in forming ions, tend to gain electrons and form negative ions. The ones on the left, the metals, tend to give up electrons, to become positive ions. The more electronegative an atom, the greater its tendency to pull the electrons in a covalent bond toward its end of the bond.

The most electronegative element in the periodic table is fluorine, in the upper right-hand corner of the table. You can use this fact as a guide. Within a period (a row) of the table, elements become more electronegative toward the right. Oxygen is more electronegative than nitrogen. Within a group (column) of the table, elements become less electronegative toward the bottom. Chlorine is less electronegative than fluorine. The comparison is not quite so straightforward when one is considering two elements that are in neither the same period nor the same group.

Hydrogen is difficult to place. It often is placed in Group IA because, like lithium, it has one electron in its outer level, but it really isn't much like the Group IA metals. Or it

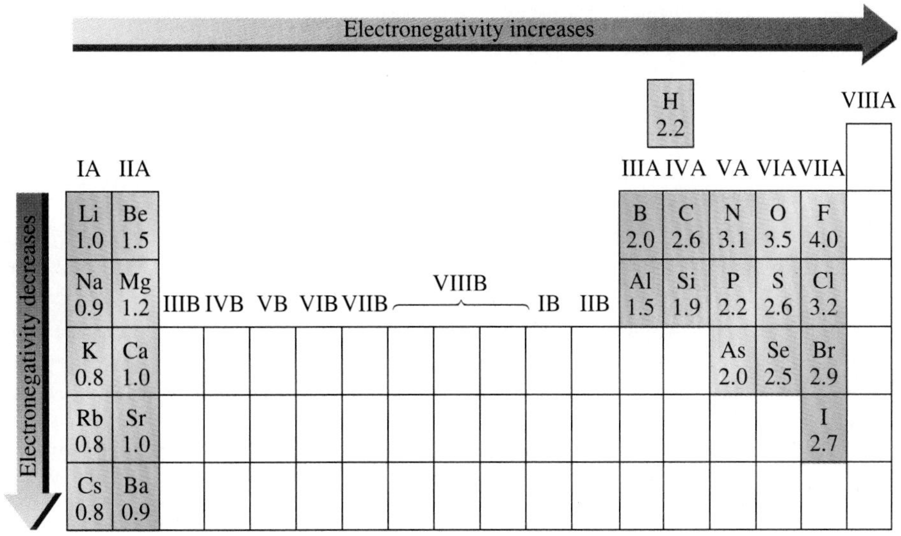

Figure 4.7
Relative electronegativities
of some common elements.

Electronegativity increases

Electronegativity decreases

	IA	IIA									IB	IIB	IIIA	IVA	VA	VIA	VIIA

H
2.2

VIIIA

IIIA IVA VA VIA VIIA

IA IIA

Li 1.0	Be 1.5
Na 0.9	Mg 1.2
K 0.8	Ca 1.0
Rb 0.8	Sr 1.0
Cs 0.8	Ba 0.9

IIIB IVB VB VIB VIIB ——— VIIIB ——— IB IIB

B 2.0	C 2.6	N 3.1	O 3.5	F 4.0
Al 1.5	Si 1.9	P 2.2	S 2.6	Cl 3.2
		As 2.0	Se 2.5	Br 2.9
				I 2.7

could be placed in Group VIIA because, like fluorine, it is just one electron short of having the configuration of a noble gas. However, hydrogen isn't nearly as electronegative as fluorine. If you look at a number of versions of the periodic table, you'll see that chemists still haven't decided what to do with it. We have placed it at the top, away from everything, in Figure 4.7. Its electronegativity fits that position. It will take electrons from an atom that gives them up readily (lithium, for example), and it will shift electrons to a more electronegative element (chlorine, for example).

4.14
RULES FOR WRITING ELECTRON-DOT FORMULAS

Recall that electrons are transferred (Section 4.5) or shared (Section 4.7) in ways that leave most atoms with an octet of electrons in their outermost energy level. In this section we describe the procedure we can follow in writing electron-dot formulas for molecules. First we must put the atoms of the molecules in their proper places.

The *skeletal structure* of a molecule tells us the order in which the atoms are attached to one another. In the absence of experimental evidence, the following rules help us to devise likely skeletal structures.

1. Hydrogen atoms form only one bond; they are always at the end of a sequence of atoms. Hydrogen often is bonded to carbon, nitrogen, or oxygen.

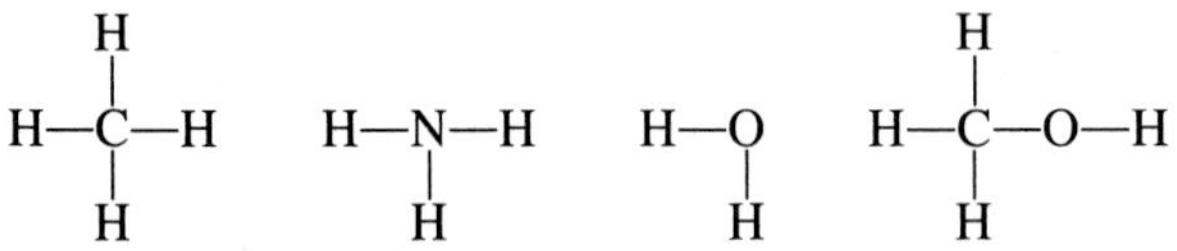

$$\begin{matrix} & H & & & & & & & H \\ & | & & & & & & & | \\ H - & C & - H & \quad H - N - H \quad & H - O & \quad H - & C & - O - H \\ & | & & | & & | & & | \\ & H & & H & & H & & H \end{matrix}$$

2. Polyatomic molecules and ions often consist of a central atom surrounded by more

electronegative atoms. (Hydrogen is an exception; it is always on the outside, even when bonded to a more electronegative element.)

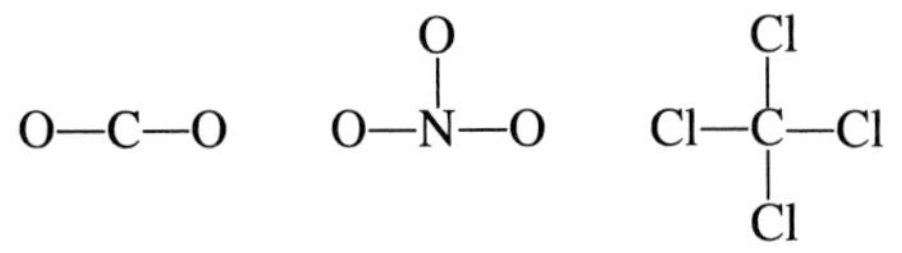

$$O-C-O \qquad O-\underset{\underset{}{|}}{\overset{\overset{O}{|}}{N}}-O \qquad Cl-\underset{\underset{Cl}{|}}{\overset{\overset{Cl}{|}}{C}}-Cl \qquad (\textit{incomplete structures})$$

After a skeletal formula for a polyatomic molecule or ion has been chosen, we can use the following steps to write an electron-dot formula.

1. Calculate the total number of valence electrons. The total for a molecule is the sum of the valence electrons for each atom. For a polyatomic anion, add the number of negative charges. For a polyatomic cation, subtract the number of positive charges.
    Examples:

    N_2O_4 has $(2 \times 5) + (4 \times 6) = 34$ valence electrons.
    NO_3^- has $5 + (3 \times 6) + 1 = 24$ valence electrons.
    NH_4^+ has $5 + (4 \times 1) - 1 = 8$ valence electrons.

2. Write the skeletal structure, and connect bonded pairs of atoms by a dash (one electron pair).
3. Place electrons about outer atoms so that each (except hydrogen) has an octet.
4. Subtract the number of electrons assigned so far from the total calculated in Step 1. Any electrons that remain are assigned in pairs to the central atom(s).
5. If a central atom has fewer than eight electrons after Step 4, a multiple bond is likely. Move one or more nonbonding pairs from an outer atom to the space between the atoms to form a double or triple bond. A deficiency of two electrons suggests a double bond, and shortage of four electrons is indicative of a triple bond or two double bonds to the central atom.

EXAMPLE 4.18

Give the electron-dot formula for methanol, CH_4O.

SOLUTION
1. The total number of valence electrons is $4 + (4 \times 1) + 6 = 14$.
2. The skeletal structure in which all the hydrogen atoms are on the outside and the least electronegative atom, carbon, is most central is

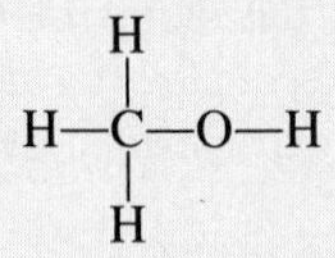

$$H-\underset{\underset{H}{|}}{\overset{\overset{H}{|}}{C}}-O-H$$

3. Now, counting each bond as two electrons gives 10 electrons. The four remaining electrons are placed (as two nonbonding pairs) on the oxygen atom.

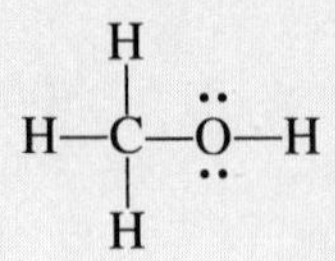

$$H-\underset{\underset{H}{|}}{\overset{\overset{H}{|}}{C}}-\overset{\cdot\cdot}{\underset{\cdot\cdot}{O}}-H$$

(The remaining steps are not necessary; both carbon and oxygen have an octet of electrons.)

Practice Exercise
Give the electron-dot formula for ethyl chloride, C_2H_5Cl.

EXAMPLE 4.19

Give the electron-dot formula for nitrogen trifluoride, NF_3.

SOLUTION
1. There are $5 + (3 \times 7) = 26$ valence electrons.
2. The skeletal structure is

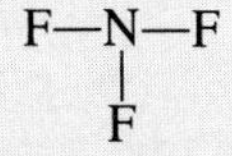

$$F-N-F$$
$$|$$
$$F$$

3. Place three nonbonding pairs on each fluorine atom.

$$:\ddot{F}-N-\ddot{F}:$$
$$|$$
$$:\ddot{F}:$$

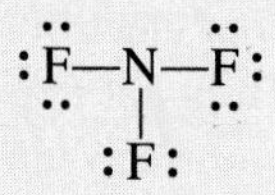

4. We have assigned 24 electrons. Place the remaining two as a nonbonding pair on the nitrogen atom.

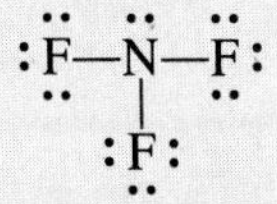

$$:\ddot{F}-\ddot{N}-\ddot{F}:$$
$$|$$
$$:\ddot{F}:$$

(Each atom has an octet; Step 5 is not needed.)

Practice Exercise
Give the electron-dot formula for oxygen difluoride, OF_2.

EXAMPLE 4.20

Give the electron-dot formula for the BF_4^- ion.

SOLUTION
1. There are $3 + (4 \times 7) + 1 = 32$ electrons.
2. The skeletal structure is

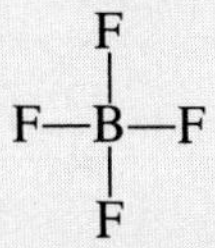

$$F$$
$$|$$
$$F-B-F$$
$$|$$
$$F$$

3. Place three nonbonding pairs on each fluorine atom.

$$:\ddot{F}:$$
$$|$$
$$:\ddot{F}-B-\ddot{F}:$$
$$|$$
$$:\ddot{F}:$$

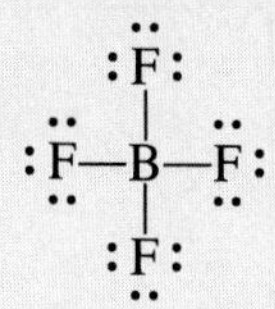

4. We have assigned 32 electrons. None remains to be assigned.

Practice Exercise
Give the electron-dot formula for PH_4^+ ion.

EXAMPLE 4.21

Give the electron-dot formula for carbon dioxide, CO_2.

SOLUTION
1. There are $4 + (2 \times 6) = 16$ valence electrons.
2. The skeletal structure is

$$O—C—O$$

3. Place three nonbonding pairs on each oxygen atom.

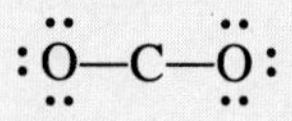

$$:\overset{..}{\underset{..}{O}}—C—\overset{..}{\underset{..}{O}}:$$

4. We have assigned 16 electrons. None remains to be placed.
5. The carbon atom has only four electrons. It needs to form two double bonds in order to have an octet. (It is not reasonable to expect that carbon would form a triple bond to one of the oxygen atoms. Why?) Move a nonbonding pair from each oxygen atom to the space between the atoms to form a double bond on each side of the carbon atom.

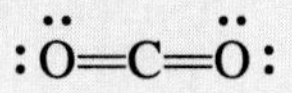

$$:\overset{..}{O}=C=\overset{..}{O}:$$

Practice Exercise
Give the electron-dot formula for nitryl fluoride, NO_2F.

The rules we have used here lead to results that are summarized and illustrated for selected elements in Table 4.5. Figure 4.8 relates the number of covalent bonds to the periodic table.

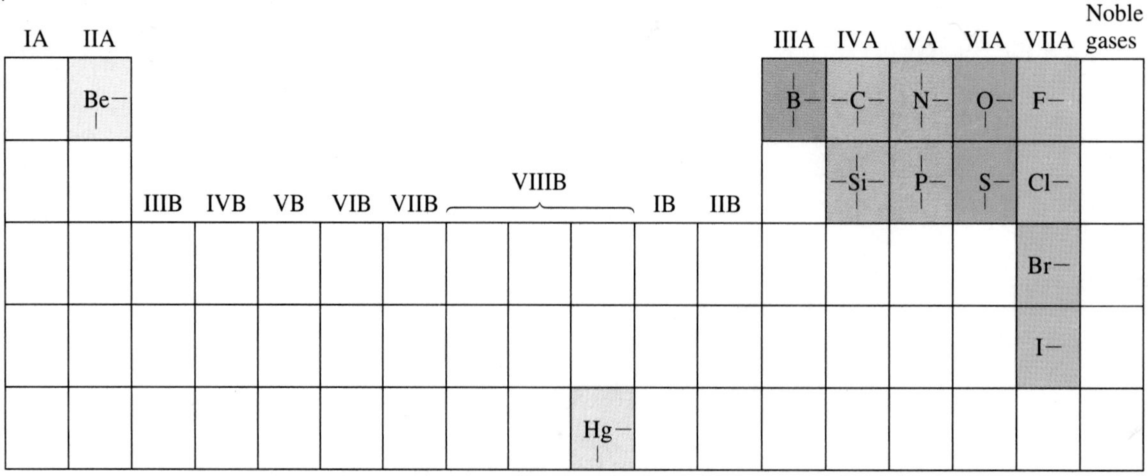

Figure 4.8
Covalent bonding of representative elements of the periodic table.

Table 4.5
Number of Bonds Formed
by Selected Elements

Electron-Dot Picture	Bond Picture	Number of Bonds	Representative Molecules	
H	H—	1	H—H H—Cl	 HCl
He:		0		
·C·	—C—	4	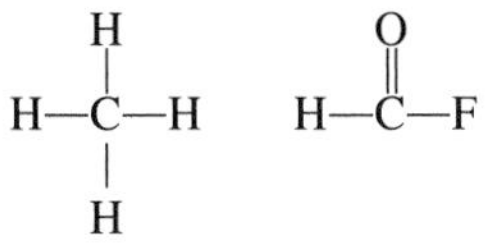 H H—C—H H—C—F H	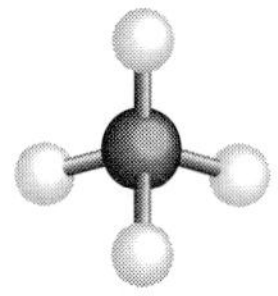 CH_4
·N·	—N—	3	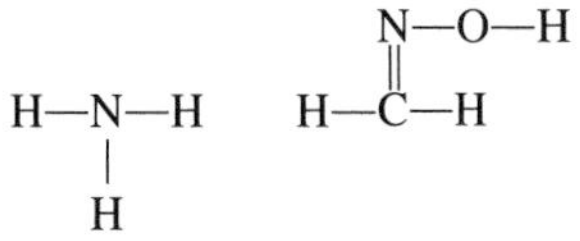 H—N—H H—C—H N—O—H H	
·O:	—O—	2	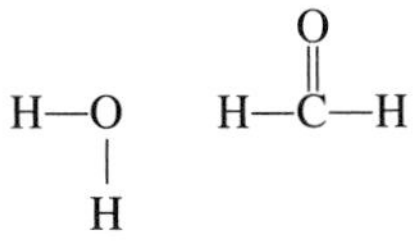 H—O H—C—H O H	 H_2O
·F:	—F	1	H—F F—F	
·Cl:	—Cl	1	Cl—Cl 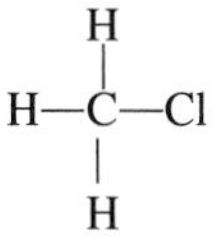 H—C—Cl H H	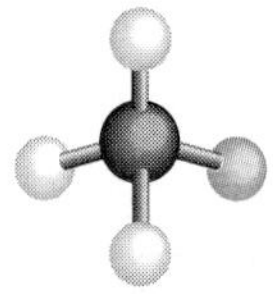 CH_3Cl

4.15
EXCEPTIONS TO THE OCTET RULE

Many molecules made of atoms of the main group elements have electron structures that follow the octet rule. There are numerous exceptions, however. The exceptions fall into three main groups. Each type is readily identified by some structural characteristic.

Molecules with an odd number of valence electrons obviously cannot satisfy the octet rule. Examples of such molecules are nitrogen monoxide (NO, also called nitric oxide), with $5 + 6 = 11$ valence electrons; nitrogen dioxide (NO_2), with 17 valence electrons; and chlorine dioxide (ClO_2), which has 19 outer electrons. Obviously one of the atoms in each of these molecules will have an odd number of electrons and therefore cannot have an octet.

Atoms and molecules with an unpaired electron are called **free radicals.** Most free radicals are highly reactive and have only transitory existence as intermediates in chemical reactions. An example is the chlorine atom that is formed from the breakdown of chlorofluorocarbons in the stratosphere and that leads to the depletion of the ozone shield (Section 13.9). Some free radicals are quite stable, however. The nitrogen oxides are

All atoms and molecules with an odd number of electrons must have one unpaired electron. Filled energy levels and sublevels have all their electrons paired, with two electrons in each orbital (Section 2.5). We need only consider valence electrons to determine whether or not an atom or molecule is a free radical. Electron-dot formulas of NO, NO_2, and ClO_2 show that one atom of each has an unpaired electron; that atom obviously does not have an octet of electrons in its outer energy level.

:N::O: :O:N::O:

:O:Cl:O:

major components of smog. Chlorine dioxide is made in vast quantities and is used for bleaching flour.

Boron atoms have three valence electrons; fluorine atoms have seven. When boron reacts with fluorine, it shares those electrons with three fluorine atoms to form boron trifluoride.

$$
\begin{array}{c}
:\ddot{\text{F}}:\\
|\\
\text{B}-\ddot{\text{F}}:\\
|\\
:\ddot{\text{F}}:
\end{array}
$$

This structure uses all 24 of the valence electrons; there are none left to put on the central boron atom. Experimental evidence indicates that this structure is consistent with the reactivity of BF_3 toward molecules with unshared pairs. For instance, BF_3 reacts readily with ammonia to form BF_3NH_3, a compound in which all atoms (except hydrogen) have an octet of electrons.

$$
\begin{array}{c}
:\ddot{\text{F}}: \quad\quad \text{H}\\
|\quad\quad\quad |\\
:\ddot{\text{F}}-\text{B} \quad + \quad :\text{N}-\text{H}\\
|\quad\quad\quad |\\
:\ddot{\text{F}}: \quad\quad \text{H}
\end{array}
\longrightarrow
\begin{array}{c}
:\ddot{\text{F}}: \text{ H}\\
|\quad\; |\\
:\ddot{\text{F}}-\text{B}-\text{N}-\text{H}\\
|\quad\; |\\
:\ddot{\text{F}}: \text{ H}
\end{array}
$$

The second period elements—carbon, nitrogen, oxygen, and fluorine—nearly always obey the octet rule. (The odd-electron compounds are obvious exceptions.) The valence electron level of the second period elements holds a maximum of eight electrons $(2s^2 2p^6)$. The third main energy level can hold up to 18 electrons $(3s^2 3p^6 3d^{10})$. Third period elements therefore can violate the octet rule by having more than eight electrons in their valence level. These so-called *expanded octets* are evident in the following compounds.

$$
\begin{array}{c}
:\ddot{\text{Cl}}: \ddot{\text{Cl}}:\\
\quad | \;/\\
:\ddot{\text{Cl}}-\text{P}\\
\quad | \;\backslash\\
:\ddot{\text{Cl}}: \ddot{\text{Cl}}:
\end{array}
\quad\quad\quad
\begin{array}{c}
:\ddot{\text{F}}: \ddot{\text{F}}:\\
\; \backslash | /\\
\quad\text{S}\\
\; / | \backslash\\
:\ddot{\text{F}}: \ddot{\text{F}}:
\end{array}
$$

Phosphorus pentachloride   Sulfur hexafluoride

Elements in the third period and beyond, then, are not limited to an octet. Yet many of their compounds still follow the octet rule.

4.16
MOLECULAR SHAPES: THE VSEPR THEORY

We have represented molecules in two dimensions on paper, but molecules have three-dimensional shapes. We can use electron-dot formulas as part of the process of predicting molecular shapes. The shapes that we consider in this book are shown in Figure 4.9.

The **valence shell electron pair repulsion theory (VSEPR)** often is used to predict the arrangement of atoms about a central atom. The basis of the VSEPR theory is that electron pairs will arrange themselves about a central atom in a way that minimizes repulsion between the like-charged particles. This means that they will get as far apart as

The shapes of molecules are of considerable importance. Biologically active molecules (Chapters 19–23) must have the right groups in the right places in order to carry out their activities.

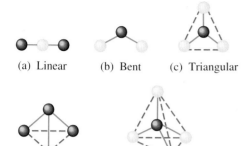

(a) Linear   (b) Bent   (c) Triangular

(d) Pyramidal   (e) Tetrahedral

Figure 4.9
Shapes of molecules. In a *linear* molecule (a), all the atoms are along a line; the bond angle is 180°. The *bent* molecule (b) has an angle less than 180°. Connecting the three outer atoms of the *triangular* molecule (c) with imaginary lines produces a triangle with an atom at the center. Imaginary lines connecting all four atoms of a *pyramidal* molecule (d) form a three-sided pyramid. Connecting the four outer atoms of the *tetrahedral* molecule (e) with imaginary lines produces a tetrahedron (a four-sided figure in which each side is a triangle) with an atom at the center.

possible. Table 4.6 gives the geometric shapes associated with the arrangement of two, three, or four entities about a central atom.

The farthest apart two substituent atoms can get are the opposite sides of the central atom at an angle of 180°. Three groups assume a triangular arrangement about the central

Table 4.6
Bonding and the Shape of Molecules

Number of Bonded Atoms	Number of NBPs[a]	Number of Sets	Molecular Shape	Examples			
2	0	2	Linear	$BeCl_2$	$HgCl_2$	CO_2	$BeCl_2$
3	0	3	Triangular	BF_3	$AlBr_3$	CH_2O	BF_3
4	0	4	Tetrahedral	CH_4	CBr_4	$SiCl_4$	CH_4
3	1	4	Pyramidal	NH_3	PCl_3		NH_3
2	2	4	Bent	H_2O	H_2S	SCl_2	H_2O
2	1	3	Bent	SO_2	O_3		SO_2

[a] NBP = nonbonding pair of electrons.

atom, forming angles of separation of 120°. Four groups form a tetrahedral array around the central atom, giving a separation of about 109.5°.

You can determine the shapes of many molecules (and polyatomic ions, Section 4.12) by following these simple rules.

1. Draw an electron-dot formula. In the structure, indicate a shared electron pair (bonding pair, BP) by a line. Use dots to indicate any nonbonding pairs (NBPs) of electrons.

2. To determine shape, count the number of atoms *and* NBPs attached to the *central atom*. Note that a multiple bond counts only as *one set*. Examples are

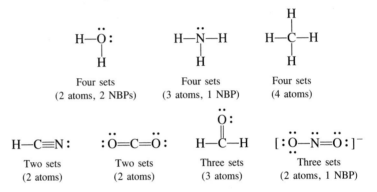

H—Ö:	H—N̈—H	H—C—H (with H above and below)
Four sets	Four sets	Four sets
(2 atoms, 2 NBPs)	(3 atoms, 1 NBP)	(4 atoms)

H—C≡N:	:Ö=C=Ö:	H—C—H (with Ö: above, double bond)	[:Ö—N̈=Ö:]⁻
Two sets	Two sets	Three sets	Three sets
(2 atoms)	(2 atoms)	(3 atoms)	(2 atoms, 1 NBP)

3. Determine the number of electron sets and draw a shape *as if* all were bonding pairs.

4. Sketch that shape, placing the electron pairs as far apart as possible (see Table 4.5). If there is *no* NBP, that is the shape of the molecule. If there *are* NBPs, remove them, leaving the BPs exactly as they were. (This may seem strange, but it stems from the fact that *all* the sets determine the geometry but only the arrangement of bonded atoms is considered in the shape of the molecule.)

EXAMPLE 4.22

What is the shape of the BH_3 molecule?

SOLUTION

1. The electron-dot formula is

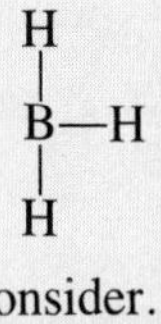

$$\begin{array}{c} H \\ | \\ B—H \\ | \\ H \end{array}$$

2. There are three electron sets to consider. The three sets get as far apart as possible, giving a triangular arrangement of the sets.

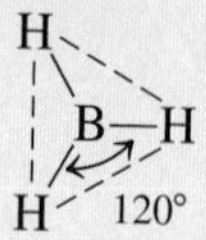

   (triangular BH₃ structure with B—H, and two H at corners, 120° angle labeled)

3. All the sets are bonding pairs; the molecular shape is the same as the arrangement of the electrons.

Practice Exercise
What is the shape of the BeH_2 molecule?

EXAMPLE 4.23

What is the shape of the SCl_2 molecule?

SOLUTION

1. The electron-dot formula is

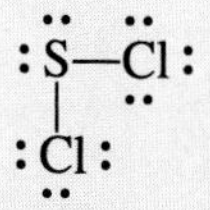

$$:\overset{\displaystyle ..}{\underset{\displaystyle }{S}}\!-\!\overset{\displaystyle ..}{\underset{\displaystyle ..}{Cl}}:$$

$$:\underset{\displaystyle ..}{Cl}:$$

2. There are four electron sets on the sulfur atom to consider, two chlorines and two NBPs. The four sets get as far apart as possible, giving a tetrahedral arrangement of the sets about the central atom.

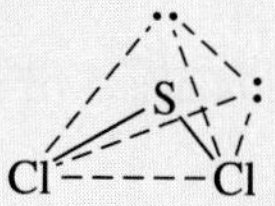

3. Remove (ignore) the NBPs; the molecular shape is bent, with a bond angle of 109.5°.

Practice Exercise

What is the shape of the PH_3 molecule?

4.17
SHAPES AND PROPERTIES: POLAR AND NONPOLAR MOLECULES

In Section 4.9 we discussed polar and nonpolar bonds. Diatomic molecules are polar if their bonds are polar and nonpolar if their bonds are nonpolar.

$$\overset{\delta+\ \ \delta-}{H\!-\!Cl} \qquad Cl\!-\!Cl$$

Polar   Nonpolar

For molecules with three or more atoms, we also must consider the orientation of the bonds to determine whether or not the molecule as a whole is polar.

We should expect the bonds in water to be polar because oxygen is more electronegative than hydrogen. (Like chlorine, oxygen is to the right in the periodic table, and electronegativity increases as one moves to the right in the table.) Just because a molecule contains polar bonds, however, does not mean that the molecule as a whole is polar. If the atoms in the water molecule were in a straight row (that is, in a linear arrangement), the two polar bonds would cancel one another.

$$\overset{\delta+\ \ \delta-\ \ \delta+}{H\!-\!O\!-\!H} \qquad (incorrect\ structure)$$

Instead of one end of the molecule being positive and the other end negative, the electrons would be pulled toward the right in one bond and toward the left in the other. Overall there

A *polar molecule* has separate centers of positive and negative charge, just as the Earth or a magnet has north and south poles.

Figure 4.10
Polar molecules are aligned
in an electric field, with the
positive end of the mole-
cule preferentially pointing
toward the negative plate
and the negative end of the
molecule directed toward
the positive plate.

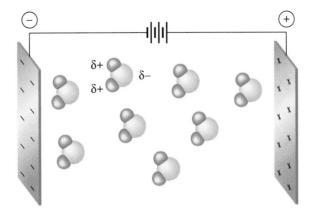

would be no net dipole. (A **dipole** consists of a positive and an equal negative charge separated by some distance.)

But water *does* act like a dipole. If you place a sample of water between two electrically charged plates, the water molecules align themselves, one end attracted toward the positive plate and the other end toward the negative plate. To act like a dipole, the molecule must be bent so that the bonds do not cancel one another out.

$$^{\delta +}\text{H} : \overset{..}{\underset{..}{\text{O}}} : {}^{\delta -} \qquad \text{or} \qquad {}^{\delta +}\text{H} \diagdown \overset{\text{O}^{\delta -}}{\underset{|}{}}$$
$$\overset{..}{\underset{..}{\text{H}}}{}^{\delta +} \qquad\qquad\qquad\qquad \text{H}^{\delta +}$$

Such molecules would align themselves between charged plates as shown in Figure 4.10.

Dipoles often are represented by an arrow with a plus at the tail end ($+\!\!\longrightarrow$).

$$\overset{+\longrightarrow}{\text{H}-\text{Cl}}$$

The positive end of the arrow is obvious; the head of the arrow indicates the negative end of the dipole. Using these arrows, we see that carbon dioxide is nonpolar despite its polar bonds.

$$\overset{\longleftarrow +\ +\longrightarrow}{\text{O}=\text{C}=\text{O}}$$

The two cancel one another out. In water, both dipoles point toward the oxygen, giving a net dipole toward that end of the molecule.

$$\nearrow \overset{\text{O}}{} \nwarrow \qquad\qquad \overset{\uparrow}{}$$
$$\times \diagup \quad \diagdown \times \qquad\qquad \overset{\text{O}}{}$$
$$\text{H} \qquad \text{H} \qquad \text{or} \qquad \text{H} \ + \ \text{H}$$

The shape of the water molecule can be accounted for by a modification of the VSEPR theory. According to VSEPR, the two bonds and two NBPs should form a tetrahedral arrangement.

$$\overset{..}{\underset{\text{H}--\text{H}}{\text{O}}} \, ..$$

Ignoring the NBPs, the molecular shape has the atoms in a bent arrangement, with a bond angle of 109.5° (the tetrahedral angle).

The predicted bond angle of 109.5° for water is a bit larger than the measured angle of 104.5°. The difference is explained by the fact that the NBPs occupy a greater volume than do the bonding pairs. These larger orbitals push the smaller BPs closer together.

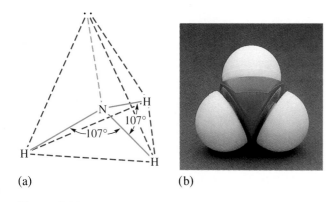

Figure 4.11

The ammonia molecule. In the drawing (a), solid lines indicate covalent bonds; the dashed lines outline a tetrahedron. The unshared pair of electrons is ignored in determining the pyramidal shape of the molecule. The photograph (b) shows a space-filling model of an ammonia molecule.

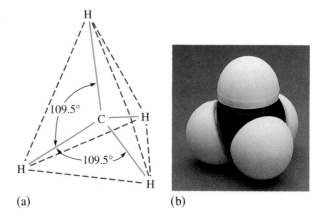

Figure 4.12

The methane molecule. In the drawing (a), solid lines indicate covalent bonds; the dashed lines outline the tetrahedron. All bond angles are 109.5°. The photograph (b) shows a space-filling model of the methane molecule.

In ammonia there are three BPs and one NBP about the nitrogen atom. The VSEPR theory predicts a tetrahedral arrangement with bond angles of 109.5°. The actual bond angles are 107°, close to the theoretical value. Presumably, the unshared pair of electrons occupies a greater volume than does a shared pair, pushing the latter slightly closer together. The arrangement is therefore that of a tripod with a hydrogen atom at the end of each leg and the nitrogen atom with its unshared pair sitting at the top (Figure 4.11). Each nitrogen–hydrogen bond is somewhat polar, making the ammonia molecule polar.

In the methane molecule there are four pairs of electrons on the central carbon atom. Using the VSEPR theory, we would expect a tetrahedral arrangement and bond angles of 109.5°. The actual bond angles are 109.5°, in perfect agreement with theory (Figure 4.12). All four electron pairs are shared with hydrogen atoms; thus, all four pairs occupy identical volumes. Each carbon–hydrogen bond is slightly polar, but the methane molecule as a whole is symmetrical. The slight bond polarities cancel out, leaving the methane molecule nonpolar.

Many of the properties of compounds such as melting point, boiling point, and solubility depend on the polarity of the molecules of the compound.

EXERCISES

1. Which group of elements in the periodic table is characterized by stable electron arrangements?
2. What is the structural difference between a sodium atom and a sodium ion?
3. How does sodium metal differ from sodium ions (e.g., in sodium chloride) in properties?
4. What is the structural difference between a sodium ion and a neon atom? What is similar?

5. What are the structural differences among a chlorine atom, a chlorine molecule, and a chloride ion? How do their properties differ?
6. Give electron-dot formulas for each of the following elements. You may use the periodic table.
   a. sodium   b. oxygen
   c. fluorine   d. aluminum
7. Give electron-dot formulas for each of the following ele-

ments. You may use the periodic table.
a. carbon   **b.** potassium
c. magnesium   **d.** chlorine
e. nitrogen

8. Using electron-dot formulas, show the formation of ion from atom for each of the following.
a. barium   **b.** bromine
c. aluminum   **d.** sulfur

9. Indicate the charges on simple ions formed from the following elements.
a. Group IIIA   **b.** Group VIA

10. Indicate the charges on simple ions formed from the following elements.
a. Group IIA   **b.** Group VIIA

11. Use electron-dot formulas to show the transfer of electrons from aluminum atoms to sulfur atoms to form ions with noble gas configurations.

12. Use electron-dot formulas to show the transfer of electrons from magnesium atoms to phosphorus atoms to form ions with noble gas configurations.

13. Use electron-dot formulas to show the sharing of electrons between two iodine atoms to form an iodine (I_2) molecule. Label all electron pairs as bonding or nonbonding.

14. Use electron-dot formulas to show the sharing of electrons between a hydrogen atom and a fluorine atom. Label ends of the molecule with symbols that indicate polarity.

15. Classify the following bonds as ionic or covalent. For those bonds that are covalent, indicate whether they are polar or nonpolar.
a. KF   **b.** IBr   **c.** MgS

16. Classify the following bonds as ionic or covalent. For those bonds that are covalent, indicate whether they are polar or nonpolar.
a. NO   **b.** CaO   **c.** NaBr

17. Classify the following bonds as ionic or covalent. For those bonds that are covalent, indicate whether they are polar or nonpolar.
a. Br_2 **b.** F_2 **c.** HCl

18. Classify the following covalent bonds as polar or nonpolar.
a. H—O   **b.** N—Cl   **c.** B—F

19. Classify the following covalent bonds as polar or nonpolar.
a. H—N   **b.** Be—F   **c.** P—Cl

20. Use the symbol $\leftrightarrow$ to indicate the direction of the dipole in each of the bonds in Exercise 18.

21. Use the symbol $\leftrightarrow$ to indicate the direction of the dipole in each of the bonds in Exercise 19.

22. The molecule BeF_2 is linear. Is it polar or nonpolar? Use the symbol $\leftrightarrow$ to explain your answer.

23. The molecule SF_2 is bent. Is it polar or nonpolar? Use the symbol $\leftrightarrow$ to explain your answer.

24. How many covalent bonds do each of the following usually form? You may refer to the periodic table.
a. H   **b.** C   **c.** O
d. F   **e.** N   **f.** Br

25. List three elements that readily form double bonds.

26. List two elements that readily form triple bonds.

27. Give electron-dot formulas for each of the following ionic compounds.
a. sodium fluoride   **b.** potassium chloride
c. potassium fluoride

28. Give electron-dot formulas for each of the following ionic compounds.
a. magnesium fluoride
b. calcium chloride
c. sodium oxide
d. potassium sulfide

29. Give electron-dot formulas for each of the following ionic compounds.
a. sodium nitride   **b.** aluminum chloride
c. magnesium nitride

30. Use electron-dot formulas to show the sharing of electrons between a silicon atom and hydrogen atoms to form a molecule in which silicon has an octet of electrons.

31. Use electron-dot formulas to show the sharing of electrons between a carbon atom and fluorine atoms to form a molecule in which each atom has an octet of electrons.

32. Use electron-dot formulas to show the sharing of electrons between a nitrogen atom and chlorine atoms to form a molecule in which each atom has an octet of electrons.

33. Name the following ions.
a. Na^+ **b.** Mg^{2+} **c.** Al^{3+}
d. Cl^- **e.** O^{2-} **f.** N^{3-}

34. Name the following ions.
a. K^+ **b.** Ca^{2+} **c.** Zn^{2+}
d. Br^- **e.** Li^+ **f.** S^{2-}

35. Name the following ions.
a. Fe^{3+} **b.** Cu^{2+} **c.** Ag^+

36. Name the following ions.
a. Fe^{2+} **b.** Cu^+ **c.** I^-

37. Give symbols for the following ions.
a. bromide ion   **b.** calcium ion
c. potassium ion   **d.** iron(II) ion

38. Give symbols for the following ions.
a. sodium ion   **b.** aluminum ion
c. oxide ion   **d.** copper(II) ion

39. Name the following ionic compounds.
a. NaBr   **b.** $CaCl_2$ **c.** $FeCl_2$
d. LiI   **e.** K_2S **f.** CuBr

40. Name the following ionic compounds.
a. KCl   **b.** $MgBr_2$ **c.** CuI_2
d. CaS   **e.** $FeCl_3$ **f.** Al_2O_3

41. Give formulas for the following covalent compounds.
a. dinitrogen monoxide
b. tetraphosphorus trisulfide
c. phosphorus pentachloride
d. sulfur hexafluoride

42. Give formulas for the following covalent compounds.
a. oxygen difluoride

b. dinitrogen pentoxide
c. phosphorus tribromide
d. tetrasulfur tetranitride

43. Name the following covalent compounds.
   a. CS_2 **b.** N_2S_4
   c. PF_4 **d.** S_2F_{10}

44. Name the following covalent compounds.
   a. CBr_4 **b.** Cl_2O_7
   c. P_4S_{10} **d.** I_2O_5

45. Use the symbols $\delta+$ and $\delta-$ to indicate partial charges, if any, on the following bonds.
   a. N—H   **b.** C—F   **c.** C—C

46. Use the symbols $\delta+$ and $\delta-$ to indicate partial charges, if any, on the following bonds.
   a. Si—Cl   **b.** Cl—Cl   **c.** O—F

47. Give electron-dot formulas that follow the octet rule, showing all valence electrons as dots, for the following covalent molecules.
   a. CH_4O **b.** NOH_3
   c. CH_5N **d.** N_2H_4

48. Give electron-dot formulas that follow the octet rule, showing all valence electrons as dots, for the following covalent molecules.
   a. NF_3 **b.** C_2H_4
   c. C_2H_2 **d.** CH_2O

49. Shared pairs of electrons can be represented by dashed lines. Give dashed-line formulas for the molecules in Exercise 47.

50. Shared pairs of electrons can be represented by dashed lines. Give dashed-line formulas for the molecules in Exercise 48.

51. Give electron-dot formulas that follow the octet rule for the following covalent molecules.
   a. COF_2 **b.** PCl_3
   c. H_3PO_3 **d.** HCN

52. Give electron-dot formulas that follow the octet rule for the following covalent molecules.
   a. SCl_2 **b.** H_2SO_4
   c. XeO_3 **d.** $HClO_4$

53. Give electron-dot formulas that follow the octet rule for the following ions.
   a. ClO^- **b.** HPO_4^{2-}
   c. ClO_2^- **d.** BrO_3^-

54. Give electron-dot formulas that follow the octet rule for the following ions.
   a. CN^- **b.** IO_4^-
   c. HSO_4^- **d.** PO_3^{3-}

55. Give an electron-dot formula for each of the following.
   a. NO   **b.** BCl_3

56. Give an electron-dot formula for each of the following.
   a. PF_5 **b.** $AlBr_3$

57. Use the valence shell electron pair repulsion (VSEPR) theory to predict the shape of each of the following molecules.
   a. silane (SiH_4)   **b.** hydrogen selenide (H_2Se)

58. Use the VSEPR theory to predict the shape of each of the following molecules.
   a. beryllium chloride ($BeCl_2$)
   b. boron chloride (BCl_3)

59. Use the VSEPR theory to predict the shape of each of the following molecules.
   a. arsine (AsH_3)
   b. carbon tetrafluoride (CF_4)

60. Use the VSEPR theory to predict the shape of each of the following molecules.
   a. oxygen difluoride (OF_2)
   b. silicon tetrachloride ($SiCl_4$)

61. Use the VSEPR theory to predict the shape of each of the following molecules.
   a. nitrogen trichloride (NCl_3)
   b. sulfur dichloride (SCl_2)

62. Use the VSEPR theory to predict the shape of each of the following molecules.
   a. phosphorus trifluoride (PF_3)
   b. dichlorodifluoromethane (CCl_2F_2)

63. Name the following ions.
   a. CO_3^{2-} **b.** HPO_4^{2-} *monohydrogen phosphate*
   c. MnO_4^- **d.** OH^-

64. Name the following ions.
   a. NO_3^- **b.** SO_4^{2-}
   c. $H_2PO_4^-$ **d.** HCO_3^-

65. Give formulas for the following ions.
   a. ammonium ion
   b. hydrogen sulfate ion
   c. cyanide ion
   d. nitrite ion

66. Give formulas for the following ions.
   a. phosphate ion   **b.** hydrogen carbonate ion
   c. dichromate ion   **d.** oxalate ion

67. Give formulas for the following ionic compounds.
   a. magnesium sulfate
   b. sodium hydrogen carbonate
   c. potassium nitrate
   d. calcium monohydrogen phosphate

68. Give formulas for the following ionic compounds.
   a. calcium carbonate
   b. potassium dihydrogen phosphate
   c. magnesium cyanide
   d. lithium hydrogen sulfate

69. Give formulas for the following ionic compounds.
   a. iron(II) phosphate
   b. potassium dichromate
   c. copper(I) iodide
   d. ammonium nitrite

70. Give formulas for the following ionic compounds.
   a. iron(III) oxalate
   b. sodium permanganate
   c. copper(II) bromide
   d. zinc monohydrogen phosphate

71. Name the following ionic compounds.
   a. KNO_2
   b. $LiCN$
   c. NH_4I
   d. $NaNO_3$
   e. $KMnO_4$
   f. $CaSO_4$

72. Name the following ionic compounds.
   a. $NaHSO_4$
   b. $Al(OH)_3$
   c. Na_2CO_3
   d. $KHCO_3$
   e. NH_4NO_2
   f. $Ca(HSO_4)_2$

73. Name the following ionic compounds.
   a. Na_2HPO_4
   b. $(NH_4)_3PO_4$
   c. $Al(NO_3)_3$
   d. NH_4NO_3

74. Name the following ionic compounds.
   a. Li_2CO_3
   b. $Na_2Cr_2O_7$
   c. $Ca(H_2PO_4)_2$
   d. $(NH_4)_2C_2O_4$

75. Fill in this table assuming that elements X, Y, and Z are all in A subgroups in the periodic table.

	Element X	Element Y	Element Z
Group number	IA	_____	_____
Electron dot formula	_____	$\cdot \dot{Y} \cdot$	_____
Charge on ion	_____	_____	2−

76. Consider the hypothetical elements X, Y, and Z with electron-dot formulas:

$$:\overset{\cdot\cdot}{\underset{\cdot\cdot}{X}}\cdot \qquad :\overset{\cdot\cdot}{Y}\cdot \qquad :\overset{\cdot}{Z}\cdot$$

   a. To which group in the periodic table would each belong?
   b. Write the electron-dot formula for the simplest compound of each with hydrogen.
   c. Write electron-dot formulas for the ions formed when X and Y react with sodium.

77. Chlorine dioxide is used to bleach flour. Give the formula for chlorine dioxide.

78. Tetraphosphorus trisulfide is used in the tips of "strike anywhere" matches. Give the formula for tetraphosphorus trisulfide.

79. The gas phosphine (PH_3) is used as a fumigant to protect stored grain and other durable produce from pests. Phosphine is generated in situ by adding water to aluminum phosphide or magnesium phosphide. Give formulas for the two phosphides.

Chapter 5
CHEMICAL REACTIONS

The "magic" of chemistry. The combination of two colorless liquids produces a yellow solid.

The complex chemical processes in the living cell involve changes in energy as well as changes in chemical composition. Some reactions provide the energy that keeps the cell alive and well. Other reactions, vital to life processes, require an input of energy.

Chemical reactions proceed at various rates. Some are explosively fast; others, exceedingly slow. Rates are affected by a number of factors. Perhaps the most important of these factors, for living organisms, are complex molecules called enzymes that accelerate reaction rates enormously. Yet, strange as it may seem, the enzymes are still there unchanged *after* doing their job (see Chapter 22).

Chemical reactions also proceed to different extents. In some, the reactants are converted entirely to products; these reactions are said to go to *completion*. In others, the products react, re-forming the original starting materials. These reactions, outside the cell, come to equilibrium. In a living cell, equilibrium would be deadly. The cellular processes must go to completion—or very nearly so. Products can become reactants in the body, but not under equilibrium conditions.

In this chapter, we will examine energy changes, reaction rates, and equilibria. For the most part, we will deal with simple, nonliving systems. The principles developed, however, will be exceedingly important in later chapters, where we will deal with the more complex chemistry of living cells.

5.1
BALANCING CHEMICAL EQUATIONS

Chemistry is a study of matter and the changes it undergoes. More than that, it is a study of the energy that brings about those changes—or the energy that is released when those changes occur. In Chapter 4, we discussed the symbols and formulas that have been invented to represent elements and compounds. Now let's look at a shorthand way of describing chemical changes—the **chemical equation.**

Carbon reacts with oxygen to form carbon dioxide. In chemical shorthand, this reaction is written

$$C + O_2 \longrightarrow CO_2$$

The plus sign (+) indicates the addition of carbon to oxygen (or vice versa) or a mixing of the two in some manner. The arrow ($\rightarrow$) is often read "yields." Substances on the left of the arrow are **reactants,** or *starting materials*. Those on the right are the **products** of the reaction. The conventions here are like those we used in writing nuclear equations (Section 3.2). Now, however, the nucleus will remain untouched. Chemical reactions involve only electrons.

Chemical equations have meaning on the atomic and molecular levels. The equation

$$C + O_2 \longrightarrow CO_2$$

means that one atom of carbon (C) reacts with one molecule of oxygen (O_2) to produce one molecule of carbon dioxide (CO_2).

Sometimes the physical states of the reactants and products are indicated. The initial letter of the state is written immediately following the formula. Thus (g) indicates a gaseous substance, (l) a liquid, and (s) a solid. The label (aq) indicates an aqueous solution, that is, a water solution. Using these labels, our equation becomes

$$C(s) + O_2(g) \longrightarrow CO_2(g)$$

Not all chemical reactions are as simply represented. Hydrogen reacts with oxygen to form water. We can write this reaction as

$$H_2 + O_2 \longrightarrow H_2O \quad \textit{(not balanced)}$$

This representation, however, is not consistent with the law of conservation of matter. There are two oxygen atoms shown among the reactants (as O_2), and only one among the products (in H_2O). For the equation to correctly represent the chemical happening, it must be balanced. To balance the oxygen atoms, we need only place the coefficient 2 in front of the formula for water.

$$H_2 + O_2 \longrightarrow 2 H_2O \quad \textit{(not balanced)}$$

This coefficient means that there are two molecules of water involved. As in the case with subscripts, a coefficient of 1 is understood. A coefficient preceding a formula multiplies everything in the formula. In the above equation, the coefficient 2 not only increases the number of oxygen atoms to two but also increases the number of hydrogen atoms to four.

But the equation is still not balanced. We balanced oxygen at the expense of unbalancing hydrogen. To balance hydrogen, we place a coefficient 2 in front of the H_2.

$$2 H_2 + O_2 \longrightarrow 2 H_2O \quad \textit{(balanced)}$$

Note that we could not balance the equation by changing the subscript for oxygen in water.

$H_2 + O_2 \longrightarrow H_2O_2$
(not correct)

The equation would be balanced, but it would not mean "hydrogen reacts with oxygen to form water." The formula H_2O_2 represents hydrogen peroxide, *not* water.

Figure 5.1
To balance the equation for the reaction of hydrogen with oxygen that forms water, the same number of each kind of atom must appear on each side (atoms are conserved). When the equation is balanced, there are four hydrogen atoms and two oxygen atoms on each side.

● Hydrogen
● Oxygen

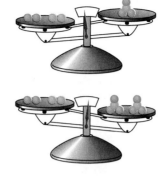

Figure 5.2
To balance the equation for the reaction of nitrogen with hydrogen that forms ammonia, the same number of each kind of atom must appear on each side of the equation. When the equation is balanced, there are two nitrogen atoms and six hydrogen atoms on each side.

● Hydrogen
● Nitrogen

Now there are enough hydrogen atoms on the left. In fact, there are four hydrogen atoms and two oxygen atoms on each side of the equation. Atoms are conserved; the equation is balanced (Figure 5.1).

EXAMPLE 5.1

Balance the following equation.

$$N_2 + H_2 \longrightarrow NH_3 \quad \textit{(not balanced)}$$

SOLUTION
For this sort of problem, we will use the concept of the least common multiple. We will balance the hydrogen first. There are two hydrogen atoms on the left and three on the right. The least common multiple of 3 and 2 is 6. Six is the smallest number of hydrogen atoms that can be evenly converted from reactants to products. We therefore need three molecules of H_2 and two of NH_3.

$$N_2 + 3 H_2 \longrightarrow 2 NH_3 \quad \textit{(balanced)}$$

We have balanced the hydrogen atoms and, in the process, the nitrogen atoms. There are two nitrogen atoms on the left and two on the right. The entire equation is balanced (Figure 5.2).

Practice Exercise
Balance the following equation.

$$P_4 + H_2 \longrightarrow PH_3$$

EXAMPLE 5.2

Balance the following equation.

$$Fe + O_2 \longrightarrow Fe_2O_3 \quad \textit{(not balanced)}$$

SOLUTION

Begin by balancing the oxygen. The least common multiple is 6. We need three molecules of O_2 and two of Fe_2O_3.

$$Fe + 3\,O_2 \longrightarrow 2\,Fe_2O_3 \quad \textit{(not balanced)}$$

We now have four atoms of iron on the right side. We can get four on the left by placing the coefficient 4 in front of Fe.

$$4\,Fe + 3\,O_2 \longrightarrow 2\,Fe_2O_3 \quad \textit{(balanced)}$$

The equation is now balanced.

Practice Exercise

Balance the following equation.

$$V + O_2 \longrightarrow V_2O_5$$

EXAMPLE 5.3

Balance the following equation.

$$CH_4 + O_2 \longrightarrow CO_2 + H_2O \quad \textit{(not balanced)}$$

SOLUTION

In this equation, oxygen appears in two different products; we will leave the oxygen for last and balance the other two elements first. Carbon is already balanced, with one atom on either side of the equation. The least common multiple of 2 and 4 is 4, so to balance hydrogen we place the coefficient 2 in front of H_2O. Now we have four hydrogen atoms on each side.

$$CH_4 + O_2 \longrightarrow CO_2 + 2\,H_2O \quad \textit{(not balanced)}$$

Now for the oxygen. There are four oxygen atoms on the right. If we place a 2 in front of O_2, the oxygen atoms balance.

$$CH_4 + 2\,O_2 \longrightarrow CO_2 + 2\,H_2O \quad \textit{(balanced)}$$

The equation is balanced.

Practice Exercise

Balance the following equation.

$$C_3H_8 + O_2 \longrightarrow CO_2 + H_2O$$

EXAMPLE 5.4

Balance the following equation.

$$H_2SO_4 + NaCN \longrightarrow HCN + Na_2SO_4 \quad \textit{(not balanced)}$$

SOLUTION

Here we have an equation that involves compounds with polyatomic ions. The SO_4 group (actually $SO_4{}^{2-}$) should be treated as a unit and balanced as a whole. The same is true of the CN group (actually CN^-). As the equation is presently written, the SO_4 groups and the CN groups are balanced, but the hydrogen atoms and the sodium atoms are not. The least common multiple for sodium is 2, so to get two sodium atoms on the left we place a 2 before NaCN. The least common multiple for hydrogen is 2, so to get two hydrogen atoms on the right we place a 2 before HCN.

$$H_2SO_4 + 2\,NaCN \longrightarrow 2\,HCN + Na_2SO_4 \quad \textit{(balanced)}$$

The same coefficients balance the CN groups and the SO_4 groups, and the entire equation is balanced.

Practice Exercise

Balance the following equation.

$$H_3PO_4 + Ca(OH)_2 \longrightarrow Ca_3(PO_4)_2 + H_2O$$

We have made the task of balancing equations deceptively easy by considering simple reactions. It is more important at this point for you to understand the principle than to be able to balance complicated equations. You should know what is meant by a balanced equation and be able to handle simple systems.

5.2
VOLUME RELATIONSHIPS IN CHEMICAL EQUATIONS

Chemists generally cannot work with individual atoms and molecules. Even the tiniest speck of matter that we can see contains billions of atoms. John Dalton postulated that atoms of different elements had different masses. Therefore, equal masses of different elements would contain different numbers of atoms. Consider the analogous situation of golf balls and Ping-Pong balls. A kilogram of golf balls contains a smaller number of balls than a kilogram of Ping-Pong balls. One could determine the number of balls in each case simply by counting them. For atoms, however, such a straightforward method is not available. It was in the experiments of a French chemist and the mind of an Italian scientist that approaches to the problem of numbering atoms were found.

In 1809 the French scientist Joseph Louis Gay-Lussac (1778–1850) announced the results of some chemical reactions that he had carried out with gases. These experiments were summarized in his **law of combining volumes,** which states that when all measurements are made at the same temperature and pressure, the volumes of gaseous reactants

Figure 5.3
Two volumes of hydrogen gas react with one volume of oxygen gas to give two volumes of steam.

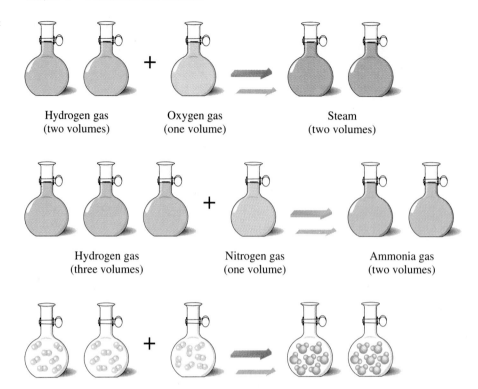

Hydrogen gas
(two volumes)

Oxygen gas
(one volume)

Steam
(two volumes)

Figure 5.4
Three volumes of hydrogen gas react with one volume of nitrogen gas to give two volumes of ammonia gas.

Hydrogen gas
(three volumes)

Nitrogen gas
(one volume)

Ammonia gas
(two volumes)

Figure 5.5
Avogadro's explanation of Gay-Lussac's law of combining volumes. Equal volumes of each of the gases contain the same number of molecules.

Hydrogen gas
(two volumes)

Oxygen gas
(one volume)

Steam
(two volumes)

and products are in a small-whole-number ratio. For example, when he allowed hydrogen to react with oxygen to form steam at 100 °C, two volumes of hydrogen united with one volume of oxygen to give two volumes of steam (Figure 5.3). The small-whole-number ratio was $2:1:2$.

In another experiment (Figure 5.4), Gay-Lussac found that if hydrogen is permitted to react with nitrogen to form ammonia, the combining volumes are three of hydrogen with one of nitrogen to give two of ammonia ($3:1:2$).

Gay-Lussac thought there must be some relationship between the *numbers* of molecules and the *volumes* of gaseous reactants and products. But it was the Italian chemist Amadeo Avogadro (1776–1856) who first explained the law of combining volumes. **Avogadro's hypothesis,** based on shrewd interpretation of experimental facts, was that equal volumes of all gases (at the same temperature and pressure) contain the same number of molecules (Figure 5.5).

The equation for the combination of hydrogen and oxygen to form water (steam) is

$$2\,H_2(g)\ +\ O_2(g)\ \longrightarrow\ 2\,H_2O(g)$$

The coefficients of the molecules are the same as the combining ratio of the gas volumes, $2:1:2$ (see Figure 5.3). Similarly, the formation of ammonia is described in the equation

$$3\,H_2(g)\ +\ N_2(g)\ \longrightarrow\ 2\,NH_3(g)$$

The coefficients are identical to the factors of the combining ratio (see Figure 5.4). The equation says that a nitrogen molecule reacts with three hydrogen molecules to produce two ammonia molecules. It also indicates that if you had 1 million nitrogen molecules, you would need 3 million hydrogen molecules to produce 2 million ammonia molecules. The equation provides the combining ratios. If identical volumes of gases contain identical numbers of molecules, then, according to the equation, one volume of nitrogen reacts with three volumes of hydrogen to produce two volumes of ammonia.

EXAMPLE 5.5

According to the equation

$$CH_4(g) \longrightarrow C(s) + 2 H_2(g)$$

what volume of hydrogen would be obtained from 1.00 L of methane (CH_4)?

SOLUTION
The equation indicates that two volumes of hydrogen are obtained from the reaction of one volume of methane. Therefore, 1.00 L of methane would yield 2.00 L of hydrogen.

Practice Exercise
Using the equation in Example 5.5, calculate what volume of hydrogen is obtained from the reaction of 25.0 L of methane.

Note that no calculations such as these can be made that involve carbon, a solid [C(s)]. The law of combining volumes applies only to gases.

EXAMPLE 5.6

Using the equation in Example 5.5, calculate how much methane must react to produce 10.0 L of H_2.

SOLUTION

$$10.0 \; \cancel{L \; H_2} \times \frac{1 \text{ L } CH_4}{2 \; \cancel{L \; H_2}} = 5.00 \text{ L } CH_4$$

5.3
AVOGADRO'S NUMBER: 6.02×10^{23}

The periodic table (inside front cover) gives *relative* atomic weights for the various elements. It isn't possible to weigh individual atoms. What we do is to weigh *equal numbers* of atoms or molecules of different substances and use the ratio of masses as relative weights. Avogadro's hypothesis, which has been verified repeatedly through the years, gives us a way to do that by measuring the masses of equal volumes of gases.

Once the relative weights are known, it is possible to plan reactions so no materials are wasted. If we could (we can't) weigh out 12 amu of carbon and 16 amu of oxygen, we would have one atom of each, and we could make one molecule of carbon monoxide (CO). If we weigh out 12 g of carbon and 16 g of oxygen, we still have the proper *ratio* of atoms. Gram quantities are easily weighed. From these amounts of reactants, we could make 28 g of CO, with none of the reactants left over. Similarly, if we wished to make

Figure 5.6
The Earth—and the Jolly Green Giant—covered with 6.02×10^{23} peas.

carbon dioxide (CO_2), we could weigh out 12 g of carbon and 32 g of oxygen to make 44 g of CO_2 with no left-over reactants.

Avogadro had no way of knowing how many molecules there were in a given volume of gas. Scientists since his time have determined the number of atoms in various weighed samples of substances. The numbers are enormously large, even for tiny samples. In defining atomic weights (Section 2.3), the mass of a carbon-12 atom is defined as exactly 12 amu. The number of carbon-12 atoms in a 12-g sample of carbon-12 is called **Avogadro's number.** The value of Avogadro's number has been determined experimentally to be 6.0221367×10^{23}. For most purposes the number is rounded to 6.02×10^{23}.

Avogadro's number is unimaginably large. If you had 6.02×10^{23} dollars, you could spend a billion dollars a second for your entire lifetime and still have used only 0.001% of your money. If carbon atoms were the size of peas, 6.02×10^{23} of them could cover the entire surface of Earth to a depth of 15 m (Figure 5.6).

Avogadro's number is a fundamental one. There are 6.02×10^{23} atoms of oxygen in 16.0 g of oxygen, 6.02×10^{23} atoms of nitrogen in 14.0 g of nitrogen, and 6.02×10^{23} atoms of sulfur in 32.0 g of sulfur. The atomic weight of any element, expressed in grams, contains 6.02×10^{23} atoms of that element.

5.4
CHEMICAL ARITHMETIC AND THE MOLE

We buy eggs by the dozen (12 eggs), pencils by the gross (144 pencils), and paper by the ream (500 sheets). A dozen is the same *number,* whether we have a dozen oranges or a dozen watermelons. A dozen oranges and a dozen watermelons do not have the same *mass,* however. Chemists use a term, *mole,* to refer to a particular number (Avogadro's number) of things. For example, a mole of magnesium atoms is Avogadro's number (6.02×10^{23}) of magnesium atoms, and a mole of sugar molecules is Avogadro's number (6.02×10^{23}) of sugar molecules. A mole of any substance has Avogadro's number (6.02×10^{23}) of particles (atoms, molecules, ions, or whatever) of that substance. A mole of magnesium atoms and a mole of sugar molecules do *not* have the same mass.

The SI definition of the **mole** (symbol *mol*) is the amount of substance containing as many elementary units as there are atoms in exactly 12 g of the carbon-12 isotope. The elementary units must be specified by a chemical formula. They may be molecules such as O_2 or CO_2 or even $C_{24}H_{50}O_4N_5Cl$. The units may be atoms such as C or O or Pu. They may be ions such as K^+ or SO_4^{2-}. They may be any kind of formula unit; if you can write a formula for it, you can describe a mole of it.

There are 6.02×10^{23} atoms of carbon-12 in 12 g of that isotope, so a mole of carbon dioxide is 6.02×10^{23} molecules of CO_2, a mole of magnesium is 6.02×10^{23} atoms of Mg, and a mole of NaCl is 6.02×10^{23} formula units of NaCl. (Note that a mole of NaCl consists of 6.02×10^{23} Na^+ ions and 6.02×10^{23} Cl^- ions.)

A **formula unit** is simply the atoms or ions specified by the formula. A formula unit of Al_2O_3 is two aluminum atoms and three oxygen atoms. The **formula weight** is the sum of all the atomic weights in the formula (expressed in amu).

The **molar mass** of a substance is the mass of 1 mol of that substance in grams. Recall that in Section 2.3 we used the term *atomic weight* instead of atomic mass. Likewise, we use the term **molecular weight** instead of molar mass. Carbon dioxide, CO_2, has a molecular weight of 44.1 g/mol. Sucrose (table sugar, $C_{12}H_{22}O_{11}$) has a molecular weight of 342.3 g/mol. Figure 5.7 is a photograph of 1 mol each of several chemical substances. By weighing out the proper molar ratio of reactants, we get the correct atomic and molecular ratios for a reaction (Figure 5.8).

Chemists and other scientists and engineers often are confronted with questions such as: "How many grams of phosphoric acid can I make from 660 g of phosphorus?" and

Figure 5.7
One mole of each of several familiar substances: salt (left), sugar (top), copper (right), and carbon (center front).

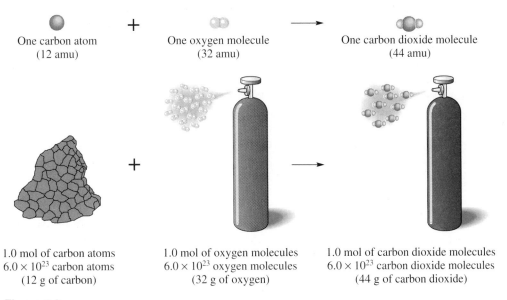

| One carbon atom (12 amu) | + | One oxygen molecule (32 amu) | → | One carbon dioxide molecule (44 amu) |

1.0 mol of carbon atoms	1.0 mol of oxygen molecules	1.0 mol of carbon dioxide molecules
6.0×10^{23} carbon atoms	6.0×10^{23} oxygen molecules	6.0×10^{23} carbon dioxide molecules
(12 g of carbon)	(32 g of oxygen)	(44 g of carbon dioxide)

Figure 5.8
We cannot weigh single atoms or molecules, but we can weigh larger numbers of these fundamental particles.

"How many grams of hydrogen peroxide do I need to convert 5.82 g of lead sulfide to lead sulfate?" It is not possible to calculate grams of one substance directly from grams of another. To answer such questions, it is necessary to convert grams of a substance to moles of that substance and to write balanced equations to determine molar ratios. Let's take it one step at a time. First, let's convert grams to moles and moles to grams. Like most other things, such calculations are best learned by practice.

EXAMPLE 5.7

Calculate the molecular weight of sulfuric acid, H_2SO_4.

SOLUTION
First, calculate the formula weight of H_2SO_4.

$$2 \times \text{atomic weight H} = 2 \times 1.0 \text{ amu} = 2.0 \text{ amu}$$
$$1 \times \text{atomic weight S} = 1 \times 32.1 \text{ amu} = 32.1 \text{ amu}$$
$$4 \times \text{atomic weight O} = 4 \times 16.0 \text{ amu} = \underline{64.0 \text{ amu}}$$
$$\text{Formula weight } H_2SO_4 = 98.1 \text{ amu}$$

The molecular weight of H_2SO_4 is therefore 98.1 g/mol.

Practice Exercise
Calculate the molecular weight of butyric acid, $C_4H_8O_2$.

EXAMPLE 5.8

Calculate the mass of 0.260 mol of sulfuric acid, H_2SO_4.

SOLUTION
The molecular weight of sulfuric acid is 98.1 g/mol (Example 5.7). The mass of 0.260 mol of sulfuric acid is therefore

$$0.260 \text{ mol } H_2SO_4 \times \frac{98.1 \text{ g } H_2SO_4}{1 \text{ mol } H_2SO_4} = 25.5 \text{ g } H_2SO_4$$

Practice Exercise
Calculate the mass of 1.37 mol of butyric acid, $C_4H_8O_2$.

EXAMPLE 5.9

How many moles of sulfuric acid are there in 19.6 g of pure H_2SO_4?

SOLUTION
Use the molecular weight from Example 5.7 and invert the conversion factor (to give an answer with the unit *moles*).

$$19.6 \text{ g } H_2SO_4 \times \frac{1 \text{ mol } H_2SO_4}{98.1 \text{ g } H_2SO_4} = 0.200 \text{ mol } H_2SO_4$$

Practice Exercise
Calculate the number of moles in each of the following.
a.   3.71 g of C_4H_{10} b.   76.0 g of H_3PO_4
c.   47.8 g of C_2H_6O d.   1.99 g of K_2CrO_4

5.5
MOLE AND MASS RELATIONSHIPS IN CHEMICAL EQUATIONS

Chemical equations not only represent ratios of atoms and molecules but also give us mole ratios. The equation

$$C + O_2 \longrightarrow CO_2$$

tells us that one atom of carbon reacts with one molecule (two atoms) of oxygen to form one molecule of carbon dioxide. The equation also indicates that 1 mol (6.02×10^{23} atoms) of carbon reacts with 1 mol (6.02×10^{23} molecules) of oxygen to give 1 mol (6.02×10^{23} molecules) of carbon dioxide. Since the molecular weight in grams per mole of a substance is numerically equal to the formula weight of the substance in atomic mass units, the equation also tells us (indirectly) that 12.0 g (1 mol) of carbon reacts with 32.0 g (1 mol) of oxygen to give 44.0 g (1 mol) of CO_2 (see Figure 5.8).

We need not use exactly 1 mol of each reactant. The important thing is to keep the ratio constant. For example, in the reaction above, the mass ratio of oxygen to carbon is 32.0:12.0 or 8.0:3.0. We could use 8.0 g of oxygen and 3.0 g of carbon to produce 11.0 g of CO_2. In fact, to calculate the amount of oxygen needed to react with a given amount of carbon, we need only multiply the amount of carbon by the factor 32.0/12.0.

EXAMPLE 5.10

Nitrogen monoxide (nitric oxide) combines with oxygen to form nitrogen dioxide according to the equation

$$2\,NO + O_2 \longrightarrow 2\,NO_2$$

State the molecular, molar, and mass relationships indicated by the equation.

SOLUTION

Molecular: 2 molecules of NO react with 1 molecule of O_2 to form 2 molecules of NO_2. Molar: 2 mol of NO react with 1 mol of O_2 to form 2 mol of NO_2. Mass: 60.0 g of NO react with 32.0 g of O_2 to form 92.0 g of NO_2.

Practice Exercise

Hydrogen sulfide burns in air to produce sulfur dioxide and water according to the equation

$$2\,H_2S + 3\,O_2 \longrightarrow 2\,SO_2 + 2\,H_2O$$

State the molecular, molar, and mass relationships indicated by the equation.

The ratio of moles of reactants and products is given by the coefficients in a balanced chemical equation.

EXAMPLE 5.11

Propane burns in air to form carbon dioxide and water.

$$C_3H_8 + 5 O_2 \longrightarrow 3 CO_2 + 4 H_2O$$

How many moles of oxygen are required to burn 0.105 mol of propane?

SOLUTION

The equation tells us that 5 mol O_2 is required to burn 1 mol C_3H_8. We can write

$$1 \text{ mol } C_3H_8 \approx 5 \text{ mol } O_2$$

where the symbol ≈ means "is chemically equivalent to." From this relationship we can construct conversion factors to relate moles from oxygen to moles of propane. The possible conversion factors are

$$\frac{5 \text{ mol } O_2}{1 \text{ mol } C_3H_8} \quad \text{and} \quad \frac{1 \text{ mol } C_3H_8}{5 \text{ mol } O_2}$$

Multiply the given quantity (0.105 mol C_3H_8) by the appropriate factor to get an answer with the asked-for units (moles of oxygen).

$$0.105 \text{ mol } C_3H_8 \times \frac{5 \text{ mol } O_2}{1 \text{ mol } C_3H_8} = 0.525 \text{ mol } O_2$$

Practice Exercise

Use the equation in Example 5.11 to answer the following questions. (a) How many moles of carbon dioxide are produced when 0.529 mol of propane is burned? (b) How many moles of water are produced when 76.2 mol of propane is burned? (c) How many moles of carbon dioxide are produced when 1.020 mol of oxygen are consumed?

Problems are seldom formulated in moles. Typically, you are given an amount of one substance in grams and are asked to calculate how many grams of another substance can be made from it. Such calculations involve several steps.

1. Write a balanced chemical equation for the reaction.
2. Determine the molecular weights of the substances involved in the calculation.
3. Write down the quantity that is given, and use the molecular weight to convert that quantity to moles.
4. Use the balanced chemical equation to convert number of moles of given substance to number of moles of desired substance.
5. Use the molecular weight to convert number of moles of desired substance to number of grams of desired substance.

The conversion process is diagrammed in Figure 5.9. It is best learned from examples and by working out exercises.

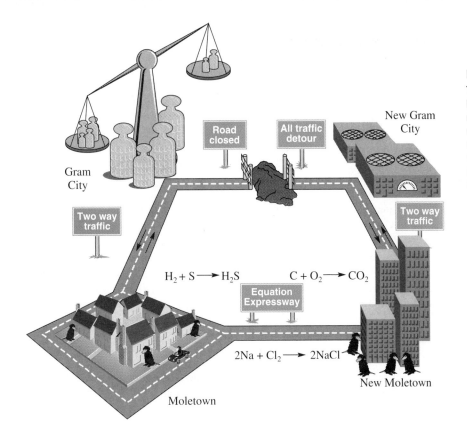

Figure 5.9
The chemical equation re-
lates moles of reactants to
moles of products. For
mass relationships between
reactants and products, we
must convert masses to
moles, then moles back to
masses.

EXAMPLE 5.12

Calculate the mass of oxygen needed to react with 10.0 g of carbon in the reaction that forms carbon dioxide.

SOLUTION

1. The balanced equation is

$$C + O_2 \longrightarrow CO_2$$

2. The molecular weight of oxygen is 32.0 g, and that of carbon is 12.0 g.

3. $10.0 \ \cancel{g \ C} \times \dfrac{1 \ mol \ C}{12.0 \ \cancel{g \ C}} = 0.833 \ mol \ C$

4. From the chemical equation, we have

$$1 \ mol \ C \approx 1 \ mol \ O_2$$

$$0.833 \ \cancel{mol \ C} \times \dfrac{1 \ mol \ O_2}{1 \ \cancel{mol \ C}} = 0.833 \ mol \ O_2$$

5. $0.833 \ \cancel{mol \ O_2} \times \dfrac{32.0 \ g \ O_2}{1 \ \cancel{mol \ O_2}} = 27.0 \ g \ O_2$

The problem also can be set up and solved in one step.

$$10.0 \ \cancel{g \ C} \times \dfrac{1 \ \cancel{mol \ C}}{12.0 \ \cancel{g \ C}} \times \dfrac{1 \ \cancel{mol \ O_2}}{1 \ \cancel{mol \ C}} \times \dfrac{32.0 \ g \ O_2}{1 \ \cancel{mol \ O_2}} = 27.0 \ g \ O_2$$

Note that the units in the denominators of the conversion factors are chosen so that each cancels the unit in the numerator of the preceding term.

Practice Exercise
Calculate the mass of oxygen needed to react with 0.334 g of sulfur in the reaction that forms sulfur dioxide. The equation is

$$S_8 + 8 O_2 \longrightarrow 8 SO_2$$

Note that we can set up conversion factors for any two compounds involved in a reaction. (Conversion factors are explained in Special Topic A.) Typical problems will ask you to calculate how much of one compound is equivalent to a given amount of one of the other compounds. All you need to do to solve such a problem is put together a conversion factor relating the two compounds. The following exercises involve additional conversion factors derived from the equation

$$C + O_2 \longrightarrow CO_2$$

EXAMPLE 5.13

Calculate the mass of carbon that can be burned to carbon dioxide in the presence of 400 g of oxygen. (*Hint:* You are given the amount of oxygen and are asked for the amount of carbon. You need a conversion factor that relates carbon to oxygen.)

SOLUTION

$$400 \text{ g } O_2 \times \frac{1 \text{ mol } O_2}{32.0 \text{ g } O_2} \times \frac{1 \text{ mol } C}{1 \text{ mol } O_2} \times \frac{12.0 \text{ g } C}{1 \text{ mol } C} = 150 \text{ g } C$$

Practice Exercise
How many grams of carbon dioxide can be obtained from 48.0 g of carbon, assuming sufficient oxygen? (*Hint:* You are given the amount of carbon and are asked for the amount of carbon dioxide. You need a conversion factor that relates carbon dioxide to carbon.)

Practice Exercise
How many grams of carbon are needed to produce 500 g of carbon dioxide? (*Hint:* You are given the amount of carbon dioxide and asked for the amount of carbon. You need a conversion factor that relates carbon to carbon dioxide.)

The examples and exercises so far have been based on an exceptionally simple reaction. In the balanced equation, there are no coefficients other than 1. That is, 1 mol of oxygen reacts with 1 mol of carbon to produce 1 mol of carbon dioxide. The following example and exercises involve balanced equations that have coefficients other than 1. We shall work out the sort of problem industrial chemists face every day. Whether chemists are making drugs, obtaining metals from their ores, synthesizing plastics, or whatever, they must use materials economically.

EXAMPLE 5.14

Ammonia (for fertilizer and other uses) is made by causing hydrogen and nitrogen to react at high temperature and pressure. The equation is

$$N_2 + H_2 \longrightarrow NH_3 \quad \textit{(not balanced)}$$

How many grams of ammonia can be made from 60.0 g of hydrogen?

SOLUTION

First, we must balance the equation.

$$N_2 + 3 H_2 \longrightarrow 2 NH_3 \quad \textit{(balanced)}$$

Next, we calculate the molecular weights of the compounds involved in the problem, that is, of hydrogen and ammonia.

H_2 has a molecular weight of 2.02 g

NH_3 has a molecular weight of 17.0 g

We are given an amount of hydrogen and are asked for a mass of ammonia. We must convert the mass given to number of moles.

$$60.0 \text{ g } H_2 \times \frac{1 \text{ mol } H_2}{2.02 \text{ g } H_2} = 29.7 \text{ mol } H_2$$

Next, we must use the coefficients given in the equation to determine the number of moles of ammonia that can be obtained from the calculated moles of hydrogen gas. The equation tells us that 3 mol of H_2 produces 2 mol of NH_3. Therefore,

$$29.7 \text{ mol } H_2 \times \frac{2 \text{ mol } NH_3}{3 \text{ mol } H_2} = 19.8 \text{ mol } NH_3$$

Finally, we must convert number of moles of ammonia to number of grams of ammonia.

$$19.8 \text{ mol } NH_3 \times \frac{17.0 \text{ g } NH_3}{1 \text{ mol } NH_3} = 337 \text{ g } NH_3$$

Or we could do it all in one step.

$$60.0 \text{ g } H_2 \times \frac{1 \text{ mol } H_2}{2.02 \text{ g } H_2} \times \frac{2 \text{ mol } NH_3}{3 \text{ mol } H_2} \times \frac{17.0 \text{ g } NH_3}{1 \text{ mol } NH_3} = 337 \text{ g } NH_3$$

Practice Exercise

Phosphorus reacts with oxygen to form tetraphosphorus decoxide. The equation is

$$P_4 + O_2 \longrightarrow P_4O_{10} \quad \textit{(not balanced)}$$

How many grams of tetraphosphorus decoxide can be made from 3.50 g of phosphorus?

EXAMPLE 5.15

How many grams of oxygen must be consumed if the following reaction is to produce 9.00 g of water?

$$H_2 + O_2 \longrightarrow H_2O \quad \textit{(not balanced)}$$

SOLUTION

The balanced equation is

$$2 H_2 + O_2 \longrightarrow 2 H_2O$$

The molecular weight of oxygen gas is 32.0 g, and the molecular weight of water is 18.0 g.

$$9.00 \text{ g } H_2O \times \frac{1 \text{ mol } H_2O}{18.0 \text{ g } H_2O} \times \frac{1 \text{ mol } O_2}{2 \text{ mol } H_2O} \times \frac{32.0 \text{ g } O_2}{1 \text{ mol } O_2} = 8.00 \text{ g } O_2$$

Practice Exercise

How many grams of oxygen can be made from 2.47 g of potassium chlorate? The equation (not balanced) is

$$KClO_3 \longrightarrow KCl + O_2$$

Practice Exercise

How many grams of carbon dioxide is formed when 73.9 g of propane is burned? The equation (not balanced) is

$$C_3H_8 + O_2 \longrightarrow CO_2 + H_2O$$

Practice Exercise

How many grams of magnesium metal is required to reduce 83.6 g of titanium(IV) chloride to titanium metal? The equation (not balanced) is

$$TiCl_4 + Mg \longrightarrow Ti + MgCl_2$$

5.6
STRUCTURE, STABILITY, AND SPONTANEITY

We saw at the beginning of Chapter 4 that some electron configurations are more stable than others. Sodium *ions* and chloride *ions,* arranged in a crystal lattice, are less reactive than sodium *atoms* and chlorine *molecules.* When sodium metal and chlorine gas are mixed, a vigorous reaction ensues.

$$2 Na + Cl_2 \longrightarrow 2 NaCl$$

A great deal of energy is produced as heat and light during the reaction. Sometimes this energy is listed as one of the products in the chemical equation.

$$2 Na + Cl_2 \longrightarrow 2 NaCl + energy$$

Figure 5.10
Coal burns with a highly exothermic reaction. The heat released can be used to convert water to steam that can turn a turbine to produce electricity.

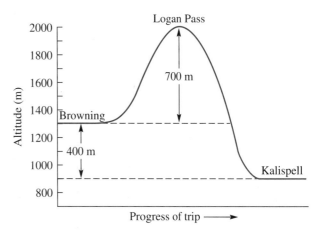

Figure 5.11
To get from Browning, Montana, to Kalispell via Going-to-the-Sun Highway, we would first have to climb to Logan Pass, even though Kalispell is 400 m lower in elevation than Browning.

Chemical reactions that result in the release of heat are said to be **exothermic.** The burning of methane is an exothermic reaction, as is the burning of gasoline or coal (Figure 5.10). In each case, chemical energy is converted into heat energy. There are other reactions, such as the decomposition of water, for which energy must be supplied. If the energy is supplied as heat, such reactions are said to be **endothermic.**

$$2 H_2O + energy \longrightarrow 2 H_2 + O_2$$

The energy released in an exothermic reaction is related to the *chemical energy* present in the reactants. This energy is a form of *potential energy*. Potential energy is released (as heat or light, for example) when reactants are brought together in a way that allows them to achieve more stable electron arrangements.

To say that a reaction is exothermic does not necessarily mean that it is instantaneous. For example, coal (carbon) has a lot of chemical energy. Carbon reacts with oxygen to form carbon dioxide with the release of considerable heat. But coal doesn't react very rapidly with oxygen at ordinary temperatures. In fact, one can store a pile of coal in air indefinitely without perceptible change. Before coal will react with oxygen to release its stored energy, it must be heated to a temperature of several hundred degrees. The coal must be supplied with a certain amount of energy (called the energy of activation; Section 5.7) before it will begin to burn steadily. Once this energy of activation is supplied, the heat evolved in the reaction will keep the coal burning brightly, and the energy eventually produced will exceed the amount of energy required to start the reaction. Overall, you get more energy out than you put in. The reaction is exothermic. In an endothermic reaction, more energy goes in than comes out.

We will return to the burning of coal in a moment, but first let's consider an analogy involving a more familiar situation. If you were in Browning, Montana (elevation 1300 m), and wished to travel to Kalispell (elevation 900 m), you could choose to drive the scenic Going-to-the-Sun Highway through Glacier National Park, which crosses the continental divide at Logan Pass (elevation 2000 m). First, you would have to climb 700 m, but then it would be downhill the rest of the way (Figure 5.11).

When heat energy is released during a chemical reaction, heat often is listed as a product in the chemical equation. When energy is supplied to keep a chemical reaction going, heat can be listed as a reactant in the equation.

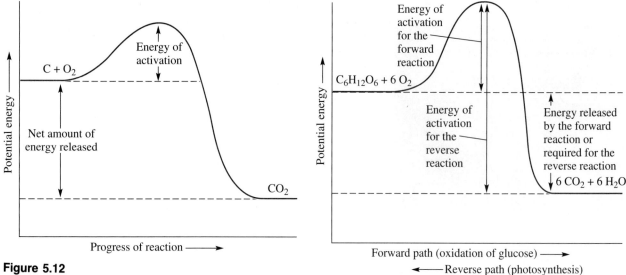

Figure 5.12
To get from reactants (carbon and oxygen) to product (carbon dioxide), we must first put some energy into the system.

Figure 5.13
Potential energy diagram for a reversible reaction.

The reaction of coal (carbon) with oxygen can be explained in much the same way. The reactants, carbon and oxygen, lie in a rather high potential energy valley (Figure 5.12). A certain amount of energy has to be put into the system for it to get over the top of the ''mountain.'' Once the reaction is under way, the heat released more than compensates for the energy needed to get the reaction going in the first place.

5.7
REVERSIBLE REACTIONS

As we shall see in Chapter 24, living cells take in oxygen and ''burn'' glucose to obtain energy.

$$C_6H_{12}O_6 \; + \; 6\,O_2 \; \longrightarrow \; 6\,CO_2 \; + \; 6\,H_2O \; + \; energy$$
Glucose

Green plants carry out the reverse reaction (an endothermic reaction called **photosynthesis**), using energy from sunlight to convert carbon dioxide and water to glucose and oxygen.

$$6\,CO_2 \; + \; 6\,H_2O \; + \; energy \; \longrightarrow \; C_6H_{12}O_6 \; + \; 6\,O_2$$

All life on our planet is based on this **reversible** reaction. But what makes it go one way in one case and the opposite way in another? One obvious answer is energy. Since energy is released when glucose is oxidized, the reactants (glucose and oxygen) must be at a higher energy level than the products (carbon dioxide and water). As we have indicated previously, this does not mean that the path is straight downhill from reactants to products. Indeed, if glucose and oxygen are mixed at ordinary temperatures outside a cell, no perceptible reaction occurs. As with the reaction of coal and oxygen, a certain amount of energy, the energy of activation, must be supplied before the reaction takes place. The potential energy diagram for this reaction (Figure 5.13) strongly resembles that for the coal and oxygen reaction (see Figure 5.12).

For reversible reactions, the energy hill, or the barrier to reaction, may be approached from either side. That is, we can read the diagram from left to right or from right to left. With the burning of glucose, we are considering an exothermic reaction in which higher-energy reactants are converted to lower-energy products. When read in the reverse direction, the diagram presents the energy changes that occur in the endothermic reaction called photosynthesis. The climb to the top of the barrier is longer from one side than from the other. The **energy of activation** (also called *activation energy*) for a chemical reaction is the minimum energy needed to get the reaction started. It is the difference in energy between the level the reactants are on and the top of the energy hill. When the reactants are in the valley at the left (as in the forward reaction), the climb to the top is not so long. When the reactants are in the valley to the right (as in the reverse reaction), the climb to the top is much longer. The potential energy diagram shown here is much simplified. Just as one seldom crosses a mountain by going straight up one side and straight down the other, so reactions seldom proceed by a smooth, one-hump potential energy change.

5.8
REACTION RATES: COLLISIONS, ORIENTATION, AND ENERGY

Before atoms, molecules, or ions can react, they must first get together: they must collide. Second, for all except the simplest particles, they must come together in the proper orientation. Third, the collision must provide a certain minimum energy, the energy of activation.

First, let's consider the frequency of collision, which is influenced by two factors, concentration and temperature. The more concentrated the reactants, the more frequently the particles will collide, simply because there are more of them in a given volume. Molecules also move faster at higher temperatures; thus, an increase in temperature also increases the frequency of collision. The effects of concentration and temperature on reaction rates will be discussed in more detail in following sections.

Now, though, let's take a closer look at the effect of orientation. It might not be obvious that orientation is important. Sometimes, though rarely, it's not. Consider a situation in which you want to knock someone down. You can tackle from the front, back, or side and still accomplish your objective. If, however, a kiss is the objective, then an orientation in which the participants are face-to-face works best. For chemical reactions, there are also a few instances in which orientation is not important. If two hydrogen atoms are to react to form a hydrogen molecule,

$$2 \text{ H} \longrightarrow \text{H}_2$$

then orientation is unimportant. The quantum mechanical view of hydrogen atoms depicts them as symmetrical (spherical) electron clouds. Front, back, top, and bottom all look the same. Therefore, all orientations of two hydrogen atoms are identical. If such atoms come into contact with one another, they can react.

For most particles, however, proper orientation is a prerequisite to reaction. Consider the molecule

$$\text{H}-\overset{\overset{\displaystyle H}{|}}{\underset{\underset{\displaystyle H}{|}}{C}}-\overset{\overset{\displaystyle H}{|}}{\underset{\underset{\displaystyle H}{|}}{C}}-\overset{\overset{\displaystyle H}{|}}{\underset{\underset{\displaystyle H}{|}}{C}}-\overset{\overset{\displaystyle H}{|}}{\underset{\underset{\displaystyle H}{|}}{C}}-\overset{}{\underset{\underset{\displaystyle H}{|}}{C}}=\overset{}{\underset{}{C}}-\text{H}$$

The two carbon atoms sharing the double bond can react with bromine (see Section 13.13) to form

$$H-\underset{\underset{H}{|}}{\overset{\overset{H}{|}}{C}}-\underset{\underset{H}{|}}{\overset{\overset{H}{|}}{C}}-\underset{\underset{H}{|}}{\overset{\overset{H}{|}}{C}}-\underset{\underset{H}{|}}{\overset{\overset{H}{|}}{C}}-\underset{\underset{H}{|}}{\overset{\overset{Br}{|}}{C}}-\underset{\underset{H}{|}}{\overset{\overset{Br}{|}}{C}}-H$$

Now, it is possible for bromine to strike the original compound anywhere along its length. However, only if the collision occurs in such a way as to bring bromine into contact with the electrons in the double bond will the reaction occur. It is even true that the bromine must approach the electrons of the double bond from a certain direction. Here is a case in which orientation is very important. We shall point out another example of proper orientation shortly.

The third factor, that collisions must provide a certain minimum energy, is a good deal more subtle. There is a temperature effect, which will be discussed in some detail in the next section. Certain substances, called catalysts, may substantially lower this activation energy, thus increasing the reaction rate. The effect of catalysts will be discussed more thoroughly in Section 5.10.

5.9
REACTION RATES: THE EFFECT OF TEMPERATURE

Reactions generally take place faster at a higher temperature. For example, coal (carbon) reacts so slowly with oxygen (from the air) at room temperature that the change is imperceptible. However, when coal is heated to several hundred degrees, it reacts at a much more rapid rate. The heat evolved in the reaction keeps the coal burning smoothly. The effect of temperature on the rates of chemical reactions is explained by the kinetic-molecular theory. We shall consider this theory in more detail in Chapter 7. For the moment, we will cite one of its postulates, which states that at high temperatures molecules move more rapidly. Thus, they collide more frequently, providing a greater chance for reaction.

At high temperatures, the rapidly moving molecules also strike one another harder. The harder these collisions are, the more energy is involved in them. These harder collisions are more likely to supply the activation energy needed to break the chemical bonds and get the reaction going. Consider the reaction that takes place between hydrogen gas and chlorine gas. In order for hydrogen and chlorine to react, bonds between hydrogen atoms and between chlorine atoms must be broken. In general, energy is absorbed to break a chemical bond, and energy is released when chemical bonds are formed.

$$H \vdots H \ + \ Cl \vdots Cl \ \longrightarrow \ 2\,H-Cl \ + \ heat$$

This is an exothermic reaction. Once it has started, the energy released by the formation of hydrogen–chlorine bonds more than compensates for that required to break H—H and Cl—Cl bonds. There is a net conversion of chemical energy to heat energy.

In endothermic reactions, the energy released by bond formation is less than that required to break the necessary bonds. For these reactions, energy must be supplied continuously from an external source, or the process will stop. A typical endothermic reaction is the decomposition of the salt potassium chlorate ($KClO_3$) to give oxygen and another salt, potassium chloride (KCl).

$$2\,KClO_3 \ + \ heat \ \longrightarrow \ 2\,KCl \ + \ 3\,O_2$$

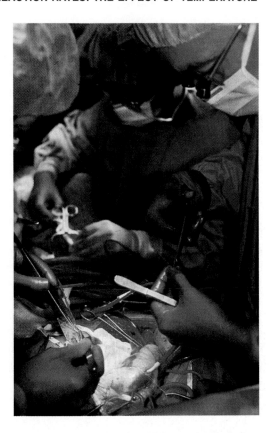

Figure 5.14
Heart surgery can be performed with the patient's temperature lowered to slow metabolic processes and thus minimize the possibility of brain damage.

The potassium chlorate must be heated continuously. If the source of heat is removed, the reaction quickly subsides.

We make use in our daily lives of our knowledge of the effect of temperature on chemical reactions. For example, we freeze foods to retard the chemical reactions that lead to spoilage. On the other hand, if we want to cook our food more rapidly, we turn up the heat. There is a general rule of thumb that reaction rates double for every temperature increase of 10 °C. (There are many exceptions to this rule, and it should be applied only if there are no other data available.) The chemical reactions that occur in our bodies generally do so at a constant temperature of 37 °C (98.6 °F). A few degrees' rise in temperature (fever) leads to an increase in respiration, pulse rate, and other physiological reactions. A drop in body temperature of a few degrees slows these same processes considerably, as is exemplified by the slowed metabolism of hibernating animals. Use is made of this phenomenon in some surgical procedures. In some cases of heart surgery, the body temperature of the patient is lowered to about 15 °C (60 °F). Ordinarily, the brain is permanently damaged when its oxygen supply is interrupted for more than 5 min. But at the lower temperature, metabolic processes slow considerably, and the brain can survive much longer periods of oxygen deprivation. The surgeon can stop the heartbeat, perform an hour-long surgical procedure on the heart, and then restart the heart and bring the patient's temperature back to normal (Figure 5.14).

It must be said that despite these examples the manipulation of reaction rates in living systems through changes in temperature is severely restricted. Increasing temperature, for example, may "kill" enzymes (Section 5.10) that mediate cellular chemistry and are essential to life. We kill germs by heat sterilization in autoclaves. For living cells, there is often a rather narrow range of optimum temperatures. Both higher and lower temperatures can be disabling, if not deadly.

5.10
REACTION RATES: CATALYSIS

Recall the endothermic reaction involving the decomposition of potassium chlorate to oxygen and potassium chloride (Section 5.9). For oxygen to be produced at a useful rate, the potassium chlorate must be heated to over 400 °C. However, if we add a small amount of manganese dioxide (MnO_2), we can get the same rate of oxygen evolution by heating the reactant to just 250 °C. Further, after the reaction is complete, the manganese dioxide can be completely recovered, unchanged. A substance that, like manganese dioxide, changes the rate of a chemical reaction without itself being changed is called a **catalyst.** In general, catalysts act by lowering the activation energy required for the reaction to occur (Figure 5.15). If activation energy is lower, then the collisions of more slowly moving molecules will be sufficient to supply that energy. The lower temperature required for a catalyzed reaction reflects this. This energy of activation is lowered because the catalyst changes the *path* of the reaction. To return to our analogy of the trip from Browning to Kalispell, it is possible to take an alternate route. U.S. Highway 2 crosses the continental divide through Marias Pass. This route involves a climb of only 300 m, compared with 700 m via Logan Pass (Figure 5.16). This alternate route is analogous to that provided by a catalyst in a chemical reaction.

Catalysts are of great importance in the chemical industry. A reaction that would otherwise be so slow as to be impractical can be made to proceed at a reasonable rate with the proper catalyst.

Catalysts are even more important in living organisms, where raising the temperature by 100 °C is not a feasible way of increasing the rate of critical reactions. If we raised our body temperature by 100 °C, we'd boil our blood, among other things, and our fatty tissue would melt (though that might not be undesirable). Biological catalysts, called **enzymes,** mediate nearly all the chemical reactions that take place in living systems. But these

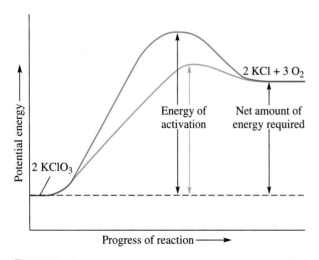

Figure 5.15
In the decomposition of potassium chlorate, a catalyst acts to lower the energy of activation (blue arrow and curve) compared with that required for the uncatalyzed reaction.

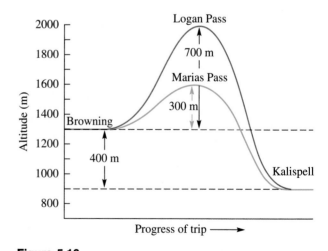

Figure 5.16
To get from Browning to Kalispell via Highway 2 involves a lower energy barrier than the route via Logan Pass.

Time from start of reaction	Rate of for... Rate of re...
0	
10	
20	
30	
40	
50	

ing back to reactants. It i...
For every reactant mole...
reverse reaction. Once e...
centrations of reactants...
their concentrations don...

Equilibrium is esta...
which external condition...
if those external conditi...
again use as an exampl...

The double arrow indic...
volved in this reaction a...
that the coefficients give...

catalysts themselves may suffer irreversible damage when living cells are subjected to too high a temperature. Once a catalyst is deactivated (rendered inactive), the reactions that require the catalyst no longer proceed at a rate that maintains life. In addition to being examples of catalysts, enzymes offer a striking illustration of the importance of orientation in chemical reactions. These compounds are huge molecules, each possessing what is called an active site. The reactants must come into contact with this active site if the enzyme is to catalyze the reaction. If the reactants do not collide at the active site, no reaction occurs. So important are enzymes to life that we discuss them more fully in Chapter 22.

5.11
REACTION RATES: THE EFFECT OF CONCENTRATION

Another factor affecting the rate of a chemical reaction is the concentration of reactants. The more reactant molecules there are in a volume of space, the more collisions will occur. The more collisions there are, the more reactions will occur. For example, you can light a wood splint and then blow out the flame. The splint will continue to glow as the wood reacts slowly with the oxygen of the air. If the glowing splint were placed in pure oxygen, the splint would burst into flame, indicating a much more rapid reaction. This more rapid reaction can be interpreted in terms of the concentration of oxygen. Air is about one-fifth oxygen. The concentration of O_2 molecules in pure oxygen is therefore about five times as great as in air. The caution against smoking in a hospital room where a patient is in an oxygen tent is not merely a concession to the sensitivity of nonsmokers. It is meant to prevent disaster (Figure 5.17).

For reactions in solution, the concentration of a reactant can be increased if more of it is dissolved. One of the first studies of reaction rate was done by Ludwig Wilhelmy in 1850. He studied the rate of reaction of sucrose (cane or beet sugar) with water. The products are two simpler sugars, glucose and fructose (see Chapter 19).

$$\text{Sucrose} \;+\; H_2O \;\xrightarrow{\text{HCl}}\; \text{Glucose} \;+\; \text{Fructose}$$

Figure 5.17
The No Smoking sign is not simply a concession to the sensitivity of the patient undergoing oxygen therapy. It is a warning about the increased danger from fires that an oxygen-rich atmosphere presents.

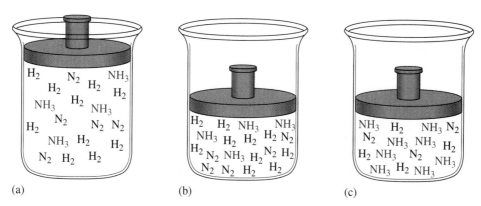

(a)   (b)   (c)

Figure 5.19

Le Châtelier's principle illustrated. (a) System at equilibrium with 10 H_2, 5 N_2, and 4 NH_3, a total of 19 molecules. (b) The same molecules forced into a smaller volume, creating a stress on the system. (c) Six H_2 and two N_2 have been converted to four NH_3. A new equilibrium has been established with 4 H_2, 3 N_2, and 8 NH_3, a total of 15 molecules. The stress is partially relieved by the reduction in the total number of molecules.

reactants occupy four volumes and the products two. In Chapter 7, we study the effects of temperature and pressure on the volume of a gas. For the moment, let us just say that increased pressure tends to reduce the volume of a gas. If the pressure of the equilibrium system we are considering is increased, the equilibrium will be disrupted, and the rate at which N_2 reacts with H_2 to form NH_3 will increase; that is, the rate of the forward reaction will increase (Figure 5.19). Increasing the rate of the forward reaction has the effect of changing H_2 and N_2, which occupy four volumes, to NH_3, which occupies only two volumes. The increased pressure therefore has the effect of changing reactants that occupy four volumes to products that occupy only two. Thus, the total volume of the system will decrease. If the pressure is held constant at the new, higher value, equilibrium will again be established. But in the new equilibrium the concentration of ammonia (NH_3) will be higher than it was before the pressure changed. We sometimes describe this change by saying that the equilibrium has been shifted to the right. Thus, for the reversible reaction of nitrogen and hydrogen, an increase in pressure shifts the equilibrium to the right.

Observations of changes that occur in an equilibrium system when factors such as concentration, pressure, or temperature are changed were made in the nineteenth century. These effects were summarized in 1884 by a French chemist, Henri Louis Le Châtelier. His rule, still called **Le Châtelier's principle,** can be stated as follows: If a stress is applied to a system in equilibrium, the system rearranges in such a direction as to minimize the stress. If heat is added to the N_2–H_2–NH_3 system, the reaction will proceed to the left, using up heat. If more nitrogen gas is added to the system, the reaction will go to the right, using up nitrogen. If additional hydrogen gas is introduced, the system will shift to the right, to use up hydrogen. Additional ammonia will cause the reaction to proceed to the left, using up ammonia.

The equilibrium can also be shifted by removal of one of the substances. Removal of ammonia will cause hydrogen and nitrogen to react, forming more ammonia. Removal of hydrogen will cause ammonia to break down and form more hydrogen (and, incidentally, more nitrogen). It should be noted, however, that a catalyst will not shift an equilibrium system. Catalysts change the rate of both the forward and reverse reactions. They do not change the position of the equilibrium, the equilibrium concentrations of reactants and products are not altered.

EXAMPLE 5.16

What effect, if any, will each of the following changes have on the equilibrium in the reaction

$$2 \, CO \; + \; O_2 \; \rightleftharpoons \; 2 \, CO_2 \; + \; heat$$

a. adding CO
b. removing O_2
c. cooling the reaction mixture
d. increasing the pressure
e. adding a catalyst

SOLUTION

a. The reaction shifts to the right to use up the added CO.
b. The reaction shifts to the left to replace the O_2 that is removed.
c. The reaction shifts to the right to replace the lost heat.
d. The reaction shifts to the right to relieve the pressure by converting three molecules (2 CO + 1 O_2) into two molecules (2 CO_2).
e. A catalyst has no effect on the position of equilibrium.

Practice Exercise

How will the addition of $N_2(g)$ to the system below (isolated, constant temperature, and at equilibrium) affect the equilibrium concentration of H_2?

$$N_2(g) \; + \; 3 \, H_2(g) \; \rightleftharpoons \; 2 \, NH_3(g)$$

In Chapter 8, we discuss equilibria between liquid and solid phases and between liquid and vapor phases. In Chapter 9, we encounter equilibria involving solutions. The concept is quite important to our study of the chemistry of life.

EXERCISES

1. Define or illustrate each of the following.
    a. molecular weight   b. mole
    c. Avogadro's number   d. exothermic
    e. endothermic   f. energy of activation
    g. catalyst   h. reversible reaction
    i. equilibrium   j. Le Châtelier's principle
2. Explain the difference between "the atomic weight of oxygen" and "the molecular weight of oxygen (gas)."
3. What is Avogadro's hypothesis? How did it explain Gay-Lussac's law of combining volumes?
4. How do the law of combining volumes and Avogadro's hypothesis indicate that hydrogen gas is composed of diatomic molecules rather than individual atoms?
5. How many oxygen molecules are there in 1.00 mol of O_2? How many oxygen atoms are there in 1.00 mol of O_2?
6. How many calcium ions and how many chloride ions are there in 1.00 mol of $CaCl_2$?
7. What is the molecular weight of CO_2? State in words how it is determined from the formula.

8. Consider the law of conservation of mass in explaining why we must work with *balanced* chemical equations.
9. What is the molar volume of each of the following gases?
    a. He   b. H_2 •   c. C_2H_6
10. What is the mass of a molar volume of each of the gases in Exercise 9?
11. How many hydrogen atoms are indicated in one formula unit of each of the following?
    a. NH_4NO_3 b. CH_3OH
    c. $CH_3CH_2CH_3$ d. C_6H_5COOH
12. How many hydrogen atoms are indicated in one formula unit of each of the following?
    a. $(NH_4)_2HPO_4$ b. $Al(C_2H_3O_2)_3$
13. How many oxygen atoms are indicated in one formula unit of each of the following?
    a. $Al_2(C_2O_4)_3$ b. $Ca_3(PO_4)_2$
14. How many atoms of each kind (Al, C, H, and O) are indicated by the notation 2 $Al(C_2H_3O_2)_3$?

15. How many atoms of each kind (N, P, H, and O) are indicated by the notation 6 $(NH_4)_2HPO_4$?

16. Indicate whether the following equations are balanced. (You need not balance the equation; just determine whether it is balanced as written.)
    a. $Mg + H_2O \rightarrow MgO + H_2$
    b. $FeCl_2 + Cl_2 \rightarrow FeCl_3$
    c. $F_2 + H_2O \rightarrow 2\ HF + O_2$
    d. $Ca + 2\ H_2O \rightarrow Ca(OH)_2 + H_2$
    e. $2\ LiOH + CO_2 \rightarrow Li_2CO_3 + H_2O$

17. Indicate whether the following equations are balanced as written.
    a. $2\ KNO_3 + 10\ K \rightarrow 6\ K_2O + N_2$
    b. $2\ NH_3 + O_2 \rightarrow N_2 + 3\ H_2O$
    c. $4\ LiH + AlCl_3 \rightarrow 2\ LiAlH_4 + 2\ LiCl$
    d. $SF_4 + 3\ H_2O \rightarrow H_2SO_3 + 4\ HF$
    e. $4\ BF_3 + 3\ H_2O \rightarrow H_3BO_3 + 3\ HBF_4$

18. Indicate whether the following equations are balanced as written.
    a. $2\ Sn + 2\ H_2SO_4 \rightarrow 2\ SnSO_4 + SO_2 + 2\ H_2O$
    b. $3\ Cl_2 + 6\ NaOH \rightarrow 5\ NaCl + NaClO_3 + 3\ H_2O$

19. Propane burns in air to form carbon dioxide and water. The equation is

$$C_3H_8 + 5\ O_2 \rightarrow 3\ CO_2 + 4\ H_2O$$

State the molecular, molar, and mass relationships indicated by the equation.

20. The poison gas phosgene reacts with water in the lungs to form hydrogen chloride and carbon dioxide. The equation is

$$COCl_2 + H_2O \rightarrow 2\ HCl + CO_2$$

State the molecular, molar, and mass relationships indicated by the equation.

21. Balance the following equations.
    a. $Cl_2O_5 + H_2O \rightarrow HClO_3$
    b. $V_2O_5 + H_2 \rightarrow V_2O_3 + H_2O$
    c. $Al + O_2 \rightarrow Al_2O_3$
    d. $Sn + NaOH \rightarrow Na_2SnO_2 + H_2$
    e. $PCl_5 + H_2O \rightarrow H_3PO_4 + HCl$

22. Balance the following equations.
    a. $TiCl_4 + H_2O \rightarrow TiO_2 + HCl$
    b. $C_4H_{10} + O_2 \rightarrow CO_2 + H_2O$
    c. $WO_3 + H_2 \rightarrow W + H_2O$
    d. $Al_4C_3 + H_2O \rightarrow Al(OH)_3 + CH_4$
    e. $Al_2(SO_4)_3 + NaOH \rightarrow Al(OH)_3 + Na_2SO_4$

23. Balance the following equations.
    a. $Na_3P + H_2O \rightarrow NaOH + PH_3$
    b. $Cl_2O + H_2O \rightarrow HClO$
    c. $CH_3OH + O_2 \rightarrow CO_2 + H_2O$
    d. $Zn(OH)_2 + H_3PO_4 \rightarrow Zn_3(PO_4)_2 + H_2O$
    e. $C_3H_8 + O_2 \rightarrow CO_2 + H_2O$

24. Balance the following equations.
    a. $Ca_3P_2 + H_2O \rightarrow Ca(OH)_2 + PH_3$
    b. $Cl_2O_7 + H_2O \rightarrow HClO_4$
    c. $MnO_2 + HCl \rightarrow MnCl_2 + Cl_2 + H_2O$
    d. $Fe + O_2 \rightarrow Fe_3O_4$
    e. $C_5H_{12} + O_2 \rightarrow CO_2 + H_2O$

25. Calculate the molecular weight of each of the following compounds.
    a. CH_4 b. AlF_3 c. UF_6

26. Calculate the molecular weight of each of the following compounds.
    a. SO_3 b. $KBrO_3$ c. $CaSO_4$

27. Calculate the molecular weight of each of the following compounds.
    a. C_6H_5Br b. $H_4P_2O_7$
    c. $K_2Cr_2O_7$ d. $Al_2(SO_4)_3$

28. Calculate the molecular weight of each of the following compounds.
    a. $(NH_4)_3PO_4$ b. $Fe(NO_3)_3$
    c. $C_2H_5NO_2$ d. $Mg(S_2O_3)_2$

29. According to the equation

$$C_3H_8 + 5\ O_2 \rightarrow 3\ CO_2 + 4\ H_2O$$

    a. how many liters of CO_2 are formed when 0.529 L of C_3H_8 are burned?
    b. how many liters of O_2 are required to burn 16.1 L of C_3H_8?

30. According to the equation

$$C_3H_8 + 5\ O_2 \rightarrow 3\ CO_2 + 4\ H_2O$$

    a. how many liters of H_2O are formed when 2.93 L of C_3H_8 are burned?
    b. how many liters of O_2 are required to form 0.370 L of H_2O?

31. Calculate the mass of 0.377 mol of sodium nitrite, $NaNO_2$.

32. Calculate the mass of 1.67 mol of potassium hydrogen oxalate, KHC_2O_4.

33. Calculate the number of moles of copper(I) sulfate, Cu_2SO_4, in 16.3 g of the compound.

34. Calculate the number of moles of octane, C_8H_{18}, in 102 g of octane.

35. Calculate the number of grams in each of the following.
    a. 0.00500 mol MnO_2
    b. 1.12 mol CaH_2
    c. 0.250 mol $C_6H_{12}O_6$
    d. 4.61 mol $AlCl_3$

36. Calculate the number of moles in each of the following.
    a. 98.6 g HNO_3 b. 9.45 g CBr_4
    c. 9.11 g $FeSO_4$ d. 11.8 g $Pb(NO_3)_2$

37. Consider the reaction for the combustion of octane.

$$2\ C_8H_{18} + 25\ O_2 \rightarrow 16\ CO_2 + 18\ H_2O$$

    a. How many moles of CO_2 are produced when 2.09 mol of octane is burned?
    b. How many moles of oxygen are required to burn 4.47 mol of octane?

38. Consider the reaction for the combustion of octane.

$$2\ C_8H_{18} + 25\ O_2 \rightarrow 16\ CO_2 + 18\ H_2O$$

a. How many moles of H_2O are produced when 2.81 mol of octane is burned?

b. How many moles of CO_2 are produced when 4.06 mol of oxygen is consumed?

39. How many grams of ammonia can be made from 440 g of H_2? The equation (not balanced) is

$$N_2 \ + \ H_2 \ \rightarrow \ NH_3$$

40. How many grams of hydrogen are needed to react completely with 892 g of N_2? The equation (not balanced) is

$$N_2 \ + \ H_2 \ \rightarrow \ NH_3$$

41. How many grams of oxygen can be prepared from 24.0 g of H_2O_2? The equation (not balanced) is

$$H_2O_2 \ \rightarrow \ H_2O \ + \ O_2$$

42. Nitric acid is used in the production of trinitrotoluene (TNT), an explosive. The equation (not balanced) is

$$\underset{\text{Toluene}}{C_7H_8} \ + \ HNO_3 \ \longrightarrow \ \underset{\text{TNT}}{C_7H_5N_3O_6} \ + \ H_2O$$

How much nitric acid is required to react with 454 g of toluene?

43. Use the equation (not balanced) in Exercise 42 to calculate the amount of TNT that can be made from 829 g of toluene.

44. What mass of quicklime (calcium oxide) can be made when 4.72 g of limestone (calcium carbonate) is decomposed by heating?

$$CaCO_3 \ \rightarrow \ CaO \ + \ CO_2$$

45. What mass of nitric acid can be made from 971 g of ammonia? The equation (not balanced) is

$$NH_3 \ + \ O_2 \ \rightarrow \ HNO_3 \ + \ H_2O$$

46. Acetylene (C_2H_2) burns in pure oxygen with a very hot flame. The equation (not balanced) is

$$C_2H_2 \ + \ O_2 \ \rightarrow \ CO_2 \ + \ H_2O$$

How much oxygen is required to react with 52.0 g of acetylene?

47. Refer to the following reaction diagram.

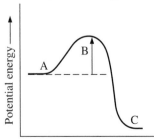

a. Which letter in the diagram refers to the products?

b. Which letter refers to the activation energy?

c. Which letter refers to the reactants?

48. Is the reaction diagrammed in Exercise 47 endothermic or exothermic?

49. If the reaction diagrammed in Exercise 47 were reversible, would the reverse reaction be endothermic or exothermic?

50. What effect does changing the temperature have on the rate of a reaction?

51. What is the effect of changing the concentration on the rate of a reaction?

52. If all other conditions were kept constant, what effect would decreasing the temperature of a reaction have on the rate of a reaction?

53. If all other conditions were kept constant, what effect would decreasing the concentration of reactants have on the rate of the reaction?

54. If all other conditions were kept constant, what effect would adding a catalyst have on the rate of the reaction?

55. Why does a catalyst increase the rate of chemical reaction?

56. According to Le Châtelier's principle, what effect will increasing the temperature have on the following equilibria?

a. $H_2 + Cl_2 \rightleftharpoons 2\ HCl + \text{heat}$

b. $2\ CO_2 + \text{heat} \rightleftharpoons 2\ CO + O_2$

c. $3\ O_2 + \text{heat} \rightleftharpoons 2\ O_3$

57. What effect would decreasing the concentration of O_2 have on the equilibrium in Exercise 56b?

58. What would increasing the concentration of O_2 do to the equilibrium in Exercise 56c?

59. The Haber process for making ammonia is given in the equation

$$N_2(g) \ + \ 3\ H_2(g) \ \rightleftharpoons \ 2\ NH_3(g) \ + \ \text{heat}$$

What would be the efffect of each of the following on the equilibrium of that reaction?

a. the removal of ammonia

b. the addition of nitrogen

c. the removal of heat

d. the addition of a catalyst

60. Explain how the orientation of reactants can influence the rate of a reaction.

61. What is the "mechanism" of a reaction?

62. Aspartame (see Table 19.1) is an artificial sweetener about 160 times as sweet as sucrose. Aspartame breaks down in acidic solution to produce (among other products) methanol, which is fairly toxic. The equation is

$$\underset{\text{Aspartame}}{C_{14}H_{18}N_2O_5} \ + \ 2\ H_2O \ \xrightarrow{H^+}$$

$$C_4H_7NO_4 \ + \ C_9H_{11}NO_2 \ + \ \underset{\text{Methanol}}{CH_4O}$$

a. A typical can of soda contains about 40 g of sucrose. How much aspartame is required to obtain the same level of sweetness?

b. How much methanol is formed by the complete breakdown of that much aspartame?

c. How many cans of soda with decomposed aspartame would you have to drink to get the approximate lethal dose of 25 g of methanol?

63. Joseph Priestley discovered oxygen in 1774 by heating "red calx of mercury," mercury(II) oxide. The calx decomposed to the elements. The equation (not balanced) is

$$HgO \rightarrow Hg + O_2$$

How much oxygen is produced by the decomposition of 10.8 g of HgO?

64. How much iron can be converted to the magnetic oxide of iron (Fe_3O_4) by 8.80 g of pure oxygen? The equation (not balanced) is

$$Fe + O_2 \rightarrow Fe_3O_4$$

65. Laughing gas (dinitrogen monoxide, N_2O, also called nitrous oxide) can be made by heating ammonium nitrate with great care. The equation (not balanced) is

$$NH_4NO_3 \rightarrow N_2O + H_2O$$

How much N_2O can be made from 4.00 g of ammonium nitrate?

66. Small amounts of hydrogen gas often are made by the reaction of calcium metal with water. The equation (not balanced) is

$$Ca + H_2O \rightarrow Ca(OH)_2 + H_2$$

How many grams of hydrogen are formed by the action of water on 0.413 g of calcium?

67. For many years the noble gases were called the "inert gases" because it was thought that they formed no chemical compounds. Neil Bartlett made the first noble gas compound in 1962. Xenon hexafluoride is made according to the equation (not balanced)

$$Xe + F_2 \rightarrow XeF_6$$

How many grams of fluorine are required to make 0.112 g of XeF_6?

68. Phosphine gas (used as fumigant to protect stored grain) is generated by the action of water on magnesium phosphide. The equation (not balanced) is

$$Mg_3P_2 + H_2O \rightarrow PH_3 + Mg(OH)_2$$

How much magnesium phosphide is needed to produce 134 g of PH_3?

Chapter 6
OXIDATION AND REDUCTION

The ammonium dichromate "volcano" is a dynamic visualization of an oxidation–reduction reaction.

Chemical reactions can be classified in several ways. In this chapter, we consider an important group of chemical reactions called *reduction–oxidation* (or *redox*) *reactions*. The two processes—oxidation and reduction—always occur together. You can't have one without the other. When one substance is oxidized, another is reduced (Figure 6.1). For convenience, however, we may choose to talk about only a part of the process—the oxidation part or the reduction part.

Our body cells obtain energy to maintain themselves by oxidizing foods. Green plants, using energy from sunlight, produce food by the reduction of carbon dioxide (photosynthesis—see Section 6.10). We win metals from their ores by reduction, then lose them again to corrosion as they are oxidized. We maintain our technological civilization by oxidizing fossil fuels (coal, natural gas, and petroleum) to obtain the chemical energy that was stored in these materials eons ago by green plants.

Reduced forms of matter—sugars, coal, gasoline—are high in energy (Figure 6.2). Oxidized forms—carbon dioxide and water—are low in energy. Let's examine the processes of oxidation and reduction in some detail, in order that we might better understand the chemical reactions that keep us alive and enable us to maintain our civilization.

137

Figure 6.1

Oxidation and reduction always occur together. Pictured here on the left is ammonium dichromate. In the reaction (center), the ammonium ion (NH_4^+) is oxidized, and the dichromate ion ($Cr_2O_7^{2-}$) is reduced. Considerable heat and light are evolved. The equation for the reaction is

$$(NH_4)_2Cr_2O_7 \longrightarrow Cr_2O_3 + N_2 + 2\,H_2O$$

The water is driven off as vapor, and the nitrogen gas escapes, leaving pure Cr_2O_3 as the visible product (right).

Figure 6.2

Reduced forms of matter, such as foods and fossil fuels, are high in energy. The energy content is released through oxidation–reduction reactions.

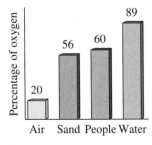

Figure 6.3

Oxygen is the most abundant element on Earth.

6.1
OXYGEN: ABUNDANT AND ESSENTIAL

It would be inaccurate to say that any one element is the most important one, for there are 20 or so elements essential to life. Nevertheless, in any list of important elements, oxygen would be at or near the top (Figure 6.3). It is an abundant element, making up about half of the outer portion of the Earth. Oxygen occurs in the **atmosphere** (the gaseous mass surrounding the Earth) as molecular oxygen (O_2). In the **hydrosphere** (the oceans, seas, rivers, and lakes of Earth), oxygen occurs in combination with hydrogen in the remarkable compound water. In the **lithosphere** (the solid portion of the Earth), oxygen occurs in combination with silicon (pure sand is largely SiO_2) and a variety of metals.

Oxygen is found in most of the compounds that are important to living organisms. Foodstuffs—carbohydrates, fats, and proteins—all contain oxygen. The human body is approximately 65% (by weight) water. Since 89% of the weight of water is due to oxygen—and many other compounds in our bodies also contain oxygen—over 60% of the weight of each one of us is oxygen.

Figure 6.4
Fuels burn more rapidly in pure oxygen than in air. Huge quantities of liquid oxygen are used to burn the fuels that blast rockets into orbit.

Elemental oxygen (O_2) accounts for about 21% (by volume) of the atmosphere. [The remainder is mainly nitrogen (N_2), which enters into few chemical reactions.] Atmospheric oxygen is said to be free or uncombined, that is, not part of a compound containing another element. Oxygen (in air) is taken into the lungs. From there it passes into the bloodstream. The blood carries oxygen to the body tissues, and the food we eat combines with this oxygen. This chemical process provides us with energy in the form of heat to maintain our body temperature. It also gives us the energy we need for mental and physical activity.

Oxygen performs many other functions. Fuels such as natural gas, gasoline, and coal need oxygen to burn and release their stored energy. Combustion of fossil fuels currently supplies over 90% of the energy that turns the wheels of civilization.

Pure oxygen is obtained by liquefying purified air, then letting the nitrogen and argon boil off. (Nitrogen boils at -196 °C, argon at -186 °C, and oxygen at -183 °C.) About 17 billion kg of oxygen of 99.5% purity is produced annually in the United States. Most of this oxygen is used directly by industry, much of it in the metals industry. About 1% is compressed into steel tanks for use in welding, in medicine, and many other purposes (Figure 6.4).

Not all that atmospheric oxygen does is immediately desirable. Oxygen causes iron to rust and copper to corrode, and it aids in the decay of wood. All these and many other chemical processes are called oxidation.

6.2
CHEMICAL PROPERTIES OF OXYGEN: OXIDATION

When iron rusts, it combines with oxygen from the atmosphere to form a reddish brown powder.

$$4\ Fe\ +\ 3\ O_2\ \longrightarrow\ 2\ Fe_2O_3$$

The chemical name for iron rust (Fe_2O_3) is iron(III) oxide. Sometimes it is called by an older name, ferric oxide. Many other metals react with oxygen to form *oxides*.

Most nonmetals also react with oxygen to form oxides. For example, carbon can react with oxygen to form carbon monoxide and carbon dioxide.

$$2\,C + O_2 \longrightarrow 2\,CO$$
$$C + O_2 \longrightarrow CO_2$$

Sulfur can combine with oxygen to form sulfur dioxide.

$$S + O_2 \longrightarrow SO_2$$

And, at high temperatures (such as those that occur in automobile engines), even nitrogen, which is ordinarily quite unreactive, combines with oxygen.

$$N_2 + O_2 \longrightarrow 2\,NO$$

The product is known as nitric oxide.

EXAMPLE 6.1

Magnesium combines readily with oxygen when ignited in air. Write the equation for the reaction.

SOLUTION

Magnesium is an element with the symbol Mg. Oxygen occurs as diatomic molecules (O_2). The two react to form MgO. The reaction is

$$Mg + O_2 \longrightarrow MgO \qquad (not\ balanced)$$

To balance the equation, we need two oxygens (2 MgO) on the right and then 2 Mg on the left.

$$2\,Mg + O_2 \longrightarrow 2\,MgO$$

Practice Exercise

When ignited, zinc burns in air with a bluish-green flame to form zinc oxide (ZnO). Write the equation for the reaction.

Oxygen also reacts with many compounds. Methane, the principal ingredient in natural gas, burns in air to produce carbon dioxide and water:

$$CH_4 + 2\,O_2 \longrightarrow CO_2 + 2\,H_2O$$

Hydrogen sulfide, a gaseous compound with a rotten-egg odor, burns, producing water and sulfur dioxide.

$$2\,H_2S + 3\,O_2 \longrightarrow 2\,H_2O + 2\,SO_2$$

In each example, oxygen combines with each of the elements in the compound to form their oxides.

EXAMPLE 6.2

Carbon disulfide is highly flammable; it combines readily with oxygen, burning with a blue flame. What products are formed? Write the equation.

SOLUTION

Carbon disulfide is CS_2. The products are carbon dioxide and sulfur dioxide. The balanced equation is

$$CS_2 + 3 O_2 \longrightarrow CO_2 + 2 SO_2$$

Practice Exercise

When heated in air, lead sulfide (PbS) combines with oxygen to form lead(II) oxide (PbO) and sulfur dioxide. Write a balanced equation for the reaction.

Oxidation: Three Definitions

The combination of elements and compounds with oxygen is called *oxidation*. The substances that combine with oxygen are said to have been *oxidized*. Originally the term *oxidation* was limited to reactions involving combination with oxygen as in those illustrated in Figure 6.5. Chemists came to recognize, though, that combination with chlorine (or bromine or other elements in the upper right of the periodic chart) was not all that different from reaction with oxygen. So they broadened the definition of oxidation. We won't concern ourselves with a formal definition. The three following working definitions should suffice.

Figure 6.5
Cooking, breathing, and burning fuel all involve oxidation.

1. *An element or a compound is oxidized when it gains oxygen atoms.* In the reactions we've encountered so far in this section, iron, carbon, sulfur, nitrogen, methane, and hydrogen sulfide all gain oxygen atoms. We say, then, that each of these substances is oxidized. Consider also the "burning" of glucose by living cells. This process, called **respiration,** is the main reaction by which cells obtain energy (see Section 25.2).

$$C_6H_{12}O_6 + 6 O_2 \longrightarrow 6 CO_2 + 6 H_2O$$

In the reaction, carbon atoms gain oxygen. Each carbon atom has one oxygen atom in glucose but two oxygen atoms in carbon dioxide. Glucose is oxidized.

EXAMPLE 6.3

In which of the following is the reactant undergoing oxidation? (These are not complete chemical equations.)

a. $Pb \rightarrow PbO_2$
c. $KClO_3 \rightarrow KCl$

b. $SnO_2 \rightarrow SnO$
d. $Cu_2O \rightarrow 2 CuO$

SOLUTION

a.   Pb gains oxygen atoms (it has none on the left and two on the right); it is oxidized.
b.   Sn *loses* an oxygen atom (it has two on the left and only one on the right); it is *not* oxidized.
c.   There are three oxygen atoms on the left and none on the right. The compound loses oxygen; it is *not* oxidized.
d.   The two copper atoms on the left share a single oxygen atom; they have half an oxygen each. On the right, each copper atom has an oxygen atom all its own. Cu has gained oxygen; it is oxidized.

Practice Exercise
In which of the following is the reactant undergoing oxidation? (These are not complete chemical equations.)

a. $3 Fe \rightarrow Fe_3O_4$
c. $Cr_2O_3 \rightarrow CrO_3$

b. $NO \rightarrow NO_2$
d. $C_3H_6O \rightarrow C_3H_6O_2$

2. *A compound is oxidized when it loses hydrogen atoms.* Methyl alcohol, when passed over hot copper gauze, forms formaldehyde and hydrogen gas.

$$\underset{\substack{\text{Methyl} \\ \text{alcohol}}}{CH_4O} \longrightarrow \underset{\text{Formaldehyde}}{CH_2O} + H_2$$

Since methyl alcohol loses hydrogen atoms, it is said to have been oxidized.

In which of the following is the reactant undergoing oxidation? (These are not complete chemical equations.)

a. $C_2H_6O \rightarrow C_2H_4O$ b. $C_2H_2 \rightarrow C_2H_6$

SOLUTION
a.   There are six hydrogen atoms in the compound on the left and only four in the one on the right. The compound loses hydrogen atoms; it is oxidized.
b.   There are two hydrogen atoms in the compound on the left and six in the one on the right. The compound *gains* hydrogen atoms; it is *not* oxidized.

Practice Exercise
In which of the following is the reactant undergoing oxidation? (These are not complete chemical equations.)

a. $C_6H_6 \rightarrow C_6H_{12}$ b. $C_3H_6O \rightarrow C_3H_4O$

3.   *An element is oxidized when it loses electrons.* When magnesium metal reacts with chlorine, magnesium ions and chloride ions are formed.

$$Mg \ + \ Cl_2 \ \longrightarrow \ Mg^{2+} \ + \ 2\,Cl^-$$

Since the magnesium atom obviously loses electrons, it has been oxidized.

In which of the following is the reactant undergoing oxidation? (These are not complete chemical equations.)

a. $Zn \rightarrow Zn^{2+}$ b. $Fe^{3+} \rightarrow Fe^{2+}$
c. $S^{2-} \rightarrow S$ d. $AgNO_3 \rightarrow Ag$

SOLUTION
a.   To form a 2+ ion, zinc loses two electrons; it is oxidized.
b.   To go from a 3+ ion to a 2+ ion, iron must *gain* an electron; it is *not* oxidized.
c.   To go from a 2− ion to an atom with no charge, sulfur loses two electrons; it is oxidized.
d.   To answer this one, you must recognize that $AgNO_3$ is an ionic compound with Ag^+ and NO_3^- ions. In going from Ag^+ to Ag, silver *gains* an electron; it is *not* oxidized.

Practice Exercise
In which of the following is the reactant undergoing oxidation? (These are not complete chemical equations.)

a. $Cu^{2+} \rightarrow Cu$ b. $MnO_4^{2-} \rightarrow MnO_4^-$
c. $Sn^{2+} \rightarrow Sn^{4+}$ d. $Cu \rightarrow CuSO_4$

EXAMPLE 6.4

The most general definition of oxidation is based on electron loss. In some equations, such as those in which a metal is converted to an ion,

$$M \longrightarrow M^+ + e^-$$

electron loss is obvious. In the reaction

$$C + O_2 \longrightarrow CO_2$$

the change is less obvious but no less real. The carbon atom in elemental carbon has four valence electrons that are all its own. In CO_2 it shares those electrons with oxygen atoms; it has only a partial share of what it originally had all to itself. In that sense carbon has lost electrons and is therefore oxidized.

EXAMPLE 6.5

Why so many different definitions of oxidation? Simply for convenience. Which do we use? Whichever is most convenient. For the reaction

$$C + O_2 \longrightarrow CO_2$$

it is convenient to see that carbon is oxidized because it gains oxygen atoms. Similarly, for the reaction

$$CH_4O \longrightarrow CH_2O + H_2$$

it is easy to see that methyl alcohol is oxidized because it loses hydrogen atoms.

6.3
HYDROGEN: OCCURRENCE, PREPARATION, AND PHYSICAL PROPERTIES

For every oxidation process, there is a complementary process—reduction. Before discussing reduction in detail, let's look at some of the chemistry of another element—hydrogen.

Hydrogen ranks low in abundance on the Earth. When we look beyond our home planet, however, hydrogen becomes much more significant. The sun, for example, is made up largely of hydrogen. In fact, in the universe, hydrogen is by far the most abundant element.

By mass, hydrogen makes up only about 0.9% of the outer portion of the Earth, so hydrogen is placed far down on the list of abundant elements. However, hydrogen atoms are quite abundant because hydrogen is the lightest element. If we considered a random sample of 10,000 atoms from the Earth's crust, 5330 would be oxygen atoms, 1590 would be silicon atoms, and 1510 would be hydrogen atoms. Unlike oxygen, hydrogen is seldom found free in nature on Earth. Most of it is combined with oxygen in water. Nearly all compounds derived from living organisms contain hydrogen. Fats, starches, sugars, and proteins all contain combined hydrogen. Petroleum and natural gas are mixtures composed mainly of *hydrocarbons,* compounds of hydrogen and carbon (Chapter 13).

Since elemental hydrogen does not occur in nature on Earth, it is necessary for humans to manufacture it. Small amounts can be made for laboratory use by the reaction of zinc with hydrochloric acid.

$$Zn + 2 HCl \longrightarrow \underset{\substack{\text{Zinc} \\ \text{chloride}}}{ZnCl_2} + H_2$$

Hydrogen gas is essentially insoluble in water, and it is readily collected by displacement of water (Figure 6.6). Commercial quantities of hydrogen usually are obtained as by-products of petroleum refining and other processes. About 200 million kg of hydrogen is produced in the United States each year. About two-thirds of this is used to make ammonia.

Hot air balloons, as their name implies, are buoyed by hot air, which is less dense than the ambient air.

Hydrogen gas is colorless and odorless. It is the least dense of all substances; its density is only 1/14 that of air under comparable conditions. For this reason it once was used in lighter-than-air craft to give them buoyancy in air (Figure 6.7). Unfortunately, a spark or flame can ignite hydrogen, causing a disastrous fire, so this use of hydrogen

Figure 6.6
Hydrogen gas in the laboratory is prepared by the reaction of zinc with hydrochloric acid. The gas bubbles from the reaction flask and is trapped in the inverted bottle. Initially the bottle was filled with water, but the hydrogen pushes the water out as it collects in the bottle.

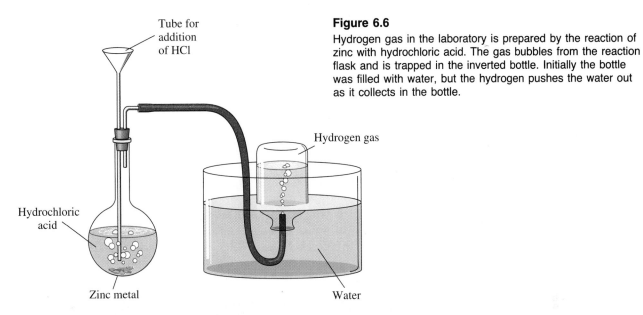

Figure 6.7
Hydrogen is the most effective buoyant gas, but it is highly flammable. Disastrous fires in hydrogen-filled dirigibles such as the German zeppelin *Hindenburg* led to the replacement of hydrogen by nonflammable helium, which buoys the Fuji blimp.

represents a considerable danger. A spectacular fire aboard the German airship Hindenburg in 1937 caused the use of hydrogen to be discontinued. Commercial use of lighter-than-air craft never recovered. The few that remain in service are filled with the noble gas helium, which, of course, is nonflammable. These airships, exemplified by the Goodyear and Fuji blimps, are used mainly in advertising.

Certain metals, such as platinum (Pt), palladium (Pd), and nickel (Ni), collect large volumes of hydrogen (in a condensed form) on their surfaces. This *adsorbed hydrogen* is a great deal more reactive than ordinary hydrogen gas. These metals often are used as catalysts (Section 5.10) for reactions of hydrogen with other substances.

6.4
CHEMICAL PROPERTIES OF HYDROGEN: REDUCTION

Chemically, a jet of hydrogen (when ignited) burns in air with an almost colorless flame. Hydrogen and oxygen can be mixed at room temperature with no perceptible reaction. When the mixture is ignited by a spark or a flame, however, a tremendous explosion results. The product in both cases is water.

$$2 H_2 + O_2 \xrightarrow{\text{spark}} 2 H_2O$$

It is interesting to note that when a piece of platinum gauze is inserted into a container of hydrogen and oxygen, the two gases react at room temperature. The platinum acts as a catalyst; it lowers the activation energy (Section 5.7) for the reaction. The platinum glows from the heat released in the initial reaction and then ignites the mixture, causing an explosion.

Hydrogen reacts with a variety of metal oxides to remove oxygen and yield the free metal. For example, when hydrogen is passed over heated copper oxide, metallic copper and water are formed.

$$CuO + H_2 \longrightarrow Cu + H_2O$$

With lead oxide, the products are metallic lead and water.

$$PbO + H_2 \longrightarrow Pb + H_2O$$

This process, in which a metal is obtained from one of its compounds (from ore), is called *reduction*.

Reduction: Three Definitions

Like oxidation, which once meant simply the combination of an element or a compound with oxygen, the term *reduction* has a broadened meaning. Instead of a formal definition, however, we use working definitions like those given for oxidation.

1. *A compound is reduced when it loses oxygen atoms.* Of the examples above, CuO and PbO obviously were reduced, since they lost oxygen. When potassium chlorate ($KClO_3$) is heated, it forms potassium chloride and oxygen gas.

$$2 KClO_3 \xrightarrow{\text{heat}} 2 KCl + 3 O_2$$

The potassium chlorate loses oxygen; therefore, it is reduced.

EXAMPLE 6.6

In which of the following is the reactant undergoing reduction? (These are not complete chemical equations.)

a. $Pb \rightarrow PbO_2$ b. $SnO_2 \rightarrow SnO$
c. $Al_2O_3 \rightarrow Al$ d. $Cu_2O \rightarrow 2 CuO$

SOLUTION

a.   Pb *gains* oxygen atoms (it has none on the left and two on the right); it is *not* reduced, rather it is oxidized (see Example 6.3).

b.   Sn loses an oxygen atom (it has two on the left and only one on the right); it is reduced.

c.   There are three oxygen atoms on the left and none on the right. The compound loses oxygen; it is reduced.

d.   The two copper atoms on the left share a single oxygen atom; they have half an oxygen each. On the right, each copper atom has an oxygen atom all its own. Cu has *gained* oxygen; it is *not* reduced, rather it is oxidized.

Practice Exercise

In which of the (partial) reactions listed in the practice exercise with Example 6.3 is the reactant undergoing reduction?

2.   *A compound is reduced when it gains hydrogen atoms.* For example, methyl alcohol is made by reaction of carbon monoxide with hydrogen.

$$CO + 2 H_2 \xrightarrow{\text{catalyst}} CH_4O$$

The carbon monoxide gains hydrogen; therefore, it is reduced.

EXAMPLE 6.7

In which of the following is the reactant undergoing reduction? (These are not complete chemical equations.)

a. $C_2H_6O \rightarrow C_2H_4O$ b. $C_2H_2 \rightarrow C_2H_6$

SOLUTION

a.   There are six hydrogen atoms in the compound on the left and only four in the one on the right. The compound *loses* hydrogen atoms. It is *not* reduced; rather, it is oxidized (see Example 6.4).

b.   There are two hydrogen atoms in the compound on the left and six in the one on the right. The compound gains hydrogen atoms; it is reduced.

Practice Exercise

In which of the (partial) reactions in the practice exercise with Example 6.4 is the reactant undergoing reduction?

The most general definition of reduction is based on electron gain. In some equations, such as those in which an ion is converted to a metal

$$M^+ + e^- \rightarrow M$$

electron gain is obvious. In the reaction

$$2 HgO \rightarrow 2 Hg + O_2$$

the change is less obvious. In HgO, mercury shares its valence electrons with oxygen atoms; it has only a partial share of them. In elemental mercury those electrons belong to Hg alone. It has all to itself the electrons that it originally had to share. In that sense mercury has gained electrons and is therefore reduced.

3.   *An atom or an ion is reduced when it gains electrons.* When an electric current is passed through a solution containing copper ions (Cu^{2+}), copper metal is plated out at the cathode.

$$Cu^{2+} + 2 e^- \longrightarrow Cu$$

The copper ion has obviously gained electrons; therefore, it has been reduced.

EXAMPLE 6.8

In which of the following is the reactant undergoing reduction? (These are not complete chemical equations.)

a. $Zn \rightarrow Zn^{2+}$ b. $Fe^{3+} \rightarrow Fe^{2+}$
c. $S^{2-} \rightarrow S$ d. $AgNO_3 \rightarrow Ag$

SOLUTION

a. To form a 2+ ion, zinc *loses* two electrons; it is *not* reduced, rather it is oxidized (see Example 6.5).
b. To go from a 3+ ion to a 2+ ion, iron must gain an electron; it is reduced.
c. To go from a 2− ion to an atom with no charge, sulfur *loses* two electrons; it is *not* reduced, rather it is oxidized.
d. To answer this one, you must recognize that $AgNO_3$ is an ionic compound with Ag^+ and NO_3^- ions. In going from Ag^+ to Ag, silver gains an electron; it is reduced.

Practice Exercise

In which of the (partial) reactions in the practice exercise with Example 6.5 is the reactant undergoing reduction?

6.5
OXIDIZING AGENTS AND REDUCING AGENTS

We said that oxidation and reduction go hand in hand (Figure 6.8). When one substance is oxidized, another is reduced. For example, in the reaction

$$CuO + H_2 \longrightarrow Cu + H_2O$$

copper oxide is reduced and hydrogen is oxidized. Further, if one substance is being oxidized, the other must be causing it to be oxidized. Here CuO causes H_2 to be oxidized. Therefore, CuO is called the **oxidizing agent.** Conversely, H_2 causes CuO to be reduced, so H_2 is the **reducing agent.** Each oxidation–reduction reaction has an oxidizing agent and a reducing agent among the reactants. The reducing agent is the substance being oxidized; the oxidizing agent is the substance being reduced.

Reduction:
copper oxide is being reduced;
CuO is the oxidizing agent.

$$CuO + H_2 \longrightarrow Cu + H_2O$$

Oxidation:
hydrogen is being oxidized;
H_2 is the reducing agent.

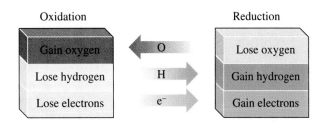

Figure 6.8
Three different definitions of oxidation and reduction.

EXAMPLE 6.9

Circle the oxidizing agents and underline the reducing agents in the following reactions.

a. $2\,C + O_2 \rightarrow 2\,CO$
b. $N_2 + 3\,H_2 \rightarrow 2\,NH_3$
c. $SnO + H_2 \rightarrow Sn + H_2O$
d. $Mg + Cl_2 \rightarrow Mg^{2+} + 2\,Cl^-$

SOLUTION
The answers are determined using the definitions previously given.

a. $\underline{C} + \boxed{O_2}$

C gains oxygen and is oxidized, so it must be the reducing agent. Therefore, O_2 is the oxidizing agent.

b. $\boxed{N_2} + \underline{H_2}$

N_2 gains hydrogen and is reduced, so it is the oxidizing agent. Therefore, H_2 is the reducing agent.

c. $\boxed{SnO} + \underline{H_2}$

SnO loses oxygen and is reduced, so it is the oxidizing agent. H_2 is the reducing agent.

d. $\underline{Mg} + \boxed{Cl_2}$

Mg loses electrons and is oxidized, so it is the reducing agent. Cl_2 is the oxidizing agent.

Practice Exercise
Circle the oxidizing agents and underline the reducing agents in the following reactions.

a. $Se + O_2 \rightarrow SeO_2$
b. $CH_3C{\equiv}N + 2\,H_2 \rightarrow CH_3CH_2NH_2$
c. $V_2O_5 + 2\,H_2 \rightarrow V_2O_3 + 2\,H_2O$
d. $2\,K + Br_2 \rightarrow 2\,K^+ + 2\,Br^-$

6.6
SOME COMMON OXIDIZING AGENTS

Oxygen is undoubtedly the most common oxidizing agent. It oxidizes the wood in our campfires and the gasoline in our automobile engines. It rusts and corrodes the metals that we get by reducing ores. It even "burns" the foods we eat to give us energy to move and to think. It is involved in the rotting of wood and the weathering of rocks. Indeed, we live in an oxidizing atmosphere. Fortunately, though, oxygen is a mild oxidizing agent; otherwise, living things would burn up quickly. As it is, it takes us about 70 years to "burn out."

Oxygen sometimes is used as a laboratory and industrial oxidizing agent. For example, acetylene (used to fuel cutting and welding torches) is made by the partial oxidation of methane.

$$4\ CH_4\ +\ 3\ O_2\ \longrightarrow\ 2\ C_2H_2\ +\ 6\ H_2O$$

Methane   Acetylene

The Breathalyzer test to detect drunk drivers makes use of the color change associated with the reduction of dichromate ion. If a suspect's breath causes the color to change to green (Cr^{3+}), this indicates his/her blood contains more than the legal level of alcohol. Further tests are then carried out at the police station to confirm the suspicion.

(Complete oxidation of methane gives carbon dioxide and water.) Often, though, it is more convenient to use other oxidizing agents. In the laboratory, potassium dichromate ($K_2Cr_2O_7$) is used frequently to oxidize alcohols to compounds called *aldehydes* and *ketones* (Chapter 15). For the oxidation of ethyl alcohol (found in alcoholic beverages) to acetaldehyde, the reaction is

$$8\ H^+\ +\ Cr_2O_7{}^{2-}\ +\ 3\ C_2H_5OH\ \longrightarrow\ 2\ Cr^{3+}\ +\ 3\ C_2H_4O\ +\ 7\ H_2O$$

Dichromate ion   Chromium(III)
(orange)   ion (green)

Another common oxidizing agent is hydrogen peroxide (H_2O_2). Pure hydrogen peroxide is a syrupy, colorless liquid. It usually is used in aqueous solution, with concentrations of 30% and 3% generally available. Some interesting uses of hydrogen peroxide are discussed in Sections 6.8 and 6.9. For now, let's look at some of its simpler chemistry.

Hydrogen peroxide oxidizes sulfides (S^{2-}) to sulfates ($SO_4{}^{2-}$). When lead-based paints are exposed to polluted air containing hydrogen sulfide (H_2S), they turn black because of the formation of lead sulfide (PbS). Hydrogen peroxide oxidizes the black sulfide to white sulfate.

Hydrogen peroxide can be used to restore the once-white areas of old paintings that have darkened from the reaction of white lead compounds (paint pigments) with sulfur compounds. The darkened pigments (black PbS) are converted to white $PbSO_4$ by the hydrogen peroxide.

$$PbS\ +\ 4\ H_2O_2\ \longrightarrow\ PbSO_4\ +\ 4\ H_2O$$

(black)   (white)

An advantage of hydrogen peroxide as an oxidizing agent is that it is converted to water, an innocuous by-product, in most reactions.

Potassium permanganate is a black, shiny, crystalline solid. It dissolves in water to give deep purple solutions. This purple color disappears as the permanganate is reduced (remember, if permanganate is an oxidizing agent, it must be reduced). So potassium permanganate is often used as a test for oxidizable substances. For example, potassium permanganate is used to oxidize iron from Fe^{2+} to Fe^{3+}. The amount of iron(II) ion in a sample can be determined by its reaction with permanganate. One can add permanganate solution, which is deep purple, to a sample of iron(II) ion. As the permanganate is reduced, it is changed to manganese(II) ion, and the purple color disappears (Figure 6.9). When all the iron(II) ion has been oxidized, the addition of more permanganate will not be accompanied by the loss of the purple color because there will be no iron(II) ion left to reduce the permanganate. Thus, one can measure just how much iron(II) ion there was in

(a)   (b)

Figure 6.9
Purple permanganate ions (MnO_4^-) are reduced to Mn^{2+} by iron(II) ions (Fe^{2+}). (a) The buret contains permanganate solution, and the flask contains a solution of iron(II) ions. (b) As the permanganate solution is added to the solution of iron(II) ions, the purple color disappears.

the sample by keeping track of how much permanganate was reduced, that is, by measuring how much permanganate was added until the purple color remained. The equation for this reaction is

$$MnO_4^- \;+\; 5\,Fe^{2+} \;+\; 8\,H^+ \;\longrightarrow\; Mn^{2+} \;+\; 5\,Fe^{3+} \;+\; 4\,H_2O$$

Permanganate   Manganese(II)
ion (purple)   ion (pale pink)

Permanganate solutions can also be used to oxidize oxalic acid (a poisonous compound found in rhubarb leaves), sulfur dioxide (SO_2), and many other compounds.

Other common oxidizing agents are the halogens—fluorine (F_2), chlorine (Cl_2), bromine (Br_2), and iodine (I_2). Chlorine, for example, oxidizes magnesium metal to magnesium ions.

$$Mg \;+\; Cl_2 \;\longrightarrow\; Mg^{2+} \;+\; 2\,Cl^-$$

We shall encounter still other interesting oxidizing agents as we go along.

6.7
SOME REDUCING AGENTS OF INTEREST

There is no single reducing agent that stands out as oxygen does among oxidizing agents. Hydrogen reduces many compounds, but it would be rather expensive for large-scale processes. Elemental carbon (as coke, obtained by driving off the volatile matter from coal) is used frequently to obtain metals from their ores. For example, tin can be obtained from tin oxide by reduction, with carbon used as the reducing agent.

$$SnO_2 \;+\; C \;\longrightarrow\; Sn \;+\; CO_2$$

Hydrogen can be used for the production of expensive metals, such as tungsten (W). First, the tungsten ore is converted to an oxide (WO_3); then it is reduced in a stream of hydrogen gas at 1200 °C.

$$WO_3 \;+\; 3\,H_2 \;\longrightarrow\; W \;+\; 3\,H_2O$$

In food chemistry, reducing agents are called *antioxidants*. Ascorbic acid (vitamin C) is used to prevent the browning of fruit (such as sliced apples) by inhibiting air oxidation. Vitamin C, which is water-soluble, presumably also retards potentially damaging oxidation of vital components of living cells. Tocopherol (vitamin E) serves as a fat-soluble antioxidant.

Figure 6.10
A photographic negative
and (below) a positive print.

Perhaps a more familiar reducing agent, by use if not by name, is the developer used in black and white photography. Photographic film is coated with a silver salt (silver bromide, Ag^+Br^-). Silver ions that have been exposed to light react rapidly with the developer, a reducing agent (such as the organic compound hydroquinone), to form metallic silver.

$$C_6H_4(OH)_2 \ + \ 2\,Ag^+ \ \longrightarrow \ C_6H_4O_2 \ + \ 2\,Ag \ + \ 2\,H^+$$
Hydroquinone   Silver
                                                    metal

Silver ions not exposed to light are not reduced as rapidly by the developer. The film is then treated with "hypo," a solution of sodium thiosulfate ($Na_2S_2O_3$), which washes out unexposed silver bromide to form the negative. That leaves the negative dark where the metallic silver has been deposited (where it was originally exposed to light) and transparent where light did not strike it. Light is then shone through the negative onto light-sensitive paper to make the positive print (see Figure 6.10).

6.8
OXIDATION AND ANTISEPTICS

Many common **antiseptics** (compounds that are applied to living tissue to kill microorganisms or to prevent their growth) are mild oxidizing agents. Hydrogen peroxide (H_2O_2), in the form of a 3% aqueous solution, finds use in medicine as a topical antiseptic. It can be used to treat minor cuts and abrasions on certain parts of the body.

Sodium hypochlorite (NaOCl), available in aqueous solution as a laundry bleach (Purex, Clorox, and the like), finds some use in the irrigation of wounds and for bladder infections. It is also used as a disinfectant and deodorizer. Solutions of iodine also are used as antiseptics. In preparation for surgery, skin may be painted with these brown antiseptic solutions.

The exact function of these oxidizing agents as antiseptics is unknown. Their action is rather indiscriminate; they attack human cells as well as microorganisms. For many purposes, these simple inorganic compounds have been replaced by organic compounds such as the phenols, which kill by other mechanisms.

Other oxidizing agents are used as disinfectants. Calcium hypochlorite [$Ca(OCl)_2$], bleaching powder, is used to disinfect clothing and bedding. Chlorine (Cl_2) is used to kill pathogenic (disease-causing) microorganisms in drinking water. Also, wastewater usually is treated with chlorine before it is returned to a stream or a lake. Swimming pools are usually disinfected by ''chlorination.'' Very large pools are sometimes sanitized with chlorine gas (transported and stored in steel cylinders as liquid chlorine). Pools chlorinated in this way soon become too acidic and must have the acidity adjusted by addition of a base (usually sodium carbonate). Smaller pools are often chlorinated with calcium hypochlorite. These pools become too alkaline, and the acidity is increased by adding hydrochloric acid. (Swimming pool water is usually maintained in the pH range of 7.2–7.8—see Chapter 10 for a discussion of acids, bases, and pH.)

Benzoyl peroxide [$(C_6H_5COO)_2$], a powerful oxidizing agent, has long been used at 5% and 10% concentrations in ointments for treating acne. In addition to its antibacterial action, benzoyl peroxide acts as a skin irritant, causing old skin to slough off and be replaced with newer, fresher looking skin. When used on areas exposed to sunlight, benzoyl peroxide is thought to promote skin cancer; such use is therefore discouraged.

6.9
OXIDATION: BLEACHING
AND STAIN REMOVAL

Bleaches are compounds that are used to remove unwanted color from fabrics, hair, or other materials. Nearly any oxidizing agent would do the job. However, some also would harm the fabrics, some would be unsafe, some would produce undesirable by-products, and some would simply be too expensive.

The most familiar laundry bleaches are aqueous solutions of sodium hypochlorite (NaOCl). Bleaching powder [$Ca(OCl)_2$] generally is preferred for large-scale bleaching operations. The paper industry uses it to make white paper, and the textile industry uses it to make whiter fabrics.

In both aqueous bleaches and bleaching powder, the active agent is the hypochlorite ion (ClO^-). Materials appear colored because loosely bound electrons are boosted to higher energy levels by absorption of visible light. Bleaching agents do their work by removing or tying down these mobile electrons. For example, hypochlorite ions (in water) take up electrons to form chloride ions and hydroxide ions. The reduction half of the complete equation is

$$2\,e^- \;+\; ClO^- \;+\; H_2O \;\longrightarrow\; Cl^- \;+\; 2\,OH^-$$

Hypochlorite bleaches are safe and effective for cotton and linen fabrics. They should not be used for wool, silk, or nylon, however.

Many stain removers are simply absorbants (e.g., cornstarch, which absorbs grease spots), solvents (e.g., amyl acetate, which removes ballpoint ink), or detergents (which remove mustard stains).

Oxalic acid also removes rust spots, but its action in that case is as a complexing agent, not a reducing agent. The oxalic acid molecules pick up the iron ions from the otherwise insoluble rust and carry them in a solution that is washed away.

Other bleaching agents include hydrogen peroxide, sodium perborate (often formulated $NaBO_2 \cdot H_2O_2$ to indicate a loose association of $NaBO_2$ and H_2O_2), and a variety of chlorine-containing organic compounds that release Cl_2 in water. When hydrogen peroxide is used to bleach hair, it acts as an oxidizing agent. In the process, the black or brown pigment in the hair (melanin) is oxidized to colorless products.

Stain removal is not nearly as simple a process as bleaching. A few stain removers are oxidizing agents or reducing agents; others have quite different chemical natures. Nearly all stains require rather specific stain removers. Hydrogen peroxide in cold water removes blood stains from cotton and linen fabrics. Potassium permanganate removes most stains from white fabrics (except for rayon); the permanganate stain then can be removed by treatment with oxalic acid.

$$5\ H_2C_2O_4 \ + \ 2\ MnO_4^- \ + \ 6\ H^+ \ \longrightarrow \ 10\ CO_2 \ + \ 2\ Mn^{2+} \ + \ 8\ H_2O$$
$$\text{Oxalic acid} \qquad \text{(purple)} \qquad\qquad\qquad\qquad \text{(colorless)}$$

Sodium thiosulfate ($Na_2S_2O_3$) readily removes iodine stains by reducing iodine to a colorless ion (I^-).

$$I_2 \ + \ 2\ Na_2S_2O_3 \ \longrightarrow \ 2\ NaI \ + \ Na_2S_4O_6$$
$$\text{(brown)} \qquad\qquad\qquad\qquad \text{(colorless)}$$

6.10
OXIDATION, REDUCTION, AND LIVING THINGS

We obtain energy for physical and mental activities by the slow, multistep oxidation of food. These processes will be discussed in detail in later chapters, but for now we will represent them schematically as

$$\text{Carbohydrates} \ + \ O_2 \ \longrightarrow \ CO_2 \ + \ H_2O \ + \ \text{energy}$$
$$\text{(sugars and starches)}$$
$$\text{Fats} \ + \ O_2 \ \longrightarrow \ CO_2 \ + \ H_2O \ + \ \text{energy}$$
$$\text{Proteins} \ + \ O_2 \ \longrightarrow \ CO_2 \ + \ H_2O \ + \ \text{Urea} \ + \ \text{energy}$$

Green plants make these foods by reducing carbon dioxide. Energy for the reaction, called **photosynthesis,** comes from the sun. The process, shown here for the formation of glucose, a simple sugar, can be written

$$6\ CO_2 \ + \ 6\ H_2O \ + \ \text{energy} \ \longrightarrow \ C_6H_{12}O_6 \ + \ 6\ O_2$$

We eat either the plants or the animals that feed on the plants, and we obtain energy from the carbohydrates, fats, and proteins of those organisms. In fact, the oxidation of carbohydrates is precisely the reverse of the photosynthesis reaction. If we take the carbohydrate to be glucose, the reaction is

$$C_6H_{12}O_6 \ + \ 6\ O_2 \ \longrightarrow \ 6\ CO_2 \ + \ 6\ H_2O \ + \ \text{energy}$$

We can see that green plants carry out the oxidation–reduction reaction that makes possible all life on Earth. Animals can only oxidize the foods that plants provide. We might therefore consider crop farming a process of reduction. Energy captured in cultivated plants, whether the plants are used directly or are fed to animals, is the basis for human life (Figure 6.11).

Figure 6.11
The food we eat is oxidized to provide energy for our activities. That energy originates in the sun and is trapped by plants through photosynthetic reactions that reduce carbon dioxide to carbohydrates.

EXERCISES

1. Define oxidation and reduction in terms of the following.
   a. oxygen atoms gained or lost
   b. hydrogen atoms gained or lost
   c. electrons gained or lost
2. List four common oxidizing agents.
3. List three common reducing agents.
4. Name some oxidizing agents used as antiseptics and disinfectants.
5. Relate the chemistry of photosynthesis to the chemistry that provides energy for your heartbeat.
6. Describe how a bleaching agent, such as hypochlorite (ClO^-), works.
7. The following ''equations'' show only part of a chemical reaction. Indicate whether the reactant shown is being oxidized or reduced.
   a. $C_2H_4O \rightarrow C_2H_4O_2$ b. $H_2O_2 \rightarrow H_2O$
8. In which of the following partial reactions is the reactant undergoing oxidation?
   a. $C_2H_4O \rightarrow C_2H_6O$ b. $WO_3 \rightarrow W$
9. In which of the following partial reactions is the reactant undergoing oxidation?
   a. $Cl_2 \rightarrow 2\ Cl^-$ b. $Fe^{3+} \rightarrow Fe^{2+}$
10. In which of the following partial reactions is the reactant undergoing oxidation?
    a. $2\ H^+ \rightarrow H_2$ b. $CO \rightarrow CO_2$
11. Circle the oxidizing agent and underline the reducing agent in these reactions.
    a. $4\ Al + 3\ O_2 \rightarrow 2\ Al_2O_3$
    b. $2\ SO_2 + O_2 \rightarrow 2\ SO_3$

12. Circle the oxidizing agent and underline the reducing agent in these reactions.
    a. $Cl_2 + 2\ KBr \rightarrow 2\ KCl + Br_2$
    b. $C_2H_4 + H_2 \rightarrow C_2H_6$
13. Circle the oxidizing agent and underline the reducing agent in these reactions.
    a. $Fe + 2\ HCl \rightarrow FeCl_2 + H_2$
    b. $CS_2 + 3\ O_2 \rightarrow CO_2 + 2\ SO_2$
14. Circle the oxidizing agent and underline the reducing agent in these reactions.
    a. $2\ AgNO_3 + Cu \rightarrow Cu(NO_3)_2 + 2\ Ag$
    b. $CuCl_2 + Fe \rightarrow FeCl_2 + Cu$
15. In the following reactions, which substance is oxidized and which is reduced?
    a. $2\ HNO_3 + SO_2 \rightarrow H_2SO_4 + 2\ NO_2$
    b. $2\ CrO_3 + 6\ HI \rightarrow Cr_2O_3 + 3\ I_2 + 3\ H_2O$
16. In the following reactions, which substance is oxidized and which is the oxidizing agent?
    a. $H_2CO + H_2O_2 \rightarrow H_2CO_2 + H_2O$
    b. $5\ C_2H_6O + 4\ MnO_4^- + 12\ H^+ \rightarrow$
    $\qquad\qquad 5\ C_2H_4O_2 + 4\ Mn^{2+} + 11\ H_2O$
17. Acetylene (C_2H_2) reacts with hydrogen to form ethane (C_2H_6). Is the acetylene oxidized or reduced? Explain your answer.
18. Unsaturated vegetable oils react with hydrogen to form saturated fats. A typical reaction is

    $$C_{57}H_{104}O_6 \ + \ 3\ H_2 \ \rightarrow \ C_{57}H_{110}O_6$$

    Is the unsaturated oil oxidized or reduced? Explain.

19. To test for an iodide ion (for example, in iodized salt), a solution is treated with chlorine to liberate iodine. The reaction is

$$2\,I^- \;+\; Cl_2 \;\rightarrow\; I_2 \;+\; 2\,Cl^-$$

Which substance is oxidized? Which is reduced?

20. Molybdenum metal, used in special kinds of steel, can be manufactured by the reaction of its oxide with hydrogen. The reaction is

$$MoO_3 \;+\; 3\,H_2 \;\rightarrow\; Mo \;+\; 3\,H_2O$$

Which substance is reduced? Which is the reducing agent?

21. Green grapes are exceptionally sour due to a high concentration of tartaric acid. As the grapes ripen, this compound is converted to glucose.

$$\underset{\text{Tartaric acid}}{C_4H_6O_2} \;\longrightarrow\; \underset{\text{Glucose}}{C_6H_{12}O_6}$$

Is the tartaric acid being oxidized or reduced?

22. The dye indigo (used to color blue jeans) is formed from indoxyl by exposure of the latter to air.

$$\underset{\text{Indoxyl}}{2\,C_8H_7ON} \;+\; O_2 \;\longrightarrow\; \underset{\text{Indigo}}{C_{16}H_{10}N_2O_2} \;+\; 2\,H_2O$$

What substance is oxidized? What is the oxidizing agent?

23. When the water pump failed in the nuclear reactor at Three Mile Island in 1979, zirconium metal reacted with the very hot water to produce hydrogen gas.

$$Zr \;+\; 2\,H_2O \;\rightarrow\; ZrO_2 \;+\; 2\,H_2$$

What substance was oxidized in the reaction? What was the oxidizing agent?

24. Vitamin C (ascorbic acid) is thought to protect our stomachs from the carcinogenic effect of nitrite ions by converting the ions to NO gas.

$$NO_2^- \;\rightarrow\; NO$$

Is the nitrite ion oxidized or reduced? Is ascorbic acid an oxidizing agent or a reducing agent?

25. In the preceding reaction (Exercise 24), ascorbic acid is converted to dehydroascorbic acid.

$$C_6H_8O_6 \;\rightarrow\; C_6H_6O_6$$

Is ascorbic acid oxidized or reduced in the reaction?

26. Give formulas for the products formed in each of the following reactions.

a. $C + O_2 \rightarrow$
b. $S + O_2 \rightarrow$
c. $CH_4 + O_2 \rightarrow$
d. $CS_2 + O_2 \rightarrow$
e. $N_2 + O_2 \rightarrow$
f. $H_2 + O_2 \rightarrow$
g. $C_3H_8 + O_2 \rightarrow$
h. $C_6H_{12}O_6 + O_2 \rightarrow$

Chapter 7
GASES

The properties of gases allow humans to fly safely in hot air balloons.

Our astronauts have seen firsthand Earth's barren, airless moon. Our spaceships have photographed the desolation of Mercury from a few kilometers up and have measured the inhospitably high temperatures of Venus. They have given us close-up portraits of the crushing, turbulent atmospheres of Jupiter, Saturn, Uranus, and Neptune. Our experiments on the harsh Martian surface failed to detect the presence of life there. As we approach the end of the twentieth century it is becoming increasingly clear that the Earth, a small island of green and blue in the vastness of space, is uniquely equipped to serve the needs of the life that inhabits it (Figure 7.1).

The life-support system of Spaceship Earth consists in part of a thin blanket of gases called the atmosphere. Although other planets in our solar system have atmospheres, the Earth's atmosphere appears to be unique in its ability to support life. The atmosphere is composed of about 5.2×10^{15} metric tons of air spread over a surface area of 5.0×10^8 km^2.

It is difficult to measure just how deep the atmosphere is. It does not end abruptly, but gradually fades as the distance from the surface of the Earth increases. It is known, though, that 99% of the atmosphere lies within 30 km of the surface of the Earth—a thin layer of air indeed (like the peel of an apple, only relatively thinner).

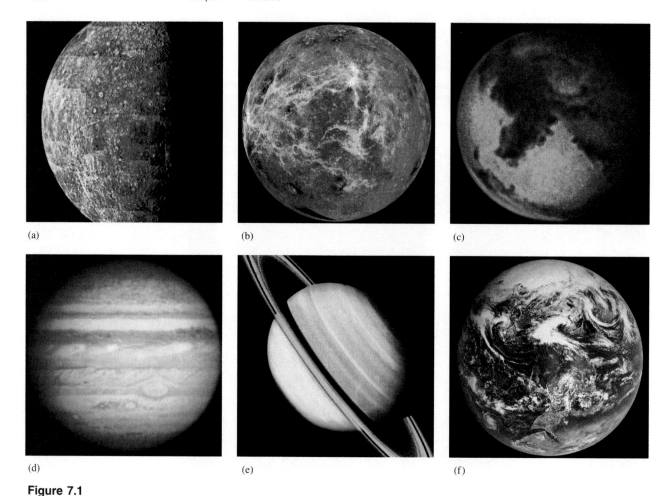

(a)   (b)   (c)

(d)   (e)   (f)

Figure 7.1

(a) Looking like the moon, the scarred planet Mercury is closest to the sun and has no atmosphere to protect its surface from meteors or intense solar radiation. (b) The crushing
atmosphere of Venus traps the sun's energy and produces temperatures at the planet's surface high enough to melt lead. (c) The dark spots are the peaks of gigantic Martian volca-
noes reaching above hazy clouds in an atmosphere too thin to support human life. (d) As
befits the largest of the planets, Jupiter possesses a turbulent atmosphere with cyclonic
storms that could swallow several planets the size of the Earth. (e) Saturn is one of the
beauties of our solar system, but its thick, poisonous atmosphere would be deadly to humans. (f) Blessed with abundant water and an atmosphere rich in oxygen, Earth offers its
inhabitants a safe refuge for as long as they respect its limits.

The atmosphere is divided into layers (Figure 7.2). The layer nearest Earth, the
troposphere, harbors nearly all living things and nearly all human activity. The next
region, the stratosphere, is where we find the ozone layer that shields living creatures
from deadly ultraviolet radiation.

Air is so familiar, and yet so nebulous, that it is difficult to think of it as matter. But
it is matter—matter in the gaseous state. All gases, air included, have mass and occupy
space. Like other forms of matter, gases obey certain physical laws. In this chapter, we
will examine some of those laws and see how they are related to certain vital processes—
such as breathing!

Table 7.1

Composition of Earth's Atmosphere (Dry): Molecules of Each
Substance per 100,000 Molecules of Air

Substance	Formula	Number of Molecules
Nitrogen	N_2	78,083
Oxygen	O_2	20,945
Argon	Ar	934
Carbon dioxide	CO_2	35
All others	—	3
Total		100,000

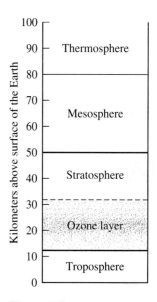

Figure 7.2
Divisions of the atmosphere
and approximate location of
the ozone layer (color).

7.1
AIR: A MIXTURE OF GASES

Air is a mixture of gases. Dry air is (by volume) about 78% nitrogen (N_2), 21% oxygen (O_2), and 1% argon (Ar). Water vapor varies from 0% up to about 4%. There are a number of minor constituents, the most important of which is carbon dioxide (CO_2). The concentration of carbon dioxide in the atmosphere is believed to have increased from 296 ppm in 1900 to its present value of more than 350 ppm. It most likely will continue to increase as we burn more and more fossil fuels (coal, oil, and gas). The composition of the atmosphere is summarized in Table 7.1.

Air is a mixture of gases, but what are gases? Perhaps they are best understood in terms of the kinetic-molecular theory.

7.2
THE KINETIC-MOLECULAR THEORY

The states of matter—solid, liquid, and gas—are obviously different from one another (Chapter 1). Chemists offer a model to explain these differences. The model is referred to as the **kinetic-molecular theory.** The basic postulates of this theory are the following.

1. All matter is composed of tiny, discrete particles called molecules.
2. Molecules are in rapid, constant motion and move in straight lines.
3. The molecules of a gas are very small compared to the distances between them.
4. There is very little attraction between molecules of a gas. /or repulsion - if its an ideal gas.
5. Molecules collide with one another, and energy is conserved in these collisions—although one molecule can gain energy at the expense of another.
6. Temperature is a measure of the *average* kinetic energy of the gas molecules.

The kinetic-molecular theory treats gases as collections of individual particles in rapid motion (*kinetic* derives from the Greek word for motion). The particles of nitrogen gas, for example, are molecules (N_2); those of argon gas (Ar) are atoms. The distances between the particles are quite large compared to the dimensions of the particles themselves. Therefore, unlike solids and liquids, gases can be readily compressed. (According to the theory, the individual particles of a solid or liquid are in contact with one another. They can't be pushed closer together because they are already touching.)

The particles of a gas are in such rapid motion and have such low densities that gravity has little effect on them. They move up and down and sideways with ease and will not fall to the bottom of a container (as a liquid will). Any container of gas is completely filled. By *filled* we do not mean that the gas is packed tightly, but rather that it is distributed throughout the container's entire volume. A particle moves along a straight path until it strikes something (another particle or the walls of the container). Then it may bounce off at an angle and travel from the point of collision along a straight path until it hits something else. These collisions are *perfectly elastic;* there is no tendency for the collection of particles to slow down and eventually stop. Two particles that are about to collide have a certain combined kinetic energy. After the collision the sum of their kinetic energies is the same. One of the particles may have been slowed down by the collision, but the other will have been speeded up just enough to compensate (Figure 7.3).

The kinetic-molecular theory explains what it is we are measuring when we measure temperature. According to the theory, temperature is just a reflection of the kinetic energy of the gas particles. The higher the kinetic energy, that is, the faster the particles are moving, the higher the temperature of the sample. On the average, the particles of a cold sample are moving more slowly than the particles of a hot sample. In any single sample, some particles are moving faster than others. Temperature reflects the *average* kinetic energy of the particles.

As a particle strikes a wall of its container, it gives the wall a little push. (If you were hit with a baseball or a brick, you would feel a push.) The sum of all these tiny pushes over a given area of the wall is what we call **pressure.**

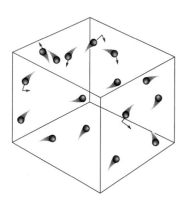

Figure 7.3
According to the kinetic-molecular theory, particles of a gas are in constant motion, occasionally bouncing off one another and off the walls of their container.

We sometimes say that the air is "thinner" at high altitudes, but we actually mean that there are fewer molecules per unit volume.

1 atm = 760 mm Hg = 76.0 cm Hg = 760 torr

1 cm H_2O = 0.74 mm Hg

1.00 atm = 29.9 in. Hg = 14.7 psi

1 Pa = 1 N/m^2 = 1.01 × 10^{-5} atm

7.3
ATMOSPHERIC PRESSURE

Molecules of air are constantly bouncing off each of us. Molecules are so tiny, though, that we don't feel the impact of individual ones. In fact, at ordinary altitudes we don't feel the molecules pushing on our skin, because there are molecules on the inside pushing out just as hard. When we increase our altitude rapidly by driving up a mountain or riding an elevator up to the top of a tall building, however, our ears pop because there are fewer molecules (because of the thinner air) on the outside pushing in than there are on the inside pushing out. Once we are at the top of the mountain or building, the pressures are soon equalized, and the popping stops.

The pressure of the atmosphere is measured by a device called a **barometer.** The simplest type of barometer is a long glass tube, closed at one end, filled with mercury, and inverted in a shallow dish containing mercury (Figure 7.4). Suppose the tube is 1 m long. Some of the mercury in the tube will drain into the dish, but *not all* of it. The mercury will drain out only until the pressure exerted by the mercury remaining in the tube exactly balances the pressure exerted by the atmosphere on the surface of the mercury in the dish. The mercury in the tube is trying to push its way out under the influence of gravity, and the air pressure is pushing it back in. At some point the competing pressures reach a stalemate. Mercury is a very heavy (dense) liquid. On the average, at sea level, a column of mercury 760 mm high balances the push of a column of air many kilometers high. The average atmospheric pressure at sea level is thus said to be 760 mm Hg (**millimeters of mercury**). This pressure is called 1 **atmosphere** (atm) or standard pressure. The unit

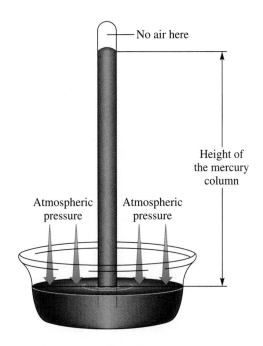

Figure 7.4
A mercury barometer.

millimeter of mercury is also called a **torr** (after the Italian physicist Evangelista Torricelli, the inventor of the mercury barometer).

If water were the fluid used in the barometer, the height of the water column required to balance 1 atm of pressure would be almost 10 m (as high as a three-story building). Respiratory therapists occasionally use a unit of pressure that does assume that water is the liquid in the barometer. The unit is **centimeters of water,** and it is particularly useful for the measurement of relatively small differences in pressure. Two other widely used units of pressure are **inches of mercury** (used in weather reports) and **pounds per square inch** (psi, used by engineers).

The SI unit for pressure is the **pascal** (Pa), a unit that is interpreted in terms of newtons (N). (A newton is the force that, when applied to a mass of 1 kg, will give that mass an acceleration of 1 m/sec^2.)

7.4
BOYLE'S LAW: PRESSURE– VOLUME RELATIONSHIP

The relationship between the volume and the pressure of a gas was first determined by Robert Boyle (Figure 7.5) in 1662. This relationship, called **Boyle's law,** states that *for a given mass of gas at constant temperature, the volume varies inversely with the pressure.* When the pressure goes up, the volume goes down; when the pressure goes down, the volume goes up (Figure 7.6). Think of gases as pictured in the kinetic-molecular theory. A sample of gas in a container exerts a certain pressure because the particles are bouncing against the walls at a certain rate with a certain force. If the volume of the container is expanded, the particles will have to travel longer distances before they strike the walls. Also, the surface area of the walls increases as the volume increases, so each unit of area is struck by fewer particles. The rate of strikes goes down, and the pressure goes down.

Figure 7.5
Robert Boyle.

Figure 7.6

The pressure–volume relationship is like a seesaw. (a) When the pressure goes down, the volume goes up. (b) When the pressure goes up, the volume goes down. (From an idea by Cindy Hill)

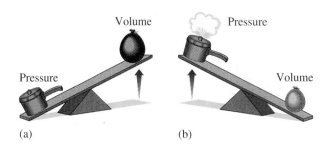

(a)   (b)

Mathematically, Boyle's law is written

$$V \propto \frac{1}{P} \quad \text{(with constant temperature and mass)}$$

where the symbol $\propto$ means "is proportional to." The inverse proportion can be changed to an equation by including a *proportionality constant, k.*

$$V = \frac{k}{P} \quad \text{or} \quad VP = k$$

Volume is inversely proportional to pressure.

$$V \propto \frac{1}{P}$$

$\propto$ means "is proportional to."

This is an elegant and precise, if somewhat abstract, way of summarizing a lot of experimental data. If the product of V times P is to be a constant, then if V goes up, P must come down. A typical Boyle's law experiment is diagrammed in Figure 7.7. Gas is enclosed in a cylinder fitted with a movable piston. At first, the gas occupies a volume of 1 L under 1 atm of pressure. When the pressure is increased to 2 atm, the volume is reduced to 0.5 L, and at 4 atm of pressure the volume is 0.25 L.

As long as we are using the *same* sample of trapped gas at a constant temperature, the product of the initial volume (V_1) times the initial pressure (P_1) is equal to the product of the final volume (V_2) times the final pressure (P_2). Thus, the following useful equation representing Boyle's law can be written.

$$V_1 P_1 = V_2 P_2$$

Boyle's law has a number of practical applications. These are perhaps best illustrated by some examples. Note that in applications of Boyle's law, any units for pressure or volume can be used as long as the use is consistent.

Figure 7.7

A diagram illustrating the effect of different pressures on the volume of a gas.

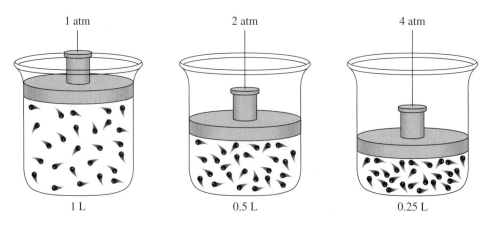

EXAMPLE 7.1

A cylinder of oxygen has a volume of 2.00 L. The pressure of the gas is 1470 psi at 20 °C. What volume will the oxygen occupy at standard atmospheric pressure (14.7 psi) assuming no temperature change?

SOLUTION

It is most helpful to first separate the initial from the final condition

Initial	Final	Change
$P_1 = 1470$ psi	$P_2 = 14.7$ psi	↓
$V_1 = 2.00$ L	$V_2 = ?$	↑

Then use the equation $V_1 P_1 = V_2 P_2$ and solve for the desired volume or pressure. In this case, we solve for V_2.

$$V_2 = \frac{V_1 P_1}{P_2}$$

$$V_2 = \frac{(2.00 \text{ L})(1470 \text{ psi})}{14.7 \text{ psi}} = 200 \text{ L}$$

This answer is reasonable because we expected the final volume to increase. Always make sure that your answer conforms to the change in volume or pressure that you expected. If the answer does not conform, you either made an arithmetic error or you transposed your numbers.

Gases are stored for use (in hospitals, for example) under high pressure, even though they will be used at atmospheric pressure. This arrangement allows much gas to be stored in a small area.

EXAMPLE 7.2

A space capsule is equipped with a tank of air that has a volume of 0.10 m^3. The air is under a pressure of 110 atm. After a space walk, during which the cabin pressure drops to zero, the cabin is closed and filled with the air from the tank. What will be the final pressure if the volume of the capsule is 11 m^3?

SOLUTION

Initial	Final	Change
$V_1 = 0.10 \text{ m}^3$	$V_2 = 11 \text{ m}^3$	↑
$P_1 = 110$ atm	$P_2 = ?$	↓

In this case we solve for P_2.

$$P_2 = \frac{V_1 P_1}{V_2}$$

$$P_2 = \frac{(0.10 \text{ m}^3)(110 \text{ atm})}{11 \text{ m}^3} = 1.0 \text{ atm}$$

As expected, the final pressure is less than the initial pressure.

> **Practice Exercise**
> A sample of nitrogen gas occupies 80.0 mL at 1.00 atm pressure. At what pressure will the nitrogen gas occupy 60.0 mL, assuming the temperature is constant?

EXAMPLE 7.3

A weather balloon is partially filled with helium gas. On the ground, where the atmospheric pressure is 740 mm Hg, the volume of the balloon is 10 m³. What will the volume be when the balloon reaches an altitude where the pressure is 370 mm Hg? Assume that the temperature is constant.

SOLUTION

Initial	Final	Change
$P_1 = 740$ mm Hg	$P_2 = 370$ mm Hg	↓
$V_1 = 10$ m³	$V_2 = ?$	↑

Since the pressure decreases, the volume increases.

$$V_2 = \frac{V_1 P_1}{P_2}$$

$$V_2 = \frac{(10 \text{ m}^3)(740 \text{ mm Hg})}{370 \text{ mm Hg}} = 20 \text{ m}^3$$

Practice Exercise
A sample of oxygen gas occupies a volume of 4.00 L at a pressure of 720 mm Hg. What will be its volume at 760 mm Hg, assuming the temperature is constant?

The pressure–volume relationship can be used to explain the mechanics of breathing. When we breathe in (inspire), the diaphragm is lowered and the chest wall is expanded (Figure 7.8). This increases the volume of the chest cavity. According to Boyle's law, the pressure inside the cavity must decrease momentarily. Outside air then enters the lungs because it is at a higher pressure than the air in the chest cavity. When we breathe out (expire), the diaphragm rises and the chest wall contracts. This decreases the volume of the chest cavity. The pressure is increased, and some of the air is forced out.

During normal inspiration the pressure inside the lungs drops about 3 mm Hg below atmospheric pressure. During expiration the internal pressure is about 3 mm Hg above atmospheric pressure. About half a liter of air is moved in and out of the lungs in this process, and this normal breathing volume is referred to as the **tidal volume.** The **vital capacity** is the maximum amount of air that can be forced from the lungs and ranges in volume from 3 to 7 L, depending on the individual. A pressure inside the lungs 100 mm Hg greater than the external pressure is not unusual during such a maximum expiration.

The lungs are never emptied completely, however. The space around the lungs is maintained at a slightly lower pressure than are the lungs themselves, causing the lungs to be kept partially inflated by the higher pressure within them. If a lung, the diaphragm, or the chest wall is punctured, allowing the two pressures to equalize, the lung will collapse.

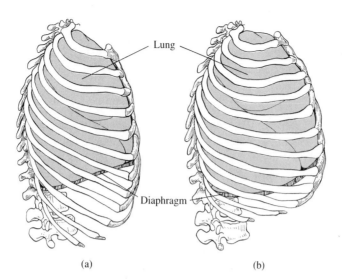

(a)   (b)

Figure 7.8
The mechanics of breathing. (a) Expiration. The diaphragm is relaxed and the rib cage is down. (b) Inspiration. The diaphragm is pulled down and the rib cage is lifted up and out, increasing the volume of the chest cavity.

Sometimes a lung will be collapsed intentionally to give it time to heal. Closing the opening reinflates the lung.

People were breathing long before Boyle formulated his law, but it is satisfying to understand how the process works. We get more than just satisfaction out of science, though. An understanding of the pressure–volume relationship has enabled us to keep people alive. When paralysis prevents people from being able to breathe, they can be kept alive by artificial respirators. The iron lung, which kept many polio victims alive during the 1950s, is a sealed chamber connected to a compressor and bellows (Figure 7.9). The pressure in the chamber is varied rhythmically. When the bellows is moved out, the pressure in the chamber is reduced. The pressure around the nose and mouth (outside the tank) is greater than the pressure on the chest (inside the tank), so air flows in and fills the lungs. When the bellows is moved in, pressure in the tank increases, and air is expelled from the lungs.

The iron lung, designed to enclose the patient completely (except for the head), is cumbersome and uncomfortable. It has been replaced by respirators that enclose the chest only. In fact, the whole area of respiratory therapy has become far more sophisticated in recent years. Specialists in this area are an indispensable part of the medical team.

Figure 7.9
An iron lung uses changes in pressure to force air into and out of the lungs.
(a) The older version enclosed the entire body except for the head. (b) A modern iron lung encloses the chest only.

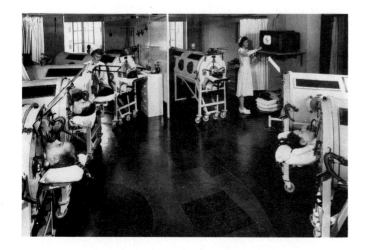

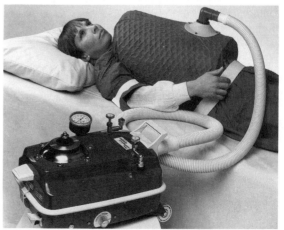

7.5
CHARLES'S LAW: TEMPERATURE–VOLUME RELATIONSHIP

In 1787, the French physicist J. A. C. Charles studied the relationship between volume and temperature of gases. When a gas is cooled at constant pressure, its volume decreases. When the gas is heated, its volume increases. Temperature and volume vary directly, that is, they rise or fall together. But this relationship requires a bit more thought. If 1 L of gas is heated from 100 °C to 200 °C at constant pressure, the volume does not double but only increases to about 1.3 L. There does not seem to be a direct proportionality.

Remember how temperature scales were defined? While zero pressure or zero volume really means zero—that is, there is zero (or no) pressure or volume to be measured, no matter what units are used—zero degrees Celsius means only the freezing point of water (0 °C). The zero point was set arbitrarily.

Charles noted a trend in the variation of volume with temperature. If you plot the change in temperature against the change in volume on a graph (Figure 7.10), you can extend the line, in theory, to the point at which the volume of the gas hits zero. Before a gas ever reaches this point, it liquefies, so this is an exercise for the imagination. From the graph you can determine the temperature at which the volume of the imaginary gas would reach zero. That temperature is −273 °C. This value was made the zero point on the absolute temperature scale, whose unit is the kelvin (K) (Section 1.9).

Charles compared changes in the volume of a gas with changes in temperature on the absolute scale. When he did this, he found that a doubling of the temperature (from 100 K to 200 K, for example) resulted in a doubling in the volume (from 1 L to 2 L, for example). Here is the simple relationship we were looking for.

Charles's law states that *the volume of a gas is directly proportional to its temperature (on the absolute scale) if the pressure is held constant.* Mathematically this relationship is expressed as

$$V \propto T$$

In the form of an equation, this becomes

$$V = kT \quad \text{or} \quad \frac{V}{T} = k \quad \text{(with constant pressure and mass)}$$

Temperature is directly proportional to volume: $T \propto V$.

Figure 7.10

The volume–temperature relationship for gases (with pressure constant).

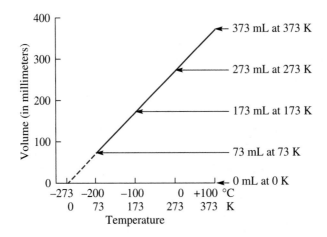

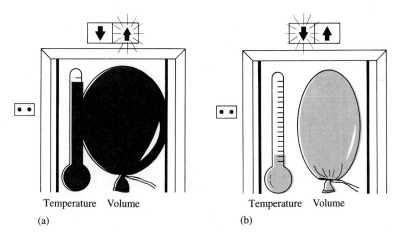

Temperature   Volume
(a)

Temperature   Volume
(b)

Figure 7.11
Temperature and volume are like passengers on an elevator. (a) When one goes up, the other must go up as well. (b) When one goes down, the other goes down too. (From an idea by Cindy Hill)

To keep V/T equal to a constant value, when the volume increases, the temperature must also increase. When volume decreases, temperature must decrease (Figure 7.11). Again let us emphasize that the values for temperature must be in kelvins, not degrees Celsius (or Fahrenheit). In the case of temperature (unlike that of volume or pressure), the units of measurement for gas law calculations are restricted.

As long as we are using the *same* sample of trapped gas at a constant pressure, the initial volume (V_1) divided by the initial *absolute* temperature (T_1) is equal to the final volume (V_2) divided by the final *absolute* temperature (T_2). The following equation is useful in solving Charles's law problems.

$$\frac{V_1}{T_1} = \frac{V_2}{T_2}$$

The kinetic-molecular model accounts for this relationship. If we heat a gas, we supply it with energy, and the particles of the gas begin moving faster. These speedier particles strike the walls of their container harder and more often. If the pressure is to remain constant, the volume of the container must increase. Increased volume means that the particles take longer to travel from one wall to another, and the increased wall area means that each unit of area will be hit less often. The pressure exerted by more slowly moving (low-temperature) particles confined within the smaller volume will be the same as the pressure of the faster moving (high-temperature) particles contained within the larger volume.

For all gas law calculations involving temperature, absolute temperatures in kelvins must be used (not °C or °F).

EXAMPLE 7.4

A balloon indoors, where the temperature is 27 °C, has a volume of 2.0 L. What will its volume be outside where the temperature is −23 °C? (Assume no change in pressure.)

SOLUTION
First, convert all temperatures to the absolute scale. The initial temperature is

27 + 273 = 300 K

The final temperature is

−23 + 273 = 250 K

Initial	Final	Change
$t_1 = 27\ °C$	$t_2 = -23\ °C$	↓
$T_1 = 300\ K$	$T_2 = 250\ K$	↓
$V_1 = 2.0\ L$	$V_2 = ?$	↓

Solving the equation

$$\frac{V_1}{T_1} = \frac{V_2}{T_2}$$

for V_2, we have

$$V_2 = \frac{V_1 T_2}{T_1}$$

$$V_2 = \frac{(2.0\ L)(250\ \cancel{K})}{300\ \cancel{K}} = 1.7\ L$$

As we expected, the volume decreased because the temperature decreased.

Practice Exercise
A sample of oxygen gas occupies a volume of 2.10 L at 25 °C. What volume will this sample occupy at 150 °C? (Assume no change in pressure.)

EXAMPLE 7.5

What would be the final volume of the balloon in Example 7.4 if it were measured where the temperature was 47 °C? (Assume no change in pressure.)

SOLUTION
The initial temperature is

$$27 + 273 = 300\ K$$

The final temperature is

$$47 + 273 = 320\ K$$

Initial	Final	Change
$t_1 = 27\ °C$	$t_2 = 47\ °C$	↑
$T_1 = 300\ K$	$T_2 = 320\ K$	↑
$V_1 = 2.0\ L$	$V_2 = ?$	↑

In this case, since the temperature increases, the volume must also increase.

$$V_2 = \frac{V_1 T_2}{T_1}$$

$$V_2 = \frac{(2.0\ L)(320\ \cancel{K})}{300\ \cancel{K}} = 2.1\ L$$

Practice Exercise
At what Celsius temperature will the initial volume of oxygen in the preceding practice exercise occupy 0.750 L? (Assume no change in pressure.)

7.6
PRESSURE–TEMPERATURE RELATIONSHIP

At about the same time that Charles was doing the experiments on which his law is based, other experiments showed that *at constant volume, pressure is directly proportional to absolute temperature*. Mathematically, this is expressed as

$$P \propto T$$

In the form of an equation, this becomes

$$P = kT \quad \text{or} \quad \frac{P}{T} = k \quad \text{(with constant volume and mass)}$$

or

$$\frac{P_1}{T_1} = \frac{P_2}{T_2}$$

At constant volume, as the temperature increases, so does the pressure. At the higher temperature, the particles move faster and hit the walls of their container harder and more often, and thus the pressure increases (Figure 7.12).

Pressure is directly proportional to temperature: $P \propto T$.

Figure 7.12
A hospital autoclave uses increased pressure to achieve the high temperatures needed to sterilize surgical instruments. The trend today is toward disposable equipment that does not require resterilization.

EXAMPLE 7.6

Automobile tires are filled to a pressure of 30 psi at 20 °C. The tires become hot from high-speed driving and reach a temperature of 50 °C. What will the pressure be at that temperature? (Assume that the volume remains constant.)

SOLUTION
The initial temperature is

$$20 + 273 = 293 \text{ K}$$

The final temperature is

$$50 + 273 = 323 \text{ K}$$

Initial	Final	Change
$t_1 = 20$ °C	$t_2 = 50$ °C	↑
$T_1 = 293$ K	$T_2 = 323$ K	↑
$P_1 = 30$ psi	$P_2 = ?$	↑

The temperature increases, so the pressure must also increase.

$$P_2 = \frac{P_1 T_2}{T_1}$$

$$P_2 = \frac{(30 \text{ psi})(323 \text{ K})}{293 \text{ K}} = 33 \text{ psi}$$

EXAMPLE 7.7

At 127 °C the pressure in an autoclave is 2.0 atm. What will the pressure be at 27 °C?

SOLUTION
The initial temperature is

$$127 + 273 = 400 \text{ K}$$

The final temperature is

$$27 + 273 = 300 \text{ K}$$

Initial	Final	Change
$t_1 = 127$ °C	$t_2 = 27$ °C	↓
$T_1 = 400$ K	$T_2 = 300$ K	↓
$P_1 = 2.0$ atm	$P_2 = ?$	↓

The temperature decreases; therefore, the pressure must also decrease.

$$P_2 = \frac{(2.0 \text{ atm})(300 \text{ K})}{(400 \text{ K})} = 1.5 \text{ atm}$$

Practice Exercise
A sample of argon gas at 25 °C exerts a pressure of 1.00 atm. What will the pressure be at 100 °C? (Assume that the volume remains constant.)

7.7
THE COMBINED GAS LAW

We have seen that the volume of a gas varies with both temperature and pressure. If we want to compare two samples, we have to do so at identical temperatures and pressures for the comparison to be meaningful. We can use a combination of Boyle's law and Charles's law to calculate the volume of a gas at any given temperature and pressure provided we know its volume at any other temperature and pressure.

The equation for the **combined gas law** is

$$\frac{V_1 P_1}{T_1} = \frac{V_2 P_2}{T_2}$$

EXAMPLE 7.8

A balloon is partially filled with helium on the ground at 27 °C and 740 mm Hg pressure. Its volume is 10 m³. What would the volume be at a higher altitude where the pressure is 370 mm Hg and the temperature is −23 °C?

SOLUTION
The temperature decreases from 27 °C to −23 °C, or from

$$27 + 273 = 300 \text{ K}$$

to

$$-23 + 273 = 250 \text{ K}$$

Initial	Final	Change
$t_1 = 27$ °C	$t_2 = -23$ °C	↓
$T_1 = 300$ K	$T_2 = 250$ K	↓
$P_1 = 740$ mm Hg	$P_2 = 370$ mm Hg	↓
$V_1 = 10$ m³	$V_2 = ?$	

Solving the combined gas law equation for V_2, we have

$$V_2 = \frac{V_1 P_1 T_2}{P_2 T_1}$$

Helpful hint: Regardless of which variable is to be determined, it is recommended that you solve the equation for the unknown *before* substituting values into the equation. Rearranging a few letters takes less time than rearranging and rewriting complex terms.

Substituting the given values (and using a calculator)

$$V_2 = \frac{(10 \text{ m}^3)(740 \text{ mm Hg})(250 \text{ K})}{(370 \text{ mm Hg})(300 \text{ K})} = 17 \text{ m}^3$$

Practice Exercise
A sample of helium occupies 38.4 mL at 40 °C and 680 mm Hg. What volume will the helium occupy at 80 °C and 720 mm Hg?

Standard condition of temperature and pressure (STP)

$$T = 273 \text{ K}$$
$$P = 1 \text{ atm}$$

Because gases are so sensitive to changes in temperature and pressure, chemists have found it convenient to define *s*tandard conditions of *t*emperature and *p*ressure (referred to as **STP**) as 0 °C (273 K) and 1 atm (760 mm Hg).

EXAMPLE 7.9

What is the volume at STP of a sample of carbon dioxide whose volume at 25 °C and 4.0 atm is 10 L?

SOLUTION

Initial	Final	Change
$t_1 = 25\ °C$		
$T_1 = 298\ K$	$T_2 = 273\ K$	↓
$P_1 = 4.0\ atm$	$P_2 = 1.0\ atm$	↓
$V_1 = 10\ L$	$V_2 = ?$	

$$V_2 = \frac{V_1 P_1 T_2}{P_2 T_1}$$

$$V_2 = \frac{(10\ L)(4.0\ \text{atm})(273\ \text{K})}{(1.0\ \text{atm})(298\ \text{K})} = 37\ L$$

Practice Exercise
A sample of hydrogen gas has a volume of 1.10 L at −40 °C and 0.520 atm. What will be its volume at STP?

Densities of gases usually are reported in the literature in grams per liter at STP. Recall (Section 5.3) that there are 6.02×10^{23} molecules in 1 mol of any gas. If we divide the molecular weight of the gas (in grams per mole) by the density at STP (in grams per liter), we will get the volume occupied by a mole of gas at STP. For example, the density of nitrogen gas (N_2) at STP is 1.25 g/L. The molecular weight of nitrogen gas is 28.0 g/mol. Dividing, we get

$$\frac{28.0\ \text{g/mol}}{1.25\ \text{g/L}} = 22.4\ \text{L/mol}$$

For oxygen (O_2) the density at STP is 1.43 g/L. The molecular weight is 32.0 g/mol. Dividing, we get

$$\frac{32.0\ \text{g/mol}}{1.43\ \text{g/L}} = 22.4\ \text{L/mol}$$

In fact, the volume occupied by a mole of most gases at STP is quite close to 22.4 L. This quantity is known as the **molar volume** of a gas. A box for a basketball has a volume of about 22.4 L. Such a box would hold 28 g of N_2 or 32 g of O_2 (Figure 7.13). Thus, Gay-Lussac's law of combining volumes (Section 5.2) is explained: No matter what the gas, a mole of it occupies about the same volume as a mole of any other gas under the same conditions.

EXAMPLE 7.10

What is the molar volume of carbon dioxide, which has a density of 1.98 g/L at STP?

SOLUTION
The molecular weight of CO_2 is 44.0 g/mol. Dividing, we get

$$\frac{44.0 \text{ g/mol}}{1.98 \text{ g/L}} = 22.2 \text{ L/mol}$$

Practice Exercise
What is the molar volume of hydrogen gas, which has a density of 0.090 g/L at STP?

7.8
THE IDEAL GAS LAW

So far we have done calculations in which the quantity of a gas does not change. As we saw above (in our discussion of molar volume), equal volumes of gases at the same temperature and pressure contain equal numbers of moles. Thus, we can write a gas law that takes into account varying quantities of gas. This relationship is called the **ideal gas equation.**

$$PV = nRT$$

In this equation, the pressure is in atmospheres, the volume in liters, and the temperature in kelvins. The number of moles of the gas is given by n. The constant R, which has a value of

$$0.082 \, \frac{\text{L} \cdot \text{atm}}{\text{mol} \cdot \text{K}}$$

is called the **universal gas constant.**

This R value is read as 0.082 liter-atmosphere per mole-kelvin.

The ideal gas law can be used to calculate any of the four quantities—P, V, n, or T—if the other three are known.

EXAMPLE 7.11

Use the ideal gas law to calculate the volume occupied by 1.0 mol of nitrogen gas at 244 K and 1.0 atm pressure.

SOLUTION

$$V = \frac{nRT}{P} = \frac{1.0 \text{ mol}}{1.0 \text{ atm}} \times \frac{0.082 \text{ L} \cdot \text{atm}}{\text{mol} \cdot \text{K}} \times 244 \text{ K} = 20 \text{ L}$$

Practice Exercise

Determine the volume occupied by 0.20 mol of nitrogen gas at 25 °C and 0.98 atm.

EXAMPLE 7.12

Use the ideal gas law to calculate the pressure exerted by 0.50 mol of oxygen in a 15-L container at 303 K.

SOLUTION

$$P = \frac{nRT}{V} = \frac{0.50 \text{ mol}}{15 \text{ L}} \times \frac{0.082 \text{ L} \cdot \text{atm}}{\text{mol} \cdot \text{K}} \times 303 \text{ K} = 0.83 \text{ atm}$$

Practice Exercise

Determine the pressure exerted by 0.033 mol of oxygen in an 18-L container at 40 °C.

7.9
HENRY'S LAW: PRESSURE–SOLUBILITY RELATIONSHIP

$PV = nRT$

Henry's law: Solubility of a gas is directly proportional to the pressure of the gas at the surface of the liquid.

In the 1760s, Joseph Priestley invented soda water by dissolving carbon dioxide gas in water. No doubt you have noticed the hissing sound and the formation of bubbles when you opened a bottle of soda. Carbon dioxide is dissolved in the liquid, and the bottle is capped under pressure. William Henry, a close friend of John Dalton, spent a great deal of time studying the solubility of gases in liquids. In 1801 he summarized his findings in the law we know as **Henry's law.** The solubility of a gas in a liquid at a given temperature, he found, is directly proportional to the pressure of the gas at the surface of the liquid. To get back to the bottle of soda: when the bottle was capped under pressure, a certain amount of carbon dioxide was dissolved in the soda. When you opened the bottle, the pressure was *reduced* (the hissing sound you heard was pressure being released), and the solubility of the carbon dioxide was *reduced* (the bubbles of gas you noticed were carbon dioxide escaping from solution).

The carbonated beverage industry is not the only group interested in Henry's law. Deep-sea divers can develop a condition called the bends (caisson disease) from nitrogen dissolved in the blood. Very little nitrogen dissolves in our blood at normal pressures. When a diver breathes air under greater pressure, appreciable amounts dissolve in the blood. If the diver comes up too rapidly, the nitrogen comes out of solution as the pressure diminishes. Tiny bubbles of nitrogen form. These cause severe pain in the arms, legs, and joints, perhaps by disrupting nerve pathways. To prevent the bends, divers sometimes use a mixture of helium and oxygen rather than air. Very little of the helium dissolves, even under increased pressure, and the problem of bubble formation as the diver ascends is minimized.

The pressure–solubility relationship is also used in therapy. In cases of carbon monoxide poisoning (see Section 28.9), the victim is placed in a hyperbaric (high-pressure) chamber. This chamber is a device that supplies oxygen at pressures of 3 or 4 atm. More oxygen is forced into the tissues at these pressures to compensate for the lack of oxygen that accompanies carbon monoxide poisoning.

Hyperbaric chambers are also used to treat infections by anaerobic bacteria (bacteria that live without oxygen). Gangrene is one such disease. The organisms that cause gangrene cannot survive in an oxygenated atmosphere. If sufficient oxygen can be forced into the diseased tissues, the infection can be arrested.

7.10
DALTON'S LAW OF PARTIAL PRESSURES

Although John Dalton is most renowned for his atomic theory (Section 2.1), he had wide-ranging interests, including meteorology. In trying to understand the weather, he did a number of experiments on water vapor in the air. In one experiment, he found that if he added water vapor at a certain pressure to dry air, the pressure exerted by the air would increase by an amount equal to the pressure of the water vapor. Based on this and other experiments, Dalton concluded that each of the gases in a mixture behaves independently of the other gases. Each gas exerts its own pressure. The total pressure of the mixture is equal to the sum of the *partial pressures* exerted by the separate gases (Figure 7.14).

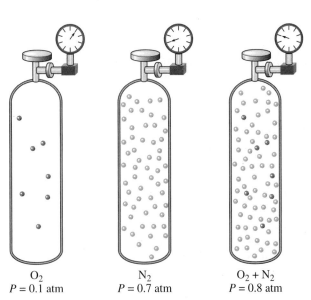

O_2
$P = 0.1$ atm

N_2
$P = 0.7$ atm

$O_2 + N_2$
$P = 0.8$ atm

Figure 7.14
Dalton's law of partial pressures states that the pressure of a mixture of gases is equal to the sum of the pressures that each gas would exert by itself.

Table 7.2

Water Vapor Pressure at
Various Temperatures

Temperature (°C)	Water Vapor Pressure (mm Hg)
0	5
10	9
20	18
30	32
40	55
50	93
60	149
70	234
80	355
90	526
100	760

Mathematically, **Dalton's law of partial pressures** is stated as

$$P_{\text{total}} = P_1 + P_2 + P_3 + \cdots$$

where the terms on the right-hand side refer to the partial pressures of gases 1, 2, 3, and so on.

Gases such as oxygen, nitrogen, and hydrogen are nonpolar. They are only slightly soluble in water and are usually collected over water by the technique of displacement. Such gases always contain water vapor, and the total pressure in the collection vessel is that of the gas plus that of water vapor.

The vapor pressure of water depends on the temperature of the water. (The **vapor pressure** of a substance is the partial pressure exerted by the molecules of the substance that are in the gas phase above the liquid phase of the substance.) The hotter the water, the higher the vapor pressure. If a gas is collected over water, we can make use of vapor pressure tables (Table 7.2) to calculate the pressure due to the gas alone. We need only subtract the vapor pressure of the water, as determined from the table, from the value for the total pressure within the collection vessel.

EXAMPLE 7.13

Oxygen is collected over water at 20 °C. The pressure inside the jar is 740 mm Hg. What is the pressure due to oxygen alone?

SOLUTION

$$P_{\text{total}} = P_{O_2} + P_{H_2O}$$

From Table 7.2 we find that the vapor pressure of water at 20 °C is 18 mm Hg. Since the total pressure is equal to 740 mm Hg, we have

$$740 \text{ mm Hg} = P_{O_2} + 18 \text{ mm Hg}$$
$$P_{O_2} = 740 - 18 = 722 \text{ mm Hg}$$

Humidity is a measure of the amount of water vapor in the air. **Relative humidity** compares the actual amount of water vapor in the air with the maximum amount the air could hold at the same temperature. If the temperature is 20 °C and the vapor pressure of water in the atmosphere is 12 mm Hg, the relative humidity is

$$\frac{12 \text{ mm Hg}}{18 \text{ mm Hg}} \times 100 = 67\%$$

The 18 mm Hg in the denominator was obtained from Table 7.2. When the relative humidity is 100%, the air is saturated with water vapor. (Note that 100% relative humidity does not mean that the air is 100% water vapor, just that it is holding as much water as it can. At 20 °C and 100% relative humidity, only about two or three molecules in every 100 molecules of air are water.)

Cool air can hold less water vapor than warm air. As the temperature falls during the night, the atmosphere may become saturated. Water vapor condenses from the air as dew.

The **heat index** relates temperature to relative humidity. Higher humidity inhibits

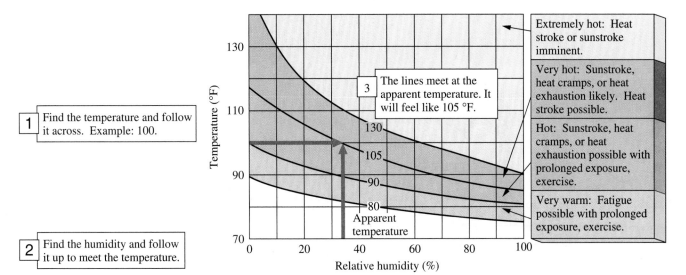

Figure 7.15
The heat index.

evaporation of sweat, which normally cools your body, thus making you now feel hotter. This apparent temperature can be found from Figure 7.15.

Respiratory therapists must concern themselves with the humidity of the gases they administer to patients. Normally, as we breathe, the inspired air is saturated with moisture as it passes through the nose and respiratory passages. Oxygen as it comes from a tank is quite dry. If oxygen is being administered over a long period of time, it must be humidified to prevent it from irritating the mucous linings of the nasal passages and the lungs. If the oxygen or mixture of gases is conducted through the nose, the therapist may assist the normal body processes by imparting about 30% humidity to the inspired gases. If the breathing mixture is conducted directly to the trachea (bypassing the nose), the therapist saturates the gas mixture with water vapor.

7.11
PARTIAL PRESSURES AND RESPIRATION

When we breathe in, the inspired air becomes moistened and warmed to our body temperature of 37 °C. The air is drawn into our lungs, where it enters a highly branched system of tubes that end in tiny air sacs called alveoli (Figure 7.16). These thin-walled pouches are surrounded by blood vessels that are part of a circulatory system serving every cell in the body.

Inspired air is rich in oxygen (P_{O_2} = 150 mm Hg) and poor in carbon dioxide (P_{CO_2} = 0.2 mm Hg). The fluid in our cells is poor in oxygen (P_{O_2} = 6 mm Hg) and rich in carbon dioxide (P_{CO_2} = 50 mm Hg). Our cells use up oxygen in metabolic reactions designed to produce energy. Carbon dioxide accumulates in the cells as a waste product of these reactions. To maintain life, we must transfer the oxygen in the inspired air to our cells. At the same time, the carbon dioxide waste in our cells must be transferred to our lungs and then exhaled to the atmosphere. The transfer of both gases occurs through the process of **diffusion.** In diffusion, gases flow from regions of higher concentration to regions of lower concentration (Figure 7.17). In our bodies, oxygen makes its way to the cells through a pressure gradient, that is, in a series of steps in which oxygen diffuses from

Diffusion—the flow of gas from a region of higher concentration to a region of lower concentration

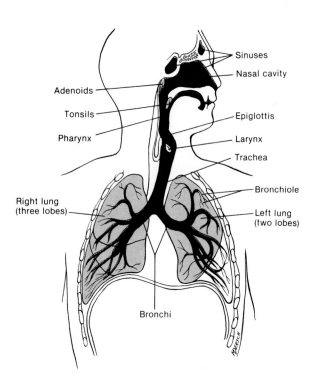

Figure 7.16

The respiratory system, showing the route of air through the nose, pharynx (throat), larynx (voice box), and trachea (windpipe) into the bronchi and bronchioles (bronchial tubes) and ending in the alveoli (air sacs).

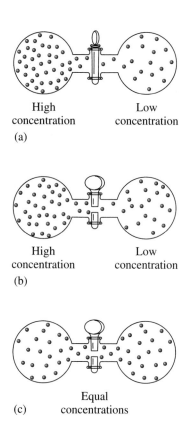

Figure 7.17

A gas (air in the lungs, for example) flows from an area of high concentration to an area of low concentration. (a) With the stopcock closed, no flow is possible. (b) With the stopcock open, there is a net flow of gases from the area of higher concentration on the left to the area of lower concentration on the right. (c) After some time, there is no net flow of gas because concentrations are equal.

areas where its concentration is higher into areas where its concentration is lower. By the same method, carbon dioxide moves in the opposite direction. It makes its way from the cells, where its partial pressure is high, to the atmosphere, where its partial pressure is low. Figure 7.18 shows the steps in the gradient for both gases. Thus, given the mechanical action of the chest and diaphragm (Section 7.4) to get air into and out of the lungs, respiration is all downhill as far as the gases are concerned.

Normally, the carbon dioxide level in the blood (not the oxygen level) acts as a trigger for the breathing process. To oversimplify, when carbon dioxide levels build up, we take a breath; if they get too low, we don't. Thus, one of the concerns of a therapist is the partial pressure of carbon dioxide in the blood. Under certain unusual conditions, it is possible for the level of carbon dioxide to fall so low that it fails to trigger the breathing

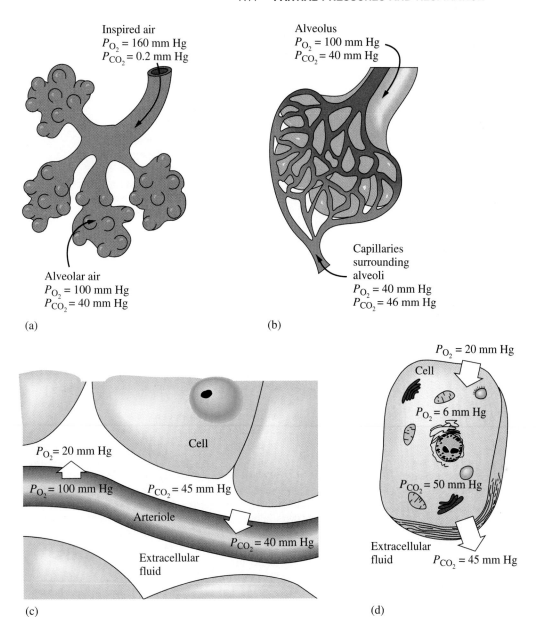

Figure 7.18

(a) Oxygen flows from the inspired air into the alveolar air. Carbon dioxide flows in the opposite direction. (b) Oxygen diffuses from the alveolus to the capillary. Carbon dioxide moves in the opposite direction. (c) Oxygen diffuses from an arteriole into the extracellular fluid. Carbon dioxide flows in the opposite direction. (d) Oxygen diffuses into a cell from the extracellular fluid. Carbon dioxide moves in the opposite direction. In each case the flow is from a region of high partial pressure to a region of low partial pressure.

mechanism. Even with access to a plentiful supply of oxygen, the person suffocates because he or she simply stops breathing.

Several commercially supplied gases are used in medicine, the majority in respiratory therapy. Table 7.3 lists these gases and some of their uses.

Table 7.3 Ten Compressed Gases Used in Medicine	Gas	Chemical Formula	Use
	Air	N_2 and O_2 (mixture)	Life support
	Carbon dioxide	CO_2	Laboratory tests, lung function tests
	Carbon dioxide–oxygen mixtures	CO_2 and O_2	Diagnosis, inhalation therapy
	Cyclopropane	C_3H_6	Anesthetic
	Helium	He	Laboratory analyses
	Helium–oxygen mixtures	He and O_2	Inhalation therapy, diagnostic tests
	Nitrogen	N_2	Diagnostic testing, inhalation therapy
	Nitrous oxide	N_2O	Anesthetic
	Oxygen	O_2	Life support, medical emergencies, adjunct to anesthetics
	Oxygen–nitrogen mixture	O_2 and N_2	Treatment of obstructive lung diseases

EXERCISES

1. Define or explain the following terms.
   a. barometer
   b. mm Hg
   c. Boyle's law
   d. vital capacity
   e. Charles's law
   f. Dalton's law
   g. combined gas law
   h. STP
   i. molar volume
   j. ideal gas law
   k. Henry's law
   l. vapor pressure
   m. relative humidity
   n. diffusion

2. List the three major gases found in dry air. Which of these are important in respiration?

3. Why is atmospheric pressure greater at sea level than on the top of a high mountain?

4. Explain how a mercury barometer works.

5. According to the kinetic-molecular theory, what change in temperature is occurring if the particles of a gas begin to move more slowly, on the average?

6. According to the kinetic-molecular theory, what change in pressure occurs when the walls of the container are struck less often by particles of a gas?

7. Container A has twice the volume but holds twice as many gas particles as container B. Using the kinetic-molecular theory as the basis for your judgment, compare the pressures in the two containers.

8. What effect will the following changes have on the volume of a gas?
   a. an increase in pressure
   b. a decrease in temperature
   c. a decrease in pressure coupled with an increase in temperature

9. What effect will the following changes have on the pressure of a gas?
   a. an increase in temperature
   b. a decrease in volume
   c. an increase in temperature coupled with a decrease in volume

10. Density is defined as mass per unit volume. For each of the following, indicate which sample you would expect to exhibit the greater density. (*Hint:* How many particles are there in equivalent volumes of the sample?)
   a. Containers A and B have the same volume and are at the same temperature, but the gas in container A is at a higher pressure.
   b. Containers A and B are at the same pressure and temperature, but the volume of container A is greater than that of container B.
   c. Containers A and B are at the same pressure and volume, but the temperature of the gas in container A is higher. (*Hint:* One can float in the air in a *hot* air balloon.)

11. Carry out the following conversions.
   a. 2.0 atm to torr   5/3   1520 ?
   b. 0.50 atm to mm Hg
   c. 0.0030 atm to cm Hg

12. Carry out the following conversions:
   a. 76 torr to atm   b. 320 torr to mm Hg
   c. 0.076 mm Hg to Pa

13. Carry out the following conversions:
   a. 10 atm to lb/in.^{2} b. 10 mm Hg to cm H_2O
   c. 3.0 in. Hg to atm

14. Carry out the following conversions:
    a. 37 °C to K   **b.** 420 K to °C
    c. −82 °C to K

15. A tank contains 500 mL of helium at 1500 mm Hg. What volume will the helium occupy at 750 mm Hg, assuming no temperature change?

16. A hyperbaric chamber with a volume of 10 m³ operates at an internal pressure of 4.0 atm. What volume would the air inside the chamber occupy at normal atmospheric pressure?

17. Oxygen used in respiratory therapy is stored under a pressure of 2200 lb/in.² in gas cylinders with a volume of 60 L.
    a. What volume would the gas occupy at normal atmospheric pressure?
    b. If oxygen flow to the patient is adjusted to 8.0 L per minute, how long will the tank of gas last?

18. The pressure within a balloon with a volume of 2.5 L is 1.0 atm. If the volume of the balloon increases to 7.5 L, what will be the final pressure within the balloon?

19. During inhalation, does the chest cavity expand or contract? Is the pressure inside the lungs decreased or increased? What happens during exhalation?

20. The cough reflex is designed to keep air passages clear. Typically, when a person coughs, he or she first inhales about 2.0 L of air. The epiglottis and the vocal cords then shut, trapping the air in the lungs. The air in the lungs is then compressed to a volume of about 1.7 L by the action of the diaphragm and chest muscles. The sudden opening of the epiglottis and vocal cords releases this air explosively. Just prior to the release, what is the pressure of the gas inside the lungs?

21. A gas at a temperature of 100 K occupies a volume of 100 mL. What will the volume be at a temperature of 10 K, assuming no change in pressure?

22. An automobile tire is inflated to 30 lb/in.² at 27 °C. What will be the pressure at 127 °C, assuming no volume change?

23. A balloon is filled with helium. Its volume is 5.0 L at 27 °C. What will be its volume at −73 °C, assuming no pressure change?

24. A gas at a temperature of 300 K and a pressure of 1.0 atm is cooled to 250 K. Assuming no change in volume, calculate the change in pressure.

25. A sealed can with an internal pressure of 720 mm Hg at 25 °C is thrown into an incinerator operating at 750 °C. What will the pressure inside the heated can be, assuming the container remains intact during incineration?

26. If a gas occupies 4.0 L at a temperature of 25 °C and a pressure of 2.0 atm, what volume will it occupy at a temperature of 200 °C and a pressure of 1.0 atm?

27. If a gas occupies 2.5 m³ at a temperature of −15 °C and a pressure of 190 mm Hg, what volume will it occupy at a temperature of 25 °C and 1140 mm Hg pressure?

28. What volume will 500 mL of a gas, measured at 27 °C and 720 mm Hg, occupy at STP?

29. If a gas has a volume of 55 cc at STP, what volume will it occupy at 100 °C and 76 mm Hg?

30. If a gas has a volume of 732 mL at 760 mm Hg and 25 °C, what volume will it occupy at 1.0 atm and 298 K?

31. A container holds oxygen at a partial pressure of 0.25 atm, nitrogen at a partial pressure of 0.50 atm, and helium at a partial pressure of 0.20 atm. What is the pressure inside the container?

32. A container is filled with equal numbers of nitrogen, oxygen, and carbon dioxide molecules. The total pressure in the container is 750 mm Hg. What is the partial pressure of nitrogen in the container?

33. Oxygen is collected over water. If the temperature is 30 °C and the collected sample has a pressure of 740 mm Hg, what is the partial pressure of the oxygen in the container? (You may refer to Table 7.2).

34. The pressure of the atmosphere on the surface of Venus is about 100 atm. Carbon dioxide makes up about 97% of the atmospheric gases. What is the partial pressure of carbon dioxide in the atmosphere of Venus?

35. Atmospheric pressure on the surface of Mars is about 6.0 mm Hg. The partial pressure of carbon dioxide is 5.7 mm Hg. What percent of the Martian atmosphere is carbon dioxide?

36. A sample of intestinal gas was collected and found on analysis to consist of 44% CO_2, 38% H_2, 17% N_2, 1.3% O_2, and 0.003% CH_4. (The percentages do not add to exactly 100% because of rounding.) What is the partial pressure of each gas if the total pressure in the intestine is 820 mm Hg?

37. What is the molar volume of methane (CH_4) gas, which has a density of 0.72 g/L at STP?

38. Calculate the approximate density of sulfur dioxide (SO_2) gas at STP.

39. Use the ideal gas law to calculate the volume occupied by 0.60 mol of a gas at a temperature of 310 K and a pressure of 0.80 atm.

40. What volume is occupied by 12 mol of hydrogen sulfide gas at a temperature of 620 K and a pressure of 18 atm?

41. What pressure is exerted by 0.010 mol of methane gas if its volume is 0.26 L at 373 K?

42. What pressure is exerted by 44 mol of carbon monoxide gas in a 36-L tank at 290 K?

43. Calculate the temperature of 0.78 mol of oxygen if its volume is 68 L and its pressure is 0.37 atm.

44. A hyperbaric chamber has a volume of 4200 L. How many moles of oxygen are needed to fill the chamber to a pressure of 3.0 atm at 290 K?

45. The interior volume of the Hubert H. Humphrey Metrodome in Minneapolis is 1.70×10^{10} L. The Teflon-coated fiberglass roof is supported by air pressure provided by 20 huge electric fans. How many moles of air are in the dome if the pressure is 1.02 atm and the temperature is 18 °C?

46. When air is inspired it becomes fully saturated with water vapor as it passes through the trachea on its way to the lungs. What is the approximate partial pressure of water vapor in the air in the alveoli? (*Hint:* At what temperature will the air be?)

47. If the P_{H_2O} in air is 12 mm Hg on a day when the temperature is 20 °C, what is the relative humidity?

48. Two flasks are connected. Flask A contains only oxygen at a pressure of 460 mm Hg. Flask B has oxygen at a partial pressure at 320 mm Hg and nitrogen at a partial pressure of 240 mm Hg.

   a. Which direction will the net flow of oxygen take?
   b. Which direction will the net flow of nitrogen take?

49. In which net direction, cells to lungs or lungs to cells, does oxygen move? What about carbon dioxide?

50. Why does oxygen flow from the alveoli to the pulmonary capillaries? Why does carbon dioxide flow in the reverse direction?

51. A person at rest breathes about 80 mL of air per second. How long does it take to breathe 22.4 L of air?

Chapter 8
LIQUIDS AND SOLIDS

Giant icebergs are frozen water. If these icebergs could be easily transported, there would be no water-shortage problems on Earth.

In Chapter 4, we considered the subject of bonding. Our primary concern then was how atoms combined to form molecules and why elements reacted to form compounds. Now we're going to expand our consideration of bonding, and this time we will be looking for an answer to this question: What makes a substance a solid rather than a liquid or a liquid rather than a gas? Some force of attraction holds the particles of solids and liquids in contact with one another. The particles of a gas fly about at random; those of liquids or solids cling together. Gas particles interact so little with one another that a collection of them retains neither a specific volume nor a specific shape. But particles of a liquid are held together with sufficient force that a collection of them has a specific volume. And particles of a solid are so rigidly held together that not only the volume but also the shape of a given sample is fixed. Are these mysterious forces of attraction important? They are if appearances are important to you. You are built of solids and liquids. If you think you've got it all together, you should thank the special forces of attraction that give shape and volume to the condensed forms of matter.

In this chapter, we're also going to consider changes in the state of matter—what happens when a solid is converted to a liquid or a liquid to a gas. Is that important? Well, consider perspiration. From the amount of advertising directed against this lowly liquid,

183

you would think it was an unnecessary annoyance. However, were it not for the conversion of this liquid to a vapor on skin surfaces, we would find it difficult just to survive in the temperate and tropical zones of our planet, let alone carry out vigorous physical activity.

8.1
STICKY MOLECULES

Gases are easily manipulated. Changes in pressure or temperature result in volume changes that are readily measured. Not so with liquids and solids. Whereas gas particles are rather far apart, allowing gases to be considerably compressed, the particles of liquids and solids are close together. The forces between rather distant gas particles are negligible, but those between particles in the liquid or solid state may be considerable.

Even in the gaseous state, there is some attraction between molecules. These forces of attraction *between* molecules are called **intermolecular forces.** This attraction may be sufficient to cause some deviation from the "ideal" behavior described in the gas laws. Generally, however, intermolecular forces are much weaker in the gaseous state than in the liquid and solid states.

Before studying the types of forces in detail, let's make some generalizations. First, all ionic compounds are solids at room temperature. It is possible to obtain them as liquids (by melting them), but generally only at high temperatures. Second, nearly all metals (mercury is a notable exception) are solids at room temperature. They, too, can be melted—some at fairly low temperatures, but others only at high temperatures.

It would be nice to have a third generalization to cover molecular substances. However, molecular substances exist in all three physical states at room temperature. Nitrogen (N_2) and carbon dioxide (CO_2) are gases, water (H_2O) and octane (C_8H_{18}) are liquids, and sulfur (S_8) and glucose ($C_6H_{12}O_6$) are solids. The physical state of a molecular substance is determined both by molecular weight (the size of the molecules) and by the type of force between molecules. If one of these two variables can be eliminated (or held constant), then simple generalizations are possible. For example, the Group VIIA elements all exist as nonpolar diatomic molecules. Intermolecular forces are similar; thus, any variation in properties can be attributed to variations in molecular weight. And we do find such a trend. Fluorine (F_2), with a molecular weight of 38 g/mol, and chlorine (Cl_2), with a molecular weight of 71 g/mol, are gases at room temperature. Bromine (Br_2), with a molecular weight of 160 g/mol, is a liquid, and iodine (I_2), with a molecular weight of 254 g/mol, is a solid. A similar trend can be found for compounds that experience similar intermolecular forces, as is the case for the following compounds of carbon and the halogens.

Compound	CF_4	CCl_4	CBr_4	CI_4
Molecular weight (g/mol)	88	154	332	520
Physical state at 20 °C	Gas	Liquid	Solid	Solid

The types of intermolecular force are usually of overriding importance if we are comparing molecules that are subject to dissimilar forces. We will consider the various types of forces in detail in the following sections.

Intermolecular forces pertain to the forces of attraction between a molecule and neighboring molecules. *Intramolecular forces* pertain to those between atoms within a molecule due to chemical bonding.

8.2
INTERIONIC FORCES

One type of force, already mentioned in Section 4.5, is the force between ions. Generally, interionic forces are the strongest of all the forces that hold solids and liquids together. The energy available to ionic compounds at room temperature is not sufficient to break the strong forces of attraction between them. Most ionic compounds melt only at extremely high temperatures.

Recall that ions are electrically charged. Those with opposite charges attract one another. The electrostatic attraction is not usually limited to just a pair of ions. Rather, each ion is attracted by several oppositely charged ions that surround it (see Figure 4.2.) These ionic bonds are exceptionally strong. Forces are greater between ions of higher charge. The interionic force between mercury(II) ion (Hg^{2+}) and sulfide ion (S^{2-}) is greater than the force between sodium ion (Na^+) and chloride ion (Cl^-). The force between aluminum ion (Al^{3+}) and nitride ion (N^{3-}) is greater still.

8.3
DIPOLE FORCES

Hydrogen chloride melts at $-112\ °C$ and boils at $-85\ °C$ (it is a gas at room temperature). The attractive forces between small molecules are not nearly as strong as the interionic forces mentioned in the preceding section. We know that covalent bonds hold the hydrogen and chlorine *atoms* together to form the hydrogen chloride *molecule,* but what makes one molecule interact with another in the solid or liquid states?

Remember that the hydrogen chloride molecule is a *dipole* (Section 4.17). It has a positive end and a negative end. Two dipoles brought close enough will attract one another. The positive end of one molecule attracts the negative end of another. Such forces may exist throughout the structure of a liquid or solid (Figure 8.1). In general, attractive forces between dipoles are much weaker than the attractive forces between ions. **Dipole interactions,** however, are stronger than the forces between nonpolar molecules of comparable size.

Attractions between dipolar molecules result from forces between centers of *partial* charge, whereas the much stronger attraction between ions involves fully charged particles.

Figure 8.1
An idealized representation of dipole forces in a liquid (a) and a solid (b). In a real liquid or solid, interactions are more complex.

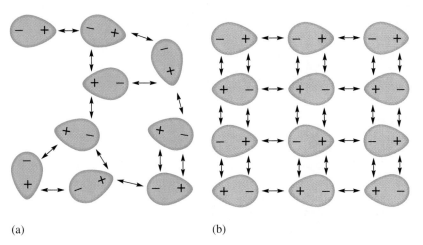

(a)   (b)

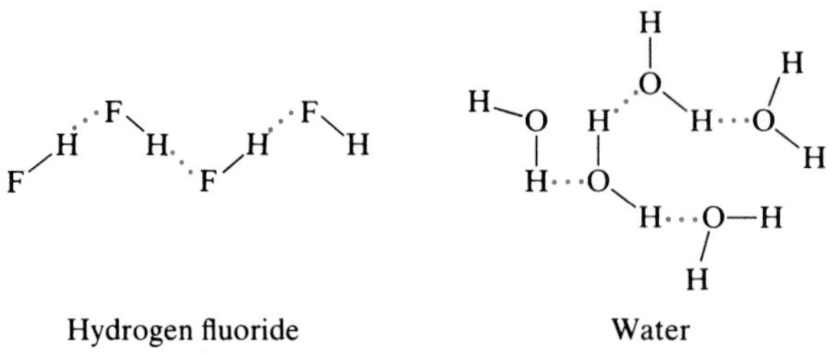

Figure 8.2
Hydrogen bonding in hydrogen fluoride and in water.

Hydrogen fluoride   Water

8.4
HYDROGEN BONDS

Certain polar molecules exhibit stronger attractive forces than would be expected on the basis of ordinary dipolar interactions. These forces are strong enough to be given a special name, the **hydrogen bond.** Note that "*hydrogen* bond" is a somewhat misleading name, as it emphasizes only one component of the interaction. Not all compounds containing hydrogen exhibit this strong attractive force; the hydrogen *must* be attached to fluorine, oxygen, or nitrogen. It is the presence of these atoms that permits us to offer an explanation for the extra strength of hydrogen bonds compared with other dipolar forces. Fluorine, oxygen, and nitrogen all have high electron-attracting power (they are very electronegative), and they are small (they are at the top of the periodic table). A hydrogen–fluorine bond, for example, is very strongly polarized, with a negative fluorine end and a positive hydrogen end. Both hydrogen and fluorine are small atoms, so the negative end of one dipole can approach very close to the positive end of a second dipole. This results in an unusually strong interaction between the two molecules—the hydrogen bond.

Hydrogen bonds are often explicitly represented by *dotted* lines to emphasize their exceptional strength compared to ordinary dipolar interactions. A *dotted* line is used to distinguish a hydrogen bond from the much stronger covalent bond, which is represented by a *solid* line (Figure 8.2).

Water has both an unusually high melting point and an unusually high boiling point for a compound with molecules of such small size. These abnormal values are attributed to water's ability to form hydrogen bonds.

The hydrogen bond may, at this point, seem merely an interesting piece of chemical theory, but its importance to life and health is immense. The structure of proteins, chemicals essential to life, is determined, in part, by hydrogen bonding. And the hereditary traits that one generation passes on to the next are dependent on an elegant application of hydrogen bonding (see Chapters 21 and 23).

Hydrogen bonds are about 5–10% as strong as covalent bonds.

8.5
DISPERSION FORCES

If one understands that positive attracts negative, then it is easy enough to understand how ions or polar molecules maintain contact with one another. But how can we explain the fact that nonpolar compounds can exist in the liquid and solid states? Even hydrogen can exist as a liquid or a solid if the temperature is low enough (its melting point is −259 °C). Some force must be holding these molecules in contact with one another in the liquid and solid states.

The answer arises from the fact that the electron cloud pictures that we talked about in Section 4.9 are only *average* positions. On the average, electrons spend more time near the fluorine atom in a fluorine–hydrogen bond, and this average, uneven sharing is what we have termed a dipole. On the average, the two electrons in the hydrogen molecule are between (and equidistant from) the two nuclei. The electrons are equally shared in this case. At any given instant, however, the electrons may be at one end of the molecule. At some other instant, the electrons may be at the other end of the molecule. Such electron motions give rise to momentary dipoles (Figure 8.3). One dipole, however momentary, can induce a similar momentary dipole in a neighboring molecule. At the instant that the electrons of one molecule are at one end, the electrons in the adjacent molecule will be repelled, prompting them to move to the far end of that molecule. Thus, at this instant there will be an attractive force between the electron-rich end of one molecule and the electron-poor end of the next. These transient, attractive forces (usually small) between molecules are called **dispersion forces.**[1]

Recall that at room temperature fluorine (F_2) and chlorine (Cl_2) are gases, bromine (Br_2) is a liquid, and iodine (I_2) is a solid. Dispersion forces are greater for larger molecules than for smaller ones. Larger atoms have larger electron clouds. Their outermost electrons are farther from the nucleus than those of smaller atoms. These remote electrons are more loosely bound and can shift toward another atom more readily than the tightly bound electrons on a smaller atom. This makes molecules with larger atoms more *polarizable* than small ones. Iodine molecules are attracted to one another more strongly than bromine molecules are attracted to one another. Bromine molecules have greater dispersion forces than chlorine molecules, and chlorine molecules, in turn, have greater dispersion forces than fluorine molecules. If you look at the periodic table, you will see that this is the order in which these elements (called, as a group, the halogens) appear in Group VIIA, with iodine the largest and fluorine the smallest. Dispersion forces, to a large extent, determine the physical properties of nonpolar compounds.

It should be noted that dispersion forces may be important even when other types of forces are present. Even though we might think of such forces as individually weaker than dipolar attractions or ionic bonds, in substances composed of large molecules the cumulative effect of dispersion forces may be considerable. For large ions, such as silver (Ag^+) and iodide (I^-), dispersion forces may play a significant role, even though interionic forces also exist. Dispersion forces play a major role in the presence of some dipolar forces. In hydrogen chloride (HCl), for example, dipolar forces may contribute as little as 15% to the intermolecular attraction; the rest is due to dispersion forces.

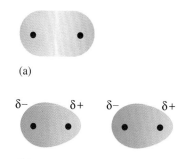

Figure 8.3
Electron cloud shapes for hydrogen molecules.
(a) Average picture with no net dipole. (b) Instantaneous pictures with momentary dipoles. The transient dipoles are constantly changing, but the net result is attraction.

Dispersion forces are weak for small, nonpolar molecules such as H_2, N_2, and CH_4. These forces can be substantial between large molecules. The properties of polymers such as polyethylene (Section 13.13) are determined to a large degree by dispersion forces between long chains of repeating $-(CH_2CH_2-)_n$ units. (In this formula, n is several hundred or even several thousand.)

8.6
THE LIQUID STATE

Molecules of a liquid are much closer together than those of a gas. Consequently, liquids can be compressed only slightly. The molecules are in constant motion, but their movements are greatly restricted by neighboring molecules. One liquid can diffuse into another, but such diffusion is much slower than in gases because of the restricted molecular motion of liquids.

The shape of the molecules that make up a liquid has an effect on one of the properties of the liquid—its **viscosity,** or resistance to flow. Liquids with low viscosity—

Viscosity—a measure of the resistance of a liquid to flow; the higher the viscosity, the slower is the liquid's rate of flow.

---

[1] Dispersion forces are sometimes called London forces, after Fritz London, professor of chemical physics at Duke University, who first treated them systematically in 1930. They are also called van der Waals forces, in honor of the Dutch physicist, Johannes D. van der Waals.

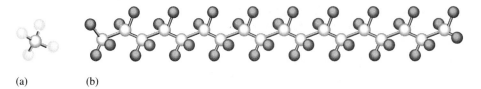

(a)   (b)

Figure 8.4

(a) Carbon tetrachloride (CCl_4) consists of small symmetrical molecules with fairly weak intermolecular forces. It has a low viscosity. (b) Octadecane ($C_{18}H_{38}$) consists of long molecules with fairly strong intermolecular forces. It has a relatively high viscosity.

Automotive oils carry a viscosity rating identified by SAE number. Oil with an SAE number of 30 is more viscous than an SAE 10 oil. The SAE abbreviation is for Society of Automotive Engineers, which sets the standards for oil viscosities.

"thin" liquids—generally consist of small, symmetrical molecules with weak intermolecular forces. Viscous liquids, on the other hand, are generally made up of large or unsymmetrical molecules with fairly strong intermolecular forces (Figure 8.4). Viscosity generally decreases with increasing temperature. Increased kinetic energy partially overcomes the intermolecular forces. Cooking oil, for example, as it pours from the bottle is thick and "oily." After it's been heated in a frying pan, it becomes thinner and more watery, that is, more like water in its consistency.

Another property of liquids is **surface tension.** A clean glass can be slightly overfilled with water before it spills over. A small needle, carefully placed, can be made to float horizontally on the surface of water—even though steel is several times denser than water is. A variety of insects can walk or skate across the surface of a pond with ease. These phenomena indicate something quite unusual about the surface of a liquid. There is a special force or tension there that resists being disrupted by the needle or water bug (Figure 8.5).

These surface forces can be explained by intermolecular forces. A molecule in the center of a liquid is pulled equally in all directions by the molecules surrounding it. A molecule on the surface, however, is attracted only by molecules at its sides and below it (Figure 8.6). There is no corresponding upward attraction. These unequal forces tend to

Figure 8.5

The surface tension of water enables this water strider to walk across the surface of a pond. Notice how the water is indented but not penetrated by the insect's legs.

Figure 8.6

Molecules in the body of a liquid are attracted equally in all directions. Those at the surface, however, are pulled downward and sideways but not upward.

pull inward at the surface of the liquid and cause it to contract, much as a stretched sheet of rubber would tend to contract. A small amount of liquid will "bead" to minimize its surface area, and a drop will be spherical for the same reason. The smaller the surface area, the smaller the number of molecules experiencing the unequal pull. Soaps and other detergents act in part by lowering surface tension, enabling water to spread out and wet a solid surface (see Section 20.5).

8.7
FROM LIQUID TO GAS: VAPORIZATION

The molecules of a liquid are in constant motion, some moving fast, some more slowly. Occasionally one of these molecules has enough kinetic energy to escape from the liquid's surface and become a molecule of vapor. If a liquid, such as water, is placed in an open container, it will soon disappear through *evaporation*. The vapor molecules disperse throughout the atmosphere, and eventually all the liquid molecules escape as they enter the vapor state. On the other hand, if the liquid is placed in a closed container, it does not go away. Some of the liquid is converted to vapor, but the vapor molecules are trapped within the container. Eventually the air above the liquid becomes saturated, and evaporation seems to stop. This vapor exerts a partial pressure (the *vapor pressure*—Section 7.10) that is constant at a given temperature.

> The rate of evaporation of a particular liquid depends on the temperature of the liquid and the amount of exposed surface area.

It may well appear that nothing further is happening, but molecular motion has not ceased. Some molecules of liquid are still escaping into the vapor state. Vapor molecules in the space above the liquid occasionally strike the liquid's surface, are captured, and thus return to the liquid state. To begin with, there are lots of liquid molecules and no vapor molecules. So, at first, conversion of liquid to vapor (**vaporization**) is taking place, but conversion of vapor to liquid (**condensation**) is not. As more molecules pass into the vapor state, the rate of condensation increases. Eventually, the rate of condensation equals the rate of vaporization. This equilibrium condition appears static but is, in fact, dynamic, as we have mentioned previously. Two opposing processes are taking place at exactly the same rate. This situation is analogous to that encountered in the case of reversible chemical reactions (Section 5.7).

At higher temperatures, more molecules of the liquid would have enough energy to escape from the liquid state. The vapor pressure would increase, but equilibrium would soon be reestablished at the higher temperature. The rate of vaporization would be greater, but so would the rate of condensation. At equilibrium, the rates would once again be equal (Figure 8.7), but the equilibrium vapor pressure would be higher at the higher temperature. Schematically, these processes may be illustrated as

$$\text{Liquid} \underset{\text{condensation}}{\overset{\text{vaporization}}{\rightleftharpoons}} \text{Vapor}$$

If a liquid is placed in an open container, the escape of molecules of the liquid is opposed by atmospheric pressure. If the liquid is heated, the vapor pressure will increase. Continued heating will eventually result in a vapor pressure equal to atmospheric pressure. At that temperature, the liquid will begin to boil. Vaporization will take place not only at the surface of the liquid but also in the body of the liquid, with vapor bubbles forming and rising to the surface. The **boiling point** of a liquid is the temperature at which its vapor pressure becomes equal to atmospheric pressure. Since the latter varies with

Figure 8.7
(a) Diagram of a liquid with its vapor at equilibrium in a closed container at a given temperature. (b) The same system at a higher temperature.

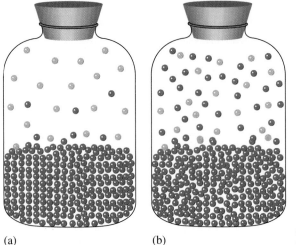

(a)   (b)

⊚ = air molecule (N$_2$, O$_2$, etc.)
⬤ = water molecule (liquid or vapor)

Table 8.1
Boiling Points of Pure Water at Various Pressures

Pressure (mm Hg)	Temperature (°C)
707	98
760	100
816	102
875	104
938	106
1004	108
1075	110
1283	115
1535	121

altitude and weather conditions, boiling points of liquids do also (Figure 8.8). The cooking of foods requires that they be supplied with a certain amount of energy. When water boils at 100 °C, an egg can be placed in the water and soft-boiled to perfection in 3 min. If, at reduced atmospheric pressure, water boils at a lower temperature, then it contains less heat energy with which to cook the egg. It would take a bit longer to prepare a soft-boiled egg on top of Mount Everest.

The boiling point is increased when external pressure is increased. Autoclaves and pressure cookers are based on this principle. We can achieve a higher temperature at the higher pressures attained in these closed vessels. (Heat added to a liquid at atmospheric pressure will merely convert liquid to vapor. No increase in temperature occurs until all the liquid has vaporized.) The chemical reactions involved in the cooking of a tough piece of meat proceed more rapidly at the temperature that can be attained in a pressure cooker. Bacteria (even resistant spores) are killed more rapidly in an autoclave, not directly by the increased pressure but rather by the higher temperatures attained. Table 8.1 gives the temperatures attainable with pure water at various pressures.

Figure 8.8
The boiling point of water at different altitudes.

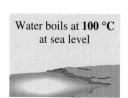

Water boils at **100 °C** at sea level

Water boils at **85 °C** at 4400 m

Pike's Peak, Colorado

Water boils at **71 °C** at 8800 m

Mount Everest, Tibet

The boiling point of a liquid is a useful physical property, often used as an aid in identifying compounds. Since boiling point varies with pressure, it is necessary to define the **normal boiling point,** that temperature at which a liquid boils under standard pressure (1 atm, or 760 mm Hg). Alternatively, one can specify the pressure at which the boiling point was determined. For example, the *Handbook of Chemistry and Physics* lists the boiling point of antipyrine (a pain reliever and fever reducer) as 319^{741}. This means that the compound boils at 319 °C under a pressure of 741 mm Hg. Table 8.2 give the normal boiling points of some familiar liquids.

Liquids can be purified by a process called **distillation.** Imagine a mixture of water and some nonvolatile material, that is, some material that will not vaporize readily. If the mixture is heated until it boils, the water will vaporize, but the nonvolatile material will not. The water vapor can then be condensed back to the liquid state and collected in a separate container. In such a distillation, the water is separated from the other component of the mixture and thereby purified.

Even if a mixture contains two or more volatile components, purification by distillation is possible. Consider a mixture of two components, one of which is somewhat more volatile than the other (e.g., water and alcohol). At the boiling point of such a mixture, both components will contribute some molecules to the vapor. The more volatile component (alcohol in this example), because it is more easily vaporized, will have more of its molecules in the vapor state than will the less volatile component. When the vapor is condensed into another container, the resulting liquid will be richer in the more volatile component than the original mixture was. Thus, a purer sample of the more volatile component will have been produced. (See a discussion of alcohol distillation in Section 15.5.) Figure 8.9 shows a typical distillation apparatus.

Heat is required for the conversion of a liquid to a vapor. A liquid evaporating at room temperature absorbs heat from its surroundings. Most of us are familiar with this cooling effect of evaporation. Even on a warm day, we feel cool after a swim, because the water evaporating from our skin removes heat. Volatile liquids, such as ethyl chloride (C_2H_5Cl), which boils at 12.5 °C, may be used as local anesthetics. Rapid evaporation

Table 8.2

Boiling Points of Various Compounds at 1 Atm

Compound	Boiling Point (°C)
Diethyl ether (anesthetic)	34.6
Acetone (solvent)	56.2
Methyl alcohol (wood alcohol)	64.5
Ethyl alcohol (grain alcohol)	78.3
Water	100.0
Mercury	356.6

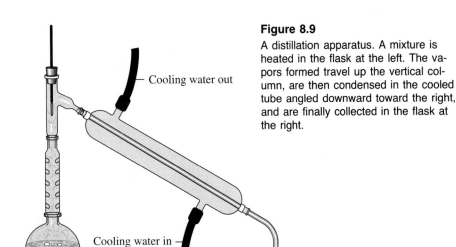

Figure 8.9

A distillation apparatus. A mixture is heated in the flask at the left. The vapors formed travel up the vertical column, are then condensed in the cooled tube angled downward toward the right, and are finally collected in the flask at the right.

Cooling water out

Cooling water in

Table 8.3
Heats of Vaporization (at the Normal Boiling Point) of Several Liquids

Compound	Molar Heat of Vaporization (cal/mol)	Heat of Vaporization (cal/g)
Diethyl ether ($C_2H_5OC_2H_5$)	6,200	84
Benzene (C_6H_6)	7,300	94
Methyl alcohol (CH_3OH)	8,400	260
Water	9,700	540
Mercury	14,200	71

from the skin removes enough heat to freeze a small area, rendering it insensitive to pain. Alcohol rubs also act to cool the skin by their evaporation.

The amount of heat required to vaporize a given amount of liquid can be measured. The quantity of heat required to vaporize 1 mol of a liquid at a constant pressure is called the **molar heat of vaporization.** The heat of vaporization is characteristic of a given liquid. It depends to a large extent on the type of intermolecular force in the liquid. Water, with molecules strongly associated through hydrogen bonding, has a heat of vaporization of 9.7 kcal/mol. Methane, with molecules held together by weak dispersion forces only, has a heat of vaporization of only 0.232 kcal/mol. Heats of vaporization of several liquids are given in Table 8.3.

Given the molar heat of vaporization, we can calculate the heat of vaporization in calories per gram.

EXAMPLE 8.1

The molar heat of vaporization of ammonia (NH_3) is 556 cal/mol. What is the heat of vaporization in calories per gram?

SOLUTION
The molecular weight of ammonia is 17.0 g/mol. Therefore, the heat of vaporization is

$$\frac{556 \text{ cal/mol}}{17.0 \text{ g/mol}} = 33.0 \text{ cal/g}$$

Practice Exercise
The molar heat of vaporization of chloroform ($CHCl_3$) is 7050 cal/mol. What is the heat of vaporization in calories per gram?

EXAMPLE 8.2

How much heat would be required to vaporize 400 g of water at its boiling point?

SOLUTION
The heat of vaporization of water is 540 cal/g.

$$\frac{540 \text{ cal}}{1.00 \text{ g}} \times 400 \text{ g} = 216{,}000 \text{ cal or } 216 \text{ kcal}$$

Practice Exercise
How much heat would be required to vaporize 400 g of chloroform at its boiling point? (Use your answer from the preceeding practice exercise.)

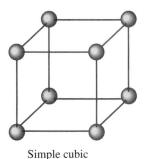
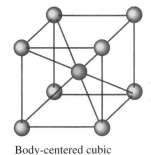
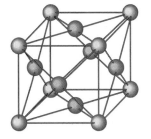

Simple cubic   Body-centered cubic   Face-centered cubic

Figure 8.10
Three types of crystal lattices based on the cube.

When a vapor condenses to a liquid, it gives up exactly the same amount of heat energy as was absorbed in converting the liquid to a vapor. A refrigerator operates by alternately vaporizing and condensing a fluid. The heat required to vaporize the fluid is drawn from the refrigerated compartment. The heat is released to the outside atmosphere when the fluid is condensed back to the liquid state.

8.8
THE SOLID STATE

Solids resemble liquids in that the particles (atoms, molecules, or ions) in them are close together, making them virtually incompressible. But these two physical states differ significantly in the motion of their particles. In the liquid state, particles are in constant (if somewhat restricted) motion. In solids, there is little motion other than vibration about a fixed point. Consequently, diffusion in solids is generally extremely slow. An increase in temperature will increase the vigor of the vibrations in a solid. If the vibrations become violent enough, the solid will melt (Section 8.9).

In most solids, the particles are arranged in regular, systematic patterns. Such solids are said to be **crystalline,** and the structure is called a **crystal lattice.** The detail of a crystal lattice can be described in terms of a small, repeating portion called a **unit cell.** Extension of these unit cells into three-dimensional space results in the plane faces and definite angles of macroscopic crystals (such as those of quartz or uncut diamonds or rock salt).

In spite of the complex appearance of the many different crystalline solids, there are relatively few fundamental types of crystal lattices. The easiest to visualize is the **simple cubic** arrangement. This cell and two others readily derived from it are shown in Figure 8.10. In one of the derived structures, there is an additional particle (atom, ion, or molecule) at the center of the cube. This arrangement is called **body-centered cubic.** The other derived structure has particles at the center of each of the six faces of the cube. This structure is described as **face-centered cubic.** Other crystal systems are based on different geometric shapes. Two of these, the orthorhombic and the hexagonal, are shown in Figure 8.11.

Crystalline solids can also be classified on the basis of the types of forces holding the particles together. The four classes are ionic, molecular, covalent network, and metallic.

Not all solids are crystalline in form. Some substances such as glass and plastic retain their shape and volume, but they do not have the ordered crystalline state. These substances are sometimes called amorphous solids or suspended liquids.

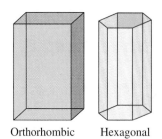

Orthorhombic   Hexagonal

Figure 8.11
Some types of crystal lattices are based on the orthorhombic and hexagonal systems.

Figure 8.12

The crystal structure of diamond, a covalent network solid. Each intersection of lines represents a carbon atom.

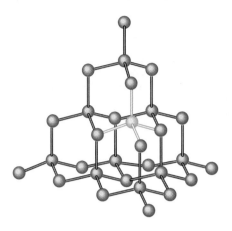

Ionic solids have ions occupying the lattice points in the crystal. A typical ionic solid is sodium chloride (NaCl), which we discussed in Section 4.4. In the lattice, each chloride ion is surrounded by six sodium ions, and each sodium ion is surrounded by six chloride ions. Interionic forces are very strong. Ionic solids consequently have high melting points and low vapor pressures and are quite hard.

Molecular crystals have discrete covalent molecules at the lattice points. These are held together by rather weak dispersion forces, as in crystalline iodine, by dispersion forces plus dipolar forces, as in iodine chloride (ICl), or by hydrogen bonds, as in ice. Molecular solids often have lower melting points than ionic solids. Ice is an exceptional molecular solid; the water molecules are strongly associated by hydrogen bonding (Section 8.10).

Covalent network crystals have atoms at the lattice points. These are joined into extensive networks by covalent bonds. Thus, each crystal is in essence one large molecule. Covalent network solids are generally extremely hard and nonvolatile and melt (with decomposition) at very high temperatures. Diamond is a familiar example. Carbon atoms occupy the lattice points. Each is covalently bonded to four other carbon atoms (Figure 8.12). Silicon carbide (SiC), also called Carborundum, is another familiar compound with an extensive network of covalent bonds.

At the lattice sites in **metallic solids** are positive ions. These ions are formed when metal atoms, such as silver (Ag), lose their outer electrons. The electrons thus released are distributed throughout the lattice, almost like a fluid. These electrons, which can move freely about the lattice, make metals good conductors of heat and electricity. Some metals, such as sodium or potassium, are fairly soft and have low melting points. Others, such as magnesium and calcium, are hard and have high melting points. The extra valence electrons in calcium and magnesium atoms seem to lead to stronger forces between atoms. As a class, metals are malleable; that is, they can be shaped under the influence of pressure or heat. They can be rolled into bars, pressed into sheets, or extruded into wire.

Table 8.4 lists some characteristics of crystalline solids.

Table 8.4
Some Characteristics of Crystalline Solids

Particles in Crystal	Principal Attractive Force Between Particles	Melting Point	Electrical Conductivity of Liquid	Characteristics of the Crystal	Examples
		IONIC CRYSTALS			
Positive and negative ions	Electrostatic attraction between ions (very strong)	High	High	Hard, brittle, most dissolve in polar solvents	NaCl, CaF$_2$, K$_2$S, MgO
		COVALENT NETWORK CRYSTALS			
Atoms	Covalent bonds (very strong)	Generally do not melt	—	Very hard, insoluble	Diamond (C), SiC, AlN
		METALLIC CRYSTALS			
Positive ions plus mobile electrons	Metallic bonds (strong)	Most are high	Very high	Most are hard, malleable, ductile, good electrical and thermal conductors, insoluble unless a reaction occurs	Cu, Ca, Al, Pb, Zn, Fe, Na, Ag
		MOLECULAR CRYSTALS			
		Hydrogen-bonded			
Molecules with H on N, O, or F	Hydrogen bonds (intermediate)	Intermediate	Very low	Fragile, soluble in other H-bonding liquids	H$_2$O, HF, NH$_3$, CH$_3$OH
		Polar			
Polar molecules (no H bonds)	Electrostatic attraction between dipoles (rather weak)	Low	Very low	Fragile, soluble in other polar and many nonpolar solvents	HCl, H$_2$S, CHCl$_3$, ICl
		Nonpolar			
Atoms or nonpolar molecules	Dispersion forces only (weak)	Very low	Extremely low	Soft, soluble in nonpolar or slightly polar solvents	Ar, H$_2$, I$_2$, CH$_4$, CO$_2$, CCl$_4$

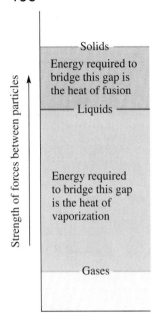

Figure 8.13

The heat of fusion for a substance is only a fraction of the heat of vaporization.

8.9
FROM SOLID TO LIQUID: MELTING (FUSION)

Solids can be changed to liquids; that is, they can be *melted*. The solid is heated, and the heat energy is absorbed by the particles of the solid. The energy causes the particles to vibrate in place with more and more vigor until, finally, the forces holding the particles in a particular arrangement are overcome. The solid has become a liquid. The temperature at which this happens is called the **melting point** of that solid. A high melting point is one indication that the forces holding a solid together are very strong.

The amount of heat required to convert 1 mol of a solid to a liquid at the melting point is called the **molar heat of fusion.** Generally, the heat of fusion is much less than the heat of vaporization (Figure 8.13). The heat of fusion is the amount of energy that will disrupt the crystal lattice but still leave the particles in contact with one another and under the influence of their mutual attraction. A much larger amount of energy is required to vaporize the liquid because, in vaporization, the attraction between particles must be completely overcome (or very nearly so). Table 8.5 gives some representative heats of fusion.

Table 8.5
Heat of Fusion (at the Melting Point) for Several Solids

Compound	Melting Point (°C)	Heat of Fusion (cal/mol)	Heat of Fusion (cal/g)
Ammonia (NH_3)	−78	1,620	95
Water (H_2O)	0	1,440	80
Benzene (C_6H_6)	−6	2,550	33
Copper (Cu)	1,083	3,110	49
Sodium chloride (NaCl)	804	7,000	120
Tungsten (W)	3,410	8,050	43

EXAMPLE 8.3

The heat of fusion of water is 80 cal/g. How much heat would be required to melt 400 g of ice?

SOLUTION

$$400 \text{ g} \times \frac{80 \text{ cal}}{1.0 \text{ g}} = 32{,}000 \text{ cal or 32 kcal}$$

Practice Exercise
The heat of fusion of sodium chloride is 120 cal/g. How much heat would be required to melt 400 g of sodium chloride?

EXAMPLE 8.4

The molar heat of fusion of naphthalene ($C_{10}H_8$) is 4610 cal/mol. What is the heat of fusion in calories per gram?

SOLUTION

The molecular weight of naphthalene is 128 g/mol. The heat of fusion is therefore

$$\frac{4610 \text{ cal/mol}}{128 \text{ g/mol}} = 36.0 \text{ cal/g}$$

8.10
WATER: A MOST UNUSUAL LIQUID

Now that we have laid something of a theoretical foundation, let's look more closely at a very special liquid, water. Next to air, water is the most familiar substance on Earth. Even so, it is a most unusual compound. At room temperature, it is the only liquid compound with a molecular weight as low as 18 g/mol. The solid form of water (ice) is less dense than the liquid, a relatively rare situation. The consequences of this peculiar characteristic for life on this planet are immense. Ice forms on the surface of lakes when the temperature drops below freezing, and this ice insulates the lower layers of water, enabling fish and other aquatic organisms to survive the winters of the temperate zones. If ice were denser than liquid water, it would sink to the bottom as it formed. This would permit the new surface water to freeze and, in its turn, sink to the bottom. The repetition of this process would eventually result in a lake frozen from top to bottom. Even the deeper lakes of the northern latitudes would freeze solid in winter. Life in the northern lakes and rivers would be very different from what it is now, if indeed there were life in those waters.

The same property—the relative density of ice and liquid water—has dangerous consequences for living cells. Since ice has a lower density than liquid water, 1 g of ice occupies a larger volume than 1 g of water. As ice crystals form in living cells, the expansion ruptures and kills the cells. The slower the cooling, the larger the crystals of ice and the more damage to the cell. The frozen food industry takes into account this property of water. Food is "flash frozen," that is, frozen so rapidly that the ice crystals are kept very small and do minimum damage to the cellular structure of the food.

Another unusual property of water is its high specific heat (Section 1.9). It takes 1 cal of heat to raise the temperature of 1 g of water 1 °C. That's 10 times as much energy as is required to raise the temperature of the same amount of iron 1 °C. The specific heats of a number of common materials are given in Table 8.6.

The reason cooking utensils are made of iron, copper, aluminum, or glass is that these materials have low specific heats. Thus, they heat up very quickly. The reason the handle of a frying pan is made of wood or plastic is that these materials have high specific heats. When they are exposed to heat, their temperatures increase more slowly.

Nearly all other solids are more dense than their liquids. Consider a lead ball and molten lead, an iron ball and molten iron, etc.

Table 8.6
Specific Heats of Some Common Substances

Substance	Specific Heat [cal/(g · °C)]
Water (liquid)	1.00
Water (solid)	0.50
Water (gas)	0.47
Ethyl alcohol	0.54
Wood	0.42
Aluminum	0.21
Glass	0.12
Iron	0.11
Copper	0.09
Silver	0.06
Gold	0.03

EXAMPLE 8.5

How much heat is required to increase the temperature of 10 g of water from 10 °C to 30 °C?

SOLUTION
The specific heat of liquid water is 1 cal/(g · °C). The temperature change is 20 °C (30 °C − 10 °C = 20 °C).

$$10 \text{ g} \times 20 \text{ °C} \times \frac{1 \text{ cal}}{(1 \text{ g})(1 \text{ °C})} = 200 \text{ cal}$$

Practice Exercise
How much heat is required to increase the temperature of 1 mol of water from 20 °C to 30 °C?

EXAMPLE 8.6

How much energy is required to change 10 g of ice at −10 °C to steam at 100 °C?

SOLUTION
This problem should be worked in parts. First calculate the energy required to raise the temperature of 10 g of ice from −10 °C to 0 °C, a change of 10 °C. The specific heat of ice is 0.5 cal/(g · °C).

$$10 \text{ g} \times 10 \text{ °C} \times \frac{0.5 \text{ cal}}{(1 \text{ g})(1 \text{ °C})} = 50 \text{ cal}$$

Next, using the heat of fusion of water (80 cal/g), calculate the energy required to melt the ice at 0 °C.

$$10 \text{ g} \times \frac{80 \text{ cal}}{1 \text{ g}} = 800 \text{ cal}$$

Then calculate the heat required to change the temperature of the water from 0 °C to 100 °C. The specific heat of water is 1 cal/(g · °C), and the temperature change is 100 °C.

$$10 \text{ g} \times 100 \text{ °C} \times \frac{1 \text{ cal}}{(1 \text{ g})(1 \text{ °C})} = 1000 \text{ cal}$$

Now calculate the amount of heat required to change the water (at 100 °C) to steam (at 100 °C). The heat of vaporization for water is 540 cal/g.

$$10 \text{ g} \times \frac{540 \text{ cal}}{1 \text{ g}} = 5400 \text{ cal}$$

Finally, total all the calculated values.

To raise the temperature of ice from $-10\ °C$ to $0\ °C$	50 cal
To change ice to liquid water	800 cal
To raise the temperature of water from $0\ °C$ to $100\ °C$	1000 cal
To change water to steam	5400 cal
Total	7250 cal

Note that almost 75% of the total energy is used in vaporizing the water.

Practice Exercise

How much energy is required to change 100 g of ice at $-10\ °C$ to liquid water at 15 °C?

The high specific heat of water means not only that much energy is required to raise the temperature of water but also that much heat is given off by water for even a small drop in temperature. The vast amounts of water on the surface of the Earth thus act to moderate daily temperature variations. We need only consider the extreme temperature changes on the surface of the waterless moon to appreciate this important property of water. The temperature of the moon varies from just above the boiling point of water (100 °C) to about $-175\ °C$, a range of 275 °C. In contrast, temperatures on Earth rarely fall below $-50\ °C$ ($-58\ °F$) or rise above 50 °C (122 °F), a range of only 100 °C.

Water also has a higher density than many other familiar liquids, including petroleum products. As a consequence, a number of liquids that are insoluble in water float on the surface of it. This situation has caused problems in recent years. Gigantic oil spills, which occur when a tanker ruptures or an offshore well gets out of control, result in a slick on the water's surface. This oil covers the feathers of waterfowl and the coats of sea animals, such as the otter and the seal. The oil is often washed onto beaches, where it does considerable ecological and aesthetic damage. If oil were denser than water, it would sink to the bottom. The problem would be of a different nature, though not necessarily less acute.

Still another way in which water is unusual is that it has a very high heat of vaporization; that is, a large amount of heat is required to evaporate a small amount of water. This is of enormous importance to animals. Large amounts of body heat, produced as a by-product of metabolic processes, can be dissipated by the evaporation of small amounts of water (perspiration) from the skin. The heat of vaporization of this water is obtained from the body, and the body is cooled. Conversely, when steam condenses, considerable heat is released. For this reason, steam causes serious burns when it contacts the skin. We previously mentioned that water's high specific heat modifies the climate. Water's high heat of vaporization also contributes to the climate-modifying effect of lakes and oceans. A large portion of the heat that would otherwise warm up the land is used instead to vaporize water from the surface of a lake or the sea. Thus, in summer it is cooler near a large body of water than in interior land areas.

All of these fascinating properties of water result from the unique structure of the water molecule (Section 4.17). Recall that the water molecule is polar. In the liquid state, water molecules are strongly associated by hydrogen bonding (see Figure 8.2). These strong attractive forces account for the high heat of vaporization of water. They must be

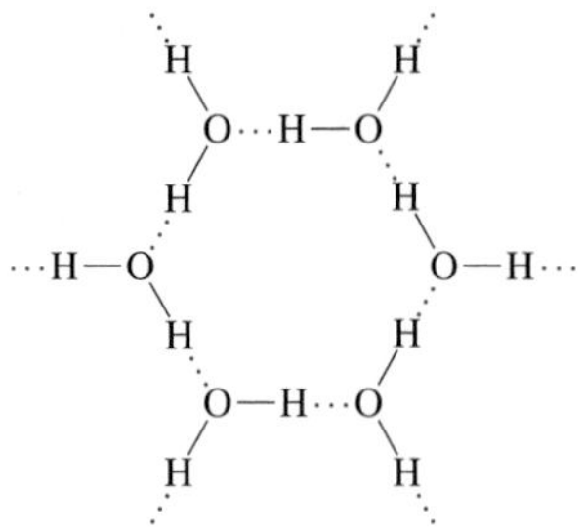

Figure 8.14
A two-dimensional representation of the structure of ice, showing a large hexagonal hole formed by six water molecules that are connected by hydrogen bonds.

overcome if vaporization is to take place, and this is why a large amount of energy must be supplied to water for it to convert from liquid to vapor.

In the liquid state, water molecules are quite close together but randomly arranged. In the solid state, water molecules are in a much more ordered arrangement. But this orderly arrangement, as illustrated in Figure 8.14, is less compact than that achieved in the liquid state. Large hexagonal holes are incorporated into the ice lattice. It is this empty space that makes ice less dense than liquid water.

Our bodies are about 65% water. Chemical reactions in living cells take place in water solutions. The importance of solutions is such that we devote the next chapter to the subject.

EXERCISES

1. Define or illustrate the following terms.
    - a. intermolecular forces
    - b. hydrogen bond
    - c. dispersion forces
    - d. viscosity
    - e. surface tension
    - f. vaporization
    - g. condensation
    - h. boiling point
    - i. normal boiling point
    - j. melting point
    - k. ionic crystal
    - l. molecular crystal
    - m. metallic solid
    - n. molar heat of fusion
    - o. covalent network crystal
    - p. molar heat of vaporization

2. How do liquids and solids differ from gases in their compressibility, spacing of molecules, and intermolecular forces?

3. In what ways are liquids and solids similar? In what ways are they different?

4. List four types of interactions between particles in the liquid and in the solid states. Give an example of each type.

5. Rank N_2, H_2O, and HCl in order of increasing strength of intermolecular interaction (weakest interaction first). What type of interaction is involved in each case?

6. Explain how oxygen can be liquefied if the temperature is lowered sufficiently, even though O_2 molecules are nonpolar.

7. What type of interaction exists between molecules of each of the following?
    - a. H_2
    - b. NO
    - c. HF

8. List three distinctive properties of water.

9. Use the kinetic-molecular theory (Section 7.2) to explain the properties of liquids and solids.

10. Describe the effect of temperature on the viscosity of a liquid.

11. Describe, in terms of intermolecular forces, the cause of the phenomenon of surface tension.

12. In which of the following compounds would hydrogen bonding be an important intermolecular force?

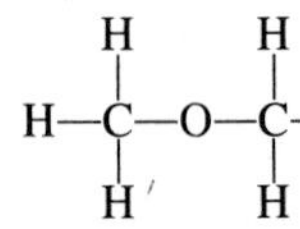

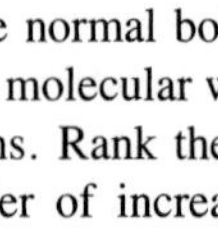

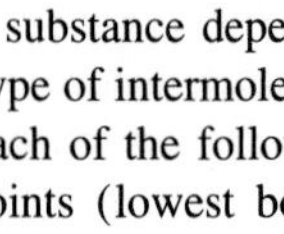

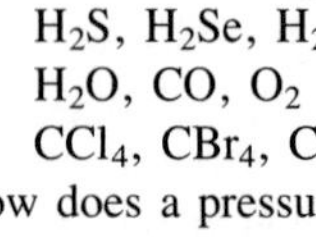

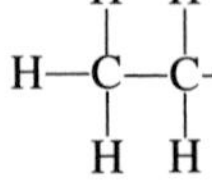

13. The normal boiling point of a substance depends on both the molecular weight and the type of intermolecular attractions. Rank the members of each of the following sets in order of increasing boiling points (lowest boiling first).
    - a. H_2S, H_2Se, H_2Te
    - b. H_2O, CO, O_2
    - c. CCl_4, CBr_4, CI_4

14. How does a pressure cooker work?

15. The heat of vaporization of bromine (Br_2) is 45 cal/g. What is the molar heat of vaporization of bromine?

16. The heat of vaporization of ammonia (NH_3) is 327 cal/g. What is the molar heat of vaporization of ammonia?

17. The molar heat of vaporization of acetic acid ($C_2H_4O_2$) is 5.81 kcal/mol. How much heat is required to vaporize 1.00 g of acetic acid?

18. The molar heat of vaporization of acetone (C_3H_6O) is 7.23 kcal/mol. How much heat is required to vaporize 5.80 g of acetone?

19. How is the heat of vaporization of a liquid related to intermolecular forces in the liquid?

20. The heat of fusion of water is 80 cal/g. How much heat would be required to melt a 15-kg block of ice?

21. The molar heat of fusion of acetone (C_3H_6O) is 2.58 kcal/mol. How much heat is required to melt 1.00 g of acetone?

22. The molar heat of fusion of water is 1.44 kcal/mol. How much heat is required to melt 40.0 g of ice?

23. There is a rule of thumb that says that for many liquids the molar heat of vaporization is approximately 21 times the normal boiling point in kelvins. Use this rule to calculate the molar heat of vaporization for benzene, which has a boiling point of 353 K. How does your calculated value compare with the experimental value given in Table 8.3?

24. The specific heat of silver is 0.06 cal/(g · °C). That of gold is 0.03 cal/(g · °C). Which metal will be hotter (i.e., will be at a higher temperature) if both absorb the same amount of heat?

25. The molar heats of vaporization of ethanol (C_2H_6O) and ethyl acetate ($C_4H_8O_2$) are 9.39 kcal/mol and 7.77 kcal/mol, respectively. Which liquid has the stronger intermolecular forces?

26. Calculate the amount of heat required to raise the temperature of 25 g of water from 20 °C to 60 °C. The specific heat of water is 1 cal/(g · °C).

27. How much energy will be expended in changing 100 g of ice at −5 °C to steam at 100 °C?

28. How much heat is required to convert 10.0 g of ice at −12.0 °C to steam at 130 °C?

29. How much heat is absorbed when 50 g of ammonia (NH_3) is vaporized? The heat of vaporization of ammonia is 327 cal/g.

30. Why does it take longer to boil an egg at high altitude than at sea level? Does it take longer to fry an egg at high altitude? Explain.

31. Why does ice float on liquid water?

32. Why does steam at 100 °C cause more severe burns than liquid water at the same temperature?

33. Which has a higher boiling point, methane (CH_4) or ethane (C_2H_6)? Why?

34. Which has a higher boiling point, ethane or methanol? Why?

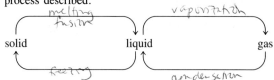

Ethane   Methanol

35. Which noble gas, neon (Ne) or xenon (Xe), has the higher boiling point? Why? (Both exist as monatomic gases.)

36. To obtain water each day, a bird in winter eats 5 g of snow at 0 °C. How many kcal (food Calories) of energy does it take to melt this snow and warm the liquid to the bird's body temperature of 40 °C?

37. To obtain water on a winter hike, a woman decides to eat snow. How many extra kcal (food Calories) of food would she have to take in each day to raise the 1500 g of snow that she needs from −10 °C to 0 °C, melt it, then raise the liquid water from 0 °C to her body temperature of 37 °C?

38. In which process is energy absorbed by the material undergoing the change of state?
   a. melting or freezing
   b. condensation or vaporization

39. Label the arrows with the term that correctly identifies the process described.

melting
fusion

vaporization

solid   liquid   gas

freezing

condensation

Chapter 9
SOLUTIONS

Fish, plants, and marine animals reside within a water solution that contains all necessary ingredients to support life.

You are a solution of sodium ions, potassium ions, calcium ions, bicarbonate ions, chloride ions, glucose, amino acids, fatty acids, glycerol, fats, proteins, acetylcholine, and lots of other goodies. Well, you aren't all in solution, or else you would wash away in the shower. But, except for a few semisolid parts, such as skin and muscle and bone, you are mostly water. The rest of you is floating around in that water.

Almost all living systems are made up of thin chemical "soups" in contact with membranes and small cellular parts called organelles. The membranes and organelles are made up of complex chemicals called lipids (fats and fatlike substances), carbohydrates (sugars, starches, and cellulose), proteins, and nucleic acids (DNA and RNA). Life processes occur in solutions and at the interfaces between solutions and semisolid organelles and membranes. Although the chemistry of these processes is now being rapidly unraveled, it is still poorly understood. Therefore, in this chapter we will deal mainly with the simpler solutions.

9.1
SOLUTIONS: DEFINITIONS AND TYPES

Put a teaspoonful of sugar in a cup of water. Stir until all the sugar is dissolved. Taste the sweetened water from one side of the cup and then from the other. Use a straw to taste it from the center and then from the bottom. If you did a proper job of mixing, the water has the same degree of sweetness throughout; that is, it is **homogeneous.** You could have added more sugar to make the water sweeter, or you could have used less sugar to make it less sweet. Further, you could boil the water away and recover the sugar. The sugar and water have not reacted chemically. This mixture of sugar in water is called a *solution,* and we say that the sugar is *dissolved* in the water. We can define a **solution** as a homogeneous mixture of two or more kinds of atoms, molecules, or ions. In a true solution, the mixture is intimate right down to the molecular level. There are no clumps of sugar molecules; single molecules are randomly distributed among the water molecules in much the same way that black marbles can be distributed among white ones (Figure 9.1).

The components of a solution are given special names. The substance being dissolved (usually present in a lesser amount) is called the **solute.** The substance doing the dissolving (usually present in a greater amount) is called the **solvent.** Water is undoubtedly the most familiar solvent, and it dissolves many familiar substances such as sugar, salt, and alcohol. There are many other solvents. Gasoline dissolves grease. Many drugs are dissolved in alcohol. Banana oil is a solvent for the glue used in making model airplanes.

A solvent need not be a liquid. Air is a solution of oxygen, argon, water vapor, and other gases in nitrogen gas. Steel is a solution of carbon (the solute) in iron (the solvent)— a solid in a solid. The most common types of solutions are summarized in Table 9.1.

In this chapter, we will deal for the most part with **aqueous solutions,** solutions in which the solvent is water.

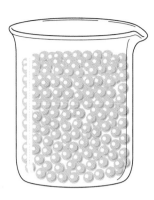

◯ = water molecule

◯ = sugar molecule

Figure 9.1

In a solution the sugar molecules are randomly distributed among the water molecules.

9.2
QUALITATIVE ASPECTS OF SOLUBILITY

We say that sugar is **soluble** in water. Just what does that mean? Can we dissolve a teaspoonful of sugar in a cup of water? Can we dissolve 10 teaspoonfuls, or 100 teaspoonfuls? We know from everyday experience that there is a limit to the amount of sugar we can dissolve in a given volume of water. Nevertheless, we still find it convenient to say that sugar is soluble in water, for we can dissolve an appreciable amount.

Solute	Solvent	Solution	Example
Gas	Gas	Gas	Air (O_2 in N_2)
Gas	Liquid	Liquid	Club soda (CO_2 in H_2O)
Liquid	Liquid	Liquid	Wine (alcohol in H_2O)
Solid	Liquid	Liquid	Saline solution (NaCl in H_2O)
Solid	Solid	Solid	14-karat gold (Ag in Au)

Table 9.1
Types of Solutions

There are a few substances that can be mixed in all proportions. Water and alcohol are familiar examples. We say that such substances are completely **miscible.** For most "soluble" substances, though, there is a limit to the amount that will dissolve in a given solvent. For others, which we call **insoluble,** that limit is near zero. Put an iron nail in a beaker of water. There is no apparent change. We say that iron is insoluble in water. Even insolubility is relative, however. If we had a method sensitive enough, we would find that some iron had dissolved. The amount might well be regarded as insignificant, and that is the sense in which the term *insoluble* is used. We will find terms such as soluble and insoluble useful, but they are imprecise and must be used with care.

Two other imprecise but sometimes useful terms are dilute and concentrated. A **dilute solution** is one that contains a little bit of solute in lots of solvent. A **concentrated solution** is one in which lots of solute is dissolved in a relatively small amount of solvent. If we dissolved a few milliliters of ethylene glycol, or antifreeze (Chapter 14), in several liters of water, the dilute solution would be quite "thin"—little changed in appearance from that of pure water. However, if we dissolved 1 L of ethylene glycol in 1 L of water, the concentrated solution would be rather syrupy—similar to pure ethylene glycol in consistency.

The terms dilute and concentrated are used in a quantitative way for solutions of acids and bases. We specify their meanings in this context in Section 11.1.

9.3
SOLUBILITY OF IONIC COMPOUNDS

In Chapter 8, we saw the unique structure of water results in relatively strong forces between water molecules. Further, we examined the different types of forces that exist between identical molecules in pure liquids. Now let's look at the types of forces that exist between the solute and solvent molecules in solutions. The solubility of a given solute depends on the relative attraction between particles in the pure substances and in the solution.

Water is a good solvent for compounds of the Group IA elements. Most lithium, sodium, and potassium salts are quite soluble in water. Examples are sodium chloride ($NaCl$), sodium sulfate (Na_2SO_4), potassium phosphate (K_3PO_4), and lithium bromide ($LiBr$). Further, nearly all nitrate salts are soluble, as are salts that incorporate the ammonium ion. Silver nitrate ($AgNO_3$), mercury(II) nitrate [$Hg(NO_3)_2$], aluminum nitrate [$Al(NO_3)_3$], ammonium chloride (NH_4Cl), ammonium sulfate [$(NH_4)_2SO_4$], and ammonium phosphate [$(NH_4)_3PO_4$] are examples. Why do these compounds dissolve in water? In essence, three things must happen. The attractive forces holding the salt ions together must be overcome. Similarly, the attractive forces holding at least some of the water molecules together must be overcome. Finally, the solute and solvent molecules must interact; that is, they must attract one another. **Hydration** is the process in which water molecules surround the solute ions (Figure 9.2).

In some solids, the forces holding the ions together are so strong that they cannot be overcome by the hydration of the ions. Many solids in which both ions are doubly or triply charged are essentially insoluble in water. Examples are calcium carbonate (Ca^{2+} and CO_3^{2-}), aluminum phosphate (Al^{3+} and PO_4^{3-}), and barium sulfate (Ba^{2+} and SO_4^{2-}). The large electrostatic forces between the ions hold the particles together despite the attraction of solvent molecules. Table 9.2 summarizes the solubilities of a variety of ionic compounds.

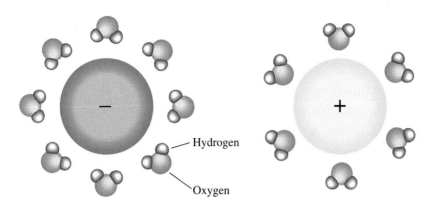

Figure 9.2
The interaction of polar water molecules with ions—hydration.

Hydrogen

Oxygen

	NO_3^-	CH_3COO^-	Cl^-	SO_4^{2-}	OH^-	S^{2-}	CO_3^{2-}	PO_4^{3-}
NH_4^+	S	S	S	S	N	S	S	S
Na^+	S	S	S	S	S	S	S	S
K^+	S	S	S	S	S	S	S	S
Ba^{2+}	S	S	S	I	S	D	I	I
Ca^{2+}	S	S	S	P	P	P, D	I	I
Mg^{2+}	S	S	S	S	I	D	I	I
Cu^{2+}	S	S	S	S	I	I	I	I
Fe^{2+}	S	S	S	S	I	I	I	I
Fe^{3+}	S	N	S	P	I	D	N	I
Zn^{2+}	S	S	S	S	I	I	I	I
Pb^{2+}	S	S	P	I	I	I	I	I
Ag^+	S	P	I	I	N	I	I	I
Hg_2^{2+}	S, D	P	I	I	N	I	I	I
Hg^{2+}	S	S	S	D	N	I	N	I

Table 9.2
Solubilities of Solid Ionic Compounds in Pure Water[a]

[a] S = is soluble in water; P = is partially soluble in water; I = is insoluble in water; D = decomposes; N = does not exist as an ionic solid.

9.4
SOLUBILITY OF COVALENT COMPOUNDS

An old but helpful rule states that *like dissolves like*. This means that nonpolar (or only slightly polar) solutes dissolve best in nonpolar solvents, and polar solutes dissolve best in polar solvents (Figure 9.3). The rule works well for nonpolar substances. Fats, oils, and greases (nonpolar or only slightly polar) dissolve well in nonpolar solvents such as toluene, C_7H_8. The forces that hold nonpolar molecules together are generally weak. Thus, the amount of energy needed to pull apart molecules of pure solute and to disrupt the attractive forces between molecules of pure solvent is small. And this energy can be balanced by the energy released through the interaction of solute and solvent molecules, although this too is slight.

The important thing for water solubility is the ability of water to form *hydrogen bonds* to the solute molecules. Thus, molecules containing a high proportion of nitrogen

Figure 9.3

(a) Lawn mowers with two-cycle engines are fueled and lubricated with a solution of nonpolar lubricating oil in nonpolar gasoline. (b) In Italian dressing, polar vinegar and nonpolar olive oil are mixed. The two liquids do not form a solution and will separate on standing. (c) Wine is a solution of polar alcohol in polar water.

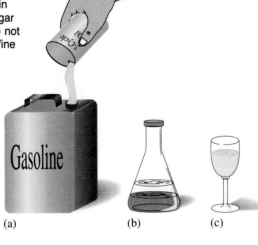

(a)   (b)   (c)

Figure 9.4

Water can form hydrogen bonds to molecules that contain nitrogen or oxygen atoms. Molecules with four or fewer carbon atoms per nitrogen or oxygen atom are usually soluble in water.

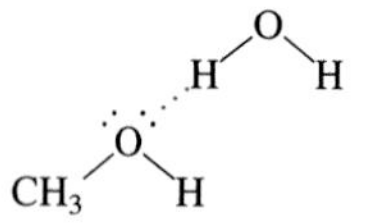

Methyl alcohol
(soluble in water)

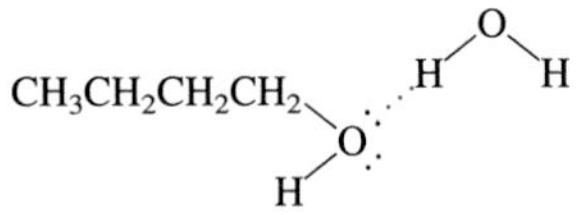

Butyl alcohol (partially soluble in water)

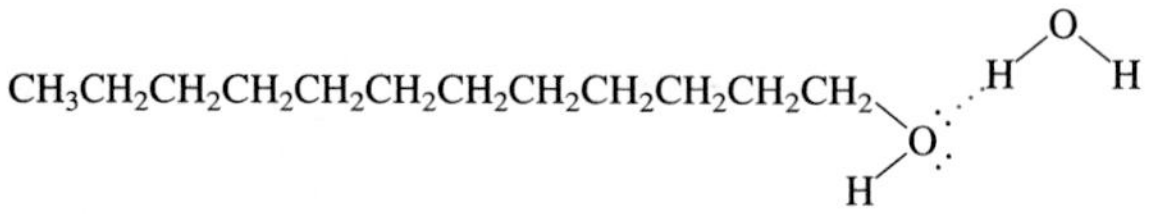

Lauryl alcohol
(essentially insoluble in water)

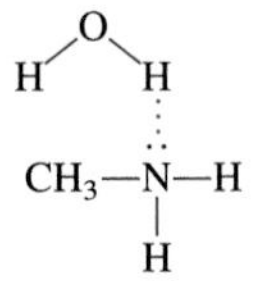

Methylamine
(soluble in water)

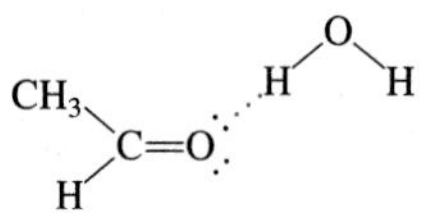

Acetaldehyde
(soluble in water)

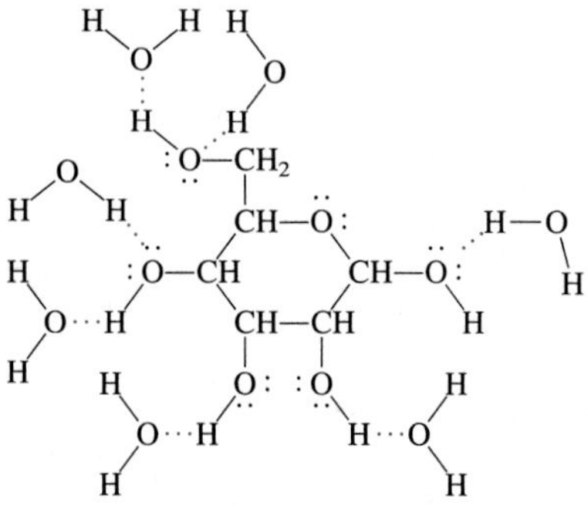

Glucose
(soluble in water)

or oxygen atoms will dissolve in water because these are the elements that can form hydrogen bonds (Section 8.4). One example is methyl alcohol (CH_3OH), which is completely miscible with water. Methylamine (CH_3NH_2) is also quite soluble in water. Figure 9.4 gives the structures of these molecules (and others mentioned in this section) and shows how they interact with water by forming hydrogen bonds.

There need not be a hydrogen atom on the oxygen (or nitrogen) atom of the solute molecules. Acetaldehyde (CH_3CHO) is completely miscible with water, even though none of the hydrogen atoms in the molecule is attached to the oxygen (see Figure 9.4). The hydrogen bonds depicted in the drawing incorporate hydrogen atoms covalently bonded to the oxygen of the water molecules. We will have many occasions to discuss the importance of hydrogen bonding in later chapters of this text.

Some fairly complex molecules, such as those of the sugars, are quite soluble in water. Glucose, or blood sugar ($C_6H_{12}O_6$), contains six carbon atoms, but its six oxygen atoms permit it to hydrogen bond to many water molecules, thereby making it water-soluble.

Hydrogen bonding, then, is the important factor in water solubility. Polarity alone is not enough. Methyl chloride (CH_3Cl) and methyl alcohol (CH_3OH) have about the same polarity, yet methyl chloride is essentially insoluble in water while methyl alcohol is completely miscible with water. Methyl chloride does not engage in hydrogen bonding, and methyl alcohol does. A few polar substances, such as hydrogen chloride (HCl), dissolve in water because they react chemically to form ions; these are discussed in subsequent chapters.

Terminology of Aqueous Systems

The importance of water as a solvent is reflected in the number of terms that have been coined to describe systems involving water. For example, the general term for the interaction of solvent with solute is **solvation,** but there is a special term for the interaction of water with a solute—*hydration*.

Certain compounds, such as calcium sulfate, tend to hold on to some water molecules even when they crystallize from solution. These compounds with their bound water molecules are called **hydrates.** The formulas for these crystals are written in such a way as to indicate the number of attached water molecules. Plaster of Paris is $(CaSO_4)_2 \cdot H_2O$ (one water molecule for every two calcium sulfate units). If more water is available, $CaSO_4 \cdot 2H_2O$ is formed (now there are two water molecules for every $CaSO_4$ unit). When a plaster cast is formed to immobilize a broken bone, the first hydrate is converted to the second. The powdery plaster of Paris changes to the rigid, protective material of the cast.

If a hydrate is heated strongly enough, the bound water can be driven off to produce the **anhydrous** compound, that is, the compound without water. Some hydrates lose their bound water simply on standing in dry air. Such compounds are said to be **efflorescent.** Other compounds form hydrates by picking up water from the atmosphere. These are described as **hygroscopic.** And finally, some compounds are so good at pulling water molecules from the air that they eventually dissolve in the water thus accumulated. These compounds are said to be **deliquescent.**

9.5
DYNAMIC EQUILIBRIA

For most substances, there is a limit to how much solute can be dissolved in a given volume of solvent. This limit varies with the nature of the solute and that of the solvent. Solubility also varies with temperature. We usually wash our clothes in hot water because most forms of ''dirt'' are more soluble at higher temperatures.[1]

Solubilities are often expressed in terms of grams of solute per 100 g of solvent. Since solubility varies with temperature, it is necessary to indicate the temperature at which the solubility is measured. For example, at 20 °C, 100 g of water will dissolve up to 109 g of sodium hydroxide (NaOH). At 50 °C, 145 g of NaOH will dissolve in 100 g of water. In a shorthand method, the solubility of sodium hydroxide is expressed as 109^{20} and 145^{50} (the 100 g of water is understood).

The solubility of sodium chloride, or common table salt (NaCl), is 36 g per 100 g of water at 20 °C. Suppose we place 40 g of NaCl in 100 g of water. What happens? Initially, many of the sodium (Na^+) ions and chloride (Cl^-) ions leave the surface of the crystals and wander about at random through the solvent. Some of the ions in their wanderings return to a crystal surface. These ions can even be trapped there, becoming once more a part of the crystal lattice. As more and more salt dissolves, the number of ''wanderers'' that return to be trapped once again in the solid state increases. Eventually (when 36 g of NaCl has dissolved), the number of ions leaving the surface of the undissolved crystals just equals the number returning. A condition of **dynamic equilibrium** has been established. The *net* amount of sodium chloride in solution remains the same despite the fact that there is still a lot of activity as ions come and go from the surface of the crystals. The net amount of undissolved crystals also remains constant (in this example, 4 g), although individual crystals may change in shape and size as ions leave one part of the crystal to enter solution while others are deposited at another part of the lattice. Some small crystals may even disappear as others grow larger, yet the net amount of undissolved salt does not change. The rate of dissolution just equals the rate of regrowth.

A solution that contains all the solute that it can at equilibrium and at a given temperature is said to be **saturated.** One that contains less than this amount is **unsaturated.** A solution with 24 g of NaCl in 100 g of water at 20 °C would be unsaturated because it could dissolve 12 g more at that temperature.

Equilibrium is established at a given temperature. If the temperature changes, more solute will dissolve or separate until equilibrium is established at the new temperature. Consider once more a sodium hydroxide solution. If we add 145 g of NaOH to 100 g of water at 20 °C, 109 g of the NaOH will dissolve, leaving 36 g as undissolved solute. If the solution is then warmed, more solute will dissolve. Finally, at 50°C all 145 g of NaOH will be in solution. The solubility of sodium hydroxide increases with increasing temperature.

Most solid compounds are increasingly soluble as the temperature is raised (Figure 9.5). This should not be surprising. As the temperature goes up, the motion of all the particles is increased. More ions are knocked loose from the lattice and go into solution. Further, it is more difficult for the crystal to recapture the ions that return to its surface, because they are moving at higher speeds. There are a few exceptions to this general rule of increased solubility at higher temperatures (see the graphs of NaCl and Na_2SO_4 in

---

[1] For some stains, such as that produced by blood, cold water is recommended. In this case, hot water causes a change in the structure of proteins in the blood that makes them more insoluble. Cold water does not change the blood in this way, and it is for this reason that cold water is recommended.

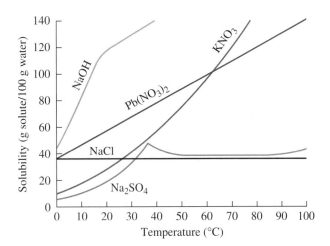

Figure 9.5
The effect of temperature on the solubility of several solids in water.

Figure 9.5). These exceptions probably involve changes in the hydration of the ions as the temperature increases.

Notice that for some substances, such as sodium hydroxide (NaOH) and lead nitrate [$Pb(NO_3)_2$], solubility increases rapidly as the temperature increases. The solubility of sodium chloride (NaCl), in contrast, changes very little over the indicated range of temperatures. The solubility of sodium sulfate (Na_2SO_4) first increases rather rapidly and then decreases, indicating a change in the hydration of the sodium (Na^+) ions and sulfate (SO_4^{2-}) ions.

If a saturated solution (with excess solid solute present) is cooled, more solute precipitates until the equilibrium is once again established at the lower temperature. For example, consider a saturated solution of lead nitrate at 90 °C. For each 100 g of water, there will be 120 g of $Pb(NO_3)_2$ dissolved. When the solution is cooled to 20 °C, the solution at equilibrium can contain only 54 g of $Pb(NO_3)_2$. The excess, 66 g, will precipitate out, increasing the amount of undissolved solute.

Now consider what would happen if one started to cool a saturated solution of lead nitrate with *no* excess solute present. Would lead nitrate precipitate? It might. Then again, it might not. There is no equilibrium—no crystals to capture the wandering ions. One might well be able to cool the solution to 20 °C without precipitation. Such a solution, containing solute in excess of that which it could contain if it were at equilibrium, is said to be **supersaturated.** This system is not stable because it is not at equilibrium. Solute may precipitate when the solution is stirred or if the inside of the container is scratched with a glass rod. Addition of a "seed" crystal of solute will nearly always result in the sudden precipitation of all the excess solute. Equilibrium is established rather rapidly when there is a crystal to which the ions can attach themselves (Figure 9.6).

Supersaturated solutions are not unknown in nature. Honey is one example; the solute is sugar. If honey is left to stand, the sugar crystallizes. We say, not very scientifically, that the honey has "turned to sugar." Supersaturated sugar solutions are fairly common in cooking. Jellies are one example. The sugar often crystallizes from jelly that has been standing for a long time.

Some wines have high concentrations of potassium hydrogen tartrate ($KHC_4H_4O_6$). When chilled, the solution becomes supersaturated. Crystals eventually form and settle out, often when the wine is stored in the consumer's refrigerator. Modern wineries solve this problem—and render the wine less acidic—by chilling the wine to −3 °C and adding tiny seed crystals of $KHC_4H_4O_6$. Precipitation is complete in 2 or 3 hr, and the crystals are filtered off.

A **precipitate** is an insoluble or nearly insoluble solid.

Figure 9.6
The addition of a seed crystal induces rapid crystallization of excess solute from a supersaturated solution.

9.6
SOLUTIONS OF GASES IN WATER

Figure 9.7
Bubbles of gas, primarily O_2 and N_2 (air), form when a beaker of water is heated.

You are no doubt familiar with a number of solutions in which gases are dissolved in water. A soft drink is a solution of carbon dioxide (and flavoring and sweetening agents) in water. Blood contains dissolved oxygen and carbon dioxide. Some familiar household cleansers contain ammonia (NH_3) dissolved in water. Formalin, used as a biological preservative, is an aqueous solution of formaldehyde (HCHO). Natural waters contain dissolved oxygen. Although only slightly soluble in water (0.0043 g of O_2 will dissolve in 100 g of water at 20 °C), this oxygen is vital to the survival of fish and other aquatic species.

Unlike most solid solutes, gases become less soluble as the temperature increases. Heat increases the molecular motion of both solute and solvent particles. In contrast to solid solutes, gaseous ones can escape from the solution when they reach the surface of a liquid in an open container.

We all heat water from time to time and know that long before the water begins to boil, bubbles of dissolved gases appear (Figure 9.7). These soon rise to the surface and escape into the atmosphere. Figure 9.8 shows the effect of temperature on the solubility of oxygen in water.

The solubility of gases in water also varies with the pressure of the gas (Section 7.9). The higher the pressure, the more gas one can dissolve in a given amount of water. Figure 9.9 shows how the solubility of oxygen in water at 25 °C varies with pressure. Soft drinks are bottled under pressure in order to dissolve large amounts of CO_2 gas. The fizz you hear when opening a can of soda is caused by the escape of the excess CO_2.

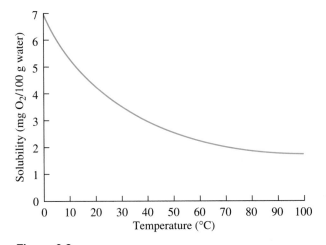

Figure 9.8

The solubility of oxygen at various temperatures at 1 atm of pressure.

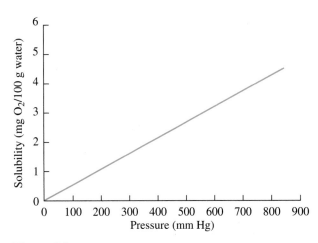

Figure 9.9

The effect of pressure on the solubility of oxygen in water at 25 °C.

9.7
MOLARITY

Most reactions of interest to us, including those in our bodies, take place in solution. A good cook may well get by with concentrations expressed as "a pinch of salt in a pint of water," but scientific work generally requires more precise measurement of amounts. We have already discussed the measurement of solubility in grams of solute per 100 g of solvent. Since substances react in small whole-number ratios of molecules or ions (Chapter 5), it is often convenient to measure the amount of solute in moles. The amount of solution is usually measured in liters or milliliters. A solution that contains 1 mol of solute per L of solution is called a 1 **molar** (or 1 M) solution. **Molarity,** then, is defined as the number of moles of solute divided by the number of liters of solution.

$$\text{Molarity (M)} = \frac{\text{moles of solute}}{\text{liters of solution}}$$

EXAMPLE 9.1

What is the molarity of a solution made by dissolving 3 mol of NaCl in enough water to make 6 L of solution?

SOLUTION

$$\frac{3 \text{ mol of solute}}{6 \text{ L of solution}} = 0.5 \text{ M}$$

Recall that the number of moles of a substance is calculated by division of mass in grams by molecular weight in grams per mole.

$$\text{Moles} = \frac{\text{mass (g)}}{\text{molecular weight (g/mol)}}$$

The molecular weight of NaOH is $23 + 16 + 1 = 40$ g/mol. In 120 g of NaOH there would be

$$120 \text{ g NaOH} \times \frac{1 \text{ mol NaOH}}{40 \text{ g NaOH}} = 3.0 \text{ mol NaOH}$$

EXAMPLE 9.2

What is the molarity of a solution of 300 g of $KHCO_3$ in enough water to make 10 L of solution?

SOLUTION
The molecular weight of $KHCO_3$ is $39 + 1 + 12 + (3 \times 16) = 100$ g/mol. The number of moles of $KHCO_3$ is

$$300 \text{ g KHCO}_3 \times \frac{1 \text{ mol KHCO}_3}{100 \text{ g KHCO}_3} = 3.00 \text{ mol KHCO}_3$$

The molarity of the solution is

$$\frac{3.00 \text{ mol}}{10 \text{ L}} = 0.30 \text{ M}$$

Practice Exercise
What is the molarity of a sodium chloride solution made by dissolving 11.7 g of NaCl in enough water to make 80 mL of solution?

Frequently we need to know how much solute is required for the preparation of a specified amount of solution of a given molarity. Recall that

$$\text{Molarity} = \frac{\text{moles of solute}}{\text{liters of solution}}$$

We can rearrange this to get

$$\text{Moles of solute} = \text{molarity} \times \text{liters of solution}$$

EXAMPLE 9.3

How many grams of NaOH (which has a molecular weight of 40 g/mol) are required for the preparation of 500 mL (0.50 L) of 6.0 M solution?

SOLUTION
We can use the rearranged equation to determine the number of *moles* of NaOH required.

$$\text{Moles of NaOH} = 6.0\text{ M} \times 0.50\text{ L} = \frac{6.0\text{ mol}}{1.0\text{ L}} \times 0.50\text{ L} = 3.0\text{ mol}$$

But we were asked for the number of grams and not moles, so we must now calculate the number of grams in 3 mol.

$$\text{Moles of NaOH} = \frac{\text{grams of NaOH}}{\text{molecular weight of NaOH}}$$

Rearranging, we get

$$\text{Grams of NaOH} = \text{moles of NaOH} \times \text{molecular weight of NaOH}$$

$$= 3.0\text{ mol} \times \frac{40\text{ g}}{1\text{ mol}} = 120\text{ g}$$

Practice Exercise
How many grams of potassium sulfate (K_2SO_4) are required to prepare 200 mL of a 0.225 M solution?

Quite often solutions of known molarity are available. How would you calculate the volume you would need in order to get a certain number of moles of solute? We can again rearrange the definition of molarity to obtain

$$\text{Liters of solution} = \frac{\text{moles of solute}}{\text{molarity}}$$

EXAMPLE 9.4

How many liters of 12 M HCl solution would one have to take to get 0.48 mol of HCl?

SOLUTION

$$\text{Liters of HCl solution} = \frac{0.48\text{ mol HCl}}{12\text{ M}} = \frac{0.48\text{ mol HCl}}{12\text{ mol HCl/L}} = 0.040\text{ L}$$

We would need 0.040 L (40 mL) of the solution to have 0.48 mol.

Practice Exercise
How many liters of 15 M aqueous ammonia (NH_3) solution do you need to get 0.45 mol of NH_3?

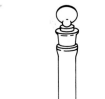

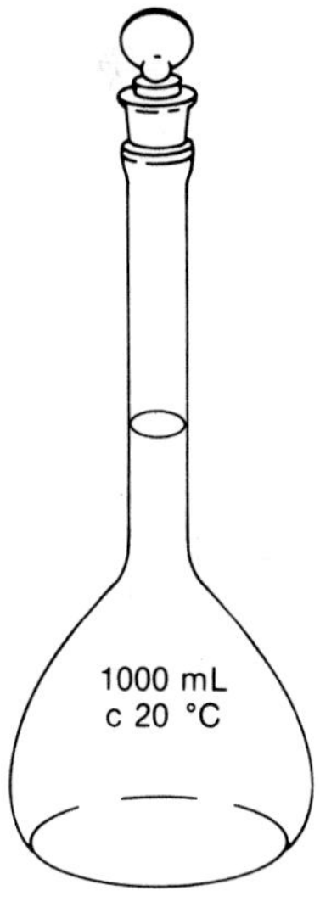

1000 mL
c 20 °C

Figure 9.10
A volumetric flask. When filled to the mark on the neck, the flask contains 1000 mL (1 L) of solution.

Remember that molarity is moles per liter of *solution,* not per liter of solvent. To make a liter of a 1 M solution, we weigh out 1 mol of solute and place it in a volumetric flask, which is a standard piece of laboratory glassware designed to contain a precisely specified volume of liquid (Figure 9.10). Enough water is added to dissolve the solute, and then more water is added to bring the volume up to the mark that indicates 1 L of *solution.* (Simply adding 1 mol of solute to 1 L of water would, in most cases, give more than 1 L of solution.)

9.8
PERCENT CONCENTRATIONS

For many purposes, it is not necessary to know the number of moles of solute but only the relative amounts of solute and solvent. One way of expressing this is in terms of the percent of solute. If both the solute and solvent are liquids, **percent by volume** is used.

$$\text{Percent by volume} = \frac{\text{volume of solute}}{\text{volume of solute + solvent}} \times 100$$

EXAMPLE 9.5

What is the percent by volume of a solution of 100 mL of alcohol dissolved in enough water to make 400 mL of solution?

SOLUTION

$$\text{Percent by volume} = \frac{100 \text{ mL alcohol}}{400 \text{ mL solution}} \times 100$$

$$= \frac{100 \text{ mL}}{400 \text{ mL}} \times 100 = 25\%$$

Practice Exercise
What is the percent by volume of a solution of 8 mL of alcohol dissolved in enough water to make 100 mL of solution?

EXAMPLE 9.6

What volumes of alcohol and water must one mix to obtain about 200 mL of a 40% by volume solution of alcohol in water?

SOLUTION

$$40\% = \frac{\text{milliliters of alcohol}}{200 \text{ mL total volume}} \times 100$$

Rearranging, we get

$$\text{Milliliters of alcohol} = \frac{40 \times 200}{100} = 80 \text{ mL of alcohol}$$

To find the amount of water required, subtract the volume of alcohol from the total volume.

$$200 \text{ mL} - 80 \text{ mL} = 120 \text{ mL of water}$$

When the 80 mL of alcohol is mixed with 120 mL of water, the total volume may not be exactly 200 mL, but it will be close.

Practice Exercise

What volumes of alcohol and water must one mix to obtain 100 mL of a 12% by volume solution of alcohol and water?

For solutions involving solutes that are solids, a more commonly used relationship is **percent by mass.**

$$\text{Percent by mass} = \frac{\text{mass of solute}}{\text{mass of solute} + \text{solvent}} \times 100$$

EXAMPLE 9.7

What is the percent by mass of a solution of 5 g of NaCl dissolved in 495 g (495 mL) of water?

SOLUTION

$$\text{Percent by mass} = \frac{5 \text{ g}}{495 \text{ g} + 5 \text{ g}} \times 100 = \frac{5 \cancel{\text{ g}}}{500 \cancel{\text{ g}}} \times 100 = 1\%$$

Recall that the density of water is 1 g/mL. Since 1 mL of water has a mass of 1 g, one can obtain 495 g of water simply by measuring 495 mL of water.

Practice Exercise

What is the percent by mass of a solution of 9 g of H_2O_2 dissolved in 300 g of water?

EXAMPLE 9.8

How would you prepare about 100 mL of a 5% NaOH solution?

SOLUTION

For dilute aqueous solutions, the density of the solution is approximately equal to the density of water. Therefore, let's assume that 100 mL of the solution will have a mass of about 100 g.

$$5\% = \frac{\text{grams of NaOH}}{100 \text{ g total}} \times 100$$

Rearranging, we get

$$\text{Grams of NaOH} = \frac{5 \times 100}{100} = 5 \text{ g}$$

Percent means "parts per hundred." To make a hundred "parts" of solution, weigh out 5 g of NaOH. Add 95 g (95 mL) of water. (The final volume might not be exactly 100 mL, but it will be close to that.)

EXAMPLE 9.9

How would you prepare about 500 mL of a 10% $NaHCO_3$ solution?

SOLUTION

$$10\% = \frac{\text{grams of } NaHCO_3}{500 \text{ g total}} \times 100$$

Rearranging, we get

$$\text{Grams of } NaHCO_3 = \frac{10 \times 500}{100} = 50 \text{ g}$$

Weigh out 50 g of $NaHCO_3$ and dissolve it in $500 - 50 = 450$ g (450 mL) of water. Again, the total volume will be about, rather than exactly, 500 mL.

Practice Exercise

How would you prepare about 100 mL of a 3% H_2O_2 solution?

Let us call attention to a special aspect of percent concentrations. If you are trying to calculate how much solute to weigh out for a particular percent concentration, it makes no difference what the solute is. A 10% solution (by mass) of NaOH contains 10 g of NaOH per 100 g of total solution. Similarly, 10% HCl and 10% $(NH_4)_2SO_4$ and 10% $C_{110}H_{190}N_3O_2Br$ each contain 10 g of the specified solute per 100 g of solution. For molar solutions, however, the mass of solute in a solution of specified molarity is different for different solutes. A liter of a 0.1 M solution requires 4 g (0.1 mol) of NaOH, 3.7 g (0.1 mol) of HCl, 13.2 g (0.1 mol) of $(NH_4)_2SO_4$, and 166 g (0.1 mol) of $C_{110}H_{190}N_3O_2Br$.

Low concentrations of solutes, such as those in blood and urine, are often expressed in milligrams per deciliter (mg/dL), which is equal to milligrams per 100 mL of solution.

$$\text{Milligrams/deciliter} = \frac{\text{milligrams of solute}}{100 \text{ mL of solution}}$$

For example, normally there is about 320–350 mg of sodium ion per 100 mL of blood plasma. Thus, the normal concentration of sodium ion in blood plasma ranges from 320 to 350 mg/dL. The use of milligrams per deciliter avoids the sometimes cumbersome use of decimal numbers. For example, the above concentrations of sodium ion would be expressed as 0.320–0.350% by mass.

For even more dilute solutions, concentrations are expressed in parts per million (ppm) or even parts per billion (ppb). For aqueous solutions, ppm is the same as milligrams per liter (mg/L), and ppb is identical to micrograms per liter (μg/L). These units are used to measure extremely low levels of toxic materials. Current concern over the purity of our drinking water centers on minute amounts of potentially dangerous compounds. For example, benzene is a compound that has been shown to produce leukemia-like symptoms in laboratory animals and humans. The Supreme Court has dealt with the question of whether the concentration of benzene in the air breathed by workers should be limited to 10 ppm or 1 ppm. The court decided that industries could not be required to lower the concentration from 10 to 1 ppm unless the higher concentration was *proven* to be dangerous.

As our technology becomes more sophisticated, our ability to detect minute quantities of materials increases. This increase in the sensitivity of analytical techniques raises questions. When a substance is first detected in the ppm or ppb range, is it a new contaminant in our environment or has it been there all along at levels that were previously undetectable?

9.9
RATIO CONCENTRATIONS

The strength of a solution is sometimes presented as a ratio, for example, 1:1000 or 1:2000. Unfortunately, there is no universal agreement as to the meaning of the ratio. The ratio 1:1000 may be used to indicate that 1 g of pure solute (perhaps a drug) is to be dissolved in 1000 mL of solvent. In the case of a liquid solute, the ratio may mean 1 mL of solute in 1000 mL of solution. Sometimes the ratio is presented as a dilution factor. Thus, 1:1000 may specify that 1 part of a sample is to be diluted with solvent until the volume of the solution is 1000 times the original volume. In specific instances, the meaning of the ratio may be clear, but because of the possibility for confusion, the use of ratio concentrations is being discouraged. Nonetheless, let us look at one example of the use of a ratio as a dilution factor.

EXAMPLE 9.10

The directions for use of a liquid drug indicate that it should be diluted 1:2000. How would 500 mL of the solution be prepared?

SOLUTION
The ratio should be read as

$$\frac{1 \text{ part solute}}{2000 \text{ parts solution}}$$

The final volume of solution required is 500 mL (500 ''parts'').

$$\text{Volume of solute} = 500 \text{ mL (parts) of solution} \times \frac{1 \text{ part solute}}{2000 \text{ parts solution}}$$

$$= 0.25 \text{ part (mL) solute}$$

To prepare the solution, 0.25 mL of the drug (as supplied) should be diluted to a total volume of 500 mL by addition of the appropriate solvent. This gives exactly the same final concentration as the dilution of 1 mL of drug to a total volume of 2000 mL.

9.10
STRENGTHS OF DRUGS

Many drug manufacturers market their products for use in strengths that are not specified by concentration units discussed above. Instead, they provide explicit instructions on how to prepare the drug for administration (referred to as **reconstitution** of the drug).

EXAMPLE 9.11

A drug label reads, "Add 5.0 mL of sterile water; each mL of solution will contain 0.50 g." If the dose prescribed for a patient is 750 mg, how many mL of the reconstituted drug should be administered?

SOLUTION

First, the drug must be reconstituted by addition of 5.0 mL of water to the container. After thorough mixing, the drug is ready for use. To calculate the dose in mL, the conversion factor given on the label (1.0 mL = 0.50 g) is used. Before that can be done, the specified dose (750 mg) must be converted to grams.

$$\text{Dose in mL} = 750 \text{ mg} \times \frac{1.0 \text{ g}}{1000 \text{ mg}} \times \frac{1.0 \text{ mL}}{0.50 \text{ g}} = 1.5 \text{ mL}$$

For some medications, the strength of the medication is reported literally in **units** (U). A unit of the substance is defined as an amount that elicits a particular effect. USP units, for example, meet standards set by the United States Pharmacopeia. The strengths of some vitamins and antibiotics, among other substances, are reported in this way.

EXAMPLE 9.12

The label on a container of crystalline penicillin G states, "Reconstitution: Add 9.6 mL diluent to provide 100,000 U per mL." If the prescribed dosage is 500,000 U, how much of the reconstituted drug should be administered?

SOLUTION

Let us assume that you have correctly reconstituted the drug. The rest is easy.

$$\text{Dose in mL} = 500,000 \text{ U} \times \frac{1 \text{ mL}}{100,000 \text{ U}} = 5 \text{ mL}$$

9.11
COLLIGATIVE PROPERTIES OF SOLUTIONS

Solutions have higher boiling points and lower freezing points than the corresponding pure solvent. The antifreeze in automobile cooling systems is there precisely because of these effects. If water alone were used as the engine coolant, it would boil away in the heat of summer and freeze in the depths of northern winters. Addition of antifreeze to the water raises the boiling point of the coolant and also prevents the coolant from freezing solid when the temperature drops below 0 °C. Salt is thrown on icy sidewalks and streets because the salt dissolves in the ice and lowers its freezing point. Therefore, the ice melts because the outdoor temperature is no longer low enough to maintain it as a solid.

The extent to which freezing points and boiling points are affected by solutes is related to the number of solute particles present in solution. The higher the concentration of solute particles, the more pronounced the effect. **Colligative properties** of solutions are those properties, like boiling point elevation and freezing point depression, that depend on the number of solute particles present in solution. For living systems, perhaps the most important colligative property is osmotic pressure. Osmosis is a phenomenon we shall discuss in detail in the next section.

Before we do that, we must first consider a rather subtle aspect of solute concentration. See if you can answer the following questions. How many solute particles are there in 1 L of a 1 M glucose, $C_6H_{12}O_6$, solution? How many in 1 L of 1 M sodium chloride, NaCl, solution? In 1 M calcium chloride, $CaCl_2$, solution? The answer "should" be 6×10^{23}, right? All of the solutions contain 1 mol of their respective solute compounds. However, the question did *not* ask for the number of *formula units;* it asked for the number of *solute particles*. Glucose is a covalent compound; its atoms are all firmly tied together in molecules. In the glucose solution, each solute particle is a glucose molecule, and there *are* 6×10^{23} of these. But in the sodium chloride solution, each formula unit of NaCl consists of a separate sodium ion (Na^+) and chloride ion (Cl^-) in solution. When sodium chloride dissolves in water, individual ions are carried off into solution by solvent molecules. So 6×10^{23} formula units of NaCl produce 12×10^{23} particles in solution— 6×10^{23} sodium ions plus 6×10^{23} chloride ions. Each calcium chloride unit produces three particles in solution—one calcium ion plus two chloride ions. Thus, the effect of a 1 M NaCl solution on colligative properties is twice that of a 1 M glucose solution. A calcium chloride solution has about three times the effect of a glucose solution of the same molarity.

When colligative properties (specifically osmotic pressure, Section 9.12) are being discussed, concentration may be reported in terms of *osmol per liter* or *osmolarity* (osmol/L). An **osmol** is a mole of solute particles. A 1 M NaCl solution contains 2 osmol of solute per liter of solution; a 1 M $CaCl_2$ solution contains 3 osmol/L. The osmolarity of a 1 M glucose solution is 1 osmol/L. The concentration of body fluids is typically reported in milliosmoles per liter (mosmol/L).

$$C_6H_{12}O_6 \xrightarrow[\text{dissociate}]{\text{does not}} C_6H_{12}O_6$$

1 M   1 osmol/L

$$NaCl \xrightarrow[\text{dissociate}]{\text{does}} Na^+ \ + \ Cl^-$$

1 M $\underbrace{\text{1 osmol/L} \quad \text{1 osmol/L}}_{\text{2 osmol/L}}$

$$CaCl_2 \xrightarrow[\text{dissociate}]{\text{does}} Ca^{2+} \ + \ 2\,Cl^-$$

1 M $\underbrace{\text{1 osmol/L} \quad \text{2 osmol/L}}_{\text{3 osmol/L}}$

9.12
SOLUTIONS AND CELL MEMBRANES: OSMOSIS

Solutions have a variety of characteristic properties. The particles are molecules, atoms, or ions. Once the solute and solvent are thoroughly mixed, the *solute does not settle out*. Molecular motion keeps the particles randomly distributed. The sugar in a can of soda, for instance, does not settle to the bottom on standing. The last drop is just as sweet as, but no sweeter than, the first. You cannot get the sugar out by passing the soda through a piece of filter paper. The sugar molecules go through the pores of the paper as readily as the water molecules. All true solutions can be filtered without removal of the solute.

Solutions may be colored, but they are *clear* and *transparent*. Copper sulfate ($CuSO_4$), dissolved in water, gives a beautiful clear blue solution. A beam of light will shine right through the solution but will not be visible in the solution. When the path of light through a solution is visible, then particles larger than those of a true solution are present (Section 9.13).

Solutions—both solute and solvent—will go through filter paper (see Figure 9.15a). Another way to describe the phenomenon is to say that the paper is **permeable** to water and other solvents and to solutions. Everyday experience tells us that other materials are **impermeable.** Water and solutions will not pass through the metal walls of cans nor through the glass walls of jars and bottles. Are there, perhaps, materials with intermediate properties? Materials that will pass solvent molecules but not solute molecules? Materials that are permeable to some solutes but not others? The answer is an emphatic yes! Many natural membranes are **semipermeable.** Cell membranes, the lining of the digestive tract, and the walls of blood vessels are all semipermeable; they allow certain substances to go through while holding others back.

The sieve model of osmosis pictures semipermeable membranes as having extremely small pores. The size of these pores is such that tiny water molecules can pass through, but larger particles, like sugar molecules, cannot. If a membrane with these characteristics separates a compartment containing pure water from one containing a sugar solution, an interesting thing happens. The volume of liquid in the compartment containing sugar increases, and the volume in the pure water compartment decreases.

Use Figure 9.11 as a reference. On both sides of the membrane all molecules are moving about at random, occasionally bumping against the membrane. If a water molecule happens to hit one of the pores, it passes through the membrane into the other compartment. When the much larger sugar molecule strikes a pore, it bounces back instead of through. The more sugar molecules there are in solution (i.e., the more concentrated the solution), the smaller the chance that a water molecule will strike a pore. Thus, in our example, there will be a *net* flow of water from the left compartment into the right compartment. This net diffusion of water through a semipermeable membrane is called **osmosis.** During osmosis, there is always a *net* flow of solvent from the more dilute (or pure solvent) area to the compartment in which the solution is more concentrated (Figure 9.12). The net diffusion of water would be *from* a 5% sugar solution into a 10% sugar solution.

As the liquid level in the right compartment builds up and that in the left compartment drops, the increased weight of fluid in the right compartment exerts pressure that makes it more difficult for additional water molecules to enter that compartment. (See

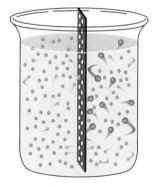

• Solvent particles
◦ Solute particles

Figure 9.11
The sieve model of osmosis holds that the semipermeable membrane has pores large enough to permit the passage of water molecules (small particles) but too small to permit the passage of larger molecules.

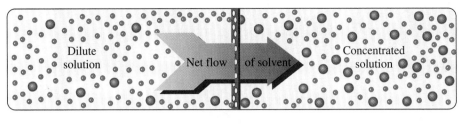

● Solute particles   ● Solvent particles

Figure 9.12
Net solvent flow through a semipermeable membrane occurs spontaneously in only
one direction, from the compartment containing dilute solution (or pure solvent) into
the compartment of concentrated solution. Remember—ordinarily the terms *dilute* and
concentrated are used to describe the concentration of *solute.* The net flow of solvent
is from where the solvent is more concentrated to where the solvent is less concentrated.

Section 7.3 on how a barometer works.) Eventually the buildup of pressure is sufficient to
prevent further net flow of water into that compartment. Things have not really come to
a standstill; the rates at which water molecules move back and forth are just equal.

Instead of waiting for the liquid level to build up and stop the new flow of water, we
can apply an external pressure to the compartment containing the more concentrated
solution and accomplish the same thing. The precise amount of pressure needed to prevent
the net flow of solvent from the dilute solution to the concentrated one is called the
osmotic pressure. The magnitude of the osmotic pressure depends only on the concentration of solute particles, that is, on the osmolarity of the solution. The more particles (the
higher the osmolarity), the greater the osmotic pressure. You can think of osmotic pressure or osmolarity as a measure of the tendency of a solution to draw solvent into itself.

Living cells can be regarded as selectively permeable bags filled with solutions of
ions, small and large molecules, and still larger cell components. Normally, the fluid
surrounding a cell has the same osmolarity as the fluid within the cell. Flow in and flow
out are about equal. If a cell is surrounded by a solution of lower osmolarity (a **hypotonic
solution**), there is a net flow of water into the cell. The cell swells. It may even burst. The
rupture of a cell by a hypotonic solution is called **plasmolysis.** If the cell is a red blood
cell, the more specific term **hemolysis** is used. What happens if a cell is placed in a
solution of higher osmolarity (a **hypertonic solution**)? The net flow of water is then out of
the cell. The cell wrinkles and shrinks (Figure 9.13). This shriveling, or **crenation,** is a
process that can also lead to the death of the cell.

Solutions that exhibit the same osmotic pressure as that of the fluid inside the cell are
said to be **isotonic.** In replacing body fluids intravenously, it is important that the fluid be
isotonic. Otherwise, hemolysis or crenation results, and the patient's well-being is seriously jeopardized. A 0.92% sodium chloride solution, called physiological saline, and a
5.5% glucose (also called dextrose or blood sugar) solution are isotonic with the fluid
inside red blood cells. The ''D5W'' so often referred to by television's doctors and
paramedics is an approximately 5% solution of dextrose (the *D*) in water (the *W*). A
0.92% sodium chloride solution is about 0.16 M. A 5.5% glucose solution is approximately 0.31 M. The osmolarity of the solutions is about the same.

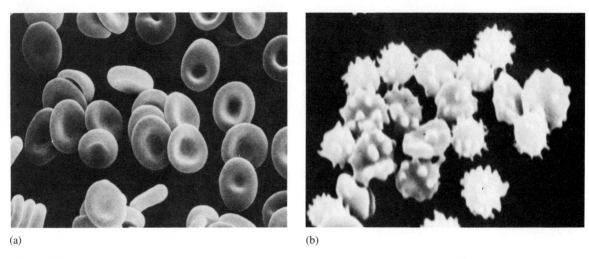

(a)   (b)

Figure 9.13
(a) Normal human red blood cells. (b) After exposure to a hypertonic solution, the cells
are wrinkled and shriveled.

EXAMPLE 9.13

What are the osmolarities of the above two isotonic solutions?

SOLUTION
The concentration of the glucose (dextrose) solution is 0.31 M. Each glucose
molecule represents one particle in solution; therefore the osmolarity of the solution is 0.31 osmol/L or 310 mosmol/L. Each formula unit of NaCl provides two
particles in solution. Therefore, the osmolarity of the 0.16 M solution is
0.32 osmol/L or 320 mosmol/L.

Isotonic solutions have their limitations. Consider the case of a patient who is to be
fed intravenously. There is a limit to the amount of water such a patient can handle in a
day—about 3 L. If an isotonic solution of 5.5% glucose were to be used, 3.0 L of the
solution would supply about 160 g of glucose. This would yield about 4.0 kcal/g to the
patient, or a total of about 640 kcal, an amount woefully inadequate. Even a resting
patient requires about 1400 kcal/day. And for a person suffering from serious burns, for
example, requirements as high as 10,000 kcal/day have been recorded. We have oversim-
plified the situation. Other vital nutrients are required by such patients, and, in fact, a
person can normally be given up to 1200 kcal through carefully formulated solutions.
This still falls short of the requirements of many seriously ill people. One answer to the
problem is to use very concentrated solutions (about six times as concentrated as isotonic
solutions). Instead of being administered through the vein of an arm or a leg, this solution
is infused directly through a tube into the superior vena cava, a large blood vessel leading
to the heart (Figure 9.14). The large volume of blood flowing through this vein quickly
dilutes the solution to levels that will not damage the blood. With this technique, patients
have been given 5000 kcal/day and have even gained weight.

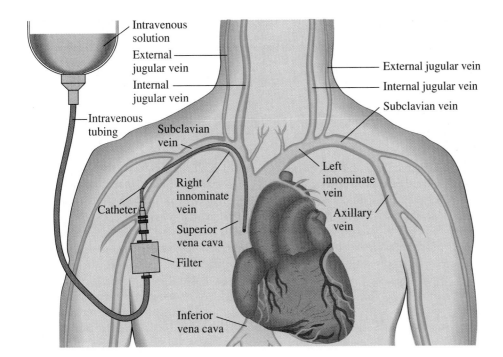

9.13
COLLOIDS

If potassium permanganate ($KMnO_4$) is dissolved in water, the molecules become intimately mixed. The solution is homogeneous; that is, it has the same properties throughout. The $KMnO_4$ cannot be filtered out by ordinary filter paper (Figure 9.15a), nor does it settle out on standing. On the other hand, if one tries to dissolve sand in water, the two substances may momentarily appear to be mixed, but the sand rapidly settles to the bottom. The temporary dispersion of sand in water is called a **suspension.** By passing the suspension through a filter paper, one can trap the sand (Figure 9.15b). The mixture is obviously heterogeneous, for part of it is clearly sand with one set of properties and part of it is water with another set of properties.

Is there nothing in between the true solution, with particles the size of ordinary molecules and ions, and suspensions, with gross chunks of insoluble matter? Yes, there is something else. And it is a highly important arrangement of matter at that—the **colloidal dispersion,** or simply **colloid.**

Colloids are defined not by the kind of matter they contain but by the *size of the particles* involved. True solutions have particles of about 0.05 to about 0.25 nanometers (nm) in diameter (1 nm = 10^{-9} m). Suspensions have particles with diameters of 100 nm or more. Particles intermediate between these are said to be colloidal.

The properties of a colloidal dispersion are different from those of a true solution and also different from those of a suspension (Table 9.3). Colloidal particles cannot be filtered by filter paper, nor do they settle on standing. Colloidal dispersions usually appear milky or cloudy. Even those that appear clear will reveal a beam of light that passes through the

(a)   (b)

Figure 9.15
(a) This aqueous solution is colored by its solute, potassium permanganate. After filtering the color remains in the solution because the filter paper is unable to stop the solute particles. (b) A suspension of sand in water can be separated by filtration. The suspended material is trapped by the filter paper, giving a colorless filtrate.

Table 9.3
Properties of Solutions,
Colloids, and Suspensions

Property	Solution	Colloid	Suspension
Particle size	0.1–1.0 nm	1–100 nm	>100 nm
Settles on standing?	No	No	Yes
Filter with paper?	No	No	Yes
Separate by dialysis?	No	Yes	Yes
Homogeneous?	Yes	Borderline	No

dispersion (Figure 9.16). This phenomenon, called the **Tyndall effect,** is not observed in true solutions. Colloidal particles, unlike tiny molecules, are large enough to scatter and reflect light off to the side. You have probably observed the Tyndall effect in a movie theater. The shaft of light that originates in the projection booth and ends at the movie screen is brought to you through the courtesy of colloidal dust particles in the air.

There are eight different kinds of colloidal dispersions, based on the physical state of the particles themselves (the dispersed phase) and that of the ''solvent'' (the dispersing phase). These are listed, with examples of each, in Table 9.4.

The most important colloids in biological systems are emulsions—either liquids or solids dispersed in water. Living matter consists of colloidal particles as well as simple ions and molecules. Substances with high molecular weights, such as starches and proteins, often form colloidal dispersions rather than true solutions in water.

In most colloidal dispersions, the particles are charged. This charge is often due to the adsorption of ions on the surface of the particle. A given colloid will preferentially adsorb only one kind of ion (either positive or negative) on its surfaces; thus, all the particles of a given colloidal dispersion bear like charges. Since like charges repel, the particles tend to stay away from one another. They cannot come together and form particles large enough to settle out. These colloids can be made to coalesce and separate out by addition of ions of opposite charge, particularly those doubly or triply charged. Aluminum chloride ($AlCl_3$), which contains Al^{3+} ions, is great for breaking up colloids in which the particles are negatively charged.

Figure 9.16
(a) A beam of light passes through a true solution without noticeable effect. The same beam passing through a colloidal solution is clearly visible because the light is scattered by the colloidal particles. This phenomenon is called the Tyndall effect. (b) Light piercing fog, which is a dispersion of water in air, is another example of the Tyndall effect.

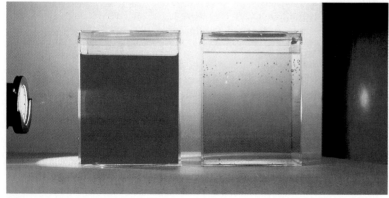

(a)   (b)

Table 9.4
Types of Colloidal
Dispersions

Type	Particle Phase	Medium Phase	Example
Foam	Gas	Liquid[b]	Whipped cream
Solid foam	Gas	Solid	Floating soap
Aerosol	Liquid	Gas	Fog, hair spray
Liquid emulsion	Liquid	Liquid	Milk, mayonnaise
Solid emulsion	Liquid	Solid	Butter
Smoke	Solid	Gas	Fine dust or soot in air
Sol[a]	Solid	Liquid	Starch solutions, jellies
Solid sol	Solid	Solid	Pearl

[a]Sols that set up in semisolid, jellylike form are called gels.
[b]By their very nature, gas mixtures always qualify as solutions. The size of particles in gas mixtures and their homogeneity fulfill the requirements of a true solution.

Some colloids are stabilized by addition of a material that provides a protective coating. Oil is ordinarily insoluble in water, but it can be emulsified by soap. The soap molecules form a negatively charged layer about the surface of each tiny oil droplet. These negative charges keep the oil particles from coming together and separating out (see Section 21.5). In a similar manner, bile salts emulsify the fats we eat, keeping them dispersed as tiny particles that can be more efficiently digested. Milk is an emulsion in which fat droplets are stabilized by a coating of casein, a protein. Casein, soap, and bile salts are examples of **emulsifying agents,** substances that stabilize emulsions.

9.14
DIALYSIS

To live, an organism must take in food and get rid of toxic wastes. The nutrients necessary to life must enter into cells, and the wastes must leave the cells. Generally, then, cell membranes must permit the passage not only of water molecules but also of other small molecules and ions. At the same time, it is important that large molecules and colloidal particles not be lost from the cell. Membranes that pass small molecules and ions while holding back large molecules and colloidal particles are called **dialyzing membranes.** The process is called **dialysis.** It differs from osmosis in that osmotic membranes pass only solvent molecules. In dialysis, molecules and ions always diffuse from areas of higher concentration to areas of lower concentration.

Dialyzing membranes are used in laboratories to purify colloidal dispersions of smaller molecules and ions. The mixture to be purified is placed in a bag made of a dialyzing membrane. The bag is placed in a container of pure water (Figure 9.17). The

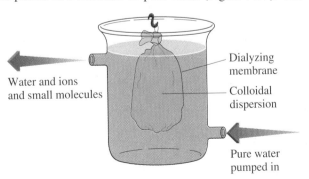

Water and ions
and small molecules

Dialyzing
membrane

Colloidal
dispersion

Pure water
pumped in

Figure 9.17
A simple apparatus for dialysis.

Figure 9.18
In renal dialysis, elaborate machinery substitutes for human kidneys that are no longer capable of cleansing the blood.

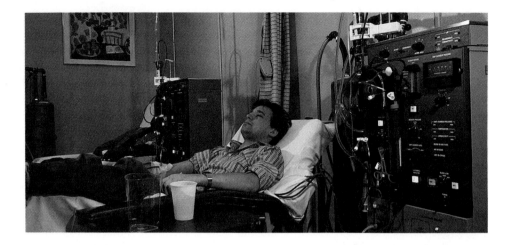

ions and small molecules pass out through the membrane, leaving the colloidal particles behind. Pure water is continuously pumped past the bag, and the unwanted small particles are carried away. The dialyzing membrane may be an animal bladder, or it may be an artificial bladder made from cellophane or from collodion, a semisynthetic plastic made by treatment of cellulose with nitric acid, alcohol, and camphor.

The kidneys are a complex dialyzing system responsible for the removal of certain potentially toxic waste products from the blood. By first gaining an understanding of the function of living kidneys, scientists have been able to construct artificial ones (Figure 9.18). Artificial kidneys are more elaborate in structure than the simple apparatus shown in Figure 9.17, but their principle of operation is the same. We discuss the operation of the kidneys in Section 28.12.

Many substances—medications, poisons, anesthetics—are thought to act by changing the permeability of membranes. Methyl mercury, a powerful poison that acts on the nervous system, is thought to act by making the membranes of nerve cells leakier than normal. A person going into shock has leaky capillaries that allow proteins and other colloidal particles as well as fluids to escape into the spaces between cells. If untreated, the cells may die from lack of oxygen and nutrients. As research increases our understanding of cellular membranes, we will undoubtedly gain a better understanding of drug action and of poisoning, health, and disease.

EXERCISES

1. Define or explain and, where possible, illustrate the following terms.

a.	solution	**b.**	solvent
c.	solute	**d.**	aqueous
e.	soluble	**f.**	insoluble
g.	miscible	**h.**	dilute
i.	concentrated	**j.**	hydration
k.	saturated	**l.**	unsaturated
m.	supersaturated	**n.**	precipitate
o.	molarity	**p.**	percent by volume
q.	percent by mass	**r.**	milligrams per deciliter
s.	ppm	**t.**	ppb
u.	USP unit	**v.**	colligative property

2. Define or explain and, where possible, illustrate the following terms.

a.	permeable	**b.**	semipermeable
c.	impermeable	**d.**	osmosis
e.	osmolarity	**f.**	osmotic pressure
g.	isotonic solution	**h.**	hypotonic solution
i.	hypertonic solution	**j.**	crenation
k.	plasmolysis	**l.**	hemolysis
m.	suspension	**n.**	colloidal dispersion
o.	Tyndall effect	**p.**	emulsifying agent
q.	dialysis	**r.**	hydrate
s.	anhydrous compound	**t.**	hygroscopic compound
u.	efflorescent compound	**v.**	deliquescent compound

3. Describe the properties of a true solution with respect to the following.
   a. size of solute particles
   b. distribution of solute and solvent particles
   c. filtration
   d. Tyndall effect
   e. color and clarity

4. Use the kinetic-molecular theory to explain why most solid solutes become more soluble with increasing temperature but gases become less soluble.

5. Fish live on oxygen dissolved in water. Would it be a good idea to thoroughly boil the water you place in a fishbowl? Explain.

6. Without referring to Table 9-2, indicate which compounds you would expect to be soluble in water. Explain each answer. You may use the periodic table.
   a. NaBr   b. $(NH_4)_2CO_3$
   c. $Ca(NO_3)_2$ d. RbCl
   e. $BaCO_3$ f. $PbSO_4$
   g. $FePO_4$ h. LiOH

7. Ethyl alcohol, CH_3CH_2OH, is soluble in water; ethyl chloride, CH_3CH_2Cl, is insoluble in water. Explain.

8. In a dynamic equilibrium, two processes are occurring at the same rate. In the equilibrium involving a saturated solution, for which processes are the rates equal?

9. A supersaturated solution is maintained at constant temperature, and precipitation is induced by the addition of a seed crystal. When no more solid appears to precipitate, is the solution saturated, unsaturated, or supersaturated?

10. What is the percent by volume concentration of a mixture containing 60 mL of water and enough alcohol to give 600 mL of solution? Which component is the solvent and which is the solute?

11. How would you prepare exactly 2.0 L of a 2.0% by volume aqueous solution of acetic acid?

12. How much acetone and water should be mixed to give about 50 mL of an approximately 20% by volume aqueous solution of acetone?

13. If 4.0 g of NaOH is dissolved in 96 g of water, what is the percent by mass concentration of the solution?

14. If 100 g of glucose is mixed with 400 g of water, what is the percent by mass concentration of the solution?

15. If 2.0 g of $NaHCO_3$ is mixed with 38 mL of water, what is the percent by mass concentration of the solution?

16. How would you prepare 100 g of a 3.0% by mass aqueous solution of $C_{12}H_{22}O_{11}$?

17. Explain how you would prepare 5.0 kg of a 10% by mass solution of NaCl in water?

18. How would you prepare about 50 mL of a 2.0% by mass aqueous solution of Na_2CO_3?

19. Arrange the following in order of increasing concentration: 1%, 1 ppb, 1 ppm, 1 mg/dL.

20. On the average, glucose makes up about 0.10% by mass of human blood. What is the concentration in mg/dL?

21. A liquid drug is to be diluted for use. How would you prepare the following amounts of the drug?
    a. 100 mL of a $1:1000$ dilution
    b. 1 L of a $1:2000$ dilution

22. If 6.0 mol HCl is dissolved in 3.0 L of solution, what is the molarity of the solution?

23. If 0.50 mol NaCl is dissolved in water to give a total of 250 ml of solution, what is the molarity of the solution?

24. If 1.0 L of solution contains 9.8 g of H_2SO_4, what is the molarity of the solution?

25. If 500 mL of solution contains 60 g NaOH, what is the molarity of the solution?

26. How would you prepare 2.0 L of 0.20 M aqueous NaOH solution?

27. Describe the preparation of 100 mL of 1.00 M aqueous NH_4Cl solution.

28. Explain how to prepare 0.40 L of 0.40 M aqueous NaCl solution.

29. Describe how you would prepare each of the following aqueous solutions. In each case, state how much solute you would measure and how much solvent you would use or how much solution you would make.
    a. exactly 100 mL of a 30% by volume alcohol solution
    b. about 100 mL of a 5.0% by mass NaOH solution
    c. exactly 10 g of a 10% by mass $NaHCO_3$ solution
    d. exactly 1.00 L of a 2.00 M NaCl solution
    e. about 500 mL of a 0.92% by mass NaCl solution
    f. exactly 500 mL of a 0.50 M $C_6H_{12}O_6$ solution

30. Reconstitution of an enzyme preparation used in the treatment of infection gives a concentration of 25,000 U/2.5 mL. How much of this solution should be administered to provide a single dose of 5000 U?

31. How many solute particles does each formula unit of the following give in solution? A quick review of covalent and ionic compounds (Chapter 4) might prove useful.
    a. KCl   b. CH_3OH c. $CaBr_2$
    d. NaOH   e. $(NH_4)_2SO_4$ f. $Al(NO_3)_3$

32. How many osmol are there in 1 mol of each of the compounds shown in Exercise 31?

33. To provide 1.0 osmol, how many moles of each of the compounds listed in Exercise 31 are required?

34. What is the osmolarity of a 0.50 M $(NH_4)_3PO_4$ solution?

35. If two containers of gas at different pressures are connected, the net diffusion of gas is from the *higher* to the *lower* pressure. The net diffusion of water in osmosis is from the *lower* to the *higher* concentration solution. How would you explain this apparent difference?

36. Which has the higher osmotic pressure: a 1% NaCl solution or a 5% NaCl solution?

37. For each pair of solutions, indicate which has the higher osmotic pressure.
    a. 0.1 M $NaHCO_3$, 0.05 M $NaHCO_3$
    b. 1 M NaCl, 1 M glucose
    c. 1 M NaCl, 1 M $CaCl_2$
    d. 1 M NaCl, 3 M glucose

38. For each pair of solutions, indicate which has the higher osmotic pressure.
    a. 1 osmol/L NaCl, 1 osmol/L CaCl$_2$
    b. 1 osmol/L NaCl, 2 osmol/L glucose
    c. 1 osmol/L NaHCO$_3$, 0.5 osmol/L NaHCO$_3$

39. If red blood cells were placed in 5.5% NaCl solution, what would be the effect on the cells? What if they were placed in 0.92% glucose solution?

40. Compare and contrast true solutions, colloidal dispersions, and suspensions.

41. Compare and contrast osmosis and dialysis with respect to the kind of particles that pass through the membrane.

42. Complete the following table.

Concentration of Solute (g/L)	Molarity (mol/L)	Molecular Weight of Solute (g/mol)
98	1.0	—
32	0.5	—
0.74	0.01	—
—	0.1	26
—	0.025	80
120	—	40
17	—	68

Know/ be able to do

43. Benzene (C$_6$H$_6$) is a nonpolar solvent. Would you expect NaCl to dissolve in benzene? Explain.

44. Motor oil is nonpolar. Would you expect it to dissolve in water? In benzene? Explain.

45. Bubbles of carbon dioxide escape when the cap is removed from a bottle of soda. Explain.

46. Alcohol that is used as a disinfectant to clean the skin prior to an injection is actually a solution of 3 parts water and 7 parts alcohol. Which component is the solvent, and which is the solute?

47. Explain why table salt (NaCl) dissolves in water.

48. You have a stock solution of 6 M HCl. How many moles of HCl are there in the following amounts of solution?
    a. 1.0 L   **b.** 100 mL   **c.** 1 mL

49. On the average, glucose (C$_6$H$_{12}$O$_6$) makes up about 0.10% by weight of human blood. How much glucose is there in 1 kg of blood?

50. A cyanide solution, made by adding 1 lb of sodium cyanide (NaCN) to 1 ton of water, is used to leach gold from its ore. What is the concentration of the cyanide solution in each of the following units of concentration?
    a. percent by mass   **b.** ppm
    c. g/kg   **d.** mol/L

Chapter 10
ACIDS AND BASES I

Most foods are either neutral or slightly acidic. We use bases to counter "acid indigestion."

Central to much of the chemistry in our everyday lives is the chemistry of two interrelated classes of compounds called *acids* and *bases*. We use many of these compounds around the house (Figure 10.1). We clean with ammonia and lye (sodium hydroxide), two familiar bases. We use vinegar (acetic acid) in cooking and as dressing on salads. We take vitamin C (ascorbic acid) and drink beverages made tart by citric acid and phosphoric acid. We treat our excess stomach acid (hydrochloric acid) with a variety of bases, including milk of magnesia (magnesium hydroxide) and baking soda (sodium bicarbonate). Our bodies produce and consume acids and bases, maintaining the delicate balance necessary to our good health and well-being.

The four tastes are related to acid–base chemistry. Acids taste *sour* and bases taste *bitter*. Salts, compounds formed when acids react with bases, taste *salty*. *Sweet* tastes are related to acid–base chemistry in a more subtle way. To taste sweet, a compound must have a hydrogen-bond donor group (acidic), a hydrogen-bond acceptor group (basic), and a nonpolar group, all arranged in just the proper geometry to fit the sweet-taste receptor.

Acid rain is an environmental problem in industrialized countries. Bitter, undrinkable alkaline (basic) water is often all that is available in areas with dry climates. Much of our understanding of air and water pollution depends on our knowledge of acids and bases.

229

Figure 10.1

Acids (left) and bases (right) found in the home.

The quantitative aspects of acid–base chemistry are discussed in Chapter 11. In this chapter, we will look at some of the more qualitative properties of acids and bases—what they are, how they are produced, how they are named, and how they react chemically.

10.1
ACIDS: THE ARRHENIUS CONCEPT

Acidic substances have been known for millennia. However, it was only in 1887 that Svante Arrhenius, a young Swedish chemist still in graduate school, proposed definitions of acids and bases that are still in common use today. **Acids** were generally recognized as substances that, in aqueous solution, would

1. Turn the indicator dye litmus from blue to red.
2. React with active metals (such as zinc, iron, and tin) to dissolve the metal and produce hydrogen gas.
3. Taste sour when diluted enough to be safely tasted.
4. React with bases to form water and ionic compounds called salts.

Litmus is a dye of plant origin isolated from certain mosses.

Concentrated or dilute acids used in the laboratory are quite corrosive; they should never be tasted.

Arrhenius proposed that these characteristic properties of acids are actually properties of the hydrogen ion (H^+) and that acids are compounds that yield hydrogen ions in aqueous solutions. Arrhenius's definition is still useful—as long as we deal only with aqueous solutions.

How can you tell when a compound is an acid? You can tell by dissolving some of the compound in water. Since tasting may be hazardous (some acids are poisonous, and nearly all are very corrosive unless they have been highly diluted), you can stick a piece of blue litmus paper into a sample of the compound. If the paper turns red, the compound is an acid.

Many foods are acidic. Vinegar contains 4–10% acetic acid. Citrus fruits and many fruit-flavored drinks contain citric acid. If a food tastes sour, most likely it contains one or more acids. Lactic acid is formed in sour milk and is responsible for the tart taste of yogurt. Phosphoric acid is used to impart tartness to beer and some forms of soft drinks.

Arrhenius's theory has been modified somewhat through the years. We know, for example, that simple hydrogen ions do not exist in water solutions. When a hydrogen atom is stripped of its only electron, all that is left is the bare nucleus, which consists of a single proton. Other positive ions, even the relatively small lithium ion (Li^+), still have their nuclei shielded by electrons remaining in the inner levels. Thus, the hydrogen ion *(proton)* is too reactive to exist as a stable ion in solution. A wealth of experimental evidence indicates that in water solution the properties of acids are due to H_3O^+. This ion is a water molecule to which a hydrogen ion (H^+) has been added. Its electron dot formula is

$$\left[H : \ddot{O} : H \atop \ddot{H} \right]^+$$

This species is called the **hydronium ion.** Today, we might well say that an acid is a substance that produces hydronium ions when dissolved in water. Even this is an oversimplification, for these ions are hydrated by other water molecules. For most purposes, though, we can use the simple hydronium ion (H_3O^+) and ignore any further hydration. The hydronium ion is quite reactive; it can readily transfer a proton to other molecules and ions. Thus, to talk about protons when we mean their source (hydronium ions) is permissible as long as we understand that we are using a simplification of the real situation.

In water, then, the properties of acids are those of the hydronium ion. It is the hydronium ion that turns litmus red, tastes sour, and reacts with active metals and bases. Nowadays, however, chemists and other scientists work with solvents other than water. They have broadened the concept of an acid to operate in these other media. Generally speaking, an **acid** is a **proton donor**—to any receptor, not just to water.

Some acids give up one hydrogen ion (proton) per molecule. These are called **monoprotic acids.** An example is hydrogen chloride. When this compound, a gas, is dissolved in water, we get one hydronium ion per molecule of hydrogen chloride.

$$HCl(g) \ + \ H_2O \ \longrightarrow \ H_3O^+(aq) \ + \ Cl^-(aq)$$

(The symbol *g* in parentheses indicates the gaseous state; *aq* in parentheses indicates an aqueous solution.)

Other acids may give up more than one hydrogen ion per molecule. Sulfuric acid (H_2SO_4) is a **diprotic acid;** it is capable of giving up two protons. Phosphoric acid (H_3PO_4) is a **triprotic acid.** A general term for acids that give up more than one proton is **polyprotic acid.** You should not assume that all the hydrogen atoms in a compound are acidic, meaning that they are released in water solutions. For example, none of the hydrogens of methane (CH_4) is given up in aqueous solution. Only one of the hydrogens in acetic acid ($C_2H_4O_2$) is acidic. For this reason, the formula for acetic acid is frequently written CH_3COOH or $HC_2H_3O_2$ to emphasize that only one proton is released. You will gain experience in determining which hydrogen atoms in a molecule are acidic as you learn more about molecular structure.

Acids are *proton* (H^+) *donors.*

10.2
STRONG AND WEAK ACIDS

When gaseous hydrogen chloride (HCl) reacts with water, it reacts completely to give hydronium ions and chloride ions.

$$HCl + H_2O \longrightarrow H_3O^+ + Cl^-$$

Similarly, the poisonous gas hydrogen cyanide (HCN) reacts with water to produce hydronium ions and cyanide ions.

$$HCN + H_2O \rightleftharpoons H_3O^+ + CN^-$$

The reaction in the latter case takes place only to a slight extent, however, and produces relatively few hydronium ions. Note the unequal double arrow, which indicates that the equilibrium strongly favors the undissociated acid and water. Most of the hydrogen cyanide remains as HCN molecules.

Acids that react completely with water are called **strong acids.** Those that react only slightly are called **weak acids.** Note that Table 10.1 classifies each acid as strong, moderately strong, or weak. This classification is a measure of the degree of reaction. *Strong acids* react completely, producing a relatively large number of hydronium ions. *Weak acids* react only partially, producing fewer hydronium ions.

Strong acids, when moderately concentrated, can cause serious damage to skin and flesh. They eat holes in clothes made of natural fibers such as cotton, silk, and wool. Strong acids also destroy most synthetic fibers such as nylon, polyester, and acrylics. Care always should be taken to prevent spills on skin and clothing. In very dilute solutions, strong acids can be harmless. Hydrochloric acid, for example, is produced in dilute solutions in our stomachs, where it aids in the digestion of certain foodstuffs. Even there, however, under certain conditions, it can cause problems—as anyone with an ulcer can attest. Concentrated weak acids can cause problems too. Although it is pleasant to eat salads with vinegar as a condiment, concentrated solutions of acetic acid are corrosive to the skin and are especially destructive to the lining of the digestive tract.

Strong acids are those that react completely with water to form hydronium ion (and an anion). Weak acids react only to a slight extent with water to form relatively few hydronium ions; most of the molecules of the weak acid remain in the molecular form (unchanged).

The term *strong* does not refer to the amount of solute in the solution. Whether an acid is strong or weak, a solution with a relatively large amount of acid solute in a given volume of solution is called a *concentrated acid.* A solution with relatively little solute in a given volume of solution is called a *dilute acid* (see Section 11.1).

Table 10.1
Some Familiar Acids

Name	Formula	Classification
Sulfuric acid	H_2SO_4	Strong
Nitric acid	HNO_3	Strong
Hydrochloric acid	HCl	Strong
Phosphoric acid	H_3PO_4	Moderately strong
Hydrogen sulfate ion	HSO_4^-	Moderately strong
Lactic acid	$CH_3CHOHCOOH$	Weak
Acetic acid	CH_3COOH	Weak
Boric acid	H_3BO_3	Weak
Carbonic acid	H_2CO_3	Weak
Hydrocyanic acid	HCN	Weak

Figure 10.2
Ammonia is a
water contains
nia molecules
base.

Ammonia is
low concentr:
once ammoni
called *ammo*
NH₃, *aqueoi*
contain amm
solutions are
ammonia.

How ca
unshared pai
from water l
electrons tha

In wate:
is a base in
and reacts w
solvents othe
acceptor. T
as ammonia
(HCO₃⁻), a
solution (se(

10.3
NAMES OF SOME COMMON ACIDS

In Chapter 4 you learned the names of some common anions (and if you didn't, now is the time). It is easy to learn the names of acids derived from these anions. Just follow these simple rules.

If the anion name ends in *-ide, -ate,* or *-ite,* then the acid name is, respectively, hydro____ic acid, ____ic acid, or ____ous acid. Let's apply these rules to a few examples.

Ion	Name	Acid	Name
Cl^-	Chloride	HCl	Hydrochloric acid
CN^-	Cyanide	HCN	Hydrocyanic acid
S^{2-}	Sulfide	H_2S	Hydrosulfuric acid
NO_3^-	Nitrate	HNO_3	Nitric acid
SO_4^{2-}	Sulfate	H_2SO_4	Sulfuric acid
PO_4^{3-}	Phosphate	H_3PO_4	Phosphoric acid
CO_3^{2-}	Carbonate	H_2CO_3	Carbonic acid
CH_3COO^-	Acetate	CH_3COOH	Acetic acid
BO_3^{3-}	Borate	H_3BO_3	Boric acid
SO_3^{2-}	Sulfite	H_2SO_3	Sulfurous acid

Note that most of the names have the same stem in the acid as in the anion. A few have extra letters: for example, the *ur* in sulfuric and sulfurous and the *or* in phosphoric. These you will have to learn as special cases if you haven't done so already.

Any acid with three or more replaceable hydrogens can exist in two forms, called **ortho** and **meta.** The meta form can be derived by subtracting two hydrogens and one oxygen (i.e., H_2O) from the ortho formula. For example, H_3PO_4 is really orthophosphoric acid, although the prefix is often dropped in everyday usage. Metaphosphoric acid is therefore HPO_3 (i.e., H_3PO_4 minus H_2O). Metaboric acid is derived in a similar manner from orthoboric acid (H_3BO_3). Subtraction of H_2O gives HBO_2. The corresponding anions are metaphosphate (PO_3^-) and metaborate (BO_2^-). Sodium metaphosphate and sodium metasilicate are used as water softeners.

EXAMPLE 10.1

The formula for phosphite ion is PO_3^{3-}. What is the formula for (ortho) phosphorous acid? For metaphosphorous acid? For metaphosphite ion?

SOLUTION
Orthophosphorous acid is H_3PO_3 (add one H^+ for each negative charge on PO_3^{3-}). Metaphosphorous acid is HPO_2 (H_3PO_3 minus H_2O). The metaphosphite ion is PO_2^-.

Practice Exercise
HNO_3 is named nitric acid. What is the name of HNO_2?

Practice Exercise
$HClO_2$ is named chlorous acid. What is the name of $HClO_3$?

Aluminum ions are found in unusually large amounts in the abnormal brain cells characteristic of Alzheimer's disease. It has not been established whether these ions cause the disease or whether they are deposited there as a result of the disease.

There is concern that antacids containing aluminum ions deplete the body of essential phosphate ions. The aluminum phosphate formed is insoluble and is secreted from the body.

$$Al^{3+} \; + \; PO_4^{3-} \; \longrightarrow \; AlPO_4(s)$$

Aluminum hydroxide is the only antacid in Amphojel. It occurs in combination in many popular products.

Magnesium compounds constitute the fourth category of antacids. These include magnesium carbonate ($MgCO_3$) and magnesium hydroxide [$Mg(OH)_2$]. Milk of magnesia is a suspension of magnesium hydroxide in water. It is sold under a variety of brand names, the best known of which is probably Phillips'. In small doses, magnesium compounds act as antacids. In large doses, they act as laxatives. Magnesium ions are poorly absorbed in the digestive tract. Rather, these small, dipositive ions draw water into the colon (large intestine), causing the laxative effect.

A variety of popular antacids combine aluminum hydroxide, with its tendency to cause constipation, and a magnesium compound, which acts as a laxative. These tend to counteract one another. Maalox and Mylanta are familiar brands.

Rolaids, a highly advertised antacid, contains the complex substance aluminum sodium dihydroxy carbonate [$AlNa(OH)_2CO_3$]. Both the hydroxide ion and the carbonate ion consume acid. Sodium-free Rolaids contains calcium carbonate and magnesium hydroxide.

Antacids interact with other medications. Anyone taking any type of medication should consult a physician before taking an antacid. Generally, antacids are safe and effective for occasional use in small amounts. All antacids are basic compounds. If you are otherwise in good health, you can choose a product on the basis of price. Anyone with severe or repeated attacks of indigestion should consult a physician. Self-medication in such cases might be dangerous.

Many me because t (baselike) nine, caffe tamines a

Claims of "fast action" are almost meaningless. All acid–base reactions are almost instantaneous. Some tablets may dissolve a little more slowly than others. You can speed their action by chewing them.

A strong ionized base is water onl producing droxide i remains lar form.

10.9
ACIDS, BASES, AND HUMAN HEALTH

Strong acids and bases cause damage on contact with living cells. Their action is nonspecific—all cells, regardless of type, are damaged. These corrosive poisons produce what are known as chemical burns. Once the offending agent is neutralized or removed, the injuries are similar to burns from heat, and they are often treated the same way.

Sulfuric acid (H_2SO_4) is by far the leading chemical product of U.S. industry. About 40 billion kg is produced annually. Most of this acid is used by industry. A major use is in the conversion of phosphate rock to soluble compounds for use as fertilizer. Only small quantities are used in or around the home. Automobile batteries contain sulfuric acid. It is also the major ingredient in one type of drain cleaner—sulfuric acid is one of the few substances that dissolves drain-clogging hair.

Sulfuric acid is a powerful dehydrating agent. It takes up water from cellular fluid, rapidly killing the cell. The sulfuric acid molecules dehydrate by reacting with water in the cells to form hydronium ions and hydrogen sulfate ions.

$$H_2SO_4 \; + \; H_2O \; \longrightarrow \; H_3O^+ \; + \; HSO_4^-$$

Hydration of these ions and other secondary reactions may also be involved in the dehydration process.

In Section 10.7 we saw how the burning of sulfur-containing coal produces aerosol mists of sulfuric acid. When this airborne pollutant comes into contact with the alveoli of the lungs, the cells are broken down. The alveoli lose their resilience, which makes it difficult for them to remove carbon dioxide. Such lung damage may contribute to pulmonary emphysema, a condition characterized by increasing shortness of breath. Emphysema is the fastest-growing cause of death in the United States. Most likely the principal factor in the rise of emphysema is cigarette smoking. Air pollution is also known to be a factor, however. For example, the incidence of the disease among smokers is three times as great in St. Louis, where air pollution is rather heavy, as in Winnipeg, Manitoba, where air pollution is rather mild.

Hydrochloric acid (also called muriatic acid) is used in homes to clean calcium carbonate deposits from toilet bowls. It is used in building construction to remove excess mortar from bricks. In industry, it is used to remove scale or rust from metals. Concentrated solutions (about 38% HCl) cause severe burns, but dilute solutions are considered safe enough for use around the home. Even dilute solutions, however, can cause skin irritation and inflammation.

Ingestion of sulfuric, hydrochloric, or any other strong acid causes corrosive damage to the digestive tract. As little as 10 mL of concentrated (98%) H_2SO_4, taken internally, can be fatal.

Sodium hydroxide (lye) is by far the most common of the strong bases used for household purposes. It is used to open clogged drains and as an oven cleaner. It destroys tissue rapidly, causing severe chemical burns. Several detergent additives, such as carbonates, silicates, and borates, form strongly basic solutions when dissolved in water. These, too, can cause corrosive damage to tissues, particularly those of the digestive tract and the eyes.

Both acids and bases, even in dilute solutions, break down the protein molecules in living cells. Generally, the fragments are not able to carry out the functions of the original proteins. In cases of severe exposure, the fragmentation continues until the tissue has been completely destroyed. And, within living cells, proteins function properly only at an optimum acidity. If the acidity changes much in either direction, the proteins can't carry out their usual functions properly (see Chapter 21).

When they are misused, acids and bases can be damaging to human health. But acids and bases affect human health in more subtle—and ultimately more important—ways. A delicate balance must be maintained between acids and bases in the blood and body fluids. If the acidity of the blood changes very much, the blood loses its capacity to carry oxygen. Fortunately, the body has a complex but efficient mechanism for maintaining proper acid–base balance (see Section 28.10).

EXERCISES

1. Define and, where possible, illustrate the following terms.

   a. acid   b. base
   c. salt   d. hydronium ion
   e. hydroxide ion

2. List four general properties of acidic solutions.
3. List four general properties of basic solutions.
4. What ion is responsible for the properties of acidic solutions (in water)?
5. What ion is responsible for the properties of basic solutions (in water)?

6. Give the formulas and the names of two strong acids and two weak acids.
7. Give the formulas and the names of two strong bases and one weak base.
8. Strong acids and weak acids both have properties characteristic of hydronium ions. How do strong and weak acids differ?
9. Give a general definition for an acid. Write an equation that illustrates your definition.
10. Give examples of a monoprotic, a diprotic, and a triprotic acid.

11. Indicate whether the acid is monoprotic, diprotic, triprotic, or tetraprotic.
    a. $H_2SO_3 + 2 H_2O \longrightarrow SO_3^{2-} + 2 H_3O^+$
    b. $CH_3CHOHCOOH + H_2O \longrightarrow$
                $CH_3CHOHCOO^- + H_3O^+$
    c. $CH_2(COOH)_2 + 2 H_2O \longrightarrow$
                $CH_2(COO^-)_2 + 2 H_3O^+$
    d. $H_3AsO_4 + 3 H_2O \longrightarrow AsO_4^{3-} + 3 H_3O^+$
    e. $H_4P_2O_7 + 4 H_2O \longrightarrow P_2O_7^{4-} + 4 H_3O^+$

12. Br^- is bromide ion. What is the name of the acid HBr?

13. Se^{2-} is selenide ion. What is the name of the acid H_2Se?

14. NO_2^- is nitrite ion. What is the name of the acid HNO_2?

15. PO_3^{3-} is phosphite ion. What is the name of the acid H_3PO_3?

16. $C_2O_4^{2-}$ is oxalate ion. What is the name of the acid $H_2C_2O_4$?

17. $C_6H_5COO^-$ is benzoate ion. What is the name of the acid C_6H_5COOH?

18. H_4SiO_4 is orthosilicic acid. Write the formulas for (a) metasilicic acid, (b) metasilicate ion, (c) sodium metasilicate.

19. H_3AsO_4 is orthoarsenic acid. Write the formulas for (a) metaarsenic acid, (b) metaarsenate ion, (c) potassium metaarsenate.

20. What is an acid anhydride?

21. What is a basic anhydride?

22. How does one brand of household ammonia differ from another?

23. Write the equation for the decomposition of carbonic acid.

24. Write the net ionic equation for the reaction of a bicarbonate salt with an acid.

25. Write the net ionic equation for the reaction of a carbonate salt with an acid.

26. Write an equation that describes the effect of acid rain on marble.

27. Use the definitions to identify the first compound in each equation as an acid or base. (*Hint:* What is *produced* by the reaction?)
    a. $C_5H_5N + H_2O \longrightarrow C_5H_5NH^+ + OH^-$
    b. $C_6H_5OH + H_2O \longrightarrow C_6H_5O^- + H_3O^+$
    c. $CH_3COCOOH + H_2O \longrightarrow CH_3COCOO^- + H_3O^+$

28. Use the definitions to identify the first compound in each equation as an acid or base.
    a. $C_6H_5SH + H_2O \longrightarrow C_6H_5S^- + H_3O^+$
    b. $CH_3NH_2 + H_2O \longrightarrow CH_3NH_3^+ + OH^-$
    c. $C_6H_5SO_2NH_2 + H_2O \longrightarrow C_6H_5SO_2NH^- + H_3O^+$

29. Give the formulas for the following acids.
    a. hydrochloric acid   b. sulfuric acid
    c. carbonic acid   d. hydrocyanic acid

30. Give formulas for the following acids.
    a. nitric acid   b. sulfurous acid
    c. phosphoric acid   d. hydrosulfuric acid

31. Give formulas for the following bases.
    a. lithium hydroxide

b. magnesium hydroxide
c. sodium hydroxide

32. Give formulas for the following bases.
    a. calcium hydroxide
    b. potassium hydroxide
    c. ammonia

33. Write the equation that shows hydrogen chloride gas reacting with water to form ions. What is the name of the acid formed?

34. Write the equation that shows how ammonia acts as a base in water. $NH_3 + H_2O \rightarrow NH_4^+ + OH^-$

35. Give the formula for the compound formed when sulfur trioxide reacts with water. Is the product an acid or a base?

36. Give the formula for the compound formed when magnesium oxide reacts with water. Is the product an acid or a base?

37. Give the formula for the compound formed when potassium oxide reacts with water. Is the product an acid or a base?

38. Give the formula for the compound formed when carbon dioxide reacts with water. Is the product an acid or a base?

39. Cesium hydroxide (CsOH) is ionic in the solid state and is quite soluble in water. Classify the compound as a strong acid, weak acid, weak base, or strong base.

40. Hydrogen iodide (HI) gas reacts completely with water to form hydronium ions and iodide ions. Classify the compound as a strong acid, weak acid, weak base, or strong base.

41. Hydrogen sulfide (H_2S) gas reacts slightly with water to form relatively few hydronium ions and hydrogen sulfide ions (HS^-). Classify the compound as a strong acid, weak acid, weak base, or strong base.

42. Methylamine (CH_3NH_2) gas reacts slightly with water to form relatively few hydroxide ions and methyammonium ions ($CH_3NH_3^+$). Classify the compound as a strong acid, weak acid, weak base, or strong base.

43. According to the equation, is phenol, C_6H_5OH, an acid or a base? Should it be classified as strong or weak?

    $C_6H_5OH + H_2O \rightleftharpoons C_6H_5O^- + H_3O^+$

44. According to the equation, is aniline, $C_6H_5NH_2$, an acid or a base? Should it be classified as strong or weak?

    $C_6H_5NH_2 + H_2O \rightleftharpoons C_6H_5NH_3^+ + OH^-$

45. Write the equation for the reaction of sodium hydroxide with hydrochloric acid.

46. Write the equation for the reaction of lithium hydroxide with nitric acid.

47. Write the equation for the reaction of 1 mol of calcium hydroxide with 2 mol of hydrochloric acid.

48. Write the equation for the reaction of 1 mol of sulfuric acid with 2 mol of potassium hydroxide.

49. Write the equation for the reaction of 1 mol of phosphoric acid with 3 mol of sodium hydroxide.

50. Write the equation for the reaction of 1 mol of sulfuric acid with 1 mol of calcium hydroxide.

51. Describe the taste and effect on litmus of a solution that has been neutralized.

52. Magnesium hydroxide is completely ionic, even in the solid state, yet it can be taken internally as an antacid. Explain why it does not cause injury as sodium hydroxide would.

53. How does magnesium hydroxide, in large doses, act as a laxative? Would other magnesium compounds have a similar effect?

54. Name some of the active ingredients in antacids.

55. Should a person who has hypertension be advised to use baking soda or milk of magnesia as an antacid? Why?

56. What is aqueous ammonia? Why is it sometimes called ammonium hydroxide?

57. What is the leading chemical product of U.S. industry?

58. What is the effect of strong acids on clothing?

59. What is the effect of strong acids and strong bases on the skin?

60. Corrosive acids and bases destroy living cells. Explain.

61. Examine the labels of at least five antacid preparations. Make a list of the ingredients in each. What kind of chemical compound is each of the active ingredients?

62. Examine the labels of at least five toilet bowl cleaners and five drain cleaners. Make a list of the ingredients in each. Look up the formulas and properties of each ingredient in a reference book. Which ingredients are acids? Which are bases?

Chapter 11
ACIDS AND BASES II

Laboratory reagent bottles of acids and bases clearly indicate whether their contents are dilute or concentrated solutions.

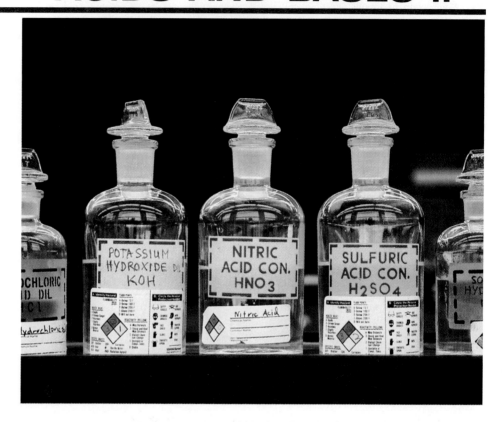

Our bodies are largely solutions—and very special solutions. Delicate balances must be maintained among the many solutes in our blood and other body fluids. And no balance is more important than the one between acids and bases. If the acidity of the blood changes very much, the blood loses its capacity to carry oxygen. Since many bodily processes produce acids, the control of acidity is—quite literally—a matter of life or death.

Our bodies have developed a marvelously complex yet efficient mechanism for maintaining the proper acid–base balance. Before we can talk about this mechanism in a meaningful way, however, we need to develop a few concepts—concepts that are more quantitative in nature than those developed in Chapter 10. It is important to know how exact concentrations of acids and bases are determined and expressed. We must understand the concept of pH, particularly as it relates to the chemistry of the blood and other body fluids. Most important, we must see just how, through substances called buffers, the level of acidity is controlled.

11.1
CONCENTRATIONS OF ACIDS AND BASES

In Chapter 9, the terms *dilute* and *concentrated* were introduced to indicate relative amounts of solute for a given amount of solvent. For solutions of acids and bases, these same terms are often used in a more precise manner. For example, in a laboratory, a stock solution of "concentrated" hydrochloric acid is one that is 12 M. A stock bottle of hydrochloric acid labeled "dilute" contains a 6 M HCl solution. A concentrated sulfuric acid solution is 18 M H_2SO_4; dilute sulfuric acid is 3 M. Table 11.1 gives a summary of concentrations of common stock acids and bases. In general, the concentrated solutions are the strongest solutions commercially available.

Sometimes it is necessary for a laboratory worker to make a more dilute solution from a more concentrated one. When a solution is diluted, the amount of solute does not change. Adding solvent changes *only* the concentration of the solution. The moles of solute in the concentrated solution must equal the moles of solute in the dilute solution. Therefore we can use the following relationship.

$$\underbrace{\begin{array}{c}\text{Volume of the} \\ \text{concentrated} \\ \text{solution}\end{array} \times \begin{array}{c}\text{concentration of the} \\ \text{concentrated} \\ \text{solution}\end{array}}_{\begin{array}{c}\text{moles of solute in} \\ \text{concentrated solution}\end{array}} = \underbrace{\begin{array}{c}\text{volume of the} \\ \text{dilute} \\ \text{solution}\end{array} \times \begin{array}{c}\text{concentration} \\ \text{of the dilute} \\ \text{solution}\end{array}}_{\begin{array}{c}\text{moles of solute} \\ \text{in dilute solution}\end{array}}$$

or

$$V_c \times M_c = V_d \times M_d$$

EXAMPLE 11.1

How much 6.0 M HCl would you use to make 1.0 L of 0.50 M HCl? How much water is required?

Table 11.1
Concentrations of Stock Acids and Bases

Name	Formula	Molarity	
		Dilute	Concentrated
Hydrochloric acid	HCl	6	12
Nitric acid	HNO_3	6	16
Sulfuric acid	H_2SO_4	3	18
Phosphoric acid	H_3PO_4	—	15
Acetic acid	CH_3COOH	6	17
Ammonium hydroxide[a]	NH_4OH	6	15
Sodium hydroxide	NaOH	6	—

[a] A better name, perhaps, is aqueous ammonia (NH_3); however, stock solutions are still labeled "ammonium hydroxide (NH_4OH)."

SOLUTION

$$V_c \times 6.0 \text{ M} = 1.0 \text{ L} \times 0.50 \text{ M}$$

$$V_c = \frac{1.0 \text{ L} \times 0.50 \text{ M}}{6.0 \text{ M}} = 0.083 \text{ L or } 83 \text{ mL}$$

$$1000 \text{ mL} - 83 \text{ mL} = 917 \text{ mL water}$$

Practice Exercise

How much 15.7 M HNO_3 would you use to make 400 mL of 6.00 M HNO_3? How much water is required?

EXAMPLE 11.2

How much 18 M H_2SO_4 and how much water would you need to make 100 mL of 3.0 M H_2SO_4?

SOLUTION

$$V_c \times 18 \text{ M} = 100 \text{ mL} \times 3.0 \text{ M}$$

$$V_c = \frac{100 \text{ mL} \times 3.0 \cancel{\text{ M}}}{18 \cancel{\text{ M}}} = 17 \text{ mL}$$

$$100 \text{ mL} - 17 \text{ mL} = 83 \text{ mL water}$$

Practice Exercise

How much 12 M CH_3COOH and how much water would you need to make 30 mL of 1.5 M CH_3COOH?

11.2
ACID–BASE TITRATIONS

Sometimes it is necessary to determine just how much acid (or base) there is in a solution of unknown concentration. This is often done by a process called **titration.**
If the unknown is acidic, a certain volume of the acid is carefully measured from a buret into a flask. The buret is a piece of laboratory glassware designed to deliver known amounts of liquid into another container. An indicator dye (Section 11.3) is added to the measured volume of acid. Then a basic solution of known concentration (a **standard base**) is added from another buret, slowly and carefully, until finally just one additional drop of base changes the color of the indicator dye (Figure 11.1). This is the **end point** of the titration.

For the determination of the amount of hydroxide ion in a base of unknown concentration, the procedure is reversed. A quantity of the basic solution is measured from one buret into a flask. Then standard acid is added from the second buret until the indicator changes color. If everything is done properly, the color change (end point) occurs when the number of moles of hydronium ion (or H^+) is just equal to the number of moles of hydroxide ion (or other H^+ acceptor).

Number of moles of H^+ = number of moles of OH^-

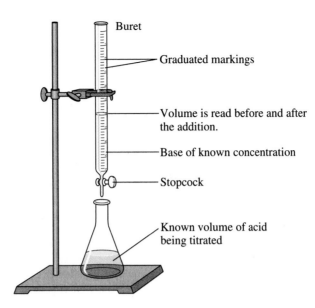

Buret

Graduated markings

Volume is read before and after
the addition.

Base of known concentration

Stopcock

Known volume of acid
being titrated

Figure 11.1
An apparatus for titration. A
sample of acid is measured
into the flask. Base is
added from a buret until an
indicator changes color.

Acid–base titrations are used frequently to determine quantitatively the concentration
of an acid or base. For example, if we know the original volume of the acid, the concentration of the base, and the volume of base needed to reach the end point, we can calculate
the concentration of the acid.

EXAMPLE 11.3

Calculate the concentration (molarity) of 20 mL of HCl that is just neutralized by
28 mL of 0.10 M NaOH solution.

SOLUTION
First write the equation.

$$HCl(aq) \ + \ NaOH(aq) \ \longrightarrow \ NaCl(aq) \ + \ H_2O$$

From the equation we see that 1 mol of HCl neutralizes 1 mol of NaOH. Next we
want to convert 28 mL of NaOH to moles of HCl. Our conversion factors are the
molarity of the NaOH and the mole ratio.

$$28 \ \text{mL NaOH} \times \frac{1 \ \text{L}}{1000 \ \text{mL}} \times \frac{0.10 \ \text{mol NaOH}}{1 \ \text{L NaOH}} \times \frac{1 \ \text{mol HCl}}{1 \ \text{mol NaOH}}$$
$$= 0.0028 \ \text{mol HCl}$$

Now that we know that 0.0028 mol of HCl is contained in 0.020 L (20 mL ×
1 L/1000 mL), we can easily calculate molarity.

$$\text{Molarity} = \frac{\text{moles}}{\text{liter}}$$

$$\text{Molarity of HCl} = \frac{0.0028 \ \text{mol HCl}}{0.020 \ \text{L HCl}} = 0.14 \ \text{M}$$

EXAMPLE 11.4

What is the concentration of an H_2SO_4 solution if 30 mL of it requires 40 mL of 0.15 M NaOH for neutralization?

SOLUTION
From the equation

$$H_2SO_4(aq) \ + \ 2\,NaOH(aq) \longrightarrow Na_2SO_4(aq) \ + \ 2\,H_2O$$

we see that 1 mol of H_2SO_4 will neutralize 2 mol of NaOH.

$$40 \text{ mL NaOH} \times \frac{1 \text{ L}}{1000 \text{ mL}} \times \frac{0.15 \text{ mol NaOH}}{1 \text{ L NaOH}} \times \frac{1 \text{ mol } H_2SO_4}{2 \text{ mol NaOH}}$$
$$= 0.0030 \text{ mol } H_2SO_4$$

$$\text{Molarity of } H_2SO_4 = \frac{0.0030 \text{ mol } H_2SO_4}{0.030 \text{ L } H_2SO_4} = 0.10 \text{ M}$$

Practice Exercise
What is the concentration of an H_3PO_4 solution if 35 mL of 0.12 M KOH is required to neutralize 50 mL of the acid?

11.3
THE pH SCALE

When we think of water, we think of H_2O molecules. But even the purest water isn't all H_2O. About 1 molecule in 500 million transfers a proton to another water molecule, giving a hydronium ion and a hydroxide ion.

$$H_2O \ + \ H_2O \ \rightleftharpoons \ H_3O^+ \ + \ OH^-$$

In other words, water is in equilibrium with hydronium ion and hydroxide ion, although the equilibrium lies far to the left. The concentration of hydronium ion in pure water is 0.0000001, or 1×10^{-7}, mol/L (M). The concentration of hydroxide ion is also 1×10^{-7} M. Pure water, then, is a neutral solution, without an excess of either hydronium ions or hydroxide ions.

The product of the hydronium ion concentration and the hydroxide ion concentration is

$$(1 \times 10^{-7})(1 \times 10^{-7}) = 1 \times 10^{-14}$$

This product, called the **ion product** of water, is sometimes represented by the symbol K_w. K_w is a constant. If you add acid to pure water, thereby increasing the concentration of hydronium ion, the concentration of hydroxide ion will fall until the product of the concentrations of the two ions equals 1×10^{-14}. Adding base to pure water will result in a similar adjustment in ion concentrations. These facts can be summarized in a single equation.

$$K_w = [H_3O^+][OH^-] = 1 \times 10^{-14}$$

The bracketed symbols are a shorthand for the concentration of the enclosed ions in moles per liter. Thus $[H_3O^+]$ should be read as the molarity of hydronium ion. The equation is valid for all aqueous solutions. And since it relates the hydronium and hydroxide ion concentrations, if the concentration of one ion is known, the concentration of the other can be calculated.

EXAMPLE 11.5

Lemon juice has a $[H_3O^+]$ of 0.01 M. What is the $[OH^-]$?

SOLUTION

In exponential form, 0.01 M is written 1×10^{-2} M. Since

$$[H_3O^+]\,[OH^-] = 1 \times 10^{-14}$$
$$[OH^-] = \frac{1 \times 10^{-14}}{1 \times 10^{-2}}$$

To divide exponential numbers, you subtract the exponent in the denominator from the exponent in the numerator, in this case, $-14 - (-2) = -14 + 2 = -12$. So

$$[OH^-] = 1 \times 10^{-12} \text{ M}$$

EXAMPLE 11.6

A sample of bile has a $[OH^-]$ of 1×10^{-6} M. What is the $[H_3O^+]$?

SOLUTION

$$[H_3O^+] = \frac{1 \times 10^{-14}}{1 \times 10^{-6}} = 1 \times 10^{-8} \text{ M}$$

It is inconvenient to use exponential numbers to express the often minute concentrations of hydronium and hydroxide ion. In 1909, S. P. L. Sørensen proposed that only the number in the exponent be used to express acidity. Sørensen's scale came to be known as the **pH scale,** from the French *pouvoir hydrogene* ("hydrogen power").

$$pH = \log \frac{1}{[H_3O^+]} = -\log [H_3O^+]$$

The reaction of most people on first encountering this definition of pH is "You call this more convenient?" Well, it really is. To determine the pH of a solution whose hydronium ion concentration $[H_3O^+] = 1.0 \times 10^{-n}$, you need only take the exponent of the hydronium ion concentration and reverse its sign. The pH of a solution whose hydronium ion concentration is 1×10^{-4} M is 4. The point is this: it is easier to say, "The pH is 4," than to say, "The hydronium ion concentration is 1×10^{-4} M." They mean the same thing.

EXAMPLE 11.7

What is the pH of the bile sample in Example 11.6?

SOLUTION

$$pH = -\log [H_3O^+]$$
$$= -\log (1 \times 10^{-8})$$

To take the logarithm of 1×10^{-8}, we need to know that

$$\log (a + b) = \log a + \log b$$

So we can write

$$\log (1 \times 10^{-8}) = \log 1 + \log 10^{-8}$$

The logarithm of 1 is 0 ($10^0 = 1$), and the logarithm of 10^{-8} is -8. So the answer can be written

$$pH = -\log 1 - \log 10^{-8}$$
$$= 0 - (-8)$$
$$= 8$$

EXAMPLE 11.8

Calculate the pH of a solution that has a hydronium ion concentration of 1×10^{-3} M.

SOLUTION

$$pH = -\log [H_3O^+]$$
$$= -\log (1 \times 10^{-3})$$
$$= -\log 1 - \log 10^{-3}$$
$$= 0 - (-3)$$
$$= 3$$

Practice Exercise

What is the pH of a solution that has a hydronium ion concentration of 1.0×10^{-11} M?

EXAMPLE 11.9

What is the pH of a solution that has a hydronium ion concentration of 4.5×10^{-3} M?

SOLUTION

We need to take the logarithm of 4.5×10^{-3}.

$$\log (4.5 \times 10^{-3}) = \log 4.5 + \log 10^{-3}$$

Using a scientific calculator with a base 10 log key, we find that the logarithm of 4.5, to two significant figures, is 0.65. Thus we have

$$\begin{aligned} pH &= -\log [H_3O^+] \\ &= -\log (4.5 \times 10^{-3}) \\ &= -\log 4.5 - \log 10^{-3} \\ &= -0.65 - (-3) \\ &= 2.35 \end{aligned}$$

EXAMPLE 11.10

Calculate the pH of a solution that has a hydronium ion concentration of 2.0×10^{-8} M.

SOLUTION

$$\begin{aligned} pH &= -\log [H_3O^+] \\ &= -\log (2.0 \times 10^{-8}) \\ &= -\log 2.0 - \log 10^{-8} \\ &= -0.30 - (-8) \\ &= 7.70 \end{aligned}$$

Practice Exercise

What is the pH of a solution that has a hydronium ion concentration of 2.7×10^{-9} M?

The pH scale has been universally adopted. The relationship between pH and $[H_3O^+]$ is given in Table 11.2. Note from the table that

$$pH + pOH = 14$$

where

$$pOH = -\log [OH^-]$$

Table 11.2
The Relationship Between pH and $[H_3O^+]$ and Between pOH and $[OH^-]$ (at 20 °C)

$[H_3O^+]$	pH	$[OH^-]$	pOH	
1×10^{0}	0	1×10^{-14}	14	↑
1×10^{-1}	1	1×10^{-13}	13	
1×10^{-2}	2	1×10^{-12}	12	
1×10^{-3}	3	1×10^{-11}	11	Acidic solutions
1×10^{-4}	4	1×10^{-10}	10	
1×10^{-5}	5	1×10^{-9}	9	
1×10^{-6}	6	1×10^{-8}	8	↓
1×10^{-7}	7	1×10^{-7}	7	Neutral solution
1×10^{-8}	8	1×10^{-6}	6	↑
1×10^{-9}	9	1×10^{-5}	5	
1×10^{-10}	10	1×10^{-4}	4	
1×10^{-11}	11	1×10^{-3}	3	Basic solutions
1×10^{-12}	12	1×10^{-2}	2	
1×10^{-13}	13	1×10^{-1}	1	
1×10^{-14}	14	1×10^{0}	0	↓

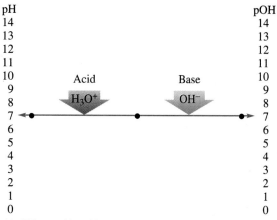

(a) When acid and base are in balance, the solution is neutral.

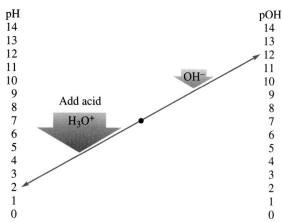

(b) When excess acid is added, pH decreases, and pOH increases.

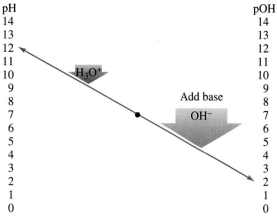

(c) When excess base is added, pH increases, and pOH decreases.

Figure 11.2

Effect of adding acid or base on the pH and pOH of a solution.

A pH of 7 represents a neutral solution. A pH lower than 7 means the solution is acidic; a pH higher than 7 indicates that the solution is basic. The *lower* the pH, the more acidic the solution; the *higher* the pH, the more basic the solution (Figure 11.2).

The cells in our bodies are bathed in solutions of fairly constant pH. Indeed, even small changes in pH may be fatal. Many chemical reactions, especially those in living cells, are quite sensitive to pH. The control of pH, by compounds called buffers, is described in Sections 11.5 and 11.6.

A variety of dyes, some synthetic and some from plant and animal sources, have the property of changing color over a range of pH values. We noted in Chapter 10 that litmus, a vegetable dye, turns red in acidic solutions and blue in basic solutions. Actually the change occurs over a range of pH from 4.5 to 8.3. Certain combinations of indicator dyes will exhibit a whole range of colors as the pH changes from strongly basic to strongly acidic (Figure 11.3). By selecting the proper indicator, we can determine the pH of almost any clear, colorless aqueous solution.

More accurate measurement of pH can be made electrically with pH meters. Gener-

A substance that has the property of changing color when acid or base is added to it is called an **indicator.**

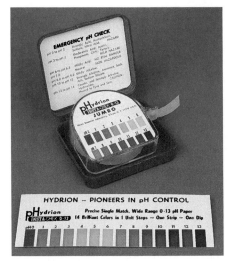

Figure 11.3
pH paper changes colors over a wide range of pH values.

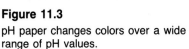

Figure 11.4
A pH meter.

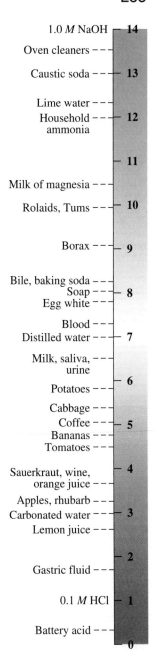

Figure 11.5
pH values of some common substances.

ally, these instruments can measure pH to a precision of about 0.01 pH unit. Also, colors and turbidity generally do not interfere with electrical measurement of pH as they do with indicator color changes. Thus, pH meters can be used on blood, urine, and other complex mixtures without prior treatment of the specimen. A typical pH meter is illustrated in Figure 11.4.

Keep in mind that the pH scale is logarithmic; for each unit change in pH, there is a 10-fold change in $[H_3O^+]$. Acid rain with a pH of 4 is ten times as acidic as rain with a pH of 5. Figure 11.5 gives the approximate pH of some substances.

11.4
SALTS IN WATER: ACIDIC, BASIC, OR NEUTRAL?

When an acid reacts with a base, the products are a salt and water. The process is called neutralization, but is the solution neutral? What if we simply take a salt and dissolve it in water? Would the solution be acidic, basic, or neutral? Although we might well expect a salt solution to be neutral, not all of them are. There are many neutral salts, including the familiar sodium chloride and others such as potassium sulfate and sodium nitrate. Some, such as ammonium nitrate and aluminum chloride, are acidic. Others, such as sodium acetate and potassium cyanide, are basic.

How do we tell if a salt is acidic, basic, or neutral? Simple. Just dissolve some of the salt in water and test the solution with indicator paper or with a pH meter. That is the experimental way, and who can argue with it? There is another way, though. We can *predict* whether a solution of a salt will be acidic, basic, or neutral by considering the relative strengths of the acid and base from which the salt was made. Just think how much more convenient it is to apply a rule than to go into a laboratory and do an experiment. The rules are as follows:

1. The salt of a strong acid and a strong base forms a neutral solution.
2. The salt of a strong acid and a weak base forms an acidic solution.
3. The salt of a weak acid and a strong base forms a basic solution.
4. The salt of a weak acid and a weak base may form an acidic, a basic, or (by chance) a neutral solution.

To apply these rules, you must be able to recognize strong acids, strong bases, weak acids, and weak bases (recall Sections 10.2 and 10.4). The following examples will illustrate the process.

EXAMPLE 11.11

Does NaCl form an acidic, basic, or neutral solution?

SOLUTION

In any aqueous solution, there will always be H^+ (actually, H_3O^+) and OH^-. NaCl dissociates in water to yield $Na^+ + Cl^-$. Therefore, in an aqueous solution of NaCl, there will be the following four ions:

$$Na^+, \quad Cl^-, \quad H^+, \quad OH^-$$

Combine the positive ion of the salt with OH^-, and the negative ion of the salt with H^+.

$$Na^+ + OH^- = NaOH$$
$$Cl^- + H^+ = HCl$$

Then examine the base and the acid that are formed. NaOH is a strong base, and HCl is a strong acid. Therefore, NaCl is a salt of a strong acid and a strong base, and (according to Rule 1) it forms a neutral solution.

Practice Exercise

Does KNO_3 form an acidic, basic, or neutral solution?

EXAMPLE 11.12

Is a solution of $(NH_4)_2SO_4$ acidic, basic, or neutral?

SOLUTION

The four ions in solution are

$$NH_4^+, \quad SO_4^{2-}, \quad H^+, \quad OH^-$$

Combining the appropriate ions, we have

$$NH_4^+ + OH^- = NH_4OH$$
$$2\,H^+ + SO_4^{2-} = H_2SO_4$$

NH_4OH (really aqueous NH_3) is a weak base, and H_2SO_4 is a strong acid. Therefore, $(NH_4)_2SO_4$ is a salt of a weak base and a strong acid, and (according to Rule 2) it forms an acidic solution.

Practice Exercise

Is a solution of NH_4Cl acidic, basic, or neutral?

EXAMPLE 11.13

Is a solution of CH_3COOK acidic, basic, or neutral?

SOLUTION
The four ions in solution are

$$K^+, \quad CH_3COO^-, \quad H^+, \quad OH^-$$

Combining the appropriate ions, we have

$$K^+ + OH^- = KOH$$
$$CH_3COO^- + H^+ = CH_3COOH$$

KOH is a strong base, and CH_3COOH (acetic acid) is a weak acid. Therefore, CH_3COOK is a salt of a weak acid and a strong base, and (according to Rule 3) it forms a basic solution.

Practice Exercise
Is a solution of Li_2CO_3 acidic, basic, or neutral? (LiOH is a strong base.)

EXAMPLE 11.14

Is a solution of CH_3COONH_4 acidic, basic, or neutral?

SOLUTION
The base, NH_4OH (aqueous NH_3), is a weak one. The acid, CH_3COOH, is also weak. There is no way to tell from the rules whether the solution would be acidic, basic, or neutral (Rule 4).

11.5
BUFFERS: CONTROL OF pH

Who cares about the pH of a salt solution anyway? You do, that's who, for some of these salts play a vital role in the control of the pH of body fluids. If they should fail to function, so will you. Our bodies are acid factories. Our stomachs produce hydrochloric acid. Our muscles produce lactic acid. Starches and sugars produce pyruvic acid when metabolized. Carbon dioxide from respiration produces carbonic acid in the blood. Our bodies must eliminate or neutralize these acids, because excess acidity in the wrong place would kill us rather quickly.

A **buffer solution** is one in which the pH remains nearly constant even if acid or base is added. Chemically, a buffer solution is one that contains a weak acid and one of its salts (or a weak base and one of its salts), usually in approximately equal concentrations. For example, 1 L of a solution that contains 0.1 mol of acetic acid (CH_3COOH) and 0.1 mol of sodium acetate (CH_3COONa) acts as a buffer at pH 4.74, an acidic value.[1] This buffer

---

[1] This may be a bit confusing because we have just spent some time considering the fact that a solution of sodium acetate is slightly basic (i.e., has a pH greater than 7). However, it is a solution containing *only* the salt that is basic. The buffer solution consists not only of a salt of acetic acid but also of some acetic acid itself. It is the presence of this acid that makes the buffer solution acidic.

can absorb significant amounts of additional acid or base without appreciable change in pH. For instance, the addition of 10 mL of 0.1 M NaOH to this buffer causes a pH change of only 0.01 pH unit. If that amount of base were added to 1 L of pure water, the pH would change by 4 full pH units. The buffer system is what prevents such a large change in pH.

How does a buffer work? It may seem strange that a solution can absorb acid or base without the pH changing appreciably. The explanation, however, is fairly simple. It follows Le Châtelier's principle (Section 5.12). The buffer solution has a large reservoir of both acid molecules and the anions from the salt. If a strong acid is added, the hydronium ions from the added acid will donate protons to the anions of the buffer to form the weak acid and water.

$$H_3O^+ + CH_3COO^- \rightleftharpoons CH_3COOH + H_2O$$

Although the reaction is reversible to a slight extent, most of the protons are removed from the solution as they are added, and the pH changes hardly at all.

When a strong base is added, the hydroxide ions will react with the hydronium ions formed in the solution by the acetic acid of the buffer.

$$OH^- + H_3O^+ \longrightarrow 2\,H_2O$$

The added hydroxide ions are tied up, and the hydronium ions removed from the solution are immediately replaced by further ionization of the acetic acid in the buffer.

$$CH_3COOH + H_2O \rightleftharpoons CH_3COO^- + H_3O^+$$

The concentration of hydronium ions returns to approximately the original value, and the pH is only slightly changed.

There are many important buffer solutions. Most biochemical reactions, whether they occur in a laboratory or in our bodies, are carried out in buffered solutions. The buffers that control the pH of our blood will be discussed in the next section. Table 11.3 lists some buffers of interest and the pH range in which they operate.

11.6
BUFFERS IN BLOOD

The pH of the blood of higher animals is held remarkably constant. In humans, blood plasma normally varies from 7.35 to 7.45 in pH. Should the pH rise above 7.8 or fall below 6.8, due to starvation or disease, the person may suffer irreversible damage to

Table 11.3
Some Important Buffers

Buffer Components	Buffer System Names	pH[a]
$CH_3CHOHCOOH/CH_3CHOHCOO^-$	Lactic acid/lactate ion	3.86
CH_3COOH/CH_3COO^-	Acetic acid/acetate ion	4.74
$H_2PO_4^-/HPO_4^{2-}$	Dihydrogen phosphate ion/ monohydrogen phosphate ion	7.20
H_2CO_3/HCO_3^-	(Carbon dioxide) carbonic acid/ bicarbonate ion	6.46[b]
NH_4^+/NH_3	Ammonium ion/ammonia	9.25

[a]The values listed are for solutions that are 0.1 M in each compound at 25 °C.
[b]This value includes dissolved CO_2 molecules as undissociated H_2CO_3. The value for H_2CO_3 alone is about 3.8.

the brain or even die. Fortunately, human blood has not one, but at least three, buffering systems. Of these, the bicarbonate/carbonic acid (HCO_3^-/H_2CO_3) buffering system is the most important.

If acids are put into the blood, hydronium ions are taken up by the bicarbonate ions to form undissociated carbonic acid and water.

$$HCO_3^- + H_3O^+ \longrightarrow H_2CO_3 + H_2O$$

As long as there is sufficient bicarbonate to take up the added acid, the pH will change little.

The carbonic acid is a weak acid and exists in equilibrium with hydronium and bicarbonate ions.

$$H_2CO_3 + H_2O \rightleftharpoons H_3O^+ + HCO_3^-$$

If bases come into the bloodstream, reacting with hydronium ions to form water, more carbonic acid molecules will ionize to replace the removed hydronium ions. Further as the carbonic acid molecules are used up, more carbonic acid can be formed from the large reservoir of dissolved carbon dioxide in the blood.

$$CO_2 + H_2O \rightleftharpoons H_2CO_3$$

Thus, bicarbonate/carbonic acid buffers the blood against either added base or added acid.

Another blood buffer is the dihydrogen phosphate/monohydrogen phosphate ($H_2PO_4^-/HPO_4^{2-}$) system. Any acid reacts with monohydrogen phosphate to form dihydrogen phosphate.

$$HPO_4^{2-} + H_3O^+ \longrightarrow H_2PO_4^- + H_2O$$

The dihydrogen phosphate is a weak acid and exists in equilibrium with hydronium ions and monohydrogen phosphate.

$$H_2PO_4^- + H_2O \rightleftharpoons H_3O^+ + HPO_4^{2-}$$

Any base that comes into the blood would react with hydronium ions to form water. However, more dihydrogen phosphate would ionize to replace these hydronium ions, leaving the pH essentially unchanged.

Proteins act as a third type of blood buffer. These complex molecules (Chapter 21) contain —COO^- groups, which, like acetate ions (CH_3COO^-), can act as proton acceptors. Proteins also contain —NH_3^+ groups, which, like ammonium ions (NH_4^+), can donate protons. If acid comes into the blood, hydronium ions can be neutralized by the —COO^- groups. (The wavy lines represent a part of a protein molecule.)

$$COO^- + H_3O^+ \longrightarrow COOH + H_2O$$

If base is added, it can be neutralized by the —NH_3^+ groups.

$$NH_3^+ + OH^- \longrightarrow NH_2 + H_2O$$

These three buffers (and perhaps others) act to keep the pH of the blood constant. Buffers can be overridden by large amounts of acid or base; their capacity is not infinite. The blood buffers can be overwhelmed if the body's metabolism goes badly amiss.

11.7
ACIDOSIS AND ALKALOSIS

Have your muscles ever hurt after prolonged physical activity? If so, you have had your blood buffers somewhat overloaded. Muscle contraction produces lactic acid. This acid ionizes somewhat more strongly than carbonic acid and thus tends to lower the pH of the blood (it tends to release more hydronium ions into the blood). Moderate amounts of lactic acid can be handled by the blood buffers. For bicarbonate, the reaction would be

$$\underset{\text{Lactic acid}}{CH_3\overset{\overset{\displaystyle OH}{|}}{CH}COOH} + HCO_3^- \longrightarrow \underset{\text{Lactate ion}}{CH_3\overset{\overset{\displaystyle OH}{|}}{CH}COO^-} + H_2CO_3$$

Excessive amounts of lactic acid overload the buffers, however, and the pH is lowered. Nerve cells respond to the increased acidity by sending a message of pain to the brain.

If the pH of the blood falls below 7.35, the condition is called **acidosis.** If the pH of the blood rises above 7.45, **alkalosis** sets in. These pathological conditions can be caused by faulty respiration or by metabolic problems. In severe cases of starvation, the body gets its energy by the oxidation of stored fats. The products of fat metabolism are acidic, and prolonged starvation leads to acidosis. Fad diets, such as those that severely limit the intake of carbohydrates, can also lead to acidosis.

The body's excretory system tries to compensate for acidosis or alkalosis by selectively excreting certain compounds. Conversely, the conditions can be brought on by kidney failure or other excretory problems. We discuss these pathological problems further in later chapters.

EXERCISES

1. Define or illustrate the following terms.
   a. end point
   b. K_w
   c. pH
   d. pOH
   e. titration
   f. indicator
   g. pH meter
   h. buffer
   i. acidosis
   j. alkalosis
2. Calculate the molarity of each of these solutions.
   a. 3 mol of H_3AsO_4 in 6 L of solution
   b. 20 g of NaOH in 0.5 L of solution
   c. 9.8 g of H_2SO_4 in 0.2 L of solution
   d. 342 g of $Ba(OH)_2$ in 10 L of solution
   e. 12 g of CH_3COOH in 100 mL of solution
3. What volume of concentrated (18 M) sulfuric acid would be required to make each of the following?
   a. 12 L of 6.0 M solution
   b. 500 mL of 0.10 M solution
4. You have a stock solution of 12 M HCl. How would you prepare the following solutions?
   a. 100 mL of 1.2 M HCl solution
   b. 5.0 L of 6.0 M HCl solution
   c. 1.50 L of 1.00 M HCl solution

5. Would 1 mol of NaOH neutralize each of the following?
   a. 1 mol HNO_3
   b. 1 mol CH_3COOH
   c. 3 mol H_3PO_4
   d. 0.5 mol H_2SO_4
6. Write an equation for the equilibrium established when CH_3COO^- is placed in water.
7. Write an equation for the equilibrium established when NH_4^+ is placed in water.
8. Classify the aqueous solution of each of these salts as acidic, basic, or neutral.
   a. KCl
   b. NaCN
   c. NH_4CN
   d. CH_3COOK
   e. $(NH_4)_2SO_4$
   f. Na_2SO_4
9. A weak acid is titrated with a strong base. Would the solution at the end point be acidic, basic, or neutral? Explain.
10. Calculate the molarity of an HCl solution if 20 mL of it requires the following for neutralization.
    a. 40 mL of 0.25 M NaOH
    b. 10 mL of 0.50 M KOH
11. A 20-mL sample of gastric fluid is neutralized by 25 mL of 0.10 M NaOH. What is the molarity of HCl in the fluid? Assume that all the acidity of the gastric fluid is due to HCl.

12. If 25.0 mL of a sulfuric acid solution requires 15.0 mL of 0.500 M sodium hydroxide to neutralize it, what is the molarity of the H_2SO_4 solution?

13. A 40.0-mL sample of a 0.150 M solution of KOH is titrated with 0.600 M HCl solution. What volume of the HCl solution is required to reach the end point?

14. A 50.0-mL sample of an acid is titrated with 0.250 M NaOH. It takes 12.5 mL of the base to reach the end point. What is the molarity of the acid?

15. A 35.0-mL sample of a base is titrated with 0.100 M HCl. It takes 42.5 mL of the acid just to neutralize the base. What is the molarity of the base?

16. When the stomach isn't being stimulated by food to make more, it produces 0.0023 mol of HCl and 30–60 mL of total juices per hour. What range of concentrations of HCl in the stomach does this represent?

17. What is the pH of each of the following solutions?
    a. 1×10^{-2} M HCl   **b.** 1×10^{-3} M HNO_3

18. What is the pH of each of the following solutions?
    a. 0.0001 M HCl   **b.** 0.1 M HBr

19. What is the pOH of each of the following solutions?
    a. 1×10^{-1} M NaOH   **b.** 1×10^{-5} M KOH

20. What is the pOH of each of the following solutions?
    a. 0.001 M NaOH   **b.** 0.01 M KOH

21. What is the pOH of each of the solutions in Exercises 17 and 18?

22. What is the pH of each of the solutions in Exercises 19 and 20?

23. Indicate whether each of the following pH values represents an acidic, basic, or a neutral solution.
    a. 11   **b.** 4   **c.** 7   **d.** 3.4

24. Answer Exercise 23, assuming that the values given are for pOH.

25. Use acetic acid and acetate ion to explain how a buffer controls pH.

26. Use ammonia and ammonium chloride to explain how a buffer controls pH.

27. Name three buffer systems operating in the blood.

28. What groups on protein buffers react with (a) added acid and (b) added base?

29. If acid is added to an unbuffered solution, will the pH increase or decrease?

30. If someone is suffering from alkalosis, is the blood pH too high or too low?

Note: Exercises 31–38 require the use of a table of logarithms or a scientific calculator with a base 10 logarithm function.

31. Calculate the pH of solutions with the following molar hydronium ion concentrations.
    a. 3×10^{-3} **b.** 5×10^{-7} **c.** 8×10^{-10}

32. Calculate the pH of a 3.6×10^{-4} M HCl solution.

33. Calculate the pH of a 8.8×10^{-2} M HNO_3 solution.

34. Calculate the pH of a 3.0×10^{-3} M NaOH solution.

35. Calculate the pH of a blood solution that has a hydronium ion concentration of 4.6×10^{-8} M.

36. Calculate the pH of a urine sample that has a hydronium ion concentration of 2.3×10^{-6} M.

37. Calculate the pH of an ammonia solution that has a hydronium ion concentration of 2.0×10^{-12} M.

38. Calculate the pH of a sample of gastric juice that has a hydronium ion concentration of 0.12 M.

Equilibrium Calculations

In Chapters 10 and 11, we described the equilibria established when a weak acid, a weak base, a salt of either, or some combination of these is dissolved in water. Our discussion was entirely qualitative. It is possible to treat these equilibria in a more quantitative fashion. We will now do that, in this special topic, for those who are interested or whose work might require it.

B.1

EQUILIBRIUM CONSTANT EXPRESSIONS

The proportions of reactants and products at equilibrium are determined by a simple relationship. Let's write a generalized reaction,

$$a\,A \; + \; b\,B \; \rightleftharpoons \; c\,C \; + \; d\,D$$

where A and B are reactants and C and D are products. The small letters are the coefficients for the substances. The relationship at a given temperature is given by the expression

$$K = \frac{[C]^c \times [D]^d}{[A]^a \times [B]^b}$$

The quantities in brackets stand for molar concentration at equilibrium. The quantity K is called the **equilibrium constant.** The entire expression is called the **equilibrium constant expression.** The coefficients in the generalized reaction become exponents in the equilibrium constant expression.

Let's consider a specific reaction.

$$H_2 \; + \; Cl_2 \; \rightleftharpoons \; 2\,HCl$$

The equilibrium constant expression for this reaction is

$$K = \frac{[HCl]^2}{[H_2] \times [Cl_2]}$$

For the reaction

$$N_2 \; + \; 3\,H_2 \; \rightleftharpoons \; 2\,NH_3$$

The equilibrium expression is

$$K = \frac{[NH_3]^2}{[N_2] \times [H_2]^3}$$

EXAMPLE B.1

Write the equilibrium constant expression for each of the following reactions.

a. $H_2 + F_2 \rightleftharpoons 2\,HF$
b. $2\,NO + O_2 \rightleftharpoons 2\,NO_2$
c. $CO + H_2O \rightleftharpoons CO_2 + H_2$

SOLUTION

a. $K = \dfrac{[HF]^2}{[H_2] \times [F_2]}$ b. $K = \dfrac{[NO_2]^2}{[NO]^2 \times [O_2]}$

c. $K = \dfrac{[CO_2] \times [H_2]}{[CO] \times [H_2O]}$

All of the above examples involve gaseous reactants and products. The same treatment is applied to substances in aqueous solution in the next section. Special cases involving a solid reactant or product are considered in Section 12.9.

B.2

IONIZATION OF WEAK ACIDS

Now let's take another look at the ionization of acetic acid. The reaction is

$$CH_3COOH \; + \; H_2O \; \rightleftharpoons \; H_3O^+ \; + \; CH_3COO^-$$

At equilibrium, the equilibrium constant expression is

$$K = \frac{[H_3O^+] \times [CH_3COO^-]}{[CH_3COOH] \times [H_2O]}$$

For dilute solutions, the molar concentration of water, $[H_2O]$, stays nearly constant at 55 M.[1] We can substitute this value into the equation:

$$K = \frac{[H_3O^+] \times [CH_3COO^-]}{[CH_3COOH] \times 55}$$

---

[1] In 1 L of dilute aqueous solution there is approximately 1 L of water, which weighs 1000 g and contains

$$\frac{1000\ g}{18\ g/mol} = 55 \text{ mol of water}$$

Thus, the concentration of water in 1 L of dilute solution is approximately 55 M.

Rearranging, we get

$$K \times 55 = \frac{[H_3O^+] \times [CH_3COO^-]}{[CH_3COOH]} = K_a$$

The product of the two constants, $K \times 55$, is a constant. This new constant, K_a, is called the **ionization constant** for the weak acid.

Sometimes the involvement of water is omitted, and the ionization of acetic acid is written

$$CH_3COOH \rightleftharpoons H^+ + CH_3COO^-$$

The ionization constant becomes

$$K_a = \frac{[H^+] \times [CH_3COO^-]}{[CH_3COOH]}$$

If you recognize that the hydronium ion is simply a hydrated proton, you will see that the two K_a expressions are the same.

The strength of an acid is related to the degree of ionization. The greater the degree of ionization, the stronger the acid. (Strong acids, such as HCl, are 100% ionized in dilute solution.) Also, the greater the degree of ionization, the larger the K_a. Table B.1 lists the K_a values for several common acids. Don't forget: The larger the K_a (the less negative the exponent), the stronger the acid.

Ionization constants can be calculated from measurements of the hydrogen ion concentration as shown in the following example.

EXAMPLE B.2

Find the K_a for an organic acid (symbolized by HOrg), if the hydrogen ion concentration of a 0.0200 M solution is 3.0×10^{-4} M. The simplified ionization reaction is

$$HOrg \rightleftharpoons H^+ + Org^-$$

and the equilibrium constant expression is

$$K_a = \frac{[H^+][Org^-]}{[HOrg]}$$

SOLUTION

Ionization of the acid produces equal concentrations of H^+ and Org^-.

$$[H^+] = [Org^-] = 3.0 \times 10^{-4} \text{ M} = 0.00030 \text{ M}$$

At equilibrium, the concentration of nonionized acid equals its original concentration minus the amount that has ionized to form H^+ and Org^-.

$$[HOrg] = 0.0200 - 0.00030 = 0.0197 \text{ M}$$
$$= 1.97 \times 10^{-2} \text{ M}$$
$$K_a = \frac{(3.0 \times 10^{-4})(3.0 \times 10^{-4})}{1.97 \times 10^{-2}} = 4.6 \times 10^{-6}$$

Usually, though, we calculate hydrogen ion concentrations from K_a values found in tables. This approach is illustrated in the following examples.

EXAMPLE B.3

Calculate the $[H^+]$ in a 0.10 M solution of acetic acid. The equation is $CH_3COOH \rightleftharpoons H^+ + CH_3COO^-$.

SOLUTION

First, write the equilibrium constant expression. (The K_a of acetic acid is given in Table B.1.)

$$K_a = \frac{[H^+][CH_3COO^-]}{[CH_3COOH]} = 1.8 \times 10^{-5}$$

Acid	Simplified Reaction	K_a
Hydrochloric acid	$HCl \longrightarrow H^+ + Cl^-$	Very large
Sulfuric acid	$H_2SO_4 \longrightarrow H^+ + HSO_4^-$	Very large
	$HSO_4^- \rightleftharpoons H^+ + SO_4^{2-}$	1.2×10^{-2}
Hydrofluoric acid	$HF \rightleftharpoons H^+ + F^-$	6.6×10^{-4}
Hydrocyanic acid	$HCN \rightleftharpoons H^+ + CN^-$	6.2×10^{-10}
Formic acid	$HCOOH \rightleftharpoons H^+ + HCOO^-$	1.8×10^{-4}
Acetic acid	$CH_3COOH \rightleftharpoons H^+ + CH_3COO^-$	1.8×10^{-5}
Benzoic acid	$C_6H_5COOH \rightleftharpoons H^+ + C_6H_5COO^-$	6.6×10^{-5}
Carbonic acid	$H_2CO_3 \rightleftharpoons H^+ + HCO_3^-$	4.2×10^{-7}
	$HCO_3^- \rightleftharpoons H^+ + CO_3^{2-}$	4.8×10^{-11}
Phosphoric acid	$H_3PO_4 \rightleftharpoons H^+ + H_2PO_4^-$	7.5×10^{-3}
	$H_2PO_4^- \rightleftharpoons H^+ + HPO_4^{2-}$	6.2×10^{-8}
	$HPO_4^{2-} \rightleftharpoons H^+ + PO_4^{3-}$	4.4×10^{-13}

Table B.1
K_a Values for Several Acids in Water at 25 °C

For every CH_3COOH that ionizes, one H^+ and one CH_3COO^- are formed. We can therefore let $x = [H^+] = [CH_3COO^-]$.

Since each H^+ formed represents one CH_3COOH ionized, the equilibrium concentration of CH_3COOH will be $0.10 - x$. Substitution gives

$$\frac{(x)(x)}{0.10 - x} = 1.8 \times 10^{-5}$$

The amount of CH_3COOH ionized is very small compared with the total amount of CH_3COOH that we started with, so let's assume that

$$0.10 - x \cong 0.10 = 1.0 \times 10^{-1}$$

We then have

$$\frac{x^2}{1.0 \times 10^{-1}} = 1.8 \times 10^{-5}$$

$$x^2 = (1.8 \times 10^{-5})(1.0 \times 10^{-1})$$
$$= 1.8 \times 10^{-6}$$
$$x = 1.3 \times 10^{-3} \text{ M} = [H^+]$$

Now let's check our assumption. Is 0.10 changed significantly by subtracting 0.0013?

$$0.10 - 0.0013 \stackrel{?}{=} 0.10$$

Certainly no significant error was introduced.

EXAMPLE B.4

Calculate the $[H^+]$ in a 0.0010 M solution of HCN.

SOLUTION

$$HCN \rightleftharpoons H^+ + CN^-$$

$$K_a = \frac{[H^+][CN^-]}{[HCN]} = 6.2 \times 10^{-10}$$

Let $x = [H^+] = [CN^-]$. Then

$$[HCN] = 0.0010 - x \cong 0.0010$$

Substituting, we have

$$\frac{(x)(x)}{1.0 \times 10^{-3}} = 6.2 \times 10^{-10}$$

$$x^2 = 6.2 \times 10^{-13} = 62 \times 10^{-14}$$

$$x = 7.9 \times 10^{-7} = [H^+]$$

Is x small compared with 0.0010? Yes!

Approximations such as those made in the preceding examples are good only if the degree of ionization is small. Such an assumption should always be checked.

B.3
EQUILIBRIA INVOLVING WEAK BASES

Equilibrium constant expressions can also be written for weak bases. For ammonia, the reaction is

$$NH_3 + H_2O \rightleftharpoons NH_4^+ + OH^-$$

and the equilibrium constant expression is

$$K = \frac{[NH_4^+][OH^-]}{[NH_3][H_2O]}$$

Since the concentration of water is quite constant for dilute solutions, we can include $[H_2O]$ with K to give a new constant, K_b.

$$K_b = \frac{[NH_4^+][OH^-]}{[NH_3]}$$

Ionization constants for several weak bases are given in Table B.2.

Table B.2
Ionization Constants for Some Weak Bases in Water at 25 °C

Base	Reaction	K_b
Ammonia	$NH_3 + H_2O \rightleftharpoons NH_4^+ + OH^-$	1.8×10^{-5}
Methylamine	$CH_3NH_2 + H_2O \rightleftharpoons CH_3NH_3^+ + OH^-$	4.2×10^{-4}
Aniline	$C_6H_5NH_2 + H_2O \rightleftharpoons C_6H_5NH_3^+ + OH^-$	4.2×10^{-10}
Phosphate ion	$PO_4^{3-} + H_2O \rightleftharpoons HPO_4^{2-} + OH^-$	4.5×10^{-2}
Carbonate ion	$CO_3^{2-} + H_2O \rightleftharpoons HCO_3^- + OH^-$	1.8×10^{-4}
Bicarbonate ion	$HCO_3^- + H_2O \rightleftharpoons H_2CO_3 + OH^-$	2.3×10^{-8}
Zinc hydroxide	$Zn(OH)_2 \rightleftharpoons ZnOH^+ + OH^-$	9.6×10^{-4}
Silver hydroxide	$AgOH \rightleftharpoons Ag^+ + OH^-$	1.1×10^{-4}

EXAMPLE B.5

Calculate the hydroxide ion concentration in a 0.010 M solution of aniline.

SOLUTION

$$K_b = \frac{[C_6H_5NH_3^+][OH^-]}{[C_6H_5NH_2]} = 4.2 \times 10^{-10}$$

Let $x = [OH^-] = [C_6H_5NH_3^+]$.

$$[C_6H_5NH_2] = 0.010 - x \cong 0.010 = 1.0 \times 10^{-2}$$

Substituting, we have

$$\frac{(x)(x)}{1.0 \times 10^{-2}} = 4.2 \times 10^{-10}$$
$$x^2 = 4.2 \times 10^{-12}$$
$$x = 2.0 \times 10^{-6} \text{ M} = [OH^-]$$

Again, we see that x is small compared with 0.010.

B.4

CALCULATIONS INVOLVING BUFFERS

Let's take another look at solutions that contain both a weak acid and a salt of that acid. We saw in Section 11.5 that such a solution acts as a buffer to resist a change in pH. Now we can calculate the hydrogen ion concentration in such a system.

If sodium acetate is added to a solution of acetic acid, ionization of the acid is considerably decreased. In effect, we are increasing the concentration of acetate ion in the system.

$$CH_3COOH \rightleftharpoons H^+ + CH_3COO^-$$

The acetate ions from sodium acetate react with the H^+, forming more un-ionized acetic acid molecules. A greater proportion of the total acetic acid is in the molecular form; ionization is decreased. This is an example of the **common ion effect**. If an ion common to those in the equilibrium is added, the degree of ionization is decreased. A new equilibrium is established with more acetate ions but fewer H^+. The value of K_a remains unchanged.

EXAMPLE B.6

What is the $[H^+]$ in a solution that is 0.01 M in acetic acid and 0.10 M in sodium acetate?

SOLUTION

Use the K_a expression for acetic acid.

$$K_a = \frac{[H^+][CH_3COO^-]}{[CH_3COOH]} = 1.8 \times 10^{-5}$$

Let $x = [H^+]$.

Sodium acetate is completely ionized; thus the concentration of acetate ion equals the sum of $[CH_3COO^-]$ from the salt and $[CH_3COO^-]$ from the ionization of acetic acid.

$$[CH_3COO^-] = 0.10 + x \cong 0.10 = 1.0 \times 10^{-1}$$

The concentration of acetic acid at equilibrium is the original concentration minus the amount ionized.

$$[CH_3COOH] = 0.10 - x \cong 0.10 = 1.0 \times 10^{-1}$$
$$\frac{(x)(1.0 \times 10^{-1})}{1.0 \times 10^{-1}} = 1.8 \times 10^{-5}$$
$$x = 1.8 \times 10^{-5} \text{ M} = [H^+]$$

Was our assumption valid that x was small enough to be ignored when added to or subtracted from 0.10? Yes, 1.8×10^{-5} is negligibly small compared with 0.10, that is, with 1.0×10^{-1}. (This buffer solution tends to keep the hydrogen ion concentration constant at a value of 1.8×10^{-5} mol/L.)

The ionization of a weak base is also decreased by the addition of a common ion. Adding ammonium chloride to a solution of ammonia decreases the amount of hydroxide ion in the system.

$$NH_3 + H_2O \rightleftharpoons NH_4^+ + OH^-$$

EXAMPLE B.7

Calculate the $[OH^-]$ and $[H^+]$ of a solution that is 0.50 M in NH_3 and 0.10 M in NH_4^+.

SOLUTION

Use the K_b expression for NH_3.

$$K_b = \frac{[NH_4^+][OH^-]}{[NH_3]} = 1.8 \times 10^{-5}$$

Let $y = [OH^-]$.

$$[NH_4^+] \cong 0.10 = 1.0 \times 10^{-1}$$
$$[NH_3] \cong 0.50 = 5.0 \times 10^{-1}$$
$$\frac{(1.0 \times 10^{-1})(y)}{5.0 \times 10^{-1}} = 1.8 \times 10^{-5}$$
$$y = 9.0 \times 10^{-5} = [OH^-]$$

Now use the relationship

$$K_w = [H^+][OH^-] = 1.0 \times 10^{-14}$$

to calculate $[H^+]$.

$$[H^+] = \frac{1.0 \times 10^{-14}}{9.0 \times 10^{-5}} = 1.1 \times 10^{-10}$$

Buffered solutions are widely used in analytical chemistry, medicine, and biochemistry and in many industrial applications such as leatherworking and dyeing. A rearrangement of the equilibrium constant expression, called the Henderson–Hasselbach equation, is available for calculating the pH of a buffer. For a weak acid, HA, the equation is

$$pH = pK_a + \log \frac{[A^-]}{[HA]}$$

In the equation, pK_a is a logarithmic term similar to pH.

$$pK_a = -\log K_a$$

If large amounts of strong acid or base are added, the capacity of a buffer is exceeded, and the pH changes rapidly. A buffer works most effectively when [HA] and $[A^-]$ are equal. Under these conditions, the buffer is said to operate at its optimum pH. If $[A^-] = [HA]$, then

$$\frac{[A^-]}{[HA]} = 1$$

The logarithm of 1 is 0, so

$$\log \frac{[A^-]}{[HA]} = 0$$

Therefore, according to the Henderson–Hasselbach equation, the optimum pH of a buffer is equal to the pK_a of the acid in the buffer.

$$pH = pK_a + \log \frac{[A^-]}{[HA]} = pK_a + 0 = pK_a$$

EXAMPLE B.8

What is the pH of a solution that is 0.2 M in H_2S and 0.2 M in HS^-? The K_a for H_2S is 1×10^{-7}. (Ignore the second ionization of H_2S.)

SOLUTION

Since $[H_2S] = [HS^-]$, the Henderson–Hasselbach equation reduces to

$$pH = pK_a$$
$$pK_a = -\log K_a = -\log (1 \times 10^{-7}) = 7$$

Therefore,

$$pH = 7$$

EXAMPLE B.9

What is the pH of a solution that is 0.50 M in HF and 0.50 M in F^-? The K_a for HF is 6.6×10^{-4}.

SOLUTION

$$pH = pK_a + \log \frac{[F^-]}{[HF]}$$
$$= -\log (6.6 \times 10^{-4}) + \log \frac{0.50}{0.50}$$
$$= -\log 6.6 - \log (10^{-4}) + \log 1$$
$$= -0.82 - (-4.00) + 0$$
$$= 3.18$$

EXAMPLE B.10

What is the pH of a solution that is 0.10 M in HCN and 0.50 M in CN^-? The K_a for HCN is 6.2×10^{-10}.

SOLUTION

$$pH = pK_a + \log \frac{[CN^-]}{[HCN]}$$
$$= -\log (6.2 \times 10^{-10}) + \log \frac{0.50}{0.10}$$
$$= -\log 6.2 - \log 10^{-10} + \log 5.0$$
$$= -0.79 - (-10.00) + 0.70$$
$$= 9.91$$

EXERCISES

1. Write the equation for the ionization of each of the following acids.
    a. HBO_2 **b.** $HClO_2$ **c.** $HC_9H_7O_4$
    d. H_2Se (first ionization only)

2. Write equations for the ionization of each of the following bases (i.e., for the reaction of the base with water to form ions).
    a. $C_4H_9NH_2$ **b.** $C_{11}H_{21}O_4N$ **c.** C_3H_5N

3. Find the K_a for the acid HZ if the $[H^+]$ of a 0.100 M solution of HZ is 0.0002 M.

4. Find K_b for the base QOH if $[OH^-]$ of a 0.0200 M solution of QOH is 0.0004 M. The ionization reaction is

    $$QOH \rightleftharpoons Q^+ + OH^-$$

5. Write an equilibrium constant expression for each of the following reactions.
    a. $HOCl \rightleftharpoons H^+ + OCl^-$
    b. $HC_6H_7O_6 \rightleftharpoons H^+ + C_6H_7O_6^-$
    c. $HCO_2H \rightleftharpoons H^+ + HCO_2^-$

6. Write an equilibrium constant expression for each of the following reactions.
    a. $C_5H_5N + H_2O \rightleftharpoons C_5H_5NH^+ + OH^-$
    b. $C_2H_5NH_2 + H_2O \rightleftharpoons C_2H_5NH_3^+ + OH^-$
    c. $HPO_4^{2-} + H_2O \rightleftharpoons H_2PO_4^- + OH^-$

7. Calculate the $[H^+]$ of each of these solutions. You may refer to Table B.1 for K_a values.
    a. 0.001 M CH_3COOH
    b. 0.02 M C_6H_5COOH
    c. 0.50 M HCN

8. Calculate the $[H^+]$ for each of these solutions. You may refer to Table B.1 for K_a values.
    a. 0.01 M HCOOH   **b.** 0.1 M HF
    c. 0.005 M HCN

9. What is $[OH^-]$ in each of the solutions in Exercise 7?

10. What is $[OH^-]$ in each of the solutions in Exercise 8?

11. Calculate $[OH^-]$ for each of these solutions. You may refer to Table B.2 for K_b values.
    a. 0.025 M NH_3 **b.** 0.10 M CH_3NH_2
    c. 0.10 M $C_6H_5NH_2$

12. Calculate the $[OH^-]$ for each of these solutions. You may refer to Table B.2 for K_b values.

    a. 0.010 M $Zn(OH)_2$ **b.** 0.15 M HCO_3^-
    c. 0.030 M CO_3^{2-}

13. What is $[H^+]$ in each of the solutions in Exercise 11?

14. What is $[H^+]$ in each of the solutions in Exercise 12?

15. What is $[H^+]$ in each of the following buffer solutions? You may refer to Table B.i for K_a values.
    a. 0.25 M HCN and 0.25 M KCN
    b. 0.50 M HF and 0.50 M NaF
    c. 0.033 M C_6H_5COOH and 0.033 M $C_6H_5COO^-$

16. What is $[H^+]$ in each of the following buffer solutions? You may refer to Table B.1 for K_a values.
    a. 0.1 M HCN and 0.2 M KCN
    b. 0.5 M HF and 0.2 M NaF
    c. 0.4 M C_6H_5COOH and 0.2 M $C_6H_5COO^-$

17. Calculate $[OH^-]$ in a solution that is 0.4 M in NH_3 and 0.4 M in NH_4^+. What is $[H^+]$ in the solution?

18. Calculate $[OH^-]$ in a solution that is 0.04 M in NH_3 and 0.02 M in NH_4^+. What is $[H^+]$ in the solution?

19. What is the pH of a buffer solution that is 0.1 M in HOC_6H_5 and 0.1 M in $NaOC_6H_5$? The K_a for HOC_6H_5 is 1×10^{-10}.

20. What is the pH of a buffer solution that is 0.05 M in $C_5H_{11}COOH$ and 0.05 M in $C_5H_{11}COO^-$? The K_a for $C_5H_{11}COOH$ is 1×10^{-8}.

Note: Exercises 21–26 require the use of a table of logarithms or a scientific calculator with a base 10 logarithm function. You may refer to Table B.1 for K_a values.

21. Calculate the pH of a solution that is 0.040 M in HCN and 0.040 M in CN^-.

22. Calculate the pH of a solution that is 0.20 M in HCOOH and 0.20 M in $HCOO^-$.

23. Calculate the pH of a solution that is 0.15 M in benzoic acid and 0.15 M in benzoate ion.

24. Calculate the pH of a solution that is 0.20 M in HF and 0.50 M in F^-.

25. Calculate the pH of a solution that is 0.15 M in HCOOH and 0.60 M in $HCOO^-$.

26. Calculate the pH of a solution that is 0.11 M in benzoic acid and 0.96 M in benzoate ion.

Chapter 12
ELECTROLYTES

Chemical reactions in solution are the source of the electric current produced by a battery.

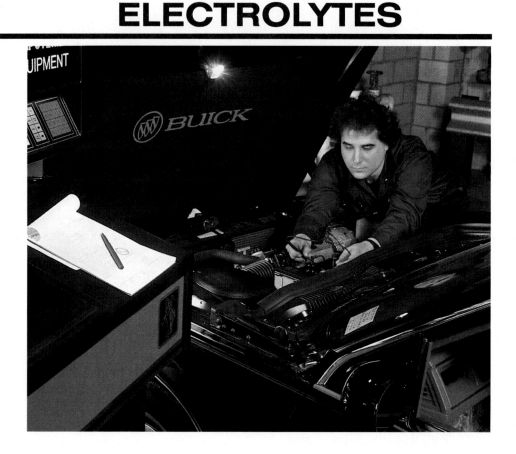

The term **ion** was introduced in Chapter 4 to describe charged particles. We have come a long way since then, and along the way we've discussed ionic bonds, ionic solids, and solutions of ions. The nineteenth-century scientists whose experiments laid the foundation for the ionic theory were working mainly for the joy of discovery and the recognition gained with success. We can only speculate as to whether they imagined the importance of ions in living systems.

The fluids in our bodies are like the saltwater of the oceans. In our fetal development, we have gill-like organs, hands that look like fins, and a shape much like that of a fish. Until we are born and breathe the air, we float in the watery darkness of the womb. Even after birth, fluids bathe our tissues and transport the materials that keep them alive. It is now well established that messages are sent to and from the brain in the form of electric signals. These messages are often carried by ions through cellular and intercellular fluids. Certain ions are essential to the proper functioning of all living organisms. They must be present in proper concentrations, however. Too few or too many can be dangerous. We will discuss some of these ions and their properties in this chapter.

12.1
EARLY ELECTROCHEMISTRY

In 1800, an Italian physicist, Alessandro Volta, invented a battery that produced an electric current. Electrical phenomena had already been subjected to considerable study before this time. (In 1752, Benjamin Franklin determined that lightning was a form of electricity by flying a kite in a thunderstorm.) The importance of Volta's invention lay in the fact that the battery provided a convenient source of electricity, one that could be used by other scientists who wished to study the interaction of matter and electricity.

Within 6 weeks of its invention, a battery was used by two English chemists, William Nicholson and Anthony Carlisle, to decompose water into hydrogen gas and oxygen gas. They accomplished the decomposition by passing an electric current through a water sample. Humphry Davy, another English chemist, used an electric current to liberate potassium metal from potassium hydroxide (KOH), sodium metal from sodium hydroxide (NaOH), and metallic magnesium, strontium, barium, and calcium from their respective compounds.

Humphry's protégé, Michael Faraday, named the process of splitting compounds by means of electricity **electrolysis,** a word of Greek origin that literally means ''releasing by electricity.'' It was Faraday who was responsible for many of the terms used in electrochemistry today. Some of these terms were introduced in Chapter 2, but we offer a brief review here. The carbon rods or metal strips that are connected to a source of electricity (the battery) and inserted into solutions under study are called **electrodes** (Figure 12.1). Electricity, we know now, is a flow of electrons. In systems used by electrochemists, electrons flow from one electrode to the other. The electrode that has lost electrons and is therefore positively charged is called the **anode.** The electrode that has gained electrons and is negatively charged is called the **cathode.** Electrolysis involves oxidation and reduction reactions. *Oxidation* occurs at the *anode,* where substances give up electrons to the positively charged electrode. *Reduction* occurs at the *cathode,* where substances can pick up electrons from the electrode, which has an excess of them.

In Faraday's studies, electrodes were placed in certain solutions and melts (Section 12.2). When this was done, the electric ''circuit'' was completed. The battery provided the driving force (the electric potential or voltage), and electric current flowed from one electrode to the other through the solution and then back to the first electrode, that is, around the circuit. Faraday hypothesized that the electric current was carried through the solution (or melt) by atoms that had electric charges. He called these charged atoms **ions,** a term we have used extensively in preceding chapters. You will recall that positively charged ions, which are attracted to the negatively charged electrode (the cathode), are called **cations,** and negatively charged ions, which are attracted to the positively charged anode, are called **anions.** (It is useful to remember that in electrolysis *anions* migrate to the *anode,* and *cations* migrate to the *cathode.*)

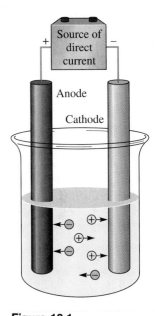

Figure 12.1
When electric current is passed through an electrolyte, positive ions move to the cathode and negative ions move to the anode. Reduction occurs at the cathode, oxidation at the anode.

12.2
ELECTRICAL CONDUCTIVITY

The electricity with which we are most familiar usually flows in metallic wires. The outer electrons of the metal atoms flow through the wire, while the nucleus and inner-level electrons of the atom remain nearly fixed in place. Most nonmetals are nonconductors. The outer-level electrons of these elements are tightly bound or are shared

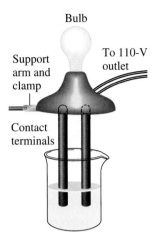

Figure 12.2

An apparatus for testing the electrical conductivity of solutions.

with neighboring atoms. They are not free to roam around. Likewise, in most covalent compounds, the electrons are firmly held and cannot conduct electricity. A few polar covalent compounds, such as hydrochloric acid, react with water to form ions.

$$HCl \; + \; H_2O \; \longrightarrow \; H_3O^+ \; + \; Cl^-$$

Solutions in which these ions are formed are then able to conduct electricity.

Solid ionic compounds such as salts, however, do not conduct electricity. The ions occupy fixed positions in the crystal lattice and are unable to move very much in an electric field. When the lattice is broken down by heat (melting) or by the process of dissolution, the ions are freed. These charged particles can then move in an electric field. Substances that conduct electricity as melts or in solutions are called **electrolytes.**

How do we know an electrolyte when we see one? Simple. Just dissolve some of the compound in water and test it with a conductivity apparatus (Figure 12.2). If a solution that conducts electricity is placed in the beaker, thus completing the circuit, electric current will flow in the apparatus, and the light bulb will glow. Compounds that, in aqueous solution, are good conductors (i.e., those that cause the bulb to glow brightly) are called **strong electrolytes.** Compounds whose aqueous solutions produce only a dim light are called **weak electrolytes.** Other compounds, whose solutions don't light the bulb at all, are called **nonelectrolytes.** These compounds produce few if any ions in solution and conduct no appreciable current.

Isn't there an easier way to determine whether a compound is an electrolyte? Can we predict electrical conductivity? Indeed we can, at least in many cases. All strong acids and strong bases (Chapter 10) and all salts that are appreciably soluble in water are strong electrolytes. All weak acids and weak bases and some slightly soluble salts are weak electrolytes. Molecular substances, such as sugar and alcohol, which are neither acids nor bases, are nonelectrolytes. *The deciding factor in all cases is the ability of the compound being considered to provide ions in aqueous solution.* Table 12.1 lists some familiar compounds and the behavior of their solutions toward electric current.

The work of Faraday and other pioneers in electrochemistry generated a great deal of interest and excitement among scientists whose concern was the fundamental structure of matter. It is worth noting, however, that the interest of the general public was excited by a purely fictional experiment involving electricity and described by a 21-year-old woman. Mary Wollstonecraft Shelley's novel *Frankenstein,* published in 1818, captured the imagination of nineteenth- and twentieth-century readers by relating electricity to life (Figure 12.3).

Table 12.1

A Selection of Strong Electrolytes, Weak Electrolytes, and Nonelectrolytes

Compound Name	Formula	Kind of Compound	Electrical Conductivity
Hydrochloric acid	HCl	Strong acid	Strong
Sulfuric acid	H_2SO_4	Strong acid	Strong
Sodium hydroxide	$NaOH$	Strong base	Strong
Sodium chloride	$NaCl$	Salt	Strong
Calcium nitrate	$Ca(NO_3)_2$	Salt	Strong
Acetic acid	CH_3COOH	Weak acid	Weak
Ammonia	NH_3	Weak base	Weak
Calcium sulfate	$CaSO_4$	Slightly soluble salt	Weak
Copper(II) sulfide	CuS	Insoluble salt	None
Sugar (sucrose)	$C_{12}H_{22}O_{11}$	Molecular solid	None
Ethyl alcohol	C_2H_5OH	Molecular liquid	None

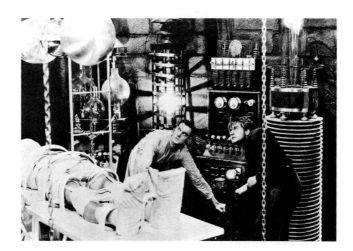

Figure 12.3
The laboratory of Frankenstein, with its apparatus for the production of electric current.

12.3
THE THEORY OF ELECTROLYTES: IONIZATION AND DISSOCIATION

In 1887, Svante Arrhenius proposed a general theory to explain the properties of electrolytes. We encountered a part of that theory in Chapter 10 in our study of acids and bases. Further, Arrhenius's theory has been modified somewhat through the years to account for new data. The modernized theory is summarized here.

1. An electrolyte, when dissolved in water, dissociates into ions, each of which carries an electric charge. For salts, this process can be summarized by an equation. For example,

$$Na_2SO_4(s) \longrightarrow 2\,Na^+(aq) + SO_4^{2-}(aq)$$
$$CaCl_2(s) \longrightarrow Ca^{2+}(aq) + 2\,Cl^-(aq)$$

For acids and some bases, the process is actually one of *ion formation* or **ionization.**

$$HCl(g) + H_2O \longrightarrow H_3O^+(aq) + Cl^-(aq)$$
$$NH_3(g) + H_2O \longrightarrow NH_4^+(aq) + OH^-(aq)$$

For strong bases, the process is usually one of **dissociation**—the separation of already existing ions.

$$NaOH(s) \longrightarrow Na^+(aq) + OH^-(aq)$$

2. When an ionic solid dissociates or when a polar molecule ionizes, the algebraic sum of all positive charges and all negative charges is 0.

$$K_3PO_4(s) \longrightarrow \underset{3(+1)}{3\,K^+(aq)} + \underset{(-3)}{PO_4^{3-}(aq)} = 0$$

Solutions as a whole are therefore electrically neutral.

3. Each ion, regardless of size, charge, or shape, has the same effect on boiling-point elevation, freezing-point depression, and osmotic pressure (Chapter 9) as an undissociated molecule would have. This assumption holds quite well for dilute solutions but is not strictly true for concentrated ones. In concentrated solutions, each ion is rather

closely surrounded by others of opposite charge. This hinders the movement of the ions; they are not completely free. This decreases the expected activity somewhat. For example, the freezing-point depression of a solution of 1 mol of NaCl in 1 kg of water is 1.8 times (almost two times) that of a nonelectrolyte at the same concentration.

4. Weak electrolytes react with water to a limited extent and hence provide only a limited number of ions in solution.
5. Nonelectrolytes exist in molecular form in solution or are insoluble salts; they produce no appreciable concentration of ions.

12.4
ELECTROLYSIS: CHEMICAL CHANGE CAUSED BY ELECTRICITY

If sodium chloride is heated until it melts, the molten salt conducts electricity, as one can see by testing the melt in the conductivity apparatus in Figure 12.2. In addition to causing the light bulb to glow, the electric current, as it passes through the melt, causes observable chemical changes. Yellow-green chlorine gas forms at the anode. At the cathode, silvery metallic sodium is formed and is rapidly vaporized by the hot melt. The molten, ionic sodium chloride is decomposed by electrical energy into elemental sodium and chlorine.

$$2\,NaCl \; + \; energy \; \longrightarrow \; 2\,Na \; + \; Cl_2$$

In crystalline form, sodium chloride does not conduct electricity. The ions occupy relatively fixed positions in the lattice and do not move very much, even under the influence of an electric potential. When sodium chloride melts, the ions are freed to move around. When a battery is connected to the melt through a pair of electrodes, the sodium ions are attracted to the electron-rich cathode, where they pick up electrons (Figure 12.4).

$$Na^+ \; + \; e^- \; \longrightarrow \; Na \qquad (at\ the\ cathode)$$

The chloride ions migrate to the electron-poor anode, where they give up electrons.

$$2\,Cl^- \; \longrightarrow \; Cl_2 \; + \; 2\,e^- \qquad (at\ the\ anode)$$

The battery in the circuit is responsible for seeing to it that this exchange of electrons does not eventually lead to neutralized electrodes. Electrons picked up by the anode from chloride ions are immediately shunted, under the influence of the battery, to the cathode, which has been losing electrons to sodium ions. As long as sodium ions and chloride ions are present in the melt, current will flow. When all of the salt has been converted to elemental sodium and chlorine, the circuit will be broken and current will cease.

This electrolytic reaction is an oxidation–reduction process. Oxidation occurs at the anode, where chloride ions lose electrons. Reduction occurs at the cathode, where sodium ions gain electrons.

Electrolysis, then, is a process of using electricity to bring about chemical change. The process is useful for the preparation and purification (refining) of many metals. Electrolysis is also used for coating one metal with another, an operation called **electroplating.** Usually the object to be electroplated, such as a spoon, is cast of a cheaper metal. It is then coated with a thin layer of a more attractive and more corrosion-resistant metal,

Electrolytic processes are used in the production of Al, Li, K, Na, and Mg and in the refining of Cu.

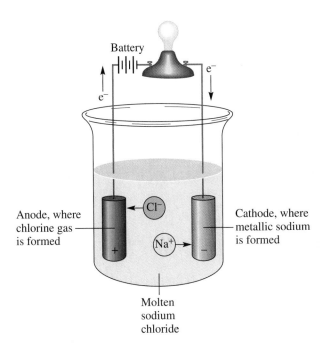

Figure 12.4

The electrolysis of molten sodium chloride.

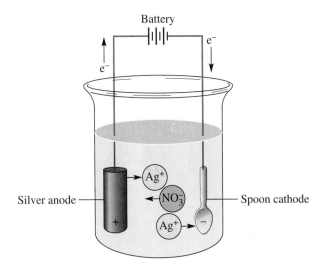

Figure 12.5

An electrolytic cell for the plating of silver.

such as gold or silver. The cost of the finished product is far less than that of a corresponding item made entirely of silver or gold. A cell for the electroplating of silver is shown in Figure 12.5. The silver is made the anode, and the spoon is made the cathode. A solution of silver nitrate is used as the electrolyte. Under the influence of the battery in the system (or any voltage source), the silver ions (Ag^+) are attracted to the cathode (spoon), where they pick up electrons and are deposited as silver atoms.

$$Ag^+ + e^- \longrightarrow Ag$$

At the anode, electrons are removed from the silver bar. Some of the silver atoms lose electrons to become silver ions.

$$Ag \longrightarrow Ag^+ + e^-$$

The net process is one in which the silver from the bar is transferred to the spoon. The thickness of the deposit can be controlled by accurate measurement of the amount of current flow and of the duration of the process.

Electrolysis finds some biological applications, including the removal of unwanted hair. In this process, a tiny wire needle is used to supply a mild electric current to the hair root. The chemical changes engendered there kill the living follicle. This is perhaps the only permanent method of hair removal. In the hands of a skilled technician, the method can be clean and safe. It is tedious, however, for each hair root must be treated individually. Similar procedures are sometimes used for the removal of warts or other growths.

Figure 12.6

The photograph on the left shows a blue solution of Cu^{2+} ions and a sample of zinc metal. When the zinc is added to the Cu^{2+} solution, the more active Zn displaces the less active copper from solution. The products of the displacement reaction (right) are a reddish brown precipitate of copper metal and a colorless solution of Zn^{2+}. The equation for the reaction is

$$Cu^{2+} + Zn \longrightarrow Cu + Zn^{2+}$$

12.5
ELECTROCHEMICAL CELLS: BATTERIES

Electricity can cause chemical change. Conversely, chemical change can produce electricity. Dry cells, storage batteries, and fuel cells all convert chemical energy to electrical energy.

When a strip of zinc metal is placed in a solution of copper(II) sulfate, the following reaction takes place.

$$Zn + Cu^{2+} \longrightarrow Zn^{2+} + Cu$$

The zinc atoms give up their outer electrons to the copper(II) ions. (The sulfate ions are not changed in the reaction and can be omitted from the equation.) The zinc metal dissolves, going into solution as Zn^{2+} ions. The copper ions precipitate from the solution as copper metal (Figure 12.6). The Zn is oxidized; the Cu^{2+} ions are reduced. Electrons are transferred directly from zinc atoms to copper(II) ions. To emphasize this transfer of electrons, the reaction can be split into two half-reactions (the oxidation portion and the reduction portion).

$$Zn \longrightarrow Zn^{2+} + 2\,e^- \quad \textit{(oxidation half-reaction)}$$
$$Cu^{2+} + 2\,e^- \longrightarrow Cu \quad \textit{(reduction half-reaction)}$$

If we place copper ions in one compartment and zinc metal in another, physically separating them (Figure 12.7), electrons have to pass through a wire to get from the zinc metal to the copper ions. This flow of electrons constitutes an **electric current,** and it can be used to run a motor or light a bulb.

Note that in Figure 12.7 the zinc compartment also contains a solution of zinc sulfate. The compartment with the copper ions also has a strip of copper metal. The metal strips serve as electrodes, conducting electrons into and out of the solutions. The strips also participate in the chemical reactions. The zinc electrode slowly dissolves as zinc atoms give up their electrons. Copper metal is plated out on the copper electrode as the

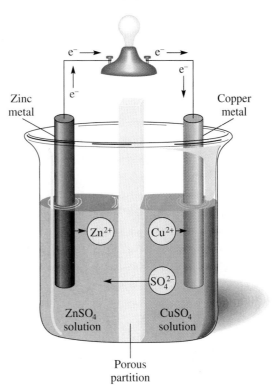

Figure 12.7
A simple electrochemical cell. A battery may consist of one or more such cells.

copper(II) ions pick up those electrons after they have traveled through the wire.

Both compartments of the apparatus in Figure 12.7 contain sulfate ions. These ions don't enter into any chemical reaction, but they do cause complications. Consider the zinc sulfate solution. During the reaction, zinc atoms are converted to zinc ions, so the solution becomes more concentrated in zinc ions. However, nothing happens to the sulfate ions. We are faced with the possibility of the solution containing more positively charged ions (Zn^{2+} ions) than negatively charged ions (SO_4^{2-} ions). This electrical imbalance is an unstable situation, and if something isn't done about it, the reaction will stop. In the other compartment the concentration of copper ions is being reduced while the concentration of sulfate ions remains constant. Thus, there is a tendency for this compartment to develop an excess of negatively charged ions. One way to deal with the problem is to connect the two solutions through a porous plate made of a material that will retard the mixing of the metal ions in solution and still permit passage of sulfate ions. When a zinc atom is converted to a zinc ion in the left compartment, a copper ion changes to a copper atom in the right, and a sulfate ion moves through the porous plate from the solution in the right compartment to the solution in the left. The charges in the two solutions stay balanced. Why bother with the porous plate? Why not just put both solutions and both electrodes in a single compartment so the sulfate ions don't have to move through pores? If we were to do that, the electrons wouldn't have to make the trip through the wire; they could just be passed from zinc atoms to copper ions, which would now be in contact with one another. And unless the electrons go through the wire, the light bulb won't glow (nor will battery-operated portable radios, hand calculators, or cardiac pacemakers work).

The net reaction for this **electrochemical cell** is

$$Zn \ + \ Cu^{2+} \ \longrightarrow \ Zn^{2+} \ + \ Cu$$

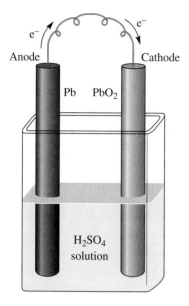

Figure 12.8
Cell of a lead storage battery.

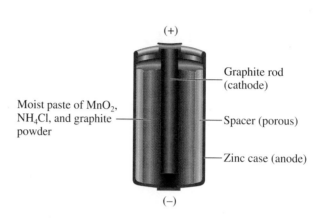

Figure 12.9
Cross section of a zinc–carbon cell.

If 1 M solutions of $ZnSO_4$ and $CuSO_4$ are used, the system will produce about 1.1 volts at 25 °C. The **volt** (V) is a measure of electrical potential, or of the tendency of the electrons in a system to flow (see Section 12.7).

A **battery** is a series of electrochemical cells. The familiar 12-V automobile storage battery consists of six cells wired in series (that is, one after the other). Each cell consists of two electrodes. The cells do not incorporate the zinc–copper couple just described but a somewhat more complicated system involving lead, lead oxide, and sulfuric acid (Figure 12.8). A distinctive feature of the lead storage battery is its capacity for being recharged. It will supply electricity (i.e., it will discharge) when you turn on the ignition to start the car or when the motor is off and the lights are on. But it can be recharged when the car is moving and an electric current is supplied *to* the battery by the mechanical action of the car. The net reaction during discharge is

$$Pb \ + \ PbO_2 \ + \ 2\,H_2SO_4 \ \longrightarrow \ 2\,PbSO_4 \ + \ 2\,H_2O$$

The net reaction during recharge is just the reverse:

$$2\,PbSO_4 \ + \ 2\,H_2O \ \longrightarrow \ Pb \ + \ PbO_2 \ + \ 2\,H_2SO_4$$

Lead storage batteries are durable, but they are heavy and contain corrosive sulfuric acid.

The carbon–zinc dry cell (Figure 12.9) is used in flashlights, tape recorders, camera flash units, and many other small devices that require a portable source of power. The case is made of zinc metal and serves as the anode. The carbon rod serves as the cathode. It is surrounded by a moist paste of graphite powder, manganese dioxide (MnO_2), and ammonium chloride. The anode reaction is simple; zinc is oxidized to zinc ions. The reaction at the cathode is more complex, but a major process is the reduction of manganese dioxide. The overall reaction is mainly

$$Zn \ + \ 2\,MnO_2 \ \longrightarrow \ ZnMn_2O_4$$

A battery is a grouping of two or more cells that are connected together to supply an electric current.

Alkaline cells are similar, but the zinc case is porous, and the carbon cathode is surrounded by moist manganese dioxide and potassium hydroxide. These cells are more expensive than the usual carbon–zinc cells, but they maintain a high voltage for a longer period of time.

12.6
FUEL CELLS

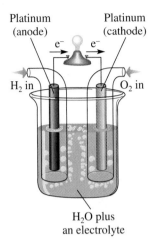

Figure 12.10
A hydrogen–oxygen fuel cell.

A **fuel cell** is a device in which chemical reactions are used to produce electricity directly. Fuel cells differ from the usual electrochemical cell in two ways. First, the fuel and oxygen are fed in continuously. Second, the electrodes are made of an inert material such as platinum that does not react during the process. The electrodes serve to conduct the electrons to and from the solution.

Consider a fuel cell that uses hydrogen and oxygen (Figure 12.10). Hydrogen is oxidized at the anode; oxygen is reduced at the cathode. The overall reaction is similar to combustion, but not all the chemical energy is converted to heat.

$$2 H_2 + O_2 \longrightarrow 2 H_2O$$

Rather, about 40–55% of the chemical energy is converted to electricity.

Fuel cells are used on spacecraft to produce electricity. The water produced can be used for drinking, so less water has to be taken aboard. Fuel cells overall have a weight advantage over storage batteries, an important consideration in a spaceship.

On Earth, more research is needed to reduce cost and to design long-lasting cells. Perhaps someday fuel cells will provide electricity to meet peak needs in large power plants. Unlike huge boilers and nuclear reactors, fuel cells can be started and stopped simply by turning the fuel on or off.

12.7
THE ACTIVITY SERIES

In Section 12.5, we saw that the zinc–copper electrochemical cell produced a voltage of about 1.1 V. This value does not depend on the size of the cell or on the size of the electrodes. The voltage measures the force with which electrons are moved around the circuit, and therefore it measures the tendency of this reaction to occur. Thus, electrochemical cells give a quantitative measure of the relative tendency of various oxidation–reduction reactions to occur.

If zinc metal is treated with hydrochloric acid, a reaction occurs to produce hydrogen gas and zinc chloride (see Figure 6.6). The net reaction is

$$Zn(s) + 2 HCl(aq) \longrightarrow ZnCl_2(aq) + H_2(g)$$

In fact, zinc will react with any acid to yield zinc ions and hydrogen gas. The reaction of zinc with sulfuric acid is

$$Zn(s) + H_2SO_4(aq) \longrightarrow ZnSO_4(aq) + H_2(g)$$

In these reactions with acids, zinc gives up electrons (is oxidized) and hydrogen ions gain electrons (are reduced).

$$Zn(s) \longrightarrow Zn^{2+} + 2 e^- \qquad \textit{(oxidation half-reaction)}$$
$$2 H^+ + 2 e^- \longrightarrow H_2(g) \qquad \textit{(reduction half-reaction)}$$

In contrast, if copper metal is treated with hydrochloric acid or sulfuric acid, no reaction occurs. Why? Because copper does not readily give up electrons to hydrogen ions. We say that zinc is *more active* than hydrogen and that copper is *less active* than hydrogen.

Zinc is one of a number of **active metals** that react with acids in this way. Not all these metals react at the same rate, however. Some, such as sodium and potassium, react violently with plain water—without any added acid. The reaction may produce enough heat to ignite the hydrogen gas, causing an explosion. With sodium, the by-product is a solution of sodium hydroxide.

$$2 Na + 2 H_2O \longrightarrow 2 NaOH(aq) + H_2$$

A similar reaction occurs with potassium.

$$2 K + 2 H_2O \longrightarrow 2 KOH(aq) + H_2$$

All the alkali metals (those in Group IA) react in this manner. Knowing the reaction for sodium, we can write the equation for the reaction of cesium with water.

$$2 Cs + 2 H_2O \longrightarrow 2 CsOH(aq) + H_2$$

We could have substituted lithium or rubidium for cesium, and we would still have been correct. Similar reactions involving members of the same chemical family are a part of the framework that helps make chemistry more understandable. We discuss the chemistry of some of the families of elements in Special Topic C. A word of caution, though. Even members of a family differ. They do react at different rates. Sometimes even the *kind* of reaction is different. But these exceptions will not concern us very much here.

Table 12.2
An Activity Series of the Metals

Metal	Reaction
Potassium Calcium Sodium	React with cold water, steam, or acids, releasing hydrogen
Magnesium Aluminum Zinc Chromium Iron	React with steam or acids, releasing hydrogen
Nickel Tin Lead	React with acids, releasing hydrogen
Hydrogen	
Copper Silver Mercury Platinum Gold	Do not react with acids to form hydrogen

In contrast to the alkali metals, metals like zinc or iron do not react appreciably when placed in water. However, if the temperature is raised and zinc is brought into contact with steam, then hydrogen gas is produced. Tin will not react with steam, but it will release hydrogen on contact with acids. Gold won't even react with acids.

It is possible to arrange metals in an **activity series** with the most reactive metals (such as potassium) at the top and the least reactive (such as gold) at the bottom (Table 12.2). The position of a metal in the table reflects its tendency to give up electrons to form ions. All of those listed above hydrogen in the series (the **active metals**) will react with acids to produce hydrogen gas. These metals will give their electrons to the hydrogen ions produced by the acids, thus becoming ions. Metals below hydrogen in the series will not give up electrons to hydrogen ions and hence will not liberate hydrogen gas from acids.

A lot of chemical information is summarized in Table 12.2. Remember that the arrangement of the table indicates how readily these metallic elements give up their electrons. When one of the metals is brought into contact with the ions of another metal, one of two things can happen. Either the metal can transfer electrons to the ions (a reaction occurs) or the ions will not accept the electrons (no reaction occurs). We can use the table to predict what will happen. A metal can transfer electrons to the ions of any metal that appears lower in the table. For example, an iron nail placed in a dilute solution of copper(II) sulfate ($CuSO_4$) gradually becomes coated with copper metal (Figure 12.11). Since iron atoms have a greater tendency than copper atoms to give up electrons, the iron atoms transfer electrons to the copper ions.

$$Fe \; + \; Cu^{2+} \longrightarrow \; Fe^{2+} \; + \; Cu$$

What we observe is the copper metal plating out of solution as it is formed. Some of the iron dissolves as iron(II) ions, but this is not as obvious to the observer.

If we use another system, it is possible to see both processes as they occur. A strip of copper metal placed in a solution of silver nitrate ($AgNO_3$) becomes coated with crystals of metallic silver (Figure 12.12). The silver ions grab electrons from the more reactive copper atoms.

$$2 \, Ag^+ \; + \; Cu \longrightarrow \; 2 \, Ag \; + \; Cu^{2+}$$

There is also visual evidence that the copper is dissolving. A solution of copper ions is blue, whereas a solution of silver ions is colorless. Thus, as the silver metal crystallizes

Figure 12.12

Copper wire in a silver nitrate solution becomes coated with silver. Copper ions replace silver ions in solution.

Iron nail

Coating of copper

$CuSO_4$ solution

Figure 12.11

An iron nail placed in a solution of copper sulfate becomes coated with copper.

out of solution, the copper ions form and slowly turn the solution blue.

Note that in one of the examples copper loses electrons and in the other it gains them. That is because there are elements both above and below copper in the activity series, and what happens depends only on the *relative tendency of the elements involved to gain or lose electrons.*

Corrosion

An oxidation–reduction reaction of particular economic importance is the corrosion of metals. It is estimated that in the United States alone, corrosion costs $70 billion a year. Perhaps 20% of all the iron and steel production in the United States each year goes to replace corroded items. Let's look first at the corrosion of iron.

In moist air, iron is oxidized, particularly at a nick or scratch.

$$Fe \longrightarrow Fe^{2+} + 2\,e^-$$

In order for iron to be oxidized, oxygen must be reduced.

$$O_2 + 2\,H_2O + 4\,e^- \longrightarrow 4\,OH^-$$

The net result, initially, is the formation of insoluble iron(II) hydroxide.

$$2\,Fe + O_2 + 2\,H_2O \longrightarrow 2\,Fe(OH)_2$$

This product is usually further oxidized to iron(III) hydroxide.

$$4\,Fe(OH)_2 + O_2 + 2\,H_2O \longrightarrow 4\,Fe(OH)_3$$

The latter, sometimes written as $Fe_2O_3 \cdot 3H_2O$, is the familiar iron rust.

The corrosion of iron appears to be an electrochemical reaction (Figure 12.13). Oxidation and reduction often occur at separate points on the

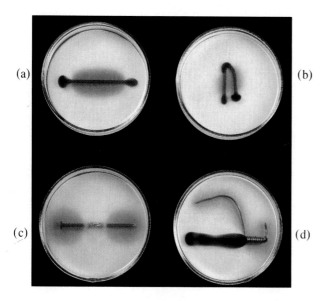

(a)   (b)

(c)   (d)

Figure 12.13

In the corrosion of an iron nail, oxidation of the Fe to Fe^{2+} occurs first at the head and tip (top left). The presence of Fe^{2+} is indicated by the formation of a blue precipitate as the Fe^{2+} reacts with $K_3[Fe(CN)_6]$. Reduction of dissolved O_2 to OH^- occurs in the middle of the nail. The presence of OH^- is indicated by the color of the acid–base indicator phenolphthalein, which is colorless in acidic solution and pink in basic solution. A bent nail (top right) oxidizes rapidly at the bend. An iron nail is protected from corrosion by the more active metal zinc (bottom left). Here no blue precipitate is formed, indicating no Fe^{2+}. Coating of an iron nail with the less active metal copper causes the exposed portions of the nail to corrode even more rapidly than otherwise.

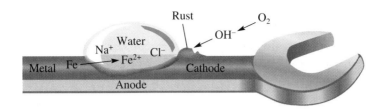

Figure 12.14
The corrosion of iron requires water, oxygen, and an electrolyte.

metal surface. Electrons are transferred through the iron metal. The circuit is completed by an electrolyte in aqueous solution. In the snow belt, this solution is often the slush from road salt and melting snow. The metal is pitted in an anodic area, where iron is oxidized to Fe^{2+}. These ions migrate to the cathodic area, where they react with the hydroxide ions formed by reduction of oxygen.

$$Fe^{2+} + 2\,OH^- \longrightarrow Fe(OH)_2$$

As indicated above, this iron(II) hydroxide is then oxidized to $Fe(OH)_3$, or rust. This process is diagrammed in Figure 12.14. Notice that the anodic area is protected from oxygen by the water film, while the cathodic area is exposed to air.

Aluminum is more reactive than iron (see Table 12.2). Therefore, we might expect it to corrode more rapidly, but billions of beer cans testify to the fact that it doesn't. How can that be? It so happens that freshly prepared aluminum quickly forms a thin, hard film of aluminum oxide on its surface. This film is impervious to air and protects the underlying metal from further oxidation.

The tarnish on silver is another example of corrosion. Silver is oxidized to silver ions.

$$Ag \longrightarrow Ag^+ + e^-$$

These silver ions react with sulfur compounds such as hydrogen sulfide (from eggs, for example) to form black silver sulfide.

$$2\,Ag^+ + H_2S \longrightarrow Ag_2S + 2\,H^+$$

You can use a silver polish to remove the tarnish, but in doing so you also lose part of the silver. An alternative method involves the use of aluminum metal to reduce the silver ions back to silver metal.

$$3\,Ag^+ + Al \longrightarrow 3\,Ag + Al^{3+}$$

This reaction also requires an electrolyte. Sodium bicarbonate (baking soda, $NaHCO_3$) is usually used. The tarnished silver is placed in contact with aluminum foil, covered with a solution of sodium bicarbonate, and heated. A precious metal is conserved at the expense of a cheaper one.

12.8
PRECIPITATION: THE SOLUBILITY PRODUCT RELATIONSHIP

We have thus far focused our attention on strong electrolytes, those compounds that supply high concentrations of ions in solution. There are also ionic compounds whose solubility in water is quite low. For example, the solubility of barium sulfate ($BaSO_4$) is 0.0002 g per 100 g of water at 25 °C. This makes barium sulfate a very weak electrolyte in aqueous solution. The fact that a salt is a weak electrolyte or a nonelectrolyte does not necessarily mean that it is unimportant in the chemistry of living systems. On the contrary, for some physiological processes, the relative insolubility of certain salts is of critical importance. To understand these processes, it will be necessary to consider first a new way of describing solubility, or, more correctly, of determining when a salt will start precipitating from solution because its solubility has been exceeded.

When solid barium sulfate is added to water, an equilibrium is established between the undissolved solute and the ions.

$$BaSO_4(s) \rightleftharpoons Ba^{2+}(aq) + SO_4^{2-}(aq)$$

An extremely small amount goes into solution. Even though barium salts are quite toxic, large amounts of barium sulfate can be swallowed or given by enema because very little will dissolve in the solutions of the body. Since barium sulfate is opaque to X-rays, technicians can use it to outline the stomach or intestines for X-ray photographs. The undissolved barium salt scattered throughout the intestine blocks the X-rays. The X-ray film is unexposed in these areas, which therefore appear white in the developed negative. The barium sulfate is later voided from the body unchanged.

Even though we say that barium sulfate is insoluble, the small amount that does dissolve can be measured (as we indicated above). The concentration of barium ions (Ba^{2+}) and sulfate ions (SO_4^{2-}) in a saturated solution of barium sulfate is 0.00001 M (or 1×10^{-5} M) each at 18 °C.

The product of the concentrations of the ions in a saturated solution

$$[Ba^{2+}][SO_4^{2-}]$$

is a constant. At 18 °C, the product is

$$(1 \times 10^{-5})(1 \times 10^{-5}) = 1 \times 10^{-10}$$

This value is called the **solubility product constant** (K_{sp}) for barium sulfate. It is this constant that we will find useful in predicting whether precipitation will occur if a solution containing barium ions is mixed with one containing sulfate ions. Precipitation will occur if the product $[Ba^{2+}][SO_4^{2-}]$ is greater than 1×10^{-10}. Barium sulfate will continue to dissociate until the product is just equal to 1×10^{-10}. At that value, the solution will be saturated. If $[Ba^{2+}][SO_4^{2-}]$ is less than 1×10^{-10}, no barium sulfate precipitate will be formed. The solution will be unsaturated in barium sulfate.

As in the case of the ion product of water, K_w (Section 11.3), it is the *product* of the ion concentrations that is important. If barium sulfate is simply placed in water, the concentration of barium ion equals the concentration of sulfate ion. It is, however, possible to prepare a solution in which the two concentrations are not equal (Example 12.1). The barium ion concentration could be higher than the sulfate ion concentration, or vice versa. Precipitation will always occur when the *product* of the two concentrations exceeds 1×10^{-10}, but precipitation will not necessarily coincide with concentrations of 1×10^{-5} for each ion.

EXAMPLE 12.1

If 0.001 mol of $BaCl_2$ and 0.0001 mol of Na_2SO_4 are added to 1 L of water, will $BaSO_4$ precipitate? (Assume no volume change.)

SOLUTION

$BaCl_2$ and Na_2SO_4 are soluble salts; each one will completely dissociate.

$$BaCl_2 \longrightarrow Ba^{2+} + 2\,Cl^-$$
$$\text{0.001 mol} \qquad \text{0.001 mol} \quad \text{0.002 mol}$$

$$Na_2SO_4 \longrightarrow 2\,Na^+ + SO_4^{2-}$$
$$\text{0.0001 mol} \qquad \text{0.0002 mol} \quad \text{0.0001 mol}$$

The 0.001 mol of $BaCl_2$ will yield 0.001 M Ba^{2+} (as well as 0.002 M Cl^-). The 0.0001 mol of Na_2SO_4 will yield 0.0001 M SO_4^{2-} (as well as 0.0002 M Na^+). The product of the barium and sulfate ion concentrations is

$$K_{sp} = [Ba^{2+}][SO_4^{2-}]$$
$$= (1 \times 10^{-3})(1 \times 10^{-4}) = 1 \times 10^{-7}$$

Since 1×10^{-7} is greater than 1×10^{-10}, $BaSO_4$ will precipitate until the product of the $[Ba^{2+}]$ and $[SO_4^{2-}]$ is just equal to 1×10^{-10}. The sodium and chloride ions in solution have no tendency to precipitate as sodium chloride salt, because sodium chloride is quite soluble in water.

Remember that the smaller negative exponent corresponds to the larger number. For example, $10^{-1} = 0.1$ and $10^{-2} = 0.01$.

The solubility product principle can be used in the preparation of certain compounds. If silver nitrate solution is mixed with sodium chloride solution, slightly soluble silver chloride will precipitate.

$$\underbrace{Ag^+ + NO_3^-}_{\text{Soluble}} + \underbrace{Na^+ + Cl^-}_{\text{Soluble}} \longrightarrow \underbrace{AgCl(s)}_{\text{Insoluble}} + \underbrace{Na^+ + NO_3^-}_{\text{Soluble}}$$

The silver chloride can be removed by filtration, that is, by pouring the mixture through a porous paper that will trap the solid silver chloride but permit passage of the solution. Further, if equimolar quantities of silver nitrate and sodium chloride are used, the filtrate (the solution going through the filter paper) can be evaporated, and the other ions can be obtained as sodium nitrate in crystalline form. The solubility product (K_{sp}) of silver chloride, $[Ag^+][Cl^-]$, is 1.6×10^{-10} at 25 °C. That means that a very small number of silver ions and chloride ions in solution will pass through the filter paper. But the vast majority of these ions will be trapped on the paper as the solid salt.

The solubility products for many salts are known. A few values are given in Table 12.3.

Precipitation is important in the formation of many minerals in nature. Geologists are not the only ones concerned with the formation of mineral deposits, however. Our teeth and bones are largely calcium phosphate salts. One such salt is $Ca_3(PO_4)_2$, sometimes called tricalcium phosphate. Teeth and bones are formed by the precipitation of calcium phosphate salts from solution. In order for this precipitation to occur, the concentrations of ions must exceed the solubility product in the immediate area of deposition. If we assume that the precipitation reaction is

$$3\,Ca^{2+} + 2\,PO_4^{3-} \rightleftharpoons Ca_3(PO_4)_2(s)$$

Table 12.3
Selected Solubility
Product Constants at
25 °C

Compound	Formula	K_{sp}
Barium carbonate	$BaCO_3$	2.0×10^{-9}
Barium sulfate	$BaSO_4$	1.1×10^{-10}
Calcium carbonate	$CaCO_3$	4.8×10^{-9}
Copper(II) sulfide	CuS	8.7×10^{-36}
Lead chromate	$PbCrO_4$	2.0×10^{-14}
Magnesium carbonate	$MgCO_3$	2.0×10^{-8}
Mercury(II) sulfide	HgS	3.0×10^{-53}
Silver acetate	$AgC_2H_3O_2$	2.0×10^{-3}
Silver bromide	$AgBr$	5.0×10^{-13}
Silver chloride	$AgCl$	1.2×10^{-10}

the solubility product is

$$[Ca^{2+}][Ca^{2+}][Ca^{2+}][PO_4^{3-}][PO_4^{3-}]$$

Each calcium phosphate unit provides three calcium ions (Ca^{2+}) and two phosphate ions (PO_4^{3-}). More simply, we can write the expression as

$$K_{sp} = [Ca^{2+}]^3[PO_4^{3-}]^2$$

At a temperature of 37 °C (normal body temperature), the solubility product constant for calcium phosphate is about 4×10^{-27}. In the blood, the concentration of free calcium ions is about 0.0012 M (1.2×10^{-3} M), and the concentration of phosphate ions is about 1.6×10^{-8} M. Plugging these values into the solubility product relationship, we get

$$(1.2 \times 10^{-3})^3 \times (1.6 \times 10^{-8})^2 = (1.7 \times 10^{-9})(2.6 \times 10^{-16}) = 4.4 \times 10^{-25}$$

(A brief review on the manipulation of exponential numbers is offered in Appendix II.) Since this value is *larger* than the solubility product (4×10^{-27}), we expect precipitation of calcium phosphate and the subsequent growth of bone and tooth.

   In growing children, the foregoing description may closely represent the actual mechanism. In adults, however, teeth and bones are no longer growing. What keeps them from getting ever larger? The pH at the area of growth is a little below 7.4 owing to metabolic processes in the cells. The hydrogen ions that are present tie up phosphate ions as hydrogen phosphate ions.

$$PO_4^{3-} + H^+ \rightleftharpoons HPO_4^{2-}$$

The phosphate ion concentration is reduced to a value that just satisfies the solubility product relationship at the constant value of 4.0×10^{-27}. Normal bone and tooth maintenance takes place, with neither growth nor diminution.

   Two pathological conditions of the mouth and teeth are worthy of mention. If the salivary glands are removed or destroyed, the teeth are no longer bathed by saliva. At the usual pH values, saliva provides just the right concentration of calcium ions and phosphate ions to prevent dissolution of the teeth. With the salivary glands gone, the concentration of the ions rapidly falls below the solubility product constant. When this happens, more calcium phosphate dissolves, and the teeth, if not removed, erode away.

   A similar situation occurs when children suffer from chronic acidosis. The blood pH may be as low as 7.1, and the pH in the immediate areas of bone formation is probably even lower. The additional hydrogen ions tie up phosphate ions, lowering the concentration of the latter. Bone growth is greatly hindered, and the child's skeleton is badly formed.

Tooth decay (caries) can also be related to solubility. A combination carbohydrate-protein called **mucin** forms a film, called plaque, on teeth. If it is not removed by brushing and flossing, buildup of plaque continues. Food and bacteria trapped in the plaque metabolize carbohydrates, producing lactic acid (see Chapter 24). Saliva does not penetrate plaque and hence cannot buffer against the buildup of the acid. The pH at the surface of the tooth may go as low as 4.5, and the concentration of phosphate ions in solution is rapidly depleted as hydrogen phosphate ions are formed. The calcium phosphate of the tooth dissolves to replenish the phosphate ion, leaving a cavity in the tooth.

Teeth are also eroded in patients suffering from **bulimia,** a condition characterized by binge eating followed by vomiting. Hydrochloric acid vomited from the stomach acts in the mouth to tie up phosphate ions. The pH may drop as low as 1.5, and erosion can be much more rapid than in caries.

Knowledge of solubility product relationships can also be used to bring slightly soluble salts into solution. If a relatively concentrated solution of calcium chloride ($CaCl_2$) is added to a solution with a fair concentration of sodium oxalate ($Na_2C_2O_4$), a precipitate of calcium oxalate (CaC_2O_4) is formed (some kidney stones are primarily calcium oxalate).

$$\underbrace{Ca^{2+} + 2\,Cl^-}_{\text{Soluble}} + \underbrace{2\,Na^+ + C_2O_4^{2-}}_{\text{Soluble}} \longrightarrow \underbrace{CaC_2O_4(s)}_{\text{Insoluble}} + \underbrace{2\,Na^+ + 2\,Cl^-}_{\text{Soluble}}$$

If hydrochloric acid is now added to the precipitate, the precipitate disappears. The solid dissolves because the oxalate ions ($C_2O_4^{2-}$) are tied up by the hydronium ions (from the HCl) to form soluble, slightly ionized oxalic acid.

$$C_2O_4^{2-} + 2\,H_3O^+ \rightleftharpoons H_2C_2O_4 + 2\,H_2O$$

This decreases the concentration of oxalate ions so that the solubility product constant for calcium oxalate is no longer exceeded. The precipitate dissolves.

Unfortunately, strong hydrochloric acid solutions cannot be used to dissolve kidney stones while they are still in the kidneys. The acid would be much too corrosive to our cells. There are chemicals, though, that have shown modest success in dissolving kidney stones. If the stones are calcium oxalate, removing foods containing oxalates (e.g., chocolate, spinach, black tea) from the diet might be of some help.

12.9
THE SALTS OF LIFE: MINERALS

A variety of inorganic compounds are necessary for the proper growth and repair of our tissues. It is estimated that such minerals represent about 4% of human body weight. Some of these, such as the chlorides (Cl^-), phosphates (PO_4^{3-}), bicarbonates (HCO_3^-), and sulfates (SO_4^{2-}), occur in the blood and other body fluids. Others, such as iron (as Fe^{2+}) in hemoglobin and phosphorus in the nucleic acids (DNA and RNA), are constituents of complex organic compounds.

Minerals essential to one or more living organisms contain the elements sodium (Na), magnesium (Mg), potassium (K), phosphorus (P), sulfur (S), chlorine (Cl), calcium (Ca), manganese (Mn), iron (Fe), copper (Cu), cobalt (Co), zinc (Zn), iodine (I), fluorine (F), silicon (Si), cadmium (Cd), lithium (Li), vanadium (V), chromium (Cr), selenium

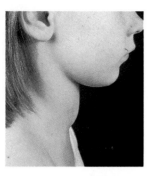

Figure 12.15
A man affected by goiter.
The swollen thyroid gland
in the neck results from a
lack of the trace element
iodine.

(Se), molybdenum (Mo), and arsenic (As). These, along with the structural elements carbon, hydrogen, nitrogen, and oxygen, make up the 26 chemical elements of life. Other elements are sometimes found in body fluids and tissues but are not known to be essential. These include barium (Ba), strontium (Sr), bromine (Br), aluminum (Al), nickel (Ni), and boron (B). It might yet be discovered that one or more of these is essential.

These minerals serve a variety of functions. Perhaps the most dramatic is that of iodine. A small amount of iodine is necessary for the proper functioning of the thyroid gland. A deficiency of iodine has serious effects, of which **goiter** is perhaps the best known (Figure 12.15). Iodine is available in seafood. To guard against iodine deficiency, a small amount of sodium iodide (NaI) is added to table salt (NaCl). Iodized salt has greatly reduced the incidence of goiter.

Iron(II) ions (Fe^{2+}) are necessary for the proper functioning of the oxygen-transporting compound hemoglobin. Without sufficient iron, there will be a shortage of oxygen supplied to the body tissues. The resulting weakened condition is called **anemia.** Foods especially rich in iron compounds include red meats and liver.

As we have seen, calcium and phosphorus are necessary for the proper development of bones and teeth. Growing children need about 1.5 g of each per day. These elements are available in plentiful quantities in milk. The need of adults for these elements is less widely known but very real. For example, calcium ions are necessary for the coagulation of blood (to stop bleeding) and for maintaining the rhythm of the heartbeat. Phosphorus is necessary for the body to metabolize carbohydrates (Chapter 24). Without phosphorus compounds, we couldn't get *any* energy from those ''quick energy'' foods. Compounds containing phosphorus play many other essential roles, also. We will encounter a number of these compounds in subsequent chapters.

Sodium chloride in moderate amounts is essential to life. It is important in the exchange of fluids between cells and plasma, for example. The presence of salt increases water retention. A high volume of retained fluids can cause swelling (edema) and high blood pressure **(hypertension).** Over 120 million prescriptions are written each year in the United States for diuretics, drugs that induce urination in an attempt to reduce the volume of retained fluids. Another 30 million prescriptions are written for potassium (K^+) supplements to replace the potassium that is washed out in the excess urine. An estimated 36 million people in the United States suffer from hypertension, and most physicians agree that our diets generally contain too much salt. The American Heart Association recommends that adults limit their salt intake to no more than 3 g (3000 mg) per day. This recommendation is at least a 50% reduction from our present levels of salt intake.

Iron, copper, zinc, cobalt, managanese, molybdenum, calcium, and magnesium are essential to the proper functioning of certain enzymes. These metalloenzymes are necessary to life. The functions of some other minerals are quite complex. Some things are known about how they operate, but a great deal remains to be learned about the role of inorganic chemicals in our bodies. Bioinorganic chemistry is a fluorishing area of research.

EXERCISES

1. Define or illustrate the following terms.
   - **a.** cathode
   - **b.** anode
   - **c.** cation
   - **d.** anion
   - **e.** electrolysis
   - **f.** strong electrolyte
   - **g.** weak electrolyte
   - **h.** nonelectrolyte
   - **i.** ionization
   - **j.** ion dissociation
   - **k.** activity series
   - **l.** solubility product constant

2. Classify the following as strong electrolytes, weak electrolytes, or nonelectrolytes in solution.
   - **a.** KCl
   - **b.** HNO_3
   - **c.** H_2CO_3
   - **d.** Na_2SO_4
   - **e.** KOH
   - **f.** CH_3OH
   - **g.** AgCl
   - **h.** CCl_4
   - **i.** $CaCl_2$

3. Why does a solution of 1 mol of NaCl in 1 kg of water freeze at a lower temperature than a solution of 1 mol of sugar in 1 kg of water? Why is the freezing-point depression for the salt solution not quite twice that for the sugar solution?

4. Hydrogen chloride is a covalent molecule. Why does a solution of the gas in water conduct electricity?

5. What new substances would be formed if electricity were passed through the following?
   a. molten KBr   **b.** molten LiCl   **c.** molten Al_2O_3

6. Both $Ba(OH)_2$ and H_2SO_4 are strong electrolytes. When equimolar solutions of the two are mixed, the resulting solution is essentially nonconducting. Write the equation for the reaction and explain the observation.

7. Write an equation that shows why hydrogen iodide (HI), a gas, forms an aqueous solution that conducts electricity.

8. Lead chromate ($PbCrO_4$) is an ionic compound, yet a saturated aqueous solution of lead chromate does not conduct electricity. Explain.

9. State the main ideas of the modernized version of Arrhenius's theory of electrolytes.

10. How does an electrolytic cell differ from an electrochemical cell?

11. Solid sodium chloride does not conduct electricity, but the molten salt does. Explain.

12. Formaldehyde (CH_2O) is a nonelectrolyte. What type of bonding exists in formaldehyde?

13. Classify each of the following electrolytes as an acid, a base, or a salt. Write equations to show what happens when each is dissolved in water.
    a. $Ca(OH)_2$ **b.** HBr   **c.** $Sr(NO_3)_2$

14. Complete the following by writing formulas for the expected products, and then balance the equations.
    a. $Ca + HCl \longrightarrow$ **b.** $Ni + HCl \longrightarrow$
    c. $Mg + HNO_3 \longrightarrow$ **d.** $Zn + H_2SO_4 \longrightarrow$
    e. $Al + H_2SO_4 \longrightarrow$ **f.** $Pb + HNO_3 \longrightarrow$

15. Complete the following by writing formulas for the expected products, and then balance the equations.
    a. $Na + H_2O \longrightarrow$ **b.** $K + H_2O \longrightarrow$
    c. $Li + H_2O \longrightarrow$ **d.** $Ca + H_2O \longrightarrow$

16. Complete the following equations for those reactions that can occur. Refer to Table 12.2.
    a. $Mg + Cu^{2+} \longrightarrow$ **b.** $Ag + Pb^{2+} \longrightarrow$
    c. $Fe + Zn^{2+} \longrightarrow$ **d.** $Al + Ni^{2+} \longrightarrow$
    e. $Cr + Na^+ \longrightarrow$ **f.** $Au + Ag^+ \longrightarrow$
    g. $Sn + K^+ \longrightarrow$ **h.** $Ca + Al^{3+} \longrightarrow$

17. Why are water, an electrolyte, and oxygen all required for the corrosion of iron?

18. Why does aluminum corrode more slowly than iron, even though aluminum is more reactive than iron?

19. How does silver tarnish? How can the tarnish be removed without the loss of silver?

20. Will a precipitate form if 0.001 mol of $MgCl_2$ and 0.001 mol of Na_2CO_3 are added to 1.0 L of water? The net ionic reaction is
$$Mg^{2+} + CO_3^{2-} \rightleftharpoons MgCO_3(s)$$

You may refer to Table 12.2 for the K_{sp} value.

21. Will a precipitate form if 0.001 mol of $AgNO_3$ and 0.0001 mol of NaCl are added to 1.0 L of water? The net ionic reaction is
$$Ag^+ + Cl^- \rightleftharpoons AgCl(s)$$

22. Will a precipitate occur if 1×10^{-6} mol of $Pb(NO_3)_2$ and 1×10^{-5} mol of Na_2CrO_4 are added to 1.0 L of water? The net ionic reaction is
$$Pb^{2+} + CrO_4^{2-} \rightleftharpoons PbCrO_4(s)$$

23. Would an object to be electroplated be made the anode or the cathode in an electrolytic cell? Why?

24. If a concentrated solution of sodium acetate is mixed with a solution of silver nitrate, a precipitate of silver acetate is formed. The precipitate dissolves readily when nitric acid is added. Write equations to explain what happens.

25. Boiler scale ($CaCO_3$) is insoluble in water, yet it readily dissolves in hydrochloric acid. Write the equation that explains what happens.

26. What is goiter? How is it prevented?

27. What is anemia? How is it prevented?

28. Describe the role of calcium and phosphorus in the proper development of bones and teeth.

29. What is hypertension?

30. What is a diuretic?

31. What causes tooth decay?

32. What is bulimia? How is it related to erosion of the teeth?

33. What happens to bone formation in children who suffer from chronic acidosis? Why?

34. What happens to the teeth when the salivary glands are removed or destroyed? Why?

35. Nineteen centuries ago, the Romans added calcium sulfate to wine. It clarifies the wine but also removes any dissolved lead. If 1 L of wine is saturated with calcium sulfate, the concentration of SO_4^{2-} is 0.014 M. What concentration of Pb^{2+} would remain in solution? The K_{sp} for lead sulfate is 1.1×10^{-8}, and the net ionic equation is
$$Pb^{2+} + SO_4^{2-} \rightleftharpoons PbSO_4(s)$$

36. Phosphates can be removed from water by precipitation using iron(III) sulfate.
$$2\ PO_4^{3-}(aq) + Fe_2(SO_4)_3 \longrightarrow$$
$$2\ FePO_4 + 3\ SO_4^{2-}(aq)$$

An estimated 7 metric tons per day of phosphates enters the East Anglian (United Kingdom) water system. How much iron(III) sulfate would be required to remove this phosphate?

37. Hard water is about 2×10^{-4} M in Ca^{2+}. Water is fluoridated with 1 g of F^- in 10^3 L of water. Will CaF_2 ($K_{sp} = 2 \times 10^{-10}$) precipitate upon fluoridation of hard water?

Inorganic Chemistry

It almost seems a contradiction of terms. **Inorganic** means nonliving and not derived from life. **Organic** means living, having the characteristics of life, or derived from life. At least, those are the everyday definitions of the terms. So how could inorganic chemicals be important to living organisms?

In the old days, chemists used the terms *organic* and *inorganic* in much the same way as everyone else. They believed that, while they could make many different inorganic chemicals, they could not hope to make organic compounds in the laboratory. They believed, with everyone else, that only living organisms, within their cells, could make organic compounds. Some mysterious **vital force,** they thought, was necessary for the synthesis of organic substances.

A series of experiments in the early 1800s led to the discarding of the vital force theory. Perhaps the most important single step was made in 1828 by Friedrich Wöhler while he was a medical student at the University of Heidelburg. He attempted to prepare ammonium cyanate by heating a mixture of two inorganic substances, lead cyanate and ammonium hydroxide. To his surprise, instead of ammonium cyanate, he obtained crystals of the well-known organic compound urea.

$$Pb(OCN)_2 \ + \ NH_4OH \ \longrightarrow$$

$$\quad \text{Lead} \qquad \text{Ammonium}$$
$$\text{cyanate} \qquad \text{hydroxide}$$

$$[NH_4OCN] \xrightarrow[\text{upon heating}]{\text{rearranges}} H_2N-\overset{\displaystyle O}{\overset{\|}{C}}-NH_2$$

$$\text{Ammonium} \qquad \qquad \text{Urea}$$
$$\text{cyanate}$$

Urea had been isolated from human urine in 1780. (Urea is synthesized in the liver, transported to the kidneys, and excreted in the urine.) Wöhler correctly concluded that ammonium cyanate is formed but then rearranges under the influence of heat to yield urea. Urea contains the same number and kinds of atoms as ammonium cyanate, but these atoms are arranged differently.

Although a few die-hard vitalists held out for several decades, the vital force theory was practically dead by the middle of the nineteenth century. **Organic chemistry** is now defined as the chemistry of the compounds of carbon. **Inorganic chemistry** is the chemistry of all the other elements. Several later chapters are devoted to organic chemistry. In this section we examine some of the properties of the various groups of elements and relate the properties to electron structure. We start with the noble gases. Then we contrast the reactivity of the elements in other groups to the inactivity of the noble gases.

The easier it is to remove the outermost electrons, the more metallic the element is. Metallic character decreases from left to right in the periodic table in parallel with ionization energies. It also increases from top to bottom with increasing atomic size. The alkali metals are all quite metallic, and the halogens are all distinctly nonmetallic. In the middle of the table we see intermediate behavior. Group IVA has carbon, a nonmetal, at the top, and two metals, tin and lead, at the bottom. In between are the *metalloids*—metal-like elements—silicon and germanium. The following brief description of the groups of elements will emphasize periodic trends.

C.1
GROUP VIIIA: THE NOBLE GASES

In the last decade of the nineteenth century a series of elements were discovered that made up an entirely new family, one completely unexpected by Mendeleev (Section 2.6) and his contemporaries. Nonetheless, this new group, called the noble gases, fit neatly between the very reactive alkali metals (Group IA) and the highly active nonmetals of Group VIIA. In the usual flat form of the periodic table the noble gases are placed to the far right in the column immediately following Group VIIA.

The six noble gases are helium, neon, argon, krypton, xenon, and radon. All are components of the atmosphere. Argon is rather abundant, making up nearly 1% by volume of the atmosphere. Xenon, on the other hand, is quite rare, making up only 91 parts per billion (by volume) of the atmosphere. Radioactive radon seeps from the ground. It makes up only a tiny portion of the atmosphere. It can be a

Figure C.1

The helium-filled Goodyear blimp is a familiar sight to those who watch sports events on television.

Figure C.2

The color of neon signs is due to changes in electronic energy levels in neon atoms.

problem, however, when it concentrates in a building with little ventilation.

The noble gases are exceptionally resistant to chemical reaction. This lack of reactivity is a reflection of their electron structure. Helium has the configuration $1s^2$, a filled first energy level. Neon has the structure $1s^2 2s^2 2p^6$, a filled second energy level. The others all have the valence configuration $ns^2 np^6$, an especially stable configuration. Other elements undergo reactions in order to achieve the more stable electron configurations of the noble gases. All the noble gases exist in elemental form as monatomic species.

Because of their stability, the noble gases were once called "inert gases." But in 1962 it was discovered that a few compounds of krypton and xenon could be formed, necessitating the change from "inert" to "noble." As yet, despite many attempts, no compounds have been made of the lighter elements helium, neon, and argon. While this family of elements is no longer called inert, its nobility is still unquestioned. More than any other family, the noble gases disdain interactions with the masses of other elements.

Helium is found in natural gas deposits, where it was formed from elements within the Earth. Helium is used to fill balloons and blimps (Figure C.1). Its lifting power is more than 90% that of hydrogen, the lightest of all the gases, and helium has the advantage of being nonflammable. Helium is also used to provide an inert atmosphere for the welding of metals that otherwise might be attacked by oxygen in the air. Liquid helium is used to achieve extremely low temperatures. It boils at only 4.2 K.

Neon is used in lighted signs for advertising (Figure C.2). A tube with electrodes is shaped into letters or symbols and is filled with neon at low pressure. An electric current is passed through the tube, causing the atoms to emit their characteristic orange glow.

Argon is the most plentiful of the noble gases. It is used to fill incandescent light bulbs. Unlike nitrogen and oxygen, it does not react with the tungsten filament. It also decreases the tendency of the filament to vaporize, thus extending the filament's life. Fluorescent lights are filled with a mixture of argon and mercury vapor.

Krypton and xenon are too expensive to have many important commercial applications, although krypton has found some use in light bulbs. Radon, although exceedingly rare in the atmosphere, can be collected from the radioactive decay of radium. This radon may be sealed in small vials and used for radiation therapy of certain malignancies. Radon accumulates in well-insulated homes; it is thought to cause about 15% of all cases of lung cancer.

C.2
GROUP IA: THE ALKALI METALS

The Group IA elements, the alkali metals, all have the outer electron configurations ns^1. There are six ele-

Figure C.4
Los Angeles on a smoggy day.

Los Angeles Smog

The word **smog** is thought to have originated in England in 1905 as a contraction of the words **smoke** and **fog.**

Automobiles fix nitrogen. Indeed, any time air is subjected to high temperatures, some of the nitrogen and oxygen combine. Unfortunately, most automobiles operate in urban areas, not on farms. And the nitrogen oxides undergo reactions with other pollutants to form a variety of noxious chemicals, not just fertilizers. This type of air pollution is called Los Angeles smog or, more properly, **photochemical smog.** In contrast to the London type, which requires cold, damp air, photochemical smog usually occurs in dry, sunny climates. The principal culprits are unburned hydrocarbons and oxides of nitrogen from automobiles. The warm, sunny climate that has drawn so many people to the Los Angeles area is also the perfect setting for photochemical smog (Figure C.4).

The combustion process in an automobile engine leads to the formation of nitric oxide (NO).

$$N_2 + O_2 \longrightarrow 2\,NO$$

Oxygen in the atmosphere slowly oxidizes the nitric oxide to nitrogen dioxide (NO_2).

$$2\,NO + O_2 \longrightarrow 2\,NO_2$$

Nitrogen dioxide is an amber-colored gas. Smarting eyes and a brownish haze are excellent indicators of Los Angeles smog. It is this nitrogen dioxide that plays a vital (villain's) role in photochemical smog. It absorbs a photon of sunlight and breaks down into nitric oxide and reactive oxygen *atoms.*

$$NO_2 + sunlight \longrightarrow NO + O$$

The oxygen atoms react with other components of automobile exhaust and of the atmosphere to produce a variety of irritating and toxic chemicals, including ozone.

The most serious environmental effect of the nitrogen oxides is that they produce smog. The gases also contribute to the fading and discoloration of fabrics, and by forming nitric acid, nitrogen oxides contribute to the acidity of rainwater. They also contribute to crop damage, although their specific effects are difficult to separate from those of sulfur dioxide and other pollutants.

Figure C.5
London smog is often highly visible, with soot and fly ash forming a dark pall of smoke.

London Smog

The two sulfur oxides, collectively called SO_x, and sulfuric acid are the major culprits of one type of smog. **London smog,** so called because of a dreadful pollution episode in that city in 1952, is a mixture of smoke, fog, soot, sulfur oxides, and often sulfuric acid (Figure C.5).

The chemistry of London smog is fairly simple. Coal is mainly carbon, but it contains as much as 3% sulfur, with varying amounts of mineral matter. The sulfur in coal burns, producing SO_2, which can react further to form SO_3 and H_2SO_4. In aerosol-mist form, sulfuric acid can damage the lungs (Section 10.9).

The harmful effect of these pollutants is considerably magnified by their interaction. A certain level of sulfur dioxide, without the presence of particulate matter (finely divided solids), might be reasonably safe. A certain level of particulate matter, without sulfur dioxide around, might be fairly harmless. But take the same levels of the two together, and the effect might well be deadly. An interaction in which the total effect of two ingredients is greater than the sum of the effects of the two ingredients taken separately is **synergistic.** Synergistic effects are quite common whenever chemicals get together. Sulfuric acid reacts with ammonia to form ammonium sulfate [$(NH_4)SO_4$].

$$2\,NH_3 \; + \; H_2SO_4 \; \longrightarrow \; (NH_4)_2SO_4$$

Sulfuric acid, in the form of minute liquid droplets, and the finely divided solid ammonium sulfate are easily trapped in the lungs. Sulfuric acid and solid ammonium sulfate also act synergistically. We will encounter synergism again in our study of the action of drugs (see Special Topics E and F).

The oxides of sulfur and the aerosol mist of sulfuric acid affect vegetation and other materials in addition to people. Leaves of plants show a bleached, splotchy effect, and both the yield and the quality of crops can be affected by exposure to oxides of sulfur. Rain in industrialized countries (and those downwind from them) has become more acidic in recent years (Section 10.7). This increase has been attributed to increased amounts of sulfur oxides released into the atmosphere. Acidic rainwater can corrode metals, eat holes in nylon stockings, decompose stone buildings and statuary, and render lakes so acidic that all life is destroyed.

C.8
GROUP VIIA: THE HALOGENS

Fluorine, chlorine, bromine, iodine, and astatine make up Group VIIA, the halogen family. All are nonmetals, and they have the electron configuration ns^2np^5. The Group IA metal sodium reacts with chlorine to form sodium chloride, the familiar table salt. Indeed, it is characteristic of the entire group that they react with metals to form salts. The word **halogen** is derived from Greek words meaning "salt former." We will limit our discussion to the first four halogens, for astatine is rare and highly radioactive.

In the elemental form all the halogens exist as diatomic molecules. The atoms can achieve a noble gas configuration by gaining an electron to form a negative ion.

$$F_2 + 2 e^- \longrightarrow 2 F^-$$
$$Cl_2 + 2 e^- \longrightarrow 2 Cl^-$$

This tendency, combined with that of many metals to give up electrons readily, is responsible for the ability of the halogens to form so many salts.

Fluoride salts, in moderate to high concentrations, are acute poisons. Small amounts of fluoride ion, however, are essential for our well-being. Concentrations of 0.7–1.0 part per million, by mass, of sodium fluoride have been added to the drinking water of many communities. Evidence indicates that such fluoridation results in a reduction in the incidence of dental caries (cavities).

Chlorine in the elemental form (Cl_2) is used to kill bacteria in water-treatment plants. Sodium hypochlorite (NaOCl) is an ingredient of common household bleaching solutions. Chlorine is present in our bodies as chloride ion (Cl^-). This ion is essential to life.

Several compounds of bromine are of importance. Silver bromide (AgBr) is sensitive to light and is used in photographic film. Sodium bromide (NaBr) and potassium bromide (KBr) have been used medicinally as sedatives. Bromide ions depress the central nervous system. Unfortunately, prolonged intake can cause mental deterioration and other problems. Bromides have been largely replaced by other, presumably safer, pain relievers.

Iodine is another essential nutrient. Compounds of iodine are necessary for the proper action of the thyroid gland. Iodine is also used as a topical antiseptic. Tincture of iodine is a solution of elemental iodine (I_2) in a mixture of alcohol and water. Iodine-releasing compounds, called iodophors, are often used as antiseptics in hospitals.

The halogens also react with hydrogen to form hydrogen halides.

$$H_2(g) + Cl_2(g) \longrightarrow 2 HCl(g)$$
$$H_2(g) + Br_2(g) \longrightarrow 2 HBr(g)$$

In water solutions these compounds form acids. Hydrochloric acid is a familiar example. This acid is present in the stomach and is involved in the digestive process.

C.9
THE B GROUPS: SOME REPRESENTATIVE TRANSITION ELEMENTS

Most of our attention so far has been focused on the A groups of the periodic table. The chemistry of the B groups is also important, if perhaps more complicated. The B groups collectively often are called the **transition elements.** All are metals in the elemental form. They conduct electricity and have a characteristic metallic luster.

The transition elements, in general, are those in which inner electron energy levels are being filled. (Group IB and IIB elements are exceptions, yet they share many properties with the other B groups and are often included as transition elements.) Recall that the third period ends with argon, which has eight electrons in its outermost energy level. The fourth period begins with potassium, which has one electron in its fourth energy level. The next element, calcium, has two electrons in its fourth energy level. Recall, however, that the $2n^2$ rule predicts a maximum of 18, not 8, electrons for the third energy level ($2n^2 = 2 \times 3^2 = 2 \times 9 = 18$). The fourth energy level has begun to fill before the third one is full. Things return to "normal" with the next element, scandium. This element skips back and continues to fill the third energy level. The first transition series (from scandium to zinc) corresponds to a filling of the $3d$ sublevel. With gallium ($Z = 31$), we return to an A group element. All inner energy levels are filled, and the next electron enters the outer energy level.

Physical properties vary widely in the transition series. For example, mercury (Hg) is a liquid at room temperature (its melting point is -38 °C). Tungsten (W), however, melts at 3410 °C. Chemically, the elements also exhibit a variety of properties. Most form more than one kind of simple ion. Iron is a familiar example. It readily forms both iron(II) and iron(III) ions (Fe^{2+} and Fe^{3+}, respectively).

Many compounds of the transition metals are colored, often brilliantly so, for a given element, the color is different for different ions. Aqueous solutions of Fe^{2+} are often pale green; those of Fe^{3+} are often yellow. With polyatomic ions, color variation often is greater still. An aqueous solution containing manganese(II) ions (Mn^{2+}) is a faint pink. Solutions of MnO_4^{2-} are an intense green, and those of MnO_4^- are deep purple.

Transition metals known to be essential to life are iron, copper, zinc, cobalt, manganese, vanadium, chromium, and molybdenum. Nickel is sometimes found in body tissues but has not been shown to be essential. Other transition metals, most notably cadmium and mercury, are dangerous poisons.

Cobalt is a component of vitamin B_{12}. A vitamin B_{12} deficiency leads to pernicious anemia. Vitamin B_{12} is found only in meat products. One hazard of a strict vegetarian diet is vitamin B_{12} deficiency.

Iron is essential for the proper functioning of hemoglobin, the red protein molecule involved in oxygen transport. This iron must be in the form of the $+2$ ion. If it is changed to the $+3$ form, the resulting compound (methemoglobin) is incapable of carrying oxygen. The oxygen-deficiency disease that results is called methemoglob-

inemia. In infants this condition is called the blue-baby syndrome.

In the fall of 1992, researchers in Finland reported the results of a five-year study of iron levels in humans. They plotted heart attack rates against iron levels and 18 other risk factors (LDL, smoking, stress, etc.) in nearly 2000 Finnish men. The results indicated that high iron levels dramatically increased the risk of heart attacks. (Only smoking was a stronger predictor of heart illness.) These findings coincided with earlier investigations that showed postmenopausal women to have a higher incidence of heart disease. The implication is that menstrual bleeding protects women's hearts by reducing the amount of iron in their bodies.

The theory requires proof, and research is under way to determine whether lowering people's iron levels will prevent heart attacks. Until the evidence is collected, it is prudent to avoid excessive amounts of iron. You should limit your consumption of meat, refrain from fortified cereals, and not use iron supplements unless prescribed by a physician.

EXERCISES

1. Define each term.
   a. inorganic chemistry b. noble gas
   c. halogen d. alkali metal
   e. alkaline earth metal f. transition element
2. Give the electron dot symbol for each of the following elements.
   a. neon b. oxygen c. fluorine
   d. potassium e. barium f. nitrogen
3. What is the outstanding chemical property of the noble gases? What structural feature accounts for this property?
4. Why is helium preferred to hydrogen for filling blimps, even though the latter has greater lifting power?
5. What function does argon serve in an electric light bulb?
6. How does a neon sign work?
7. What similarities in properties are shared by the alkali metals? What structural feature do atoms of these elements share?
8. What similarities in properties are shared by the alkaline earth metals? What structural feature do atoms of these elements share?
9. Which alkaline earth metal is most different from the others? What are some of those differences?
10. What alkali metal compound is used to treat manic-depressive psychoses?
11. What is the most abundant metal in the Earth's crust?
12. Which Group IIIA element has the greatest tendency to form covalent compounds (rather than ionic compounds)? Why?

13. Draw electron-dot symbols for gallium (Ga) and gallium ion.
14. Why is aluminum replacing steel wherever possible in automobiles?
15. Draw the electron dot-symbol for each of the following.
    a. fluoride ion b. iodide ion
    c. oxide ion d. sulfide ion
    e. potassium ion f. strontium ion
16. Name one use for each element or compound.
    a. chlorine b. iodine c. NaF
    d. AgBr e. NaOCl f. KCl
17. What is the effect on tooth enamel of small amounts of fluoride?
18. What are basic oxides? Use calcium oxide to illustrate your answer.
19. What are acidic oxides? Use sulfur trioxide to illustrate your answer.
20. Complete and balance the following equations.
    a. $Li + O_2 \longrightarrow$ b. $Ca + O_2 \longrightarrow$
    c. $S + O_2 \longrightarrow$ d. $CaO + H_2O \longrightarrow$
    e. $SO_2 + H_2O \longrightarrow$ f. $Ca + S \longrightarrow$
21. What is photochemical smog? What chemical compound starts the formation of photochemical smog by absorbing sunlight?
22. What is synergism?
23. What are allotropes? Name two sets of allotropes.
24. How do oxygen atoms, oxygen molecules, and ozone differ in structure and properties?

25. How does the burning of sulfur-containing coal lead to acid rain?
26. What is London smog? How is it formed?
27. How do weather conditions differ in episodes of Los Angeles smog as compared to London smog?
28. What is the greenhouse effect?
29. What was the vital force theory? How was it overthrown?
30. What is the origin of helium in natural gas wells?
31. Why are the noble gases no longer called the "inert gases"?
32. What are the halogens? Why are they so called?
33. What is nitrogen fixation? Why is it important?
34. What condition(s) lead(s) to formation of carbon monoxide during combustion?
35. How does carbon monoxide exert its posionous effect?
36. What two alkali metal ions play a major role in maintaining fluid balance in the body?
37. What important molecule in plants incorporates magnesium?
38. Name two functions of calcium ions in the body.
39. What is hard water? How does it affect the action of soaps?
40. List three distinguishing characteristics of transition metals.
41. List four transition metals essential to life. Indicate their function in the body.

Chapter 13
HYDROCARBONS

Offshore drilling for natural gas and crude oil. Petroleum is the chief raw material for the synthesis of plastics, pesticides, drugs, detergents, and other important commodities.

Scientists of the eighteenth and nineteenth centuries studied compounds isolated from rocks and ores, from the atmosphere and oceans, and labeled them *inorganic* because they were obtained from nonliving systems. Compounds obtained from plants and animals were called *organic* because they were isolated from organized (living) systems. The early chemists believed that only living organisms could synthesize organic compounds. Although by the middle of the nineteenth century a number of "organic" compounds had been prepared using ordinary laboratory techniques, the labels *organic* and *inorganic* remained (Figure 13.1).

Today, **organic chemistry** is defined simply as the chemistry of the compounds of carbon.[1] It may seem strange that we divide chemistry into two branches, one of which considers the chemistry of one element while the other handles the chemistry of the more than 100 remaining elements. However, this division seems more reasonable when one discovers that, of the 10 million or more compounds that have been characterized, the overwhelming majority contain carbon.

---

[1] This definition is not adhered to strictly; several of the following compounds of carbon often are considered to be inorganic chemicals.

Carbon monoxide (CO)	Carbonates (e.g., Na_2CO_3)	Thiocyanates (e.g., NaSCN)
Carbon dioxide (CO_2)	Bicarbonates (e.g., $NaHCO_3$)	Cyanates (e.g., KOCN)
Carbon disulfide (CS_2)	Cyanides (e.g., KCN)	Carbides (e.g., CaC_2)

Figure 13.1
The word *organic* has different meanings. Organic fertilizer is organic in the original sense—it is derived from living organisms. Organic foods generally mean foods grown without pesticides or synthetic fertilizers. Organic chemistry is the chemistry of compounds of carbon.

In January 1990 the Chemical Abstract Service of the American Chemical Society recorded the 10 millionth known chemical compound. Over 600,000 new compounds are added to the list each year. About 95% of all the compounds are organic substances.

What is so special about carbon that differentiates it from all of the other elements in the periodic table? Carbon has a unique ability to form stable, covalent bonds with itself and with other elements in infinite variations. The molecules thus produced may contain only one or over a million carbon atoms. So complex is the chemistry of carbon that we shall approach its study by dividing its millions of compounds into families. We'll study one family at a time and begin by concentrating on the simpler members of each family. Eventually we will move to a consideration of those molecules that deserve to be called organic in the old sense—complex, carbon-containing molecules that determine the form and function of living systems.

We might pause to ponder how far science has come since Wöhler's synthesis of urea in 1828 (see page 288). Before that year, scientists believed they could not synthesize even the simplest organic molecule. In 1980, the U.S. Supreme Court ruled that new life forms, created in a laboratory, could be patented. The patent under consideration described a new organism, designed to consume spilled oil and made by changing the genetic heritage of an existing life form. Was this "creation of life in a test tube"? Not quite, but each new experiment in genetic engineering (see Section 23.9) brings that achievement closer.

13.1
ORGANIC VERSUS INORGANIC

Organic compounds, like inorganic ones, obey all the natural laws. Often there is no clear distinction in chemical or physical properties between organic and inorganic molecules. Nevertheless, it may be useful to compare and contrast typical members of each class. Table 13.1 contrasts the general properties of organic and inorganic compounds. (It must be understood, however, that there are exceptions to every entry in this table.) Table 13.2 lists a variety of properties of the inorganic compound sodium chloride ($NaCl$, common table salt) and the organic compound benzene (C_6H_6, a solvent once widely used to strip furniture for refinishing).

Organic ~nonpolar	Inorganic
1. Low melting points	1. High melting points
2. Low boiling points	2. High boiling points
3. Low solubility in water; high solubility in nonpolar solvents	3. High solubility in water; low solubility in nonpolar solvents
4. Flammable	4. Nonflammable
5. Solutions are nonconductors of electricity	5. Solutions are conductors of electricity
6. Chemical reactions are usually slow	6. Chemical reactions are rapid
7. Exhibit isomerism	7. Isomers are limited to a few exceptions (e.g., the transition elements)
8. Exhibit covalent bonding	8. Exhibit ionic bonding
9. Compounds contain many atoms	9. Compounds contain relatively few atoms

Table 13.1
Some Contrasting Properties of Organic and Inorganic Compounds

	Benzene	Sodium Chloride
Formula	C_6H_6	NaCl
Solubility in H_2O	Insoluble	Soluble
Solubility in gasoline	Soluble	Insoluble
Flammable?	Yes	No
Melting point	5.5 °C	801 °C
Boiling point	80 °C	1413 °C
Density	0.88 g/cm^3	2.7 g/cm^3
Bonding	Covalent	Ionic

Table 13.2
Comparison of an Inorganic and an Organic Compound

13.2
ALKANES: SATURATED HYDROCARBONS

single bonds

$C_n H_{(2n+2)}$

Before you can even begin to understand the large, complex molecules on which life is based, you need to learn something about simpler molecules. We will start with organic compounds containing only two elements, carbon and hydrogen—the **hydrocarbons.** In Sections 4.10 and 4.17, we encountered compounds called methane (CH_4) and ethane (C_2H_6). Models of molecules of these two compounds are shown in Figure 13.2. These are the first two members of a series of related compounds called **alkanes** or **saturated hydrocarbons.** *Saturated,* in this case, means that each carbon atom is bonded to four other atoms; there are no double or triple bonds in the molecules. Structurally, methane and ethane are generally represented as shown below.

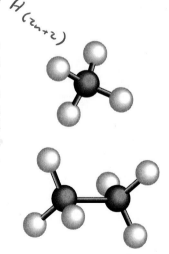

Figure 13.2
Ball-and-stick models of methane and ethane.

```
      H              H   H
      |              |   |
  H—C—H          H—C—C—H
      |              |   |
      H              H   H

   Methane          Ethane
```

These representations make no attempt to accurately portray bond angles in particular or molecular geometry in general. You should keep the models in mind when you look at

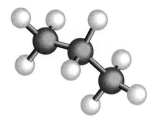

Figure 13.3

Ball-and-stick model of propane.

these flat pictures. (The bond angles and molecular geometry of methane are described in detail in Section 4.17.)

The three-carbon hydrocarbon (C_3H_8) is called propane. A ball-and-stick model of this compound is shown in Figure 13.3. In two dimensions, the structure is generally written as

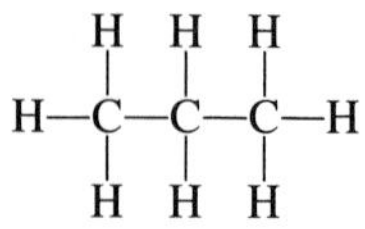

$$\begin{array}{c} \text{H}\quad\text{H}\quad\text{H} \\ |\quad\;\;|\quad\;\;| \\ \text{H}-\text{C}-\text{C}-\text{C}-\text{H} \\ |\quad\;\;|\quad\;\;| \\ \text{H}\quad\text{H}\quad\text{H} \end{array}$$

Propane

Notice that methane, ethane, and propane form a series in which the adjacent members differ by one carbon atom and two hydrogen atoms, that is, by a CH_2 unit (Figure 13.4). Compounds related in this way are said to make up a **homologous series.** Such a series has properties that vary in a regular and predictable manner. The principle called **homology** gives meaning to organic chemistry in much the same way that the periodic table gives organization and meaning to the chemistry of the elements. Instead of a bewildering array of individual carbon compounds, we can study a few members of a homologous series (called **homologs**), and from them we can deduce the properties of other compounds in the series.

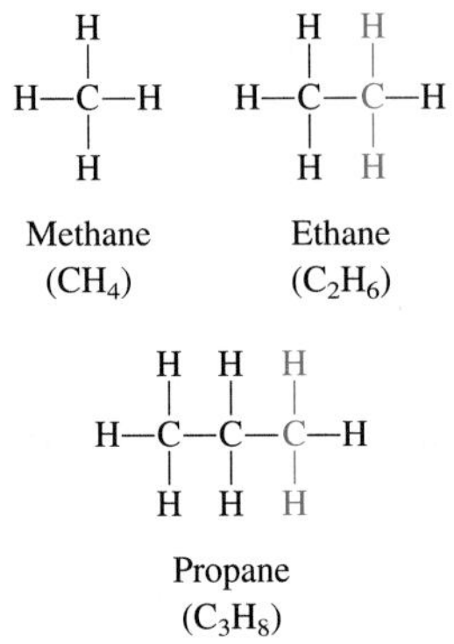

Continuing the homologous series, we add another carbon atom and a pair of hydrogens to get C_4H_{10}. It is rather easy to write a structure corresponding to this formula. We merely string four carbon atoms in a row,

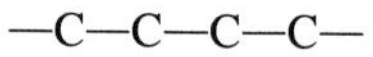

$$-\text{C}-\text{C}-\text{C}-\text{C}-$$

and then we add enough hydrogen atoms to give each carbon atom four bonds.

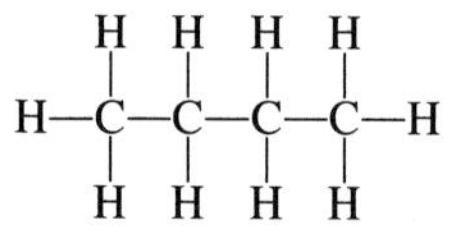

$$\begin{array}{c} \text{H}\quad\text{H}\quad\text{H}\quad\text{H} \\ |\quad\;\;|\quad\;\;|\quad\;\;| \\ \text{H}-\text{C}-\text{C}-\text{C}-\text{C}-\text{H} \\ |\quad\;\;|\quad\;\;|\quad\;\;| \\ \text{H}\quad\text{H}\quad\text{H}\quad\text{H} \end{array}$$

There is a compound called butane that has this structure. But there is another way to put four carbons and 10 hydrogens together. String out three of the four carbon atoms, and then branch the other one off the middle carbon of the chain.

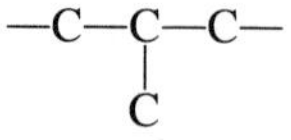

$$\begin{array}{c} -\text{C}-\text{C}-\text{C}- \\ |\;\; \\ \text{C} \end{array}$$

Now we add enough hydrogen atoms to give each carbon four bonds.

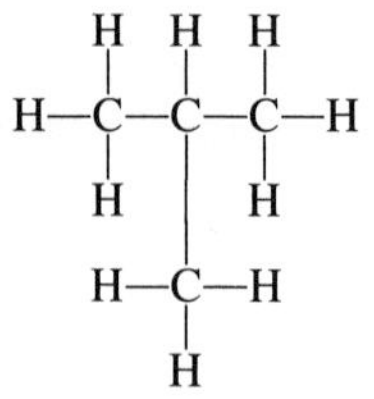

$$\begin{array}{c} \text{H}\quad\text{H}\quad\text{H} \\ |\quad\;\;|\quad\;\;| \\ \text{H}-\text{C}-\text{C}-\text{C}-\text{H} \\ |\quad\;\;|\quad\;\;| \\ \text{H}\quad\;\;|\quad\;\;\text{H} \\ |\; \\ \text{H}-\text{C}-\text{H} \\ |\; \\ \text{H} \end{array}$$

And, fortunately for the sake of our structural theory, there is another gaseous hydrocarbon that corresponds to this structure. There are two compounds, then, that have the same molecular formula, C_4H_{10}. One boils at $-1\,°C$; the other at $-12\,°C$. Different compounds with the same molecular formula are called **isomers.** To give the two butanes unique names, we call the one with the continuous chain of carbon atoms *butane.* The one

$$\begin{array}{cc} \begin{array}{c} \text{H} \\ |\; \\ \text{H}-\text{C}-\text{H} \\ |\; \\ \text{H} \end{array} & \begin{array}{c} \text{H}\quad\text{H} \\ |\quad\;\;| \\ \text{H}-\text{C}-\text{C}-\text{H} \\ |\quad\;\;| \\ \text{H}\quad\text{H} \end{array} \\ \text{Methane} & \text{Ethane} \\ (CH_4) & (C_2H_6) \end{array}$$

$$\begin{array}{c} \text{H}\quad\text{H}\quad\text{H} \\ |\quad\;\;|\quad\;\;| \\ \text{H}-\text{C}-\text{C}-\text{C}-\text{H} \\ |\quad\;\;|\quad\;\;| \\ \text{H}\quad\text{H}\quad\text{H} \end{array}$$

Propane
(C_3H_8)

Figure 13.4

Members of a homologous series. Each succeeding formula incorporates one carbon and two hydrogens more than the previous formula.

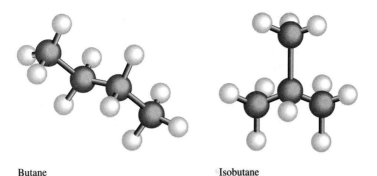

Butane   Isobutane

Figure 13.5
Ball-and-stick models of
butane and isobutane.

with the branched chain is called *isobutane*. Figure 13.5 shows ball-and-stick models of
the two isomeric butanes.

The three smaller homologs of this series do not exist in isomeric forms. There is
only one methane, one ethane, and one propane. Why? Because there is only one way to
arrange the carbon atoms and hydrogen atoms in each of their formulas so that each
carbon atom has four bonds and each hydrogen atom has one bond. When we come to
butane, there are two (and only two) ways to arrange the carbon and hydrogen atoms.
Butane is a continuous four-carbon chain. This chain may be twisted or bent, but it is still
one continuous chain. Isobutane is a true isomer of butane. The formula of isobutane
shows a continuous chain of three carbon atoms only, with the fourth carbon atom attached as a branch off the middle carbon of the continuous chain. If our figures do not
make this point clear, you should make models to convince yourself. (Borrow ball-and-
stick models or create some of your own from marshmallows and toothpicks.)

A continuous chain of carbon
atoms is also referred to as a
straight chain.

One more step up the homologous series gets us to C_5H_{12}. There are three compounds with this molecular formula. Collectively, these compounds are called *pentanes*.
We can name compound I pentane because it has all five carbons in a continuous chain.
Compound II can be called isopentane because, like isobutane, it has a single carbon atom
branched off the second carbon of the continuous chain. But what shall we call compound
III? Let's name it the way the chemists did when it was discovered in 1870. Since the
other two pentanes were characterized first, this one was called neopentane (from the
Greek *neos,* new).

Table 13.3
Stems That Indicate the
Number of Carbon Atoms
in Organic Molecules

Stem	Number
Meth-	1
Eth-	2
Prop-	3
But-	4
Pent-	5
Hex-	6
Hept-	7
Oct-	8
Non-	9
Dec-	10

Hydrocarbon chemistry seems quite complex, but there is some system to it. For
example, the first part of the name indicates the number of carbon atoms in the molecule
(Table 13.3). For five or more carbon atoms, the stems are derived from Greek or Latin
names for the numbers.

The ending of the name also has meaning. Note that all end in *-ane*. This indicates
that all are alkanes. The name heptane, then, means an alkane with seven carbon atoms.
To draw the structure of heptane, we merely write out a string of seven carbon atoms,

—C—C—C—C—C—C—C—

Name	Molecular Formula	Condensed Structural Formula	Number of Possible Isomers
Methane	CH_4	CH_4	1
Ethane	C_2H_6	CH_3CH_3	1
Propane	C_3H_8	$CH_3CH_2CH_3$	1
Butane	C_4H_{10}	$CH_3(CH_2)_2CH_3$	2
Pentane	C_5H_{12}	$CH_3(CH_2)_3CH_3$	3
Hexane	C_6H_{14}	$CH_3(CH_2)_4CH_3$	5
Heptane	C_7H_{16}	$CH_3(CH_2)_5CH_3$	9
Octane	C_8H_{18}	$CH_3(CH_2)_6CH_3$	18
Nonane	C_9H_{20}	$CH_3(CH_2)_7CH_3$	35
Decane	$C_{10}H_{22}$	$CH_3(CH_2)_8CH_3$	75

and then attach enough hydrogens to give each carbon a valence of 4. This requires three hydrogens on each end carbon and two on each of the others.

In this manner we can readily extend our list of structures of continuous-chain hydrocarbons to decane. These alkanes are shown in Table 13.4. Notice that each molecular formula in that table differs from the one preceding it by a CH_2 unit, illustrating the principle of homology. Notice also that the number of isomers increases rapidly with increasing carbon number. There are five hexanes, nine heptanes, and 18 octanes. Even more striking is the calculation that there are 366,319 possible isomers of the alkane whose molecular formula is $C_{20}H_{42}$ and there are 62,491,178,805,831 possible iosmers of the alkane whose molecular formula is $C_{40}H_{82}$. Fortunately, not all of those have been isolated or characterized. We've already run into a problem of giving unique names to the five hexanes, let alone the large numbers of isomers of higher alkanes.

13.3
CONDENSED STRUCTURAL FORMULAS

The general formula for the alkane homologs is C_nH_{2n+2}. Formulas such as we have used so far, showing all the carbon and hydrogen atoms and how they are attached to one another, are called **structural formulas.** These structures convey much more information than simple chemical formulas. For example, the molecular formula C_4H_{10} only expresses the actual number of atoms in the molecule. It doesn't tell us whether we are dealing with butane or with isobutane. The structural formulas

identify the specific isomers by the order of attachment of the various atoms.

Unfortunately, structural formulas are difficult to type (even with a word processor) and take up a lot of space. Chemists often use **condensed structural formulas** to alleviate these problems. The condensed structures show the hydrogen atoms right next to the carbon atoms to which they are attached. The two isomeric butanes become

$$CH_3-CH_2-CH_2-CH_3 \quad \text{and} \quad CH_3-\underset{\underset{\displaystyle CH_3}{|}}{CH}-CH_3$$

Sometimes these are further simplified by omitting some (or all) of the bond lines.

$$CH_3CH_2CH_2CH_3 \quad \text{and} \quad CH_3\underset{\underset{\displaystyle CH_3}{|}}{CH}CH_3 \quad [\text{or} \quad CH_3CH(CH_3)_2 \quad \text{or} \quad (CH_3)_3CH]$$

13.4
ALKYL GROUPS

The group of atoms that results when a hydrogen atom is removed from an alkane is called an **alkyl group.** *Thus, the general formula for an alkyl group is* C_nH_{2n+1}. *The group is named by replacing the -ane suffix of the parent hydrocarbon with -yl* (Table 13.5). It is important to note that alkyl groups are not independent molecules. Rather, they exist as parts of molecules.

Notice that two alkyl groups can be derived from propane. These are called propyl and isopropyl, respectively. Remember that there is only one alkane named propane, but a chain of three carbon atoms can be attached to a longer chain in two different ways. One has the attachment through an end carbon of the three-carbon chain; the other, through the middle carbon. There are many other alkyl groups. The ones that we are most likely to encounter are listed in Table 13.5.

For convenience, the organic chemist often represents an alkyl group by the symbol R. This symbol is used whenever it is desirable to represent a general class of compounds. Thus, the symbol for the family of saturated hydrocarbons is R—H. The R stands for any alkyl group—methyl, ethyl, propyl, isopropyl, or any larger group.

Table 13.5
Common Alkyl Groups

Parent Hydrocarbon		Alkyl Group		Condensed Formula
Methane	H—C—H (with H above and below)	Methyl	H—C— (with H above and below)	CH_3-
Ethane	H—C—C—H (with H's)	Ethyl	H—C—C— (with H's)	CH_3CH_2- or C_2H_5-
Propane	H—C—C—C—H (with H's)	Propyl	H—C—C—C— (with H's)	$CH_3CH_2CH_2-$ or C_3H_7-
		Isopropyl	H—C—C—C—H (with H's)	CH_3CHCH_3 or $(CH_3)_2CH-$

13.5
THE UNIVERSAL LANGUAGE: IUPAC NOMENCLATURE

To bring order to the chaotic naming of newly discovered compounds, an international meeting of chemists was held at Geneva, Switzerland, in 1892. The meeting resulted in a simple, unequivocal system for naming organic compounds. The system was modified by the International Union of Chemists and is kept up to date by its successor, the International Union of Pure and Applied Chemistry (IUPAC). What has evolved is a set of rules known as the **IUPAC System of Nomenclature.** The rules for the alkanes are as follows.

1. Saturated hydrocarbons are named according to the longest continuous chain of carbons in the molecule (rather than the total number of carbon atoms). This is the parent compound.
2. The suffix *-ane* indicates that the molecule is a saturated hydrocarbon.
3. The name of the parent hydrocarbon is modified by noting what alkyl groups are attached to the chain.
4. The chain is numbered, and the position of each substituent alkyl group is indicated by the number of the carbon atom to which it is attached. The chain is numbered in such a way that the substituents occur on the carbon atoms with the lowest numbers.
5. Names of the substituent groups are placed in alphabetical order before the name of the parent compound.
6. If the same alkyl group appears more than once, the numbers of all the carbons to which it is attached are expressed. If the same group appears more than once on the same carbon, the number of the carbon is repeated as many times as the group appears.
7. Hyphens are used to separate numbers from names of substituents; numbers are separated from each other by commas. The number of identical groups is indicated by the Greek prefixes *di-, tri-, tetra-*, etc. These prefixes are NOT considered in determining the alphabetical order of the substituents (e.g., ethyl precedes dimethyl).
8. The last alkyl group to be named is prefixed to the name of the parent alkane forming one word.

The term **substituent** refers to any atom or group of atoms that substitutes for a hydrogen atom in an organic molecule.

Table 13.6 contains some examples of the IUPAC system for naming organic compounds. The best way to learn how to name alkanes is by working out examples, not just by memorizing rules. It's easier than it sounds. Try the following.

EXAMPLE 13.1

Name the compound

$$CH_3—CH_2—\underset{\underset{CH_3}{|}}{CH}—\underset{\underset{CH_3}{|}}{CH}—CH_3$$

SOLUTION

The longest continuous chain has five carbon atoms. There are two methyl groups attached to the second and third carbon atoms (not the third and fourth; use the lowest combination of numbers, counting from the right-hand end in this case). The correct name is 2,3-dimethylpentane.

Table 13.6
Examples of IUPAC Nomenclature

Condensed Structural Formula	Rewritten and Numbered	IUPAC Name
$CH_3CH_2CH(CH_3)CH_2CH(CH_3)_2$	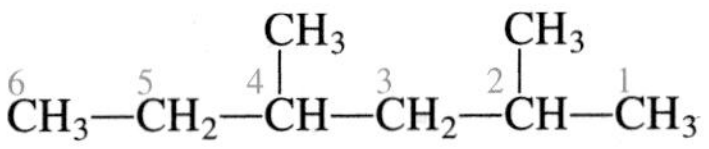 $\overset{6}{C}H_3-\overset{5}{C}H_2-\overset{4}{C}H-\overset{3}{C}H_2-\overset{2}{C}H-\overset{1}{C}H_3$ with CH_3 on carbon 4 and CH_3 on carbon 2	2,4-Dimethylhexane NOT 3,5-Dimethylhexane
$(C_2H_5)_2CHCH(CH_3)CH_2CH_2CH_3$	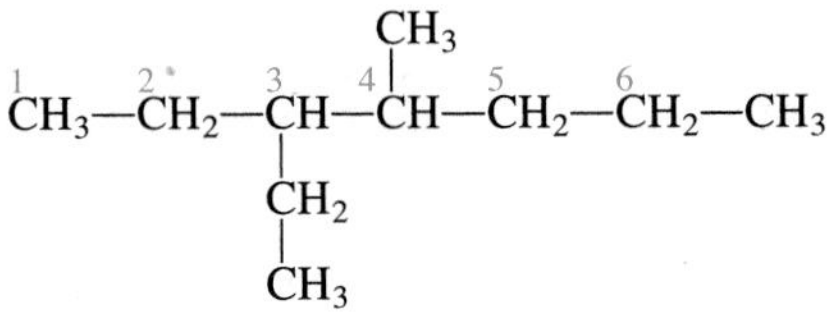 $\overset{1}{C}H_3-\overset{2}{C}H_2-\overset{3}{C}H-\overset{4}{C}H-\overset{5}{C}H_2-\overset{6}{C}H_2-CH_3$ with CH_3 on carbon 4 and CH_2-CH_3 on carbon 3	3-Ethyl-4-methylheptane NOT 4-Methyl-5-ethylheptane
$(CH_3)_2CHCH(C_3H_7)_2$	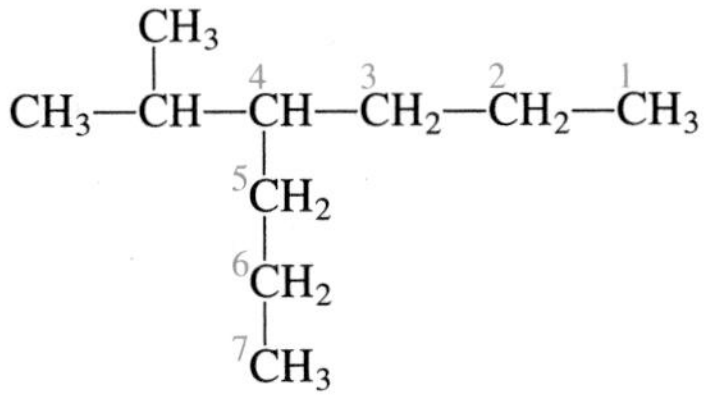 $CH_3-CH-\overset{4}{C}H-\overset{3}{C}H_2-\overset{2}{C}H_2-\overset{1}{C}H_3$ with CH_3 on first carbon and $\overset{5}{C}H_2-\overset{6}{C}H_2-\overset{7}{C}H_3$ chain	4-Isopropylheptane NOT 2-Methyl-3-propylhexane
$(CH_3)_2CHCH_2CH_2C(CH_3)_3$	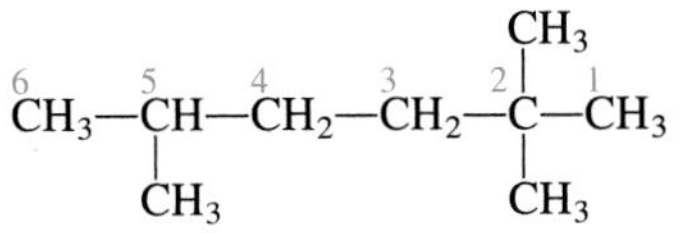 $\overset{6}{C}H_3-\overset{5}{C}H-\overset{4}{C}H_2-\overset{3}{C}H_2-\overset{2}{C}-\overset{1}{C}H_3$ with CH_3 groups on carbons 5, 2, and 2	2,2,5-Trimethylhexane NOT 2,5-Trimethylhexane
$(C_2H_5)_2CHC(CH_3)(C_2H_5)_2$	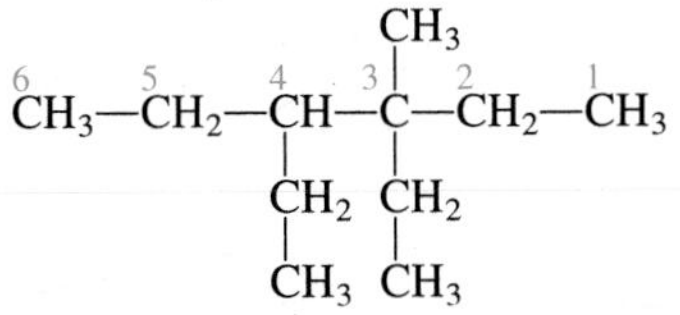 $\overset{6}{C}H_3-\overset{5}{C}H_2-\overset{4}{C}H-\overset{3}{C}-\overset{2}{C}H_2-\overset{1}{C}H_3$ with CH_3 on carbon 3 and CH_2-CH_3 groups on carbons 4 and 3	3,4-Diethyl-3-methylhexane NOT 3,4-Ethyl-3-methylhexane

EXAMPLE 13.2

Name the compound

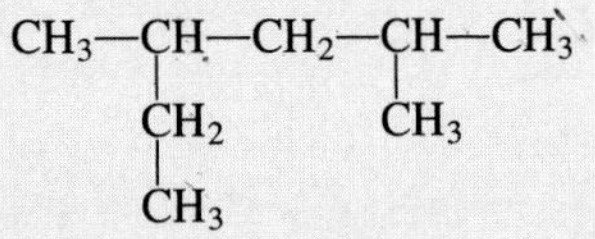

$$CH_3-CH-CH_2-CH-CH_3$$
with CH_2 then CH_3 below the second carbon, and CH_3 below the fourth carbon

SOLUTION

The correct name is 2,4-dimethylhexane, NOT 2-ethyl-4-methylpentane. This is a fooler. The parent compound is the longest continuous chain, not necessarily the chain drawn straight across the page. In this compound, the longest chain contains six, not five, carbon atoms.

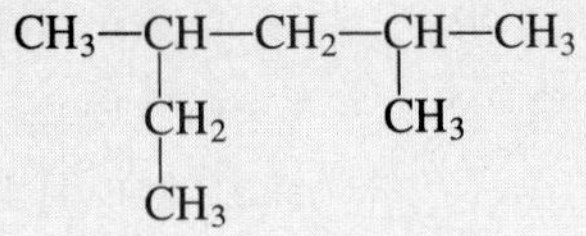

$$CH_3-CH-CH_2-CH-CH_3$$
with CH_2 then CH_3 below the second carbon, and CH_3 below the fourth carbon

EXAMPLE 13.3

Name the compound

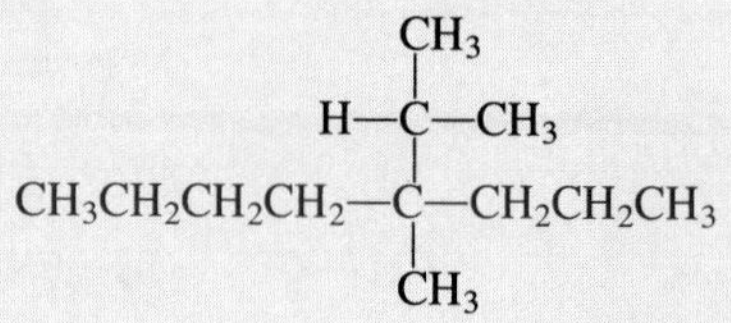

$$CH_3CH_2CH_2CH_2-\underset{\underset{CH_3}{|}}{\overset{\overset{CH_3}{|}}{\overset{H-\underset{CH_3}{\overset{|}{C}}-CH_3}{\underset{}{C}}}}-CH_2CH_2CH_3$$

SOLUTION

The correct name is 4-isopropyl-4-methyloctane.

Practice Exercise
Name the following compounds.

a. 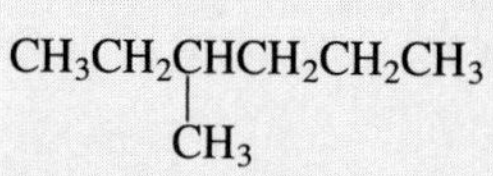 $CH_3CH_2\underset{\underset{CH_3}{|}}{C}HCH_2CH_2CH_3$

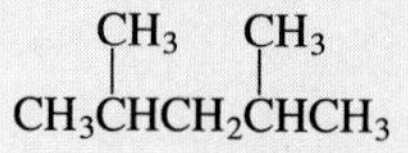

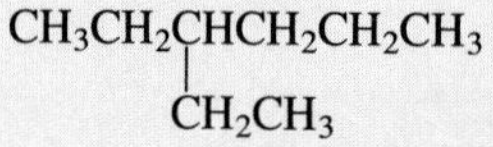

b. $CH_3\underset{\underset{CH_3}{|}}{C}HCH_2\underset{\underset{CH_3}{|}}{C}HCH_3$

c. $CH_3CH_2\underset{\underset{CH_2CH_3}{|}}{C}HCH_2CH_2CH_3$

d. $CH_3CH_2CH_2\underset{\underset{H-\underset{CH_3}{\overset{|}{C}}-CH_3}{|}}{C}HCH_2CH_2CH_3$

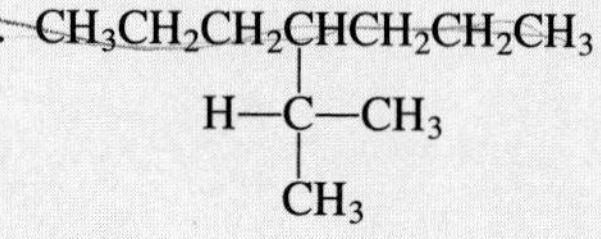

EXAMPLE 13.4

Draw the structural formula for 4-isopropyl-2-methylheptane.

SOLUTION

In drawing structural formulas for a compound, always start with the parent chain, heptane in this case.

$$-C-C-C-C-C-C-C-$$

Then add the groups at their proper positions. You can number the parent chain from either direction as long as you are consistent (don't change directions in the middle of the problem). First place the additional carbons.

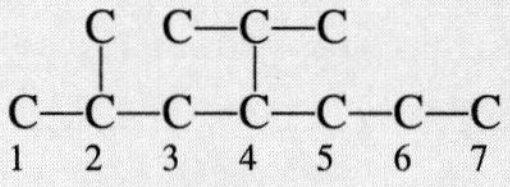

$$\begin{array}{ccccccc} & C & C-C-C \\ & | & | \\ C-C-C-C-C-C-C \\ 1 & 2 & 3 & 4 & 5 & 6 & 7 \end{array}$$

Finally, fill in all the hydrogens (each carbon atom must have four bonds).

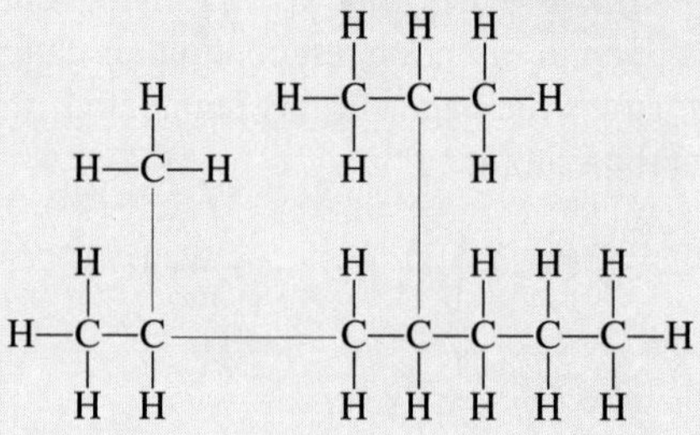

You can condense this formula by writing the hydrogens right next to the carbons to which they are attached.

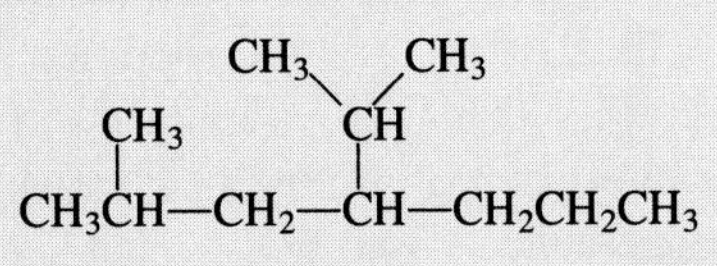

$$CH_3CH-CH_2-CH-CH_2CH_2CH_3$$

with substituents: CH_3 (on first CH), CH_3 and CH branch bearing two CH_3 groups (on third CH)

Practice Exercise

Draw the structural formulas for the following compounds.

a. 4-propylheptane
b. 2,2-dimethylbutane
c. 3-ethyl-2-methylpentane
d. 3-isopropyl-3-methyloctane

13.6
PHYSICAL PROPERTIES OF ALKANES

The alkanes are nonpolar molecules. They are therefore insoluble in water, but they are soluble in organic solvents such as toluene and methylene chloride. Alkanes themselves are good solvents; they dissolve many organic substances of low polarity, such as fats, oils, and waxes. For that reason, mixtures of alkanes are frequently used as organic solvents.

The hydrocarbons are generally less dense than water, their densities being less than 1.0 g/mL. They are colorless and tasteless, and many of them are odorless. The odor associated with natural gas is not from methane or any of the other alkanes, but rather from an odorant (usually CH_3SH) purposely added to the gas in sufficient quantities to allow for the detection of leaks. Table 13.7 summarizes the properties of the first 10 straight-chain alkanes.

All organic chemistry textbooks include many tables of physical properties. The figures in these tables are not meant to be memorized but are given in order to present a

Name	Molecular Formula	Melting Point (°C)	Boiling Point (°C)	Density (g/mL)	Normal State
Methane	CH_4	−182	−164	—	Gas
Ethane	C_2H_6	−183	−89	—	Gas
Propane	C_3H_8	−190	−42	—	Gas
Butane	C_4H_{10}	−138	−1	—	Gas
Pentane	C_5H_{12}	−130	36	0.63	Liquid
Hexane	C_6H_{14}	−95	69	0.66	Liquid
Heptane	C_7H_{16}	−91	98	0.68	Liquid
Octane	C_8H_{18}	−57	125	0.70	Liquid
Nonane	C_9H_{20}	−51	151	0.72	Liquid
Decane	$C_{10}H_{22}$	−30	174	0.73	Liquid

Table 13.7
Physical Properties of Some Alkanes

numerical description of the physical characteristics of organic molecules and to serve as standards of purity. The physical states of substances at room temperature are obtained directly from a knowledge of their melting points and boiling points. Room temperature is considered to be approximately 68 °F or about 20 °C. If the melting point of a substance is above 20 °C, the substance will exist as a solid at room temperature. (For example, the first member of the alkanes series that is a naturally occurring solid is *n*-octadecane, $C_{18}H_{38}$, mp 28 °C.) If the boiling point of a substance is above 20 °C, the substance will exist as a liquid at room temperature. (Note that pentane, bp 36 °C, is the first liquid alkane.) Similarly, knowledge of the density of a substance indicates whether the substance is lighter or heavier than water. (The density of water is 1.00 g/mL.) We observe that oil and grease do not mix with water but rather float on the surface. They do not mix with water because they are *insoluble* in water; they float on top of the water because they are *less dense* than water. Densities are also useful in calculations that call for the conversion of a certain mass of a liquid into the corresponding volume of that liquid, or vice versa. (Recall that density equals the mass of a substance divided by its volume.)

As indicated in Table 13.7, the first four members of the alkane series are gases. A comparison of the densities of these gases with the density of air allows us to predict whether or not they will rise in the air. Natural gas is composed chiefly of methane, which has a density of about 0.65 g/L. The density of air is about 1.29 g/L. Natural gas, then, is much lighter than air, and it will rise in a room in which there is a natural gas leak. Once the leak is detected and eliminated, the gas can be removed from the room by opening an upper window. On the other hand, the three constituents of bottled gas are much heavier than air. Propane has a density of 1.6 g/L, and the butane isomers have densities of about 2.0 g/L. If bottled gas escapes into a room, it collects near the floor. This presents a much more serious fire hazard than a natural gas leak because it is more difficult to rid the room of the heavier gas.

As shown in Table 13.7, the boiling points of the straight-chain alkanes increase with increasing molecular weight. This general rule holds true within all the families of organic compounds whenever the straight-chain homologs of the family are considered. Larger molecules are able to wrap around and interact with one another, and thus more energy is required to separate them. In general, the straight-chain isomers have higher boiling points than the branched-chain isomers. The former compounds can be likened to strands of spaghetti. The molecules can be very closely packed together, resulting in relatively strong intermolecular van der Waals forces of attraction (Section 8.5). Van der Waals forces depend on the total area of contact available between two molecules. The greater this area of contact, the greater the attractive force. Branched-chain hydrocarbons are more compact than their straight-chain isomers. Consequently, there is less surface area to interact. The van der Waals forces between molecules are weaker, and the molecules can more easily escape from the liquid. Table 13.8 lists the melting points and boiling points of the isomers of butane and pentane.

Bottled gas consists chiefly of propane (with some butane and isobutane) that, through pressure, has been condensed to a liquid (liquefied petroleum gas, LPG) so that it can be compactly stored and shipped in metal containers.

Table 13.8
Physical Properties of the Isomers of Butane and Pentane

IUPAC Name	Condensed Structural Formula	Melting Point (°C)	Boiling Point (°C)
Butane	$CH_3(CH_2)_2CH_3$	−138	−1
2-Methylpropane	$CH_3CH(CH_3)_2$	−159	−12
Pentane	$CH_3(CH_2)_3CH_3$	−130	36
2-Methylbutane	$CH_3CH_2CH(CH_3)_2$	−160	30
2,2-Dimethylpropane	$(CH_3)_4C$	−17	9

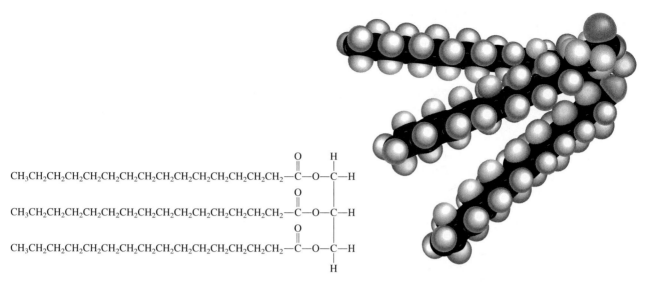

$$CH_3CH_2CH_2CH_2CH_2CH_2CH_2CH_2CH_2CH_2CH_2CH_2CH_2CH_2CH_2 \overset{O}{\underset{\|}{C}} - O - \overset{H}{\underset{|}{C}} - H$$

$$CH_3CH_2CH_2CH_2CH_2CH_2CH_2CH_2CH_2CH_2CH_2CH_2CH_2CH_2CH_2 \overset{O}{\underset{\|}{C}} - O - \overset{|}{\underset{|}{C}} - H$$

$$CH_3CH_2CH_2CH_2CH_2CH_2CH_2CH_2CH_2CH_2CH_2CH_2CH_2CH_2CH_2 \overset{O}{\underset{\|}{C}} - O - \overset{|}{\underset{|}{C}} - H$$
$$H$$

Figure 13.6
Tripalmitin, a typical fat molecule.

Notice that there is no obvious trend that enables us to predict melting points. In general, the more symmetrical isomers tend to have higher melting points than the less symmetrical isomers (because symmetrical molecules can pack closer together in the crystal lattice). Contrast 2,2-dimethylpropane with the less symmetrical pentanes in Table 13.8. Contrast, also, the very symmetrical octane isomer $(CH_3)_3CC(CH_3)_3$ with the straight-chain octane. The former is a solid and melts at 101 °C; the latter is a liquid whose melting point is −57 °C.

An extensive review of the physical properties of the alkanes has been given here, not because these compounds are so important in and of themselves, but rather because of their contributions to the structures of the other families of organic and biological compounds. A knowledge of alkane properties is also vital to an understanding of the functions of lipids because large portions of their structures consist of segments of alkane groups (Figure 13.6). One of the major functions of phospholipids (Section 20.7) and glycolipids (Section 20.8) is to serve as structural components of living tissues. Their biological importance here is dependent upon the presence of both polar and nonpolar groups, which enable them to bridge the gap between water-soluble and water-insoluble phases. This is a requisite in maintaining the selective permeability of cell membranes.

13.7 - Don't Read
PHYSIOLOGICAL PROPERTIES
OF ALKANES

The physiological properties of alkanes vary in a regular way as we proceed through the homologous series. Methane appears to be totally physiologically inert. We could breathe a mixture of 80% methane and 20% oxygen without ill effect. Such a mixture would be flammable, however, and no fire or spark of any kind could be permitted in an atmosphere consisting of methane. Breathing an atmosphere of pure methane (the ''gas'' of a gas stove) can lead to death not so much because of the presence of methane but because of the absence of oxygen (asphyxia). The other gaseous alkanes (and vapors of volatile liquid alkanes) act as anesthetics in high concentrations. They can also produce asphyxiation by excluding oxygen.

Liquid alkanes, such as those in gasoline, have varied effects on the body, depending on the part exposed. On the skin, alkanes dissolve body oils. Repeated contact may cause dermatitis. Swallowed, alkanes do little harm while in the stomach. However, in the lungs, alkanes cause "chemical pneumonia" by dissolving fatlike molecules from the cell membranes in the alveoli. The cells become less flexible, and the alveoli are no longer able to expel fluids. The buildup of fluids is similar to that which occurs in bacterial or viral pneumonia. People who swallow gasoline, petroleum distillates, or other liquid alkane mixtures should not be made to vomit. That would increase the chance of their getting the alkane into the lungs.

Heavier liquid alkanes, when applied to the skin, act as emollients (skin softeners). Such alkane mixtures as mineral oil can be used to replace natural skin oils washed away by frequent bathing or swimming. Petroleum jelly (Vaseline is one brand) is a semisolid mixture of hydrocarbons that can be applied as an emollient or simply as a protective film. Water and water solutions (e.g., urine) will not dissolve such a film, which explains why petroleum jelly protects a baby's tender skin from diaper rash.

13.8
CHEMICAL PROPERTIES: LITTLE AFFINITY

The alkanes are the least reactive of all organic compounds. They are generally unreactive toward strong acids (such as sulfuric acid), strong bases (such as sodium hydroxide), most oxidizing agents (such as potassium dichromate), and most reducing agents (such as sodium metal). In fact, the alkanes undergo so few reactions that they are sometimes called **paraffins** (from the Latin *parum affinis*, little affinity). Paraffin wax is a mixture of solid alkanes that is used as a seal for homemade preserves.

The alkanes do undergo a few important reactions. When mixed with oxygen at room temperature, alkanes give no apparent reaction. However, when a match flame or spark supplies sufficient energy to get things started (the energy of activation), an exothermic (heat-producing) reaction proceeds vigorously. The reaction, which is called **combustion,** is illustrated for methane.

$$CH_4 \;+\; 2\,O_2 \;\longrightarrow\; CO_2 \;+\; 2\,H_2O \;+\; \text{heat}$$

If the reactants are adequately mixed and there is sufficient oxygen, the products are carbon dioxide, water, and the all-important heat (for cooking foods, heating homes, and drying clothes). Such ideal conditions are rarely met, however, and products other than carbon dioxide, water, and heat are frequently formed. When the oxygen supply is limited, carbon monoxide is a by-product.

$$2\,CH_4 \;+\; 3\,O_2 \;\longrightarrow\; 2\,CO \;+\; 4\,H_2O$$

This reaction is responsible for dozens of deaths each year from unventilated or improperly adjusted gas heaters. (Similar reactions with similar results occur with kerosene heaters.) At the high temperatures achieved in some combustion reactions, some nitrogen (which makes up 80% of the atmosphere) is converted to oxides. Further, petroleum usually contains some sulfur compounds, and combustion reactions involving sulfur will yield sulfur dioxide. Thus, the same combustion reactions that heat our homes, power our industries, and propel our automobiles also produce most of our air pollution.

13.9
HALOGENATED HYDROCARBONS

In addition to undergoing combustion, alkanes react with chlorine and bromine in the presence of ultraviolet light or at a high temperature to yield organochlorine and organobromine compounds. Fluorine combines explosively with most hydrocarbons, whereas iodine is relatively unreactive. In general, halogenated hydrocarbons are compounds in which one or more hydrogens of a hydrocarbon have been replaced by halogen atoms. The following are examples of halogenated hydrocarbons:

$$\begin{array}{ccc}
\text{H} & \text{H Br Br} & \text{F H Cl} \\
\text{H—C—I} & \text{H—C—C—C—H} & \text{H—C—C—C—H} \\
\text{H} & \text{H H H} & \text{H CH}_3\text{ H}
\end{array}$$

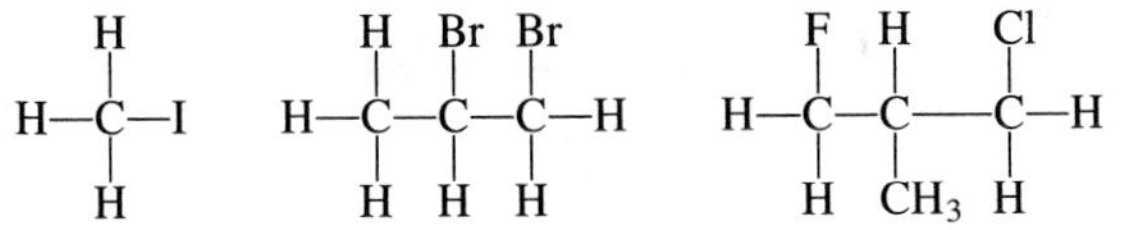

More specifically, replacement of one hydrogen atom of an alkane with a halogen atom gives an **alkyl halide.** These compounds are given common names that consist of two parts. The first is the name of the alkyl group; the second is the stem of the name of the halogen, with the ending -ide.

EXAMPLE 13.5

Give the common name for the compound with the formula

$$CH_3CH_2Br$$

SOLUTION

The alkyl group (CH_3CH_2—) is ethyl. The halogen is bromine. The compound is therefore ethyl bromide.

EXAMPLE 13.6

Give the common name for the compound

$$(CH_3)_2CHCl$$

SOLUTION

The alkyl group is isopropyl. The halogen is chlorine. The compound is named isopropyl chloride.

The IUPAC system differs from the common nomenclature in that halogen substituents are indicated by the prefixes fluoro-, chloro-, bromo-, and iodo-. The prefix is used with the name of the parent alkane, with numbers to indicate the position of the halogen if necessary.

EXAMPLE 13.7

Give the IUPAC name for

$$\overset{1}{C}H_3\overset{2}{C}H\overset{3}{C}H_2\overset{4}{C}H_2\overset{5}{C}H_3$$
$$|$$
$$Cl$$

SOLUTION

The parent alkane is pentane. The name of the compound is 2-chloropentane.

EXAMPLE 13.8

Give the IUPAC name for

$$\overset{1}{C}H_3\overset{2}{C}H\overset{3}{C}H_2\overset{4}{C}H\overset{5}{C}H_2\overset{6}{C}H_3$$
$$|\qquad\qquad|$$
$$CH_3\quad Br$$

SOLUTION

The parent alkane is hexane. The compound is named 4-bromo-2-methylhexane.

Methyl chloride   Methylene chloride

Chloroform   Carbon tetrachloride

More than one hydrogen on a hydrocarbon molecule can be replaced by halogen atoms. This leads to a wide variety of interesting and often useful compounds. Consider methane (CH_4), the simplest hydrocarbon. One hydrogen can be replaced by chlorine, yielding methyl chloride (CH_3Cl). Methyl chloride is a refrigerant. It is also used in the manufacture of silicones and synthetic rubber, and as a general methylating agent. A second hydrogen can be substituted to yield dichloromethane, or methylene chloride (CH_2Cl_2). This compound is a common solvent. It has been used, for example, to extract most of the caffeine from coffee to make the decaffeinated brands. It boils at 40 °C; hence, it can easily be removed by distillation if one wishes to recover the solute.

A third methane hydrogen can be replaced by chlorine, giving chloroform—trichloromethane ($CHCl_3$). Chloroform was one of the first anesthetics and was once widely used. It has been largely replaced by safer, less toxic chemicals for that use, but it is still an important commercial and industrial solvent.

Replacing all four of methane's hydrogens with chlorine gives carbon tetrachloride (CCl_4), also called tetrachloromethane. Carbon tetrachloride has been used as a dry-cleaning solvent and in fire extinguishers. It is no longer recommended for either use. Exposure to carbon tetrachloride (or most of the other chlorinated hydrocarbons, for that matter) can cause severe damage to the liver. Even the vapor, breathed in small amounts, can cause serious illness if the exposure is prolonged. Use of a carbon tetrachloride fire extinguisher in conjunction with water to put out a fire can be deadly. Carbon tetrachloride reacts with water at high temperatures to form phosgene ($COCl_2$), an extremely poisonous gas. In fact, phosgene was used in poison gas warfare during World War I.

With ethane, from one to six hydrogen atoms can be replaced by chlorine. Replacement of only one hydrogen gives ethyl chloride. Ethyl chloride is used as an external local anesthetic. When it is sprayed on the skin, it begins to evaporate, and this cools the area enough to be insensitive to pain. Replacement of two hydrogens can give either of two isomers. 1,2-Dichloroethane is commonly used as a solvent, particularly for rubber.

Replacement of three of the hydrogen atoms of ethane with chlorine atoms also gives two isomers. Both of the trichloroethanes are used as commercial and industrial solvents. 1,1,1-Trichloroethane is used for cleaning molds that are used in the fabrication of plastic items.

Replacement of four hydrogens of ethane with chlorine also leads to isomers. Only one is of importance, however—1,1,2,2-tetrachloroethane. This compound is widely used but highly toxic. It dissolves rubber, cellulose acetate, and other complex organic materials. It is also used to sterilize soil and as a component of weed killers and insecticide preparations.

Ozone Depletion

Carbon compounds containing fluorine, as well as chlorine, have been used as the dispersing gases in aerosol cans and as refrigerants. Properly called **chlorofluorocarbons (CFCs),** they are also known as *freons,* from their Du Pont trade name. Structures of three of the CFCs are shown below.

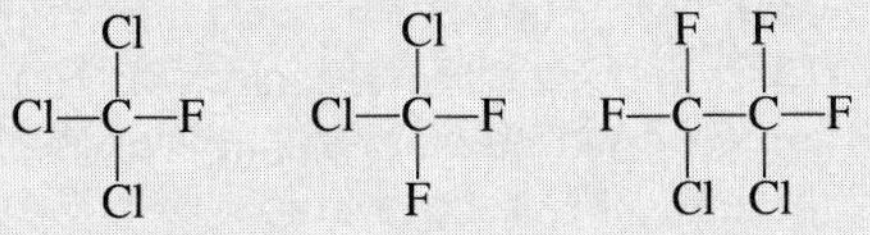

Freon 11   Freon 12   Freon 114

At room temperature, the chlorofluorocarbons are gases or liquids with low boiling points. They are essentially insoluble in water and inert toward most other substances. These properties make them ideal propellants for use in aerosol cans of deodorants, hair sprays, and food products. Unfortunately, the inertness of these compounds allows them to persist in the environment.

Chlorofluorocarbons diffuse into the stratosphere, where they are broken down by ultraviolet radiation. Chlorine atoms formed in this process break down the ozone that protects the Earth from harmful ultraviolet radiation.

$$O_3 + \text{ultraviolet light} \longrightarrow O_2 + O$$
$$CF_2Cl_2 + \text{ultraviolet light} \longrightarrow CF_2Cl\cdot + Cl\cdot$$
$$Cl\cdot + O_3 \longrightarrow ClO\cdot + O_2$$
$$\cdot ClO + O \longrightarrow Cl\cdot + O_2$$

Note that the last step results in the formation of another chlorine atom that can break down another molecule of ozone. The third and fourth steps are repeated many times; thus, the decomposition of one molecule of chlorofluorocarbon can result in the destruction of many molecules of ozone.

The U.S. National Research Council predicts a 2–5% increase in skin cancer for each 1% depletion of the ozone layer. Even a 1% increase in

Ozone in the upper atmosphere acts as a shield against ultraviolet radiation from the sun, which can cause skin cancer and harm some plants. The ozone holes are large areas of intense ozone depletion over the Antarctic continent and northern Europe that occur from late August through early October and break up in mid-November.

ultraviolet radiation can lead to a 5% increase in melanoma, a deadly form of skin cancer.

In 1984, the NRC predicted a 2–4% depletion by late in the twenty-first century. Worries escalated in 1987 with the discovery that ozone concentrations over Antarctica had plummeted to a record low. This thinning or "hole" in the ozone layer has been getting worse since 1979. A similar but less dramatic thinning has been detected in the Arctic.

Why are there ozone holes at the poles? In the polar winter a vortex of intensely cold air with ice crystals develops. Reactions occur on the surface of the ice crystals where hydrogen chloride (HCl) and chlorine nitrate ($ClONO_2$) are adsorbed. These molecules react to form chlorine and nitric acid.

$$HCl + ClONO_2 \longrightarrow Cl_2 + HNO_3$$

The Cl_2 molecules are readily dissociated by light into chlorine atoms.

$$Cl_2 \xrightarrow{\text{light}} 2 \; \overset{..}{\underset{..}{Cl}} \cdot$$

These chlorine atoms then enter into the ozone-destroying cycle.

Winter and early spring stratospheric ozone depletion at middle to high latitudes is estimated to be 2–10% over the last 20 years. Overall satellite data show that the ozone layer has been thinning at a rate of 0.5% per year since 1978. In the fall of 1991, scientists reported that the ozone hole over the Antarctic covered nearly 8 million square miles and was nearly as severe as the 1987 record level.

Chlorofluorocarbons have been banned in the United States from most aerosol preparations, but they still are used as refrigerants. They still escape into the atmosphere from refrigerators and air conditioners. As propellants in aerosol products, CFCs were replaced by methylene chloride and flammable hydrocarbons such as isobutane. Methylene chloride won't burn, but it in turn has been banned as a possible carcinogen.

A worldwide agreement, called the Montreal Protocol, was signed by the industrialized nations and implemented in 1989. The protocol calls for reduction in the production and use of ozone-depleting substances, with a nearly total ban by the year 2000. Chlorofluorocarbons that contain hydrogen atoms (HCFCs) and fluorinated hydrocarbons (HFCs) are under development as replacements for the CFCs.

Because of the immediacy of concern, the timetable for the Montreal Protocol has been moved up to 1995.

Other compounds, such as the nitrogen oxides from agricultural activities and aircraft that fly in the stratosphere (supersonic transport planes), also lead to depletion of the ozone shield.

An HFC   An HCFC

The HCFCs break down in the lower atmosphere more rapidly than CFCs. Only a tiny proportion will reach the stratosphere. They will therefore have little impact on the ozone layer. Since it is chlorine atoms from CFCs that destroy stratospheric ozone, the HFCs, which contain no chlorine, will have no effect on the ozone shield.

13.10
CYCLOALKANES

The hydrocarbons we have encountered so far have been composed of open-ended chains of carbon atoms. Carbon and hydrogen atoms can also hook up in other arrangements, some of them quite interesting. There is a synthetic hydrocarbon that has the formula C_3H_6. The three carbons are joined in a **ring** or **cycle.** The compound is called cyclopropane (Figure 13.7).

In addition to having an interesting structure, cyclopropane has some intriguing properties. It is a potent, quick-acting anesthetic with few undesirable side effects. It is seldom used in surgery, however, because it forms explosive mixtures with air at nearly all concentrations.

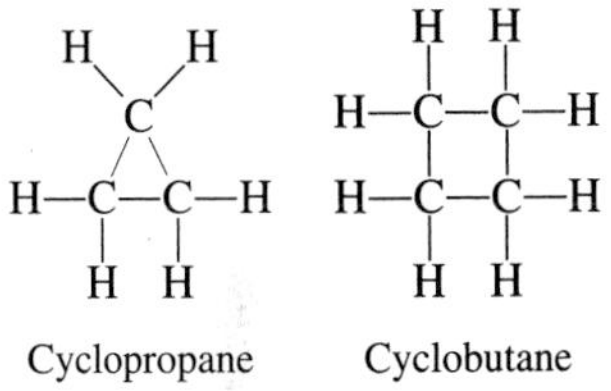

Names of cycloalkanes are formed by addition of the prefix *cyclo-* in front of the name of the open-chain compound with the same number of carbon atoms as are in the ring. Thus, the name for the cyclic compound C_4H_8 is cyclobutane. Names and structures of several cycloalkanes are given in Figure 13.8. Note that the carbon atoms of each compound form a regular geometric figure. For example, the three carbon atoms of cyclopropane form a triangle. Therefore, a triangle is frequently used to represent cyclopropane. Similarly, the five carbons of cyclopentane form a pentagon, and we use a pentagon to represent cyclopentane. It is understood that the carbon atoms occur at the angles of the particular figure and that each carbon atom is attached to sufficient hydrogens to give the carbon atom four bonds.

Cyclic compounds may also contain attached substituent groups (see Figure 13.8). If there is only one substituent on the cyclohexane ring, the substituent does not have to be numbered, since all the positions of the ring are equivalent. When there is more than one substituent, numbers are required. The ring carbons are numbered so that the carbons bearing the substituents have the lowest numbers.

The major distinction between the cyclic and noncyclic alkanes arises from the differences in the geometric configuration of the two types of compounds. There is free rotation about all carbon–carbon bonds of open-chain alkanes. This is not the case with the cyclohexanes. Here the carbons are held rigidly in place. Free rotation is impossible without disruption of the ring structure. For these compounds, the position of any substituent relative to the ring becomes extremely important. (The substituent will be situated

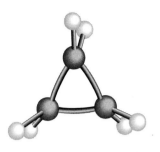

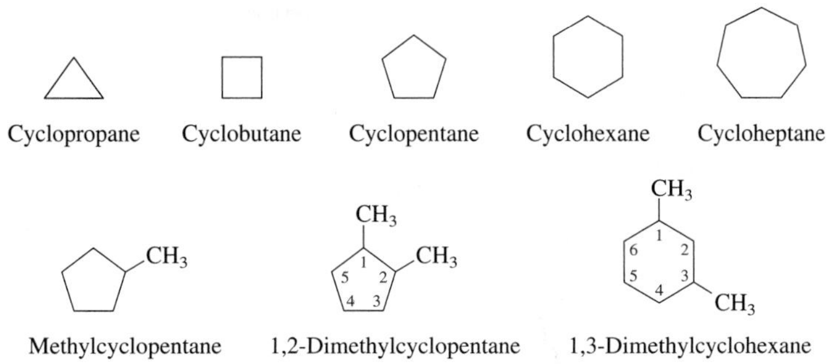

Figure 13.7
Ball-and-stick model of cyclopropane.

Cyclopropane   Cyclobutane

Figure 13.8
Some cycloalkanes.

Cyclopropane   Cyclobutane   Cyclopentane   Cyclohexane   Cycloheptane

Methylcyclopentane   1,2-Dimethylcyclopentane   1,3-Dimethylcyclohexane either above or below the ring.) A new form of isomerism is possible, which we discuss in Section 18.4. For now, let it suffice to say that the compounds at the left are not the same, but are isomers.

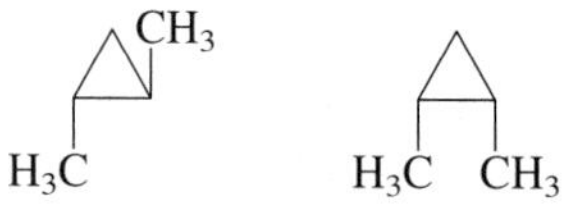

EXAMPLE 13.9

Draw structures for the following cyclic hydrocarbons.

a. cyclooctane
b. ethylcyclohexane
c. 1,1,2-trimethylcyclobutane
d. 1-ethyl-1,2,5,5-tetramethylcycloheptane

SOLUTION

a. 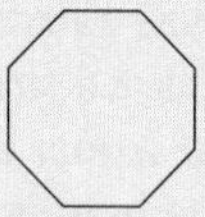   b. 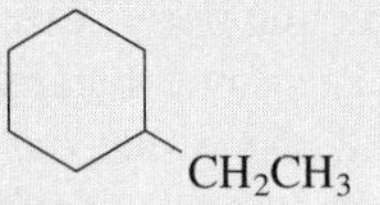   c. 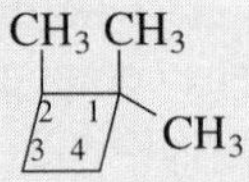 CH_3 CH_3 d.

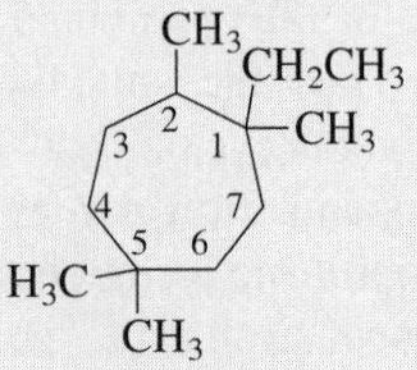

Practice Exercise
Draw structures for (a) cyclopropane and (b) 1-ethyl-2-methylcyclopentane.

The physical, chemical, and physiological properties of these cyclic hydrocarbons are generally quite similar to those of the corresponding open-chain compounds. Cycloalkanes (with the exception of cyclopropane, which has a highly strained ring) act very much like ordinary alkanes. Cyclic hydrocarbons with five- and six-membered rings occur in petroleum from certain areas; California crude, for instance, is particularly rich in these compounds. Like all other hydrocarbons, cyclic hydrocarbons burn.

13.11
ALKENES: STRUCTURE AND NOMENCLATURE

Not all hydrocarbons are as resistant to reaction as the alkanes. In fact, the next family that we will consider is quite reactive. This family, called **alkenes** (note the **-ene** ending), is characterized by the presence of a carbon–carbon double bond. Names, structures, and physical properties of a few representative alkenes are given in Table 13.9. Like all hydrocarbons, the alkenes are insoluble in water but soluble in organic solvents.

We have used only condensed structural formulas in this table. Thus, $CH_2{=}CH_2$ stands for

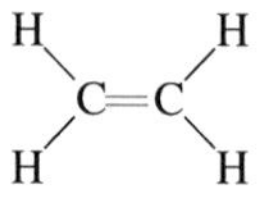

IUPAC Name	Molecular Formula	Condensed Structure	Melting Point (°C)	Boiling Point (°C)
Ethene	C_2H_4	$CH_2{=}CH_2$	−169	−104
Propene	C_3H_6	$CH_3CH{=}CH_2$	−185	−47
1-Butene	C_4H_8	$CH_3CH_2CH{=}CH_2$	−185	−6
1-Pentene	C_5H_{10}	$CH_3CH_2CH_2CH{=}CH_2$	−138	30
1-Hexene	C_6H_{12}	$CH_3(CH_2)_3CH{=}CH_2$	−140	63
1-Heptene	C_7H_{14}	$CH_3(CH_2)_4CH{=}CH_2$	−119	94
1-Octene	C_8H_{16}	$CH_3(CH_2)_5CH{=}CH_2$	−102	121

Table 13.9
Physical Properties of Some Selected Alkenes

The double bond is shared by the two carbon atoms and does not involve the hydrogens, although the condensed formula does not make this point obvious.

Compare the molecular formulas of the alkenes in Table 13.9 with those for the alkanes in Table 13.4. The molecular formula for ethane is C_2H_6. The formula for the two-carbon alkene is C_2H_4. Each alkene has two fewer hydrogens than the corresponding alkane. Compared with the carbon atoms in an alkane, each of two carbon atoms in the alkene must bond to one less hydrogen atom in order to form the double bond. Since these carbons are sharing two of their bonds with one atom (and thus bonding to fewer different atoms), the compounds are called **unsaturated hydrocarbons.** Alkenes with one double bond have the general formula C_nH_{2n}.

The first two alkenes of Table 13.9, ethene and propene, are most often called by their common names, ethylene and propylene, respectively. Ethylene ($CH_2{=}CH_2$) is a major commercial chemical. The U.S. chemical industry produces about 13 billion kg of ethylene annually, making it the most important of all synthetic organic chemicals. Over one-third of this ethylene goes into the manufacture of polyethylene, one of the most familiar plastics. Another one-sixth is converted to ethylene glycol, the principal component of most brands of antifreeze for automobile radiators.

Propylene ($CH_3CH{=}CH_2$) is also an important industrial chemical. It is converted to plastics, isopropyl alcohol (Chapter 14), and a variety of other end products and intermediates for synthesis.

Before we consider the rest of the alkenes in Table 13.9, let's pause to discuss the nomenclature of alkenes. Although there is only one alkene with the formula C_2H_4 (ethene) and only one alkene with the formula C_3H_6 (propene), there are several alkenes with the formula C_4H_8. Common names are hardly helpful in naming the many isomers of the higher alkenes. For the most part, the IUPAC system is used for these. Some of the IUPAC rules for alkenes are as follows:

1. All have names ending in -*ene*.
2. The longest chain of atoms *containing the double bond* is the parent compound. The name has the same stem as the corresponding alkane (i,e., the alkane with the same number of carbon atoms), but the ending is changed from -*ane* to -*ene*. Thus, the compound $CH_3CH{=}CH_2$, with three carbon atoms, is named *propene*.
3. When it is necessary to indicate the position of the double bond, the first carbon of the two that are doubly bonded is given the lowest possible number (i.e., the carbons are counted from the end of the chain nearer the first carbon of the double bond). The compound $CH_3CH{=}CHCH_2CH_3$, for example, has the double bond between the second and third carbon atoms. Its name is 2-pentene.

4. Substituent groups are named as usual. Their position is indicated by a number. Thus,

$$CH_3CHCH_2CH{=}CHCH_3$$
$$|$$
$$CH_3$$

5 methyl-1-2-hexene

is 5-methyl-2-hexene. Note that the numbering of the parent chain is always done in such a way as to give the double bond the lowest number, even if that forces a substituent to have a higher number. We say that the double bond has priority in numbering.

Table 13.10 contains some examples of the application of the IUPAC rules of nomenclature to the alkenes and cycloalkenes. Study the table and then try the following examples.

EXAMPLE 13.10

Name the compound

$$CH_3CH{=}CHCH_2CH{-}CHCH_3$$
$$|\quad\ \ |$$
$$CH_3\ CH_3$$

SOLUTION
The longest continuous chain has seven carbon atoms. To give the first carbon of the double bond the lowest number, we start numbering from the left.

$$\overset{1}{C}H_3\overset{2}{C}H{=}\overset{3}{C}H\overset{4}{C}H_2\overset{5}{C}H{-}\overset{6}{C}\overset{7}{H}CH_3$$
$$|\quad\ \ |$$
$$CH_3\ CH_3$$

The name of the compound is 5,6-dimethyl-2-heptene.

EXAMPLE 13.11

Name the compound

$$CH_2{=}C{-}CH_2CH_3$$
$$|$$
$$CH_2CH_3$$

SOLUTION
The name is 2-ethyl-1-butene. The longest continuous chain in the molecule contains five carbon atoms. However, the longest continuous chain *containing the double bond* incorporates only four carbon atoms, and this four-carbon chain serves as the parent compound.

Practice Exercise
Name the following compounds.

a. $CH_3C{=}CHCHCH_2CH_3$
     $\quad\ \ |\qquad\ |$
     $\quad\ CH_3\ \ CH_2CH_3$

b.

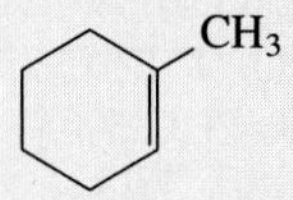

Table 13.10
Application of the IUPAC Rules to the Naming of
Alkenes and Cycloalkenes

Condensed Structural Formula	Rewritten and Numbered	IUPAC Name
$CH_3(CH_2)_2CH=CH_2$		1-Pentene
$CH_3CH=CHCH_2CH_3$		2-Pentene
$CH_3CH_2CH=C(CH_3)_2$		2-Methyl-2-pentene
$(CH_3)_2CHC(CH_3)=CHCH_3$		3,4-Dimethyl-2-pentene
$CH_3CH_2CH(CH_3)C(C_3H_7)=CH_2$		3-Methyl-2-propyl-1-pentene
		Cyclopentene
		1,3-Dimethylcyclopentene
		4-Methylcyclohexene
		1,2,3,3-Tetramethylcyclohexene
		3,4-Diethylcyclobutene

EXAMPLE 13.12

Draw the structural formula for 3,4-dimethyl-2-pentene.

SOLUTION

To draw the formula for this compound, first write down the parent chain of five carbons.

$$C-C-C-C-C$$

Then add the double bond between the second and third carbons (this is 2-pentene).

$$\overset{1}{C}-\overset{2}{C}=\overset{3}{C}-\overset{4}{C}-\overset{5}{C}$$

Now add the groups at their proper positions.

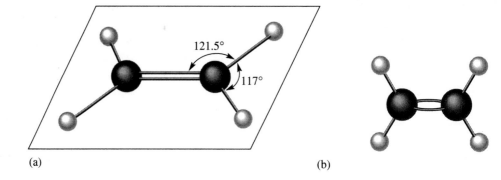

$$\left(\text{or}\quad CH_3-CH=C\overset{\displaystyle CH_3}{\underset{}{}}\overset{\displaystyle CH_3}{\underset{}{}}CH-CH_3\right)$$

Practice Exercise

Draw structural formulas for the following compounds.

a. 3-ethyl-2-methyl-1-hexene   b. 3-isopropylcyclopentene

Before leaving the subject of alkene nomenclature, one additional aspect must be mentioned. The double bond of alkene molecules, like the ring structure of cycloalkanes, imposes geometric restrictions on the molecules. Recall that according to VSEPR theory (Section 4.16), each doubly bonded carbon atom lies in the center of an equilateral triangle (to minimize the repulsive forces among the three regions of electron density). The carbon atoms of a double bond and the two atoms bonded to each carbon all lie in a single plane (Figure 13.9). Free rotation about doubly bonded carbon atoms is *not* possi-

Figure 13.9

(a) Planar configuration of the double bond. (b) Ball-and-spring model of ethylene.

121.5°

117°

(a)   (b)

ble without rupturing the bond. Therefore, the relative positions of substituent groups located above or below the double bond become very significant. The nomenclature employed in situations such as this, as well as the consequences of restricted rotation, are discussed in Chapter 18.

13.12
ALKENES AND LIVING THINGS

The physiological properties of the alkenes are also similar to those of the alkanes. Ethylene has found some use as an inhalation anesthetic. Like the gaseous alkanes, ethylene can cause unconsciousness and even death by asphyxiation. Large amounts of liquid and solid (or mixtures of liquid and solid) alkenes are seldom encountered. They would probably act on or in our bodies much as the alkanes do.

Alkenes occur widely in nature. Ripening fruits and vegetables give off ethylene, which triggers further ripening. Fruit processors artificially introduce ethylene to hasten the normal ripening process; 1 kg of tomatoes can be ripened by exposure to as little as 0.1 mg of ethylene for 24 hours. Unfortunately, the tomatoes don't taste much like those that ripen on the vine.

Other alkenes that occur in nature include 1-octene, a constituent of lemon oil, and octadecene ($C_{18}H_{36}$), found in fish liver. Dienes (which have two double bonds) and polyenes (which have many double bonds) are also common. Butadiene (CH_2=CH—CH=CH_2) is found in coffee. A hexadecadiene ($C_{16}H_{30}$) occurs in olive oil. Lycopene and the carotenes are isomeric polyenes ($C_{40}H_{56}$) that give the attractive red, orange, and yellow colors to watermelons, tomatoes, carrots, and other vegetables and fruits. Vitamin A, essential to good vision, is derived from a carotene (see Special Topic G). The world would be a much darker place without the chemistry of the alkenes.

13.13
CHEMICAL PROPERTIES OF ALKENES

Like the alkanes—and all other hydrocarbons—the alkenes burn. While this is a hazard that you should remember when working with alkenes, these compounds are not commercially important as fuels. We can write an equation for the combustion of ethene that is similar to the one for the combustion of an alkane.

$$C_2H_4 \ + \ 3\,O_2 \ \longrightarrow \ 2\,CO_2 \ + \ 2\,H_2O \ + \ heat$$

The typical reactions of the alkenes are **addition reactions.** One of the bonds in the double bond is broken, permitting each of the involved carbon atoms to bond to an additional atom or group. The originally doubly bonded carbons are still attached by the remaining single bond. Perhaps the simplest addition reaction is that which occurs with hydrogen in the presence of a nickel (Ni), platinum (Pt), or palladium (Pd) catalyst.

Ethene   Hydrogen   Ethane

The product of this reaction is an alkane with the same carbon skeleton as the original alkene. **Hydrogenation** was once widely used in industry in the conversion of unsaturated vegetable oils into saturated fats. Hydrogenation of the liquid oil yields a solid fat and makes vegetable shortening resemble the more familiar animal fat, lard. The difference in the liquid oil and the solid fat is due to a difference in the number of double bonds present; there are more in the unsaturated oils and fewer in the saturated fats. Vegetable oils are now more readily accepted and, in fact, are frequently preferred by the consumer, but some oils are still hydrogenated so that the vegetable product will resemble an animal product. Margarine, for example, resembles butter by virtue of hydrogenation (see Chapter 20).

Alkenes readily add halogen molecules. Indeed, the reaction with bromine is often used to test for alkenes. Solutions of bromine are brownish-red. When an alkene is added to such a solution, the color disappears because the alkene reacts with the bromine.

$$\underset{\text{Ethene}}{\overset{\text{H}}{\underset{\text{H}}{}}C=C\overset{\text{H}}{\underset{\text{H}}{}} \;+\; \underset{\substack{\text{Bromine}\\(\text{brownish red})}}{\text{Br—Br}} \;\longrightarrow\; \underset{\substack{\text{1,2-Dibromoethane}\\(\text{colorless})}}{H-\overset{\overset{\text{H}}{|}}{\underset{\underset{\text{Br}}{|}}{C}}-\overset{\overset{\text{H}}{|}}{\underset{\underset{\text{Br}}{|}}{C}}-H}$$

Another important addition reaction of the alkenes is that with water. This reaction, called **hydration,** requires the presence of a mineral acid, such as sulfuric acid (H_2SO_4), as a catalyst.

$$\underset{\text{Ethene}}{\overset{\text{H}}{\underset{\text{H}}{}}C=C\overset{\text{H}}{\underset{\text{H}}{}} \;+\; \underset{\text{Water}}{\text{H—OH}} \;\xrightarrow{\;H_2SO_4\;}\; \underset{\text{Ethyl alcohol}}{H-\overset{\overset{\text{H}}{|}}{\underset{\underset{\text{H}}{|}}{C}}-\overset{\overset{\text{H}}{|}}{\underset{\underset{\text{OH}}{|}}{C}}-H}$$

Vast quantities of ethyl alcohol, for use as an industrial solvent, are made from ethylene. This alcohol is structurally identical to that used in alcoholic beverages. However, federal law requires that all drinking alcohol be produced by the natural process called fermentation. (See Section 14.4 for a more extensive discussion of the hydration reaction.)

EXAMPLE 13.13

Write equations for the reaction of $CH_3CH=CHCH_3$ with each of the following.

a. H_2 (Ni catalyst)   b. Br_2 c. H_2O (H_2SO_4 catalyst)

SOLUTION

In each reaction, the reagent adds across the double bond. The answers are

a. $CH_3CH=CHCH_3 \;+\; H_2 \;\xrightarrow{\;Ni\;}\; CH_3\underset{\underset{\text{H}}{|}}{C}H-\underset{\underset{\text{H}}{|}}{C}HCH_3 \;$ or $\; CH_3CH_2CH_2CH_3$

b. $CH_3CH=CHCH_3 \;+\; Br_2 \;\longrightarrow\; CH_3\underset{\underset{\text{Br}}{|}}{C}H-\underset{\underset{\text{Br}}{|}}{C}HCH_3$

c. $CH_3CH=CHCH_3 \;+\; H_2O \;\xrightarrow{\;H_2SO_4\;}\; CH_3\underset{\underset{\text{H}}{|}}{C}H-\underset{\underset{\text{OH}}{|}}{C}HCH_3 \;$ or $\; CH_3CH_2\underset{\underset{\text{OH}}{|}}{C}HCH_3$

13.14
POLYMERIZATION

The most important alkene reaction of all, perhaps, is **polymerization.** Chemists call giant molecules assembled from much smaller building blocks **polymers** (from the Greek *poly,* many, and *meros,* parts). In a laboratory (and in living systems, for that matter), the preparation of polymers involves the hooking together of many smaller molecules. The small molecules, the building blocks, are called **monomers** (from the Greek *monos,* one and *meros,* parts). The polymer is as different from the monomer as long strips of spaghetti are from the particles of flour that spaghetti is made of. For example, polyethylene, the familiar solid, waxy "plastic" of plastic bags, is a polymer prepared from a monomer (ethylene) that is a gas.

There are two general types of polymerization reactions. One is called addition polymerization; the other is condensation polymerization (see pages 400 and 406).

In **addition polymerization,** the building blocks or monomers add to one another in such a way that the polymeric product contains all the atoms of the starting monomers. Under high pressure and temperature and in the presence of a catalyst, ethylene molecules are made to join together in long chains. The polymerization can be represented by the reaction of a few monomer units.

$$\cdots + \underset{H}{\overset{H}{>}}C=C\underset{H}{\overset{H}{<}} + \underset{H}{\overset{H}{>}}C=C\underset{H}{\overset{H}{<}} + \underset{H}{\overset{H}{>}}C=C\underset{H}{\overset{H}{<}} + \cdots \longrightarrow \cdots \overset{H}{\underset{H}{C}}-\overset{H}{\underset{H}{C}}-\overset{H}{\underset{H}{C}}-\overset{H}{\underset{H}{C}}-\overset{H}{\underset{H}{C}}-\overset{H}{\underset{H}{C}} \cdots$$

The dotted lines in the formula of the product are like etc.'s: they indicate that the structure is extended for many units in each direction. Notice that the two carbon atoms and four hydrogen atoms of each monomer molecule are incorporated into the polymer structure. Polyethylene is an addition polymer. A truer picture of the polymer is given in Figure 13.10a, which shows a ball-and-stick model of a short segment. Still better is the space-filling representation of Figure 13.10b. Both models are deficient in that they con-

Figure 13.10
(a) Ball-and-stick model of a segment of a polyethylene molecule. (b) Space-filling model of a short segment of a polyethylene molecule.

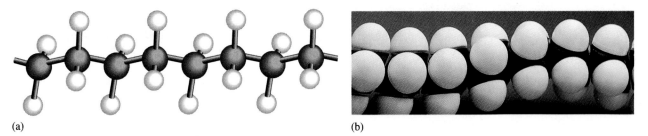

(a)   (b)

tain too few atoms. Real polyethylene molecules have varying numbers of carbon atoms—from a few hundred to several thousand.

Polyethylene is the simplest of the synthetic polymers, and it's the cheapest, too. It is familiar today in the plastic bags used for packaging fruit and vegetables, in garment bags for dry-cleaned clothing, in garbage-can liners, and in many other items.

Polyethylene was invented shortly before the start of World War II. Before long, it was used for insulating cables in a top-secret invention—radar—which helped British pilots spot enemy aircraft before the aircraft became visible to the naked eye. Polyethylene proved to be tough and flexible and an excellent electrical insulator. It could withstand both high and low temperatures. Without polyethylene, the British could not have had effective radar, and without radar the Battle of Britain might have been lost. The invention of this simple plastic may have changed the course of history.

Polymers are readily distinguishable from the small molecules from which they are synthesized. Besides the fundamental differences in molecular weight, structure, and bonding, polymers usually exhibit different physical and chemical properties. The small molecules tend to be volatile liquids of low viscosity, or even gases at room temperature. Polymers, on the other hand, tend to be nonvolatile, highly viscous liquids, glasses, or solids that soften only at high temperatures. Natural materials such as proteins (Chapter 21), natural rubber, cellulose, starch (Chapter 19), and complex silicate minerals are polymers. Artificial materials such as fibers, films, plastics, semisolid resins, and synthetic rubbers are also polymers. Synthetic polymers represent more than half of the compounds produced by the chemical industry (Table 13.11).

13.15
THE ALKYNE SERIES

In alkenes, carbon atoms in the double bond share two pairs of electrons. Carbon atoms can also share three pairs of electrons, forming triple bonds. Compounds containing such bonds are called **alkynes.** The common name of the simplest alkyne is acetylene (C_2H_2—Figure 13.11). Its structure is

$$H-C\equiv C-H$$

Figure 13.11
Ball-and-stick model of acetylene.

About 10% of all acetylene produced is used in oxyacetylene torches for cutting and welding metals. The flame from such a torch can attain a very high temperature. Most acetylene, however, is converted to chemical intermediates that are in turn used to make vinyl and acrylic plastics, fibers, and resins and a variety of other chemical products.

The alkynes are similar to the alkenes in both physical and chemical properties. For example, they undergo many of the typical addition reactions of alkenes. Like ethene, acetylene has been used as an anesthetic for surgery. At higher concentrations, it causes narcosis and asphyxia. The IUPAC nomenclature for alkynes parallels that of the alkenes, except that the family ending is -*yne* rather than -*ene*. The official name for acetylene is ethyne. Notice that the ending of the common name, acetylene, is deceptive. It sounds very much like ethylene or propylene. Remember that acetylene is an alkyne and that ethylene and propylene are alkenes.

Table 13.11
Some Addition Polymers

Monomer	Polymer	Polymer Name	Some Uses
$H_2C=CH_2$ Ethylene	$\left[\begin{array}{c} H\ \ H \\ -C-C- \\ H\ \ H \end{array}\right]_n$	Polyethylene	Plastic bags, bottles, toys, electrical insulation
$H_2C=CH-CH_3$ Propylene	$\left[\begin{array}{c} H\ \ H \\ -C-C- \\ H\ \ CH_3 \end{array}\right]_n$	Polypropylene	Indoor-outdoor carpeting, bottles, biomedical implants
$H_2C=CH-$⬡ Styrene	$\left[\begin{array}{c} H\ \ H \\ -C-C- \\ H\ \ ⬡ \end{array}\right]_n$	Polystyrene	Simulated wood furniture, styrofoam insulation and packing materials
$H_2C=CH-Cl$ Vinyl chloride	$\left[\begin{array}{c} H\ \ H \\ -C-C- \\ H\ \ Cl \end{array}\right]_n$	Poly(vinyl chloride), PVC	Plastic wrap, simulated leather (Naugahyde), phonograph records, garden hoses, rainwear, floor covering
$H_2C=CCl_2$ 1,1-Dichloroethene (Vinyiidene chloride)	$\left[\begin{array}{c} H\ \ Cl \\ -C-C- \\ H\ \ Cl \end{array}\right]_n$	Poly(vinylidene chloride), Saran	Food wrap, seatcovers
$F_2C=CF_2$ Tetrafluoroethylene	$\left[\begin{array}{c} F\ \ F \\ -C-C- \\ F\ \ F \end{array}\right]_n$	Polytetrafluoroethylene, Teflon	Nonstick coating for cooking utensils, electrical insulation, lubricant, bearings
$H_2C=CH-CN$ Cyanoethylene	$\left[\begin{array}{c} H\ \ H \\ -C-C- \\ H\ \ CN \end{array}\right]_n$	Polyacrylonitrile, Orlon, Acrilan, Creslan, Dynel	Yarns, wigs
$H_2C=CH-O-\overset{\overset{\displaystyle O}{\|}}{C}-CH_3$ Vinyl acetate	$\left[\begin{array}{c} H\ \ H \\ -C-C- \\ H\ \ O-C-CH_3 \\ \|\!|\ O \end{array}\right]_n$	Poly(vinyl acetate), PVA	Adhesives, textile coatings, chewing gum resin, paints
$H_2C=\overset{\overset{\displaystyle CH_3}{\|}}{\underset{\underset{\displaystyle O}{\|}}{C}}-C-O-CH_3$ Methyl methacrylate	$\left[\begin{array}{c} H\ \ CH_3 \\ -C-C- \\ H\ \ C-O-CH_3 \\ \|\!|\ O \end{array}\right]_n$	Poly(methyl methacrylate), Lucite, Plexiglas	Glass substitute, bowling balls

Figure 13.12

A variety of replacement parts for the human body. Many of the parts are made from synthetic polymers.

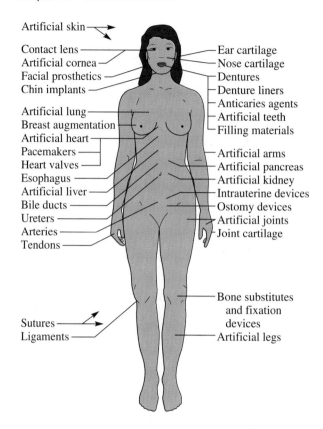

Artificial skin

Contact lens
Artificial cornea
Facial prosthetics
Chin implants

Artificial lung
Breast augmentation
Artificial heart
Pacemakers
Heart valves
Esophagus
Artificial liver
Bile ducts
Ureters
Arteries
Tendons

Sutures
Ligaments

Ear cartilage
Nose cartilage
Dentures
Denture liners
Anticaries agents
Artificial teeth
Filling materials

Artificial arms
Artificial pancreas
Artificial kidney
Intrauterine devices
Ostomy devices
Artificial joints
Joint cartilage

Bone substitutes and fixation devices
Artificial legs

Biomedical Polymers

HO—CH₂—C with =O and OH

Glycolic acid

HO—CH—C with CH₃, =O, and OH

Lactic acid

One of the most interesting uses of polymers has been in replacements for diseased, worn out, or missing parts of the human body (Figure 13.12). Artificial ball-and-socket hip joints made of steel (the ball) and plastic (the socket) are now being installed at a rate of 25,000 a year. People crippled by arthritis are not only freed from pain but are also given much more freedom of movement. Patients with heart and circulatory problems can enter a hospital for a "valve job" or the replacement of worn-out or damaged parts. Pyrolytic carbon heart valves derived from a polymer are widely used. Knitted Dacron tubes replace arteries blocked or damaged by atherosclerosis. Plastic implants reconstruct breasts removed because of cancer.

On contact with most foreign substances, blood begins to clot. To prevent such a reaction, most synthetics must be chemically treated and coated with heparin, a natural anticoagulant, before being used as implants. Dacron can be used for artificial arteries because it is relatively inert in this regard. Another approach to the problem of substituting synthetic materials for body parts is to use naturally occurring substances to construct the biomedical polymers. For example, polymers of glycolic and lactic acids have been used in synthetic films for covering burn wounds. Ordinarily, burns have to be covered with human skin from donors or with specially treated pigskin to prevent infection and excessive loss of fluids. Frequent changes of these covers are required because of the body's

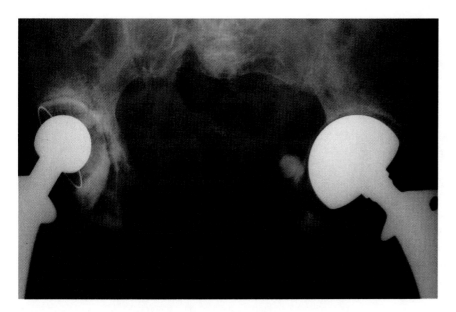

Figure 13.13
X-ray photo showing replacement hip joints (above); (right)
artificial hip joint made of metal alloy and polymer.

tendency to reject these foreign tissues. In contrast, the synthetic film is absorbed and metabolized rather than rejected.

The development of biomedical polymers has barely begun. In the future lies the prospect of the replacement of entire parts of the body. Artificial hearts made of synthetic elastomers are used to keep people alive who would otherwise die of heart failure, and artificial joints for hip and knee replacement are widely available for persons with arthritis or other degenerative joint disease (Figure 13.13).

13.16
BENZENE

In 1825 Michael Faraday isolated a hydrocarbon from a sample of illuminating gas made from whale oil. He determined that the compound had an empirical formula of CH_2 (based upon the then current belief that the atomic weight of carbon was 6), and he named the compound "bicarburet of hydrogen." Later the compound was given the name benzene because it could be obtained by distilling benzoic acid with calcium oxide. During the ensuing decades the concept of atomic weight became more clearly defined, and vapor density measurements established the molecular formula of benzene to be C_6H_6.

One of the many possible structural formulas corresponding to this molecular formula is 1,3,5-cyclohexatriene. The compound would be a slightly lopsided hexagon, having three double bonds in a six-membered ring (Figure 13.14a). The high degree of unsaturation would imply a very high reactivity. Chemists soon discovered, however, that

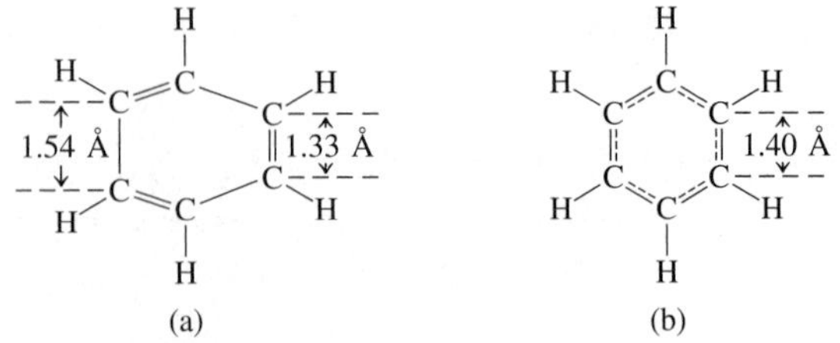

Figure 13.14

(a) The hypothetical 1,3,5-cyclohexatriene would be a slightly lopsided hexagon because double bonds are shorter than single bonds. (b) The actual benzene molecule is a regular hexagon with all sides 1.40 Å.

(a)   (b)

benzene is chemically stable and that it behaves more like an alkane than an alkene. It is unreactive with respect to the bromine test for unsaturation.

In 1865, the German chemist Friedrich August Kekulé proposed a structure that could account for all of the known chemical properties of the compound. His theory was that the benzene molecule consisted of a cyclic, hexagonal, planar structure of six carbon atoms with alternate single and double bonds. Each carbon atom was bonded to only one hydrogen atom. He accounted for the equivalence of all six carbon atoms by suggesting that the double bonds are not static but rather are mobile, and that they oscillate from one position to another.

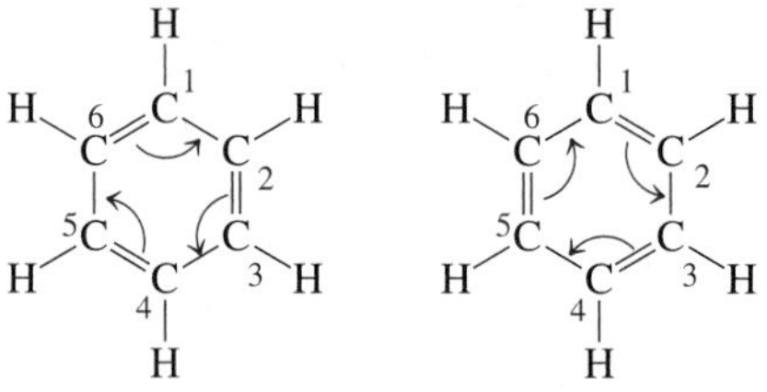

I   II

Structures I and II differ from each other only in the positions of the double and single bonds, and although they satisfy the requirements for equivalent hydrogens, they do not explain why benzene does not behave like an unsaturated hydrocarbon. Furthermore, X-ray-diffraction measurements of the bond lengths indicate that all the carbon–carbon bonds in benzene are the same length, 1.40 Å (Figure 13.14b). This value falls between the length of a carbon–carbon double bond (1.33 Å) and that of a carbon–carbon single bond (1.54 Å).

To accommodate these discrepancies, chemists have postulated that benzene exhibits resonance. **Resonance** is a word used to describe the phenomenon in which no single classical Lewis structure adequately accounts for the experimentally observed properties of a molecule (such as bond energies and bond distances). According to the resonance theory, the actual benzene molecule is a resonance hybrid of structures I and II. That is, neither of the two structures actually exists, but taken together they represent the true structure of the molecule. We depict this situation with a double-headed arrow ($\leftrightarrow$) that connects the contributing structures (Figure 13.15a). This resonance arrow should be clearly distinguished from the pair of half-headed arrows ($\rightleftharpoons$) that are used to indicate an equilibrium condition.

Some authors combine the two contributing forms into a single structure as depicted in Figure 13.15b. The inner circle indicates that the valence electrons are shared equally by all six carbon atoms (that is, the electrons are *delocalized,* or spread out, over all the carbon atoms). This method is a valid shortcut device for writing benzene, but it does not adequately account for all the electrons in the molecule. The representation of benzene that contains alternating single and double bonds (*resonance* or *Kekulé forms*) is the best model for keeping track of the electrons within the molecule. The circle within a hexagon

1 angstrom (Å) = 1 × 10⁻⁸ cm

Although the term *resonance* is much used in organic chemistry, it is not unique to organic compounds. Many inorganic compounds (e.g., O_3, SO_2, NO_3^-) also exhibit resonance.

better describes the molecule by indicating the equal sharing of the electrons and the identical bond lengths within the molecule. In this text, we shall use the hexagon with inscribed circle. Remember that when either representation is used, it is understood that each corner of the hexagon is occupied by one carbon atom. Attached to each carbon atom is one hydrogen atom, and all the atoms lie in the same plane. Whenever any other atom, or group of atoms, is substituted for a hydrogen atom, it must be shown to be bonded to a particular corner of the hexagon.

One final note: taken literally, the term resonance is a misnomer. It indicates oscillation from one structure to another or the oscillation of an electron pair from one bond to another. *There is no oscillation.* A mule, the offspring of a male donkey and a mare, can be considered a hybrid of a donkey and a horse. However, it is not a horse at one instant and a donkey at another; it is always a mule. Likewise, any molecule that exhibits resonance has *one* real structure, which is never any of the rather unsophisticated Lewis structures used to describe it. The difficulty arises because we cannot adequately portray the molecule, not because of the molecule itself.

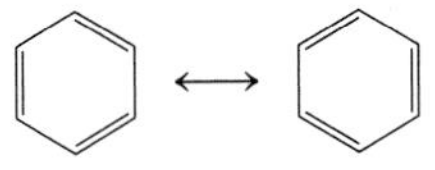

(a)

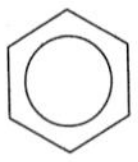

(b)

Figure 13.15
Two methods of representing the structure of benzene: (a) Kekulé structures; (b) hexagon with inscribed circle.

Kekulé's Dream

There is a story told that one evening Kekulé fell asleep while sitting in front of a fire. He dreamed about chains of atoms having the forms of twisting snakes. Suddenly one of the snakes caught hold of its own tail, forming a whirling ring. He awoke, freshly inspired, and spent the remainder of the night working on his now famous hypothesis. Kekulé is said to have written "Let us learn to dream, gentlemen, and then perhaps we shall learn the truth."

13.17
AROMATIC HYDROCARBONS: STRUCTURE AND NOMENCLATURE

There are many compounds chemically similar to benzene. Several of those discovered in the early days had pleasant odors and were known as aromatic compounds. The label *aromatic* is also used today: an **aromatic compound** simply means a compound that has, like benzene, an unusually stable ring of electrons. (All the nonaromatic hydrocarbons that we have considered—the alkanes, alkenes, etc.—are referred to collectively as **aliphatic compounds** to distinguish them from aromatic compounds. "Aliphatic" originally meant that the source of the compound was a fat. Today, however, it simply means "not aromatic.")

There is more than one system for naming aromatic hydrocarbons. Both common names and systematic names are encountered. Some aromatic compounds are referred to exclusively as derivatives of benzene, whereas others are more frequently denoted by their common names. Note that in the following structures, it is immaterial whether the substituent is written at the top, side, or bottom of the ring: a hexagon is symmetrical, and all positions are equivalent.

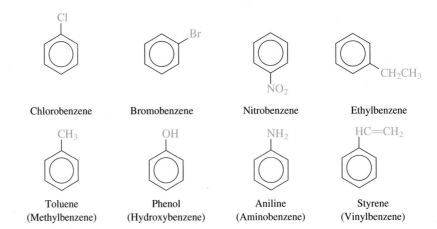

Chlorobenzene   Bromobenzene   Nitrobenzene   Ethylbenzene

Toluene   Phenol   Aniline   Styrene
(Methylbenzene)   (Hydroxybenzene)   (Aminobenzene)   (Vinylbenzene)

A complication arises when there is more than one substituent attached to a benzene ring. When this occurs, all the positions on the hexagon are no longer equivalent, and the relative positions of the substituents must be designated. In the case of a disubstituted benzene, one nomenclature system uses the prefixes **ortho** (*o-*), **meta** (*m-*), and **para** (*p-*). Ortho designates 1,2-disubstitution, meta designates 1,3-disubstitution, and para designates 1,4-disubstitution.

o-Chloronitrobenzene   *m*-Dibromobenzene   *p*-Fluoroiodobenzene

Alternatively, the ring is numbered and the substituent names are listed in alphabetical order. The first substituent is given the lowest number. When a common name is used, the carbon atom that bears the group responsible for the name is considered to be carbon number 1 (C-1).

m-Xylene   *m*-Chloroethylbenzene   *o*-Bromotoluene   *p*-Nitrophenol
(1,3-Dimethylbenzene)   (1-Chloro-3-ethylbenzene)   (2-Bromotoluene)   (4-Nitrophenol)

If more than two substituents occur, the numbering is determined by the requirement that the numbers give the smallest possible sum (substituents are listed in alphabetical order).

2-Bromo-4-fluoro-1-nitrobenzene   2,5-Dichlorophenol   3-Ethyl-4-iodotoluene   2,4,6-Trinitrotoluene
(TNT—an explosive)

Occasionally, an aromatic group is a substituent that is bonded to an aliphatic compound or to another aromatic ring. The general designation for an aromatic (**aryl**) group as a substituent is Ar just as R is used to represent an alkyl group. The most common aryl group is the one derived from benzene (C_6H_5—). It was given the name *phenyl*, derived from *pheno*, an old name for benzene.[2]

$$CH_3—CH—CH_2—CH_2—CH_2—CH_2—CH_3$$

Phenyl group   2-Phenylheptane

Mention should be made of some common aromatic hydrocarbons that are not substituted benzenes but are condensed benzene rings.

Naphthalene
mp 80 °C
bp 218 °C

Anthracene
mp 218 °C
bp 342 °C

Phenanthrene
mp 101 °C
bp 340 °C

These three substances are colorless, crystalline solids that are obtained from coal tar. Naphthalene has a pungent odor and is commonly used in mothballs and in moth flakes. Anthracene is an important starting material in the manufacture of certain dyes. A large group of naturally occurring substances, the steroids, contain the hydrogenated phenanthrene structure (see Section 20.10).

These polycyclic aromatic compounds do not exist in coal itself but are formed by the intense heating involved in the distillation of coal tar. For many years, it has been known that workers in coal-tar refineries are susceptible to a type of skin cancer known as tar cancer. Investigation has shown that a number of these polycyclic aromatic hydrocarbons have the ability to cause cancer when applied to the skin. Such compounds are called **carcinogens** (cancer producers). One of the most active carcinogenic compounds, benzpyrene, occurs in coal tar to the extent of 1.5%, and it has been isolated from cigarette smoke, automobile exhaust gases, and charcoal-broiled steaks. It is estimated that more than 1000 tons of benzpyrene is emitted into the air over the United States each year. Only a few milligrams of benzpyrene is required to induce cancer in experimental animals.

The mechanism by which these compounds cause cancer has not yet been elucidated. One hypothesis is that the active carcinogens are not the polycyclic hydrocarbons themselves but one or more of their metabolites. (As we shall learn in Chapter 24, metabolites are the products of chemical transformations in living cells.) Figure 13.16 indicates the

---

[2] This terminology is confusing because it would seem that the group derived from benzene should be called benzyl. The problem is compounded by the fact that another group is called benzyl. Replacement of one of the methyl hydrogens of toluene gives the *benzyl* group, $C_6H_5CH_2$—.

—CH_2—   —CH_2Br

Benzyl group   Benzyl bromide

Figure 13.16
Benzpyrene can be metabolized in the body to produce the active carcinogen.

Benzpyrene   A diolepoxide

Figure 13.16
Benzpyrene can be metabolized in the body to produce the active carcinogen.

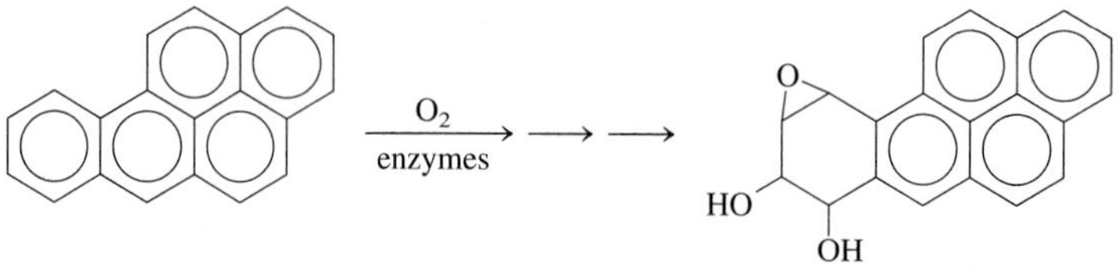

conversion of benzpyrene via a multistep oxidation sequence to yield a highly carcinogenic diolepoxide metabolite. To a certain extent, fused polycyclic hydrocarbons are formed whenever organic molecules are heated to high temperatures. It is the current belief that lung cancer is caused by the formation of carcinogenic compounds in the burning of cigarettes.

The aromatic hydrocarbons containing only one benzene ring are generally liquids, whereas the polyring aromatics are generally solids. They are all insoluble in water but soluble in organic solvents. Aromatic hydrocarbons are readily combustible. Unlike aliphatic hydrocarbons, which burn with a relatively clean flame, the aromatic compounds burn with a very sooty flame.

13.18
USES OF BENZENE AND BENZENE DERIVATIVES

Benzene is a component of gasoline, with unleaded gasolines having about 2% benzene. Exposure of most people to benzene comes from pumping gasoline, living in a house with an attached garage, or being around a cigarette smoker. For smokers, cigarettes overwhelm all other sources of benzene.

Most of the benzene used commercially comes from petroleum. Benzene is employed industrially as a starting material for the production of many other products (e.g., detergents, drugs, dyes, insecticides, plastics). Benzene was once widely used as an organic solvent, but we now know it to be a poisonous substance with both short- and long-term effects. Inhalation of large concentrations of benzene can cause nausea and even death due to respiratory or heart failure. Repeated exposure to benzene leads to a progressive disease in which the ability of the bone marrow to make new blood cells is eventually destroyed. This results in a condition called *aplastic anemia,* in which there is a decrease in the numbers of both the red and white blood cells.

Because of these hazards, many chemical laboratories have replaced benzene with toluene as a general solvent.[3] Toluene is used in the production of dyes, drugs, and explosives and as a solvent in lacquers. It is commonly used as a preservative for urine specimens, and it is added to fuels to improve their octane numbers. Trinitrotoluene (TNT), unlike nitroglycerin (see Section 14.7), is not sensitive to shock on jarring and must be exploded by a detonator.

Nitrobenzene is used extensively in the manufacture of aniline, the parent compound of many dyes and drugs. Phenol containing a small amount of water is a liquid, and in this form is referred to as carbolic acid. It is a good antiseptic and germicide, but its use is limited because of its toxicity (see Section 14.8).

---

[3]The maximum allowed concentration for an 8-hour exposure to toluene is 100 ppm, whereas the allowable exposure to benzene for the same period of time is 1 ppm. There are no toxic symptoms attributable to toluene until concentrations reach 200 ppm. Care must be taken in using toluene because most commercial toluene contains benzene as an impurity. If the concentration of benzene in toluene is less than 1%, the toluene is safe to use. *However,* toluene has been shown to cause birth defects, so pregnant women should avoid breathing toluene vapors. Prolonged inhalation of toluene should be avoided by everyone.

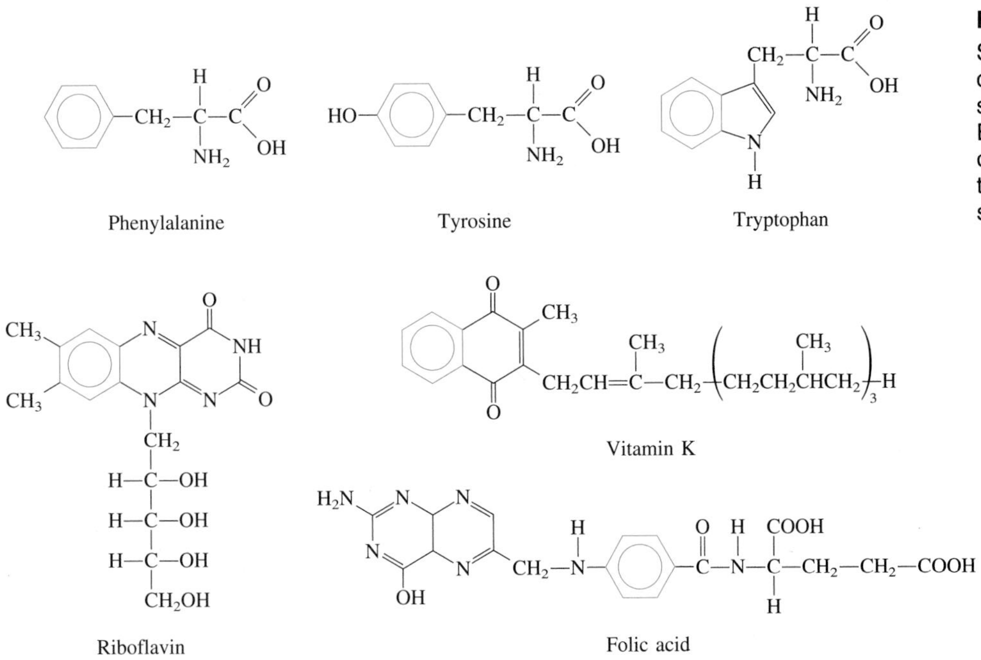

Figure 13.17
Some biologically important compounds that cannot be synthesized by animals. Each of these compounds contains a benzene ring that must be supplied from substances in the diet.

Phenylalanine

Tyrosine

Tryptophan

Vitamin K

Riboflavin

Folic acid

The xylenes are good solvents for grease and oil and are used for clearning slides and optical lenses of microscopes. *p*-Xylene is oxidized to terephthalic acid, which is then utilized in the production of Dacron. The xylenes are also used to raise the octane number of unleaded gasolines.

Substances containing the benzene ring are commonly found in both the animal and plant kingdoms, although they are more abundant in the latter. Plants have the ability to synthesize the benzene ring from carbon dioxide, water, and inorganic materials. Animals, on the other hand, are incapable of this synthesis but are dependent on benzenoid compounds for their survival. Therefore, the animal must obtain the compounds from the food that it ingests. Included among the aromatic compounds necessary for animal metabolism are the amino acids phenylalanine, tyrosine, and tryptophan and certain vitamins such as vitamin K, riboflavin, and folic acid (Figure 13.17). In addition, a great majority of drugs contain the benzene ring (see Special Topics E and F).

Xylene is the common name for dimethylbenzene. Hence, there are three xylene isomers: *o*-, *m*-, and *p*-dimethylbenzene.

EXERCISES

1. List three ways in which a "typical" organic compound differs from a "typical" inorganic one.

2. Define, illustrate, or give an example for each of the following terms.

   a. hydrocarbon
   b. alkane
   c. paraffin
   d. saturated
   e. unsaturated
   f. substituent
   g. alkene
   h. alkyne
   i. alkyl group
   j. isomers
   k. polymer
   l. aromatic compound
   m. aliphatic compound
   n. phenyl group
   o. alkyl halide
   p. chlorofluorocarbon
   q. homologous series
   r. carcinogen

3. Classify the following compounds as organic or inorganic.

   a. C_6H_{10}
   b. $CoCl_2$
   c. $C_{12}H_{22}O_{11}$
   d. CH_3NH_2
   e. $NaNH_2$
   f. $Cu(NH_3)_6Cl_2$

4. Which member of each pair has a higher melting point?

   a. CH_3OH and $NaOH$
   b. CH_3Cl and KCl
   c. $C_{20}H_{42}$ and $C_{40}H_{82}$
   d. CH_4 and LiH

5. You find a jar without a label containing a solid material. The substance melts at 48 °C. It ignites readily and burns cleanly. The substance is insoluble in water and floats on the surface of the water. Is the substance likely to be organic or inorganic?

6. How many carbon atoms are there in each of the following?
   a. ethane   **b.** heptane
   c. butane   **d.** nonane

7. Indicate whether the structures in each set represent the same compound or isomers.

   a. CH_3CH_3 and CH_3 with CH_3

   b. $CH_3\overset{CH_3}{\underset{|}{C}}H_2$ and $CH_3CH_2CH_3$

   c. $CH_3CH_2\overset{CH_3}{\underset{|}{C}}HCH_2CH_3$ and $CH_3\overset{CH_3}{\underset{|}{C}}HCH_2CH_2CH_3$

   d. $CH_3\overset{CH_3}{\underset{|}{C}}HCH_2CH_3$ and $CH_3CH_2\overset{CH_3}{\underset{|}{C}}H{\underset{CH_3}{|}}$

   e. $CH_3CH_2\overset{CH_3}{\underset{|}{C}}H-CH_2$ and $\overset{CH_3}{\underset{|}{C}}H_2CH_2\overset{CH_3}{\underset{|}{C}}HCH_3$

8. Draw the structural formulas of the four-carbon alkanes (C_4H_{10}). Identify butane and isobutane, and give the IUPAC name for the latter.

9. Write structures for the five isomeric hexanes (C_6H_{14}). Name each by the IUPAC system.

10. Draw the following alkyl groups.
   a. ethyl   **b.** isopropyl

11. Classify each of the following compounds as saturated or unsaturated.
   a. $CH_3\overset{}{\underset{|}{C}}=CH_2$ with CH_3 **b.** $CH_3C{\equiv}CCH_3$

   c. $CH_3-\overset{CH_3}{\underset{CH_3}{\overset{|}{\underset{|}{C}}}}-CH_3$ **d.** ⬠

12. Indicate whether the structures in each set represent the same compound or isomers.
   a. $CH_3CH{=}CHCH_3$ and $CH_3CH_2CH{=}CH_2$

   b. $CH_3\overset{CH_3}{\underset{|}{C}}{=}\overset{CH_3}{\underset{|}{C}}CH_3$ and $CH_3C{=}CCH_3$ with $\underset{CH_3}{|}$ $\underset{CH_3}{|}$

   c. $CH_3CH_2CH_2CH{=}\overset{CH_3}{\underset{|}{C}}CH_3$ and $CH_3\overset{CH_3}{\underset{|}{C}}HCH{=}CHCH_2CH_3$

   d. $CH_2{=}\overset{CH_3}{\underset{|}{C}}CH_2CH_3$ and $CH_2{=}\overset{CH_2CH_3}{\underset{|}{C}}CH_3$

e.
$$\underset{CH_3CH_2}{\overset{CH_3}{\diagdown}}C{=}C\underset{CH_2CH_3}{\overset{CH_2CH_3}{\diagup}} \text{ and}$$
$$\underset{CH_3}{\overset{CH_3CH_2}{\diagdown}}C{=}C\underset{CH_2CH_3}{\overset{CH_2CH_3}{\diagup}}$$

13. Give structural formulas for the following.
   a. heptane   **b.** 3-methylpentane
   c. 2,2,5-trimethylhexane   **d.** 4-ethyl-3-methyloctane

14. Write structures for the two isomers that have the molecular formula C_3H_7Br. Give the common name and the IUPAC name of each.

15. Write structures for the four isomers that have the molecular formula C_4H_9Br. Give the common name and the IUPAC name of each.

16. Draw structural formulas for the following.
   a. methylene chloride   **b.** chloroform
   c. carbon tetrachloride   **d.** ethyl chloride

17. Draw structural formulas for the following.
   a. acetylene   **b.** 3-ethyl-2-pentene
   c. 3-isopropyl-1-hexene   **d.** 2,3-dimethyl-2-butene

18. Draw structural formulas for the following.
   a. 1,1-dimethylcyclobutane
   b. cyclobutene

19. Name the following compounds.
   a. ⬡   **b.** (cyclohexene)
   c. (cyclopentane)$-CH_3$ **d.** CH_2CH_3 (benzene)

20. Which is the molecular formula for each of the following compounds?
   a. (cyclohexene)   **b.** (cyclopentene)CH_2CH_3
   c. △CH_3 **d.** (benzene)CH_3

21. Indicate whether each compound is aromatic or aliphatic.
   a. (benzene)   **b.** (cyclohexane)
   c. (cyclohexadiene)   **d.** (naphthalene)

22. Identify each substitution pattern as meta, ortho, or para.

a.   **b.**   **c.**

23. Write structural formulas for the following.
 a. 2-methylpentane
 b. 4-ethyl-2-methylhexane
 c. 2,2,3,3-tetramethylbutane
 d. 4-ethyl-3-isopropyloctane

24. Write structural formulas for the following.
 a. 2-methyl-2-pentene
 b. 5-methyl-1-hexene
 c. 2-ethyl-1-butene
 d. 2,4,6,6-tetramethyl-2-heptene

25. What is wrong with each of the following names? Give the structure and the correct name for each compound.
 a. 2-dimethylpropane
 b. 2,3,3-trimethylbutane
 c. 2,4-diethylpentane
 d. 3,4-dimethyl-5-propylhexane

26. What is wrong with each of the following names? Give the structure and correct name for each compound.
 a. 2-methyl-4-heptene
 b. 2-ethyl-3-hexene
 c. 2,2-dimethyl-3-pentane
 d. 4-bromocyclobutene

27. Draw structural formulas for the following compounds.
 a. toluene   **b.** *m*-diethylbenzene
 c. naphthalene   **d.** *p*-dichlorobenzene
 e. 2,4-dinitrotoluene   **f.** 1,2,4-trimethylbenzene

28. Draw structural formulas for the following compounds.
 a. cyclohexane   **b.** cyclopentene

29. Name the following compounds by the IUPAC system.
 a. $CH_3CH_2CH(CH_3)CH_2CH_3$
 b. $(CH_3)_2CHCH(CH_3)_2$
 c. $CH_3CH(CH_2CH_3)_2$
 d. $(CH_3)_3CCH_2C(CH_3)_2CH_2CH_3$

30. Name the following compounds by the IUPAC system.
 a. $CH_2{=}C(CH_3)CH_2CH_2CH_3$
 b. $CH_3CH_2CH{=}C(CH_3)_2$
 c. $(CH_3)_2C{=}CHCH_2CH(CH_3)_2$
 d. $(CH_3)_3CCH{=}C(CH_3)CH_2CH_3$

31. Name the following compounds by the IUPAC system.

 a. —CH₃   **b.** H₃C———CH₂CH₃

 CH₂CH₃

 c. ▢   **d.** ———CH₂CH₃

32. Name the following compounds by the IUPAC system.
 a. $CH_3CHBrCHBrCH_3$
 b. $CH_3CHClCH_2CH{=}CHCH_2CHCl_2$

c.   Cl   **d.**   Br

 Cl   CH₃

33. Name the following compounds by the IUPAC system.
 a. CH_2CH_3 **b.** $CH(CH_3)_2$

 c. NO_2 **d.** CH_3

 O_2N NO_2 NO_2

34. Give the reagents required for the following transformations.
 a. $H_2C{=}CHCH_3 \longrightarrow CH_3CH_2CH_3$

 b.   ⬡ → ⬡ OH

 c. $CH_3CH{=}CHCH_3 \longrightarrow CH_3\overset{OH}{\underset{}{C}}HCH_2CH_3$

 d. $CH_2{=}CHCH_3 \longrightarrow \overset{Cl\ Cl}{CH_2CHCH_3}$

35. What starting materials are required to complete the transformations shown?

 a. ? $\xrightarrow{H_2 / Ni}$ ⬡

 b. ? $\xrightarrow{Cl_2}$ H—C—C—H (with Cl, Cl top; H, H bottom)

 c. ? $\xrightarrow[H_2SO_4]{H_2O}$ CH_3CHCH_3 (OH)

 d. ? $\xrightarrow[H_2SO_4]{H_2O}$ ⬡—OH

36. What compounds contain fewer carbons than propane and are homologs of propane?

37. Which member of each pair will have the higher boiling point?
 a. pentane or butane
 b. $(CH_3)_2CHCH(CH_3)_2$ or $CH_3(CH_2)_4CH_3$
 c. cyclopentane or cyclohexane
 d. $CH_3(CH_2)_5CH_3$ or $CH_3(CH_2)_7CH_3$

38. Which member of each pair will have the higher melting point?
 a. pentane or butane

   b. neopentane or pentane

   c. cyclopentane or cyclohexane

   d. $CH_3(CH_2)_5CH_3$ or $CH_3(CH_2)_7CH_3$

39. Using appropriate examples, distinguish between the terms in the following pairs.

   a. alkyl group and alkane

   b. alkyl group and aryl group

   c. common name and IUPAC name

   d. propyl group and isopropyl group

   e. straight-chain alkane and branched-chain alkane

40. Distinguish between the terms in each pair.

   a. natural gas and bottled gas

   b. physical properties and chemical properties

   c. exothermic reaction and endothermic reaction

41. Complete the following equations.

   a. $(CH_3)_2C{=}CH_2 + Br_2 \longrightarrow$

   b. (ring)$-CH_3 + Cl_2 \longrightarrow$

   c. $CH_2{=}C(CH_3)CH_2CH_3 + H_2 \xrightarrow{Ni}$

   d. $CH_2{=}CHCH{=}CH_2 + 2\ H_2 \xrightarrow{Ni}$

   e. $(CH_3)_2C{=}C(CH_3)_2 \xrightarrow[H_2SO_4]{H_2O}$

   f. (ring)$-CH_2CH_3 \xrightarrow[H_2SO_4]{H_2O}$
     CH_2CH_3

42. Three isomeric pentenes, X, Y, and Z, can be hydrogenated to 2-methylbutane. Addition of chlorine to Y gives 1,2-dichloro-3-methylbutane, and 1,2-dichloro-2-methylbutane is obtained from Z. Write the structural formulas for the three isomers.

43. Pentane and 1-pentene are both colorless, low-boiling liq-

uids. Give a simple test that would distinguish the two compounds. Indicate what you would observe.

44. What is the danger in swallowing a liquid alkane?

45. Distinguish between lighter and heavier liquid alkanes in terms of their effect on the skin.

46. Describe a physiological effect of some polycyclic aromatic hydrocarbons.

47. What are some of the hazards associated with the use of benzene? Why is toluene thought to be less toxic?

48. What physiological effect is shared by ethene and cyclopropane?

49. Write equations for the complete combustion of each of the following.

   a. natural gas (methane)

   b. a typical petroleum hydrocarbon (such as octane)

50. Write equations for the incomplete combustion that forms carbon monoxide from each of the substances in Exercise 49.

51. The complete combustion of benzene forms carbon dioxide and water.

$$C_6H_6 + O_2 \longrightarrow CO_2 + H_2O$$

Balance the equation. What weight of carbon dioxide is formed by the complete combustion of 39 g of benzene?

52. The density of gasoline is 0.69 g/mL. On the basis of the complete combustion of octane, calculate the amount of carbon dioxide and water formed per gallon (3.78 L) of gasoline used in an automobile.

53. List several uses of polyethylene.

54. What is addition polymerization?

55. What structural feature usually characterizes molecules used as monomers in addition polymerization?

56. Give the structure of the monomer from which polystyrene is made (see Table 13.11).

Petroleum

D.1
NATURAL GAS AND GASOLINE

Compounds of carbon are the basis of all life on our planet. They are essential to all life processes and constitute by far our major energy source. They are the basis of many of our structural building materials and of nearly all our medicines. Some organic compounds are still obtained from plants and animals, but most come ultimately from the fossilized carbon materials coal and petroleum. Petroleum is a complex mixture of hydrocarbons produced by the decomposition of animal and vegetable matter that has been entrapped in the earth's crust for long periods of time. Most of the **petrochemicals** so vital to our modern economy are derived from these alkanes. A portion of the remaining petrochemicals, the aromatics are derived from coal.

Petroleum, as it comes from the ground, is of limited use. So that it will better suit our needs, we separate it into fractions by boiling it in a distillation column (Figure D.1). The lighter molecules, those with one to four carbon atoms each, come off the top of the column. The next fraction contains, for the most part, molecules having from five to twelve carbon atoms. These and other fractions are listed in Table D.1.

Crude oil and natural gas are the liquid and gaseous components of petroleum. Natural gas is about 80% methane and 10% ethane, and the remaining 10% is a mixture of the higher alkanes. Methane is a product of the bacterial decay of plant and marine organisms that have become buried beneath the earth's surface. It is also produced by the microbial decomposition of organic matter in sewage treatment plants. Because methane was first isolated in marshes, it was given the name "marsh gas." Methane is used primarily as a cooking fuel (and in bunsen burners). Natural gas is the cleanest of the fossil fuels because it contains the least amount of sulfur compounds. There is very little sulfur dioxide produced when natural gas is combusted.

Propane and the butanes are familiar fuels. They are usually supplied under pressure in tanks. Although they are gases at ordinary temperatures and under normal atmospheric pressure, they are liquefied under pressure and are sold as liquefied petroleum (LP) gas. Butane, liquefied

Figure D.1
The fractional distillation of petroleum.

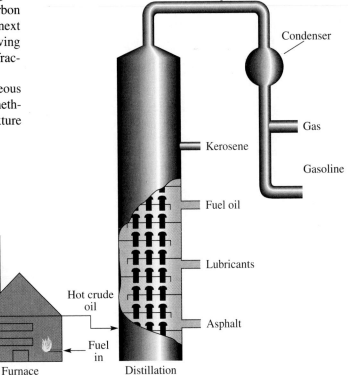

339

Table D.1
Typical Petroleum
Fractions

Fraction	Typical Range of Hydrocarbons	Approximate Range of Boiling Point (°C)	Typical Uses
Natural gas	CH_4 to C_4H_{10}	Less than 40	Fuel, starting materials for plastics
Gasoline	C_5H_{12} to $C_{12}H_{26}$	40–200	Fuel, solvents
Kerosene	$C_{12}H_{26}$ to $C_{16}H_{34}$	175–275	Diesel fuel, jet fuel, home heating; cracking to gasoline
Heating oil	$C_{15}H_{32}$ to $C_{18}H_{38}$	250–400	Industrial heating, cracking to gasoline
Lubricating oil	$C_{17}H_{36}$ and up	Above 300	Lubricants
Residue	$C_{20}H_{42}$ and up	Above 350 (some decomposition)	Paraffin, asphalt

under pressure, can be seen in disposable butane cigarette lighters. When the release lever is pressed on these lighters, the butane comes under atmospheric pressure, and some of it vaporizes and is ignited by a spark.

Since gasoline is generally the fraction most in demand, the fractions with higher boiling points are often in excess supply. These can be converted to gasoline by heating in the absence of air. This process, called **cracking,** breaks the big molecules apart. The process is illustrated in Figure D.2, where $C_{14}H_{30}$ is used as an example. Not only does cracking convert some of the molecules into those in the gasoline range (those having five to twelve carbon atoms), but it also results in a variety of useful by-products. The unsaturated hydrocarbons are starting materials for the manufacture of many plastics, detergents, and drugs—indeed a whole host of petrochemicals. Present and future shortages of petroleum will mean a great deal more than just scarce, high-priced gasoline.[1]

The cracking process described here is a crude, yet illustrative, example of how chemists modify nature's materials to meet our needs and desires. Starting with coal tar or petroleum, the chemist can create a dazzling array of substances with a wide variety of properties. These include plastics, painkillers, antibiotics, stimulants, depressants, and detergents, to name just a few. Many of these materials are discussed in this text.

Gasoline, like the petroleum from which it is derived, is a mixture of hydrocarbons. The typical alkanes in gasoline have formulas ranging from C_5H_{12} to $C_{12}H_{26}$. Since there are many isomeric forms, particularly for the higher members of the group, we see that gasoline is an exceedingly complex mixture of alkanes. There are also small amounts of other kinds of hydrocarbons present, and even some sulfur- and nitrogen-containing compounds. The gasoline fraction of petroleum as it comes from the distillation column is called **straight-run gasoline.** It doesn't burn very well in modern, high-compression automobile engines. Chemists have learned how to modify it in a variety of ways to make it burn better.

Figure D.2
Formulas of some of the products formed when $C_{14}H_{30}$ (a typical molecule in kerosene) is cracked. You need only note that a great variety of hydrocarbons with fewer carbon atoms are formed.

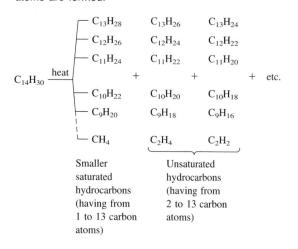

[1]In today's energy-conscious world, the unit of commerce on the oil market is the **barrel.** One barrel of crude oil equals 42 gallons. Of this, approximately 45% is converted into gasoline, 30% into fuel oil for heating purposes, and 10% into jet fuel. Only about 5% of this precious liquid is used in the petrochemical industry to manufacture detergents, dyes, fertilizers, pesticides, plastics, etc. The remainder is sold as aviation gasoline, lubricating oils, greases, and asphalt.

D.2
THE OCTANE RATING OF GASOLINE

Early in the development of the automobile engine, scientists learned that some types of hydrocarbons burned more evenly and were less likely to ignite prematurely than others. Ignition before the piston was in proper position led to a knocking in the engine. Scientists soon were able to correlate good performance with a branched-chain structure in hydrocarbon molecules. An arbitrary performance standard, called the **octane rating,** was established in 1927. The best performer in a laboratory test engine was found to be a compound that became known as isooctane (it was one of 18 isomeric octanes tested). Isooctane was assigned a value of 100 octane. An unbranched-chain compound, heptane, was found to cause a very bad knock. It was given an octane rating of 0. A gasoline rated 90 octane was one that performed the same as a mixture that was 90% isooctane and 10% heptane.

$$CH_3-\underset{\underset{CH_3}{|}}{\overset{\overset{CH_3}{|}}{C}}-CH_2-\underset{\overset{|}{CH_3}}{CH}-CH_3 \qquad CH_3CH_2CH_2CH_2CH_2CH_2CH_3$$

Isooctane   Heptane

During the 1930s, chemists discovered that the octane rating of gasoline could be improved by heating gasoline in the presence of catalysts such as sulfuric acid (H_2SO_4) and aluminum chloride ($AlCl_3$). This increase in octane rating was attributed to a conversion (**isomerization**) of a part of the unbranched structures to highly branched molecules. For example, heptane molecules could be isomerized to a branched structure.

$$CH_3CH_2CH_2CH_2CH_2CH_2CH_3 \xrightarrow[\text{heat}]{H_2SO_4}$$

$$CH_3CH_2-\underset{\underset{CH_3}{|}}{\overset{\overset{CH_3}{|}}{CH}}-CH-CH_3$$

Chemists also are able to combine small hydrocarbon molecules (below the gasoline range) into larger ones more suitable for use as fuel.

Certain chemical substances were discovered that, when added in small amounts, substantially improved the antiknock quality of gasoline. Chief among these additives was tetraethyllead. This compound, when added in amounts as small as 1 mL per liter of gasoline (about 1 part per thousand), would increase the octane rating from approximately 55 to 90 or more.

$$C_2H_5-\underset{\underset{C_2H_5}{|}}{\overset{\overset{C_2H_5}{|}}{Pb}}-C_2H_5$$

Tetraethyllead

Lead is toxic, however.[2] Large amounts of it have entered the environment through the combustion of leaded gasoline in automobiles. Further, lead fouls the catalytic converters used in modern automobiles. Tetraethyllead has been (or soon will be) phased out as an octane booster in most of the industrialized nations.

Scientists have developed a number of ways to get high octane ratings in unleaded fuels. For example, petroleum refineries use **catalytic reforming** to convert low-octane alkanes to high-octane aromatic compounds. For example, hexane (with an octane number of 25) is converted to benzene (with an octane number of 106).

$$CH_3CH_2CH_2CH_2CH_2CH_3 \xrightarrow[\text{heat}]{\text{catalyst}} \bigcirc + \ 4\ H_2$$

           (C_6H_{14})   (C_6H_6)

Octane boosters to replace tetraethyllead also have been developed. Methyl *tert*-butyl ether (Section 14.9) is perhaps the most important. Methanol, ethanol, and *tert*-butyl alcohol also are used. None of these is nearly as effective as tetraethyllead in boosting the octane rating. They must therefore be used in fairly large quantities. The amount that can be used in gasoline is limited by solubility problems. Methanol in excess of 5% and ethanol in excess of 10% tend to separate from the gasoline, especially if moisture gets into the fuel.

To recapitulate briefly, by World War II, American scientists were able to convert parts of nearly all petroleum fractions into gasoline. Furthermore, they were able to improve octane ratings substantially by rearranging the molecules and adding tetraethyllead. It was this high-performance fuel, as much as or more than superior machines, that made possible the Allied victory in the air war during World War II.

---

[2] Lead is especially toxic to the brain. Even small amounts can cause learning disabilities in children.

Chapter 14
ALCOHOLS, PHENOLS, AND ETHERS

Alcohol is obtained by the fermentation process. Grapes are the principal source of the alcohol found in wine.

The families of organic compounds discussed in this chapter occur widely in nature. The human race has been quick to adapt these materials to its own use. The earliest written histories record the isolation and use by primitive peoples of the compound known as alcohol. According to Genesis, Noah planted a vineyard after the flood, drank wine from its grapes, and became drunk.

Human ingenuity may have reached some sort of peak in finding sources of *aqua vitae,* the water of life. Alcohol has been obtained from the fermentation of fruits, grains, potatoes, rice, and even cacti. It was prescribed as medicine in the twelfth century but has been most frequently used without such justification. What we know as alcohol is actually only one member of a family known by the same name. The family includes among its members such familiar substances as cholesterol and the carbohydrates.

The name of another family of organic compounds considered in this chapter, the ethers, has become almost synonymous with anesthesia. And anyone who has ever been in a hospital would recognize the pungent, antiseptic odor of phenol, the simplest member of the third family to be introduced in this chapter.

Why are we taking up three different families in this one chapter? We do so because each can be considered an organic derivative of water. Water is by far the most important inorganic compound we have studied. It should not be surprising, therefore, that organic compounds derived from water are also of critical importance to life and health. Some, like the carbohydrates, deserve to be, and are, considered in a separate chapter (Chapter 19). For now, we shall, as usual, deal with the simpler members of each family.

14.1
THE FUNCTIONAL GROUP

Alcohols, phenols, and ethers can be viewed as organic derivatives of water. Consider the water molecule.

$$\underset{H}{\overset{O}{\diagdown}}_{H}$$

It is a bent molecule with the central oxygen attached to two hydrogen atoms. If one of these hydrogens were removed and replaced with an alkyl group (R—), we would have

$$\underset{R}{\overset{O}{\diagdown}}_{H}$$

This is the general formula for the **alcohol** family. The alkyl group may be methyl, ethyl, isopropyl, or an aliphatic group too complicated to have a simple name. As long as the carbon attached to the **hydroxyl group** (—OH) is aliphatic, the compound is an alcohol.

If the hydroxyl group is attached directly to an aromatic ring, a different family of compounds is produced. Compounds in which an aryl group (Ar—) is attached to a hydroxyl group are called **phenols.**

$$\underset{Ar}{\overset{O}{\diagdown}}_{H}$$

The chemistry of the phenols is sufficiently different from that of the alcohols to justify treating the two classes of compounds as separate, if closely related, families. Nonetheless, for both families, the chemistry is largely determined by the hydroxyl group. Nearly all the characteristic reactions of alcohols and many of those of the phenols take place at the hydroxyl group. Even the physical properties are determined to a large extent by the presence of the hydroxyl group. Such a group of atoms, which confers characteristic chemical and physical properties on a family of organic compounds, is called a **functional group.** The hydroxyl group is only one functional group. We have already encountered others. The carbon–carbon double bond (C=C) in alkenes and the carbon–carbon triple bond in alkynes (C≡C) are functional groups. In both instances, these structural features confer on the members of the families a particular chemical reactivity. For example, alkenes and alkynes tend to undergo addition reactions. The halogens in halogenated hydrocarbons are functional groups, although we did not consider in detail the particular reactions associated with these groups. The alkanes are characterized by their *lack* of a distinct functional group. Other functional groups will serve as unifying concepts for the next three chapters. Some of the more important functional groups are listed in Table 14.1. For ready reference, this table is also reproduced on the inside back cover.

Table 14.1
Selected Organic
Functional Groups

Name of Class	Functional Group	General Formula(s) of Class
Alkane	None	R—H
Alkene	—C=C—	R—C=C—R (with R groups on the double-bond carbons)
Alkyne	—C≡C—	R—C≡C—R
Alcohol	—C—O—H	R—O—H
Ether	—C—O—C—	R—O—R
Aldehyde	O‖—C—H	O‖ R—C—H
Ketone	O‖—C—	O‖ R—C—R
Amine	—C—N—	H‖ R—N—H R—N—R (H) R—N—R (R)
Carboxylic acid	O‖—C—O—H	O‖ R—C—O—H
Ester	O‖—C—O—C—	O‖ R—C—O—R
Amide	O‖—C—N—	O‖ R—C—N—H (H) R—C—N—R (H) R—C—N—R (R)

14.2
CLASSIFICATION AND NOMENCLATURE OF ALCOHOLS

The properties of alcohols depend on the structural arrangement of the carbon atoms in the molecule. Alcohols can be grouped into three classes that serve to distinguish these different structural arrangements from one another. The classes are known as primary, secondary, and tertiary, and an alcohol is classified according to the type of carbon atom to which the hydroxyl group is attached.

1. A **primary (1°) carbon atom** is a carbon atom that is attached directly to only one other carbon.

Example

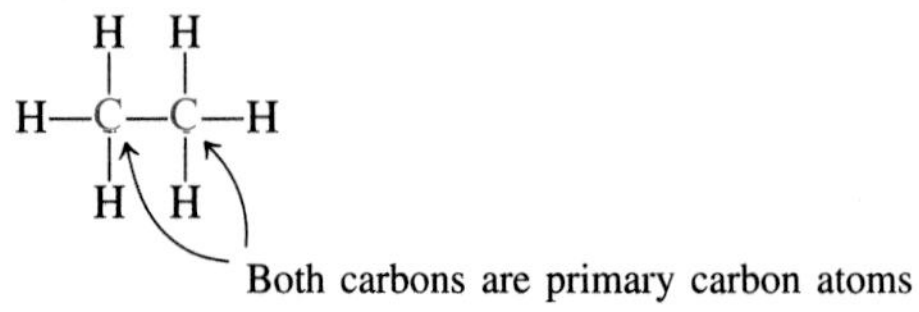

Both carbons are primary carbon atoms

2. A **secondary (2°) carbon atom** is one that is attached directly to two other carbon atoms.

Example

A secondary carbon atom

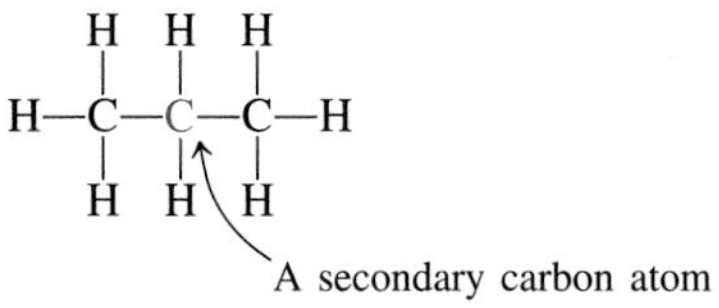

3. A **tertiary (3°) carbon atom** is one that is attached directly to three other carbon atoms.

Example

A tertiary carbon atom

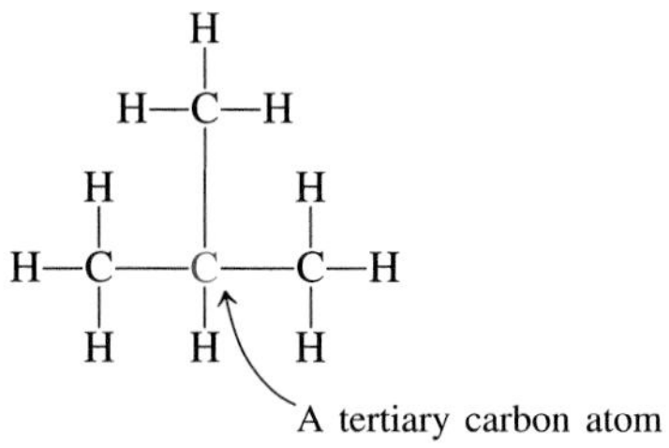

Therefore, a ***primary alcohol*** is one in which the hydroxyl group is attached to a primary carbon atom, a ***secondary alcohol*** is one whose hydroxyl group is located on a secondary carbon atom, and a ***tertiary alcohol*** has its hydroxyl group bonded to a tertiary carbon. Table 14.2 presents the nomenclature and classification of some of the simpler alcohols.

Table 14.2
Classification and Nomenclature of Some Alcohols

Structural Formula	Type of Alcohol	Common Name	IUPAC Name
CH_3OH	Primary	Methyl alcohol	Methanol
CH_3CH_2OH	Primary	Ethyl alcohol	Ethanol
$CH_3CH_2CH_2OH$	Primary	Propyl alcohol	1-Propanol
$CH_3\overset{\text{OH}}{\underset{\vert}{C}}HCH_3$	Secondary	Isopropyl alcohol	2-Propanol
$CH_3CH_2CH_2CH_2OH$	Primary	Butyl alcohol	1-Butanol
$CH_3CH_2\overset{\text{OH}}{\underset{\vert}{C}}HCH_3$	Secondary	sec-Butyl alcohol	2-Butanol
$CH_3\overset{\text{CH}_3}{\underset{\vert}{C}}HCH_2OH$	Primary	Isobutyl alcohol	2-Methyl-1-propanol
$CH_3-\overset{\text{OH}}{\underset{\underset{\text{CH}_3}{\vert}}{\overset{\vert}{C}}}-CH_3$	Tertiary	tert-Butyl alcohol	2-Methyl-2-propanol
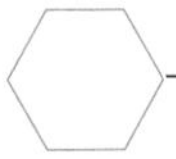 ⬡—OH	Secondary	Cyclohexyl alcohol	Cyclohexanol
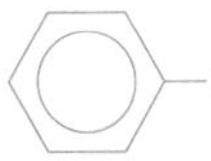 ⬡—CH_2OH	Primary	Benzyl alcohol	Phenylmethanol

As shown in the table, the common names of the lower members of the alcohol series are formed in a manner similar to that used for the naming of the alkyl halides. The name of the alkyl group is followed by the word *alcohol* to indicate the presence of the hydroxyl group. The IUPAC system is used in naming the higher homologs. In the IUPAC system, the designations primary, secondary, and tertiary are unnecessary and have no significance. Alcohols are named under the IUPAC system as follows.

$$\underset{\text{2-Methyl-2-butanol}}{CH_3CH_2\overset{\displaystyle OH}{\underset{\displaystyle CH_3}{C}}CH_3}$$

$$\underset{\text{3,5-Dimethyl-3-hexanol}}{\underset{6\quad5\quad4\quad3|2\quad1}{CH_3\overset{\displaystyle CH_3}{C}HCH_2\overset{\displaystyle OH}{\underset{\displaystyle CH_3}{C}}CH_2CH_3}}$$

$$\underset{\text{6-Methyl-3-heptanol}}{\overset{7\quad6\quad5\quad4\quad3\quad2\quad1}{CH_3\underset{\displaystyle CH_3}{C}HCH_2CH_2\underset{\displaystyle OH}{C}HCH_2CH_3}}$$

1. The longest continuous chain of carbons containing the —OH group is taken as the parent compound.
2. The chain is numbered from the end closer to the hydroxyl group. The number that then appropriately indicates the position of the hydroxyl group is prefixed to the name of the parent hydrocarbon.
3. The *-e* ending of the parent alkane is replaced by the suffix *-ol*, the generic ending for the alcohols.
4. If more than one hydroxyl group appears in the same molecule (polyhydroxy alcohols), the suffixes *-diol*, *-triol*, etc., are used. In these cases, the *-e* ending of the parent alkane is retained.

2-Bromo-4-chlorocyclopentanol

$$\underset{\text{(Ethylene glycol)}}{\underset{\text{1,2-Ethanediol}}{H-\overset{\displaystyle OH}{\underset{\displaystyle H}{C}}-\overset{\displaystyle OH}{\underset{\displaystyle H}{C}}-H}}$$

$$\underset{\text{(Glycerol)}}{\underset{\text{1,2,3-Propanetriol}}{H-\overset{\displaystyle H}{\underset{\displaystyle OH}{C}}-\overset{\displaystyle H}{\underset{\displaystyle OH}{C}}-\overset{\displaystyle H}{\underset{\displaystyle OH}{C}}-H}}$$

EXAMPLE 14.1

Give the IUPAC name for

$$\overset{10\quad9\quad8\quad7\quad6\quad5\quad4\quad3\quad2\quad1}{CH_3CH_2\underset{\displaystyle CH_3}{C}HCH_2\underset{\displaystyle CH_3}{C}HCH_2CH_2\underset{\displaystyle OH}{C}HCH_2CH_3}$$

SOLUTION

The carbons are numbered from the end closer to the —OH group. The name is 6,8-dimethyl-3-decanol (not 3,5-dimethyl-8-decanol).

Practice Exercise

Give the IUPAC name for

14.3
PHYSICAL PROPERTIES OF ALCOHOLS

It was stated earlier that the alcohols can be considered to be derivatives of water. This relation becomes particularly apparent, especially for the lower homologs, in a discussion of the physical and chemical properties of the alcohols. Remember that methane, ethane, and propane are gases and are insoluble in water. In contrast, methanol,

Table 14.3
Comparison of Boiling Points and Molecular Weights

Formula	Name	MW	Boiling Point (°C)
CH_4	Methane	16	-164
HOH	Water	18	100
C_2H_6	Ethane	30	-89
CH_3OH	Methanol	32	65
C_3H_8	Propane	44	-42
CH_3CH_2OH	Ethanol	46	78
C_4H_{10}	Butane	58	-1
$CH_3CH_2CH_2OH$	1-Propanol	60	97

ethanol, and propanol are liquids and are completely miscible with water. Therefore, replacement of a single hydrogen atom with a hydroxyl group brings about marked changes in solubility and physical state of the molecules. These differences result from the hydrogen-bonding capabilities of the alcohols. *Hydrogen bonding* (Section 8.4) is responsible for the intermolecular attractions between alcohol molecules. Even the lightest homolog of the series exists as a liquid at room temperature. Alcohols can form hydrogen bonds with water molecules as well, and so the lower homologs of the series are water-soluble. Table 14.3 lists the molecular weights and the boiling points of some common compounds. The table shows that substances that have similar molecular weights do not always have similar boiling points.

In the case of the alcohols, the relatively high boiling points are a direct result of strong intermolecular attractions. Recall that a boiling point is a rough measure of the amount of energy necessary to separate a liquid molecule from its nearest neighbors. If the nearest neighbors of a molecule are attached to that molecule by means of hydrogen bonds, a considerable amount of energy must be supplied to break those bonds. Only then can an individual molecule escape from the liquid into the gaseous state.

Figure 14.1 illustrates hydrogen bonding in water and in the alcohols. This schematic representation reveals why water boils at a higher temperature than methyl alcohol, even though water is a lighter molecule. The oxygen atom and both hydrogen atoms of the water molecule participate in hydrogen bonding to three, or even four, adjacent water molecules. Alcohols result from the replacement of one hydrogen atom of water with an alkyl group. The alkyl group does not participate in hydrogen bonding, and so the alcohol is associated to only two other alcohol molecules. More energy is required to disrupt three or four intermolecular bonds than two, and thus greater energy is needed to vaporize the water. (The energy required to break a hydrogen bond is about 5 kcal/mol. Although this is distinctly less than the energy required to break any of the intramolecular bonds in water or alcohol, it is still an appreciable amount of energy. Its significance is evidenced by the boiling point differences.)

Polarity and hydrogen bonding are significant factors in the water solubility of alcohols. A common expression among chemists is "like dissolves like," implying that polar solvents will dissolve polar solutes, and nonpolar solvents will dissolve nonpolar solutes. Care must be taken, however, not to apply this generalization haphazardly to all cases. All alcohol molecules are polar, yet not all alcohols are water-soluble. On the other hand, all alcohols are soluble in most of the common nonpolar solvents (methylene chloride, ether, hexane). Only the lower homologs of the series have an appreciable solubility in water. As the length of the carbon chain increases, water solubility decreases.

The differences in water solubility can be explained in the following manner. The hydroxyl group confers polarity and water solubility upon the alcohol molecule. The alkyl group confers nonpolarity and water insolubility. Whenever the hydroxyl group comprises

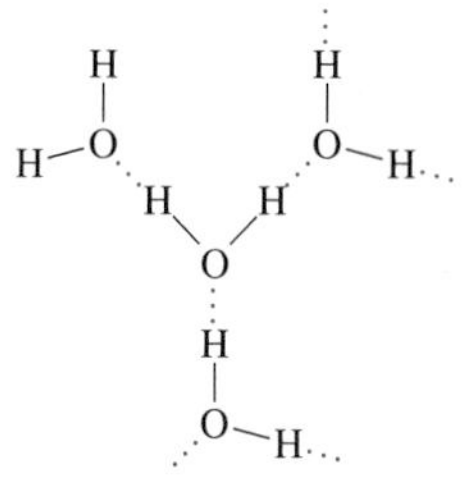

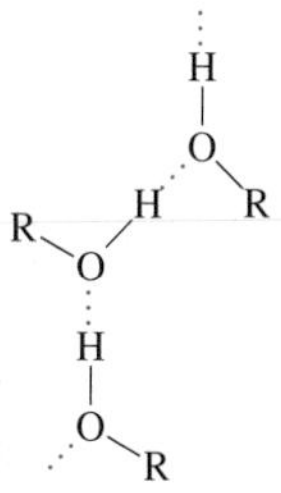

Figure 14.1
Intermolecular hydrogen bonding (a) in water and (b) in alcohol.

H—O··H—O···H—O
  H   CH₃   H

a substantial portion of a molecule, the molecule will be water-soluble. (The hydroxyl group can be thought of as dragging the remainder of the molecule into the water structure.) As the size of the alkyl group increases, however, the alcohols become more like alkanes (they become more insoluble in water) and less like water itself. Decyl alcohol ($CH_3CH_2CH_2CH_2CH_2CH_2CH_2CH_2CH_2CH_2OH$) is insoluble in water. The hydroxyl group's ability to form hydrogen bonds is almost totally overshadowed by the lack of attraction between water molecules and the long alkane portion of the molecule (Figure 14.2).

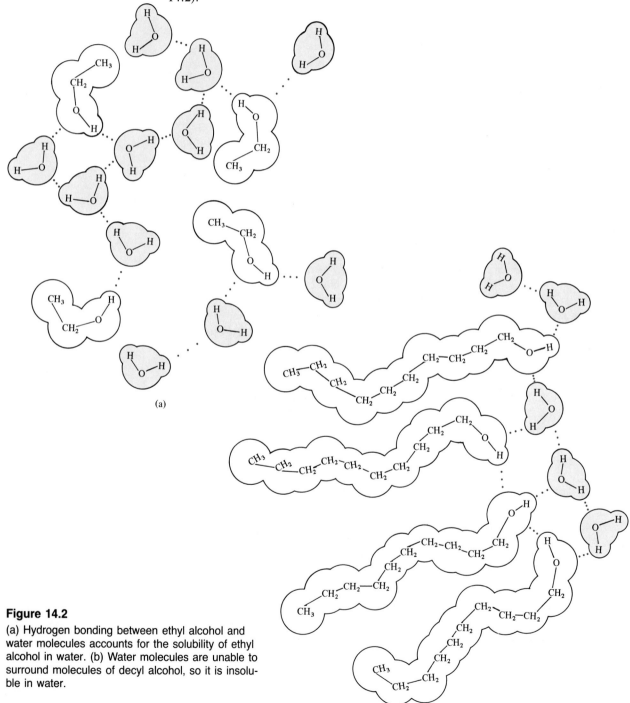

(a)

(b)

Figure 14.2

(a) Hydrogen bonding between ethyl alcohol and water molecules accounts for the solubility of ethyl alcohol in water. (b) Water molecules are unable to surround molecules of decyl alcohol, so it is insoluble in water.

Table 14.4
Solubilities of the Butyl
Alcohols in Water

Alcohol	Formula	Solubility (g/100 g H_2O)
Butyl	$CH_3CH_2CH_2CH_2OH$	8
Isobutyl	$(CH_3)_2CHCH_2OH$	11
sec-Butyl	$CH_3CH_2CH(OH)CH_3$	12.5
tert-Butyl	$(CH_3)_3COH$	Completely soluble

Consider Table 14.4, which lists the solubilities of the butyl alcohols in water. The large discrepancies in the water solubilities of these isomeric alcohols cannot be attributed to differences in molecular weight. The differing solubilities are a result of the different geometric shapes of the alcohols. The very compact *tert*-butyl alcohol molecules experience weaker intermolecular van der Waals attractions and are more easily surrounded by water molecules. Hence, *tert*-butyl alcohol has a lower boiling point (83 °C) than any of its isomers (all of which boil above 100 °C) and a higher solubility in water.

In summary, solubility considerations involve the balance of polar and nonpolar groups within a molecule, as well as molecular shape. The more polar a molecule and the more compact its shape, the greater its water solubility. Molecules that can effectively form hydrogen bonds to water will dissolve in water. Each functional group, such as the hydroxyl, that can form hydrogen bonds to water can carry along into solution an alkyl group of up to four or five carbon atoms. Thus, we will frequently find that the borderline of water solubility in a family of organic compounds occurs at four or five carbon atoms.

14.4
PREPARATION OF ALCOHOLS

Most simple alcohols are made by the hydration of alkenes (the addition of water to the double bond—Section 13.13). Since alkenes are made by the cracking of petroleum, our industrial supply of most alcohols (especially ethyl alcohol and isopropyl alcohol) is dependent to a large extent on the availability of oil.

Ethanol is made by the hydration of ethylene in the presence of sulfuric acid.

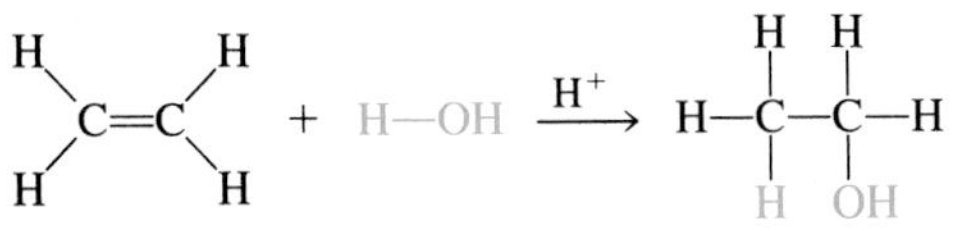

In a similar manner, isopropyl alcohol is produced by the addition of water to propylene.

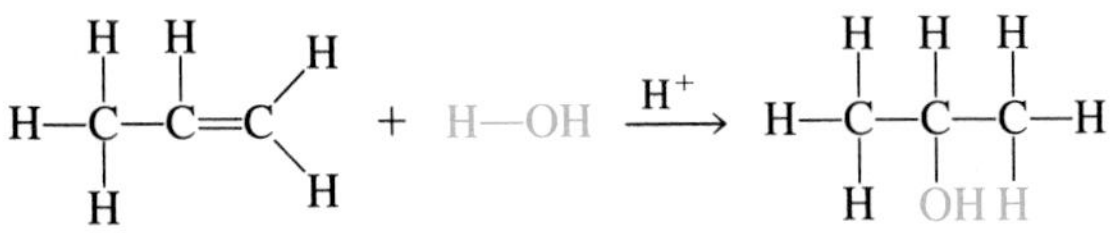

With 2-methylpropene, the product is 2-methyl-2-propanol.

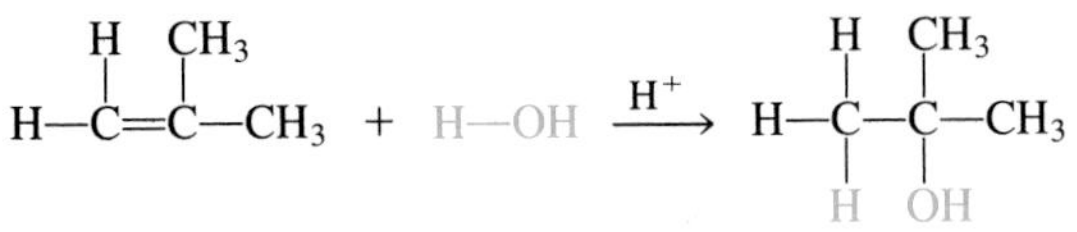

Note that in the last two reactions the hydrogen goes on the carbon atom (of the two involved in the double bond) that has the most hydrogens already bonded to it. The hydroxyl group goes on the carbon with fewer hydrogens. Thus, addition of water to propylene always gives isopropyl alcohol, never propyl alcohol.

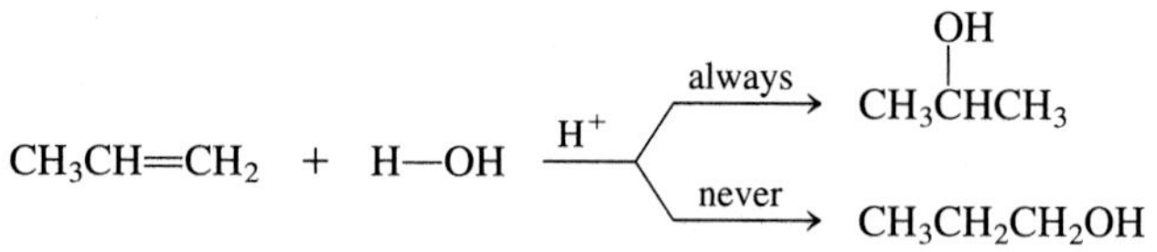

$$CH_3CH{=}CH_2 \;+\; H{-}OH \xrightarrow{H^+} \begin{array}{l} \xrightarrow{always} \underset{\substack{|\\OH}}{CH_3CHCH_3} \\ \xrightarrow{never} CH_3CH_2CH_2OH \end{array}$$

The above rule, in a more general form, was first formulated in 1870 by Vladimir V. Markovnikov, a Russian chemist. It is widely known as **Markovnikov's rule.** Sometimes the rule is stated (somewhat facetiously) as ''the rich get richer'' or ''them that has, gets.''

EXAMPLE 14.2

What alcohols are formed by the hydration of (a) 2-methyl-1-pentene and (b) 1-methylcyclopentene?

SOLUTION

First write out the structural formulas and count the number of hydrogens that are directly bonded to each of the carbons of the double bond.

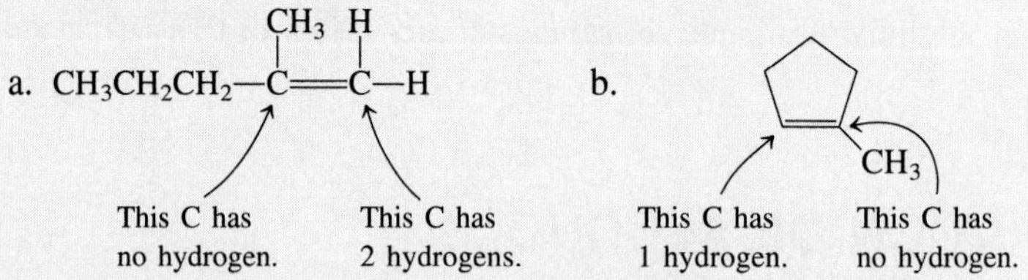

According to Markovnikov's rule, the hydrogen goes to the carbon that has the most hydrogens. The hydroxyl group goes to the other carbon.

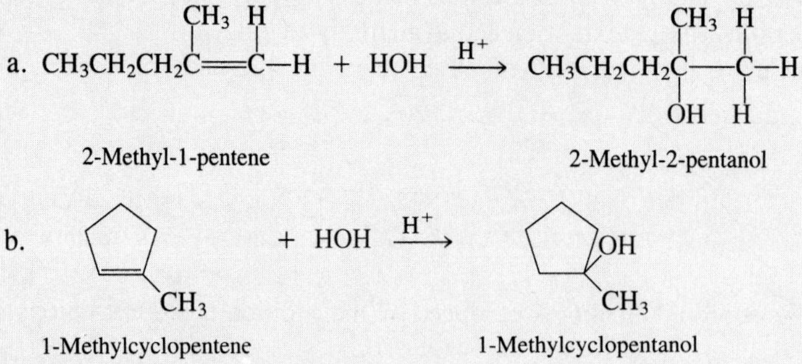

a. 2-Methyl-1-pentene → 2-Methyl-2-pentanol

b. 1-Methylcyclopentene → 1-Methylcyclopentanol

Practice Exercise

What alcohols are formed by the hydration of (a) 2-methyl-2-pentene and (b) 1-ethylcyclobutene?

Although the direct hydration of alkenes is an important industrial process for the preparation of alcohols, it is seldom used as a laboratory procedure because other synthetic methods are more convenient. The hydration reaction, however, is very common in

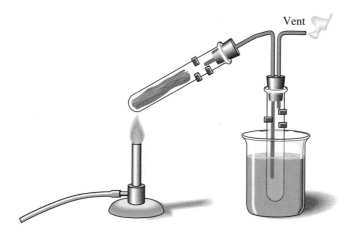

Figure 14.3
An apparatus for the de-
structive distillation of
wood. The wood is heated
in an enclosed tube, and
alcohol is condensed in the
second tube by the cold
water in the beaker. Gases
formed in the process can
be burned as they emerge
through the vent tube.

biochemistry. Many of the hydroxy compounds that occur in living systems are formed in this manner. The following reaction, for example, occurs as one of the steps in the Krebs cycle (see Figure 25.1).

$$\underset{\text{Fumaric acid}}{\text{HOOC}-\overset{\text{H}}{\underset{\text{H}}{\text{C}}}=\overset{}{\underset{}{\text{C}}}-\text{COOH}} \;+\; \text{HOH} \;\underset{}{\overset{\text{enzyme}}{\rightleftharpoons}}\; \underset{\text{Malic acid}}{\text{HOOC}-\overset{\text{H}}{\underset{\text{H}}{\text{C}}}-\overset{\text{OH}}{\underset{\text{H}}{\text{C}}}-\text{COOH}}$$

Prior to 1923, methanol was prepared by the destructive distillation of wood—hence its common name, *wood alcohol*. When wood is heated to a temperature of 450 °C in the absence of air, it decomposes to charcoal and a volatile fraction (Figure 14.3). Two to three percent of this fraction is methanol, and it can be separated from the other components (acetic acid and acetone) by fractional distillation. On the average, 1 ton of wood produces about 35 lb of the alcohol. Presently, methanol is prepared more economically by combining hydrogen and carbon monoxide under conditions of high temperature and pressure in the presence of a zinc oxide–chromium oxide catalyst.

$$2\,H_2 \;+\; CO \;\xrightarrow[\text{ZnO, Cr}_2\text{O}_3]{\text{200 atm, 350 °C}}\; CH_3OH$$

The production of alcoholic spirits is one of the oldest known chemical reactions. Even in biblical times ethanol was prepared by the fermentation of sugars or starch from various sources (potatoes, corn, wheat, rice, etc.) Biochemical investigations have shown that the fermentation process is catalyzed by enzymes found in yeast and that it proceeds by an elaborate multistep mechanism. These steps are considered in detail in Chapter 24. The equation for the overall process can be written as follows.

$$\underset{\text{Starch}}{(C_6H_{10}O_5)_x} \;\xrightarrow{\text{enzymes}}\; \underset{\text{Glucose}}{C_6H_{12}O_6} \;\xrightarrow{\text{enzymes}}\; \underset{\text{Ethanol}}{2\,C_2H_5OH} \;+\; 2\,CO_2$$

On an industrial scale, either molasses from sugarcane or starches from various types of grain are fermented by yeast to ethanol (Figure 14.4). It is ethanol that most people are referring to when they say "alcohol," meaning liquor or what is drunk. The greatest use of ethanol is as a beverage. Wines contain about 12% ethanol by volume; champagnes contain 14–20%; beers and ciders contain about 4%; and whiskey, gin, and brandy con-

Figure 14.4
Alcohol can be made by the fermentation of nearly any type of starchy or sugary material: *(left to right)* wine from rice, vodka from potatoes, aperitif from artichoke, raki from raisins, and wine from grapes.

tain 40–50%. The alcoholic content of a beverage is indicated by a measure known as *proof spirit*.[1] The proof value is twice the alcoholic content by volume, and thus whiskey that is 50% alcohol is said to be 100 proof. Fermented alcohol can be concentrated to as much as 95% by distillation. Such grain alcohol is frequently used as a solvent for drugs meant for internal consumption.

14.5
PHYSIOLOGICAL PROPERTIES OF ALCOHOLS

The simple alcohols are poisonous to some degree. In an attempt to quantify the degree of toxicity, scientists use the term LD_{50} to indicate the *l*ethal *d*ose of a chemical to 50% of a population of test animals. Like humans, individual animals respond differently to various poisons. Some are killed by amounts much smaller than the LD_{50}; others survive considerably larger amounts. The LD_{50} term, then, is only approximate for animals. Extrapolation to human toxicities can introduce even larger errors. The dosage usually is expressed as the amount of tested substance per kilogram of body weight of the test animal. The smaller the LD_{50} value, the smaller the quantity of the substance required to kill the animal, and therefore the more toxic the substance. Table 14.5 lists LD_{50} values for alcohols administered orally to rats. Note that no LD_{50} is given for methanol. Although its short-term toxicity is not terribly high, methanol can cause permanent blindness or death, even in small concentrations. Each year many accidents are attributed to this alcohol, which is frequently mistaken for its less harmful relative ethanol. Methanol should never be applied to the body, nor should its vapors be inhaled, because it is readily absorbed through the skin and respiratory tract.

Ingestion of as little as 15 mL of methanol can cause blindness; 30 mL (1 fluid ounce) can cause death.

---

[1] The term has its origin in an old seventeenth-century method for testing whiskey. Dealers were perhaps too often tempted to increase profits by adding water to the booze. The whiskey to be tested was poured over a portion of gunpowder and the mixture was ignited. Since an ethanol–water solution will ignite when the alcohol concentration is about 50%, this solution scored 100 in the test, "proof" of the spirit content in the whiskey. If the powder did not burn, it was due to the presence of too much water in the whiskey.

Table 14.5
Lethal Oral Doses (in
Rats) for Some Alcohols

Alcohol	Structure	Boiling Point (°C)	LD_{50} (g/kg body weight)	Uses
Methyl alcohol	CH_3OH	64	[a]	Solvent, fuel additive
Ethyl alcohol	CH_3CH_2OH	78	7.06	Solvent, beverages
Propyl alcohol	$CH_3CH_2CH_2OH$	97	1.87	Solvent
Isopropyl alcohol	$CH_3CHOHCH_3$	82	5.8	Solvent, body rubs
Butyl alcohol	$CH_3CH_2CH_2CH_2OH$	118	4.36	Solvent
Hexyl alcohol	$CH_3(CH_2)_4CH_2OH$	156	4.59	—
Ethylene glycol	$HOCH_2CH_2OH$	198	8.54	Antifreeze
Glycerol	$HOCH_2CHOHCH_2OH$	290 (dec)[b]	>25	Moisturizer

[a]No LD_{50} is given for methyl alcohol. Its acute (short-term) toxicity is not terribly high. However, it is metabolized to formaldehyde (HCHO) in the body, so that the chronic (long-term) toxicity is quite high. The LD_{50} for formaldehyde administered orally to rats is 0.70 g/kg of body weight. The LD_{50} for acetaldehyde, the metabolite of ethyl alcohol, is 1.9 g/kg.
[b]Glycerol decomposes.
Source: Susan Budavari (Ed.), *The Merck Index,* 11th ed. Rathway, NJ: Merck and Co., 1989.

The reason methanol is so dangerous[2] is that humans and other primates have liver enzymes that oxidize primary alcohols to compounds called aldehydes (Section 14.6). Ethanol, for example, is oxidized to acetaldehyde.

$$CH_3CH_2OH \xrightarrow{\text{liver enzymes}} \underset{CH_3}{\overset{H}{\diagdown}} C=O$$

Ethanol   Acetaldehyde

The acetaldehyde is in turn oxidized to acetic acid, a normal constituent of cells. The acetic acid can then be oxidized to carbon dioxide and water.

Similarly, methanol is oxidized to formaldehyde.

$$CH_3OH \xrightarrow{\text{liver enzymes}} \underset{H}{\overset{H}{\diagdown}} C=O$$

Methanol   Formaldehyde

Formaldehyde reacts rapidly with the components of cells. It causes proteins to be coagulated, in much the same way that an egg is coagulated by cooking. It is this property of formaldehyde that accounts for the great toxicity of methanol. The LD_{50} for formaldehyde administered orally to rats is 0.070 g per kilogram of body weight. For acetaldehyde under the same conditions, LD_{50} is 1.9 g per kilogram of body weight. Thus, formaldehyde is about 27 times as toxic to rats as acetaldehyde. Indeed, the antidote for methanol poisoning has long been ethanol, administered intravenously. The ethanol preferentially loads up the liver enzymes in humans and other primates. If the enzymes are tied up oxidizing ethanol to acetaldehyde, they cannot catalyze the oxidation of the methanol to the dangerously toxic formaldehyde. Thus, the unoxidized methanol is gradually excreted from the body.

[2]It should be noted that methanol is not particularly toxic to horses, rats, and some other animals. These animals are deficient in the enzymes that oxidize alcohols to aldehydes. Toxicity studies in other animals cannot always be extrapolated to humans. In many cases, however, trends in toxicities can be judged from animal studies.

Despite its toxicity, methanol is a valuable industrial solvent. The largest use of methanol is as the starting material for the commercial synthesis of formaldehyde. It is also used in car windshield washer fluids, and it is employed commercially as a solvent for paint, gum, and shellac. Methanol, along with ethanol (see following), is mixed with gasoline and sold for use as a motor fuel.

Ethyl alcohol is potentially toxic. Rapid ingestion of 1 pint of pure alcohol would kill most people. Alcohol freely crosses into the brain, where it depresses the respiratory control center, resulting in failure of the respiratory muscles in the lungs and hence suffocation. Alcohol is believed to act on the nerve cell membranes, causing a diminution in speech, thought, cognition, and judgment. Excessive ingestion over a long period of time leads to deterioration of the liver (cirrhosis) and loss of memory and may lead to strong physiological addiction. Addiction to alcohol (alcoholism) is the most serious drug problem in the United States. It has been estimated that there are about 40 times as many alcoholics (about 10 million) as there are heroin addicts in the United States. If the alcohol is diluted (as in alcoholic beverages) and is consumed in small quantities, it is relatively safe. The body possesses enzymes that have the capacity to metabolize ethyl alcohol to carbon dioxide and water (see Chapter 24).

Alcohol not intended for beverage purposes is commercially prepared by the direct hydration of ethylene, which is a by-product of the petroleum industry. The alcohol so produced is 95% alcohol and 5% water. The water that remains in this mixture cannot be removed by ordinary distillation procedures because 95% ethanol is a constant-boiling mixture (an **azeotrope**). This 95% alcohol is used in chemical laboratories as a solvent. If 100% alcohol is needed, special procedures must be employed to prepare it. One method is to dry the alcohol over calcium oxide for several hours. The calcium oxide has a great affinity for water but will not combine with ethanol. The remaining alcohol is then distilled. Alcohol so prepared is known as **absolute** (100%) **alcohol.**

> An azeotropic mixture is a constant-boiling mixture of two components that are present in a fixed ratio. A mixture of 95% ethanol and 5% water is an azeotropic mixture that boils at 78 °C.

Ethanol is used as a solvent for perfumes, medicinal formulations (tinctures), lacquers, varnishes, and shellacs. Alcohol denatures enzymes in bacteria (see Section 21.14) and for this reason it is widely used as an antiseptic in mouthwashes and aerosol disinfectants. Ethanol is also employed in the synthesis of other organic compounds. When ethanol is used for such industrial purposes it is not subject to a federal tax. (The tax is more than $20/gallon in most states.) To ensure the legitimate use of tax-free alcohol, the government treats it with certain additives that make it unfit to drink. Such alcohol is known as **denatured alcohol.** Common denaturants are methanol and 2-propanol. These compounds are toxic but do not interfere with the solvent properties of the alcohol.

The use of ethanol (and methanol) as a blend component of gasoline began in the United States in 1979. Favorable results have been achieved with mixtures containing 80–90% unleaded gasoline and 10–20% alcohol. Since ethanol can be obtained easily from grain (and methanol from coal, wood chips, or municipal refuse), this represents one method of augmenting our dwindling fuel supplies. Brazil, which has an abundance of sugarcane (and no oil), is producing cars that are designed to burn only alcohol as a motor fuel. In addition to its use as a fuel extender, alcohol also increases the octane rating of the gasoline with which it is blended.

A 70% solution of isopropyl alcohol is commonly referred to as rubbing alcohol. It has a high vapor pressure, and its rapid evaporation from the skin produces a cooling effect that helps to lower fever. Rubbing alcohol, like ethanol, denatures bacterial enzymes and is used as an antiseptic to cleanse the skin before a blood sample is taken or an injection is given. Isopropyl alcohol is toxic when ingested, but, unlike methanol, it is not readily absorbed through the skin. Though more toxic than ethanol, it less often causes

Contrary to popular belief, alcohol is a depressant of the central nervous system, not a stimulant. The illusionary stimulation comes from its effect of depressing brain areas responsible for judgment. The resulting lack of inhibitions and restraints may cause one to feel "stimulated." People under the influence of alcohol suffer from diminished control of their judgment and actions and hence may endanger themselves and/or others (especially if they are driving). In most states, a blood alcohol concentration (BAC) of 0.1% (100 mg of alcohol in 100 mL of blood) is legal evidence of intoxication. A BAC of 0.5–1% leads to coma and death (Table 14.6).

Table 14.6
Approximate Relationship Between Drinks Consumed, Blood Alcohol Concentration, and Effect for a 70-kg (154-lb) Moderate Drinker

Number of Drinks[a]	Blood Alcohol Concentration (% by volume)	Effect[b]
2	0.05	Mild sedation; tranquillity
4	0.10	Lack of coordination
6	0.15	Obvious intoxication
10	0.30	Unconsciousness
20	0.50	Possible death

[a] Rapidly consumed 30-mL (1-oz) "shots" of 90 proof whiskey, 360-mL (12-oz) bottles of beer, or 150-mL (5-oz) glasses of wine.
[b] An inexperienced drinker would be affected more strongly, or more quickly, than one who is ordinarily a moderate drinker. Conversely, a habitual *heavy* drinker would be affected less.

Alcohol in the Blood

Alcohol causes blood vessels to dilate. The resulting increased flow of blood through the capillaries beneath the skin imparts a feeling of warmth and a reddish hue to the skin.

fatalities. Instead, it can induce vomiting; it doesn't stay down long enough to kill you. Other higher alcohols behave in a similar manner. Much of the production of isopropyl alcohol is for the manufacture of acetone (see Section 15.5) and to introduce the isopropyl group into organic molecules.

14.6
CHEMICAL PROPERTIES OF ALCOHOLS

The reactions of the alcohols occur mainly at the functional group. They may, however, involve hydrogen atoms attached to the carbon bearing the hydroxyl group or even those on an adjacent carbon. We will discuss three major kinds of reactions of the alcohols. Dehydration and oxidation are considered here. Esterification is covered in Section 16.8.

Dehydration of Alcohols

Dehydration (removal of water) is usually accomplished by adding concentrated sulfuric acid to the alcohol and heating the resulting mixture. The hydroxyl group is removed from the alcohol carbon, and a hydrogen atom is removed from an adjacent carbon, giving an alkene.

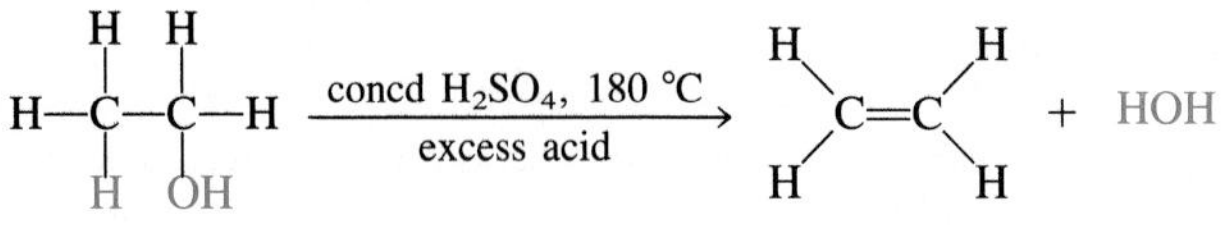

$$ \text{H-C-C-H} \xrightarrow[\text{excess acid}]{\text{concd } H_2SO_4,\ 180\ °C} \text{C=C} + HOH $$

Ethanol   Ethylene

Under the proper conditions, it is possible to perform a dehydration involving two molecules of alcohol. In this case, the hydroxyl group of one alcohol is removed, and only the hydrogen of the hydroxyl group of the second alcohol molecule is removed. The two organic groups remaining combine to form an ether molecule (see Section 14.9).

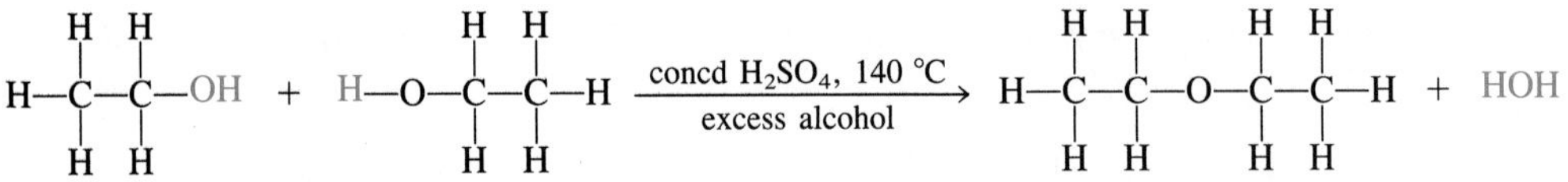

$$ \text{H-C-C-OH} + \text{H-O-C-C-H} \xrightarrow[\text{excess alcohol}]{\text{concd } H_2SO_4,\ 140\ °C} \text{H-C-C-O-C-C-H} + HOH $$

Two molecules of ethanol   Diethyl ether

Thus, depending on conditions, one can prepare either alkenes or ethers by dehydration of alcohols. At 180 °C and with an excess of H_2SO_4, dehydration of ethanol gives ethylene as the main product. At 140 °C and with an excess of ethanol, the main product of dehydration of ethanol is diethyl ether.

Dehydration (and its reverse, hydration) reactions occur continuously in cellular metabolism. In these biochemical dehydrations, enzymes serve as catalysts instead of acids, and the reaction temperature is 37 °C instead of the elevated temperatures required in the laboratory. The following reaction occurs in the Embden–Meyerhof pathway (see Figure 24.12).

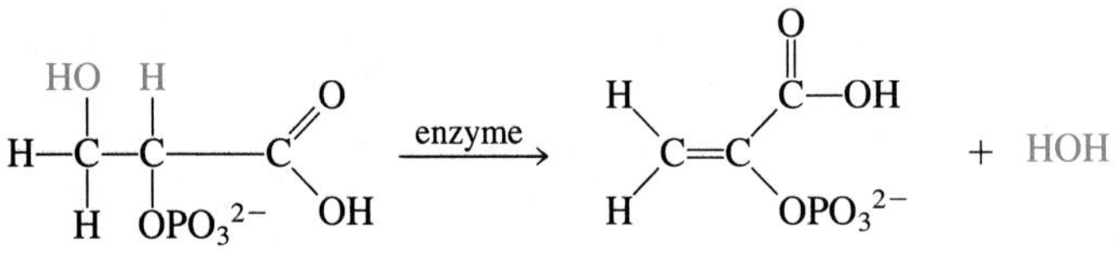

$$ \text{H-C-C-C} \xrightarrow{\text{enzyme}} \text{C=C} + HOH $$

2-Phosphoglyceric acid   Phosphoenolpyruvic acid

Look carefully at this equation. The compounds involved are more complex than the ethanol and ethylene we used in our previous example, but the reaction is not. Ignore all other functional groups that are in the molecule. In this reaction, these other functional groups remain unchanged in the product. The only thing that has happened is that a hydrogen and hydroxyl group have been eliminated from the starting material, and the product contains a double bond. The point to be made is that if you know the chemistry of a particular functional group, you know the chemistry of a thousand or a hundred thousand different individual compounds. Alcohols have a potential for undergoing dehydration, and you will find that big ones, little ones, and ones that incorporate other functional groups all dehydrate if conditions are right.

Dehydration to form simple ethers in biological systems is perhaps less common than the reaction to form an alkene. However, many important reactions, such as the formation of glycosides from sugars (Chapter 19), are at least technically dehydrations leading to etherlike compounds.

Oxidation of Alcohols

Primary and secondary alcohols are readily oxidized. We just saw how methanol and ethanol are oxidized by liver enzymes to form aldehydes. Such reactions can also be carried out in the laboratory with chemical oxidizing agents. For example, in acid solution, potassium dichromate oxidizes ethyl alcohol to acetaldehyde. The reaction is

$$8\ H^+\ +\ \underset{\substack{\text{Dichromate ion}\\ \text{(orange)}}}{Cr_2O_7^{2-}}\ +\ 3\ C_2H_5OH\ \longrightarrow\ \underset{\substack{\text{Chromium(III)}\\ \text{ion (green)}}}{2\ Cr^{3+}}\ +\ 3\ C_2H_4O\ +\ 7\ H_2O$$

Similarly, propyl alcohol is oxidized to propionaldehyde. The balanced equation for this reaction is quite complicated, even if we write only the net ionic equation.

$$3\ CH_3CH_2CH_2OH\ +\ 8\ H^+\ +\ Cr_2O_7^{2-}\ \longrightarrow\ 3\ CH_3CH_2CHO\ +\ 2\ Cr^{3+}\ +\ 7\ H_2O$$

In situations like this, organic chemists have a tendency to simplify everything until only the change involving the organic molecules is shown. Thus, the above reaction would be simplified to

$$\underset{\text{Propyl alcohol}}{CH_3CH_2CH_2OH}\ \xrightarrow[H^+]{K_2Cr_2O_7}\ \underset{\text{Propionaldehyde}}{CH_3CH_2\overset{\displaystyle O}{\overset{\displaystyle \|}{C}}{-}H}$$

The required inorganic reagents are written either above or below the arrow. The inorganic by-products are ignored (they're still formed, but we just ignore them in this form of the equation). In this way, all attention is focused on the organic starting material and product, and less time is spent balancing the frequently complicated equations.

The abbreviated form of this particular equation indicates that a primary alcohol is oxidized to an aldehyde. We shall see, in Chapter 15, that aldehydes are even more easily oxidized than alcohols and yield as products carboxylic acids. If one wishes to isolate the aldehyde initially formed in the oxidation of the alcohol, it is necessary to remove it from contact with the oxidizing agent. This can be done by distilling the aldehyde from the reaction mixture as it forms.

Secondary alcohols are oxidized to compounds called *ketones* (see Chapter 15). Oxidation of isopropyl alcohol by dichromate gives acetone.

$$\underset{\substack{\text{Isopropyl alcohol}\\ \text{(a secondary alcohol)}}}{CH_3{-}\overset{\displaystyle OH}{\overset{\displaystyle |}{C}H}{-}CH_3}\ \xrightarrow[H^+]{K_2Cr_2O_7}\ \underset{\substack{\text{Acetone}\\ \text{(a ketone)}}}{CH_3{-}\overset{\displaystyle O}{\overset{\displaystyle \|}{C}}{-}CH_3}$$

Unlike aldehydes, ketones are relatively resistant to further oxidation and special precautions to isolate the product of this reaction are not necessary.

As we saw in the preceding section, oxidation of alcohols is important in living organisms. Indeed, enzyme-controlled oxidation reactions provide the energy whereby cells can do useful work. One step in the metabolism of carbohydrates (Chapter 25) involves the oxidation of the secondary alcohol group in isocitric acid to a ketone group.

$$\begin{array}{ccc}
CH_2{-}COOH & & CH_2{-}COOH \\
| & & | \\
CH{-}COOH & \xrightarrow{\text{enzyme}} & CH{-}COOH \\
| & & | \\
HO{-}CH{-}COOH & & O{=}C{-}COOH \\
\text{Isocitric acid} & & \text{Oxalosuccinic acid}
\end{array}$$

Again note that the overall reaction is identical to that of the conversion of isopropyl alcohol to acetone. The complications of structure that distinguish isocitric acid in no way interfere with the characteristic reaction of its secondary alcohol group.

Tertiary alcohols are resistant to oxidation. Ordinary oxidizing agents, such as dichromate, bring about no change in this class of alcohols, which lack a hydrogen on the carbon atom bonded to the hydroxyl group.

$$\begin{array}{c}
CH_3 \\
| \\
CH_3{-}C{-}OH \xrightarrow[H^+]{K_2Cr_2O_7} \text{No reaction} \\
| \\
CH_3
\end{array}$$

tert-Butyl alcohol

The oxidation reactions we have described involve the formation of a carbon–oxygen double bond. Thus, the carbon atom bonded to the oxygen in the alcohol must be able to release one of the atoms attached to it so it can form the double bond with oxygen. Hydrogen readily leaves the carbon under the conditions of oxidation. Both primary and secondary alcohols have a hydrogen on the carbon holding the hydroxyl group, and both these classes of alcohols are readily oxidized. The tertiary alcohol lacks such a hydrogen and is not easily oxidized. An analogous structural feature explains why aldehydes tend to oxidize further whereas ketones don't. The aldehydes still have a hydrogen left on the carbon holding the oxygen; the ketones do not. The following equations summarize the differences in the oxidation (indicated by the symbol [O]) of the various classes of alcohols.

$$\begin{array}{ccccc}
O{-}H & & O & & O \\
| & & \| & & \| \\
R{-}C{-}H & \xrightarrow{[O]} & R{-}C{-}H & \xrightarrow{[O]} & R{-}C{-}OH \\
| & & & & \\
H & & & &
\end{array}$$

Primary alcohol   Aldehyde   Carboxylic acid

$$\begin{array}{ccc}
OH & & O \\
| & & \| \\
R{-}C{-}R & \xrightarrow{[O]} & R{-}C{-}R \\
| & & \\
H & &
\end{array}$$

Secondary alcohol   Ketone

$$\begin{array}{c}
OH \\
| \\
R{-}C{-}R \xrightarrow{[O]} \text{No reaction} \\
| \\
R
\end{array}$$

Tertiary alcohol

EXAMPLE 14.3

Write equations for the reactions of the following alcohols with $K_2Cr_2O_7$ and H_2SO_4.

a. 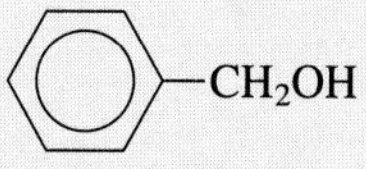 [benzene ring]—CH_2OH b. 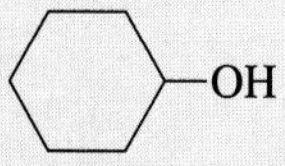 [cyclohexane ring]—OH c. 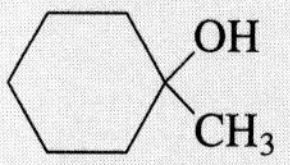 [cyclohexane ring with OH and CH_3]

SOLUTION

First recognize the class of each alcohol. (a) is a primary alcohol, (b) is a secondary alcohol, and (c) is a tertiary alcohol. Therefore,

a. 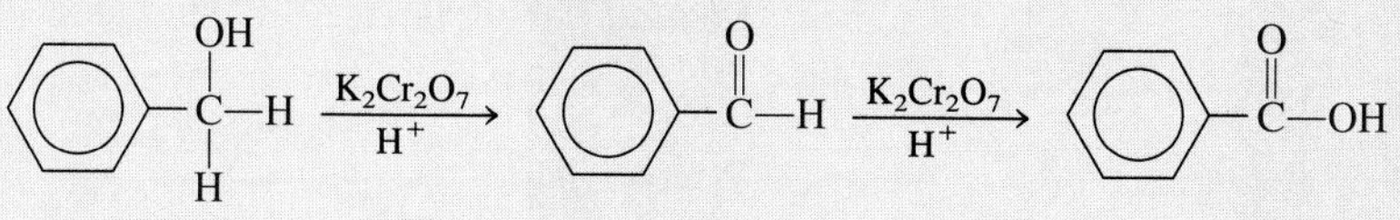 [structures showing oxidation of benzyl alcohol to aldehyde then carboxylic acid, with $\xrightarrow[H^+]{K_2Cr_2O_7}$]

b. 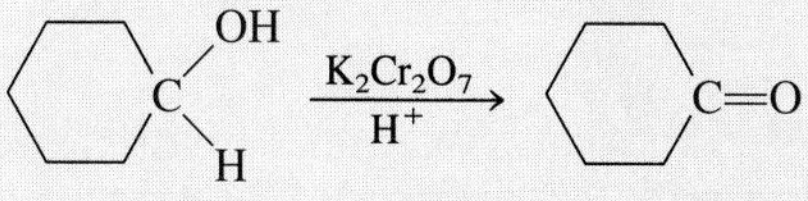 [cyclohexanol structure] $\xrightarrow[H^+]{K_2Cr_2O_7}$ [cyclohexanone, C=O]

c. [tertiary alcohol structure] $\xrightarrow[H^+]{K_2Cr_2O_7}$ No reaction

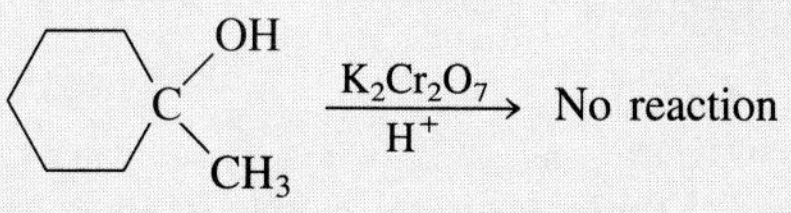

Practice Exercise

Write equations for the reactions of the following alcohols with $K_2Cr_2O_7$ and H_2SO_4.

a. $CH_3CH_2CH_2CH_2CH_2OH$ b. 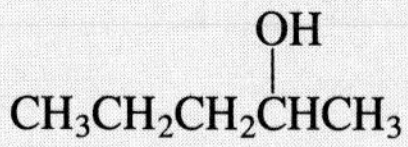 $CH_3CH_2CH_2\overset{OH}{\underset{}{C}}HCH_3$ c. $CH_3CH_2\overset{OH}{\underset{CH_3}{C}}CH_3$

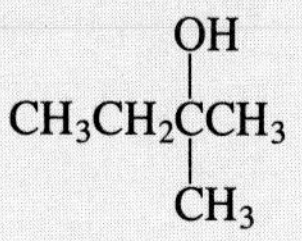

14.7
MULTIFUNCTIONAL ALCOHOLS: GLYCOLS AND GLYCEROL

The simple alcohols that we have met so far contain only one hydroxyl group each. They are called **monohydric** alcohols. Several important compounds that are frequently encountered contain more than one hydroxyl group per molecule. They are called **polyhydric** alcohols. Those with two such groups are said to be **dihydric.** Substances with three hydroxyl groups are called **trihydric** alcohols.

Dihydric alcohols are often called **glycols.** The most important of these is ethylene glycol. This compound is the main ingredient in permanent antifreeze mixtures for automobile radiators (Figure 14.5). Ethylene glycol is a sweet, somewhat viscous liquid. With two hydroxyl groups, extensive intermolecular hydrogen bonding exists. Thus, ethylene glycol has a high boiling point (198 °C) and does not boil away when used as antifreeze. It is also completely miscible with water. A solution of 60% ethylene glycol in water will not freeze until the temperature falls to −49 °C (−56 °F). The color of most of the

Figure 14.5
Ethylene glycol is the principal active ingredient in permanent antifreeze solutions.

commercial antifreezes is due to the additives. Ethylene glycol is also used in the manufacture of polyester fiber (Dacron) and film (Mylar) used in tapes for recorders and computers.

Ethylene glycol is quite toxic. As with methanol, its toxicity is due to a metabolite. Liver enzymes oxidize the ethylene glycol to oxalic acid.

$$\underset{\text{Ethylene glycol}}{\overset{\overset{\displaystyle OH \quad OH}{|\qquad\quad|}}{CH_2-CH_2}} \quad\xrightarrow{\text{liver enzymes}}\quad \underset{\text{Oxalic acid}}{\overset{\overset{\displaystyle O \quad\; O}{\|\quad\;\|}}{HO-C-C-OH}}$$

This compound crystallizes as its calcium salt, calcium oxalate (CaC_2O_4), in the kidneys, leading to renal damage. Such injury can lead to kidney failure and death. As with methanol poisoning, the usual treatment for ethylene glycol poisoning is ethanol, administered to load up and thus block the liver enzymes from catalyzing the conversion of ethylene glycol to oxalic acid.

Another common dihydric alcohol is propylene glycol. The physical properties of this compound are quite similar to those of ethylene glycol. Its physiological properties, however, are quite different.

Propylene glycol is essentially nontoxic, and it can be used as a solvent for drugs. It is also used as a moisturizing agent for foods. Like other alcohols, propylene glycol can be oxidized by liver enzymes.

$$\underset{\text{Propylene glycol}}{\overset{\overset{\displaystyle OH \quad OH}{|\qquad\quad|}}{CH_3-CH-CH_2}} \quad\xrightarrow{\text{liver enzymes}}\quad \underset{\text{Pyruvic acid}}{\overset{\overset{\displaystyle O \quad\; O}{\|\quad\;\|}}{CH_3-C-C-OH}}$$

In this case, however, the product is pyruvic acid, a normal intermediate in carbohydrate metabolism (Chapter 24).

Glycerol (glycerin) is the most important trihydric alcohol. It is a sweet, syrupy liquid. Essentially nontoxic, it is a product of the hydrolysis of fats and oils. Glycerol has widespread industrial use; the following list is not all-inclusive.

1. Preparation of hand lotions and cosmetics
2. Additive in inks, tobacco products, and plastic clays to prevent dehydration (glycerol is hygroscopic)
3. Constituent of glycerol suppositories
4. Sweetening agent and solvent for medicines
5. Lubricant
6. A source for production of plastics, surface coatings, and synthetic fibers
7. Source of nitroglycerin

The equation for the preparation of nitroglycerin shows that three molecules of nitric acid are required for every molecule of glycerin. The glycerin must be very pure to ensure stability of the product.

$$
\begin{array}{c}
\text{H} \\
| \\
\text{H}-\text{C}-\text{OH} \\
| \\
\text{H}-\text{C}-\text{OH} \\
| \\
\text{H}-\text{C}-\text{OH} \\
| \\
\text{H}
\end{array}
\; + \; 3\,\text{HONO}_2
\;\xrightarrow[\text{10–20 °C}]{\text{H}_2\text{SO}_4}\;
\begin{array}{c}
\text{H} \\
| \\
\text{H}-\text{C}-\text{ONO}_2 \\
| \\
\text{H}-\text{C}-\text{ONO}_2 \\
| \\
\text{H}-\text{C}-\text{ONO}_2 \\
| \\
\text{H}
\end{array}
\; + \; 3\,\text{H}_2\text{O}
$$

Glycerol (Glycerin) Glycerol trinitrate (Nitroglycerin)

Nitroglycerin

Nitroglycerin was first prepared in 1846 by the Italian chemist Sobrero, who was lucky that he lived to tell of his discovery. Sobrero mixed nitric acid and glycerin, and the ensuing explosion nearly killed him. It was not until 15 years later that the famed Swedish chemist and inventor Alfred Nobel (Figure 14.6) discovered a method to prepare and transport the compound safely. He found that a type of diatomaceous earth, a claylike material, was capable of absorbing the nitroglycerin, thus rendering it insensitive to shock. The stabilized mixture was referred to as *dynamite*. Unless exploded by means of a percussion cap or by a detonator containing mercuric fulminate [Hg(ONC)$_2$], it was quite stable. The production of dynamite was a major breakthrough. With dynamite, the construction of canals, dams, highways, mines, and railroads became much easier. Its use as a weapon in warfare greatly disturbed the Nobel family, however, and by Alfred Nobel's will a trust fund was established to provide an annual award for an outstanding contribution toward peace. (A trust fund was also set up to offer annual awards for contributions in the fields of chemistry, physics, literature, and medicine or physiology.)

It is surprising that a compound that is so sensitive to shock is also used as a drug to relieve the sharp chest pains called angina pectoris. Nitroglycerin is administered in tablet form (mixed with nonactive ingredients), as an alcoholic solution (spirit of glyceryl trinitrate), or in the form of a patch from which the drug is absorbed through the skin. Nitroglycerin functions as a vasodilator. It relaxes cardiac muscle and smooth muscle in the smaller blood vessels, thus increasing the supply of blood (and hence oxygen) to the heart.

Figure 14.6
Alfred Nobel, inventor of dynamite and founder of the famed Nobel prizes.

Nitroglycerin is a pale yellow, oily liquid that detonates upon slight impact. The explosive power arises from the extremely rapid conversion of a small volume of liquid into a large volume of hot, expanding gases.

$$4 \ C_3H_5(ONO_2)_3 \longrightarrow 6 \ N_2(g) + 12 \ CO_2(g) + 10 \ H_2O(g) + O_2(g)$$

The reaction produces temperatures of over 3000 °C and pressures above 2000 atm. The explosion wave caused by such temperatures and pressures is enormous, accounting for the damaging effect of the detonation.

Phenol
(Carbolic acid)

4-Hexylresorcinol

o-Phenylphenol

o-Benzyl-*p*-chlorophenol

14.8
PHENOLS

Compounds with a hydroxyl group attached directly to an aromatic ring are called **phenols.** The parent compound, C_6H_5OH, is itself called phenol. Phenol is a white crystalline compound with a distinctive (''hospital smell'') odor. Other compounds may be named as derivatives of phenol, but most of those of interest to us are best known by special, nonsystematic names.

The phenols generally are solids with low melting points or oily liquids at room temperature. Most are only sparingly soluble in water. They have found wide use as germicides or antiseptics (substances that kill microorganisms on living tissue) and as disinfectants (substances intended to kill microorganisms on furniture, fixtures, floors, and around the house in general).

The first widely used antiseptic was phenol itself, which was also called carbolic acid. Joseph Lister used it for antiseptic surgery in 1867. Unfortunately, phenol doesn't kill only undesirable microorganisms. It kills all types of cells. Applied to the skin, it can cause severe burns. In the bloodstream, it is a systemic poison, that is, one that is carried to and affects all parts of the body. Its severe side effects led to searches for safer antiseptics, a number of which have been found.

One of the most active phenolic antiseptics is 4-hexylresorcinol. It is much more powerful than phenol as a germicide and has fewer undesirable side effects. Indeed, it is safe enough to be used as the active ingredient in some mouthwashes and in antiseptic throat lozenges such as Sucrets.

Prominent among disinfectants are the compounds *o*-phenylphenol and *o*-benzyl-*p*-chlorophenol. These compounds are the main active ingredients in preparations such as Lysol. The methyl derivatives of phenols are called *cresols.* They are important ingredients in the wood preservative creosote.

o-Cresol   *m*-Cresol   *p*-Cresol

Unlike the three cresol isomers, the three dihydroxybenzenes have been given individual names. They all have commercial significance, and two of them are important components of biochemical molecules. Hydroquinone occurs in a coenzyme (Section

25.2), and catechol forms part of the structure of certain neurotransmitters termed cate-cholamines (see Sections F.2 and F.3).

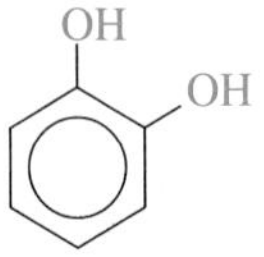

Catechol 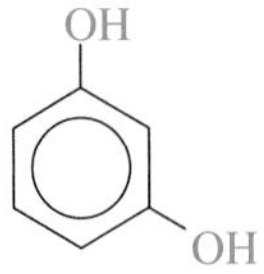   Resorcinol 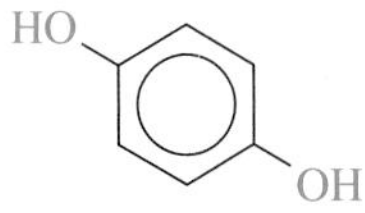   Hydroquinone

Hexachlorophene was once widely used in germicidal cleaning solutions (pHisohex) and as an ingredient in deodorant soaps and other cosmetics. In the United States, products contained at most 3% hexachlorophene. The compound was generally considered a safe and effective antibacterial agent. In 1972, however, the picture changed rapidly. An outbreak of neurological disease among infants in northeastern France was traced to a baby powder called Bébé that contained over 20% hexachlorophene. Over 30 infants died. The U.S. Food and Drug Administration acted quickly. Hexachlorophene was banned from all products intended for over-the-counter sales. It is still available for prescription use and for use in hospitals—in concentrations not to exceed 3%.

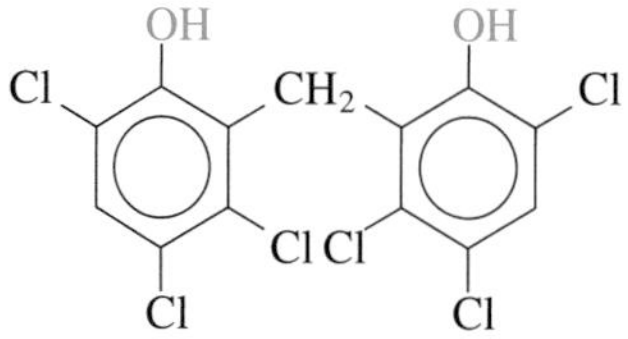

The most important commercial reaction of phenols is the condensation with formaldehyde to yield phenolic polymers (Bakelite). Bakelite was used initially as an electrical insulator and later to form plastic parts for the automotive and radio industries. Phenol is also used in the production of phenolphthalein, an acid–base indicator.

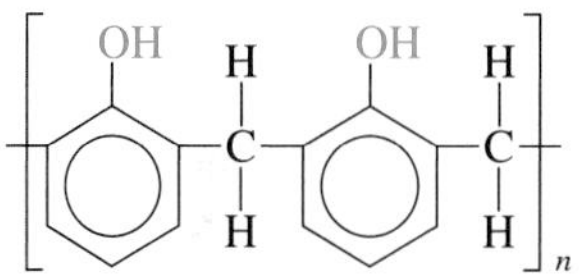

One of the major distinguishing characteristics between phenols and alcohols is that phenols are slightly acidic ($K_a \simeq 10^{-10}$), whereas alcohols are neutral. Phenols can be neutralized by strong bases, but they are too weakly acidic to react with weak bases such as aqueous sodium bicarbonate. The latter reaction serves to distinguish the phenols from the carboxylic acids (Chapter 16), which do react with $NaHCO_3$.

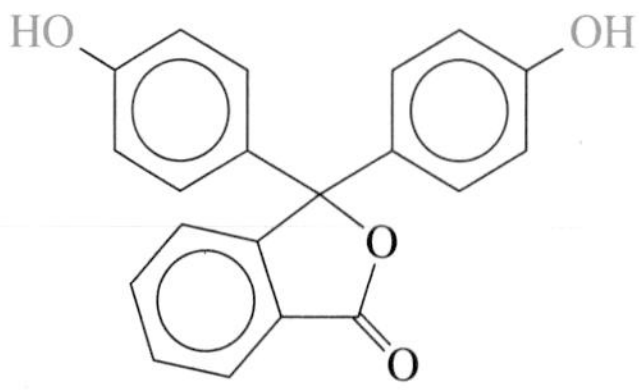

Hexachlorophene

Bakelite

Phenolphthalein

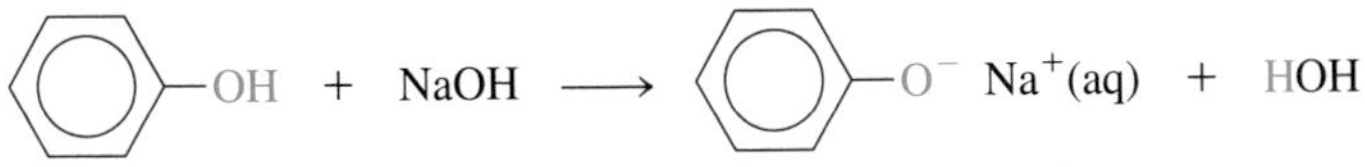

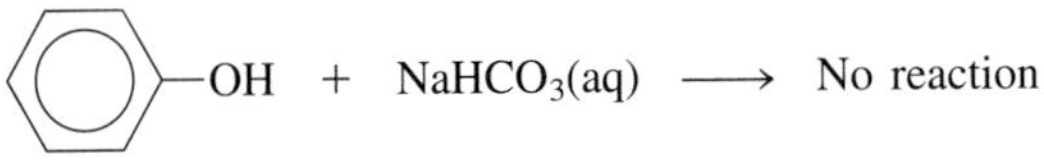

14.9
ETHERS

Ethers may be considered to be derivatives of water in which both hydrogen atoms have been replaced by alkyl or aryl groups. They may also be considered as derivatives of an alcohol in which the hydroxyl hydrogen has been replaced by an organic group.

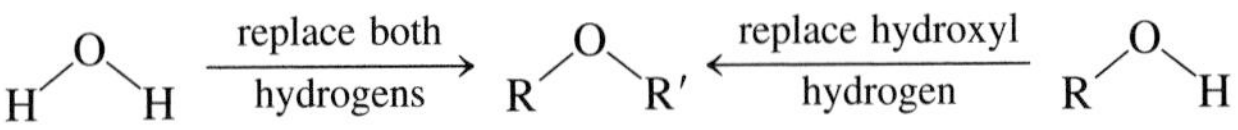

Table 14.7
Comparison of Boiling
Points of Alkanes,
Alcohols, and Ethers

Formula	Name	MW	Boiling Point (°C)
$CH_3CH_2CH_3$	Propane	44	−42
CH_3OCH_3	Dimethyl ether	46	−25
CH_3CH_2OH	Ethyl alcohol	46	78
$CH_3CH_2CH_2CH_2CH_3$	Pentane	72	36
$CH_3CH_2OCH_2CH_3$	Diethyl ether	74	35
$CH_3CH_2CH_2CH_2OH$	Butyl alcohol	74	117

The general formula for the ethers is R—O—R′. When both R groups are the same, the compound is a *symmetrical* ether. When R and R′ are different, the ether is said to be an *unsymmetrical* ether.

Symmetrical Ether   *Unsymmetrical Ether*

R—O—R H_3C—O—CH_3 R—O—R′ H_3C—O—CH_2CH_3

CH_3—O—CH_3
Dimethyl ether

CH_3—O—CH_2CH_3
Ethyl methyl ether

CH_3CH_2—O—CH_2CH_3
Diethyl ether

R—O⋯⋯⋯⋯⋯H—O
    \   \
     R   H

Simple ethers are simply named. Just name the groups attached to oxygen and then add the generic name *ether*. For symmetrical ethers the group name should be preceded by the prefix *di-*, although the prefix is sometimes dropped in common usage. The names methyl ether and dimethyl ether refer to the same compound, but the latter is preferred.

Ether molecules have no hydrogen atom on oxygen. Therefore ether molecules in the pure liquid are incapable of intermolecular hydrogen bonding. Given their molecular weight, then, the ethers have quite low boiling points. Indeed, ethers have boiling points about the same as those of alkanes of comparable molecular weight and much lower than those of the corresponding alcohols (Table 14.7).

Ether molecules do have an oxygen atom, however. They can participate in hydrogen bonding with water molecules. Consequently, the ethers have about the same water solubilities as their isomeric alcohols. (For example, methyl ether and ethanol are completely soluble in water, whereas diethyl ether and 1-butanol are soluble to the extent of 8 g/ 100 mL of water.)

Chemically, the ethers are quite inert. Like the alkanes, they do not react with the usual oxidizing agents, reducing agents, or bases. The inertness of ethers makes them excellent solvents for organic materials. Often called simply ''ether,'' diethyl ether is often used in the extraction of organic compounds from plant and animal materials or from mixtures of organic and inorganic substances. The volatile ether is then easily removed by evaporation, and the desired organic components are left behind.[3]

Ether is extremely hygroscopic. A freshly opened can of ether will immediately pick up about 1–2% water from the moisture in the air. Special techniques (dry box, nitrogen atmosphere) are employed in handling the compound when reaction conditions call for anhydrous ether.

Use of diethyl ether in the laboratory produces unusual hazards. The compound is quite volatile and extremely flammable. The vapors form an explosive mixture with air. They are also heavier than air and can travel long distances along a tabletop or the floor to

[3]The extracting power of ether has made it the solvent of choice among cocaine users. ''Freebasing'' first involves separating the cocaine from other substances by extracting it into ether. Enormous quantities of ether are used in extracting cocaine from the coca plant. Drug enforcement officials in the United States have been able to apprehend some cocaine manufacturers by keeping track of the shipments of large quantities of ether.

reach a flame or spark and set off an explosion. Hence, open flames are not permitted in a laboratory in which ether is being used. Ether fires cannot be extinguished with water because the ether is less dense than water and will float on top of it. The use of carbon dioxide fire extinguishers is recommended.

Diethyl ether should not be stored in an ordinary refrigerator. Even at low temperatures, it has sufficient vapor pressure to form an explosive mixture with air. A spark can ignite the vapors. Special explosion-proof refrigerators, with sealed electrical equipment to prevent contact between spark and flammable vapors, are available for safe storage of volatile, flammable liquids.

Still another hazard with ethers is that, upon standing, they react with oxygen from the air to form peroxides.

$$CH_3CH_2\!-\!O\!-\!CH_2CH_3 \; + \; O_2 \; \longrightarrow \; CH_3\underset{\underset{\text{O}-\text{O}-\text{H}}{|}}{CH}\!-\!O\!-\!CH_2CH_3$$

Diethyl ether   A peroxide

These peroxides are less volatile than the ether and are concentrated in the residue left behind during a distillation or evaporation. But these concentrated peroxides are highly explosive and are sensitive to both shock and heat. Hospitals avoid these problems by buying only those amounts of ether sufficient for immediate use and by keeping containers tightly closed and away from strong light, which catalyzes peroxide formation. Ether suspected of containing peroxides should be treated with a reducing agent, such as alkaline ferrous sulfate solution, before it is used.

Anesthesia

A **general anesthetic** acts on the brain to produce unconsciousness as well as insensitivity to pain. A local anesthetic (see Section F.5) renders one part of the body insensitive to pain yet leaves the patient conscious.

$$CH_3CH_2\!-\!O\!-\!CH_2CH_3$$

Diethyl ether

Diethyl ether was the first general anesthetic. It was introduced into surgical practice in 1846 by a Boston dentist, William Morton. Inhalation of ether vapor produces unconsciousness by depressing the activity of the central nervous system. Ether is relatively safe. There is a fairly wide gap between the effective level of anesthesia and the lethal dose. The disadvantages are its high flammability and its side effect, nausea.

$$N\!=\!N\!-\!O$$

Nitrous oxide
(Dinitrogen monoxide)

Nitrous oxide, or laughing gas (N_2O), was tried by Morton without success before he tried ether. Nitrous oxide was discovered by Joseph Priestley in 1772. Its narcotic effect was noted, and it soon came to be used widely at laughing gas parties among the nobility. Nitrous oxide, mixed with oxygen, finds some use in modern anesthesia. It is quick acting but not very potent. Concentrations of 50% or greater must be used to be effective. When nitrous oxide is mixed with ordinary air instead of oxygen,

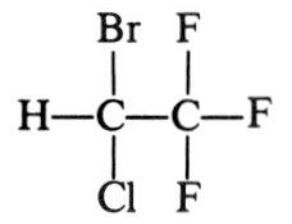

Halothane

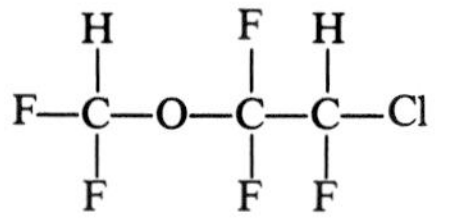

Enflurane

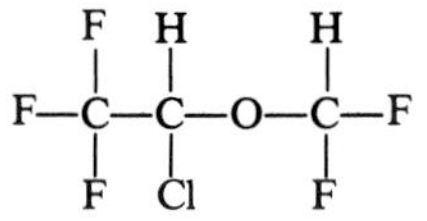

Isofluorane

Figure 14.7
Three modern general anesthetics.

Curare is the arrow poison used by South American Indian tribes. Large doses of curare kill by causing a complete relaxation of all muscles. Death occurs because of respiratory failure.

not enough oxygen gets into the patient's blood, and permanent brain damage can result.

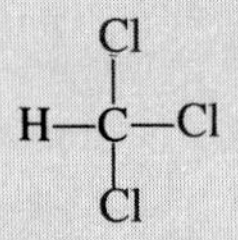

Chloroform
(Trichloromethane)

Chloroform ($CHCl_3$) was introduced as a general anesthetic in 1847. Its use quickly became popular after Queen Victoria gave birth to her eighth child while anesthetized by chloroform in 1853. Chloroform was used widely for years. It is nonflammable and produces effective anesthesia, but it has a number of serious drawbacks. For one, it has a narrow safety margin; the effective dose is close to the lethal dose. It also causes liver damage, and it must be protected from oxygen during storage to prevent the formation of deadly phosgene gas.

Modern anesthetics include fluorine-containing compounds such as halothane, enflurane, and methoxyflurane (Figure 14.7). These compounds are nonflammable and relatively safe for the patient. Their safety, particularly that of halothane, for operating-room personnel, however, has been questioned. For example, female operating room workers suffer a higher rate of miscarriages than the general population.

Modern surgical practice has moved away from the use of a single anesthetic. Generally, a patient is given an intravenous anesthetic such as thiopental (Section F.6) to produce unconsciousness. The gaseous anesthetic then is administered to provide insensitivity to pain and to keep the patient unconscious. A relaxant, such as curare, also may be employed. Curare and related compounds produce profound relaxation; thus, only light anesthesia is required. This practice avoids the hazards of deep anesthesia.

The potency of an anesthetic is related to its solubility in fat. General anesthetics seem to work by dissolving in the fatlike membranes of nerve cells. This changes the permeability of the membranes, and the conductivity of the neurons is depressed.

EXERCISES

1. What is a functional group?
2. Give the structure of and name the functional group in each of the following groups of compounds.
   a. alkenes   **b.** alcohols   **c.** ethers
3. Give structural formulas for the eight isomeric pentyl alcohols ($C_5H_{12}O$). (*Hint:* Three are derived from pentane, four are derived from isopentane, and one is derived from neopentane.) Name the alcohols by the IUPAC system.
4. Classify the alcohols in Exercise 3 as primary, secondary, or tertiary.
5. Draw and name the isomeric ethers with the formula $C_5H_{12}O$.
6. Name the following compounds.
   a. $CH_3CH_2CH_2CH_2CH_2CH_2OH$
   b. $CH_3CH_2CH_2CH_2CHOHCH_3$
   c. $(CH_3)_2CHCH_2OH$
   d. $CH_3CH_2CHOHC(CH_3)_3$
7. Name the following compounds.
   a. $CH_3CHOHCH_2CHCl_2$
   b. $(CH_3)_2COHCHBr_2CH_3$

c.

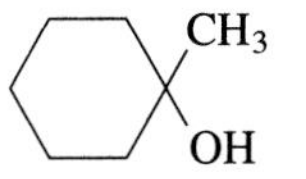

d. $CH_3CH_2CH_2COH(CH_2CH_3)_2$

8. Name the following compounds.
  a. $CH_3CH_2CH_2OCH_2CH_2CH_3$
  b. $CH_3CH_2OCH(CH_3)_2$

  c. 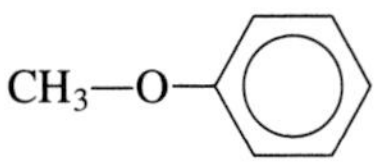 CH_3-O-⟨benzene ring⟩   **d.** 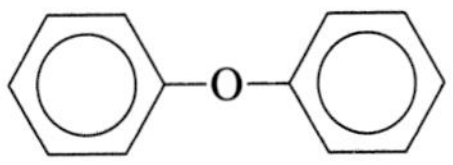 ⟨benzene⟩$-O-$⟨benzene⟩

  e.   **f.** 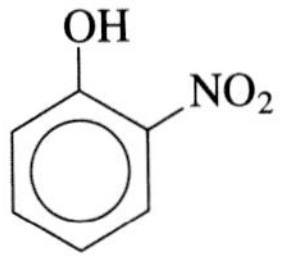 ⟨OH, NO₂ benzene⟩

9. Give structural formulas for the following alcohols.
  a. 3-hexanol   **b.** 3,3-dimethyl-2-butanol
  c. cyclopentanol   **d.** 4-methyl-2-hexanol

10. Give structural formulas for the following alcohols.
  a. 4,5-dimethyl-3-heptanol
  b. 2-ethyl-1-phenyl-1-butanol
  c. 2-bromo-2-chlorocyclobutanol
  d. 3-phenylcyclopentanol

11. Give structural formulas for the following.
  a. propylene glycol   **b.** glycerol

12. Give structural formulas for the following ethers.
  a. ethyl methyl ether   **b.** diisopropyl ether
  c. phenyl benzyl ether   **d.** cyclopropyl propyl ether

13. Give structural formulas for the following.
  a. *m*-iodophenol
  b. *p*-methylphenol (*p*-cresol)
  c. 2,4,6-trinitrophenol (picric acid)
  d. 3,5-diethylphenol

14. Give the IUPAC names for the compounds referred to by the following names.
  a. grain alcohol   **b.** wood alcohol
  c. rubbing alcohol   **d.** carbolic acid

15. **a.** What is denatured alcohol?
  b. Why is some alcohol denatured?

16. **a.** Menthol is one of the ingredients in mentholated cough drops and nasal sprays. It produces a cooling, refreshing sensation when rubbed on the skin and so is used in shaving lotions and cosmetics. What is its IUPAC name?

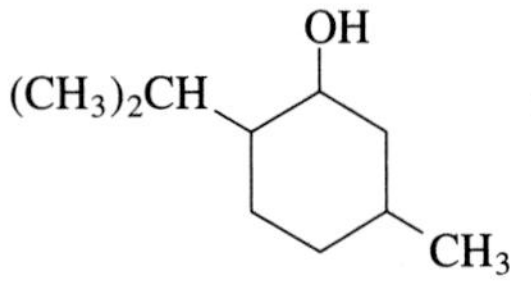

      Menthol 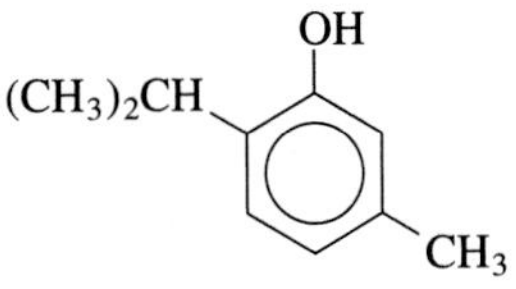   Thymol

  b. It is interesting to note that the aromatic equivalent of menthol is thymol, the flavoring constituent of thyme. Give two names for thymol.

17. Benzyl alcohol and *p*-cresol are isomers. Write the formulas for each. Compare their solubilities in **(a)** water and **(b)** an aqueous solution of NaOH.

18. Why is methanol so much more toxic to humans than ethanol?

19. Why is ethylene glycol so much more toxic to humans than propylene glycol?

20. What chemical compound is used in the treatment of acute methanol or ethylene glycol poisoning? How does it work?

21. Why is ethyl alcohol the only primary alcohol that can be prepared by the hydration of an alkene?

22. Classify each of the following conversions as oxidation, dehydration, or hydration. (Only the organic starting material and product are shown.)

  a. $CH_3OH \longrightarrow H-\overset{\overset{\displaystyle H}{|}}{C}=O$

  b. $CH_3\overset{\overset{\displaystyle OH}{|}}{C}HCH_3 \longrightarrow CH_3CH=CH_2$

  c. $CH_3\overset{\overset{\displaystyle OH}{|}}{C}HCH_3 \longrightarrow CH_3\overset{\overset{\displaystyle O}{||}}{C}CH_3$

  d. $HOOCCH=CHCOOH \longrightarrow HOOCCH_2\overset{\overset{\displaystyle OH}{|}}{C}HCOOH$

  e. $2\ CH_3OH \longrightarrow CH_3OCH_3$

23. Each of the butyl alcohols is treated with potassium dichromate in acid. Draw the product (if any) expected from each of the four isomeric alcohols.

24. Write an equation for the dehydration of 2-propanol **(a)** to yield an alkene and **(b)** to yield an ether.

25. Draw the structural formula of the ether that would form from the *intra*molecular dehydration of $HOCH_2CH_2CH_2CH_2CH_2OH$.

26. Draw the alkene that would form from the dehydration of cyclohexanol.

27. Without consulting tables, arrange the following compounds in order of increasing boiling points: ethanol, 1-propanol, methanol.

28. Without consulting tables, arrange the following compounds in order of increasing boiling points: butane, ethylene glycol, 1-propanol.

29. Without consulting tables, arrange the following compounds in order of increasing boiling points: diethyl ether, propylene glycol, 1-butanol.

30. Without consulting tables, arrange the following compounds in order of increasing solubility in water: methanol, 1-butanol, 1-octanol.

31. Without consulting tables, arrange the following compounds in order of increasing solubility in water: pentane, propylene glycol, diethyl ether.

32. State Markovnikov's rule.

33. What is the product of each of the following reactions?
  a. $CH_2=CHCH_2CH_3 \xrightarrow{H^+,\ H_2O}$

b. $CH_3CHOHCH_3 \xrightarrow[\text{H}^+]{\text{KMnO}_4}$

c. ⬠—OH $\xrightarrow[\text{180 °C}]{\text{concd H}_2\text{SO}_4}$

d. ⬡—$CH_2OH \xrightarrow[\text{H}^+]{\text{K}_2\text{Cr}_2\text{O}_7}$

34. What reagents are necessary to carry out the following conversions?

a. $CH_3CH{=}CH_2 \xrightarrow{?} CH_3\overset{\text{OH}}{\underset{}{C}}HCH_3$

b. $CH_2{=}\overset{}{\underset{CH_3}{C}}{-}CH_3 \xrightarrow{?} CH_3\overset{\text{OH}}{\underset{CH_3}{C}}{-}CH_3$

c. $CH_3CH_2OH \xrightarrow{?} CH_2{=}CH_2$

b. $CH_3CH_2CH_2OH \xrightarrow{?} CH_3CH_2\overset{\text{O}}{\underset{}{C}}{-}H$

e. $CH_3\overset{}{\underset{OH}{C}}HCH_3 \xrightarrow{?} CH_3\overset{\text{O}}{\underset{}{C}}CH_3$

f. $2\ CH_3CH_2OH \xrightarrow{?} CH_3CH_2OCH_2CH_3$

35. In the preparation of diethyl ether from ethanol, why is it so critical to maintain the reaction temperature between 130 and 150 °C?

36. Methanol is not particularly toxic to rats. If methanol were newly discovered and tested for toxicity in laboratory animals, what would you conclude about its safety for human consumption?

37. In addition to ethanol, the fermentation of grain produces other organic compounds collectively called fusel oils (FO). The four principal FO components are 1-propanol, isobutyl alcohol, 3-methyl-1-butanol, and 2-methyl-1-butanol. Draw a structural formula for each of these alcohols. (FO is quite toxic and accounts in part for hangovers.)

38. Tetrahydrocannabinol (THC) is the principal active ingredient in marijuana. What functional groups are present in the THC molecule?

Tetrahydrocannabinol
(THC)

39. Give the structure of the alkene from which each of the following alcohols is made by reaction with water in acidic solution:

a. $CH_3\overset{}{\underset{OH}{C}}HCH_3$ **b.** CH_3CH_2OH

c. $CH_3\overset{CH_3}{\underset{CH_3}{C}}{-}OH$ **d.** $CH_3\overset{}{\underset{OH}{C}}HCH_2CH_3$

e. ⬡—OH

40. Write an equation for the reaction (if any) of phenol with aqueous
a. NaOH   **b.** $NaHCO_3$

41. What is a polyhydric alcohol?

42. What is a glycol?

43. What precautions must be taken when using diethyl ether as a solvent in a laboratory experiment?

44. Ethyl alcohol, like rubbing alcohol, is often used for sponge baths. What property of alcohols makes them useful for this purpose?

45. Give the name and one use for each of the following compounds.
a. CH_3OH **b.** CH_3CH_2OH
c. $CH_3CHOHCH_3$ **d.** CH_2OHCH_2OH
e. $CH_2OHCHOHCH_2OH$ **f.** ⬡—OH

46. Define the following terms.
a. absolute alcohol   **b.** general anesthetic
c. antiseptic   **d.** disinfectant
e. 86 proof   **f.** LD_{50}
g. azeotrope   **h.** gasohol

47. Write the equation for the production of ethyl alcohol by the addition of water to ethylene. How much ethyl alcohol can be made from 14 kg of ethylene?

48. The label on a bottle of light wine indicates that 100 mL of the wine furnishes 70 Cal (food calories), 0.2 g of protein, 5.77 g of carbohydrates, and 0.0 g of fat. Assuming that carbohydrates and proteins furnish 4 Cal/g each, that alcohol furnishes 7 Cal/g, and that no other caloric nutrients are present, answer the following questions.
a. How many Calories are provided by the alcohol in a 100-mL serving of the wine? What percentage of the total Calories is provided by alcohol?
b. How many grams of alcohol are there in each 100-mL serving?
c. The density of alcohol is 0.789 g/mL. How many mL of alcohol are there in each 100-mL serving? What is the percent alcohol by volume?

Chapter 15
ALDEHYDES AND KETONES

Aldehydes are responsible for the odor and flavor of cinnamon as well as many other foods and spices.

What do certain hormones, vanilla flavor, a biological tissue preservative, and fresh cucumbers have in common? The answer: a carbonyl functional group. The carbonyl group is characteristic of aldehydes and ketones, the families we shall consider in this chapter. As the opening list indicates, this functional group and these two families of compounds are found in a most diverse company of products. Both the tempting aromas associated with cinnamon, vanilla, and fresh, buttered baked goods and the sickeningly sweet smell of some rancid foods are associated with the carbonyl group.

The aldehydes and ketones offer us an opportunity to study the carbonyl group in its simplest surroundings. In Chapter 16, we consider somewhat more complicated functional groups incorporating the carbonyl group. And we ultimately find ourselves running into this ubiquitous grouping of atoms in carbohydrates, fats, proteins, nucleic acids, hormones, vitamins, and the host of organic compounds critical to the functioning of living systems. But first things first. Let's begin by focusing on the carbonyl group in aldehydes and ketones.

Table 15.2
Nomenclature of Ketones

Molecular Formula	Condensed Structural Formula	Common Name	IUPAC Name
C_3H_6O	$CH_3\overset{\overset{\displaystyle O}{\|\|}}{C}CH_3$	Acetone (dimethyl ketone)	Propanone
C_4H_8O	$CH_3\overset{\overset{\displaystyle O}{\|\|}}{C}CH_2CH_3$	Ethyl methyl ketone	Butanone
C_4H_6O	$CH_2{=}CH\overset{\overset{\displaystyle O}{\|\|}}{C}CH_3$	Methyl vinyl ketone	3-Buten-2-one (NOT 1-Buten-3-one)
$C_5H_{10}O$	$CH_3CH_2\overset{\overset{\displaystyle O}{\|\|}}{C}CH_2CH_3$	Diethyl ketone	3-Pentanone
$C_5H_{10}O$	$CH_3CH_2CH_2\overset{\overset{\displaystyle O}{\|\|}}{C}CH_3$	Methyl propyl ketone	2-Pentanone
$C_5H_{10}O$	$CH_3\overset{\overset{\textstyle }{}}{\underset{\underset{\displaystyle CH_3}{\|}}{CH}}\overset{\overset{\displaystyle O}{\|\|}}{C}CH_3$	Isopropyl methyl ketone	3-Methyl-2-butanone (NOT 2-Methyl-3-butanone)
$C_6H_{10}O$	(cyclohexanone ring =O)	Cyclohexanone	Cyclohexanone
C_8H_8O	(phenyl)$-\overset{\overset{\displaystyle O}{\|\|}}{C}-CH_3$	Acetophenone (Methyl phenyl ketone)	Phenylethanone
$C_{13}H_{10}O$	(phenyl)$-\overset{\overset{\displaystyle O}{\|\|}}{C}-$(phenyl)	Benzophenone (Diphenyl ketone)	Diphenylmethanone

EXAMPLE 15.4

Give the IUPAC name for   O=⬠ (cyclopentanone ring with CH₃)
CH₃

SOLUTION
There are five carbon atoms in the ring, and the methyl group is on the third carbon.

O=⬠ (ring numbered 1, 5, 4, 2, 3)
CH₃

The name is 3-methylcyclopentanone.

Practice Exercise
Give the IUPAC name for   (cyclohexanone ring with Cl at top, O at right)

EXAMPLE 15.5

What is the IUPAC name for the following ketone?

$$\overset{1}{C}H_3\overset{2}{C}H\!-\!\overset{3}{C}\!-\!\overset{4}{C}H\overset{5}{C}H_3$$
$$\qquad\;\;\; \underset{CH_3}{|}\;\; \overset{\|}{O}\;\; \underset{CH_3}{|}$$

SOLUTION

The name is 2,4-dimethyl-3-pentanone.

15.4
PHYSICAL PROPERTIES OF ALDEHYDES AND KETONES

The carbon and oxygen of the carbonyl group share two pairs of electrons, but they do not share them equally. The electronegative oxygen has a much greater attraction for the bonding pairs. Thus, the electron density is greater at the oxygen end of the bond and less at the carbon end. The carbon is left with a partial positive charge; the oxygen, with a partial negative charge. (Polar bonds were encountered earlier in Chapter 4.)

$$\overset{\delta+}{\underset{/}{\overset{\backslash}{C}}}=\overset{\delta-}{O}$$

The polarity of the carbon–oxygen double bond is greater than that of the carbon–oxygen single bond. Indeed, double-bond polarity is great enough to affect the boiling points of aldehydes and ketones, whereas the polar single bonds in ethers have little effect on boiling points (Table 15.3). Such dipolar forces, however, are still not comparable to the hydrogen bonding that exists between molecules of an alcohol.

With the exception of the gaseous formaldehyde, the majority of the aldehydes are liquids. (The physical state of acetaldehyde, bp 20 °C, depends on the temperature of the laboratory; in warm rooms acetaldehyde exists as a gas.) Although the lower members of the series have pungent odors, many other aldehydes have pleasant odors and are used in making perfumes and artificial flavorings. The hydrogens of water molecules can form hydrogen bonds with the carbonyl oxygen; thus the solubility of aldehydes is about the same as that of alcohols and ethers. Formaldehyde and acetaldehyde are soluble in water; as the carbon chain increases, water solubility decreases. The borderline of solubility occurs at about four carbon atoms per oxygen atom. All aldehydes are soluble in organic solvents and, in general, are less dense than water.

Compound	MW	Type of Intermolecular Forces	Boiling Point (°C)
$CH_3CH_2CH_2CH_3$	58	Dispersion only	−1
$CH_3OCH_2CH_3$	60	Weak dipole	6
$CH_3CH_2\overset{O}{\overset{\|}{C}}H$	58	Strong dipole	49
$CH_3CH_2CH_2OH$	60	Hydrogen bonding	97

Table 15.3

Boiling Points of Compounds with Similar Molecular Weights and Different Types of Intermolecular Forces

Compound	Formula	Boiling Point (°C)	Solubility in Water (g/100 g H_2O)
Formaldehyde	HCHO	−21	Miscible
Acetaldehyde	CH_3CHO	20	Miscible
Propionaldehyde	CH_3CH_2CHO	49	16
Butyraldehyde	$CH_3CH_2CH_2CHO$	76	7
Valeraldehyde	$CH_3CH_2CH_2CH_2CHO$	103	Slightly soluble
Benzaldehyde	C_6H_5CHO	178	0.3
Acetone	CH_3COCH_3	56	Miscible
Ethyl methyl ketone	$CH_3COCH_2CH_3$	80	26
Methyl propyl ketone	$CH_3COCH_2CH_2CH_3$	102	6.3
Diethyl ketone	$CH_3CH_2COCH_2CH_3$	101	5

The physical properties of the ketones are almost identical to those of the corresponding aldehydes. Acetone has a pleasant odor, and it is the only ketone that is completely soluble in water. The higher homologs are colorless liquids, are slightly soluble in water, and, unlike the aldehydes, have rather bland odors. Table 15.4 lists some physical constants for several of the aldehydes and ketones.

15.5
PREPARATION OF ALDEHYDES AND KETONES

In Section 14.6 the oxidation of primary and secondary alcohols to form aldehydes and ketones, respectively, was mentioned. However, in aqueous solutions the product aldehyde forms a hydrate that is further oxidized to a carboxylic acid. Therefore, organic solvents are used in the preparation of aldehydes from alcohols. The reagent of choice is chromic oxide in combination with pyridine, methylene chloride, and HCl. In organic solvents, chromium(VI) ions are mild oxidizing agents that can oxidize primary alcohols to aldehydes, without further oxidizing the aldehydes to acids. We shall see in Chapters 24 and 25 that the enzyme-catalyzed oxidation of alcohols to aldehydes and ketones is of great significance in biological systems.

R—CH₂OH →[O]→ An aldehyde

A primary alcohol

$3\ CH_3CH_2CH_2CH_2OH\ +\ CrO_3\ +\ 3\ HCl\ \xrightarrow[CH_2Cl_2]{pyridine}\ 3\ CH_3CH_2CH_2CHO\ +\ CrCl_3\ +\ 3\ H_2O$

1-Butanol   (orange)   Butanal   (green)

Benzyl alcohol →[CrO₃, H⁺ / pyridine/CH₂Cl₂]→ Benzaldehyde

Like the aldehydes, ketones are obtained by the oxidation of an alcohol. However, the alcohol must be a secondary alcohol, and no special reagents are needed because the ketone is not susceptible to further oxidation. Although many oxidants are used, the most commonly employed are chromium(VI) compounds and sulfuric acid.

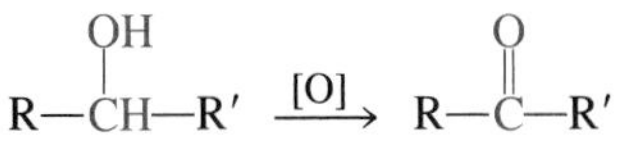

$$
\underset{\substack{\text{A secondary}\\\text{alcohol}}}{R-\overset{\overset{\displaystyle OH}{|}}{C}H-R'} \xrightarrow{[O]} \underset{\text{A ketone}}{R-\overset{\overset{\displaystyle O}{\|}}{C}-R'}
$$

$$
\underset{\text{Isopropyl alcohol}}{CH_3\overset{\overset{\displaystyle OH}{|}}{C}HCH_3} \xrightarrow[\text{H}_2\text{SO}_4]{\text{K}_2\text{Cr}_2\text{O}_7} \underset{\text{Acetone}}{CH_3\overset{\overset{\displaystyle O}{\|}}{C}CH_3}
$$

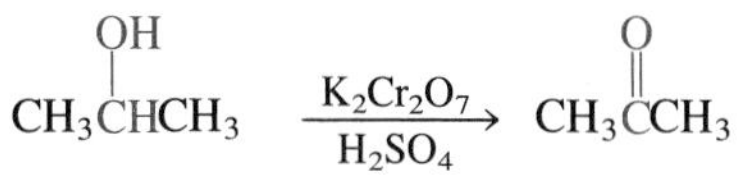

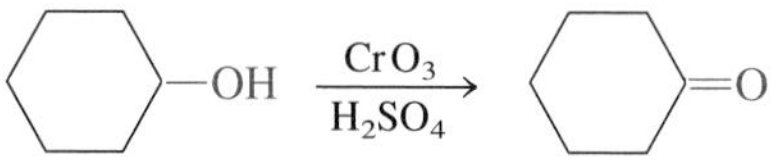

$$
\underset{\text{Cyclohexanol}}{\bighexagon\!-OH} \xrightarrow[\text{H}_2\text{SO}_4]{\text{CrO}_3} \underset{\text{Cyclohexanone}}{\bighexagon\!=O}
$$

As we shall see in Chapters 24 and 25, the electrons that are released during the oxidation of alcohols to carbonyl compounds are converted into a form of energy that can be utilized by the cell. These biochemical oxidation reactions are carried out at body temperature (~37 °C) and are catalyzed by enzymes. Notice that in the following reaction the enzyme selectively catalyzes the oxidation of the secondary alcohol group to a ketone, but it does not oxidize the primary alcohol group to an aldehyde. We discuss enzyme specificity in Section 22.4.

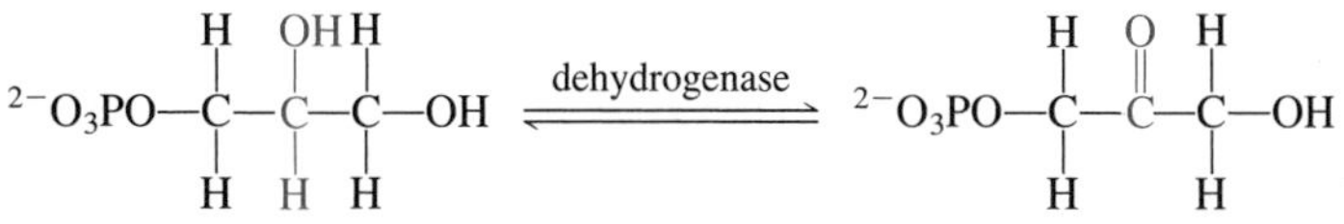

$$
\underset{\text{Glycerol 3-phosphate}}{^{2-}O_3PO-\overset{\overset{\displaystyle H}{|}}{\underset{\underset{\displaystyle H}{|}}{C}}-\overset{\overset{\displaystyle OH}{|}}{\underset{\underset{\displaystyle H}{|}}{C}}-\overset{\overset{\displaystyle H}{|}}{\underset{\underset{\displaystyle H}{|}}{C}}-OH} \underset{\xrightarrow{\text{dehydrogenase}}}{\rightleftharpoons} \underset{\text{Dihydroxyacetone phosphate}}{^{2-}O_3PO-\overset{\overset{\displaystyle H}{|}}{\underset{\underset{\displaystyle H}{|}}{C}}-\overset{\overset{\displaystyle O}{\|}}{C}-\overset{\overset{\displaystyle H}{|}}{\underset{\underset{\displaystyle H}{|}}{C}}-OH}
$$

15.6
CHEMICAL PROPERTIES OF ALDEHYDES AND KETONES

Oxidation

Aldehydes are readily oxidized to carboxylic acids. Ketones resist oxidation.

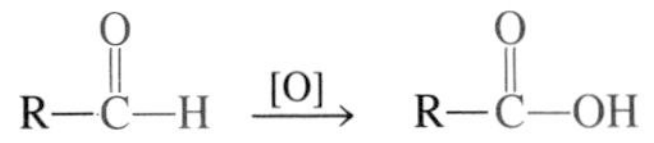

$$
\underset{\text{An aldehyde}}{R-\overset{\overset{\displaystyle O}{\|}}{C}-H} \xrightarrow{[O]} \underset{\text{Carboxylic acid}}{R-\overset{\overset{\displaystyle O}{\|}}{C}-OH}
$$

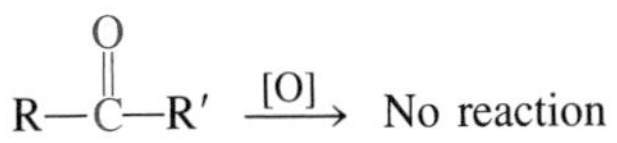

$$
\underset{\text{A ketone}}{R-\overset{\overset{\displaystyle O}{\|}}{C}-R'} \xrightarrow{[O]} \text{No reaction}
$$

The aldehydes are, in fact, among the most easily oxidized of organic compounds, and this fact helps chemists identify them. Through the use of oxidizing agents, aldehydes can be distinguished not only from ketones but also from alcohols if the reagent is gentle enough. One such test reagent was invented by Professor Bernhard Tollens (1841–1918) at the University of Göttingen in Germany. *Tollens's reagent* employs silver ion as the mild oxidizing agent. In order for the silver ion to be kept in solution, it must be complexed by two ammonia molecules.

$$H_3N\!-\!Ag^+\!-\!NH_3$$

When Tollens's reagent oxidizes an aldehyde, the silver ion is reduced to free silver.

$$H\!-\!\overset{\displaystyle O}{\underset{\displaystyle H}{C}} \;+\; 2\,Ag(NH_3)_2^+ \;+\; 3\,OH^- \;\longrightarrow\; H\!-\!\overset{\displaystyle O}{\underset{\displaystyle O^-}{C}} \;+\; 2\,Ag(s) \;+\; 4\,NH_3 \;+\; 2\,H_2O$$

Silver mirror

The silver, when deposited on a clean glass surface, produces a beautiful mirror. Indeed, mirrors are often silvered by means of the Tollens reaction. The reducing agent of choice is often the sugar glucose (which contains an aldehyde functional group) rather than a simple aldehyde. Ordinary ketones do not react with Tollens's reagent.

Two other test reagents, Benedict's and Fehling's, make use of alkaline solutions of copper(II) ion (Cu^{2+}). The source of the ion is copper(II) sulfate. Because Cu^{2+} forms an insoluble hydroxide in basic solution, another reagent must be added to the solution to keep the copper ion from precipitating out as the hydroxide. In *Benedict's solution*, sodium citrate is employed for this purpose; copper remains in solution as the copper(II) citrate ion. The additional reagent in *Fehling's solution* is sodium potassium tartrate (Rochelle salt) with which copper forms the water-soluble copper(II) tartrate ion. The blue color of these solutions is due to the presence of copper(II) ion complexes. A positive test for the aldehyde group is evidenced by a color change to brick red, indicating the presence of the copper(I) oxide.

The copper(II) ion (+2 oxidation state) is the oxidizing agent and therefore must be the substance that is reduced, in this case to copper(I) oxide (+1 oxidation state).

$$CH_3\overset{\displaystyle O}{\underset{\displaystyle H}{C}} \;+\; 2\,Cu^{2+} \;+\; 5\,OH^- \;\longrightarrow\; CH_3\overset{\displaystyle O}{\underset{\displaystyle O^-}{C}} \;+\; Cu_2O(s) \;+\; 3\,H_2O$$
(blue)   (red)

Although ketones are resistant to oxidation by ordinary laboratory oxidizing agents, it is possible to force their oxidation. And, in particular, it should be recognized that both aldehydes and ketones will undergo combustion, that is, will burn. Acetone is a common organic solvent. Neither it nor any other volatile, flammable organic solvent should be used around open flames, heating elements, or other sources of possible ignition.

Reduction

Aldehydes and ketones are readily reduced to the corresponding primary and secondary alcohols, respectively. A wide variety of reducing agents may be used.

$$R\!-\!\overset{\displaystyle O}{\underset{\displaystyle H}{C}} \;\xrightarrow{2[H]}\; R\!-\!\overset{\displaystyle H}{\underset{\displaystyle H}{C}}\!-\!OH \qquad \text{and} \qquad R\!-\!\overset{\displaystyle O}{C}\!-\!R' \;\xrightarrow{2[H]}\; R\!-\!\overset{\displaystyle OH}{\underset{\displaystyle H}{C}}\!-\!R'$$

Carbonyl compounds can be reduced to alcohols by hydrogen gas in the presence of a metal catalyst (catalytic hydrogenation). However, this method suffers from the disadvantages that many of the catalysts (Pt, Pd, Ru) are expensive and that other functional groups are also reduced.

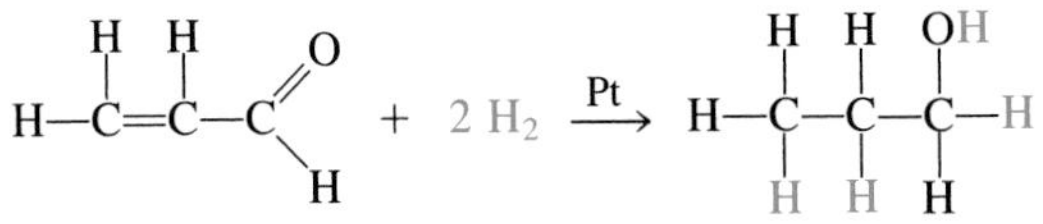

Acrolein
(2-Propenal)
1-Propanol

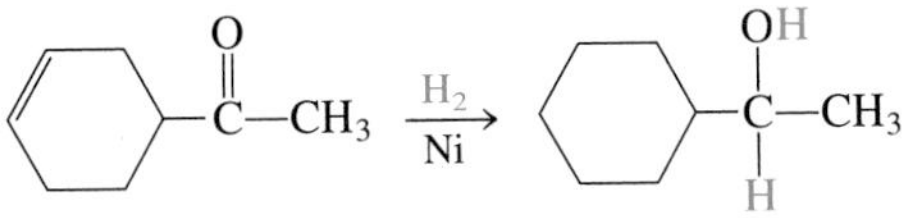

Two extremely important biochemical carbonyl reduction reactions are discussed in Chapter 24. They are the reduction of acetaldehyde to ethyl alcohol and the reduction of pyruvic acid to lactic acid. In each case, the enzyme that catalyzes the reaction contains the coenzyme NADH, which is the reducing agent. (See Table I.1 for the structure of the coenzyme.)

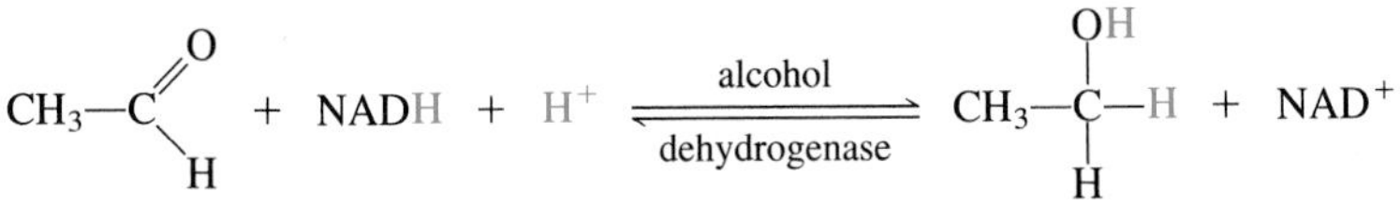

Acetaldehyde
Ethyl alcohol

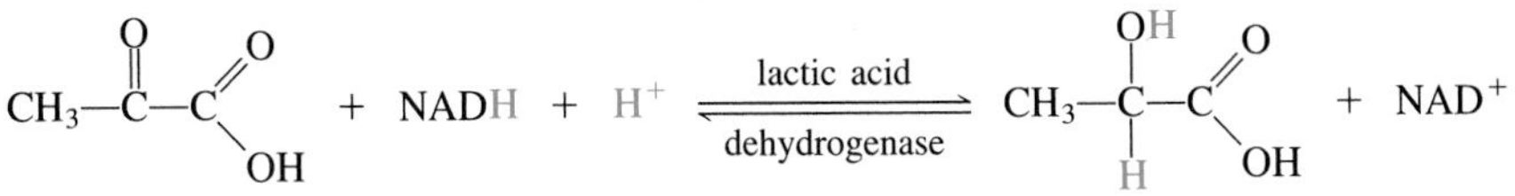

Pyruvic acid
Lactic acid

Hydration of Carbonyl Compounds

Formaldehyde is a gas at room temperature, yet it dissolves readily in water. In fact, formaldehyde actually *reacts* with water.

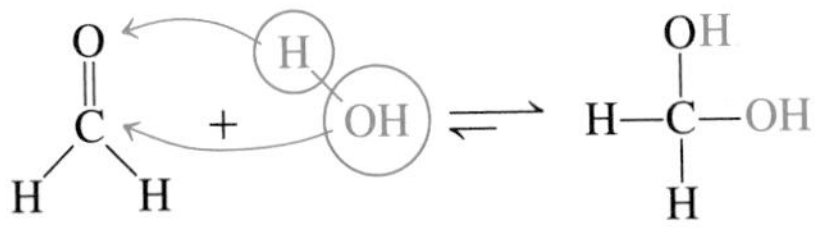

The process is an addition reaction, analogous to the hydration of the carbon–carbon double bond of an alkene. The net result is that a hydrogen from water is added to the carbonyl oxygen and a hydroxyl group from water becomes attached to the carbonyl carbon. The product is called a **hydrate.** It readily breaks down to re-form formaldehyde and water. At equilibrium at 20 °C, the hydrate predominates. Indeed, only 1 molecule in 10,000 exists as free formaldehyde. The other 9999 are in the form of the hydrate.

Acetaldehyde is also hydrated in aqueous solution, but to a lesser extent than formaldehyde.

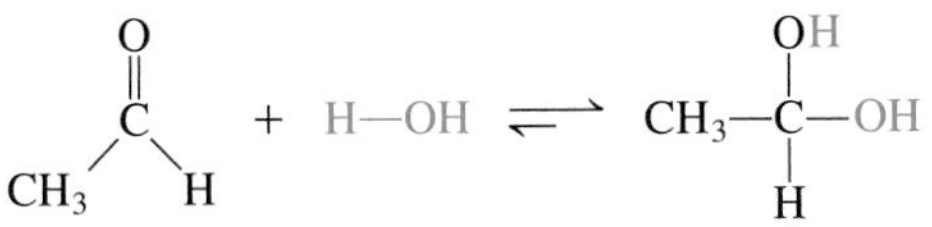

Out of 10,000 molecules, about 4200 are in the form of the free aldehyde at equilibrium. Still, that leaves 5800 in the hydrated form. Generally, higher aldehydes and ketones are even less hydrated, existing primarily in the free aldehyde (or ketone) form at equilibrium in water.

In most cases, it is impossible to isolate the hydrates from solution. Attempts to do so result in loss of water and regeneration of the carbonyl compound. An exception is the hydrate of trichloroacetaldehyde (chloral).

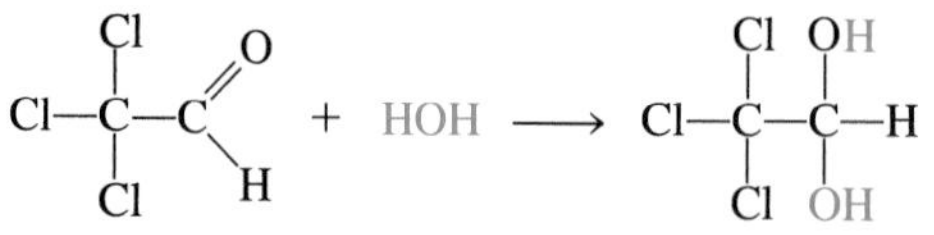

Chloral   Chloral hydrate

The product, called chloral hydrate, is a stable solid, soluble in water, and one of the very few organic compounds that possess two hydroxyl groups on the same carbon atom. It is a powerful sedative and soporific (sleep-inducing drug). Chloral hydrate has had wide use in medicine. It is perhaps even better known in fictional mystery stories. Slipped into someone's drink, the mixture is called a "Mickey Finn" or "knockout drops." Such combinations of alcohol and chloral hydrate—two "downers"—are exceedingly dangerous. A little too much, and the unfortunate victim may be put to sleep permanently.

The Addition of Alcohols

Alcohols add to the carbonyl group of aldehydes and ketones in much the same way as water. The addition of 1 mol of an alcohol to 1 mol of an aldehyde or ketone yields a **hemiacetal** or **hemiketal,** respectively. In the presence of an anhydrous acid catalyst, equilibrium is rapidly established, and the equilibrium favors the carbonyl compounds (the reactants). As with the hydrates, simple hemiacetals and hemiketals are generally not sufficiently stable to be isolated.

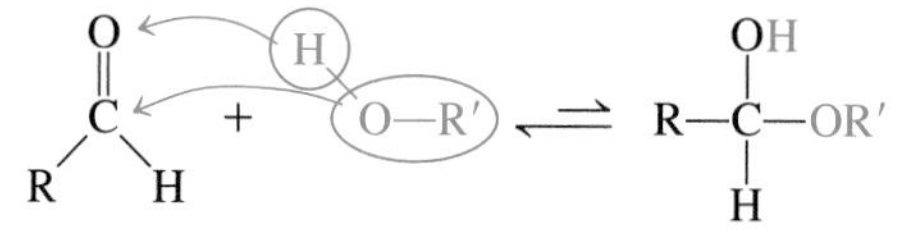

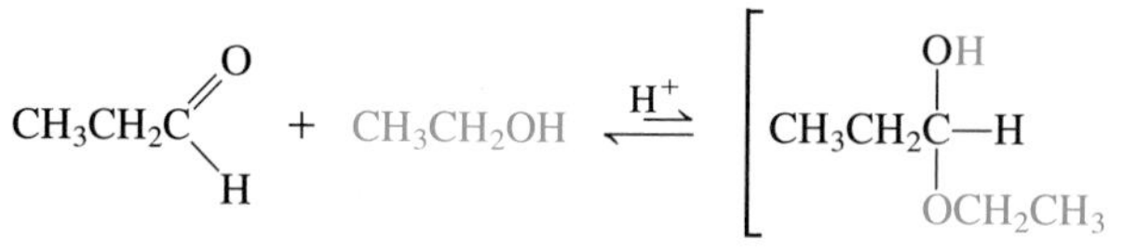

Unstable hemiacetal

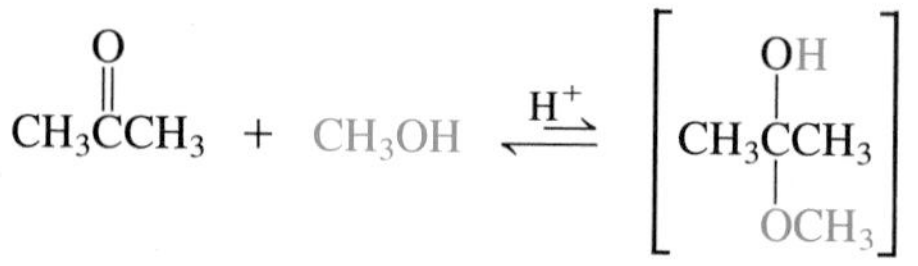

Unstable hemiketal

If, however, the alcohol and carbonyl groups occur within the same molecule, then the equilibrium favors the formation of cyclic hemiacetals and hemiketals. These cyclic compounds result from *intramolecular* interaction between the —OH and C=O groups. The cyclization reactions are of particular significance in our discussion of the structures of monosaccharides (see Section 19.5).

5-Hydroxypentanal   Cyclic hemiacetal

Hemiacetals can be made to react further with alcohols. If dry hydrogen chloride gas is bubbled into a solution of aldehyde in excess alcohol, an **acetal** is formed. The reaction for acetaldehyde and methanol is

A hemiacetal   An acetal
(unstable)   (stable)

Unlike hemiacetals and hydrates, acetals are stable. They can be isolated in pure form. First an alcohol molecule adds to the double bond of the aldehyde to form the hemiacetal. Then the hydroxyl group of the hemiacetal and the hydrogen from the hydroxyl group of a second alcohol molecule are eliminated, and the two remaining pieces combine to form the acetal. We shall see in Section 19.7 that the naturally occurring disaccharides (maltose, lactose, and sucrose) are acetals.

Acetal, or ketal (from a ketone), formation is often used to ''protect'' the functional group of aldehydes or ketones while other chemical operations are performed on the molecules, because acetals are resistant to oxidation whereas aldehydes themselves are not. An aldehyde is converted to an acetal, an oxidation reaction is then carried out on another part of the molecule, and finally the aldehyde is regenerated from the acetal. The carbonyl group is easily regenerated by aqueous acid.

In an interesting application, the antibiotic chloramphenicol is treated with acetone, and a protective cyclic ketal is formed that masks the bitter taste of the drug. Both of the alcohol hydroxyl groups required to form the ketal are attached to a single molecule in this product.

Chloramphenicol (bitter)   Acetone

A cyclic ketal (not bitter)

The cyclic ketal is converted back to the free chloramphenicol by acids in the digestive system. Chloramphenicol is a powerful, but hazardous, antibiotic. It is used only when other, less dangerous, drugs are ineffective. In about 1 person in 20,000–40,000 (depending on dosage), chloramphenicol causes fatal aplastic anemia.

EXAMPLE 15.6

Complete the following equations.

a. $\text{CH}_3\text{CH}_2\text{C}\overset{O}{\underset{H}{}}$ + 2 $\text{CH}_3\text{CH}_2\text{OH}$ $\xrightarrow{\text{H}^+}$

b. $\text{CH}_3\overset{O}{\overset{\|}{\text{C}}}\text{CH}_3$ + 2 CH_3OH $\xrightarrow{\text{H}^+}$

SOLUTION

Realize that each reaction involves two steps. First the addition of 1 mol of alcohol to 1 mol of the carbonyl compound.

a. $\text{CH}_3\text{CH}_2\text{C}\overset{O}{\underset{H}{}}$ + $\text{CH}_3\text{CH}_2\text{OH}$ $\underset{\longleftarrow}{\xrightarrow{\text{H}^+}}$ $\text{CH}_3\text{CH}_2\overset{OH}{\underset{OCH_2CH_3}{\overset{|}{\text{C}}}}\text{—H}$

b. $\text{CH}_3\overset{O}{\overset{\|}{\text{C}}}\text{CH}_3$ + CH_3OH $\underset{\longleftarrow}{\xrightarrow{\text{H}^+}}$ $\text{CH}_3\overset{OH}{\underset{OCH_3}{\overset{|}{\text{C}}}}\text{CH}_3$

This is followed by the interaction of the hemiacetal and hemiketal with a second mole of the alcohol.

a. $\text{CH}_3\text{CH}_2\overset{OH}{\underset{OCH_2CH_3}{\overset{|}{\text{C}}}}\text{—H}$ + $\text{CH}_3\text{CH}_2\text{OH}$ $\xrightarrow{\text{H}^+}$ $\text{CH}_3\text{CH}_2\overset{OCH_2CH_3}{\underset{OCH_2CH_3}{\overset{|}{\text{C}}}}\text{—H}$ + HOH

A hemiacetal   An acetal

b. $\text{CH}_3\overset{OH}{\underset{OCH_3}{\overset{|}{\text{C}}}}\text{CH}_3$ + CH_3OH $\xrightarrow{\text{H}^+}$ $\text{CH}_3\overset{OCH_3}{\underset{OCH_3}{\overset{|}{\text{C}}}}\text{CH}_3$ + HOH

A hemiketal   A ketal

Practice Exercise

Complete the following equation.

$\bigcirc\!\!-\!\text{C}\overset{O}{\underset{H}{}}$ + $\text{CH}_3\text{CH}_2\text{CH}_2\text{OH}$ $\longrightarrow$

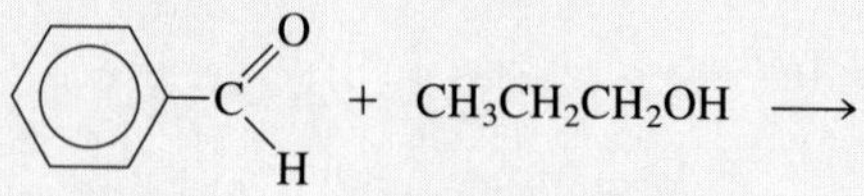

15.7
SOME COMMON CARBONYL COMPOUNDS

Formaldehyde is the simplest and industrially the most important member of the aldehyde family. It is manufactured from methanol and oxygen in the air by passing methanol vapors over a copper or silver catalyst at temperatures above 300 °C. Formaldehyde is a colorless gas with an extremely irritating odor. Because of its reactivity, formaldehyde cannot be handled easily in the gaseous state and is therefore dissolved in water and sold as a 37–40% aqueous solution (such a solution is called *formalin*).

The largest use of formaldehyde is as a reagent for the preparation of many other organic compounds and for the manufacture of polymers such as Bakelite, Formica, and Melmac. Formaldehyde can denature proteins (see Section 21.14), rendering them insoluble in water and resistant to decay bacteria. For this reason it is used in embalming solutions and in the preservation of biological specimens. Formalin is also used as a general antiseptic in hospitals to sterilize gloves and surgical instruments. However, its use as an antiseptic, preservative, and embalming fluid has declined because formaldehyde is suspected of being carcinogenic.

Acetaldehyde is an extremely volatile, colorless liquid. It is prepared by the catalytic (Ag) oxidation of ethyl alcohol or by the catalytic ($PdCl_2$) oxidation of ethylene. It serves as the starting material for the preparation of many other organic compounds such as acetic acid, ethyl acetate, and chloral. Acetaldehyde is formed as a metabolite in the fermentation of sugars and in the detoxification of alcohol in the liver (see Chapter 24).

Acetone is the simplest and most important of the ketones. It is produced in large quantities by the catalytic (Ag) oxidation of isopropyl alcohol. Because it is miscible with water as well as with most organic solvents, acetone finds its chief use as an industrial solvent (e.g., for paints and lacquers). It is the chief ingredient in some brands of nail polish remover. Acetone is also an important intermediate for the preparation of chloroform, iodoform, dyes, methacrylates, and many other complex organic compounds.

Acetone is formed in the human body as a by-product of lipid metabolism. Normally it does not accumulate to an appreciable extent because it is subsequently oxidized to carbon dioxide and water. The normal concentration of acetone in the human body is less than 1 mg/100 mL of blood. In the case of certain abnormalities, such as diabetes mellitus, the acetone concentration rises above this level. The acetone is then excreted in the urine, where it can be easily detected. In severe cases, its odor can be noted on the breath (see Section 26.6).

Many other familiar substances contain aldehydes or ketones as the active principles (Figure 15.1). Benzaldehyde is the major component of oil of bitter almond. Cinnamaldehyde is oil of cinnamon. Biacetyl contributes to the aroma and taste of fresh butter. Camphor is a bicyclic ketone. Irone is a ketone with the odor of violets; it is used in many perfumes. Vanillin is the active principle of vanilla flavoring; it is produced synthetically for use in imitation vanilla. Muscone, formed in special glands of the musk deer, is used in perfumes.

Even the odor of green leaves is due in part to carbonyl compounds. Most green leaves contain *cis*-3-hexenal. The compound *trans*-2-*cis*-6-nonadienal has a cucumber odor. These and other carbonyl compounds (with related acetals, ketals, and alcohols) impart a ''green'' herbal odor to shampoos and other cosmetics.

Formaldehyde is present in wood smoke and is one of the compounds responsible, by killing bacteria, for the preservative effect of the smoking of foods.

Figure 15.1

Some interesting aldehydes and ketones. Benzaldehyde is an oil found in almonds. Cinnamaldehyde is oil of cinnamon. 2,3-Butanedione is responsible for the flavor of butter. Irone is responsible for the odor of violets. Vanillin, of course, gives vanilla its flavor. Muscone is musk oil, a common ingredient of perfumes. *cis*-3-Hexenal provides an herbal odor. *trans*-2-*cis*-6-Nonadienal gives a cucumber odor.

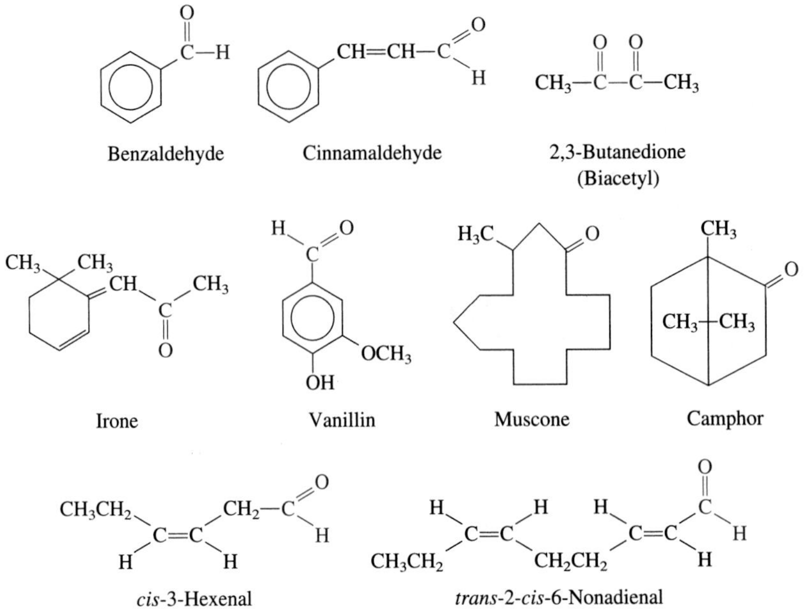

Benzaldehyde   Cinnamaldehyde   2,3-Butanedione
                                       (Biacetyl)

Irone   Vanillin   Muscone   Camphor

cis-3-Hexenal   *trans*-2-*cis*-6-Nonadienal

Several of the steroid hormones (Special Topic H) have the carbonyl functional group as an integral part of their structure. Progesterone is a hormone secreted by the ovaries. It stimulates the growth of cells in the wall of the uterus, preparing the uterine wall for attachment of the fertilized egg. Testosterone is the main male sex hormone. These (and other) sex hormones affect our development and our lives in most fundamental ways.

EXERCISES

1. Draw structures and give common and IUPAC names for the four isomeric aldehydes having the formula $C_5H_{10}O$.
2. Draw structures and give common and IUPAC names for the three isomeric ketones having the formula $C_5H_{10}O$.
3. Name the following.

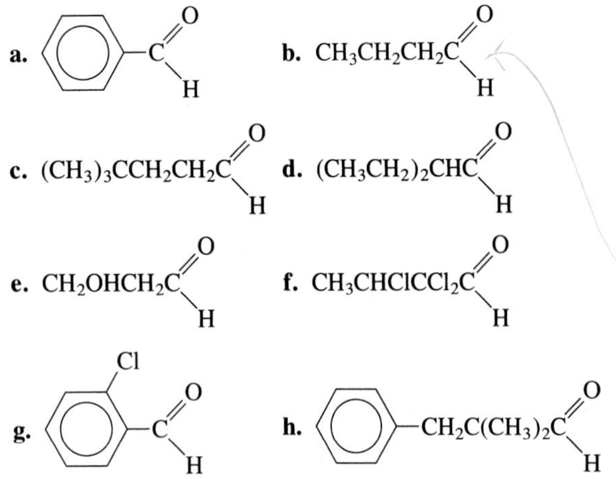

   a.

   b. $CH_3CH_2CH_2C\overset{O}{\underset{H}{\big\backslash}}$

   c. $(CH_3)_3CCH_2CH_2C\overset{O}{\underset{H}{\big\backslash}}$

   d. $(CH_3CH_2)_2CHC\overset{O}{\underset{H}{\big\backslash}}$

   e. $CH_2OHCH_2C\overset{O}{\underset{H}{\big\backslash}}$

   f. $CH_3CHClCCl_2C\overset{O}{\underset{H}{\big\backslash}}$

   g.

   h. $\bigcirc$—$CH_2C(CH_3)_2C\overset{O}{\underset{H}{\big\backslash}}$

4. Name the following.

   a. $CH_3CH_2\overset{O}{\overset{\|}{C}}CH_2CH(CH_3)_2$ b.

   c. $CH_3\overset{O}{\overset{\|}{C}}CH_2CH_2CH_3$ d. $(CH_3)_3C\overset{O}{\overset{\|}{C}}CHBrCH_3$

   e. $(CH_3)_2CH\overset{O}{\overset{\|}{C}}CHCl_2$ f. $CH_3CH_2CH(CH_3)\overset{O}{\overset{\|}{C}}CH_3$

   g. $\bigcirc$—$CH(CH_3)\overset{O}{\overset{\|}{C}}CH_3$ h. CH_3—

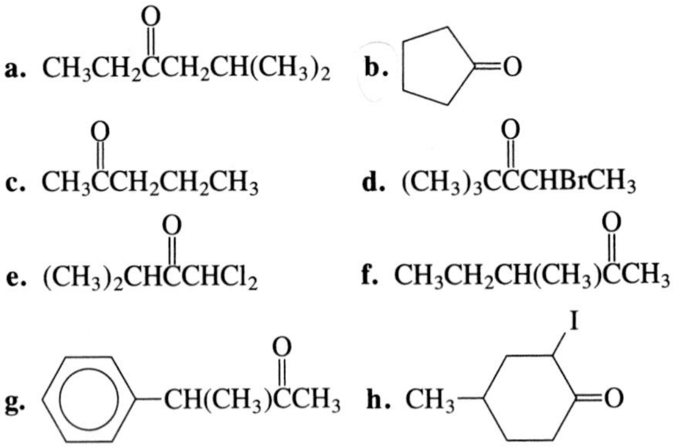

5. Give structural formulas for the following.
   a. butyraldehyde
   b. 3-methylheptanal
   c. *p*-nitrobenzaldehyde
   d. 2-chloropropanal
   e. 5-ethyloctanal
   f. 2,5-dimethylhexanal

6. Give structural formulas for the following.
   a. 2-hexanone
   b. 3-bromo-2-heptanone
   c. 4-methylcyclohexanone
   d. 1-phenyl-2-butanone
   e. 2-iodo-2-methyl-4-octanone
   f. 2-hydroxy-3-pentanone

7. As we shall see in Chapter 19, 2,3-dihydroxypropanal and 1,3-dihydroxyacetone are important carbohydrates. Draw their structural formulas.

8. Glutaraldehyde (pentanedial) is a germicide that is replacing formaldehyde as a sterilizing agent. It is less irritating to the eyes, nose, and skin. Draw the structural formula of glutaraldehyde.

9. Which compound has the higher boiling point: acetone or 2-propanol?

10. Which compound has the higher boiling point: butanal or 1-butanol?

11. Which compound has the higher boiling point: dimethyl ether or acetaldehyde?

12. Give the structures of the alcohols that could be oxidized to the following aldehydes or ketones.
   a. 4-methylcyclohexanone
   b. 2,2-dimethylpropanal
   c. 3-bromopentanal
   d. 2-pentanone
   e. phenylethanal
   f. o-methylbenzaldehyde

13. Write the equation for the reaction of acetaldehyde with each of the following.
   a. 1 mol of CH_3OH
   b. 2 mol of CH_3OH, with dry HCl present
   c. 1 mol of $HOCH_2CH_2OH$, with dry HCl present
   d. Cu^{2+}
   e. $K_2Cr_2O_7$
   f. Hydrogen gas with a nickel catalyst

14. Write the equation for the reaction, if any, of acetone with the reagents in Exercise 13.

15. Indicate whether Tollens's reagent could be used to distinguish between the compounds in each set. Explain your reasoning.
   a. 1-pentanol and pentanal
   b. 2-pentanol and 2-pentanone
   c. pentanal and 2-pentanone
   d. pentanal and pentane
   e. 2-pentanone and pentane

16. Assume that a *stronger* oxidizing agent, such as $K_2Cr_2O_7$, could be used as a test for distinguishing among compounds. For each set in Exercise 15, indicate whether this reagent would distinguish between the two compounds. Explain your reasoning.

17. What reagent would you use to distinguish between 2-pentanone and 2-pentanol? What would be observed when the reagent was added?

18. Account for the fact that the oxidation of primary alcohols usually gives poorer yields of aldehydes than the oxidation of secondary alcohols gives of ketones.

19. What is the effective chemical reagent in each of the following?
   a. Benedict's solution
   b. Fehling's solution
   c. Tollens's reagent

20. Which of the compounds in Figure 15.1 would give a positive Benedict's test?

21. Name three aldehydes or ketones that serve as active principles in flavors or aromas.

22. List the reagents necessary to carry out the following conversions.

a. $$CH_3CH_2C\overset{O}{\underset{H}{\big\langle}} \xrightarrow{?} CH_3CH_2C\overset{O}{\underset{O^-}{\big\langle}} + Ag(s)$$

b. $$CH_3CH_2CH_2CH_2OH \xrightarrow{[O]\;?} CH_3CH_2CH_2C\overset{O}{\underset{H}{\big\langle}}$$

c. $$(CH_3)_2CHCHCH_3 \xrightarrow{?} (CH_3)_2CHCCH_3$$
   (OH) → (O)

d. $$CCl_3-\overset{O}{\underset{H}{C}} \xrightarrow{?} CCl_3-\overset{OH}{\underset{H}{C}}-OH$$

e. [cyclopentanone] $\xrightarrow{?}$ [cyclopentanol with OH]

f. $$CH_3-\overset{O}{\underset{CH_3}{C}} \xrightarrow{?} CH_3-\overset{OCH_3}{\underset{CH_3}{C}}-OCH_3$$

23. Draw the hemiacetal formed from the intramolecular reaction of

$$HOCH_2CH_2CH_2CH_2C\overset{O}{\underset{H}{\big\langle}}$$

24. Name the three functional groups on the vanillin molecule (Figure 15.1).

25. Name the three functional groups on the testosterone molecule.

Testosterone

26. Chloral (CCl_3CHO), which forms a stable hydrate, also forms a stable hemiacetal. Give the structure of the hemiacetal formed by the reaction of chloral with methanol.

27. Which of the following compounds are hemiacetals?

    a. $CH_3CH_2CHOCH_3$ *hemi*
                   OH *acetal*

    b. $CH_3CH_2CHOCH_3$
                   OCH_3

    c. CH_3CH_2CHOH
                 OH

    d. ⟨benzene ring⟩—$CHOCH_2CH_3$ *hemiacetal*
                       OH

    e. ⟨benzene ring⟩—CH_2CH_2CHOH
                            OH

    f. ⟨benzene ring⟩—$CHOCH_2CH_3$
                      OCH_2CH_3

28. Which of the substances in Exercise 27 are acetals?

29. Which of the substances in Exercise 27 are hydrates?

Chapter 16
CARBOXYLIC ACIDS AND DERIVATIVES

The perfume industry owes its existence to the fragrance of esters.

Organic acids were known long before the inorganic acids were isolated. We studied some inorganic acids (HCl and H_2SO_4) first; however, primitive tribes were more familiar with organic acids, such as the acetic acid they obtained when their fermentation reactions went awry and produced vinegar instead of alcohol. Naturalists of the seventeenth century knew that the sting of a red ant's bite was due to an organic acid which that pest injected into the wound. And it was long recognized that the crisp, tart flavor of citrus fruits was produced by an organic compound appropriately called citric acid. The acetic acid of vinegar, the formic acid of red ants, and the citric acid of fruits all belong to the same family of compounds, the carboxylic acids.

A number of derivatives of carboxylic acids are also important. The amides, of which proteins (Chapter 21) are perhaps the most spectacular example, and the esters, which include fats (Chapter 20), are two classes of acid derivatives that we shall consider most carefully. Two synthetic fibers are also classed within these two families of derivatives. Nylon, like silk and wool, is a polyamide. Dacron is a polyester.

In this chapter, we shall look at simple carboxylic acids and at esters and amides, two kinds of acid derivatives. The more complex worlds of lipids and proteins we shall save for later chapters.

16.1
ACIDS AND THEIR DERIVATIVES: THE FUNCTIONAL GROUPS

The carboxyl group

We spoke of the carbonyl group in Chapter 15, and there we noted that it was this functional group that determined the chemistry of the aldehydes and ketones. The carbonyl group is also incorporated in carboxylic acids and the derivatives of carboxylic acids. However, in these compounds, the carbonyl group is only one part of the functional group that characterizes these families.

The functional group of the **carboxylic acids** is the **carboxyl** group. This group can be considered a combination of the *carb*onyl group ($>C=O$) and the hyd*roxyl* group (—OH), but it has characteristic properties of its own.

The **amide** functional group has nitrogen attached to the carbonyl group. The properties of the amide functional group are different from those of the simple carbonyl group and those of simple nitrogen-containing compounds, called amines (Chapter 17).

The functional group of the **esters** looks a little like that of an ether and a little like that of a carboxylic acid. As you should now suspect, compounds in this group react neither like acids nor like ethers, but rather like a distinctive family of compounds.

We keep talking about the derivatives of carboxylic acids. All of the families we shall discuss in this chapter, excluding the carboxylic acids themselves, are regarded as derived from the acid. In each case, the hydroxyl group of the acid's functional group is replaced with some other group in the derivative. Table 16.1 gathers all of these functional groups in one location to permit you to compare and contrast the various groups more easily. The table also offers an example (with common and IUPAC names) for each type of compound. We shall consider nomenclature in more detail as we take up each of these families separately.

Table 16.1
Carboxylic Acid Derivatives

Family	Functional Group	Example	Common Name	IUPAC Name
Carboxylic acid	$\overset{O}{\overset{\|}{-C}}$—OH	$CH_3C\overset{O}{\diagdown}_{OH}$	Acetic acid	Ethanoic acid
Amide	$\overset{O}{\overset{\|}{-C}}$—N—	$CH_3C\overset{O}{\diagdown}_{NH_2}$	Acetamide	Ethanamide
Ester	$\overset{O}{\overset{\|}{-C}}$—O—$\overset{\|}{C}$—	$CH_3C\overset{O}{\diagdown}_{OCH_3}$	Methyl acetate	Methyl ethanoate

16.2
SOME COMMON ACIDS: STRUCTURES AND NAMES

Most of the organic acids that we will consider are derived from natural sources. The acids most frequently encountered are known by their common names. Many of these are based upon Latin and Greek names and are related to the source of the acid.

The simplest carboxylic acid is formic acid. It was first obtained by the destructive distillation of ants (Latin *formica,* ant). The bite of an ant smarts because the ant injects formic acid as it bites. The stings of wasps and bees also contain formic acid (as well as other poisonous materials).

Acetic acid can be made by the aerobic fermentation of a mixture of cider and honey. This produces a solution (vinegar) that contains about 4–10% acetic acid, plus a number of other compounds that give vinegar its flavor. Acetic acid is probably the most familiar *weak* acid used in educational and industrial chemistry laboratories (Figure 16.1).

The third member of the homologous series of acids, propionic acid, is seldom encountered in everyday life. The fourth member is more familiar, at least by its odor. If you've ever smelled rancid butter, you probably wish you hadn't—but you know what butyric acid smells like. It is one of the most foul-smelling substances imaginable. Butyric acid can be isolated from butterfat or synthesized in the laboratory. It is one of the ingredients of body odor. Extremely small amounts of this and other chemicals enable bloodhounds to track fugitives.

The acid with a carboxyl group attached directly to a benzene ring is called benzoic acid. In general, carboxylic acids can be represented by the formula RCOOH. (R can be an alkyl or an aryl group.)

Table 16.2 lists several members of the family of carboxylic acids and the derivation of their common names. When using common names, substituted acids are named by locating the position of the substituent group by means of the Greek letters α, β, γ, δ, etc., rather than numbers. These letters refer to the position of the carbon atom in relation to the carboxyl carbon, as illustrated.

$$\underset{\delta}{C}-\underset{\gamma}{C}-\underset{\beta}{C}-\underset{\alpha}{C}-\overset{O}{\underset{OH}{C}}$$

$$CH_3-\underset{H}{\overset{CH_3}{C}}-\overset{O}{\underset{OH}{C}}$$
α-Methylpropionic acid

$$CH_3-\underset{OH}{\overset{H}{C}}-CH_2-\overset{O}{\underset{OH}{C}}$$
β-Hydroxybutyric acid

In the IUPAC system, the parent hydrocarbon is taken to be the one that corresponds to the longest continuous chain containing the carboxyl group. The *-e* ending of the parent alkane is replaced by the suffix *-oic*, and the word *acid* follows. As in the case of the aldehydes, the carboxyl carbon is understood to be carbon number 1. If substituents are attached to the parent chain of the acid, Greek letters are used with common names; numbers are used with IUPAC names.

$$H-\overset{O}{\underset{OH}{C}}$$
Formic acid

$$CH_3-\overset{O}{\underset{OH}{C}}$$
Acetic acid

$$CH_3CH_2-\overset{O}{\underset{OH}{C}}$$
Propionic acid

$$CH_3CH_2CH_2-\overset{O}{\underset{OH}{C}}$$
Butyric acid

Benzoic acid

Figure 16.1
Acetic acid is a familiar laboratory weak acid. It is also the active ingredient in vinegar.

Table 16.2
Some Common Aliphatic
Acids

Condensed Formula	IUPAC Name	Common Name	Derivation of Common Name
$HCOOH$	Methanoic acid	Formic acid	Latin *formica*, ant
CH_3COOH	Ethanoic acid	Acetic acid	Latin *acetum*, vinegar
CH_3CH_2COOH	Propanoic acid	Propionic acid	Greek *protos*, first, and *pion*, fat
$CH_3CH_2CH_2COOH$	Butanoic acid	Butyric acid	Latin *butyrum*, butter
$CH_3(CH_2)_3COOH$	Pentanoic acid	Valeric acid	Latin *valere*, powerful
$CH_3(CH_2)_4COOH$	Hexanoic acid	Caproic acid ⎫	
$CH_3(CH_2)_6COOH$	Octanoic acid	Caprylic acid ⎬	Latin *caper*, goat
$CH_3(CH_2)_8COOH$	Decanoic acid	Capric acid ⎭	
$CH_3(CH_2)_{10}COOH$	Dodecanoic acid	Lauric acid	Laurel tree
$CH_3(CH_2)_{12}COOH$	Tetradecanoic acid	Myristic acid	*Myristica fragrans* (nutmeg)
$CH_3(CH_2)_{14}COOH$	Hexadecanoic acid	Palmitic acid	Palm tree
$CH_3(CH_2)_{16}COOH$	Octadecanoic acid	Stearic acid	Greek *stear*, tallow

EXAMPLE 16.1

Give the common and IUPAC names for

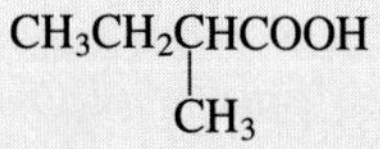

$$CH_3CH_2\overset{\underset{\displaystyle |}{}}{C}HCOOH$$
$$CH_3$$

SOLUTION
The longest continuous chain contains four carbon atoms; the compound is therefore named as a substituted butyric (or butanoic) acid. The methyl substituent is at the alpha carbon in the common system, or at the number 2 carbon in the IUPAC system. The compound is α-methylbutyric acid or 2-methylbutanoic acid.

Practice Exercise
Give the IUPAC name for $CH(CH_3)_2CH_2COOH$.

EXAMPLE 16.2

Draw the structural formula for α,β-dichloropropionic acid.

SOLUTION
Propionic acid contains three carbons.

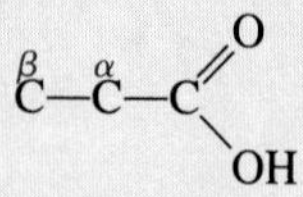

Two chlorine atoms must be attached to the parent chain, one at the alpha carbon and one at the beta carbon.

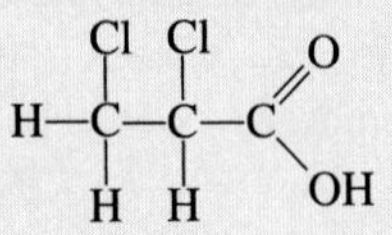

Practice Exercise
Draw a structural formula for 4-bromo-5-methylhexanoic acid.

Derivatives of the aliphatic dicarboxylic acids are very important in biological systems. These acids are almost always referred to by their common names; a mnemonic for remembering these names is given in Table 16.3.

Table 16.3
Aliphatic Dicarboxylic Acids

Formula	Common Name	Mnemonic
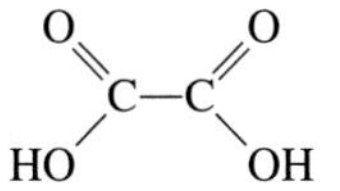 $\begin{array}{c}O \qquad O \\ \diagdown \quad \diagup \\ C-C \\ \diagup \quad \diagdown \\ HO \qquad OH \end{array}$	Oxalic acid	Oh
$HOOC-CH_2-COOH$	Malonic acid	My
$HOOC-(CH_2)_2-COOH$	Succinic acid	Such
$HOOC-(CH_2)_3-COOH$	Glutaric acid	Good
$HOOC-(CH_2)_4-COOH$	Adipic acid	Apple
$HOOC-(CH_2)_5-COOH$	Pimelic acid	Pie

16.3
PREPARATION OF CARBOXYLIC ACIDS

Few carboxylic acids occur free in nature. Many aliphatic acids, particularly those with an even number of carbon atoms, are found combined with glycerol in fats (Chapter 20). Some of the acids are thus available from the hydrolysis of fats.

Carboxylic acids are the final oxidation products of aldehydes and/or primary alcohols. The acid produced contains the same number of carbon atoms as did the precursor aldehyde or alcohol.

General Equation

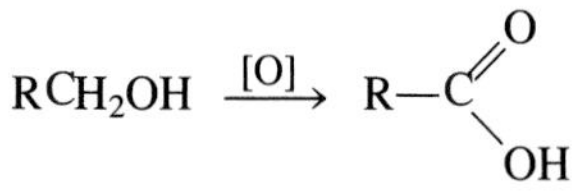

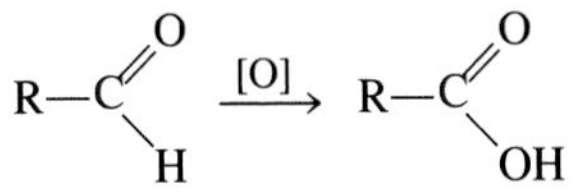

$$RCH_2OH \xrightarrow{[O]} R-C\overset{\displaystyle O}{\underset{\displaystyle OH}{\diagup}}$$

$$R-C\overset{\displaystyle O}{\underset{\displaystyle H}{\diagup}} \xrightarrow{[O]} R-C\overset{\displaystyle O}{\underset{\displaystyle OH}{\diagup}}$$

Specific Equation

$$CH_3CH_2OH \xrightarrow[H_2SO_4]{K_2Cr_2O_7} CH_3COOH$$

Ethanol   Acetic acid

Our bodies accomplish this same oxidation of alcohols to acids whenever we drink alcoholic beverages. The liver contains enzymes that convert the ethyl alcohol to acetic acid. Acetic acid is utilized to provide energy (Section 25.1), or it is converted into fat

(Section 26.2). Excess alcohol that is not oxidized in the liver continues to circulate in the blood and eventually causes intoxication.

$$CH_3-\underset{\underset{\displaystyle H}{|}}{\overset{\overset{\displaystyle H}{|}}{C}}-OH \xrightarrow[\text{dehydrogenase}]{\text{alcohol}} CH_3-C\overset{\displaystyle O}{\underset{\displaystyle H}{\diagup}} \xrightarrow[\text{dehydrogenase}]{\text{acetaldehyde}} CH_3-C\overset{\displaystyle O}{\underset{\displaystyle OH}{\diagup}}$$

Ethanol   Acetaldehyde   Acetic acid

16.4
PHYSICAL PROPERTIES OF CARBOXYLIC ACIDS

The first nine members of the carboxylic acid series are colorless liquids with very disagreeable odors. The odor of vinegar is that of acetic acid; the odor of rancid butter is primarily that of butyric acid. Caproic acid is present in the hair and the secretions of goats. The acids from C_5 to C_{10} all have "goaty" odors (odor of Limburger cheese). These acids are produced by the action of skin bacteria on perspiration oils; hence the odor of unaired locker rooms ("essence of old gym sneakers"). The acids above C_{10} are waxlike solids, and because of their low volatility they are practically odorless.

Carboxylic acids are highly polar and exhibit strong intermolecular hydrogen bonding. Consequently, these compounds have higher boiling points than even the alcohols of comparable molecular weights. Ethyl alcohol (with a molecular weight of 46 g/mol) boils at 78 °C, while formic acid (with the same molecular weight) boils at 100 °C. Similarly, propyl alcohol (with a molecular weight of 60 g/mol) boils at 97 °C, while acetic acid (with the same molecular weight) boils at 118 °C.

There is good evidence that, even in the vapor phase, some of the hydrogen bonds between acid molecules are not broken. The particular structure of the carboxyl group permits two molecules to hydrogen-bond very strongly to one another.

$$R-C\underset{\underset{\displaystyle O-H\cdots\cdots O}{}}{\overset{\overset{\displaystyle O\cdots\cdots H-O}{}}{}}C-R$$

In many situations, the interaction is so strong that the **dimer** (the two-molecule unit) acts as a single particle. Osmotic pressures, freezing-point depressions, and other properties are frequently less than one would expect when carboxylic acids are the solute. That is because we are counting as two separate molecules a combination that really is behaving as one piece.

The carboxyl groups readily hydrogen-bond to water molecules. The acids having one to four carbon atoms are colorless liquids that are completely miscible with water. Solubility decreases with increasing number of carbon atoms. Hexanoic acid ($C_6H_{12}O_2$) is soluble only to the extent of 1.0 g per 100 g of water. Palmitic acid ($C_{16}H_{32}O_2$) is essentially insoluble. The aromatic acids are odorless solids and are sparingly soluble in water. All the acids are soluble in organic solvents such as alcohol, toluene, methylene chloride, and ether.

Table 16.4 contains physical constants for the first 10 members of the aliphatic carboxylic acid family. Notice that the melting points show no regular increase with

Table 16.4
Physical Constants of
Carboxylic Acids

Formula	Name of Acid	Melting Point (°C)	Boiling Point (°C)	Solubility (g/100 g H_2O)	K_a (25 °C)
HCOOH	Formic	8	100	Miscible	1.8×10^{-4}
CH_3COOH	Acetic	17	118	Miscible	1.8×10^{-5}
CH_3CH_2COOH	Propionic	−22	141	Miscible	
$CH_3(CH_2)_2COOH$	Butyric	−5	163	Miscible	
$CH_3(CH_2)_3COOH$	Valeric	−35	187	5	
$CH_3(CH_2)_4COOH$	Caproic	−3	205	1	1.5×10^{-5}
$CH_3(CH_2)_5COOH$	Enanthic	−8	224	0.24	
$CH_3(CH_2)_6COOH$	Caprylic	16	238	0.07	
$CH_3(CH_2)_7COOH$	Pelargonic	14	254	0.03	
$CH_3(CH_2)_8COOH$	Capric	31	268	0.02	

increasing molecular weight. Pure acetic acid freezes at 16.6 °C. As this is only slightly below normal room temperature (about 20 °C), acetic acid solidifies when cooled only slightly. In the poorly heated laboratories of a century or so ago in northern North America and Europe, acetic acid often froze on the reagent shelf. For that reason, pure acetic acid (sometimes referred to as concentrated acetic acid) came to be known as **glacial acetic acid,** a name that survives to this day.

16.5
CHEMICAL PROPERTIES OF CARBOXYLIC ACIDS: NEUTRALIZATION

In Chapter 10, we defined an acid as a compound that (1) turns blue litmus red, (2) neutralizes bases, (3) reacts with active metals to give off hydrogen, and (4) tastes sour. Of the four properties of acids listed, you are probably most familiar with the last—sour taste. Vinegar is sour because it contains acetic acid. Grapefruits and lemons are sour because they contain citric acid. Sour milk contains lactic acid. The acids we eat are, for the most part, organic acids. The strong acids that we encountered in Chapter 10, such as hydrochloric acid (HCl), nitric acid (HNO_3), and sulfuric acid (H_2SO_4), are made from minerals and hence are called **mineral acids.** Historically, the first organic acids came from plant or animal matter, that is, from organisms. Now many organic acids are synthesized in the laboratory from petroleum products.

Those carboxylic acids that are water-soluble form moderately acidic solutions. They will change litmus from blue to red. Carboxylic acids, whether water-soluble or not, will react with aqueous solutions of sodium hydroxide, sodium carbonate, and sodium bicarbonate to form salts.

$$RCOOH \ + \ NaOH(aq) \ \longrightarrow \ RCOO^- Na^+(aq) \ + \ H_2O$$
$$2\,RCOOH \ + \ Na_2CO_3(aq) \ \longrightarrow \ 2\,RCOO^- Na^+(aq) \ + \ H_2O \ + \ CO_2$$
$$RCOOH \ + \ NaHCO_3(aq) \ \longrightarrow \ RCOO^- Na^+(aq) \ + \ H_2O \ + \ CO_2$$

In these reactions, the carboxylic acids act as typical acids; they neutralize basic compounds. With solutions of carbonate and bicarbonate ions they also form carbon dioxide gas.

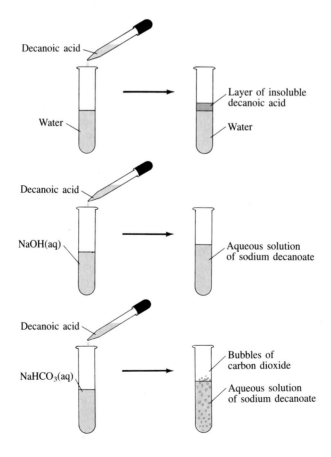

The carboxylic acids are weak acids. They tend to ionize only slightly in aqueous solution.

$$RCOOH + H_2O \rightleftharpoons RCOO^- + H_3O^+$$

Acetic acid is one of the weak acids listed in Chapter 10. We need to distinguish between degrees of weakness. We can order some organic compounds (and some inorganic ones) according to their relative acidities.

Strongest acid   Weakest acid

$$H_2SO_4, HNO_3, HCl > RCOOH > H_2CO_3 > ArOH > H_2O > ROH > RH$$

| Mineral acids | Carboxylic acids | Carbonic acid | Phenols | Water | Alcohols | Alkanes |

Water is frequently used as a dividing line for acidity and basicity. Because we live in a water world, we quite naturally regard water as neutral. In aqueous solutions, if something is more acidic than water, it is treated as an acid (as are phenols and carboxylic acids). If a compound is less acidic than water (as are alcohols and alkanes), it is not regarded as an acid. Neither alcohols nor alkanes affect the pH of aqueous solutions.

Because of the difference in relative acidities among organic compounds, solubility behavior is often an identifying feature of carboxylic acids. Carboxylic acids that are insoluble in water dissolve in aqueous hydroxide, carbonate, or bicarbonate because the insoluble acids react to form ionic salts that are water-soluble. The behavior of decanoic acid is illustrated in Figure 16.2. Solution in aqueous sodium bicarbonate, with the formation of carbon dioxide bubbles, is characteristic of carboxylic acids.

Figure 16.2

Decanoic acid is insoluble in water but soluble in aqueous sodium hydroxide or sodium bicarbonate.

EXAMPLE 16.3

Write equations for the reactions of decanoic acid with (a) NaOH and (b) $NaHCO_3$.

SOLUTION

a. $CH_3(CH_2)_8C\overset{O}{\underset{OH}{}}$ + NaOH $\longrightarrow$ $CH_3(CH_2)_8C\overset{O}{\underset{O^- Na^+}{}}$ + HOH

b. $CH_3(CH_2)_8C\overset{O}{\underset{OH}{}}$ + $NaHCO_3$ $\longrightarrow$ $CH_3(CH_2)_8C\overset{O}{\underset{O^- Na^+}{}}$ + HOH + CO_2

Practice Exercise

Write equations for the reactions of benzoic acid with (a) NaOH and (b) $NaHCO_3$.

Organic salts are named in the same manner as the inorganic salts. The name of the cation is followed by the name of the organic anion. The name of the anion is obtained by dropping the *-ic* ending of the acid name and replacing it with the suffix *-ate*. This applies whether we are using common names or IUPAC names. Following are some examples.

$CH_3C\overset{O}{\underset{O^- Li^+}{}}$ $CH_3CH_2CH_2C\overset{O}{\underset{O^- K^+}{}}$ $C_6H_5-C\overset{O}{\underset{O^- Na^+}{}}$

Lithium acetate   Potassium butyrate   Sodium benzoate
(Lithium ethanoate)   (Potassium butanoate)

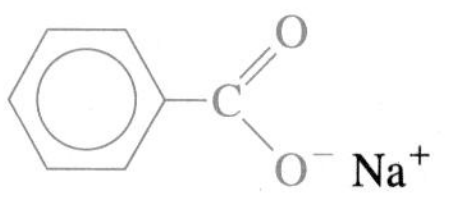

The sodium or potassium salts of long-chain carboxylic acids are called soaps. We discuss the chemistry of soaps in considerable detail in Section 20.5.

$CH_3CH_2CH_2CH_2CH_2CH_2CH_2CH_2CH_2CH_2CH_2CH_2CH_2CH_2CH_2CH_2CH_2-C\overset{O}{\underset{O^- Na^+}{}}$

Sodium stearate (a soap)

Other salts that have commercial significance are those used as preservatives. They act to prevent spoilage by inhibiting the growth of bacteria and fungi. Calcium and sodium propionate are added to processed cheese and bakery goods; sodium benzoate is used as a preservative in cider, jellies, pickles, and syrups; and sodium and potassium sorbate are added to fruit juices, sauerkraut, soft drinks, and wine. (Look for these salts on ingredient labels the next time you are shopping.)

$\left(CH_3CH_2C\overset{O}{\underset{O^-}{}}\right)_2 Ca^{2+}$ $CH_3CH=CHCH=CHC\overset{O}{\underset{O^- K^+}{}}$

Calcium propionate   Potassium sorbate

16.6
AN ESTER BY ANY OTHER NAME . . .

Esters are perhaps the most important class of derivatives of the carboxylic acids. The general formula for an ester is RCOOR′, where R may be a hydrogen, an alkyl group, or an aryl group and R′ may be alkyl or aryl, but *not* hydrogen.

Esters are widely distributed in nature. Their occurrence is particularly important in fats and vegetable oils, which are esters of long-chain fatty acids and glycerol (Chapter 20). Esters of phosphoric acid are of the utmost importance to life, and they are enumerated in Section 16.10.

Esters are named in a manner similar to that used for naming organic salts. The group name of the alkyl or aryl portion is given first and is followed by the name of the acid portion. In both common and IUPAC nomenclature, the *-ic* ending of the parent acid is replaced by the suffix *-ate*. Table 16.5 illustrates the nomenclature of several esters.

Table 16.5
Nomenclature of Esters

Formula	Common Name	IUPAC Name
$H-C(=O)-O-CH_3$	**Methyl** formate	**Methyl** methanoate
$CH_3-C(=O)-O-CH_3$	**Methyl** acetate	**Methyl** ethanoate
$CH_3-C(=O)-O-CH_2CH_3$	**Ethyl** acetate	**Ethyl** ethanoate
$CH_3CH_2-C(=O)-O-CH_2CH_3$	**Ethyl** propionate	**Ethyl** propanoate
$CH_3CH_2CH_2-C(=O)-O-CH_2CH_2CH_3$	**Propyl** butyrate	**Propyl** butanoate
$CH_3CH_2CH_2-C(=O)-O-CHCH_3(CH_3)$	**Isopropyl** butyrate	**Isopropyl** butanoate
$C_6H_5-C(=O)-O-CH_2CH_3$	**Ethyl** benzoate	**Ethyl** benzoate
$CH_3C(=O)-O-C_6H_5$	**Phenyl** acetate	**Phenyl** ethanoate

EXAMPLE 16.4

What are the common and IUPAC names for this ester?

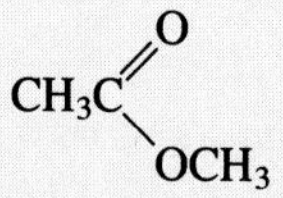

$$CH_3C \underset{OCH_3}{\overset{O}{\diagup}}$$

SOLUTION

The alkyl group attached to oxygen is methyl. The

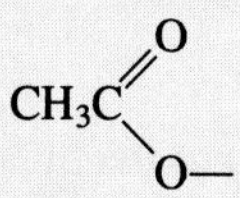

$$CH_3C \underset{O-}{\overset{O}{\diagup}}$$

is derived from acetic acid (which has two carbons). Its name is acetate. The compound is methyl acetate. The IUPAC name for the two-carbon acid unit is ethanoate. The IUPAC name for the ester is methyl ethanoate.

EXAMPLE 16.5

What are the common and IUPAC names for this ester?

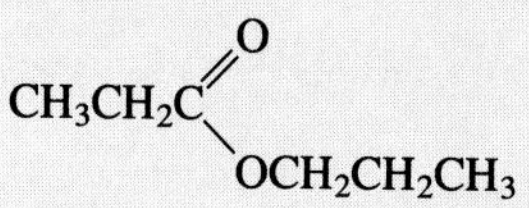

$$CH_3CH_2C \underset{OCH_2CH_2CH_3}{\overset{O}{\diagup}}$$

SOLUTION

The alkyl group (attached directly to oxygen) is propyl. The part of the molecule derived from the acid,

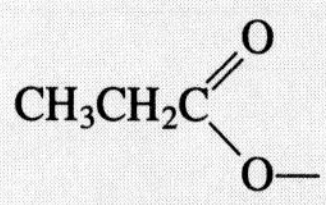

$$CH_3CH_2C \underset{O-}{\overset{O}{\diagup}}$$

has three carbon atoms. It ıs thus called propionate or, by IUPAC terminology, propanoate. The common name of the ester is therefore propyl propionate, and the IUPAC name is propyl propanoate.

Practice Exercise

Give the IUPAC name for

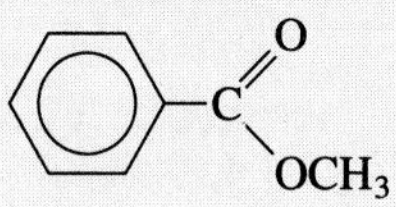

$$\text{(phenyl)} C \underset{OCH_3}{\overset{O}{\diagup}}$$

EXAMPLE 16.6

What is the name of this ester?

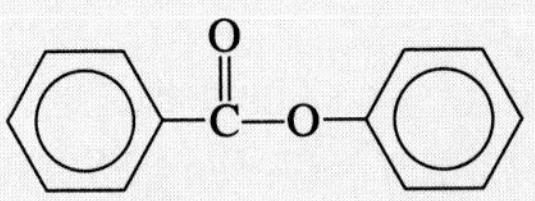

$$\text{(phenyl)} \overset{O}{\underset{}{C}} - O - \text{(phenyl)}$$

SOLUTION

The group attached by a single bond to oxygen is phenyl. The acid portion corresponds to the benzoate group (from benzoic acid). Therefore, the compound is phenyl benzoate.

EXAMPLE 16.7

Draw the structure for ethyl pentanoate.

SOLUTION
It is easier to start with the acid portion. Draw the pentanoate (five-carbon) group first.

$$CH_3CH_2CH_2CH_2C\begin{matrix}O\\\\O-\end{matrix}$$

Then simply attach the ethyl group to the bond that ordinarily holds the hydrogen in the free acid.

$$CH_3CH_2CH_2CH_2C\begin{matrix}O\\\\O-CH_2CH_3\end{matrix}$$

Practice Exercise
Draw the structure for phenyl butanoate.

16.7
PHYSICAL PROPERTIES OF ESTERS

Unlike the carboxylic acids from which they are derived, the esters generally have pleasant odors and are often responsible for the characteristic fragrances of fruits and flowers. Once a flower or fruit has been chemically analyzed, flavor chemists can attempt to duplicate the natural odor or taste. They are seldom completely successful, but they often get close enough for practical purposes. Esters are used in the manufacture of perfumes and as flavoring agents in the confectionery and soft drink industries. (A mixture of nine esters is used to produce an artificial raspberry flavor.)

The molecules of an ester are polar but are incapable of forming intermolecular hydrogen bonds with one another. Esters thus have considerably lower boiling points than the isomeric carboxylic acids. As one might expect, the boiling points of esters are about intermediate between those of ketones and ethers of comparable molecular weight. Ester molecules are capable of forming hydrogen bonds to water molecules, so esters of low molecular weight are somewhat water-soluble. Borderline solubility occurs in those molecules that have three to five carbon atoms. Table 16.6 lists the physical properties of some common esters.

Esters find their most important use as industrial solvents. Ethyl acetate is commonly used to extract organic solutes from an aqueous solution. It is also used as a nail polish remover and is a major constituent of some paint removers. Cellulose nitrate is dissolved in ethyl acetate and butyl acetate to form lacquers. The solvent evaporates as the lacquer "dries," leaving a thin protective film on the surface to which the lacquer was applied. Esters with high boiling points are used as softeners (plasticizers) for brittle plastics.

Formula	Name	MW	Melting Point (°C)	Boiling Point (°C)	Aroma
$HCOOCH_3$	Methyl formate	60	−99	32	
$HCOOCH_2CH_3$	Ethyl formate	74	−80	54	Rum
CH_3COOCH_3	Methyl acetate	74	−98	57	
$CH_3COOCH_2CH_3$	Ethyl acetate	88	−84	77	
$CH_3CH_2COOCH_3$	Methyl propionate	88	−88	80	
$CH_3CH_2COOCH_2CH_3$	Ethyl propionate	102	−74	99	
$CH_3CH_2CH_2COOCH_3$	Methyl butyrate	102	−85	102	Apple
$CH_3CH_2CH_2COOCH_2CH_3$	Ethyl butyrate	116	−101	121	Pineapple
$CH_3COO(CH_2)_4CH_3$	Amyl acetate	130	−71	148	Banana
$CH_3COOCH_2CH_2CH(CH_3)_2$	Isoamyl acetate	130	−79	142	Pear
$CH_3COOCH_2C_6H_5$	Benzyl acetate	150	−51	215	Jasmine
$CH_3CH_2CH_2COO(CH_2)_4CH_3$	Amyl butyrate	158	−73	185	Apricot
$CH_3COO(CH_2)_7CH_3$	Octyl acetate	172	−39	210	Orange

Table 16.6
Physical Properties of Esters

16.8
PREPARATION OF ESTERS: ESTERIFICATION

Some esters are prepared by direct esterification of the carboxylic acid. This is accomplished by heating a carboxylic acid with an alcohol in the presence of a mineral acid catalyst.

An alcohol molecule condenses with an acid molecule, splitting out water to form an ester. The reaction is reversible, and it soon comes to equilibrium. Experimental data reveal that, for any esterification reaction, the composition of the equilibrium mixture is governed by the value of the equilibrium constant. Consideration of the law of chemical equilibrium and Le Châtelier's principle (Section 5.11) allows predetermination of reaction conditions that produce a maximum yield of the desired ester.

If the reaction involves an inexpensive alcohol, such as methanol (CH_3OH), excess alcohol can be used. This will drive the reaction toward completion. In the preparation of methyl benzoate, for example, 10 mol of CH_3OH may be used for each mole of benzoic acid.

Forcing of the reaction in this way can give a 75% yield of methyl benzoate.

Similarly, if the acid is cheap, it can be employed in excess. In the preparation of butyl acetate, acetic acid is used in a molar ratio of 2:1 (or greater).

$$CH_3C\overset{O}{\underset{OH}{\big\langle}} + CH_3CH_2CH_2CH_2OH \; \rightleftharpoons \; CH_3C\overset{O}{\underset{OCH_2CH_2CH_2CH_3}{\big\langle}} \; + \; H_2O$$

(excess)

A third method of driving the reaction toward completion involves removal of the product water as it is formed. This is easily accomplished if the acid, the alcohol, and the ester all boil at temperatures well above 100 °C.

$$CH_3CH_2CH_2C\overset{O}{\underset{OH}{\big\langle}} + CH_3CH_2CH_2CH_2OH \; \xrightarrow{H^+} \; CH_3CH_2CH_2C\overset{O}{\underset{OCH_2CH_2CH_2CH_3}{\big\langle}} \; + \; H_2O$$

| Butyric acid (bp 164 °C) | Butyl alcohol (bp 118 °C) | Butyl butyrate (bp 165 °C) | Distilled from mixture (bp 100 °C) |

In general, the reaction between an acid and an alcohol is extremely slow, and a catalyst must be employed to speed it up. Sulfuric acid is commonly used because it is both an acid and a good dehydrating agent. Thus it serves both to catalytically increase the rate of reaction and to shift the equilibrium to the right by reacting with water to form hydronium ions, H_3O^+, thus increasing the yield of the ester.

A commercially important esterification reaction occurs between a dicarboxylic acid and a dialcohol. Such a reaction yields an ester that contains a free carboxyl group and a free alcohol group on either end. Hence further condensation reactions can occur to produce polyesters (polymers). The most significant polyester, Dacron, is made from terephthalic acid and ethylene glycol monomers. Dacron polyester is found in permanent press garments, carpets, tires, and many other products. Dacron is inert and is used in surgery (in the form of a mesh) to repair or replace diseased sections of blood vessels.

$$n\,HO{-}CH_2CH_2{-}OH \; + \; n\,HO{-}\overset{O}{\underset{}{C}}{-}\!\!\bigcirc\!\!{-}\overset{O}{\underset{}{C}}{-}OH \; \longrightarrow \; {+}O{-}CH_2CH_2{-}O{-}\overset{O}{\underset{}{C}}{-}\!\!\bigcirc\!\!{-}\overset{O}{\underset{}{C}}{\}_n} \; + \; n\,H_2O$$

| Ethylene glycol | Terephthalic acid | Dacron |

16.9
CHEMICAL PROPERTIES OF ESTERS: HYDROLYSIS

Esters are neutral compounds, unlike the acids from which they are formed. As neutral compounds they exhibit neither acidic nor basic properties. Esters typically undergo chemical reactions in which the alkoxy (—OR′) group is replaced by another group. One such reaction is **hydrolysis,** or splitting with water. Hydrolysis is catalyzed by either acid or base. Acidic hydrolysis is simply the reverse of the esterification reaction. The ester is refluxed with a large excess of water containing a strong acid catalyst. However, the equilibrium is unfavorable for ester hydrolysis, and so the reaction never goes to completion.

General Equation

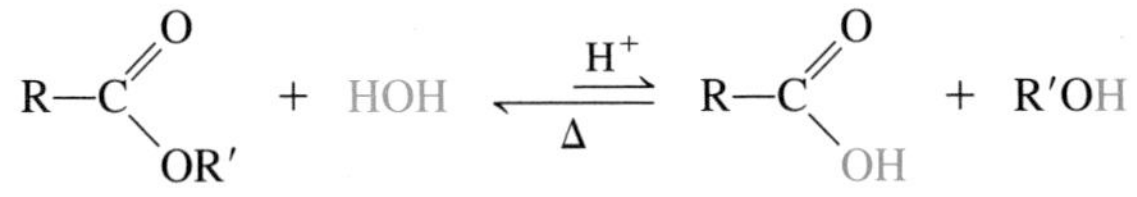

$$R-\overset{\overset{\displaystyle O}{\|}}{C}-OR' \ + \ HOH \ \underset{\Delta}{\overset{H^+}{\rightleftharpoons}} \ R-\overset{\overset{\displaystyle O}{\|}}{C}-OH \ + \ R'OH$$

Specific Equation

$$\bigcirc\!\!\!\!-\overset{\overset{\displaystyle O}{\|}}{C}-OCH_3 \ + \ HOH \ \underset{\Delta}{\overset{H^+}{\rightleftharpoons}} \ \bigcirc\!\!\!\!-\overset{\overset{\displaystyle O}{\|}}{C}-OH \ + \ CH_3OH$$

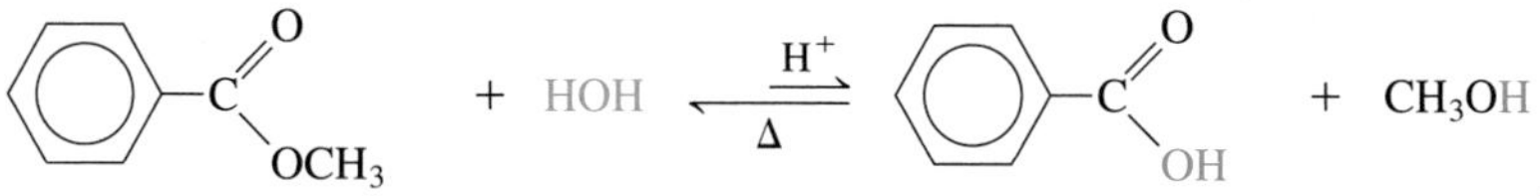

Methyl benzoate   Benzoic acid   Methanol

EXAMPLE 16.8

Write an equation for the acid hydrolysis of ethyl acetate.

SOLUTION

Remember that in acid hydrolysis, water splits the ester bond. The H of water joins to oxygen, and the OH of water joins to carbon.

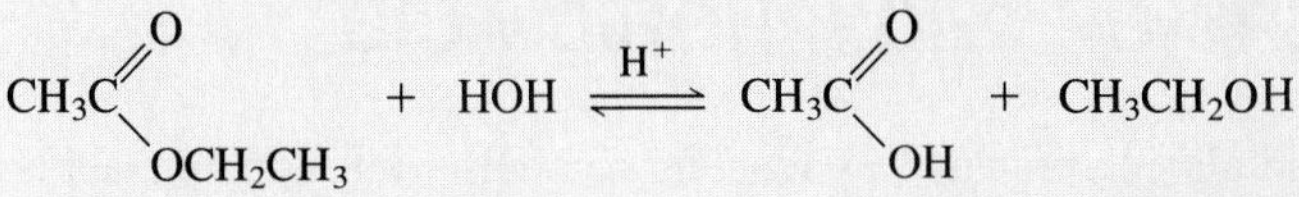

$$CH_3\overset{\overset{\displaystyle O}{\|}}{C}-OCH_2CH_3 \ + \ HOH \ \overset{H^+}{\rightleftharpoons} \ CH_3\overset{\overset{\displaystyle O}{\|}}{C}-OH \ + \ CH_3CH_2OH$$

Ethyl acetate   Acetic acid   Ethanol

Practice Exercise

Write an equation for the acid hydrolysis of isopropyl propanoate.

When a base (such as sodium hydroxide or potassium hydroxide) is used to hydrolyze an ester, the reaction goes to a completion because the carboxylic acid is removed from the equilibrium by its conversion to a salt. Organic salts do not react with alcohols, so the reaction is essentially irreversible. Accordingly, ester hydrolysis is usually carried out in basic solution. Because soaps are prepared by the alkaline hydrolysis of fats and oils, the term **saponification** (Latin *sapon*, soap; *facere*, to make) is used to describe the alkaline hydrolysis of all esters (see also Section 20.5). Note that in the saponification reaction, the base is a reactant and thus is not a catalyst.

General Equation

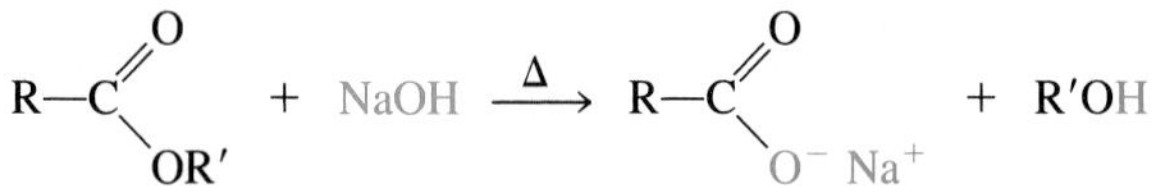

$$R-\overset{\overset{\displaystyle O}{\|}}{C}-OR' \ + \ NaOH \ \overset{\Delta}{\longrightarrow} \ R-\overset{\overset{\displaystyle O}{\|}}{C}-O^- \ Na^+ \ + \ R'OH$$

Specific Equation

$$CH_3-\overset{\overset{\displaystyle O}{\|}}{C}-OCH_2CH_3 \ + \ NaOH \ \overset{\Delta}{\longrightarrow} \ CH_3-\overset{\overset{\displaystyle O}{\|}}{C}-O^- \ Na^+ \ + \ CH_3CH_2OH$$

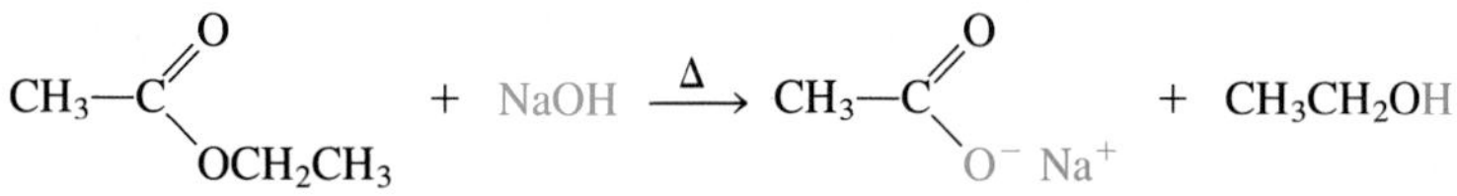

Ethyl acetate   Sodium acetate   Ethanol

EXAMPLE 16.9

Write an equation for the hydrolysis of methyl benzoate in a potassium hydroxide solution.

SOLUTION

In basic hydrolysis, the molecule of base splits the ester linkage. The hydrogen of the base joins the oxygen, and the oxy cation of the base joins the carbon.

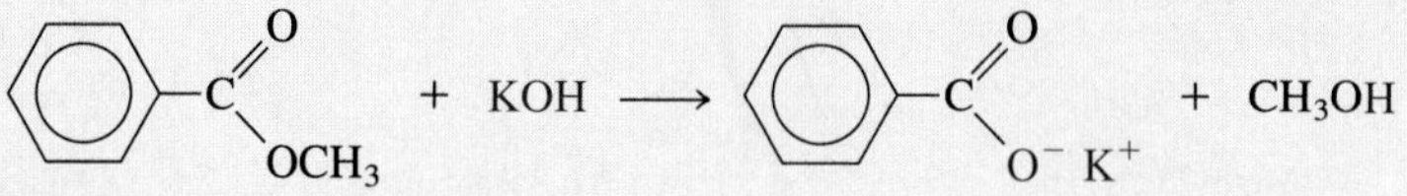

Practice Exercise

Write an equation for the hydrolysis of phenyl methanoate in a sodium hydroxide solution.

16.10
ESTERS OF PHOSPHORIC ACID

Esters can also be prepared by reacting alcohols with inorganic acids (e.g., HNO_3, H_2SO_4, H_3PO_4). In Section 14.7 we mentioned nitroglycerin (glycerol trinitrate), the ester of glycerol and nitric acid. Perhaps the most important inorganic esters in biochemistry are those of phosphoric acid and two of its anhydrides, pyrophosphoric acid[1] and triphosphoric acid. Phosphate or pyrophosphate esters are present in every plant and animal cell. They are biochemical intermediates in the transformation of food into usable energy (in the form of ATP). Organic phosphates are also important structural constituents of phospholipids (Section 20.7), nucleic acids (Chapter 23), coenzymes (Table I.1), and insecticides.

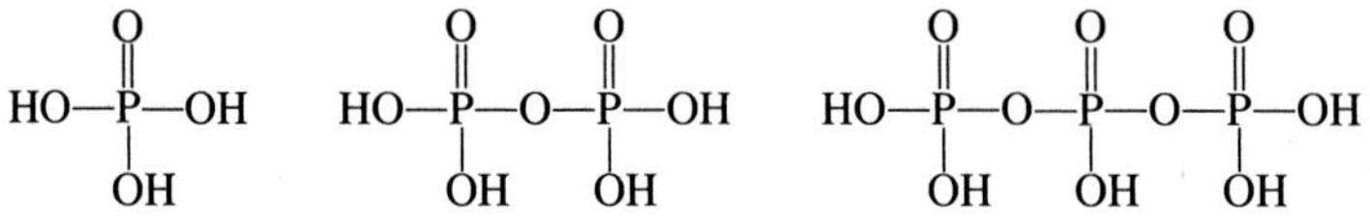

Phosphoric   Pyrophosphoric   Triphosphoric acid
acid   acid

Phosphoric acid is triprotic and can form monoalkyl, dialkyl, and trialkyl esters as one or more hydrogen atoms are replaced by alkyl groups.

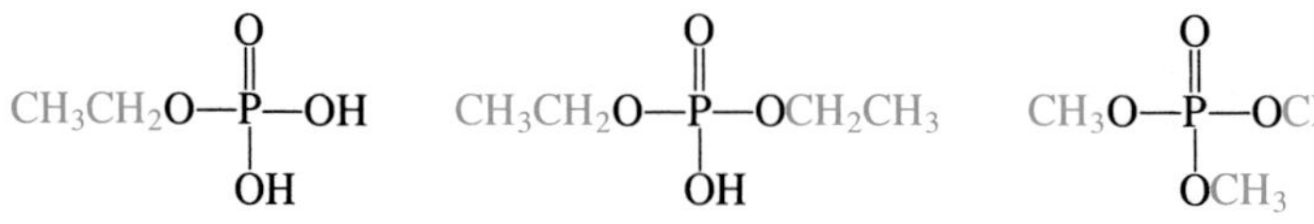

Ethyl dihydrogen   Diethyl hydrogen   Trimethyl phosphate
phosphate   phosphate

[1] Pyrophosphoric acid is an anhydride that can be considered as derived from phosphoric acid by the removal of one molecule of water from two molecules of acid.

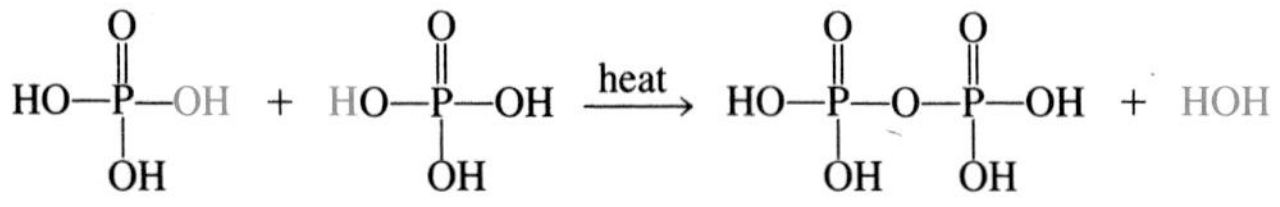

A wide variety of phosphate esters occur naturally and are compounds of central importance in metabolism. The majority of the substances obtained from our food must be converted to phosphate esters before they can be used by the cells. Many of them are formed by phosphorylation reactions—the interaction of alcohols with anhydrides of phosphoric acid.

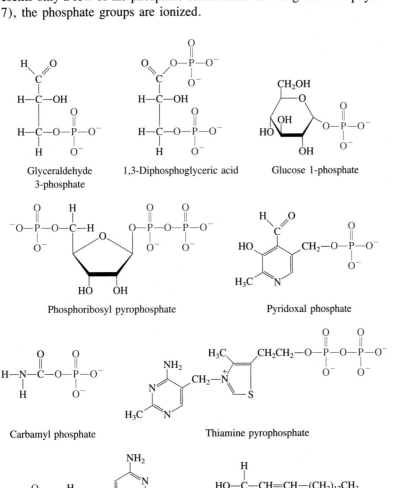

A polyphosphate   Glucose   Glucose 6-phosphate

Figure 16.3 presents only a few of the phosphate esters found in living cells. At physiological pH (≈ 7), the phosphate groups are ionized.

Figure 16.3
Some phosphate compounds of biological importance.

Glyceraldehyde 3-phosphate

1,3-Diphosphoglyceric acid

Glucose 1-phosphate

Phosphoribosyl pyrophosphate

Pyridoxal phosphate

Carbamyl phosphate

Thiamine pyrophosphate

Cytidine nucleotide

Sphingomyelin

16.11
AMIDES: STRUCTURES AND NAMES

In the amide functional group, a nitrogen is attached to a carbonyl group. If the two remaining bonds to nitrogen are attached to hydrogen atoms, the compound is called a simple amide. If one or both of the two remaining bonds to nitrogen are attached to alkyl or aryl groups, the compound is called a substituted amide. The carbonyl carbon–nitrogen bond is referred to as the **amide linkage.** This bond is very stable and is found in the repeating units of protein molecules (Chapter 21), in nylon, and in many other industrial polymers.

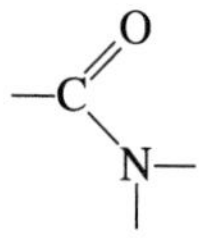

The amide group 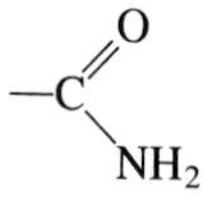   A simple amide 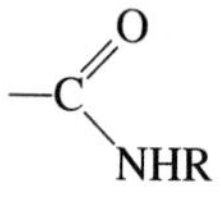   A substituted amide 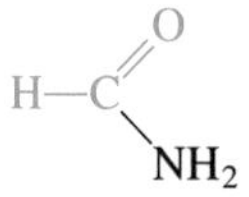   Formamide (Methanamide)

Amides are named as derivatives of organic acids. The -ic ending of the common name or the -oic ending of the IUPAC name is replaced with the suffix -amide.

EXAMPLE 16.10

Name the compound

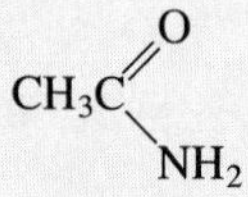

SOLUTION

This amide is derived from acetic acid. Drop the -ic suffix, attach the ending -amide, and you have the name: acetamide (or ethanamide in the IUPAC system).

EXAMPLE 16.11

Name the compound

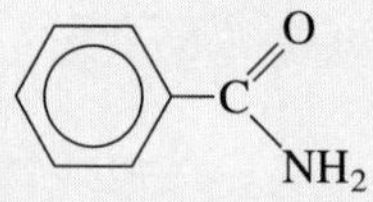

SOLUTION

This amide is derived from benzoic acid. Drop the -oic, add -amide, and you have it: benzamide.

Practice Exercise

Draw a structural formula for pentanamide.

In substituted amides, alkyl groups attached to the nitrogen atom are named as substituents. Instead of using a Greek letter or a number to specify location, chemists indicate the group's attachment to nitrogen by a capital letter *N*. If the substituent on nitrogen is phenyl, the compound is named as an anilide. The *-ic* or *-oic* ending of the acid name is replaced with *-anilide* instead of *-amide*.

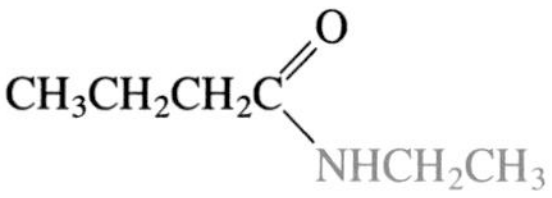

$CH_3CH_2CH_2C$
O
$NHCH_2CH_3$

N-Ethylbutyramide

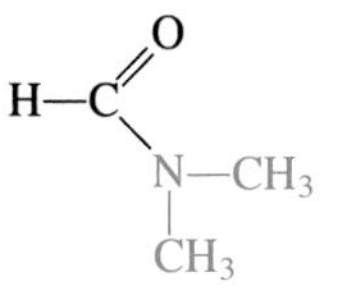

$H-C$
O
$N-CH_3$
CH_3

N,N-Dimethylformamide

CH_3C
O
$NH-$⬡

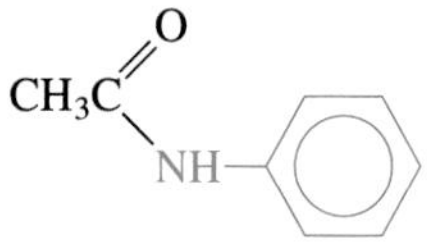

Acetanilide

EXAMPLE 16.12

Name the compound

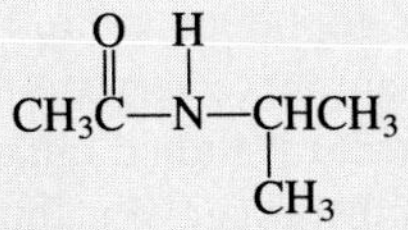

$CH_3C-N-CHCH_3$
O H
CH_3

SOLUTION
The acid portion of the molecule (the portion incorporating the carbonyl group and to one side of the nitrogen) contains two carbon atoms and is derived from acetic acid. The compound will therefore be named as a substituted acetamide. The substituent attached directly to the nitrogen is an isopropyl group. The name of the compound is *N*-isopropylacetamide. (The IUPAC name is *N*-isopropylethanamide.)

Practice Exercise
Draw a structural formula for *N,N*-dimethylpropanamide.

16.12
PHYSICAL PROPERTIES OF AMIDES

smell terrible when heated

With the exception of formamide, which is a liquid, all unsubstituted amides are solids (Table 16.7). Most amides are colorless and odorless. The lower members of the series are soluble in water, with borderline solubility occurring in those that have five or six carbon atoms.

Table 16.7
Physical Constants of Some Unsubstituted Amides

Formula	Name	Melting Point (°C)	Boiling Point (°C)	
$HCONH_2$	Formamide	2	193	
CH_3CONH_2	Acetamide	82	222	Soluble in water
$CH_3CH_2CONH_2$	Propionamide	81	213	
$CH_3CH_2CH_2CONH_2$	Butyramide	115	216	
$C_6H_5CONH_2$	Benzamide	132	290	Insoluble

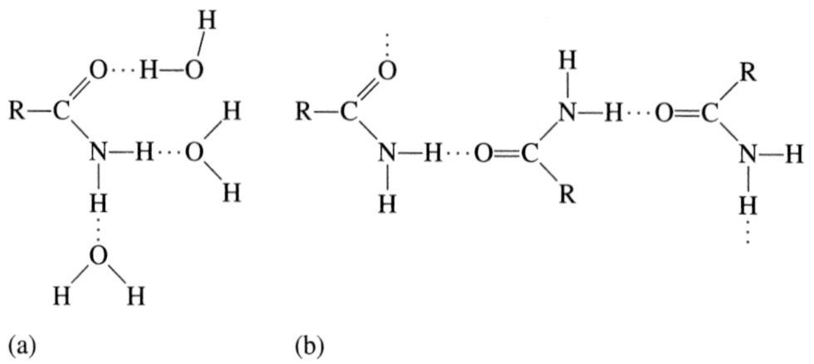

Figure 16.4
(a) Hydrogen bonding of
amides with water mole-
cules. (b) Intermolecular
hydrogen bonding in
amides.

Figure 16.4
(a) Hydrogen bonding of amides with water molecules. (b) Intermolecular hydrogen bonding in amides.

(a)   (b)

The amides have abnormally high boiling points and melting points. This phenomenon, as well as the water solubility of the amides, is a result of the polar nature of the amide group and the formation of hydrogen bonds (Figure 16.4). Similar hydrogen bonding plays a critical role in determining the structure and properties of proteins, DNA, RNA, and other giant molecules so important to life processes. Electrostatic forces and hydrogen bonding combine to account for the very strong intermolecular attractions found in the amides. Note, however, that disubstituted amides have no hydrogens bonded to nitrogen and thus are incapable of hydrogen bonding. N,N-Dimethylacetamide has a melting point of $-20\ °C$, which is about $100\ °C$ lower than the melting point of acetamide.

16.13
SYNTHESIS
OF AMIDES

The addition of ammonia to the free acid will result in the formation of an amide, but the reaction is very slow at room temperature. The ammonium salt of the acid is formed first; then water can be split out if the reaction temperature is maintained above $100\ °C$. The second step is reversible; the equilibrium favors salt formation. Continuous removal of the water shifts the equilibrium to the right.

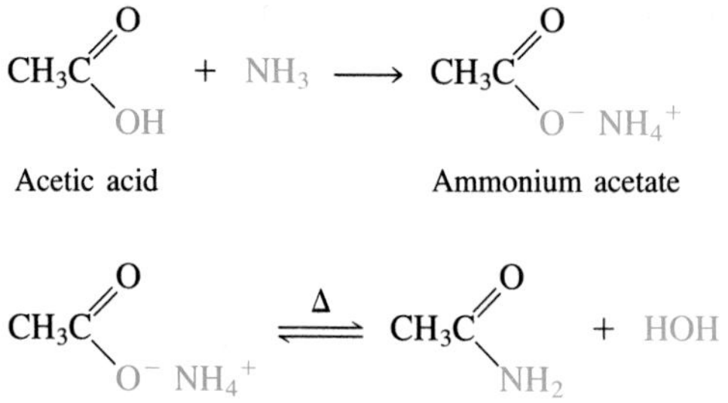

Acetic acid   Ammonium acetate

In Section 16.8, we showed the condensation polymerization reaction that forms polyesters. An even more important reaction is one that yields polyamides. Again, two difunctional monomers are employed, usually adipic acid and 1,6-hexanediamine. The monomers condense by splitting out water to form a new product, which is still difunctional and thus can react continuously to yield a polymer.

$$n \text{ H}-\underset{\overset{|}{H}}{N}CH_2CH_2CH_2CH_2CH_2CH_2\underset{\overset{|}{H}}{N}-\text{H} \ + \ n \text{ HO}-\overset{\overset{O}{\|}}{C}CH_2CH_2CH_2CH_2\overset{\overset{O}{\|}}{C}-\text{OH} \ \xrightarrow[\text{10 atm}]{270\ °C}$$

1,6-Hexanediamine   Adipic acid

$$-\underset{\overset{|}{H}}{N}\left[\overset{\overset{O}{\|}}{C}-(CH_2)_4-\overset{\overset{O}{\|}}{C}-\underset{\overset{|}{H}}{N}-(CH_2)_6-\underset{\overset{|}{H}}{N}\right]_n\overset{\overset{O}{\|}}{C}- \ + \ n \text{ HOH}$$

Nylon 6,6

Nylon is the collective name of several different synthetic polyamide fibers. Nylon 6,6 is the most common; the 6,6 designation refers to the number of carbon atoms in each of the monomers. It is possible to make other nylons by varying the number of carbons in either the dicarboxylic acid or the diamine. Nylon is a remarkable polymer. It is stable in dilute acids or bases, has a high melting point (260 °C), and is very strong. Nylon is one of the most widely used synthetic fibers; for example, it is used in rope, sails, carpets, clothing, tires, brushes, and parachutes. It also can be molded into blocks for use in electrical equipment, gears, bearings, and valves.

16.14
CHEMICAL PROPERTIES OF AMIDES: HYDROLYSIS

Generally, the amides are neutral compounds, showing neither appreciable acidity nor significant basicity in water. Furthermore, the amides resist hydrolysis in plain water, even upon prolonged heating. In the presence of added acid or base, however, hydrolysis proceeds at a moderate rate. Acidic hydrolysis of a simple amide gives a carboxylic acid and an ammonium salt.

$$CH_3CH_2C\overset{\overset{O}{\|}}{\underset{NH_2}{\diagdown}} \ + \ HCl(aq) \ + \ H_2O \ \longrightarrow \ CH_3CH_2C\overset{\overset{O}{\|}}{\underset{OH}{\diagdown}} \ + \ NH_4Cl(aq)$$

Basic hydrolysis gives a salt of the carboxylic acid and ammonia.

$$CH_3CH_2C\overset{\overset{O}{\|}}{\underset{NH_2}{\diagdown}} \ + \ NaOH(aq) \ \longrightarrow \ CH_3CH_2C\overset{\overset{O}{\|}}{\underset{O^-\ Na^+}{\diagdown}} \ + \ NH_3$$

It may be easier to see why the products of the two reactions differ if we consider the hydrolysis products that would form if the reaction could be carried out in the absence of added acid or base.

$$CH_3CH_2C\overset{\overset{O}{\|}}{\underset{NH_2}{\diagdown}} \ + \ H_2O \ \longrightarrow \ CH_3CH_2C\overset{\overset{O}{\|}}{\underset{OH}{\diagdown}} \ + \ NH_3$$

The products of this reaction are an acid (the carboxylic acid) and a base (ammonia). If we actually carry out the hydrolysis in the presence of hydrochloric acid, some of the hydrochloric acid will react with the basic product ammonia to form ammonium chloride. If, instead, we use sodium hydroxide to speed the reaction, some of this base will react with the carboxylic acid product to form the sodium salt of the carboxylic acid. There are two points to be made here. One is that the several hydrolysis reactions discussed in this chapter are closely related and should be considered as variations on a theme rather than as separate reactions. The second point is that chemical principles are not confined within chapters. If acids react with bases to form salts in Chapter 10, they will also react in Chapter 16. If a reaction under consideration happens to produce an acidic product, that product will always exhibit those properties we have previously ascribed to acids.

EXAMPLE 16.13

Write an equation for the acid (HCl) hydrolysis of butyramide.

SOLUTION
Remember that in the acid hydrolysis of amides, the products are the organic acid and an ammonium salt.

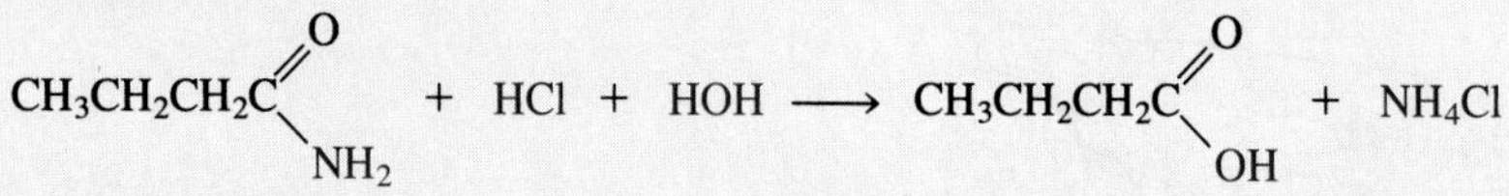

$$CH_3CH_2CH_2\overset{\displaystyle O}{\underset{\displaystyle NH_2}{C}} + HCl + HOH \longrightarrow CH_3CH_2CH_2\overset{\displaystyle O}{\underset{\displaystyle OH}{C}} + NH_4Cl$$

Practice Exercise
Write an equation for the acid (HCl) hydrolysis of benzamide.

EXAMPLE 16.14

Write an equation for the basic (KOH) hydrolysis of butyramide.

SOLUTION
In basic hydrolysis of amides, the products are the salt of the carboxylic acid and ammonia.

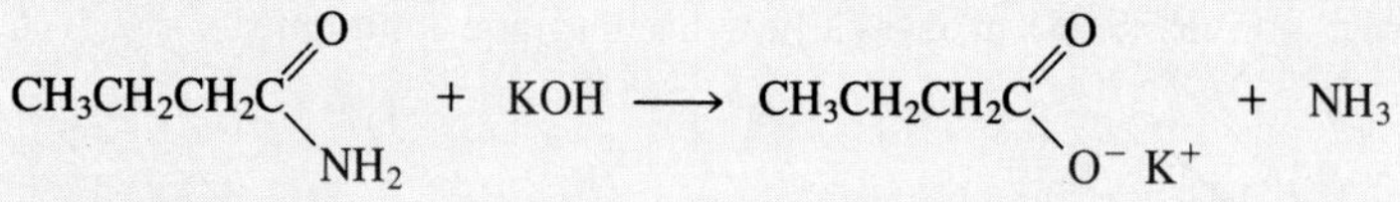

$$CH_3CH_2CH_2\overset{\displaystyle O}{\underset{\displaystyle NH_2}{C}} + KOH \longrightarrow CH_3CH_2CH_2\overset{\displaystyle O}{\underset{\displaystyle O^- K^+}{C}} + NH_3$$

Practice Exercise
Write an equation for the basic (NaOH) hydrolysis of benzamide.

The hydrolysis of amides is of more than theoretical interest. Digestion of proteins (Section 27.1) involves the hydrolysis of amide bonds. Clothes made of nylon have been known to disintegrate in air polluted with sulfuric acid mist. The acid catalyzes the hydrolysis of the amide bonds that hold the long chains of the nylon molecules together. Perhaps it is worth some time to consider what the same polluted air does to the proteins of our lungs.

EXERCISES

1. Draw the functional groups in
  a. aldehydes   **b.** ketones
  c. carboxylic acids   **d.** esters
  e. ethers   **f.** amides

2. Give common names for the straight-chain carboxylic acids containing the given number of carbon atoms.
  a. 1   **b.** 2   **c.** 3   **d.** 4

3. Draw structural formulas for the following compounds.
  a. heptanoic acid
  b. 3-methylbutanoic acid
  c. 3-chloropentanoic acid
  d. 3-phenylpropanoic acid

4. Draw structural formulas for the following compounds.
  a. *o*-nitrobenzoic acid
  b. *p*-chlorobenzoic acid
  c. 2,3-dibromobenzoic acid
  d. *m*-isopropylbenzoic acid

5. Draw structural formulas for the following compounds.
  a. oxalic acid
  b. β-hydroxybutyric acid
  c. α-chloropropionic acid
  d. phenylacetic acid

6. Name the following compounds.
  a. $(CH_3)_2CHCH_2COOH$
  b. $(CH_3)_3CCH(CH_3)CH_2COOH$
  c. $CH_2OHCH_2CH_2COOH$
  d. $(CH_3)_2CHCH_2CH(CH_3)COOH$

7. Name the following compounds.
  a. $CH_3(CH_2)_8COOH$
  b. $(CH_3)_2CHCCl_2CH_2CH_2COOH$
  c. $CH_3CHOHCH(CH_2CH_3)CHICOOH$

  d. Br—⟨benzene ring⟩—COOH

8. Draw structural formulas for the following salts.
  a. potassium acetate
  b. calcium propanoate

9. Name each of the following salts.

  a. ⟨benzene ring⟩—$\overset{\displaystyle O}{\overset{\|}{C}}$—$O^- Li^+$

  b. $CH_3CH_2CH_2\overset{\displaystyle O}{\overset{\|}{C}}$—$O^- NH_4^+$

10. Draw structural formulas for the following esters.
  a. methyl acetate   **b.** phenyl acetate

11. Draw structural formulas for the following esters.
  a. ethyl pentanoate
  b. ethyl 3-methylhexanoate

12. Draw structural formulas for the following esters.
  a. ethyl benzoate   **b.** phenyl benzoate

13. Draw structural formulas for the following esters.
  a. ethyl butyrate
  b. isopropyl propionate

14. Name the following esters.

  a. ⟨benzene ring⟩—$\overset{\displaystyle O}{\overset{\|}{C}}$—$O$—$CH_3$

  b. CH_3—O—$\overset{\displaystyle O}{\overset{\|}{C}}$—$H$

  c. $CH_3CH_2\overset{\displaystyle O}{\overset{\|}{C}}$—$O$—$CH_2CH_3$

  d. $CH_3CH_2CH_2O$—$\overset{\displaystyle O}{\overset{\|}{C}}$—$CH_3$

  e. $CH_3CH_2\overset{\displaystyle O}{\overset{\|}{C}}$—$OCH_2CH_2CH_3$

  f. $CH_3CH_2CH_2\overset{\displaystyle O}{\overset{\|}{C}}$—$O$—⟨benzene ring⟩

15. Draw structural formulas for the following amides.
  a. butanamide   **b.** hexanamide

16. Draw structural formulas for the following amides.
  a. formamide   **b.** propionamide

17. From what alcohol might each acid be prepared via oxidation with acidic dichromate (or liver enzymes)?
  a. CH_3CH_2COOH **b.** $HOOCCOOH$
  c. $HCOOH$ **d.** $(CH_3)_2CHCH_2COOH$

18. Arrange in order of increasing acidity, with the least acidic compound first.
  a. toluene
  b. benzoic acid
  c. benzyl alcohol ($C_6H_5CH_2OH$)
  d. phenol

19. Draw structural formulas for the following amides.
  a. *N*-methylacetamide
  b. *N*,*N*-dimethylbenzamide

20. Draw structural formulas for and name all the isomeric amides that have the molecular formula C_4H_9NO.

21. Name the following amides.

  a. ⟨benzene ring⟩—$\overset{\displaystyle O}{\overset{\|}{C}}$—$NH_2$

  b. $CH_3CH_2CH(CH_3)\overset{\displaystyle O}{\overset{\|}{C}}$—$NH_2$

  c. $CH_3\overset{\displaystyle O}{\overset{\|}{C}}$—$NH_2$

  d. Cl—⟨benzene ring⟩—$\overset{\displaystyle O}{\overset{\|}{C}}$—$NH_2$

e. $CH_3CH_2\overset{\overset{\displaystyle O}{\|}}{C}-\overset{\underset{\displaystyle CH_3}{|}}{N}-CH_3$

f. $CH_3CH_2CH_2\overset{\overset{\displaystyle O}{\|}}{C}-NH-\langle\bigcirc\rangle$

22. Which compound would have the higher boiling point?

$CH_3CH_2CH_2-O-CH_2CH_3$ $CH_3CH_2CH_2\overset{\overset{\displaystyle O}{\|}}{C}-OH$

          I   **II**

23. Which compound would have the higher boiling point?

$CH_3CH_2CH_2CH_2CH_2OH$ $CH_3CH_2CH_2\overset{\overset{\displaystyle O}{\|}}{C}-OH$

           I   **II**

24. Which compound would have the higher boiling point?

$CH_3CH_2CH_2\overset{\overset{\displaystyle O}{\|}}{C}-NH_2$ $CH_3\overset{\overset{\displaystyle O}{\|}}{C}-O-CH_2CH_3$

       I   **II**

25. Which compound would have the higher boiling point?

$CH_3CH_2CH_2\overset{\overset{\displaystyle O}{\|}}{C}-OH$ $CH_3CH_2\overset{\overset{\displaystyle O}{\|}}{C}-O-CH_3$

       I   **II**

26. Without consulting tables, arrange the following compounds in order of increasing boiling point.
   a. butyl alcohol   **b.** propionic acid
   c. pentane   **d.** methyl acetate

27. Which compound is more soluble in water?

$CH_3\overset{\overset{\displaystyle O}{\|}}{C}-OH$ $CH_3CH_2CH_2CH_3$

    I   **II**

28. Which compound is more soluble in water?

$CH_3CH=CHCH_3$ $CH_3\overset{\overset{\displaystyle O}{\|}}{C}-NH_2$

    I   **II**

29. Which compound is more soluble in water?

$CH_3\overset{\overset{\displaystyle O}{\|}}{C}OCH_3$ $CH_3CH_2CH_2CH_2\overset{\overset{\displaystyle O}{\|}}{C}OCH_2CH_3$

    I   **II**

30. Which compound is more soluble in water?

$\langle\bigcirc\rangle-\overset{\overset{\displaystyle O}{\|}}{C}-OH$ $\langle\bigcirc\rangle-\overset{\overset{\displaystyle O}{\|}}{C}-O^-\ Na^+$

      I   **II**

31. Of the families of compounds discussed in this chapter, which has members with characteristically unpleasant odors? Which group has characteristically pleasant aromas?

32. Write an equation for the reaction of butyric acid with aqueous NaOH and with aqueous $NaHCO_3$.

33. Write an equation for the reaction of benzoic acid with aqueous NaOH and with aqueous $NaHCO_3$.

34. Benzoic acid is insoluble in water. If the reactions described in Exercise 33 were carried out in test tubes, what would you observe?

35. Write an equation for the acid-catalyzed hydrolysis of ethyl acetate.

36. Write an equation for the base-catalyzed hydrolysis of ethyl acetate.

37. Write an equation for the acid-catalyzed hydrolysis of benzamide.

38. Write an equation for the base-catalyzed hydrolysis of benzamide.

39. Complete the following equations.

a. $CH_3CH_2\overset{\overset{\displaystyle O}{\|}}{C}-OH \xrightarrow{\text{NaOH}}$

b. $\langle\bigcirc\rangle-COOH \xrightarrow{\text{KOH}}$

c. $HOOC-COOH \xrightarrow{\text{excess NaOH}}$

d. $\langle\overset{\displaystyle COOH}{\underset{\displaystyle COOH}{\bigcirc}}\rangle \xrightarrow{\text{excess } NaHCO_3}$

40. Complete the following equations.

a. $CH_3\overset{\overset{\displaystyle O}{\|}}{C}-OH + CH_3CH_2CH_2OH \xrightarrow{H^+}$

b. $HO-\overset{\overset{\displaystyle O}{\|}}{C}CH_2\overset{\overset{\displaystyle O}{\|}}{C}-OH + 2\ CH_3OH \xrightarrow{H^+}$

41. Complete the following equations.

a. $CH_3CH_2CH_2O-\overset{\overset{\displaystyle O}{\|}}{C}-\langle\bigcirc\rangle + H_2O \xrightarrow{H^+}$

b. $(CH_3)_2CH-\overset{\overset{\displaystyle O}{\|}}{C}-O-CH_2CH_3 + H_2O \xrightarrow{H^+}$

42. Complete the following equations.

a.

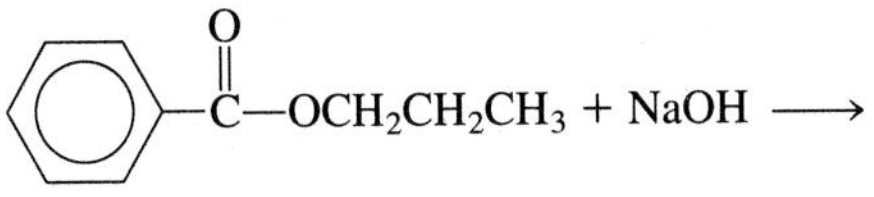

b. $CH_3\overset{O}{\underset{\|}{C}}{-}OCH(CH_3)_2 + KOH \longrightarrow$

43. Complete the following equations.

a.

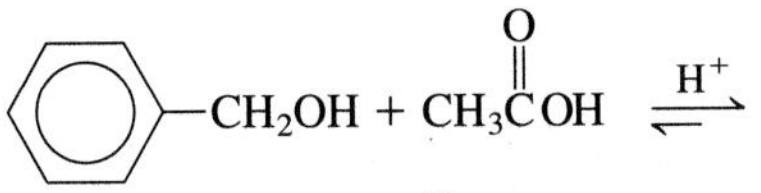

b.

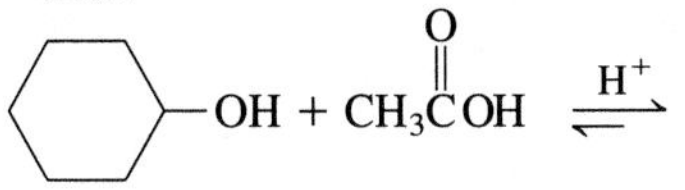

44. Complete the following equations.

a. $CH_3\overset{O}{\underset{\|}{C}}{-}NH_2 + HCl + H_2O \longrightarrow$

b.

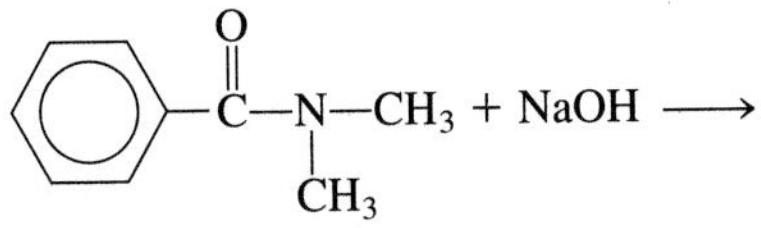

45. List the reagents necessary to carry out the following conversions.

a. $CH_3CH_2\underset{CH_3}{\underset{|}{C}H}CH_2OH \overset{?}{\longrightarrow} CH_3CH_2\underset{CH_3}{\underset{|}{C}H}\overset{O}{\underset{\|}{C}}{-}OH$

b. $CH_3\overset{O}{\underset{\|}{C}}{-}H \overset{?}{\longrightarrow} CH_3\overset{O}{\underset{\|}{C}}OH$

c.

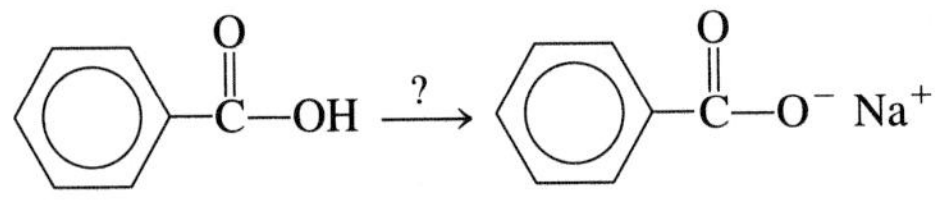

46. List the reagents necessary to carry out the following conversions.

a.

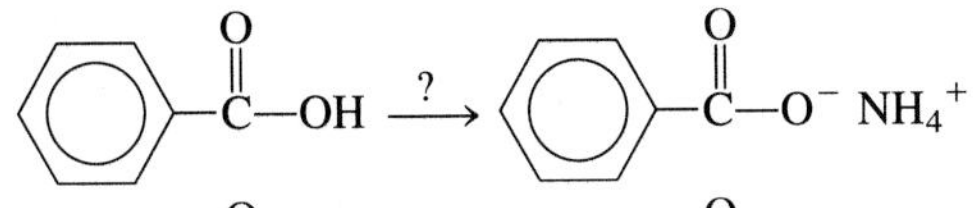

b.

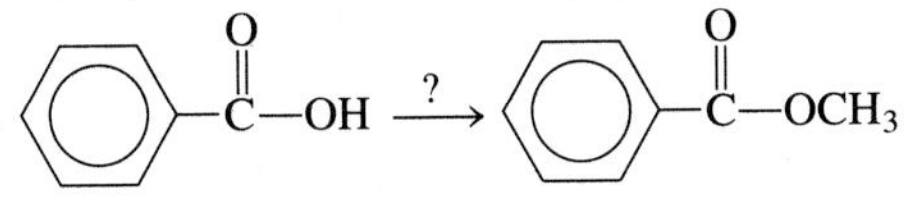

c. $CH_3CH_2CH_2CH_2OH \overset{?}{\longrightarrow} CH_3CH_2CH_2CH_2O\overset{O}{\underset{\|}{C}}CH_3$

d. $CH_3CH_2CH_2\overset{O}{\underset{\|}{C}}{-}OCH_3 \overset{?}{\longrightarrow}$
$CH_3CH_2CH_2\overset{O}{\underset{\|}{C}}{\underset{O^- Li^+}{\diagdown}} + CH_3OH$

47. List the reagents necessary to carry out the following conversions.

a.

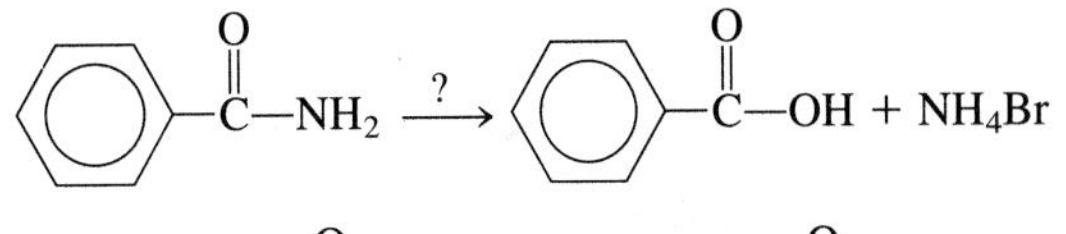

b. $CH_3CH_2{-}\overset{O}{\underset{NH_2}{\overset{\|}{C}}} \overset{?}{\longrightarrow} CH_3CH_2{-}\overset{O}{\underset{O^- K^+}{\overset{\|}{C}}} + NH_3$

48. A lactone is a cyclic ester. What product is formed in each of the following reactions?

a. 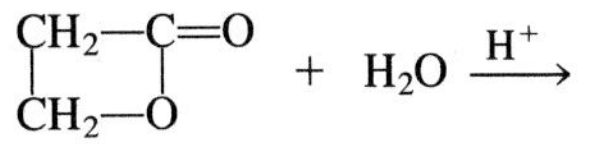 $\underset{CH_2{-}O}{\overset{CH_2{-}C=O}{|}} + H_2O \overset{H^+}{\longrightarrow}$

b.

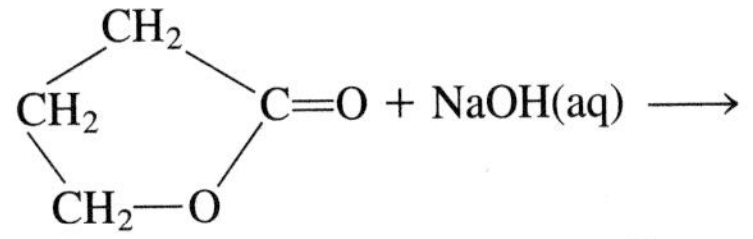

49. A lactam is a cyclic amide. What product is formed in each of the following reactions?

a. 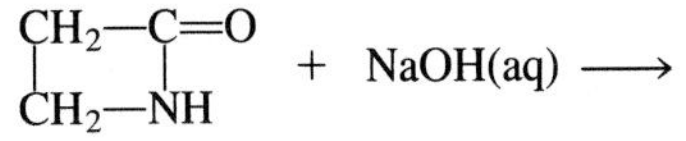 $\underset{CH_2{-}NH}{\overset{CH_2{-}C=O}{|}} + NaOH(aq) \longrightarrow$

b.

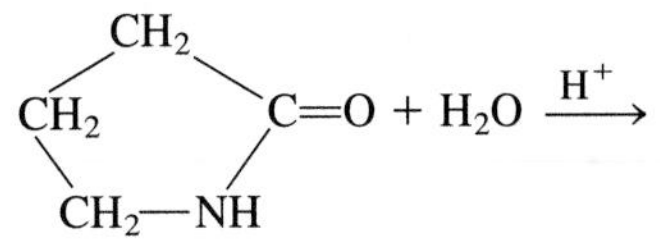

50. Pellagra is a vitamin deficiency disease. Corn contains the antipellagra factor nicotinamide, which is not readily absorbed in the digestive tract. When corn is treated with quicklime (calcium hydroxide) to make hominy, the vitamin is rendered more available. What sort of reaction takes place? Write the equation. Nicotinamide is

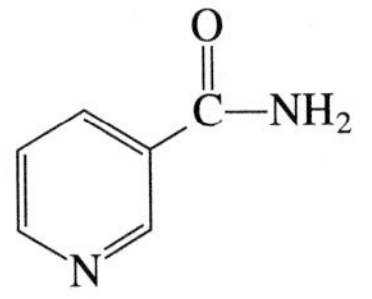

51. The following compounds are isomers. Circle and name the functional groups in each.

a. $CH_3CH_2CH_2\overset{O}{\underset{\|}{C}}OH$ **b.** $CH_3CH_2\overset{O}{\underset{\|}{C}}CH_2OH$

c. $CH_3\overset{O}{\underset{\|}{C}}CH_2CH_2OH$ **d.** $H\overset{O}{\underset{\|}{C}}CH_2CH_2CH_2OH$

e. $CH_3OCH_2CH_2\overset{O}{\underset{\|}{C}}H$ **f.** $CH_3CH_2OCH_2\overset{O}{\underset{\|}{C}}H$

g. $CH_3CH_2CH_2O\overset{\overset{\displaystyle O}{\|}}{C}H$

h. $CH_3OCH_2\overset{\overset{\displaystyle O}{\|}}{C}CH_3$

i. $CH_3CH_2O\overset{\overset{\displaystyle O}{\|}}{C}CH_3$

j. $CH_3CH_2\overset{\overset{\displaystyle O}{\|}}{C}OCH_3$

k. $CH_3\overset{\overset{\displaystyle O}{\|}}{C}-\overset{\overset{\displaystyle OH}{|}}{C}HCH_3$

l. $CH_3\overset{|}{C}HCH_2\overset{\overset{\displaystyle O}{\|}}{C}H$
$\quad\ \ \underset{OH}{|}$

52. Draw structural formulas for the following.
  a. diethyl hydrogen phosphate
  b. methyl dihydrogen phosphate
  c. triphosphoric acid

53. Name the following compounds.

a. $HO-\overset{\overset{\displaystyle O}{\|}}{\underset{\underset{\displaystyle OH}{|}}{P}}-O-\overset{\overset{\displaystyle O}{\|}}{\underset{\underset{\displaystyle OH}{|}}{P}}-OH$

b. $CH_3CH_2O-\overset{\overset{\displaystyle O}{\|}}{\underset{\underset{\displaystyle OH}{|}}{P}}-OH$

54. Define and illustrate the following terms.
  a. neutralization
  b. esterification
  c. acid hydrolysis
  d. saponification

55. Offer an explanation for the following.
  a. Methyl acetate has a lower boiling point (57 °C) than either methyl alcohol (65 °C) or formic acid (100 °C), even though it has a higher molecular weight.
  b. Sodium benzoate is soluble in water, whereas benzoic acid is insoluble.
  c. The alkaline hydrolysis of esters is irreversible, whereas the acidic hydrolysis of esters is reversible.
  d. Both acidic hydrolysis and alkaline hydrolysis of amides are irreversible.

56. A compound, $C_7H_{14}O_2$, was treated with potassium hydroxide to yield ethanol and a potassium salt. Write four possible structural formulas of the original compound.

57. An ester, $C_6H_{12}O_2$, was hydrolyzed in aqueous acid to yield an acid (Y) and an alcohol (Z). Oxidation of the alcohol with potassium permanganate resulted in the identical acid (Y). What is the structural formula of the ester?

58. 5.1 g of a monocarboxylic acid was required to neutralize 125 mL of a 0.4 M NaOH solution. Write all possible structural formulas for the acid.

59. If 3.0 g of acetic acid reacted with excess methanol, how many grams of methyl acetate could be formed?

60. How many milliliters of a 0.10 M barium hydroxide solution would be required to neutralize 0.50 g of dichloroacetic acid?

Drugs: Some Esters and Amides

People have long sought relief from pain and from discomfort. Alcohol, opium, cocaine, and marijuana have been used as medicines for centuries. Often they were used for their pleasurable effects, not just for relief of pain. According to the broadest definition, a **drug** is any chemical substance that affects an individual in such a way as to bring about physiological, emotional, or behavioral change. In this special topic, we will discuss several important drugs that are either esters or amides, types of compounds discussed in the preceding chapter.

E.1
ASPIRIN AND OTHER SALICYLATES

Soon after acetic anhydride[1] became available, in the nineteenth century, chemists began to acetylate a variety of physiologically active compounds. Such structural modifications often change the properties of drugs to enhance their effectiveness or to minimize undesirable side effects. Two such cases will be described here, that of aspirin and that of heroin.

The first successful synthetic pain relievers were derivatives of salicylic acid (Figure E.1). Salicylic acid was first isolated from willow bark in 1860, although an English clergyman named Edward Stone had reported to the Royal Society as early as 1763 that an extract of willow bark was useful in reducing fever. Salicylic acid is itself a good **analgesic** (pain reliever) and **antipyretic** (fever re-

[1] Acetic anhydride can be considered as derived from acetic acid by the removal of one molecule of water from two molecules of acid.

$$CH_3\overset{O}{\overset{\|}{C}}-OH + HO-\overset{O}{\overset{\|}{C}}CH_3 \longrightarrow CH_3\overset{O}{\overset{\|}{C}}-O-\overset{O}{\overset{\|}{C}}CH_3 + HOH$$

ducer), but it is sour and irritating when taken by mouth. Chemists sought to modify the structure of the molecule to remove this undesirable property while retaining (or even improving) the desirable properties.

The first modification was simple neutralization of the acid. The salt sodium salicylate was first used in 1875. It was less unpleasant to swallow, but it proved to be highly irritating to the lining of the stomach. Phenyl salicylate (salol) was introduced in 1886. Salol is a widely used intestinal antiseptic. It is not hydrolyzed by acids and therefore passes through the stomach unchanged. In the alkaline medium of the intestines, hydrolysis occurs to yield phenol and the salicylate ion. Salol is also employed as a coating for some medicinal pills in order to permit them to pass through the stomach intact but disintegrate in the intestines.

Methyl salicylate is an oil found in numerous plants and has a fragrance associated with wintergreen. Commercially, it is used in perfumes and for flavoring candy. It finds widespread use as the pain-relieving ingredient in liniments such as Ben-Gay. When rubbed on the skin, this ester has the unusual ability of penetrating the surface. Hydrolysis then occurs, liberating salicylic acid, which relieves the soreness.

Acetylsalicylic acid (**aspirin**) was first introduced in 1899 and soon became the largest-selling drug in the world. Over 55 billion tablets are produced annually in the United States (that's over 200 tablets for every person in the country). Acetylsalicylic acid is made by treating salicylic acid with acetic anhydride. In this reaction, the hydroxyl group of the phenol reacts exactly like that of an alcohol.

Since aspirin is the most widely used drug in the world, let's take a close look at it. Aspirin is a chemical

Figure E.1
Salicylic acid and some of its derivatives.

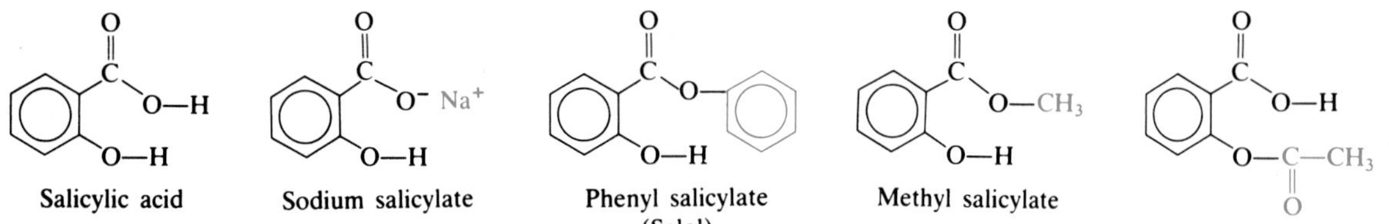

Salicylic acid   Sodium salicylate   Phenyl salicylate (Salol)   Methyl salicylate   Acetylsalicylic acid

compound. Like other compounds, its properties are invariant. Each aspirin *tablet* usually contains 325 mg of the compound, held together with an inert binder (usually starch). Aspirin has been tested extensively. The conclusions of impartial studies are invariably the same: the only significant difference between brands is price. In fact, nearly all of the 14 million kg of aspirin produced annually in the United States is made by two companies, Dow Chemical and Monsanto. Only two drug companies, Sterling Drug (which makes Bayer) and Norwich-Eaton, make their own product.

"Buffered" aspirin contains antacids but is not truly buffered. A **buffer** reacts with either acid or base and keeps the pH of a solution essentially constant. The antacids neutralize acid only; they do not buffer. The experts who evaluated Bufferin, the most highly advertised brand of aspirin, for the FDA, found that there was no basis for the advertising claim that Bufferin is "twice as fast as aspirin." Neither did they find evidence that Bufferin "helps prevent the stomach upset often caused by aspirin." Some people experience mild stomach irritation when they take aspirin on an empty stomach. Eating a little food first or drinking a full glass of water with aspirin is just as effective as taking "buffered" tablets.

"Extra-strength" formulations simply have 500 mg of aspirin rather than the usual 325 mg. They have no other active ingredients. Simple arithmetic tells us that three plain aspirin tablets are equal to two extra-strength tablets in dosage, and they are usually much lower in price.

Aspirin relieves minor aches and pains and suppresses inflammation. We are beginning to understand how aspirin accomplishes its extraordinary effects. It was found that aspirin acetylates, and thus inhibits, an enzyme (cyclooxygenase) necessary for the synthesis of prostaglandins (see Section H.6). Among their many functions, prostaglandins are involved in inflammation, increased blood pressure, and the contraction of smooth muscle. Elevated concentrations of prostaglandins appear to activate pain receptors in the tissues, making the tissues more sensitive to any pain stimulus. Generally speaking, prostaglandins enhance inflammatory effects, whereas aspirin diminishes them. Aspirin often is the initial drug of choice for treatment of arthritis, a disease characterized by the inflammation of joints and connective tissues.

Aspirin also acts as an **anticoagulant**—it inhibits the clotting of blood. Studies indicate that there is some intestinal bleeding every time aspirin is ingested. The blood loss is usually minor (0.5–2.0 mL per two-tablet dose), but it may be substantial in some cases. Aspirin should not be used by people facing surgery, childbirth, or other haz-

Table E.1

Acute Toxicities of Chemicals Presently or Formerly Used in Over-the-Counter Drugs

Chemical Compound	LD_{50}[a]
Acetaminophen	338[b]
Acetanilide	800
Aspirin	1500
Caffeine	355 (246)
Diphenhydramine	500
Ibuprofen	(1050)
Methyl salicylate	887
Naproxen	534
Phenacetin	1650
Piroxicam	360[b]

[a]LD_{50} values are for oral administration of the drug in rats in milligrams per kilogram of body weight (unless otherwise noted). Values in parentheses are for male rats.
[b]Orally in mice.
Source: Susan Budavari (Ed.), *The Merck Index*, 11th ed. Rathway, NJ: Merck and Co., 1989.

ards involving the possible loss of blood (nonuse should start a week before the hazard). On the other hand, small daily doses seem to lower the risk of coronary heart attack and stroke, presumably by the same anticoagulant action that causes bleeding in the stomach. For people with heart problems, many doctors now prescribe one aspirin tablet to be taken every other day. (Some doctors are prescribing daily doses of one baby aspirin, 81 mg.)

Fevers are induced by substances called **pyrogens.** These compounds are produced by and released from leukocytes and other circulating cells. Pyrogens usually use prostaglandins as secondary mediators. Fevers therefore can be reduced by aspirin and other prostaglandin inhibitors.[2]

Our bodies try to fight off infections by elevating the temperature. Mild fevers (those below 39 °C or 102 °F) usually are therefore best left untreated. High fevers, however, can cause brain damage and require immediate treatment.

Aspirin is not effective for severe pain—for example, pain from a migraine headache. Prolonged use of aspirin, as for arthritic pain, can lead to gastrointestinal disorders. Like all drugs, aspirin is somewhat toxic. It is the drug most often involved in the accidental poisoning of children. The toxicities of aspirin and other drugs are listed in Table E.1.

[2] Some pyrogens do not work through prostaglandins. Aspirin doesn't affect those pyrogens.

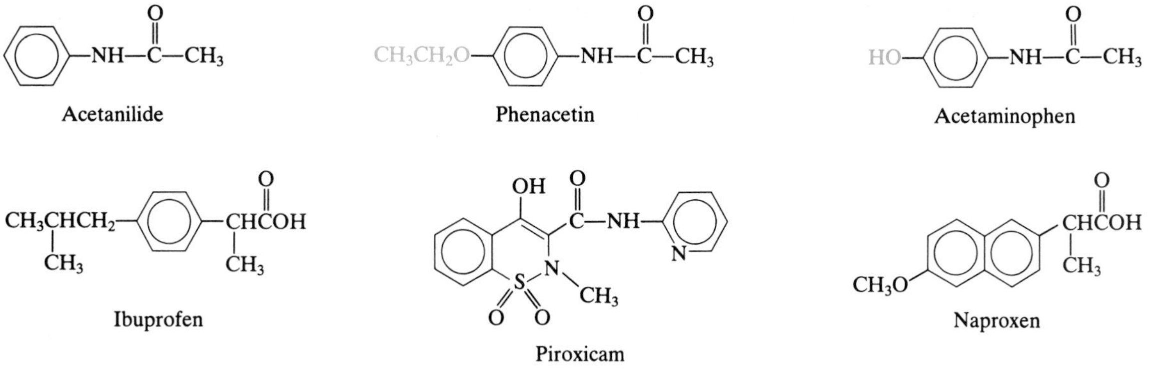

Figure E.2
Some aspirin substitutes. Phenacetin and acetaminophen are derivatives of acetanilide.
Ibuprofen and naproxen are derivatives of propionic acid. Acetanilide and phenacetin are
no longer used. Naproxen and piroxicam are under review for over-the-counter use.

Use of aspirin to treat children feverish with flu or chicken pox is associated with **Reye's syndrome.** This syndrome is characterized by vomiting, lethargy, confusion, and irritability. Fatty degeneration of the liver and other organs can lead to death unless treatment is begun promptly. Just how aspirin use enhances the onset of Reye's syndrome is not known, but the correlation is quite strong. Aspirin products bear a warning not to use aspirin to treat children with fevers.

Still another hazard associated with aspirin use is allergic reaction. In some people, an allergy to aspirin can cause skin rashes, asthmatic attacks, and even loss of consciousness. Some doctors claim that the allergic reaction may be delayed 3–5 hours, so the victim may not associate the reaction with aspirin. Susceptible individuals must be careful to avoid aspirin by itself or in any combination with other drugs.

E.2
ASPIRIN SUBSTITUTES AND COMBINATION PAIN RELIEVERS

Some people who are allergic to aspirin may safely take substitute medicines (Figure E.2). The most common is acetaminophen. This compound gives relief of pain and reduction of fever comparable to the action of aspirin. Unlike aspirin, however, it is not effective against inflammation, so it is of limited use to people with arthritis. But neither does it induce bleeding, so it is often used to relieve the pain that follows minor surgery. The fact that

acetaminophen does not promote bleeding probably accounts for its more frequent use than aspirin in hospitals. Acetaminophen usually costs more than aspirin, especially in the form of highly advertised brands such as Tylenol, Panadol, and Anacin-3. Regular acetaminophen tablets are 325 mg; "extra-strength" forms are 500 mg. Overuse of acetaminophen has been linked to liver and kidney damage.

Ibuprofen, an anti-inflammatory drug available for decades by prescription under the trade name Motrin, has been available for over-the-counter use since 1984.[3] Brand names are Advil, Nuprin, Mediprin, and Motrin IB. Ibuprofen also is available in generic form. The usual nonprescription dosage is 200 mg. Ibuprofen is perhaps superior to aspirin in its action against inflammation. It also relieves mild pain and reduces fevers but is unlikely to be any better than aspirin for those purposes. It usually is much more expensive than aspirin, and many people who are sensitive to aspirin are also sensitive to ibuprofen.

Much of what we spend on analgesics is spent on combinations of aspirin and other drugs. Is our money well spent? What are those other drugs? Do they really add anything to the effectiveness of the medication? Must the consumer be at the mercy of misleading advertising? These are questions that can be answered rather simply. Let's look first at the chemical compounds in some of the more familiar "combination pain relievers."

[3] Two other prescription anti-inflammatory drugs, naproxen (Naprosyn) and piroxicam (Feldene), are under review by the U.S. FDA for over-the-counter sale. They are quite similar to ibuprofen in their properties and mode of action.

For many years, the most familiar combination was aspirin, phenacetin, and caffeine (APC). This combination was available under a variety of trade names, or it could be purchased as APC tablets USP, usually at a lower price than the proprietary medications. Phenacetin has about the same effectiveness as aspirin in reducing fever and relieving minor aches and pains. It has been implicated in damage to the kidneys, in blood abnormalities, and as a likely carcinogen, however. The U.S. FDA banned further use of phenacetin in 1983.

Anacin, which was once an APC formulation, now contains only aspirin and caffeine. Caffeine is a mild stimulant found in coffee, tea, and cola syrup. There is no reliable evidence that caffeine significantly enhances the effect of aspirin. In fact, evidence indicates that for fever reduction, caffeine *counteracts* the action of aspirin. Combinations containing caffeine are therefore *less effective* than plain aspirin for this use. A tablet of Anacin does contain a little more aspirin (400 mg) than a regular (325-mg) aspirin tablet. The usual two-tablet dose of Anacin provides 800 mg of aspirin. You could get the same dose—at a lower price—from two and a half regular aspirin tablets.

Caffeine

Another combination drug, Excedrin, contains both aspirin and acetaminophen. Excedrin also contains caffeine. Many other combination products are available. Brands and formulations change frequently. Extensive studies show repeatedly that, for most people, plain aspirin is the cheapest, safest, and most effective product.

E.3
ALKALOIDS OF OPIUM

An **alkaloid** is any nitrogen-containing compound that is obtained from plants and has physiological activity (nicotine, atropine, quinine, etc.). The opium alkaloids, (Figure E.3) sometimes referred to as opiates or opioids, originally meant drugs obtained from the opium poppy (*Papaver somniferum*—not the common garden-variety poppy). Today the term includes all the synthetic compounds that have morphinelike activity. That is, when

administered in medicinal doses, they relieve pain (analgesics) and they induce sleep **(hypnotics).**

Morphine, which has been used extensively for many years, is still the most important sedative-analgesic. Morphine remains the standard against which new analgesics are measured. Although morphine can be synthesized in the laboratory, it is still obtained from opium. When pure, morphine is an odorless, white crystalline solid with a bitter taste, and it is insoluble in water.

Commercial opium is the dried juice from the unripened seed pod of the poppy plant. Opium is a complex mixture containing more than 20 different compounds. The principal alkaloid, morphine, makes up about 10% of the weight of raw opium. Raw opium was used in many of the patent medicines of the nineteenth century. Ayer's Cherry Pectoral, Jayne's Expectorant, Pierce's Golden Medical Discovery, and Mrs. Winslow's Soothing Syrup were but a few.

Morphine

Morphine was first isolated in 1805 by Friedrich Sertürner, a German pharmacist. With the invention of the hypodermic syringe (Figure E.4) in the 1850s, a new method of administration became available. Injection of morphine directly into the bloodstream was more effective for the relief of pain, but it also seriously escalated the problem of addiction. Morphine was used widely during the American Civil War (1861–1865). It was effective for relief of pain caused by battle wounds. One side effect of morphine use is constipation. Noting this, soldiers came to use morphine as a treatment for that other common malady of men on the battlefront—dysentery. During their wartime service, over a hundred thousand soldiers became addicted to morphine. The affliction was so common among veterans that it came to be known as ''soldier's disease.''

Morphine and other narcotics were placed under the federal government's control by the Harrison Act of 1914. Morphine is still used by prescription for relief of severe pain. It also induces lethargy, drowsiness, confusion, euphoria, chronic constipation, and depression of the respiratory system. It is strongly addictive if administered in

Figure E.3
Opium poppy flower (a) and seed pod (b); (c) heroin; (d) morphine.

Figure E.4
An old-fashioned hypodermic syringe.

amounts greater than the prescribed doses or for a period longer than the prescribed time.

Slight changes in the molecular architecture of morphine produce altered physiological properties. Replacement of one of the —OH groups by an —OCH$_3$ group produces codeine. Actually, codeine occurs in opium to an extent of about 0.5%. It is usually synthesized, however, by methylating the more abundant morphine molecules. About 55,000 kg of codeine is produced each year in the United States, enough for 16 doses of 15 mg for every person in the country.

Codeine is similar to morphine in its physiological action, except that it is less potent and has less tendency to induce sleep. It is also thought to be less addictive. In amounts of less than 2.2 mg/mL, codeine is exempt from

CH₃

Codeine
(Methylmorphine)

the stringent narcotics regulations and is used in cough syrups.

In the laboratory, reaction of morphine with acetic anhydride produces heroin. This morphine derivative was first prepared by chemists at the Bayer Company of Germany in 1874. It received little attention until 1890, when it was proposed as an antidote for morphine addiction. Shortly thereafter, Bayer was widely advertising heroin as a sedative for coughs, often in the same ads as aspirin (Figure E.5). It soon was found, however, that heroin induced addiction more quickly than morphine and that heroin addiction was harder to cure.

CH₃

Heroin
(Diacetylmorphine)

The physiological action of heroin is similar to that of morphine, except that heroin seems to produce a stronger feeling of euphoria for a longer period of time. Heroin is not legal in the United States, even by prescription. It has, however, been advocated for use in relief of pain in terminal cancer patients and has been so used in Britain.

Addiction probably has three components: psychological dependence, physical dependence, and tolerance. **Psychological dependence** is evident in the uncontrollable desire for the drug. **Physical dependence** is shown by acute withdrawal symptoms such as convulsions. The body cells are conditioned to the drug and cannot function normally without it. The difference in physical and psychological dependence may be only one of degree, not kind. Both are likely to be biochemical in origin. **Toler-**

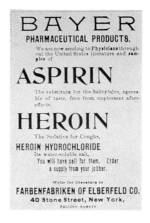

Figure E.5
Heroin was advocated as a safe medicine in 1900. It was widely marketed as a sedative for coughs.

ance for the drug is evidenced by the increasing dosages required to produce in the addict the same degree of narcosis and analgesia.

Deaths from heroin usually are attributed to overdose, but the situation is not altogether clear. The problem seems to be a matter of quality control. As an illustration, the office of the chief medical examiner for New York City analyzed 132 samples of drugs, supposedly heroin, that had been confiscated on the streets (Figure E.6). Twelve contained no heroin at all. The remaining 120 varied from 1 to 77% heroin. A user could think he or she was getting a dose of 1 unit, when he or she was actually getting 77 times as much—a catastrophic overdose.

Figure E.6
Forms of heroin as found on the street.

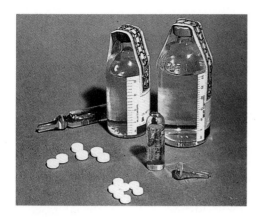

Figure E.7
Dosage forms of meperidine (Demerol), a synthetic narcotic.

E.4
SYNTHETIC NARCOTICS: ANALGESIA AND ADDICTION

Much research has gone into developing a drug that would be as effective as morphine for the relief of pain but that would not be addictive. Perhaps the best known of the synthetic narcotics is meperidine (Demerol) (Figure E.7). Meperidine is somewhat less effective than morphine, but it has the advantage that it does not cause nausea. Repeated use, unfortunately, does lead to addiction.

Meperidine
(Demerol)

Another synthetic narcotic is methadone. This drug has been widely used to treat heroin addiction. Like heroin, methadone is highly addictive. However, when taken orally, it does not induce the sleepy stupor characteristic of heroin intoxication. Unlike a heroin addict, a person on methadone maintenance usually is able to hold a productive job. Methadone is available free in clinics. If an addict who has been taking methadone reverts to heroin, the methadone in his or her system effectively blocks the eu-

phoric rush normally given by heroin and so reduces the addict's temptation to use heroin.

Methadone maintenance is not a perfect answer. Perhaps it is not even a good one. When injected into the body, methadone gives an effect similar to that of heroin, and methadone has been diverted for illegal use in this manner. And an addict on methadone is still an addict. All the problems of tolerance (and cross-tolerance with heroin and morphine) still exist.

Methadone

Chemists have synthesized thousands of morphine analogs. Only a few have shown significant analgesic activity. Most are addictive. Morphine acts by binding to receptors in the brain. Molecules that mimic the action of a drug are called **agonists.** An **antagonist** blocks the action of a drug. Morphine antagonists are molecules that block the action of morphine, most likely by blocking the receptors. Some molecules have both agonist and antagonist activity. These show great promise as analgesics. An example is pentazocine (Talwin). It is less addictive than morphine and yet it is effective for relief of pain. There is some hope that an effective analgesic could be developed that is not addictive, but to date the two effects seem inseparable.

Pentazocine
(Talwin)

Naloxone

Pure antagonists such as naloxone are of value in treating opiate addicts. Overdosed addicts can be brought back from death's door by an injection of naloxone. Long-acting antagonists can block the action of heroin for as much as a month, thus aiding an addict in overcoming his or her addiction.

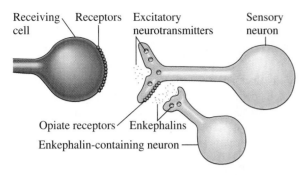

Figure E.8
A proposed mechanism of opiate activity. The enkephalins bind to the "opiate" receptors and prevent the release of neurotransmitters that convey the pain message to the brain.

E.5

A NATURAL HIGH: THE BRAIN'S OWN OPIATES

Morphine acts by fitting specific receptor sites in the brain (Figure E.8). These morphine receptors were first demonstrated in 1973 by Solomon Snyder and Candace Pert at Johns Hopkins University School of Medicine.

Why should the human brain have receptors for a plant-derived drug like morphine? There seemed to be no good reason, so several investigators started a search for morphinelike substances produced by the human body. Not one, but several such substances, called **endorphins** ("endogenous morphines"), were soon found. Each was a short peptide chain composed of amino acid units (Chapter 21). Those with five amino acid units are called **enkephalins.** There are two enkephalins, and they differ only in the amino acid at the end of the chain. *Leu*-enkephalin has the sequence Tyr-Gly-Gly-Phe-Leu, and *Met*-enkephalin is Tyr-Gly-Gly-Phe-Met.[4] Other substances with chains of 30 amino acids also were found.

Some of the enkephalins have been synthesized and shown to be potent pain relievers. Their use in medicine is quite limited, however, because, after being injected, they are rapidly broken down by the enzymes that hydrolyze proteins. It is hoped, though, that analogs more resistant to hydrolysis can be employed as morphine substitutes for the relief of pain. Unfortunately, both the natural enkepha-

[4]The three-letter symbols are abbreviations for amino acids; see Table 21.1.

lins and the analogs, like morphine, seem to be addictive.

It appears that endorphins are released as a response to pain deep in the body. Bruce Pomeranz of the University of Toronto has collected evidence that indicates that acupuncture anesthetizes by stimulating the release of the brain "opiates." The long needles stimulate deep sensory nerves that cause the release of the peptides that then block the pain signals. Endorphin release also has been used to explain other phenomena once thought to be largely psychological. A soldier wounded in battle feels no pain until the skirmish is over. His body has secreted its own pain-killer. The production of these compounds during strenuous athletic activity also may explain the "high" reported by distance runners.

E.6

LSD: HALLUCINOGENIC DRUG

One of the most interesting amides of all is the N,N-diethylamide of lysergic acid, better known as LSD (from the German *lysergsaure diethylamid*) or "acid." The physiological properties of this compound were discovered quite accidentally by Albert Hofmann in 1943. Hofmann, a chemist at the Sandoz Laboratories in Switzerland, unintentionally ingested some LSD. He later took 250 μg, which he considered a small dose, to verify that LSD had caused the symptoms he had experienced. Hofmann had a very rough time for the next few hours, exhibiting such symptoms as visual disturbance and schizophrenic behavior.

Lysergic acid dietnylamide
(LSD)

Lysergic acid is obtained from ergot, a fungus that grows on rye.[5] It is converted to the diethylamide by treatment with thionyl chloride ($SOCl_2$) followed by diethyl-

[5]Several useful drugs are obtained from the ergot fungus. Ergotamine shrinks blood vessels in the brain; it is used to treat migraine headaches. Ergonovine causes small blood vessels to contract. It is used to induce uterine contractions and thus reduce bleeding after childbirth. Both compounds, like LSD, are amides of lysergic acid.

amine. Note that a part of the LSD structure resembles that of serotonin (see Section F.2). LSD seems to act as a serotonin agonist.

LSD is a potent drug, as indicated by the small amount required for a person to experience its fantastic effects. The usual dose is probably about 10–100 μg. No wonder Hofmann had a bad time with 250 μg! To give you an idea of how small 10 μg is, let's compare that amount of LSD to the amount of aspirin in one tablet—one aspirin tablet contains 325,000 μg of aspirin.

Is LSD a dangerous drug? A few facts are known, but most are disputed. In 1967, Maimon Cohen of the State University of New York at Buffalo reported that LSD damages chromosomes, especially those of the leukocytes (white blood cells). The report received wide publicity. Fear of damage to germ cells—and the subsequent birth of deformed babies—caused a decline in the use of LSD. Additional studies produced mixed results. Some seemed to confirm Cohen's findings. For example, in hamsters, LSD administered to pregnant females caused gross fetal deformities. Other studies, however, seemed to exonerate LSD as a cause of chromosomal damage. The question still has not been resolved.

The great concern over the problem is due, in part, to the thalidomide tragedy of the late 1950s and early 1960s. Thalidomide was a completely legal, amidelike drug used as a tranquilizer.

Figure E.9
Frances O. Kelsey, M.D., of the U.S. Food and Drug Administration, refused to approve the tranquilizer drug thalidomide for marketing in the United States. Her action prevented many mothers from facing the tragedy of giving birth to a "thalidomide baby." She received the Distinguished Federal Civilian Service Award in 1962.

Thalidomide

Thalidomide was considered so safe, based on laboratory studies, that it often was prescribed for pregnant women. In Germany it was available without a prescription. It took several years for the human population to provide evidence that laboratory animals had not. The drug had a disastrous effect on developing human embryos. Women who had taken the drug during the first 12 weeks of pregnancy had babies that suffered from phocomelia, a condition characterized by shortened or absent arms and legs and other physical defects. The drug was used widely in Germany and Great Britain, and these two countries bore the brunt of the tragedy. The United States escaped relatively unscathed because an official of the FDA had believed that there was evidence to doubt the drug's safety and therefore had not approved it for use in the United States (Figure E.9).

E.7
MARIJUANA: SOME CHEMISTRY OF *CANNABIS*

Although marijuana is neither an ester nor an amide, we discuss it here because of its significance. Many books have been written about marijuana, yet all we know for certain about the drug would fill only a few pages. Let's look at some chemistry of the weed and at some of the ways chemists are involved with the marijuana problem.

The weed *Cannabis sativa* (Figure E.10) has long been useful. The stems yield tough fibers for making ropes. *Cannabis* has been used as a drug in tribal religious rituals. Marijuana also has a long history as a medicine, particularly in India. In the United States, marijuana is second only to alcohol as an intoxicant.

Figure E.10
The marijuana plant.

Figure E.11
Retail forms of marijuana.

The term **marijuana** refers to a preparation made by gathering the leaves, flowers, seeds, and small stems of the plant (Figures E.11 and E.12). These are generally dried and smoked and referred to as "pot." They contain a variety of chemical substances, many of them still unidentified. The principal active ingredient, however, is tetrahydrocannabinol (THC). Actually, there are several active cannabinoids in marijuana; only one is shown here.

$$CH_3$$

Tetrahydrocannabinol
(THC)

Figure E.12
Marijuana dealers were early converts to the metric system, Shown are kilo bricks of marijuana.

Marijuana plants vary considerably in THC content. Most of the marijuana sold in North America has a THC content of about 1%. Plants native to the United States have a low THC content, usually about 0.1%. Potency depends on the genetic variety of plant, not to any significant extent on the climate or the soil where it is grown.

More potent preparations sometimes are made from marijuana. By selecting only the flowering tops and tender top leaves, you get a stronger product called *ganja*. (The ordinary marijuana is called *bhang* in India.) Jamaican ganja has a THC content of between 4 and 8%. Indian ganja is generally somewhat less potent. By collecting only the resinous secretions of the flowering parts, you get a product called hashish, or "hash" (known as *charas* in India). Hash has a THC content of between 5 and 12%.

Liquid hash and hash oil are probably solvent extracts of marijuana.

The effects of marijuana are difficult to measure, partly because of the variable THC content of different preparations. A variety of standard potency is now grown and supplied for controlled clinical studies. With this standard product, some of the effects of marijuana can be measured in reproducible experiments. Smoking cannabis increases the pulse rate, distorts the sense of time, and impairs some complex motor functions. These effects can be measured easily. Other results also have been noted widely, if less quantitatively. Marijuana smoking sometimes induces a euphoric sense of lightness—a floating sensation. Sometimes it causes a feeling of anxiety. Often, the user has an impression of brilliance, although studies have shown no mind-expanding effects. Users sometimes experience hallucinations, although these are much less frequent than with LSD. Marijuana seems to heighten one's enjoyment of food, with users relishing beans as much as they normally would enjoy steak.

The long-term effects of marijuana use are more difficult to evaluate. Some people claim that smoking marijuana leads to the use of harder drugs. There is little objective evidence of this. More heroin addicts start drug use with alcohol than with marijuana.

There is some evidence—both direct and indirect—that marijuana causes brain damage. Studies in rats indicate that marijuana causes brain lesions. Heavy users often are lazy, passive, and mentally sluggish. It is difficult, however, to prove that these are effects of marijuana use. There are millions of users, and even if the observed brain damage is due to marijuana, it is less extensive than that caused by alcohol.[6]

Some people also have claimed that excessive use of marijuana leads to psychoses. It has long been known to induce short-term psychotic episodes in those already predisposed and in others who take excessive amounts. Long-term psychoses, however, occur at the same rate among regular marijuana users as among the general population.

One of the more interesting reports has come from two surgeons at the Harvard Medical School. Menelaos Aliapoulios and John Harman claim to have treated 13 young men for gynecomastia (enlarged breasts). All were heavy marijuana users. Their breasts also discharged a white milky liquid. The painful swelling receded in three of the men after they stopped smoking cannabis, but three others needed surgery. The two doctors are convinced that marijuana contains a "feminizing ingredient." There is a slight structural similarity of THC to the female hormones

(Special Topic H). Some studies indicate that THC binds weakly to estrogen receptors.

Estradiol

Despite its apparent feminizing properties, THC in both high and low doses causes an initial rise in testosterone levels in men. With high doses, however, the slight increase is followed by a rapid fall to below-normal testosterone levels.

Chemists have isolated the active components and synthesized them. They can monitor the THC content of marijuana. They can monitor THC in the bloodstream and identify the products of its breakdown. They cannot, however, tell how it changes the body chemistry or what its long-term effects are.

Perhaps the most significant harm from marijuana comes through its impairment of complex motor functions—such as those used in driving an automobile. In addition to dodging drunks, we have to look out for potheads. Incidentally, chemists have developed a THC detector similar to the one used to test blood alcohol. Let's hope the road is cleared of all intoxicated people, no matter what they are intoxicated with.

Unlike alcohol, THC persists in the bloodstream for several days because it is soluble in fats. The products of its breakdown remain in the blood for as long as 8 days. The persistence of these chemicals inside the body indicates that some of a given dose may still be active in the body at the time another dose is taken. This might account for the fact that an experienced pot smoker can get high on a dose that doesn't affect a novice.

Marijuana has some legitimate medical uses. It reduces pressure in the eyes of people who have glaucoma. If not treated, the buildup of pressure eventually causes blindness. Also, marijuana relieves the nausea that afflicts cancer patients undergoing radiation treatment and chemotherapy.

---

[6]The gene for a THC receptor was cloned in 1990 (see Section 23.9). The receptors are found in movement control centers, thus explaining the loss of coordination in those intoxicated with the drug. The memory and cognition areas of the brain also are rich in THC receptors. This explains why marijuana users often do poorly on tests. There are few receptors in the brainstem where breathing and heartbeat are controlled. This correlates with the fact that it is difficult to get a lethal dose of marijuana.

EXERCISES

1. Define, explain, or give an example of the following.
   a. drug
   b. analgesic
   c. antipyretic
   d. pyrogen
   e. Reye's syndrome
   f. hypnotic
   g. opium
   h. psychological dependence
   i. physical dependence
   j. tolerance
   k. agonist
   l. antagonist
   m. endorphins
   n. enkephalins

2. a. How does heroin differ in structure from morphine?
   b. How does it differ in physiological properties?

3. Examine the labels of at least five "combination" pain relievers (e.g., Excedrin, Empirin, Anacin). Make a list of the ingredients in each. Look up the properties (medical use, dosage, side effects, toxicity, etc.) in a reference work such as *The Merck Index*.

4. Aspirin is a chemical compound.
   a. What is its structure and chemical name?
   b. In what ways may one brand of aspirin differ from another?
   c. In what ways must brands of aspirin be the same?

5. Do a cost analysis on at least five brands of plain aspirin, calculating the cost per gram.

6. To what family of organic compounds does ibuprofen belong?

7. What alkyl group is attached to the *para* position of the benzene ring of the ibuprofen molecule? Can you see how the generic name ibuprofen was derived?

8. List the reagents necessary to carry out the following conversions.

a.

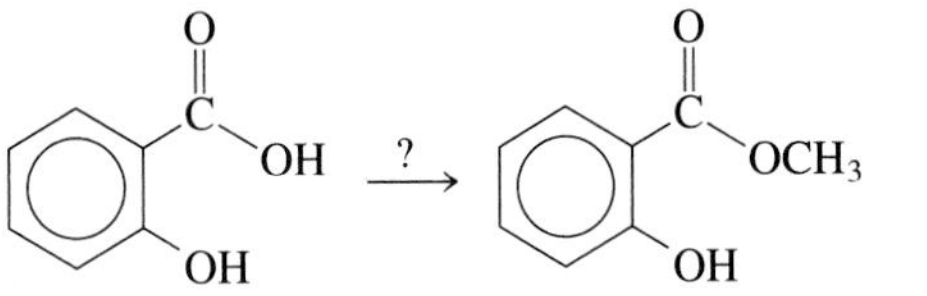

b.

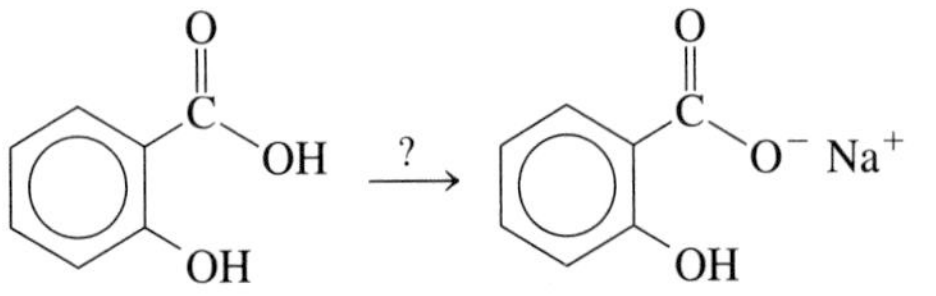

9. List the reagents necessary to carry out the following conversions.

a.

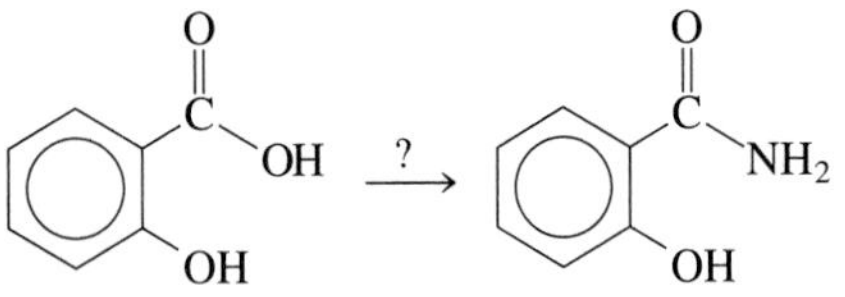

b.

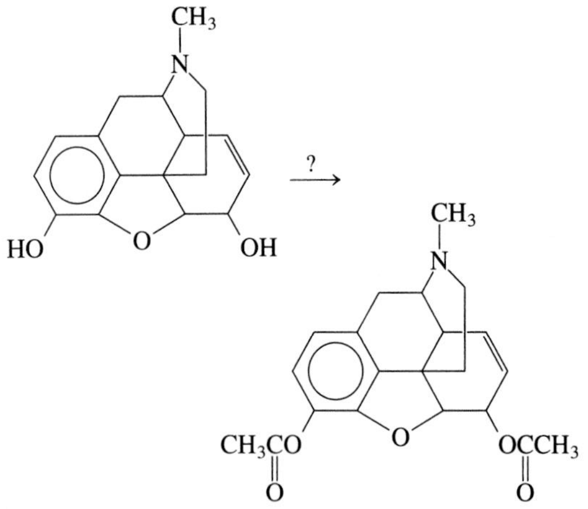

10. "New Maximum Strength" Bayer aspirin is "1000 mg of pure . . . aspirin per [2 tablet] dose." How many "regular" tablets of 325 mg aspirin each would you take to get about the same amount of aspirin?

11. What is an alkaloid? Name several alkaloids.

12. What are the effects of a hallucinogenic drug?

13. a. How does methadone maintenance work?
   b. What is your position on the use of methadone maintenance to treat heroin addiction?

14. How are endorphins related to each of the following?
   a. the anesthetic effect of acupuncture
   b. the absence of pain in a badly wounded soldier

15. What are some of the problems in the clinical evaluation of LSD?

16. What are some of the problems in the clinical evaluation of marijuana?

17. What is the role of the chemist in the marijuana controversy?

18. How might marijuana have a feminizing effect on males?

19. Why is tetrahydrocannabinol retained in the body for several days?

20. In what areas of the brain are THC receptors found? How does the location of these receptors explain some of the effects of THC as a drug?

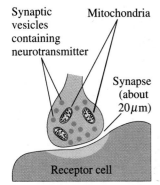

Synaptic vesicles containing neurotransmitter

Mitochondria

Synapse (about 20μm)

Receptor cell

Figure F.3

Schematic diagram of a synapse. When an electric signal reaches the nerve ending, neurotransmitter molecules are released from the vesicles. They then migrate across the narrow gap (synapse) and move to the receptor cell, where they fit specific sites.

F.2
BRAIN AMINES

We all have our ups and downs. These moods probably result from multiple causes, but it appears likely that a variety of chemical compounds formed in the brain are involved. Before we consider these ups and downs, though, let's take a look at epinephrine, an amine formed in the adrenal glands.

Commonly called *adrenaline,* epinephrine is secreted by the adrenal glands. A tiny amount of epinephrine causes a great increase in blood pressure. When a person is under stress or is frightened, the flow of adrenaline prepares the body for fight or flight. Because culturally imposed inhibitions prevent fighting or fleeing in most modern situations, the adrenaline-induced supercharge is not used. This sort of frustration has been implicated in some forms of mental illness.

Epinephrine
(Adrenaline)

Norepinephrine

One simplified biochemical theory of mental illness involves two brain amines. One is norepinephrine (NE), a relative of epinephrine. NE is a neurotransmitter formed in the brain. When formed in excess, NE causes the person to be elated—perhaps even hyperactive. In large excess, NE induces a manic state. A deficiency of NE, on the other hand, could cause depression.

Serotonin

The other brain amine is serotonin. A neurotransmitter, serotonin also seems to play a role in mental illness. Serotonin is involved in sleep, sensory perception, and the regulation of body temperature. Its exact role in mental illness is not clear. A metabolite of serotonin, 5-hydroxyindoleacetic acid (5-HIAA), is found in unusually *low* levels in the spinal fluid of violent suicide victims.[1] This indicates that abnormal serotonin metabolism may play a role in depression.

Our cells have at least six different receptors that are activated by NE and related compounds. NE agonists (drugs that enhance or mimic its action) are stimulants. NE antagonists (drugs that block the action of NE) slow down various processes. Drugs called beta blockers reduce the stimulant action of epinephrine and NE on various kinds of cells. Propranolol (Inderal) is used to treat cardiac arrhythmias, angina, and hypertension by lessening slightly the force of the heart beat. Unfortunately, it also causes lethargy and depression. Metoprolol (Lopressor) acts selectively on cells in the heart. It can be used by hypertensive patients who have asthma because it does not act on receptors in the bronchi.

Serotonin agonists are used to treat depression and anxiety. Agonist drugs are being used experimentally to treat obsessive-compulsive disorder. Serotonin antagonists are used to treat migraine headaches and to relieve the nausea caused by cancer chemotherapy.

Richard Wurtman of the Massachusetts Institute of Technology found a relationship between diet and serotonin levels in the brain. Serotonin is produced in the body from the amino acid tryptophan (Figure F.4). The synthesis involves several steps, each catalyzed by an enzyme. Wurtman found that diets high in carbohydrates lead to high levels of serotonin. A large amount of protein lowers the serotonin concentration. That may seem strange, for

---

[1]Levels of 5-HIAA also are low in murderers and other violent offenders. Levels of this serotonin metabolite are higher than normal in persons with obsessive-compulsive disorders, sociopaths, and people who have guilt complexes.

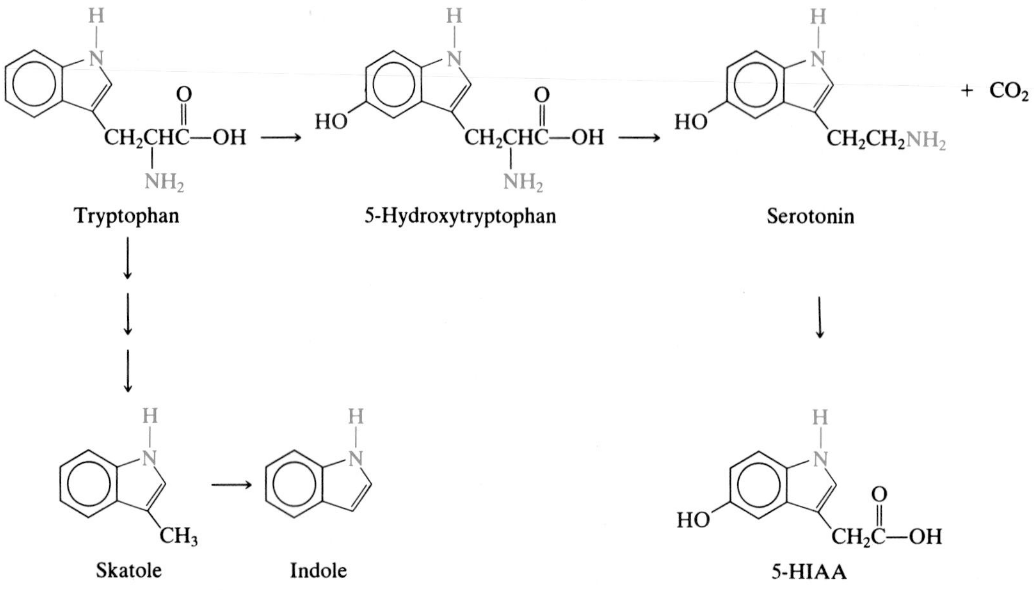

Figure F.4
Serotonin is produced in the brain from tryptophan. A variety of other interesting compounds are also produced from tryptophan. One route leads (through several intermediates) to skatole, the principal odoriferous ingredient in human feces, and to indole. Indole is used in perfumes.

protein has lots of tryptophan and carbohydrates have little. But, Wurtman says, protein is only 1% tryptophan. In the presence of all those other amino acids little tryptophan reaches the brain. With a carbohydrate meal, the hormone insulin lowers the level of the other amino acids in the blood, allowing relatively high levels of tryptophan to reach the brain.

Norepinephrine also is synthesized in the body from an amino acid. It is derived from tyrosine. The synthesis is complex and proceeds through several intermediates (Figure F.5). Each step is catalyzed by one or more enzymes. The intermediate compounds also have physiological activity: dopa has been used successfully in the treatment of Parkinson's disease, and dopamine has been employed to

Figure F.5
The biosynthesis of norepinephrine from phenylalanine via tyrosine. The skin pigment melanin also is produced from tyrosine (Chapter 21). Dopamine is a neurotransmitter with a stimulant effect on the brain. The antipsychotic tranquilizers (Section F.9) are dopamine antagonists.

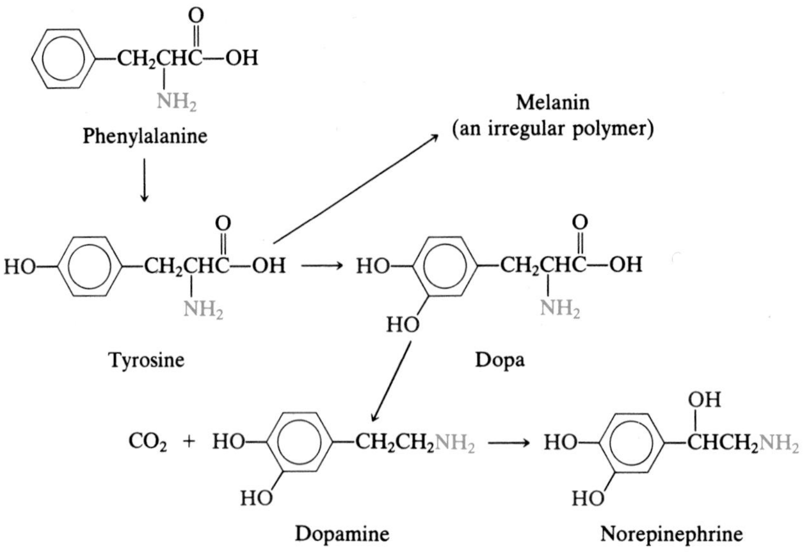

treat low blood pressure (see Section 27.5). Since tyrosine is also a component of our diets, it may well be that our mental state depends to a fair degree on our diet.

Nearly one out of every ten people in the United States suffers from mental illness. Over half the patients in hospitals are there because of mental problems. When the biochemistry of the brain is more fully understood, mental illness may be cured (or at least alleviated) by administration of drugs. In subsequent sections, we see just how far we have already come in learning to control our moods with drugs. As is true for so many things, the potential for good that such compounds represent is matched by a potential for abuse.

F.3
STIMULANT DRUGS: AMPHETAMINES

Among the more widely known stimulant drugs are a variety of synthetic amines related to phenylethylamine (Figure F.6). Note the similarity of these molecules to those of epinephrine and norepinephrine; all are derived from the basic phenylethylamine structure. The amphetamines probably act as stimulants by mimicking the natural brain amines.

Amphetamine[2] and methamphetamine have been widely abused. Amphetamine has been extensively used for weight reduction. It has also been employed for treating mild depression and narcolepsy, a rare form of sleeping sickness. Amphetamine induces excitability, restless-

ness, tremors, insomnia, dilated pupils, increased pulse rate and blood pressure, hallucinations, and psychoses. It is no longer recommended for weight reduction. It was found that, generally, any weight loss was only temporary. The greatest problem, however, was the diversion of vast quantities of amphetamines into the illegal drug market. Amphetamines are inexpensive. Armed forces personnel, truck drivers, and college students have been among the heavy users.

Methamphetamine has a more pronounced psychological effect than amphetamine. Generally, the "speed" that abusers inject into their veins is methamphetamine.[3] Such injections, at least initially, are said to give the abuser a euphoric rush. Shooting methamphetamine is quite dangerous, though, because the drug is rather toxic (see Table F.1).

Another amphetamine derivative, phenylpropanolamine, is widely used as an over-the-counter appetite suppressant. Like its relatives, this compound is a stimulant. Studies show that it is at best marginally effective as a diet aid, and it poses a threat to people with hypertension. Nevertheless, sales of phenylpropanolamine are 1 billion tablets, or $150 million, each year.

One controversial use of amphetamines has been their employment in the treatment of hyperactivity in children. The drug of choice is often methylphenidate (Ritalin). Although it is a stimulant, the drug seems to calm kids who otherwise can't sit still. This use has been criticized as "leading to drug abuse" and as "solving the teacher's problem, not the kid's."

---

[2] Amphetamine is not a single compound, but a mixture of two stereoisomers (see Section 18.5) that has been marketed under the trade name Benzedrine.

[3] Like other amine drugs, the amphetamines normally are used in the form of salts called hydrochlorides (Section 17.4). A freebase form of methamphetamine is used like crack cocaine (Section F.4) for smoking. This form is called "ice" because it exists in a clear crystalline form that resembles the solid form of water.

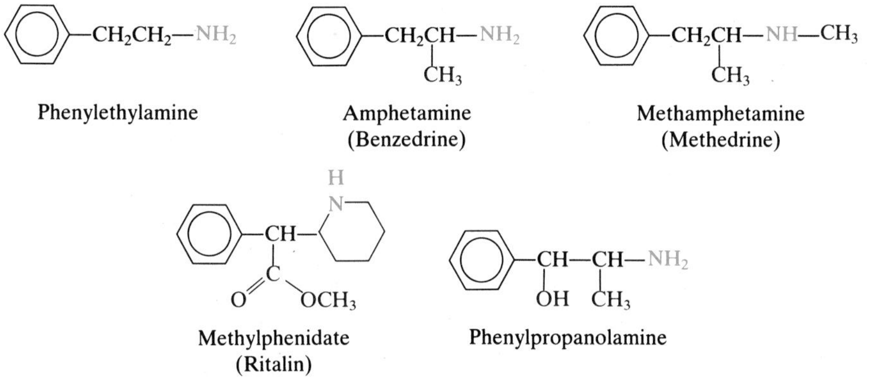

Figure F.6
Phenylethylamine and related compounds.

Phenylethylamine

Amphetamine (Benzedrine)

Methamphetamine (Methedrine)

Methylphenidate (Ritalin)

Phenylpropanolamine

Table F.1
Toxicities of Various Drugs[a]

Drug	LD$_{50}$ (mg/kg body weight)	Method of Administration	Experimental Animal
Local anesthetics			
Lidocaine	292	Oral	Mice
Procaine	45	Intravenous	Mice
Cocaine	17.5	Intravenous	Rats
Barbiturates			
Barbital	600	Oral	Mice
Pentobarbital	118	Oral	Rats
Phenobarbital	162	Oral	Rats
Amobarbital	212	Subcutaneous	Mice
Thiopental	149	Intraperitoneal	Mice
Narcotics			
Morphine	500	Subcutaneous	Mice
Heroin	21.8	Intravenous	Mice
Meperidine	170	Oral	Rats
Stimulants			
Caffeine	355	Oral	Rats
Nicotine	230	Oral	Mice
	0.3	Intravenous	Mice
Amphetamine	180	Subcutaneous	Rats
Methamphetamine	70	Intraperitoneal	Mice
Mescaline	370	Intraperitoneal	Rats

[a]Comparisons of toxicities in different animals—and extrapolation to humans—are at best crude approximations. The method of administration can have a profound effect on the observed toxicity. *Source:* Susan Budavari (Ed.), *The Merck Index,* 11th ed. Rahway, NJ: Merck and Co., 1989.

F.4

CAFFEINE, NICOTINE, AND COCAINE

The beverages coffee and tea and some soft drinks[4] contain the mild stimulant caffeine. The effective dose of caffeine is about 200 mg, corresponding to about two cups of strong coffee or tea. Caffeine is also available in tablet form as a stay-awake or keep-alert type of drug. The best-known brands are probably No-Doz and Vivarin. No-Doz contains about 100 mg of caffeine per tablet; each Vivarin tablet has 200 mg.

Caffeine

[4]Each year a million kilograms of caffeine is added to food in the United States. Most of it goes into soft drinks.

Is caffeine addictive? A 1992 study indicated that people who consume as little as one or two cups of coffee a day experienced withdrawal symptoms (headaches, depression, fatigue, drowsiness) when they abruptly avoided caffeine. There is also evidence that caffeine may be involved in chromosome damage. To be safe, females in their childbearing years should avoid large quantities of caffeine.

Another common stimulant is nicotine. This drug is taken by smoking or chewing tobacco. Nicotine is highly toxic to animals (see Table F.1). It is especially deadly when injected; the lethal dose for a human is estimated to be about 50 mg. Nicotine is used widely in agriculture as a contact insecticide. Nicotine seems to have a rather transient effect as a stimulant. This initial response is followed by depression.

Nicotine

Figure F.7
Coca leaves and illicit forms of cocaine.

Cocaine hydrochloride, the form used by those who snort cocaine, is readily absorbed through the watery mucous membrane of the nose. Those who smoke cocaine use the freebase (crack), which readily vaporizes at the temperature of a burning cigarette. When smoked, cocaine reaches the brain in 15 seconds. Cocaine acts by preventing the neurotransmitter dopamine from being taken back up after it is released by nerve cells. High levels of dopamine are therefore available to stimulate the pleasure centers of the brain. After the binge, dopamine is depleted in less than an hour. This leaves the user in a pleasureless state and (often) craving more cocaine.

Use of cocaine increases stamina and reduces fatigue, but the effect is short-lived. Stimulation is followed by depression. Once quite expensive and limited to use mainly by the wealthy, cocaine is now available in cheap and potent forms. Hundreds, including several well-known athletes, have died from cocaine overdose.

Is nicotine addictive? Casual observation of a person trying to quit smoking seems to indicate that it is.[5] There is also evidence of the development of tolerance. It is difficult, however, to separate all the social factors involved in smoking from the physiological effects.

Cocaine (Figure F.7), first used as a local anesthetic (Section F.5), also serves as a powerful stimulant. The drug is obtained from the leaves of a shrub that grows almost exclusively on the eastern slopes of the Andes Mountains. Many of the Indians living in and around the area of cultivation chew coca leaves—mixed with lime and ashes—for their stimulant effect. Cocaine used to arrive in the United States as the salt cocaine hydrochloride. Now much of it comes in the form of broken lumps of the freebase. This form is called *crack cocaine*.[6]

Cocaine

F.5
LOCAL ANESTHETICS

An **anesthetic** is any substance that causes either unconsciousness or insensitivity to pain. **Local anesthetics** are drugs that block transmission of nerve signals when applied to nerve tissue. They act on all kinds of nerves and on all parts of the nervous system.

For dental work and minor surgery, it is usually desirable to deaden the pain in one part of the body only. The first local anesthetic to be used successfully was cocaine. Its structure was determined by Richard Willstätter in 1898. Even before Willstätter's work, there were attempts to develop synthetic compounds with similar properties.

Certain esters of *p*-aminobenzoic acid act as local anesthetics (Figure F.8). The ethyl and butyl esters are used to relieve the pain of burns and open wounds. These are applied as ointments, usually in the form of picrate salts.

More powerful in their anesthetic action are a series of derivatives with a second nitrogen atom in the alkyl group of the ester. Perhaps the best known of these is procaine (Novocaine), first synthesized by Alfred Einhorn in 1905. Procaine can be injected as a local anesthetic, or it can be injected into the spinal column to deaden the entire lower portion of the body. Local anesthetics work by blocking nerve impulses to the brain. When the block involves the spinal cord, messages of pain from the lower parts of the body are prevented from reaching the brain.

[5]Clonidine, a drug used to treat high blood pressure, reduces nicotine withdrawal symptoms.
[6]Crack is the same chemical as the freebase that occasionally made the news several years ago. When cocaine hydrochloride was the main street form of the drug, some people made freebase by reacting the salt with a strong base. The person then used a solvent to extract the freebase (pure cocaine) from solution. Some freebasers were badly burned when the solvent (such as diethyl ether) was accidentally ignited.

Figure F.8

Some local anesthetics. Three (b) are derived from *para*-aminobenzoic acid (a). The other two (c) have amide functions. All often are used in the form of the hydrochloride salt, which is more soluble in water than the free base.

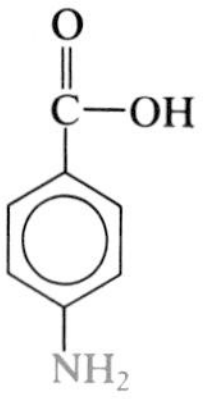

para-Aminobenzoic acid
(a)

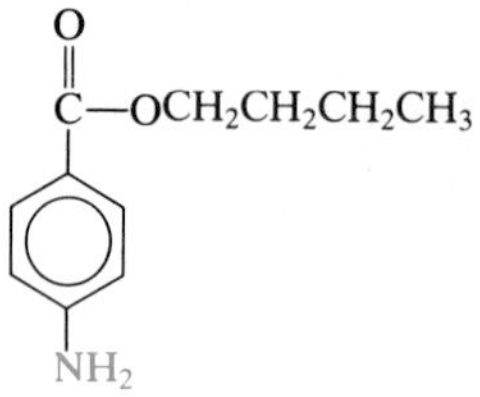

Butyl *para*-aminobenzoate
(Butesin)
(b)

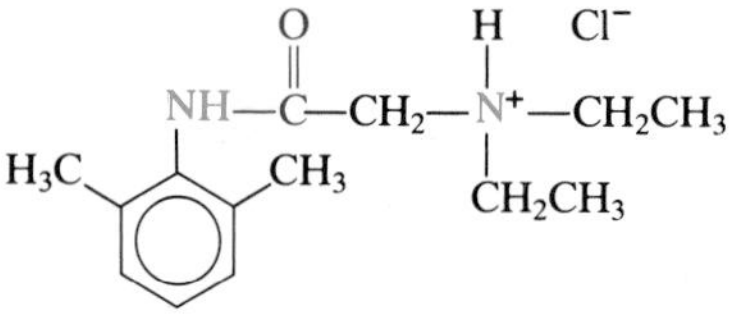

Lidocaine (Xylocaine)
(c)

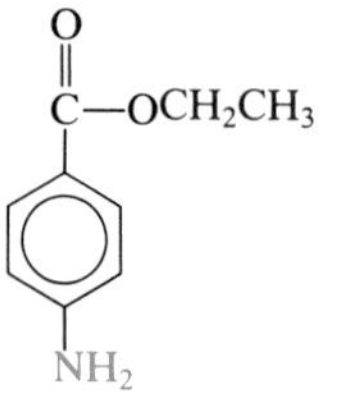

Ethyl *para*-aminobenzoate
(Benzocaine)
(b)

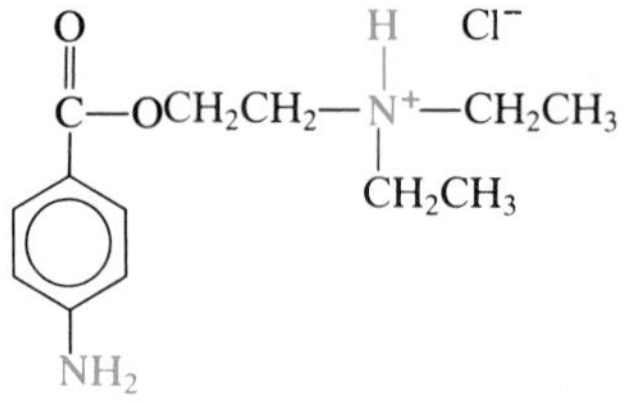

Procaine
(Novocaine)
(b)

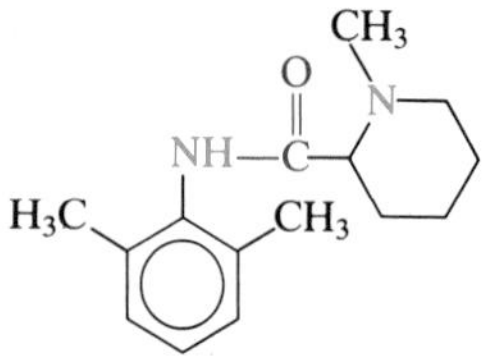

Mepivicaine
(c)

The local anesthetic of choice nowadays is often lidocaine or mepivicaine. Each compound is highly effective and yet has a fairly low toxicity (Table F.1). Note that lidocaine and mepivicaine are not derivatives of *p*-aminobenzoic acid, but they do share some structural features with the compounds that are.

F.6

BARBITURATES: SEDATION, SLEEP, AND SYNERGISM

As a family of related compounds, the barbiturates display a wide variety of properties. They can be employed to produce mild sedation, deep sleep, and even death.

Barbituric acid was first synthesized in 1864 by Adolph von Baeyer, a young student of August Kekulé (Chapter 13). He made it from urea, which occurs in urine, and malonic acid, which occurs in apples. The term **barbiturates,** according to Willstätter, came about because, at the time of the discovery, von Baeyer was infatuated with a girl named Barbara. The word comes from *Barbara* and *urea*. Curiously, in the United States, the names of the barbiturates end in *-al* even though they are ketones rather than aldehydes. The British spelling uses the suffix *-one*.

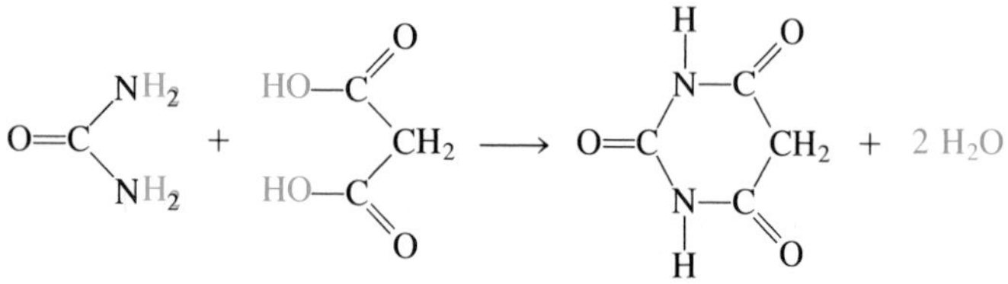

The medicinal value of the barbiturates was discovered in 1903 by Joseph von Mering. A derivative called barbital (Figure F.9) was found to be useful in putting dogs to sleep. Several thousand barbiturates have been synthesized through the years, but only a few have found widespread use in medicine. Pentobarbital (Nembutal) is employed as a short-acting hypnotic drug. Before the discovery of the modern tranquilizers, pentobarbital was used widely to calm anxiety and other disorders of psychic origin.

Phenobarbital (Luminal) is a long-acting drug. It, too, is a hypnotic and can be used as a sedative. Phenobarbital is employed widely as an anticonvulsant for epileptics and brain-damaged people. The action of amobarbital (Amytal) is intermediate in duration. Thiopental (Pentothal),[7] a

---

[7]Thiopental has been investigated as a possible truth drug. It does seem to help psychiatric patients recall traumatic experiences. It also helps uncommunicative individuals talk more freely. It does not, however, prevent one from withholding the truth or even from lying. No true truth drug exists.

Urea   Malonic acid   Barbituric acid

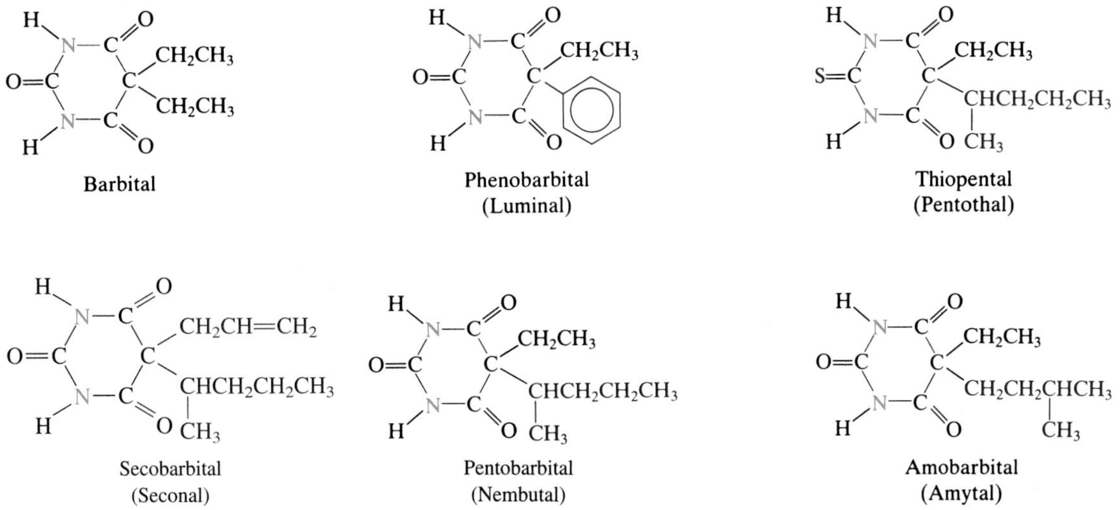

Barbital

Phenobarbital
(Luminal)

Thiopental
(Pentothal)

Secobarbital
(Seconal)

Pentobarbital
(Nembutal)

Amobarbital
(Amytal)

Figure F.9
Some barbiturate drugs.

compound that differs from pentobarbital only in that an oxygen atom on the ring has been replaced by a sulfur atom, is used widely in anesthesia (Figure F.9).

The barbiturates were once used in small doses as sedatives. The dose for sedation was generally a few milligrams. In larger doses (about 100 mg), barbiturates induce sleep. They were once the sleeping pills so widely used—and abused—by middle-class, often middle-aged, people. The lethal dose is in the vicinity of 1500 mg (1.5 g). Barbiturates are the drug of choice for many suicides—news reports list the cause of death as "an overdose of sleeping pills." There is also potential for accidental overdose. After a couple of tablets, the person becomes groggy. If the person is unable to remember taking the sleeping pills, he or she may take more pills.

The barbiturates are especially dangerous when ingested along with ethyl alcohol. This combination produces an effect much more drastic than just the sum of the effects of two depressants. The effect of a barbiturate is enhanced by as much as 200-fold when it is taken with an alcoholic beverage. This effect of one drug enhancing the action of another is called a **synergistic effect.**

Synergism has probably led to many deaths. According to news reports, an autopsy showed that a well-known Hollywood gossip columnist ingested only about 200 mg of barbiturate and 2 oz of alcohol. Either alone would have produced only mild sedation; the combination killed her. Synergistic effects are not limited to alcohol–barbiturate combinations. You should never take two

drugs at the same time without competent medical supervision.

The barbiturates are strongly addictive. Habitual use leads to the development of a tolerance to the drugs, and ever-larger doses are required to get the same degree of intoxication. Barbiturates are legally available by prescription only, but they are a part of the illegal drug scene also. They are known as "downers" because of their depressant, sleep-inducing effects.

The side effects of barbiturates are similar to those of alcohol. Abuse leads to hangovers, drowsiness, dizziness, and headaches. Withdrawal symptoms are often severe, accompanied by convulsions and delirium. In fact, some medical authorities now say that withdrawal from barbiturates is more dangerous—that is, more likely to cause death—than withdrawal from heroin.

Barbiturates are cyclic amides. Notice, however, that the barbiturate ring resembles that of thymine, one of the bases found in nucleic acids. Evidence indicates that barbiturates may act by substituting for thymine (or cytosine or uracil) in nucleic acids, thus interfering with protein synthesis (see Chapter 23).

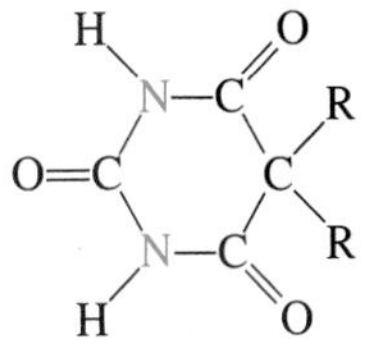

A barbiturate 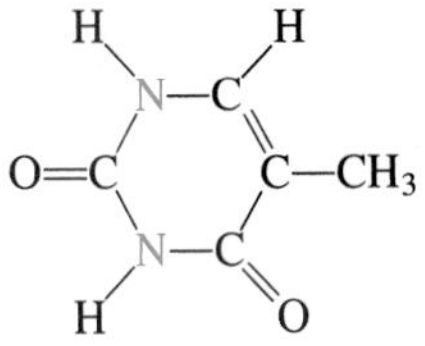   Thymine

F.7

DISSOCIATIVE ANESTHETICS: KETAMINE AND PCP

Ketamine, like thiopental an intravenous anesthetic, is called a **dissociative anesthetic**—it induces hallucinations similar to those reported by people who have had near-death experiences. They seem to remember observing their rescuers from a vantage point above it all, or moving through a dark tunnel toward a bright light. Unlike thiopental, ketamine seems to affect associative pathways before it hits the brainstem.

Ketamine

Little is known of the action of ketamine at the molecular level. If it acts by fitting receptors in the body, we can assume that our bodies produce their own chemicals that fit those receptors. These compounds may be synthesized or released only in extreme circumstances—such as in near-death experiences. Is it possible that we are on the threshold of the discovery of the chemistry of "life after death"?

Closely related to ketamine is phencyclidine (PCP), known on the street as "angel dust" or "crystal." PCP is soluble in fat and has no appreciable water solubility. It is stored in fatty tissue and released when the fat is metabolized; this accounts for the "flashbacks" commonly experienced by users.

Phencyclidine
(PCP)

PCP is an important part of the illegal drug scene. It is cheap and easily prepared. It was tested and found to be too dangerous for human use, but it has been used as an animal tranquilizer and marketed for this purpose under the trade name Sernylan. Many users experience bad "trips" with PCP. About 1 in 1000 develops a severe form of schizophrenia. Laboratory tests show that PCP depresses the immune system. This could lead to in-

creased risk of infection. Despite these well-known problems, every few years a new crop of young people appears on the scene to be victimized by this hog tranquilizer.

From a pharmacological viewpoint, the compound is of interest because it possesses a variety of actions. It can both stimulate and depress the central nervous system; it is a hallucinogen and an analgesic. There are few drugs that seem to induce so wide a range of effects. The immediate danger is that PCP is an extremely toxic substance, particularly when mixed with alcohol (it triggers violent, psychotic behavior). PCP is considered more dangerous than heroin because medically nothing is known about its effects, nor are there drugs available to counteract it. It is very easy to overdose on PCP to the point of convulsions.

F.8

ANTIANXIETY AGENTS

The hectic pace of life in the modern world has driven people to seek rest and relaxation in chemicals. Ethyl alcohol is undoubtedly the most widely used tranquilizer. The drink before dinner—to "unwind" from the tensions of the day—is very much a part of the American way of life. Many people, however, seek their relief in other chemical forms.

Several over-the-counter drugs—Cope, Vanquish, and Compoz among others—claim to be able to help us cope with or vanquish our problems or at least to compose ourselves in the face of minor adversity. Such products usually contain a little aspirin plus an antihistamine. The latter has a side effect of making one drowsy. These products have come under attack by consumer groups for being worthless at best and perhaps even dangerous.

Another class of widely used antianxiety drugs is the benzodiazepines, compounds that feature seven-member heterocyclic rings (Figure F.10). Of these, perhaps the best known are diazepam (Valium), lorazepam (Ativan), alprazolam (Xanax), and chlorodiazepoxide (Librium). These drugs are among the most widely prescribed medications in the United States.[8]

The benzodiazepine derivatives once were called "minor tranquilizers." They often are used to treat anxiety. Antianxiety agents make people feel better simply by

---

[8] Certain benzodiazepines, such as flurazepam (Dalmane) and triazolam (Halcion), are used to treat insomnia. They do help a person fall asleep, but the kind of sleep achieved is not restful. These pills may be useful to help someone get through tough times, but they do not cure insomnia or correct the conditions that cause it.

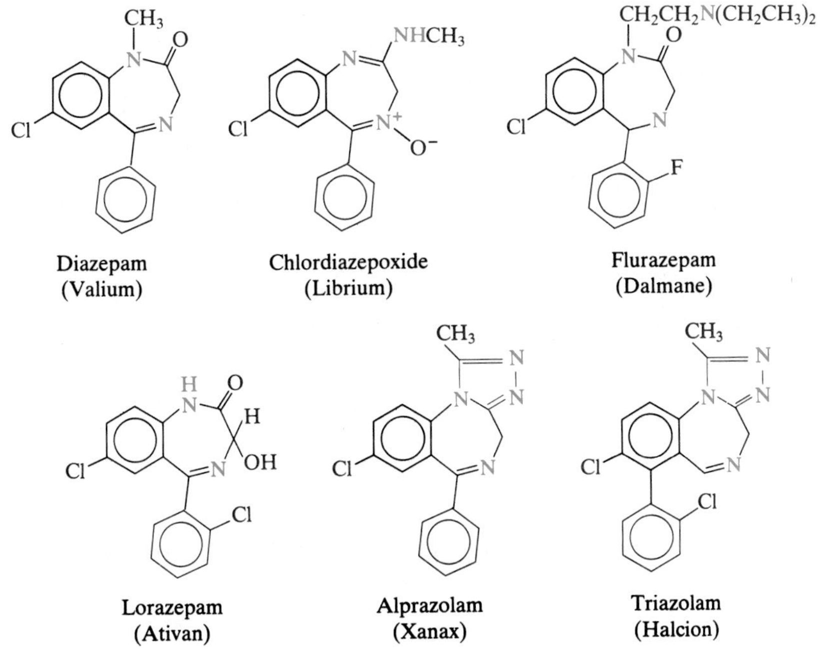

Figure F.10
Some benzodiazepine drugs.

Diazepam
(Valium)

Chlordiazepoxide
(Librium)

Flurazepam
(Dalmane)

Lorazepam
(Ativan)

Alprazolam
(Xanax)

Triazolam
(Halcion)

making them feel dull and insensitive. They do not solve any of the underlying problems that cause anxiety.

Like most other mind-altering drugs, the benzodiazepines act by fitting specific receptors. Presumably our bodies produce compounds that fit these receptors. To date, no such compound has been found. Rather, scientists have found compounds, called β-carbolines, that act on the brain's anxiety receptors to produce *terror*. There is yet so much to learn about the chemistry of the brain.

Anyway, what price tranquility? After 20 years of use, benzodiazepines were found to be addictive. People trying to get off the drugs after prolonged use go into painful withdrawal.

F.9
ANTIPSYCHOTIC AGENTS

For centuries, the people of India used the snakeroot plant, *Rauwolfia serpentina,* to treat a variety of ailments including fever, snakebite, and other poisonings and—most important—to treat maniacal forms of mental illness. Western scientists became interested in the plant near the middle of the twentieth century—after disdaining such remedies as quackery for many generations.

In 1952, rauwolfia was introduced into American medical practice as a hypertensive (blood-pressure-reducing) agent by Robert Wilkins of Massachusetts General Hospital. In the same year, Emil Schlittler of Switzerland isolated an active alkaloid, which he named reserpine, that has an impressive (intimidating?) structure.

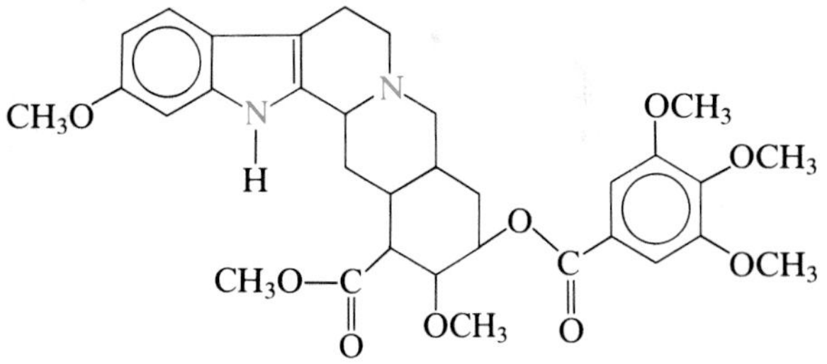

Reserpine

Rauwolfia was found not only to reduce blood pressure but also to bring about sedation. The latter finding attracted the interest of psychiatrists, who found reserpine so effective that by 1953 it had replaced electroshock therapy for 90% of psychotic patients.

Also in 1952, chlorpromazine (Thorazine) was tried as a tranquilizer on psychotic patients in the United States. The drug had been tested in France as an antihistamine. Medical workers there noted that it calmed mentally ill people who were being treated for allergies. Chlorpromazine was found to be quite effective in controlling the symptoms of schizophrenia.

Halcion

Halcion is the most widely prescribed sleeping pill in the world. It is a prescription drug manufactured by Upjohn that is intended for use as a sleeping pill, but only for short-term use. It is marketed in more than 90 countries; annual sales of Halcion amount to about $250 million. In the United States, pharmacists fill about a half million Halcion prescriptions every month. The FDA declared the drug safe and effective in 1982; however, controversy arose almost immediately.

Critics have called Halcion a "defective drug." They say that it is more dangerous than the other benzodiazepine drugs. Halcion is more likely to cause such nervous system disturbances as amnesia, anxiety, delusions, hallucinations, hostility, and paranoia. Many violent acts have been committed by people who were taking Halcion.

Supporters say that drug-induced violence has never been documented in controlled clinical studies. Many researchers found Halcion to be as safe as other sleeping pills. Users like the fact that it puts them to sleep quickly and causes less morning grogginess than the other benzodiazepines. The NIH recommends intermittent use for no longer than three weeks. The agency concludes that "sleeping pills should be thought of as short-term aid when insomnia is due to situational stress."

In October 1991, after reviewing previously undisclosed Upjohn studies, British health officials concluded that the risks of treatment outweighed the benefits and imposed a ban on Halcion. Based in part on the British findings, various groups in the United States asked the FDA to take Halcion off the U.S. market. In the spring of 1992, an FDA advisory panel reaffirmed "that Halicon is a safe and effective sleeping pill." But the panel said the label should contain a stronger warning on the side effects of the drug. Upjohn insists that "there is absolutely no scientific or medical evidence that warrants withdrawal of Halcion." Because of the continuing controversy, it is predicted that U.S. doctors will now prescribe Halcion less often.

Chlorpromazine is one of a group of related compounds called phenothiazines (Figure F.11). Several of these compounds are used in medicine. Promazine itself is a tranquilizer, but it is much less potent than chlorpromazine. Thioridazine (Mellaril) is one of the more potent compounds.

It is interesting to note that slight changes in structure can result in profound changes in properties. Replacing the sulfur atom of promazine by a $-CH_2CH_2-$ group produces imipramine (Tofranil), a compound that is not a tranquilizer at all. Rather, it is an antidepressant. Another common tricyclic (three-ring) antidepressant drug is amitriptylene (Elavil), which is further modified by replacement of the ring nitrogen atom by a carbon atom (see Figure F.11).

The phenothiazines are dopamine antagonists. Dopamine (see Figure F.5) is important in the control of detailed motion (such as grasping small objects), in memory and emotions, and in exciting the cells of the brain. Some researchers think that schizophrenic patients produce too much dopamine; others believe that they have too many dopamine receptors. In either case blocking the action of dopamine relieves the symptoms of schizophrenia. The tricyclic antidepressants work by an entirely different method. They inhibit the enzymes (called monoamine oxidases, MAO) that catalyze the deactivation of norepinephrine. This leaves more of this excitatory neurotransmitter to activate the receptor nerve cells. If a person is depressed, the MAO inhibitor thus would raise the level of nerve activity and relieve the depression.

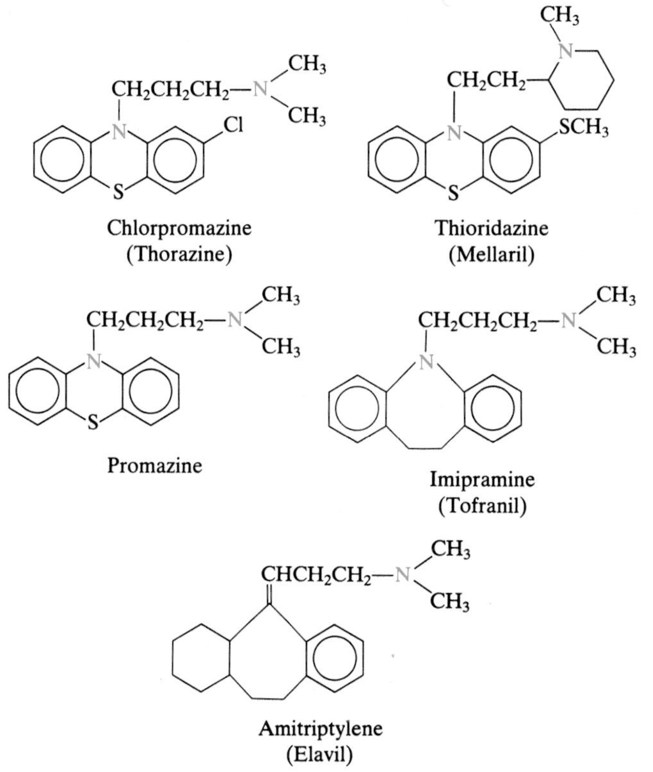

Figure F.11
Three phenothiazine tranquilizers (chlorpromazine, thorazine, and promazine) and two tricyclic
antidepressants (imipramine and amitriptylene).

The antipsychotic drugs have been one of the real tri-
umphs of chemical research. They have served to greatly
reduce the number of patients confined to mental hospitals
by controlling the symptoms of schizophrenia to the extent
that 95% of all schizophrenics no longer need hospitaliza-
tion. These drugs are not cures. We can only hope that
continued research will ascertain the causes of schizophrenia. At that time perhaps a real cure—or, better yet, a
preventive—can be found.

F.10
DESIGNER DRUGS

So-called designer drugs are analogs of compounds
that have some proven pharmacological activity.
These compounds first surfaced in California in the early
1980s. They are synthesized in clandestine laboratories
and then sold on the street as heroin. (PCP, which was
synthesized in the 1950s, can be considered the forerunner
of the designer drugs.) Like PCP, the designer drugs do
not fit into any one drug category; rather, they possess a
wide diversity of pharmacological effects.

The term ''designer drugs'' was first used to describe
analogs of fentanyl (Figure F.12). Fentanyl is a powerful
narcotic that is marketed under the trade name Sublimaze.
It is a short-acting anesthetic, and it has been used in about
70% of all surgical procedures in the United States since
the early 1970s. It is just as addictive as heroin, and its
euphoric effects are similar to those of heroin although of
shorter duration. By making slight modifications in the

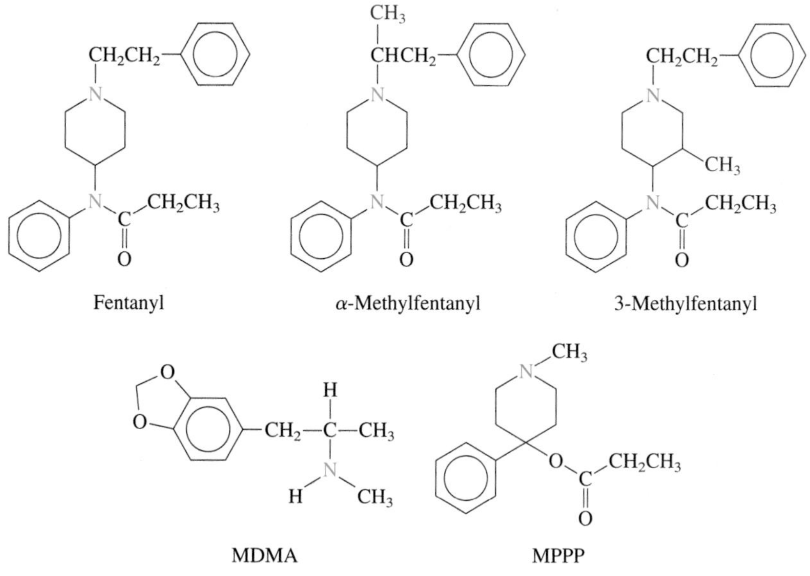

Figure F.12
Designer drugs.

structure of fentanyl, the underground chemists have developed more potent, longer lasting drugs that give a greater feeling of euphoria. For example, α-methylfentanyl (''China White'') is purported to be 20–40 times more potent as a euphoric drug than heroin. 3-Methylfentanyl is currently the most common fentanyl analog on the street. It is an extremely potent narcotic, and its duration of action and euphoriant effects are indistinguishable from those of heroin.

MDMA (3,4-methylenedioxymethamphetamine)—also called the ''Yuppie drug,'' ''Ecstasy,'' and ''Adam''—has been classified as a designer drug because it appeared on the drug scene at about the same time as the other drugs. It was first synthesized in 1914 as an appetite suppressant but was never marketed. The interesting aspect of this compound is that although it bears a close structural similarity to methamphetamine (Figure F.6), it bears little pharmacological relationship to that drug. MDMA does not have the stimulating effects of methamphetamine. Instead, it seems to enhance open communication between people and to ease psychic trauma. Some psychiatrists have been using MDMA since the 1970s in counseling sessions as an adjunct to psychotherapy. They report that under clinical conditions MDMA has few negative side effects and can act to ease psychic trauma and increase communicativeness.

To recreational users, MDMA is a pleasant way of raising one's consciousness without experiencing the hallucinating properties of LSD. However, a report of the National Institute on Drug Abuse (NIDA) calls MDMA ''a nationwide problem as well as a serious health threat.'' The report states that MDMA users experience problems similar to those associated with the use of amphetamines and cocaine. The Drug Enforcement Administration (DEA) outlawed the use of MDMA because research suggested that it may cause permanent brain damage to its users (mostly college students and young professionals).

A third class of designer drugs comprises analogs of meperidine (Demerol). The principal analog that has appeared on the street is MPPP (1-methyl-4-phenyl-4-propionoxypiperidine). MPPP is about three times as potent as morphine and 25 times as potent as Demerol. An impurity in the synthesis of MPPP has caused several cases of irreversible Parkinson's disease among the addicts who used it.

The lack of quality control in the production of designer drugs is a major problem. The drugs are all contaminated with impurities, and the only animal testing is done on the humans who buy the drugs. It is therefore no wonder that designer drugs are often quite lethal. Over 100 drug overdose deaths in California in recent years have been attributed to the use of designer drugs. During one weekend in February 1991, over a dozen people died and more than 100 others were hospitalized in three northeastern states during a wave of abuse of fentanyl derivatives.

EXERCISES

1. Define or identify each of the following terms:
   a. neuron   b. synapse
   c. neurotransmitter

2. Which two naturally occurring amines are presently considered to play major roles in the biochemistry of mental health? What are their proposed roles?

3. a. Which amino acids serve as precursors for the amines of Exercise 2?
   b. How may our mental state in part be related to our diet?

4. How do amphetamines exert a stimulant effect?

5. How does cocaine exert a stimulant effect?

6. When administered intravenously to rats, the LD_{50} of procaine is 50 mg/kg and that of cocaine is 17.5 mg/kg. Which drug is more toxic?

7. What is a local anesthetic? How does a local anesthetic work?

8. What is a general anesthetic? How does a general anesthetic work?

9. Name two dissociative anesthetics. How do they work?

10. a. What do we mean when we say amphetamines are ''uppers''?
    b. Why are barbiturates called ''downers''?

11. a. What is the basic structure common to all barbiturate molecules?
    b. How is the basic structure modified to change the properties of individual barbiturate drugs?

12. What is synergism?

13. Examine the structure of the reserpine molecule and identify the following.
    a. five ether functional groups
    b. two amine functional groups
    c. two ester functional groups

14. Acetbutolol is used as a drug for the treatment of heart

disease (angina and arrhythmias) and hypertension. There are five functional groups in the compound. Name the five families of organic compounds to which acetbutolol could be assigned.

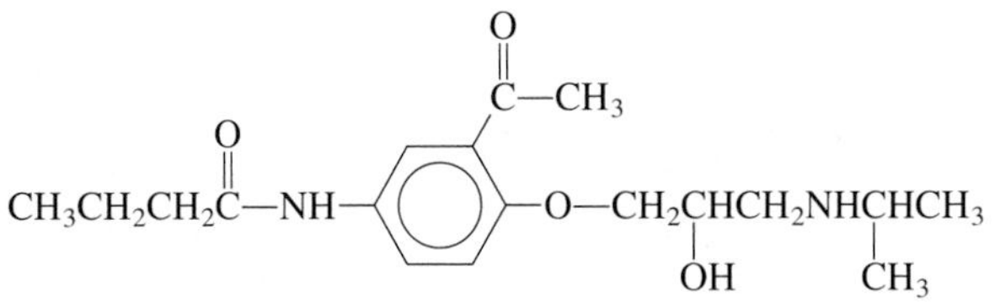

$$CH_3CH_2CH_2C-NH-\underset{}{\overset{O}{\parallel}}$$

Acetbutolol

15. Labetalol is used as a drug for the treatment of angina and hypertension. Circle the four functional groups in the molecule, and name the families of organic compounds that incorporate these functional groups.

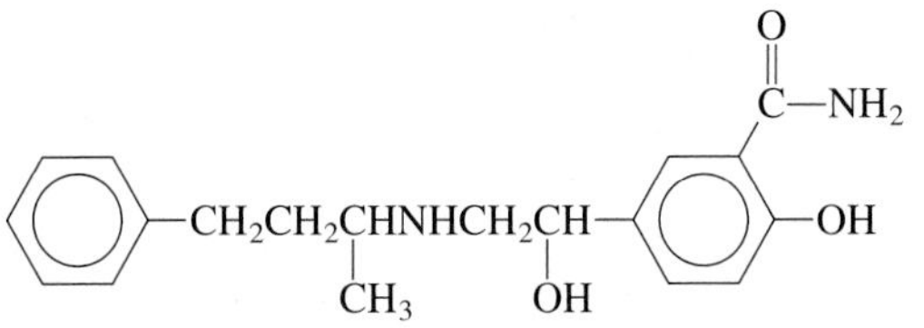

Labetalol

16. If the minimum lethal dose (MLD) of amphetamine is 5 mg per kilogram of body weight, what would be the MLD for a 70-kg person? Can toxicity studies on animals always be extrapolated to humans?

17. Cocaine is usually used in the form of the salt cocaine hydrochloride and sniffed up the nose, where it is readily absorbed through the watery mucous membranes. Some prefer to take their cocaine by smoking it (mixed with to-

bacco, for example). Before smoking, the cocaine hydrochloride must be converted back to the freebase (that is, to the molecular form). Explain the choice of dosage form for each route of administration.

18. Overdoses of phencyclidine (PCP) are treated by intravenous administration of ammonium chloride. The ammonium ion presumably converts the PCP to a salt that is somewhat more water-soluble and thus more readily excreted.

$$C_{17}H_{25}N + NH_4^+ \rightleftharpoons C_{17}H_{25}NH^+ + NH_3$$

a. Which is the stronger acid, NH_4^+ or $C_{17}H_{25}NH^+$?
b. Which is the stronger base, $C_{17}H_{25}N$ or NH_3?

19. Drugs such as lithium carbonate and reserpine block the release of norepinephrine. How might these be useful for treating manic patients?

20. Electroconvulsive therapy (shock treatment) induces the release of norepinephrine. What sort of mental problems are treated with this therapy?

21. Haloperidol is one of the most widely prescribed antipsychotic drugs. What five functional groups are present in the molecule?

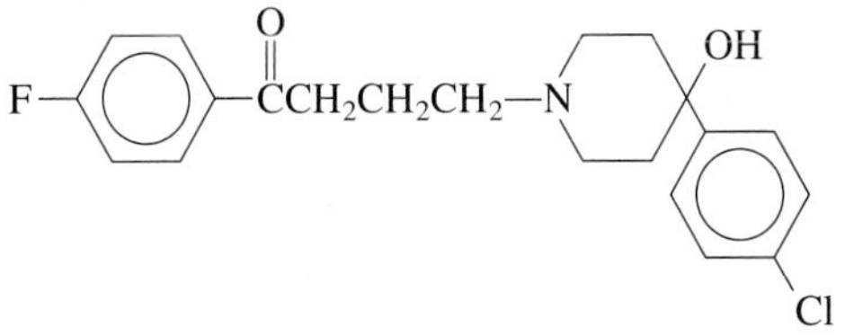

Haloperidol

Chapter 18
STEREOISOMERISM

A space-filling model and its mirror image. We shall discover that the three-dimensional shapes of molecules are vital to life.

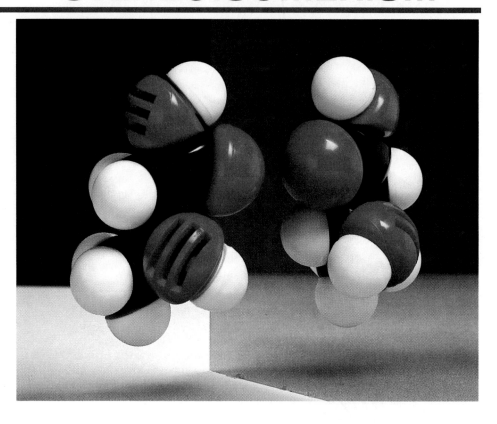

Isomerism has been defined as the phenomenon whereby two or more *different* compounds are represented by *identical* molecular formulas. Isomeric molecules have different physical and chemical properties, and these differences are attributed to the existence of different structural formulas.

Two types of **structural isomers** have been mentioned—positional isomers and functional group isomers. *Positional isomers* result from the presence of an atom or a group of atoms at different positions on the carbon chain. Examples of such isomers were discussed in the sections on alkanes, alkyl halides, and alcohols (Table 18.1).

Two molecules that have the same molecular formula but contain different functional groups are *functional group isomers*. Examples of functional group isomers of alcohols and ethers, aldehydes and ketones, and acids and esters are given in Table 18.2.

The second major category of isomerism is **stereoisomerism,** or space isomerism. Unlike structural isomers, stereoisomers have the identical order of atoms and identical functional groups. They differ only with respect to the spatial arrangement of atoms or groups of atoms within the molecule. Therefore, **stereoisomers** are isomers that have the same structural formulas but differ in the arrangement of atoms in three-dimensional space. Much of what we know about stereoisomers comes from their effect on plane-polarized light. Let's take a look at the nature of this light before we examine the molecules that act upon it.

Table 18.1
Examples of Positional Isomers

ALKANES	
$CH_3CH_2CH_2CH_3$	$CH_3CH(CH_3)_2$
Butane	Isobutane

ALKYL HALIDES	
CH_3CHCl_2	CH_2ClCH_2Cl
1,1-Dichloroethane	1,2-Dichloroethane

ALCOHOLS	
$CH_3CH_2CH_2OH$	$CH_3CHOHCH_3$
1-Propanol	2-Propanol

18.1
POLARIZED LIGHT AND OPTICAL ACTIVITY

First of all, let's establish a convention for drawing waves. A wave is something that goes up and down or increases and decreases or varies regularly in some such way. Here's a wave coming toward you and moving to the right and left across the page.

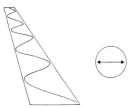

The arrow in the circle is our convention for indicating the same thing. When you see such an arrow, you are supposed to imagine a wave moving out of the page toward you and vibrating back and forth in the direction indicated by the arrow.

Here's another example.

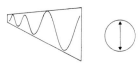

This wave is coming toward you in an up-and-down motion. The vertical double-headed arrow stands for the same motion.

Ordinary light may be described as exhibiting a wave motion (i.e., it has characteristics that are associated with wavelike motion). A beam of ordinary light can be pictured as a bundle of waves, some of which move up and down, some sideways, and others at all conceivable angles (Figure 18.1). While such light can be described as ordinary, a more scientific term is **nonpolarized.** Which brings us to what we wanted to talk about all along—polarized light.

The waves of **polarized light** vibrate in a single plane. Both of the beams of light shown in Figure 18.2 are polarized. The two polarized beams differ only in the angle of the plane of polarization (represented in our drawings by the orientation of the double-headed arrows).

Figure 18.1
The beam of light in this illustration is not polarized.

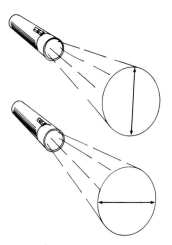

Figure 18.2
Both of the illustrated light beams are polarized. The planes of polarization differ.

Table 18.2
Examples of Functional Group Isomers

ALCOHOLS AND ETHERS

CH_3CH_2OH CH_3OCH_3

Ethanol   Dimethyl ether

ALDEHYDES AND KETONES

$$\overset{O}{\overset{\|}{CH_3CH_2CH}} \qquad \overset{O}{\overset{\|}{CH_3CCH_3}}$$

Propanal   Acetone

ACIDS AND ESTERS

$$CH_3C\overset{O}{\underset{OH}{\diagup}} \qquad H-C\overset{O}{\underset{OCH_3}{\diagup}}$$

Acetic acid   Methyl formate

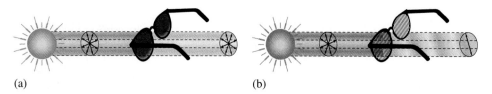

(a)   (b)

Figure 18.3

Ordinary sunglasses (a) dim light by preventing some light from passing through the lenses. But these lenses do not discriminate among light waves vibrating at different angles—all are cut down. The light reaching the eyes through ordinary sunglasses is nonpolarized. Sunglasses with Polaroid lenses (b) selectively pass light waves vibrating in a single plane. Light waves vibrating in the other planes are rejected, not allowed to pass through. The light reaching the eyes through Polaroid sunglasses is plane-polarized.

Sunlight, in general, is not polarized, nor is the light from an ordinary light bulb, nor the beam of light from an ordinary flashlight. One way to get polarized light is to pass ordinary light through Polaroid sheets, such as those used for the lenses of some sunglasses. These lenses are made by carefully orienting organic compounds in plastic to produce a material that permits only light vibrating in a single plane to pass through (Figure 18.3). To the eye, polarized light doesn't ''look'' any different than nonpolarized light. We can detect polarized light, however, by using a second sheet of polarizing material (Figure 18.4).

Certain substances act on polarized light by rotating the plane of vibration. Such substances are said to be **optically active.** The extent of optical activity is measured in an instrument called a **polarimeter** (Figure 18.5). The device consists of two polarizing lenses, called the **polarizer** and the **analyzer.** With the sample tube empty or containing distilled water, maximum light reaches the observer's eye when both the polarizer and the analyzer are aligned so that both pass light vibrating in the same plane (the pointer on the analyzer indicates 0° difference between the alignment of the analyzer and the polarizer).

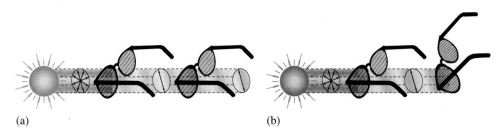

(a)   (b)

Figure 18.4

(a) The light that passes through the first Polaroid lens is polarized. The second pair of glasses is oriented like the first; therefore its Polaroid lens passes the polarized light. (b) The second pair of glasses is oriented 90° to the first. Thus, the plane of polarization of the light that made it through the first lens is oriented incorrectly for the second lens. No light gets through. If the second pair of sunglasses were not made with Polaroid lenses, some light would get through no matter how they were oriented because such glasses do not distinguish among various orientations of the plane of polarization.

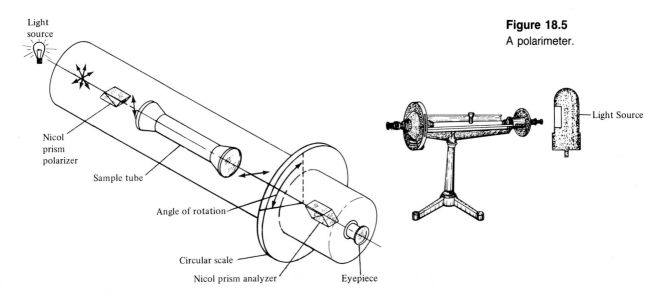

Figure 18.5
A polarimeter.

When an optically active substance[1] is placed in the sample tube, the plane of polarization is rotated. The polarized light emerging from the sample tube is vibrating in a different direction than when it entered the tube. To see the maximum amount of light when the sample is in place, the observer must rotate the analyzing lens to accommodate this change in the plane of polarization. The angle of rotation, indicated by the pointer on the analyzing lens, corresponds to the change in the plane of polarization caused by the sample.

The size of the angle depends not only on the structure of the optically active material but also on the length of the sample tube, the concentration of the solution, even the color of light used and the temperature of the experiment. However, scientists have agreed to certain standard conditions for reporting the angle of rotation. When a rotation is calculated and reported with these conditions taken into account, the value is referred to as the *specific rotation* (symbol $[\alpha]$) and is a physical constant as characteristic of the material as the melting point, boiling point, density, or solubility. For example, the specific rotation of an aqueous solution of sucrose is $+66.5$.

Some substances rotate the plane of polarized light to the right (clockwise from the observer's point of view). These compounds are said to be **dextrorotatory** (Latin *dexter*, right); substances that rotate light to the left (counterclockwise) are **levorotatory** (Latin *laevus*, left) (Figure 18.6). To denote the direction of rotation, a positive sign (+) is given

---

[1] The sample may be a pure gas, a pure liquid, a pure crystalline solid, or a solute dissolved in an appropriate solvent.

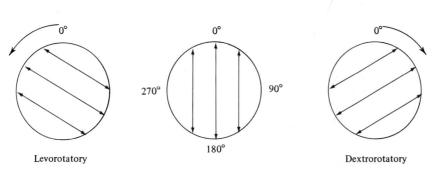

Levorotatory   Dextrorotatory

Figure 18.6
Direction of rotation of analyzer.

to dextrorotatory substances and a negative sign $(-)$ to levorotatory substances. Sucrose is said to be dextrorotatory because it rotates plane-polarized light $66.5°$ in the clockwise direction, and it is designated as $(+)$-sucrose.

18.2
CHIRAL CARBON ATOMS

So far, the properties of optically active compounds have been described, and certain terms have been defined for use in dealing with them. Certain fundamental questions have yet to be answered.

1. Why do some compounds exhibit optical activity?
2. What are the spatial arrangements of these compounds?
3. How do they differ from one another and from compounds that are not optically active?

The background to the discovery and explanation of optically active isomers is one of the most interesting sagas in the history of chemistry. You may wish to consult a more comprehensive organic chemistry textbook or, for full enlightenment, the original papers of Pasteur (1848), of Wislicenus (1873), and of van't Hoff and Le Bel (both 1874). The key to the their explanation of optical activity in organic compounds was the *tetrahedral carbon atom*.

Recall that the configuration of the methane molecule is tetrahedral. If any or all hydrogen atoms are substituted by other atoms or groups of atoms, the tetrahedral arrangement about the central carbon atom is still retained (Figure 18.7). If additional carbon atoms are added and if they all are joined by single bonds, the configuration about each of the carbon atoms will be tetrahedral.

Figure 18.7

Tetrahedral configuration of methane and a trisubstituted compound.

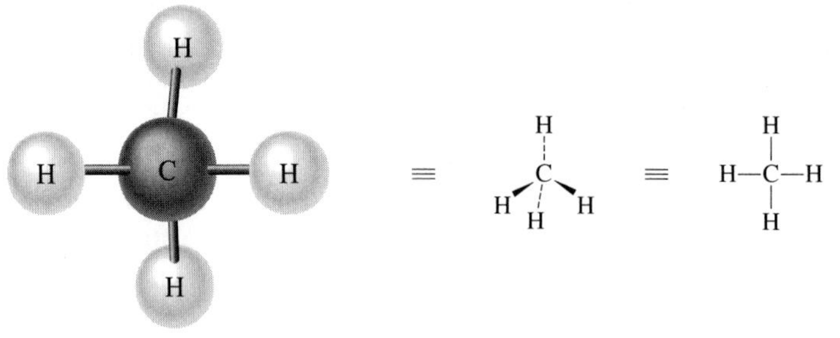

(a)

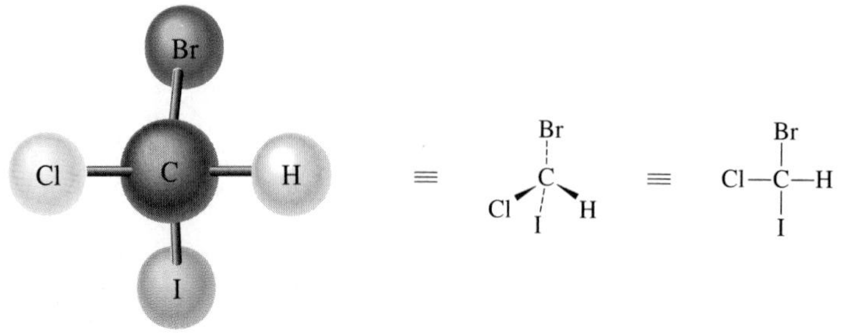

(b)

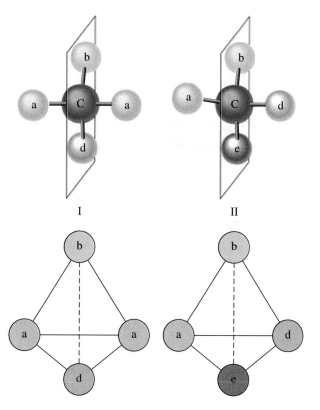

Figure 18.8
Compound I contains two similar and two dissimilar substituents. Compound II contains four dissimilar substituents.

Consider the two generalized molecules shown in Figure 18.8. Compound I contains two identical substituents and two different substituents bonded to the central carbon atom, and compound II contains four dissimilar substituents. Compound I is a symmetrical molecule; that is, a plane of symmetry passes through b, d, and the central carbon atom. Compound II is a nonsymmetrical molecule; you can't draw a plane of symmetry through the molecule.

To understand better why compound I is classified as a symmetrical molecule and compound II as a nonsymmetrical molecule, we can place both compounds in front of a mirror and attempt to impose the mirror images upon the original molecules (Figure 18.9). A symmetrical molecule is *superimposable* on (matches) its mirror image, whereas a

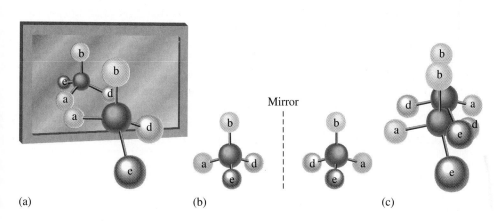

(a) (b) (c)

Figure 18.9
(a) and (b) Mirror images of a nonsymmetric molecule. (c) A nonsymmetric molecule cannot be superimposed on its mirror image.

Figure 18.10

The left and right hands are nonsuperimposable mirror images.

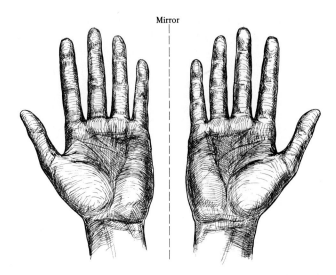

Mirror

nonsymmetrical molecule cannot be superimposed upon its mirror image. The bonds can be twisted and turned, but none of them can be broken.[2]

A useful analogy can be drawn between *nonsuperimposable* (nonmatching) mirror-image compounds and your right and left hands (Figure 18.10). Regardless of how you twist and turn them, you cannot superimpose your right hand upon your left, or vice versa. This is because, although your right and left hands are nearly perfect mirror images of each other, they are *not* identical. This difference becomes immediately apparent if you try to place your right hand in a left-hand glove. The general property of "handedness" is called **chirality** (Greek *cheir*, hand). An object that is not superimposable upon its mirror image is said to be **chiral.** An object that is superimposable upon (that is, identical to) its mirror image is **achiral.**

A **chiral carbon** is defined as a carbon atom that is bonded to four different groups. If a compound contains only one chiral atom, the molecule is always chiral. However, as we shall see in Section 18.3, sometimes a molecule may be achiral if it contains more than one chiral atom.

Molecules that are nonsuperimposable (nonidentical) mirror images are called **enantiomers** (Greek *enantios,* opposite). Enantiomers have identical physical and chemical properties, and they rotate plane-polarized light the same number of degrees—*but in opposite directions.*

The word *chiral* (pronounced KYE-ral) is now widely accepted and has displaced the earlier terms dissymmetric and asymmetric.

$$CH_3CHC\overset{\displaystyle O}{\underset{\displaystyle OH}{\diagdown}}$$
$$\underset{\displaystyle OH}{|}$$

Lactic acid

Let's look at some real molecules that have chiral carbons. A usual representation of lactic acid is given in the margin. Examination of the structure reveals that the second carbon atom has four different groups attached—a hydrogen atom, a carboxyl group, a hydroxyl group, and a methyl group. This carbon, then, is a chiral center. There should be *two* lactic acids. Are there? Yes! (Why, it's enough to give you faith in chemical theory!) One lactic acid is found in sour milk. It is levorotatory and can be designated as (−)-lactic acid. Another lactic acid is found in muscle tissue, particularly after exercise. It is dextrorotatory and is called (+)-lactic acid. How are the two related? They are enantiomers—

[2] These molecules, like any others, can be turned upside down or spun about or tipped forward or backward, just as you can stand on your feet or on your head or lie on your back, etc. If you had a mole on your right arm, however, no matter what position you assumed, the mole would still be on your right arm. Your mirror image would always have a left-arm mole. No amount of spinning or turning will cause the mole to change from your right to your left arm. The molecules pictured in Figure 18.9 also have such a distinguishing arrangement. If, for these molecules, we call atom d the *head* and the side to which atom e is attached the *front,* then one compound always has atom d on its right side and the other always has d on its left side.

one is the mirror image of the other. Using perspective formulas, the enantiomers can be represented as follows.

Mirror

Enantiomers can be drawn using "flat" formulas, for example,

but such drawings require more of *you*. You have to visualize that the horizontal bonds project above the plane of the paper and the vertical bonds project below. Because they are easier to draw, we shall use these "flat" formulas (often called **Fischer projections**) to represent stereoisomers.

In what ways do the actual isomers differ? In many respects, they seem more alike than different. All of their physical properties are identical save one, the direction in which they rotate the plane of polarized light. One has a specific rotation of $+2.6$; the other, -2.6. Only the sign is different. Simple chemical properties are also the same. Both form acidic solutions. Both neutralize bases. Both form esters. Indeed, when reacting with *achiral* molecules, the two enantiomers exhibit identical chemical properties. Such common reagents as water, hydroxide ion, and ethyl alcohol do not contain chiral centers and are achiral. It is only when the lactic acid isomers react with other chiral molecules that they behave differently. They may react at different rates and to different extents. The products formed will have different properties. Are these differences really important? In living cells reactions are controlled by enzymes, and enzymes are chiral. That means that enantiomers may behave quite differently in living cells. Enantiomers may have different tastes and smells. One may be an effective drug and the other worthless. One may be essential to health and the other toxic. Quite literally, we may be talking about differences between life and death.

When lactic acid is made from pyruvic acid in the laboratory, it shows *no* optical activity.

Pyruvic acid   Racemic lactic acid
[50%$(+)$-lactic acid and 50%$(-)$-lactic acid]

How can this be? The lactic acid has a chiral center. The answer: in syntheses of this sort, the $(+)$ and $(-)$ forms are formed in exactly equal amounts. Such a mixture of enantiomers is called a **racemic mixture.** It shows no optical activity because it contains equal amounts of molecules with equal but opposite rotatory power. Everything cancels out. Racemic lactic acid is designated $(\pm)$. The racemic mixture may exhibit different physical properties than the pure enantiomers (Table 18.3).

Table 18.3
Properties of Lactic Acids

Form	Melting Point (°C)	Specific Rotation	pK_a
$(+)$	53	$+2.6$	3.8
$(-)$	53	-2.6	3.8
$(\pm)$	16.8	0	3.8

Figure 18.11
Enantiomers are nonsuper-
imposable mirror images.

"Mirror, mirror on the wall, who is the enantiomerest of them all?"

In summary, so far we have mentioned those compounds that contain only one chiral carbon atom. For such compounds there always exist two isomers that are nonsuperimposable mirror images (enantiomers—Figure 18.11), a dextrorotatory form and a levorotatory form. A convenient method of drawing the enantiomer of an optically active compound is to maintain the position of two of the substituents (usually the larger ones) about the chiral center and invert the positions of the other two.

Table 18.4 lists some optically active and inactive compounds from several classes of organic compounds. A worthwhile exercise is to draw the enantiomers for each of the optically active compounds listed.

EXAMPLE 18.1

2-Methyl-1-butanol exists in two optically active forms. The specific rotation of one is +5.756, and for the other it is −5.756. Draw structural formulas of the enantiomers.

SOLUTION
Write the structural formula for one enantiomer and identify the chiral center. (To allow for valid comparisons, the convention is to draw the carbon chain vertically.) Then generate the other enantiomer by interchanging the positions of two of the groups about the chiral center while maintaining the positions of the other two groups.

$$
\begin{array}{cc}
CH_3 & CH_3 \\
| & | \\
CH_2 & CH_2 \\
| & | \\
H-C-CH_3 & H_3C-C-H \\
| & | \\
CH_2OH & CH_2OH \\
I & II
\end{array}
$$

Table 18.4
Optically Active and
Inactive Organic
Compounds

Family	Optically Active	Optically Inactive
Alkane	$\underset{\underset{\displaystyle CH_2CH_3}{\mid}}{\overset{\overset{\displaystyle H}{\mid}}{CH_3-C-CH_2CH_2CH_3}}$	$\underset{\underset{\displaystyle CH_3}{\mid}}{\overset{\overset{\displaystyle H}{\mid}}{CH_3-C-CH_2CH_2CH_2CH_3}}$
Alkyl halide	$\underset{\underset{\displaystyle Cl}{\mid}}{\overset{\overset{\displaystyle H}{\mid}}{CH_3-C-CH_2CH_3}}$	$\underset{\underset{\displaystyle Cl}{\mid}}{\overset{\overset{\displaystyle H}{\mid}}{H-C-CH_2CH_2CH_3}}$
Alcohol	$\underset{\underset{\displaystyle CH_3}{\mid}}{\overset{\overset{\displaystyle OH}{\mid}}{CH_3CH_2-C-H}}$	$\underset{\underset{\displaystyle H}{\mid}}{\overset{\overset{\displaystyle H}{\mid}}{CH_3CH_2CH_2-C-OH}}$
Ether	$\underset{\underset{\displaystyle CH_3}{\mid}}{\overset{\overset{\displaystyle H}{\mid}}{CH_3CH_2-C-OCH_3}}$	$\underset{\underset{\displaystyle H}{\mid}}{\overset{\overset{\displaystyle CH_3}{\mid}}{CH_3CH_2O-C-CH_3}}$
Aldehyde	$\underset{\underset{\displaystyle H}{\mid}}{\overset{\overset{\displaystyle CH_3}{\mid}}{CH_3CH_2-C-C\overset{O}{\underset{H}{\diagdown}}}}$	$\underset{\underset{\displaystyle CH_3}{\mid}}{\overset{\overset{\displaystyle CH_3}{\mid}}{CH_3-C-C\overset{O}{\underset{H}{\diagdown}}}}$
Ketone	$\underset{\underset{\displaystyle CH_3}{\mid}}{\overset{\overset{\displaystyle O\ \ H}{\parallel\ \ \mid}}{CH_3C-C-CH_2CH_3}}$	$\underset{\underset{\displaystyle H}{\mid}}{\overset{\overset{\displaystyle O\ \ H}{\parallel\ \ \mid}}{CH_3C-C-CH_2CH_2CH_3}}$
Amine	$\underset{\underset{\displaystyle CH_3}{\mid}}{\overset{\overset{\displaystyle H}{\mid}}{CH_3CH_2-C-NH_2}}$	$\underset{\underset{\displaystyle H}{\mid}}{\overset{\overset{\displaystyle H}{\mid}}{CH_3CH_2CH_2-C-NH_2}}$
Hydroxy acid	$\underset{\underset{\displaystyle OH}{\mid}}{\overset{\overset{\displaystyle H}{\mid}}{CH_3-C-CH_2C\overset{O}{\underset{OH}{\diagdown}}}}$	$\underset{\underset{\displaystyle OH}{\mid}}{\overset{\overset{\displaystyle H}{\mid}}{H-C-CH_2CH_2C\overset{O}{\underset{OH}{\diagdown}}}}$
Amino acid	$\underset{\underset{\displaystyle NH_2}{\mid}}{\overset{\overset{\displaystyle H}{\mid}}{CH_3-C-C\overset{O}{\underset{OH}{\diagdown}}}}$	$\underset{\underset{\displaystyle NH_2}{\mid}}{\overset{\overset{\displaystyle H}{\mid}}{H-C-CH_2C\overset{O}{\underset{OH}{\diagdown}}}}$
Ester	$\underset{\underset{\displaystyle CH_3}{\mid}}{\overset{\overset{\displaystyle H}{\mid}}{CH_3CH_2-C-C\overset{O}{\underset{OCH_3}{\diagdown}}}}$	$\underset{\underset{\displaystyle CH_3}{\mid}}{\overset{\overset{\displaystyle H}{\mid}}{CH_3-C-C\overset{O}{\underset{OCH_2CH_3}{\diagdown}}}}$
Amide	$\underset{\underset{\displaystyle H}{\mid}}{\overset{\overset{\displaystyle CH_3}{\mid}}{\bigcirc-C-C\overset{O}{\underset{NH_2}{\diagdown}}}}$	$\underset{\underset{\displaystyle H}{\mid}}{\overset{}{CH_3-\bigcirc-C-C\overset{O}{\underset{NH_2}{\diagdown}}}}$

It is not possible to tell which isomer is dextrorotatory and which is levorotatory merely by inspection of the structural formulas. The distinction can be made only by measuring the optical rotation of each compound in a polarimeter.

Practice Exercise
Draw structural formulas for the enantiomers of 2-butanol.

18.3
MULTIPLE CHIRAL CENTERS

A molecule may have more than one chiral center. Indeed, simple carbohydrate molecules generally have several each. Giant molecules, such as starch, cellulose, and the proteins, may have several hundred or even several thousand chiral centers. Let us look first, though, at molecules with just two.

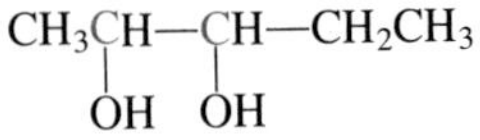

Consider 2,3-pentanediol. There are *four ways* in which the groups can be arranged about the two chiral centers (Figure 18.12). Note that structures I and II are enantiomers; they are nonsuperimposable mirror images of one another. Compounds III and IV make up another set of enantiomers. What is the relationship, though, between structures II and III? They are stereoisomers because they differ only in their spatial arrangement. They are not enantiomers because they are not mirror images. Such pairs of isomers are called **diastereomers.** Note that II and IV are also diastereomers, as are I and III and I and IV. Diastereomers generally have *different* physical properties (like boiling point and solubility, as well as specific rotation). Unlike enantiomers, they can be separated by distillation or fractional crystallization.

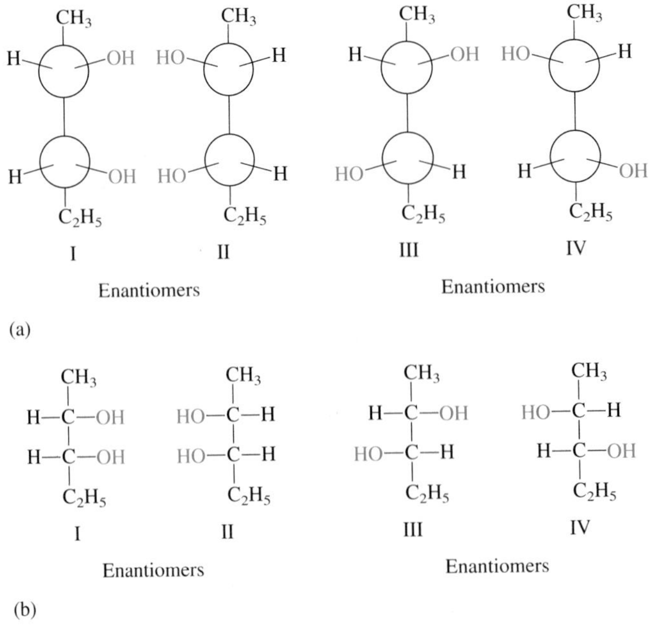

$$CH_3CH-CH-CH_2CH_3$$
$$OHOH$$

2,3-Pentanediol

Figure 18.12
The four stereoisomeric 2,3-pentanediols. (a) Perspective drawings. (b) Flat projection formulas.

The first chemist to postulate the existence of multiple stereoisomeric forms was van't Hoff (first Nobel prize in chemistry, 1901). He formulated a statement (**van't Hoff's rule**) that makes it possible to predict the total number of possible stereoisomers for a molecule with more than one chiral center. *The maximum number of different configurations is 2^n, where n is the number of chiral carbon atoms.* The rule is best illustrated with a specific example.

EXAMPLE 18.2

Draw all the possible stereoisomers of 2-methyl-1,3-butanediol.

SOLUTION

1. First draw the structural formula and note the number of chiral carbon atoms. Using the formula 2^n, calculate the number of possible stereoisomers.

$$CH_3-\underset{\underset{H}{|}}{\overset{\overset{OH}{|}}{C}}-\underset{\underset{H}{|}}{\overset{\overset{CH_3}{|}}{C}}-CH_2OH$$

There are two chiral carbons, and therefore there are 2^2 or 4 possible stereoisomers.

2. Since two configurations are possible for each chiral carbon, there will be two sets of mirror images.

Mirror   Mirror

```
        CH3                 CH3               CH3               CH3
     H—C—OH    |   HO—C—H       H—C—OH   |   HO—C—H
     H—C—CH3   |   H3C—C—H      H3C—C—H  |   H—C—CH3
       CH2OH   |     CH2OH        CH2OH  |     CH2OH
```

Practice Exercise

Draw all the possible stereoisomers of 2,3-dibromobutanal.

Figure 18.13
Louis Pasteur, a French chemist who invented the process, now called pasteurization, of heating milk to destroy pathogenic bacteria and slow the fermentative action. Pasteur's work also led to the germ theory of disease, to immunization procedures, and to the original discovery of the stereoisomerism associated with enantiomers.

$$HOOC-\underset{\underset{OH}{|}}{CH}-\underset{\underset{OH}{|}}{CH}-COOH$$

Tartaric acid

Let us consider one last set of compounds before we move on to the next section. These compounds, the tartaric acids, were involved in the earliest studies relating structure and optical activity. The investigator was Louis Pasteur (Figure 18.13), and the compounds he studied also included a new type of stereoisomer.

The formula for tartaric acid, like that of 2,3-pentanediol, contains two chiral carbon atoms. There is one notable difference between tartaric acid and 2,3-pentanediol. In tartaric acid, each of the two chiral carbons is attached to the same four different groups, namely, to —COOH, —OH, —H, and —CH(OH)COOH. In 2,3-pentanediol, one chiral carbon is attached to a methyl group, whereas the other is attached to an ethyl group. This difference is significant, as we shall see.

Writing out the perspective formulas of tartaric acid (Figure 18.14), we see a pair of enantiomers (I and II). The other apparent pair (III and IV), though, are not enantiomers; they are not even isomers. They are, in fact, the same compound. The structures can be superimposed by rotating one of them 180° in the plane of the paper (Figure 18.15). It is

Figure 18.14
The three stereoisomeric tartaric acids. (a) Perspective drawings. (b) Flat projection formulas. Note that the meso form has an internal plane of symmetry.

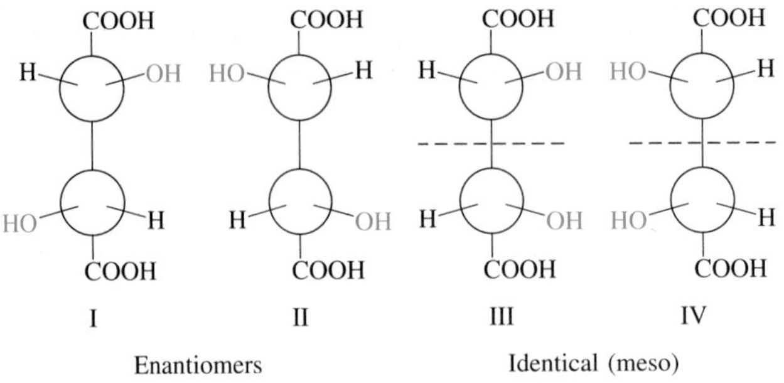

(a)

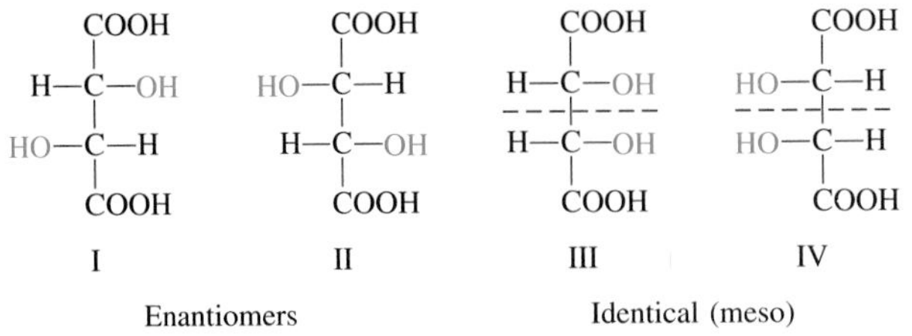

(b)

Figure 18.15
meso-Tartaric acid is superimposable on its mirror image. It does not exist as a pair of enantiomers, but as a single compound only. (a) Perspective drawings. (b) Flat projection formulas.

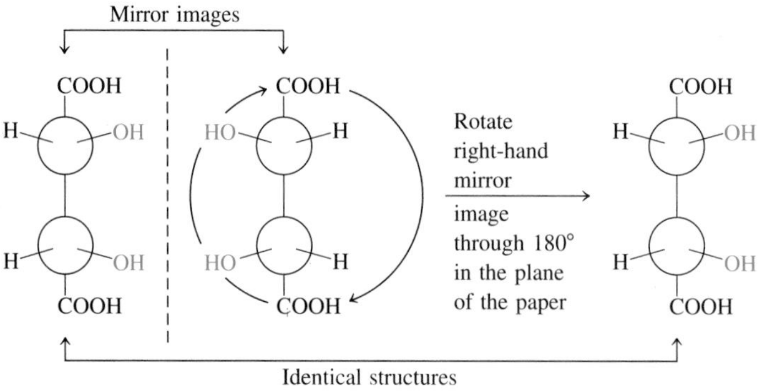

(a)

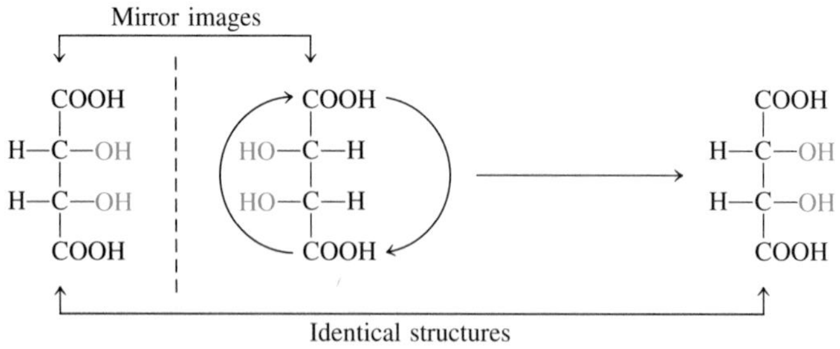

(b)

important to emphasize again that enantiomers are not simply mirror images of one another, but *nonsuperimposable* mirror images. Structures III and IV are mirror images, but they are also superimposable and therefore identical. The corresponding structures for 2,3-pentanediol (Figure 18.12) are not superimposable because the CH_3 group and the C_2H_5 group are different.

Form	Melting Point (°C)	Specific Rotation	pK_a	Solubility (g/100 g H_2O)
(+)	168–170	+12.0	2.93	133
(−)	168–170	−12.0	2.93	133
(±)	206	0	2.96	21
Meso	140	0	3.11	125

Table 18.5
Properties of the Tartaric Acids

The single compound represented by both structures III and IV in Figure 18.14 is termed a **meso** compound. It is diastereomer of compound I and compound II. All meso compounds are optically inactive.[3] Meso compounds contain at least two chiral centers, but have an internal symmetry plane; in Figure 18.14 we have indicated this mirror plane in structures III and IV by a dashed line. The meso molecule as a whole is *not* chiral. Table 18.5 lists all the forms of tartaric acid.

Let us summarize here the various conditions that result in a lack of optical activity. A compound (like ethanol, CH_3CH_2OH) may contain no chiral center. It is not chiral (i.e., it is achiral) and therefore not optically active. A meso compound contains chiral centers, but it is not chiral because it also contains an internal mirror plane. It, too, is optically inactive. A racemic mixture contains chiral molecules but is not optically active because the dextrorotatory molecules cancel out the effects of the levorotatory ones.

18.4
GEOMETRIC ISOMERISM (CIS–TRANS ISOMERISM)

It has been mentioned that molecules that have a cyclic structure or a carbon–carbon double bond have certain restrictions placed upon them. In Section 13.11 three isomers of butene, C_4H_8, were identified.

$$CH_3CH_2-\overset{H}{\underset{}{C}}=\overset{H}{\underset{}{C}}-H \qquad CH_3-\overset{H}{\underset{}{C}}=\overset{H}{\underset{}{C}}-CH_3 \qquad CH_3-\overset{H_3C}{\underset{}{C}}=\overset{H}{\underset{}{C}}-H$$

$$\text{1-Butene} \qquad\qquad \text{2-Butene} \qquad\qquad \text{2-Methylpropene}$$
$$\text{I} \qquad\qquad\qquad \text{II} \qquad\qquad\qquad \text{III}$$

Experimental evidence has shown, however, that there are four different butene molecules, all having distinctly different physical properties. The fourth isomer also has the

---

[3] A meso compound is superimposable on its mirror image even though it contains chiral carbon atoms. For this reason it is incorrect to say that all molecules that contain chiral carbon atoms are chiral and thus optically active. Another example of a meso compound is ribitol, whose structural formula is

$$\begin{array}{c}
CH_2OH \\
| \\
H-C-OH \\
| \\
H-C-OH \quad\text{---- Plane of symmetry} \\
| \\
H-C-OH \\
| \\
CH_2OH
\end{array}$$

Structures that have meso forms will have fewer stereoisomers than predicted by van't Hoff's rule. Tartaric acid has two chiral carbon atoms but exists in only three stereoisomeric forms, including one that is meso.

Figure 18.16
cis-2-Butene and *trans*-2-butene.

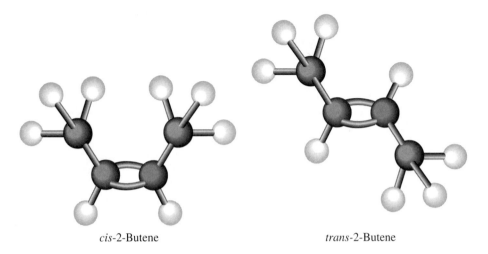

cis-2-Butene   *trans*-2-Butene

structure that we have designated as 2-butene, and therefore these two isomeric 2-butenes, although they are structural isomers of compounds I and III, are not structural isomers of one another. Our knowledge of stereoisomerism leads us to the conjecture that these two molecules might differ in the spatial configuration of their atoms. Because they have different physical properties, however, they cannot be enantiomers; a different type of configurational explanation must be sought to account for this phenomenon. The explanation is based upon the geometric arrangement of the carbon–carbon double bond.

Recall that the two carbon atoms of a C=C double bond and the four atoms attached to them are all in the same plane and that rotation around the double bond is prevented. (This is in sharp contrast to the free rotation enjoyed by carbon atoms that are linked to one another by single bonds.) Construction of ball-and-stick models indicates that there are two possible ways to arrange the atoms of 2-butene that are in keeping with its structural formula (Figure 18.16). These three-dimensional ball-and-stick models are more simply represented as follows.

H   H   CH₃   H
  \   /   \   /
   C=C   C=C
  /   \   /   \
CH₃   CH₃   H   CH₃

IIa   IIb

cis-2-Butene   *trans*-2-Butene
mp −139 °C   mp −106 °C
bp 4 °C   bp 1 °C

In structure IIa, both methyl groups lie on the same side of the molecule, and this compound is the **cis** isomer (Latin *cis,* on this side). The methyl groups of structure IIb are on opposite sides of the molecule; it is the **trans** isomer (Latin *trans,* across). Because of the restriction on free rotation about the double bond, the structures are clearly nonsuperimposable and hence not identical. *cis*-2-Butene and *trans*-2-butene are geometric isomers of each other.

Geometric isomers are compounds that have different configurations because of the presence of a rigid structure in the molecule. Geometric isomers are diastereomers because they are stereoisomers that are not enantiomers. For alkenes there are *only two* geometric isomers that correspond to each double bond (cis and trans).

We can draw two *seemingly* different propenes (structures IV and V).

However, the second structure is not really different from the first. If you could pick it up from the page and flip it over, you would see that the two formulas are identical.

The same thing *cannot* be done with cis and trans isomers. If we start with drawings of the two 2-butenes (VI and VII),

and then flip the trans isomer over, we see that the resulting structure is still clearly different from the cis isomer.

As the case of propene proves, the mere presence of a double bond is not the only criterion (nor is it a necessary one) for the occurrence of geometric isomerism. The requirements for such isomerism are (1) that rotation be restricted in the molecule and (2) that there be two nonidentical groups on *each* of the doubly bonded carbons. For 2-butene ($CH_3CH{=}CHCH_3$), the doubly bonded carbon on the left has a hydrogen group and a methyl group (two different groups), and the doubly bonded carbon on the right has a hydrogen group and a methyl group (two different groups). Thus, 2-butene exists as a cis and a trans isomer. Propene ($CH_3CH{=}CH_2$) has a doubly bonded carbon with two hydrogens (two identical groups) attached. The second requirement for geometric isomerism is not fulfilled, therefore, and this compound does *not* exist as cis and trans isomers. One of the doubly bonded carbons in propene does have two different groups attached, but the rules require that *both* carbons have two different groups.

In general, when two identical (or nearly identical) substituents are on the same side of the double bond, the compound is the cis isomer. The trans isomer is the one in which similar groups are on opposite sides of the double bond. Following are some examples of geometric isomers.

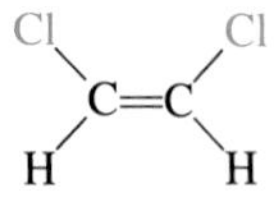

cis-1,2-Dichloroethene
mp $-80\ °C$
bp $60\ °C$

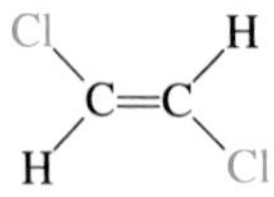

trans-1,2-Dichloroethene
mp $-50\ °C$
bp $48\ °C$

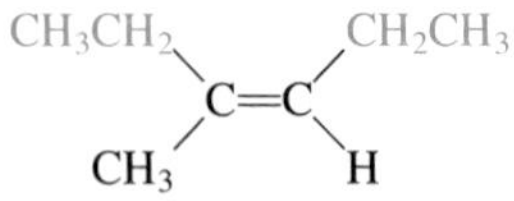

cis-3-Methyl-3-hexene

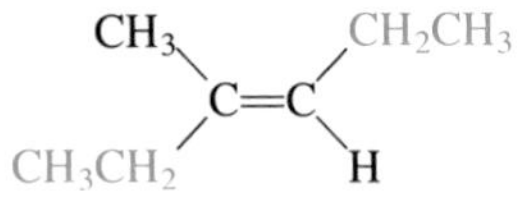

trans-3-Methyl-3-hexene

EXAMPLE 18.3

Draw all alkenes with the formula C_5H_{10} and indicate which ones exist as cis and trans isomers. Give the IUPAC names for the isomers.

SOLUTION

First we'll draw the various possible carbon skeletons incorporating a double bond.

$$CH_2{=}CHCH_2CH_2CH_3 \qquad CH_3CH{=}CHCH_2CH_3$$

1-Pentene
VIII

2-Pentene
IX

$$CH_2{=}\overset{\underset{\displaystyle |}{CH_3}}{C}{-}CH_2CH_3 \qquad CH_3{-}\overset{\underset{\displaystyle |}{CH_3}}{C}{=}CH{-}CH_3 \qquad CH_2{=}CH{-}\overset{\underset{\displaystyle |}{CH_3}}{CH}{-}CH_3$$

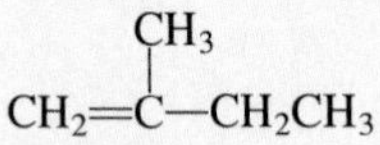

2-Methyl-1-butene
X

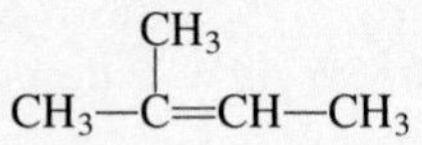

2-Methyl-2-butene
XI

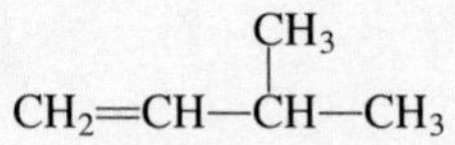

3-Methyl-1-butene
XII

Of these, only IX exists as cis and trans isomers.

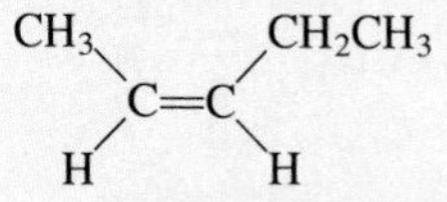

cis-2-Pentene

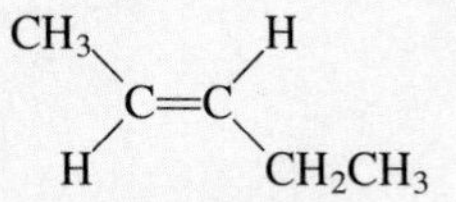

trans-2-Pentene

Structures VIII, X, and XII each have two hydrogens on one of their doubly bonded carbon atoms, and structure XI has two methyl groups on one of its doubly bonded carbons.

Practice Exercise

Draw and name all alkenes with the formula $C_3H_4Br_2$, and indicate which ones exist as cis and trans isomers.

Maleic and fumaric acids are classic examples of geometric isomers that have widely different chemical and physical properties (Figure 18.17). Because of the proximity of its

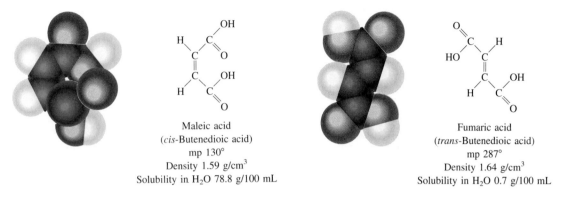

Maleic acid
(*cis*-Butenedioic acid)
mp 130°
Density 1.59 g/cm^3
Solubility in H$_2$O 78.8 g/100 mL

Fumaric acid
(*trans*-Butenedioic acid)
mp 287°
Density 1.64 g/cm^3
Solubility in H$_2$O 0.7 g/100 mL

Figure 18.17
Space-filling models, structural formulas, and properties of maleic acid and fumaric acid.

carboxyl groups, maleic acid readily loses water to form an anhydride upon gentle heating.

Maleic acid   Maleic anhydride

Fumaric acid is incapable of anhydride formation under the same reaction conditions. If it is heated to high temperatures (~300 °C), fumaric acid rearranges to form maleic acid, which then loses water to form the anhydride. We shall see in Section 25.1 that fumaric acid is the isomer produced and utilized by enzymes in living cells.

Recall from Section 13.10 that the nature of the bonding in the cycloalkanes also imposes geometric constraints on the groups bonded to the ring carbon atoms. Common to all ring structures is the inability of groups to rotate about any of the ring carbon–carbon bonds. Therefore, groups can either be on the same side of the ring (cis) or on opposite sides of the ring (trans). For our purposes here, we represent all cycloalkanes as planar structures, but we clearly indicate the positions of the groups, either above or below the plane of the ring.

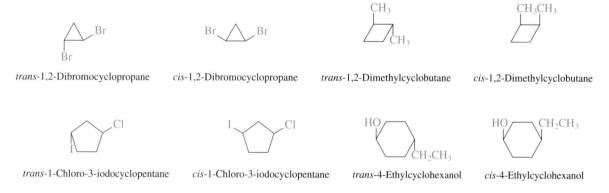

trans-1,2-Dibromocyclopropane   *cis*-1,2-Dibromocyclopropane   *trans*-1,2-Dimethylcyclobutane   *cis*-1,2-Dimethylcyclobutane

trans-1-Chloro-3-iodocyclopentane   *cis*-1-Chloro-3-iodocyclopentane   *trans*-4-Ethylcyclohexanol   *cis*-4-Ethylcyclohexanol

18.5
BIOCHEMICAL SIGNIFICANCE

Molecular configurations are of the utmost importance in biochemistry. The example of the two enantiomers of lactic acid has already been cited in Section 18.2. The dextrorotatory form is isolated from muscle tissue, and the levorotatory isomer is found in yeast and some bacteria. Laboratory synthesis of lactic acid from either acetaldehyde or pyruvic acid produces a racemic mixture of (±)-lactic acid; it is impossible to synthesize chemically either of the chiral forms using achiral reagents.

The obvious question, then, is how do muscle cells synthesize only (+)-lactic acid, and yeast only the (−)-isomer? The explanation arises many times during the study of biochemistry. Enzymatic control is the answer.

$$\begin{array}{ccc}
\underset{\text{(−)-Lactic acid}}{\begin{array}{c}\text{HO}\quad\text{O}\\\diagdown\;\diagup\\\text{C}\\|\\\text{H—C—OH}\\|\\\text{CH}_3\end{array}} & \xleftarrow[\substack{\text{lactic acid}\\\text{dehydrogenase}\\\text{in yeast}}]{} \underset{\text{Pyruvic acid}}{\begin{array}{c}\text{HO}\quad\text{O}\\\diagdown\;\diagup\\\text{C}\\|\\\text{C=O}\\|\\\text{CH}_3\end{array}} \xrightarrow[\substack{\text{lactic acid}\\\text{dehydrogenase}\\\text{in muscle}}]{} & \underset{\text{(+)-Lactic acid}}{\begin{array}{c}\text{HO}\quad\text{O}\\\diagdown\;\diagup\\\text{C}\\|\\\text{HO—C—H}\\|\\\text{CH}_3\end{array}}
\end{array}$$

Enzymes are biological catalysts that are themselves chiral organic compounds. Similarly, almost every organic compound that occurs in living organisms is one enantiomer of a pair. Foods and medicines must have the proper molecular configurations if they are to be beneficial to the organism. For example, the popular flavoring agent Accent is levorotatory monosodium glutamate.[4] The dextrorotary form of this salt would not enhance the flavor of meat because our taste buds could not recognize it (Figure 18.18). Similarly, the natural form of epinephrine is levorotatory. It has a physiological activity about 15–20 times greater than that of its dextrorotatory enantiomer.

[4] L-Monosodium glutamate (MSG) does not itself impart any taste, but it somehow enhances the flavor of foods that it combines with. Some people seem to be sensitive to MSG and display symptoms such as dizziness, hot flashes, numbness, sweating, and swelling of the hands and feet.

Figure 18.18
(a) The levorotatory stereoisomer of monosodium glutamate forms a precise fit with the bonding sites on the receptor protein of our taste buds.
(b) The dextrorotatory isomer does not fit and therefore cannot bind to the receptor sites.

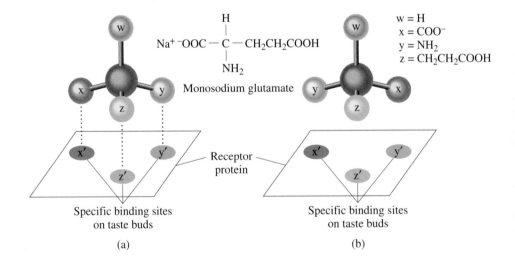

$$\text{Na}^+\,{}^-\text{OOC—}\underset{\underset{\text{NH}_2}{|}}{\overset{\overset{\text{H}}{|}}{\text{C}}}\text{—CH}_2\text{CH}_2\text{COOH}$$

Monosodium glutamate

w = H
x = COO⁻
y = NH₂
z = CH₂CH₂COOH

Receptor protein

Specific binding sites on taste buds

(a)   (b)

In Section F.3 we mentioned that the stimulant drug amphetamine is not a pure compound. Rather, it is a mixture of two enantiomers.

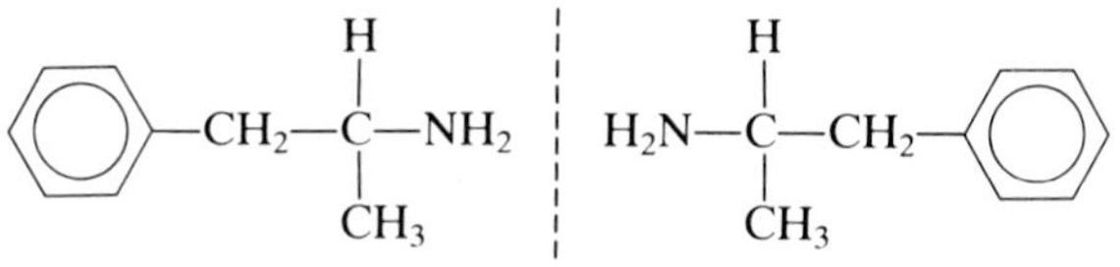

The dextrorotatory form is a stronger stimulant than its levorotatory isomer. Dexedrine is the trade name for the pure dextro isomer. Benzedrine is the trade name for a mixture of the two isomers in equal amounts. Dexedrine is two to four times as active as Benzedrine.

Because of the dangers of pesticides, scientists have been searching for alternative methods to control insects. One method that has made a promising start involves the use of chemicals known as pheromones. **Pheromones** are chemicals that are used for communication between members of the same species of insects. Insects emit pheromones for a variety of purposes, such as sending an alarm, social regulation, attracting a mate, trail marking, and territorial marking.

The most important pheromone for insect control is the sex attractant. The females of many insect species depend upon an attractant to lure males for mating. These chemicals are remarkably powerful; a few drops can attract males within a range of 2 miles. Traps baited with the sex attractant, and also containing an insecticide, can be used to lure all the males of that species to their deaths. One such compound is trimedlure, which has been found to be strongly attractive to the male Mediterranean fruit fly. Trimedlure has eight possible stereoisomers, and they differ considerably in attraction for the insect. The fly is most strongly drawn to the isomer in which the methyl and ester groups are trans to each other.

In a few insect species, including the boll weevil, the male emits the pheromone.

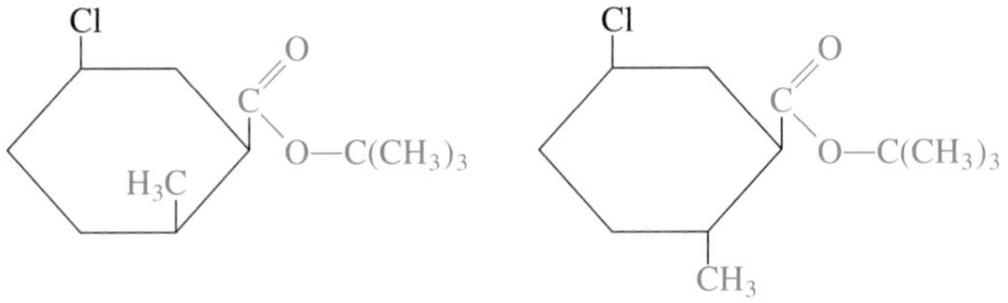

Methyl and ester groups are cis.   Methyl and ester groups are trans.

Sex attractants of more than 30 insects have been identified. In most cases, just one of the possible stereoisomers in physiologically active (Figure 18.19).

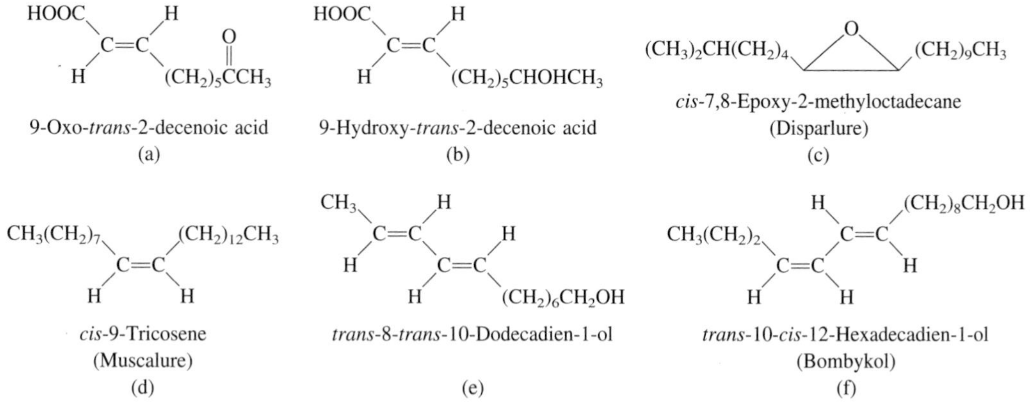

9-Oxo-*trans*-2-decenoic acid
(a)

9-Hydroxy-*trans*-2-decenoic acid
(b)

cis-7,8-Epoxy-2-methyloctadecane
(Disparlure)
(c)

cis-9-Tricosene
(Muscalure)
(d)

trans-8-*trans*-10-Dodecadien-1-ol
(e)

trans-10-*cis*-12-Hexadecadien-1-ol
(Bombykol)
(f)

Figure 18.19
Sex attractants of some female insects. (a, b) Queen honeybee, (c) gypsy moth, (d) common house fly, (e) codling moth, (f) silkworm moth.

The very subtle differences in structural configurations of organic molecules are of primary importance to life. We will deal with the vitally important stereoselectivity of enzymes only after examining the compositions of the three major classes of biochemical compounds—the carbohydrates, the lipids, and the proteins. It is necessary to observe strictly the proper configurational formulas of these compounds. If enzymes can recognize such subtle differences of shape and structure, so must we.

EXERCISES

1. Define the following terms.
   a. chiral center   b. enantiomers
   c. polarimeter   d. optically active
   e. specific rotation   f. geometric isomers
   g. polarized light   h. diastereomers
   i. stereoisomers   j. meso form
   k. racemic mixture   l. levorotatory
   m. dextrorotatory   n. pheromones

2. Which of the carbon atoms shown in color are chiral?

   a. —CH$_2$CH—NH$_2$ / CH$_3$

   b. H$_2$NCOCH$_2$CCH$_2$OCNH$_2$ / CH$_2$CH$_2$CH$_3$ / CH$_3$ (with C=O groups)

3. Are these structures mirror images? Are they superimposable?

   CH$_3$ CH$_3$
   H—C—OH   HO—C—H
   CH$_3$ CH$_3$

4. Are these structures mirror images? Are they superimposable?

   H   H
   CH$_3$CH$_2$—C—CH$_3$ CH$_3$—C—CH$_2$CH$_3$
   OH   OH

5. Circle each chiral carbon atom in the following molecules.

   a. HO— —CHCH$_2$—NH—CH$_3$ / OH / HO

   Epinephrine

   b. O=C—CH$_2$—CH$_2$—CH—C=O / HO / NH$_2$ / O$^-$ Na$^+$

   MSG

6. Circle each chiral carbon atom in the following molecules.

   a. H$_2$N—CH—C—NH—CH—C—OCH$_3$ / CH$_2$ / COOH / CH$_2$ (with two C=O groups and benzene ring)

   Aspartame

   b. CH$_3$—N$^\pm$—CHCH$_2$—C / Cl$^-$ / CH$_3$ / H CH$_3$ / O CH$_2$CH$_3$ (with benzene rings)

   Methadone

7. Circle each chiral carbon atom.
   a. CH$_3$CHCH$_2$OH   b. CH$_3$CHCOOH
        OH   NH$_2$
   c. C$_6$H$_5$CH$_2$CHCH$_3$ d. CH$_3$CHCH$_2$CH$_3$
             NH$_2$ Br
   e. CH$_3$CHCHO   f. CH$_3$CH—CHCH$_3$
        OH   OH OH

8. Which of the following can exist in the meso form?
   a. CH$_3$CH—CHCH$_3$
        OH OH
   b. CH$_3$CH—CHCH$_2$CH$_3$
        Br Br
   c. HOOCCH—CHCOOH
         OH OH

9. Which of the following can exist in the meso form?
   a. CH$_2$OH b. CHO   c. COOH   d. CH$_3$
      CHOH   CHOH   CHOH   CHCl
      CHOH   CHOH   CHOH   CHCl
      CH$_2$OH   CH$_2$OH   CH$_2$OH   CH$_3$

10. Indicate whether the compound is chiral.

  a. 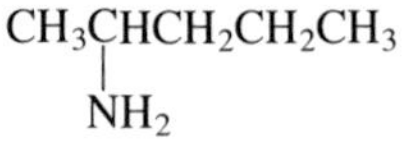 $CH_3CHCH_2CH_2CH_3$
       |
       NH_2

  b. 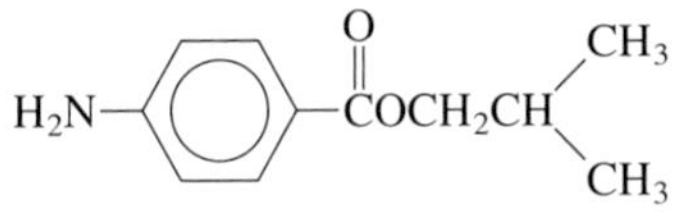 H_2N—⟨benzene ring⟩—$\overset{\overset{O}{\|}}{C}OCH_2CH\overset{CH_3}{\underset{CH_3}{<}}$

  c.

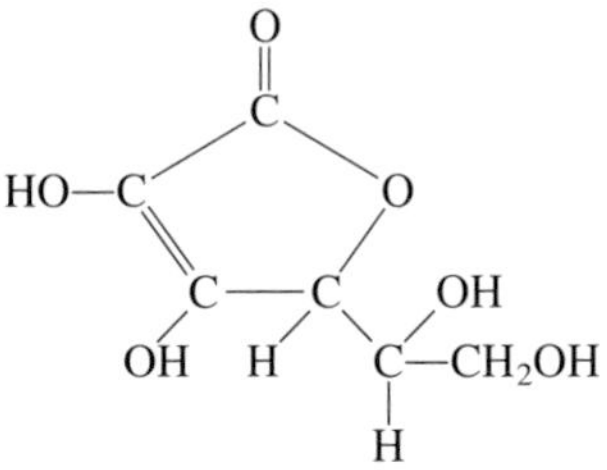

  d. 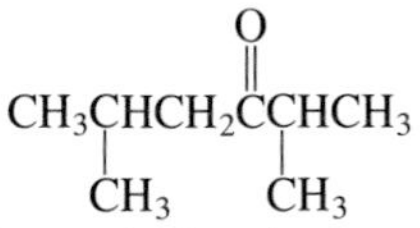 $CH_3CHCH_2\overset{\overset{O}{\|}}{C}CHCH_3$
        | |
       CH_3 CH_3

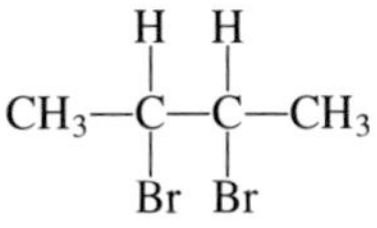

11. Indicate whether the compound is chiral.

  a. 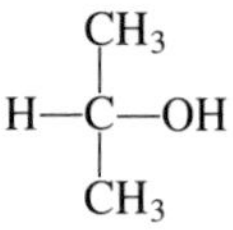 $H—\overset{\overset{CH_3}{|}}{\underset{\underset{CH_3}{|}}{C}}—OH$

  b. $CH_3—\overset{\overset{H}{|}}{\underset{\underset{Br}{|}}{C}}—\overset{\overset{H}{|}}{\underset{\underset{Br}{|}}{C}}—CH_3$

  c. $CH_3—\overset{\overset{H}{|}}{\underset{\underset{Br}{|}}{C}}—\overset{\overset{Br}{|}}{\underset{\underset{H}{|}}{C}}—CH_3$

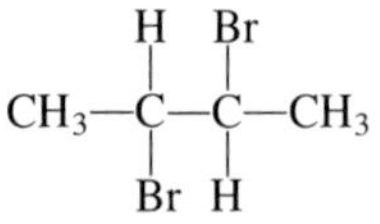

  d. 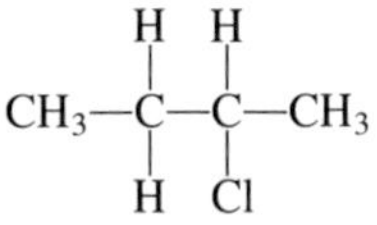 $CH_3—\overset{\overset{H}{|}}{\underset{\underset{H}{|}}{C}}—\overset{\overset{H}{|}}{\underset{\underset{Cl}{|}}{C}}—CH_3$

12. Compare (+)-lactic acid and (−)-lactic acid with respect to
  a. boiling point
  b. melting point
  c. specific rotation
  d. solubility in H_2O
  e. reaction with ethanol
  f. reaction with (+)-sec-butylamine

13. How does a meso compound differ from a racemic mixture?

14. Draw projection formulas for (+), (−), and meso forms of 2,3-butanediol. Which is meso? Can you tell which is (+) and which is (−)?

15. In Chapter 14, Exercise 3, you are asked to draw structures for the eight isomeric pentyl alcohols. Which of these could exist in enantiomeric forms? (That is, which molecules have chiral carbon atoms?) Draw projection formulas for each pair of enantiomers.

18. Draw Fischer projections (flat) for the stereoisomers of bromochlorofluoromethane.

17. Draw Fischer projections (flat) for 2,3-dichlorobutanal. Label pairs of enantiomers.

18. How many stereoisomers are there for each of the following?

  a. 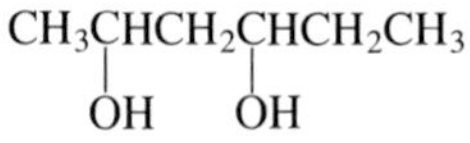 $CH_3CHCH_2CHCH_2CH_3$
        | |
       OH OH

  b. 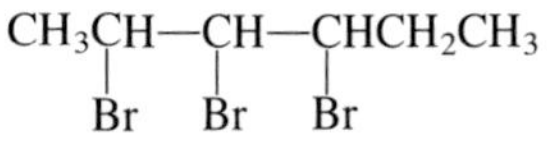 $CH_3CH—CH—CHCH_2CH_3$
        | | |
       Br Br Br

19. How many stereoisomers are there for the following structure?

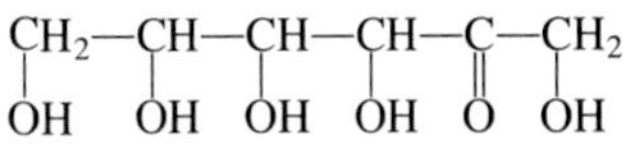

$$CH_2—CH—CH—CH—\overset{\overset{}{C}}{}—CH_2$$
$$\underset{OH}{|}\quad\underset{OH}{|}\quad\underset{OH}{|}\quad\underset{OH}{|}\quad\underset{O}{\|}\quad\underset{OH}{|}$$

20. (−)-Menthol melts at 43 °C, boils at 212 °C, and has a density of 0.890 g/cm³ and a specific rotation of −50. List the corresponding properties of (+)-menthol.

21. Would (+)-nicotine and (−)-nicotine react the same way (a) with HCl? (b) with lactic acid? (c) with tartaric acid?

22. 1-Butene reacts with HCl to form 2-chlorobutane. Does the reactant have a chiral center? Does the product have a chiral center? Would 2-chlorobutane formed in this manner show optical activity? Why or why not?

23. Write formulas for pairs of isomers that represent the following.
  a. positional isomers
  b. functional group isomers
  c. enantiomers
  d. diastereomers
  e. noncyclic geometric isomers
  f. cyclic geometric isomers

24. In what ways do enantiomers resemble each other? How do they differ?

25. Draw the configurational isomers for the remaining three stereoisomers of 3-phenyl-2-butanol. Given the following properties for the stereoisomer shown, predict as many properties as you can for the other three. Include a statement as to why you were not able to predict some of the properties.

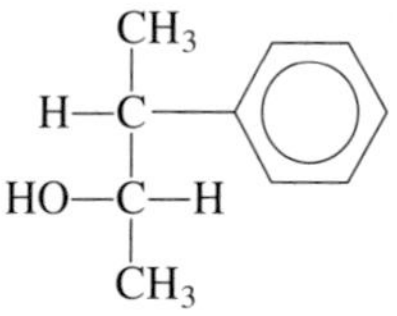

bp (25 mm Hg) 118 °C
[α] +30.9

26. Write the formulas of the geometric isomers for each of the following compounds. Label them cis and trans. If there are no geometric isomers, write None.
  a. 2-hydroxy-2-pentene
  b. 3-hexene
  c. 4-methoxy-2-pentyne
  d. 1,1-dibromo-1-butene
  e. 2-butenoic acid
  f. 4-methyl-2-pentene
  g. 2,3-dimethyl-2-pentene
  h. 1,1-dibromo-2-ethylcyclopropane
  i. 1,2-dibromocyclohexene
  j. 2-chlorocyclohexanol

k. 1-bromo-3-chlorocyclobutane

l. 1,2,3-trimethylcyclopropane

27. 1,2-Dimethylcyclobutane exists as cis and trans isomers. One isomer is chiral, the other is achiral. Draw mirror images for both sets of geometric isomers, and identify which isomers are identical and which isomers are enantiomers. Are *cis*- and *trans*-1,2-dimethylcyclobutane diastereomers?

28. For 1,2-dibromocyclopropane, there are three stereoisomers. Draw structures for the three stereoisomers.

29. Draw the other geometric isomers for all of the pheromones shown in Figure 18.19.

30. Urushiol is an unsaturated phenolic compound that is the active agent in poison ivy and poison sumac. Draw all of the geometric isomers for urushiol.

31. If the four bonds of carbon were directed toward the corners of a square, how many isomers of CH_2BrCl would exist? Draw them.

32. Draw and name all the geometric isomers (both noncyclic and cyclic) corresponding to the molecular formula C_5H_{10}.

33. Only one of the isomers of pentanal is optically active. What is its formula? Draw the two enantiomers of this compound.

34. Give the formula of the smallest noncyclic alkane that could be optically active.

35. An alcohol has the molecular formula $C_4H_{10}O$.

a. Write all the possible structural formulas for the alcohol.

b. If the alcohol can be separated into two optically active forms, which of the formulas in **a** is correct?

36. Draw the enantiomers of each of the following compounds.

a. 2-butanol

b. 2-methylpropanoic acid

c. 2,3-dihydroxypropanal

d. 2-bromobutanoic acid

37. β-Hydroxybutyric acid occurs in the urine of diabetics in amounts up to 30 g/day. Draw the structure of both enantiomers. (*Hint:* Place the carbon atoms in a vertical column with the carboxyl group at the top.)

38. Draw structures for the two enantiomeric forms of 3-chloro-1,2-propanediol. (*Hint:* Place the carbon atoms in a vertical column with the CH_2OH at the bottom).

39. Benzedrine is a racemic mixture of (+)- and (−)-amphetamine. The pure dextrorotatory enantiomer, Dexedrine, has a much greater physiological activity than either the racemic mixture or the pure levorotatory form. Draw the structural formulas for the enantiomers of amphetamine (see Figure F.6).

40. Lysergic acid diethylamide (LSD) contains two chiral carbon atoms (see Section E.6). Identify which carbon atoms are chiral and then draw structural formulas of the four stereoisomers of LSD. [Only one of these four isomers, (+)-lysergic acid diethylamide, is physiologically active.]

41. Examine the labels of some common household products (detergents, foods, drugs, sprays, cosmetics). Write structural formulas for the chemical compounds contained in these products.

Molecules to See and to Smell

The retina of the human eye contains two kinds of receptor cells, rods and cones. 11-*cis*-Retinal forms a complex with various proteins, and these complexes are the photosensitive chemicals found in the rods and cones. Rhodopsin, a complex of 11-*cis*-retinal and a protein called opsin, is the visual pigment found in the cones. When light strikes the retina of the eye, a complex series of reactions is initiated by a cis–trans isomerization. This reaction is termed a **photochemical isomerization** (photoisomerization) because the energy of light causes the geometric change to occur. The only function of light in vision is to alter the shape of the absorbing molecule, 11-*cis*-retinal, to the trans configuration (Figure G.1). This primary event in vision occurs within a few picoseconds (1 ps = 10^{-12} s). Proteins in the eye are altered by this single photochemical act, and some energy is released that triggers a nerve impulse (Figure G.2). The impulse is transmitted via the optic nerve to the brain. Then an enzyme converts *trans*-retinal back to *cis*-retinal so it can bind to opsin to await the next exposure to light. Notice that *trans*-retinal is analogous to vitamin A in all respects except that it contains a carbonyl group, whereas the vitamin contains a primary alcohol group. Vitamin A is the reduced form of *trans*-retinal (or *trans*-retinal is the oxidized form of vitamin A), and the oxidation–reduction interconversion are essential to the chemical events of vision (Figure G.3). It is interesting to note that of all the tissues in the body, the retina has the highest respiration rate. You may think of seeing as a rather passive activity, but vision is metabolically demanding.

When light strikes rhodopsin, 11-*cis*-retinal is isomerized to the all-trans isomer. This change in structure is accompanied by a change in electrical potential. The photoisomerization of the retinal triggers a nerve impulse. The rhodopsin complex then splits into opsin and the free aldehyde. The all-trans retinal is converted by enzymes back

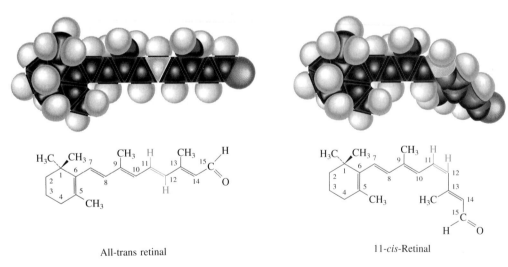

All-trans retinal

11-*cis*-Retinal

Figure G.1

Fundamental molecule of vision is retinal ($C_{15}H_{28}O$), also known as retinene, which combines with proteins called opsins to form visual pigments. Because the nine-member carbon chain in retinal contains an alternating sequence of single and double bonds, it can assume a variety of bent forms. Two isomers of retinal are depicted here. In the space-filling models carbon atoms are dark except C-11, which is colored; hydrogen atoms are light. The red atom attached to C-15 is oxygen. When tightly bound to opsin, retinal is in the bent and twisted form known as 11-*cis*-retinal. When struck by light, it straightens out into the all-trans configuration. This simple photochemical event provides the basis for vision. [From "Molecular Isomers in Vision" by Ruth Hubbard and Allen Kropf. Copyright © 1967 by Scientific American, Inc. All rights reserved.]

In contrast, most animal tissue contains a comparatively small amount of carbohydrate (less than 1% in humans). Plants use carbohydrates both as a source of energy and as supporting tissue. Plants are able to synthesize their own carbohydrates from the carbon dioxide of the air and water taken from the soil. Animals are incapable of this synthesis, and therefore they are dependent upon the plant kingdom as a source of these vital compounds. We humans use carbohydrates not only for our food (about 60–65% by mass of the average diet) but also for our clothing (cotton, linen, rayon), shelter (wood), fuel (wood), and paper (wood).

19.1
CARBOHYDRATES: DEFINITIONS AND CLASSIFICATIONS

Carbohydrates are compounds of carbon, hydrogen, and oxygen. They include the starches, the sweet-tasting compounds called sugars, and structural materials such as cellulose (Figure 19.1). The term *carbohydrate* has its origin in a misinterpretation of the molecular formulas of many of these substances. For example, the formula for blood sugar is $C_6H_{12}O_6$, but we could also represent it as that of a ''carbon hydrate'' $(C \cdot H_2O)_6$. These compounds are not hydrates of carbon, however. They are alcohols—they all contain the hydroxyl (—OH) functional group. Most contain a real or latent carbonyl (C=O) group. By a *latent* carbonyl group we mean a functional group such as a hemiacetal or an acetal (Section 15.6) that can be more or less readily converted to a carbonyl group. Consequently, some carbohydrates give reactions as aldehydes or ketones even though the ''carbonyl'' may exist primarily in a hemiacetal form. Complex carbohydrates containing the acetal function can be hydrolyzed like other acetals to simpler compounds that give reactions typical of aldehydes or ketones.

Figure 19.1

The principal sugars in our diets are sucrose (cane or beet sugar), glucose (corn syrup), and fructose (fruit sugar, often in the form of high-fructose corn syrup).

Carbohydrates are often called saccharides (from the Latin *saccharon,* sugar). In fact, William Prout, an English physician who first recognized the three general classes of foodstuffs (now called carbohydrates, fats, and proteins), suggested that they be called the saccharine, the oily, and the albuminous. Simple carbohydrates, those that cannot be further hydrolyzed, are called **monosaccharides.** Carbohydrates than can be hydrolyzed to two monosaccharide units are called **disaccharides,** and carbohydrates that can be hydrolyzed to many monosaccharide units are called **polysaccharides.**

Carbohydrate molecules contain several —OH groups each. The molecules can form an extensive network of intermolecular hydrogen bonds. The monosaccharides and disaccharides are sweet-tasting, white crystalline solids at room temperature. They have relatively high melting points and often will char before melting. Carbohydrate molecules also can form hydrogen bonds to water molecules. The simpler ones are readily soluble in water. For example, 100 g of glucose will dissolve in 100 mL of water at 25 °C. Polysaccharides are frequently tasteless, insoluble, amorphous compounds with exceedingly high molecular weights.

19.2
MONOSACCHARIDES: GENERAL TERMINOLOGY

The general names for the monosaccharides are obtained in a manner analogous to the naming of organic compounds by the IUPAC system. The number of carbon atoms in the molecule is denoted by the appropriate stem and *-ose* is the generic designation for any sugar. For example, the terms triose, tetrose, pentose, and hexose signify three-, four-, five-, and six-carbon monosaccharides, respectively. In addition, those monosaccharides that contain an aldehyde group are called **aldoses;** those containing a ketone group are **ketoses.** By combining these terms, both the type of carbonyl group and the number of carbon atoms in the molecule are easily expressed. Thus, monosaccharides are generally referred to as aldotetroses, aldopentoses, ketopentoses, ketoheptoses, etc. Glucose and fructose are specific examples of an aldose and a ketose, respectively.

Glucose
(an aldohexose)

Fructose
(a ketohexose)

19.3
STEREOCHEMISTRY

The simplest sugars are the trioses. Two trioses derived from glycerol by oxidation are important intermediates in metabolism. Dihydroxyacetone is a ketotriose, whereas glyceraldehyde is an aldotriose. Dihydroxyacetone does not contain a chiral center. Glyceraldehyde possesses a chiral carbon atom and thus can exist in two optically active forms. Except for the direction in which they rotate plane-polarized light, they have identical physical properties. One form has a specific rotation of +8.7, the other a rotation of −8.7. The great German chemist Emil Fischer (Nobel prize, 1902) initiated the convention of projecting the formulas onto a two-dimensional plane so that the aldehyde group is written at the top, with the hydrogen and hydroxyl written to the right and left. (Formulas of chiral molecules represented in this manner are referred to as Fischer projections, Fischer models, or Fischer configurations.) Arbitrarily, Fischer then decided that

Dihydroxyacetone

the formula of glyceraldehyde in which the hydroxyl group is positioned to the right of the chiral carbon atom represents the dextrorotatory isomer. He assigned the letter D as its prefix. The levorotatory isomer, in which the —OH group is positioned to the left of the chiral carbon atom, was accordingly assigned the letter L as its prefix.[1]

D-Glyceraldehyde   L-Glyceraldehyde

The two forms of glyceraldehyde are especially important because the more complex sugars can be considered to be derived from them. They serve as a reference point for designating and drawing all other monosaccharides. Sugars whose Fischer projections terminate in the same configuration as D-glyceraldehyde are designated as **D sugars;** those derived from L-glyceraldehyde are designated as **L sugars.** We cannot obtain energy from L carbohydrates (though beings on some other planet, with enzymes that are mirror images of ours, might be able to use only L forms). Since we will encounter only D carbohydrates, we will not emphasize this form of isomerism here.

The letters D and L often mislead the beginning student. It must be emphasized that these prefixes serve only to signify the absolute configuration of a molecule. A **D sugar** is one that has the same configuration about the **penultimate** carbon atom as D-glyceraldehyde has. The letters do not in any way refer to the optical rotation of the molecule. D-Glyceraldehyde just happens to be dextrorotatory. It is not to be expected, however, that all compounds derived from it will have the same optical rotation, for such compounds will have additional chiral carbon atoms.

For example, there are two aldopentoses of interest to us. One is D-ribose, the sugar unit that occurs in ribonucleic acids (RNA). Related to D-ribose, but not an isomer of it (its molecular formula is different), is D-2-deoxyribose, the sugar unit that occurs in deoxyribonucleic acids (DNA). As the 2-*deoxy*- implies, this sugar is "missing" an oxygen on the second carbon atom. Both of these D sugars are levorotatory.

The direction of rotation of plane-polarized light is a specific property of each optically active molecule. It is not at all dependent upon the configuration about the penultimate carbon. The symbols plus $(+)$ and minus $(-)$ are used to denote the optical rotation of the monosaccharides: $(+)$ indicates a clockwise, or dextrorotatory, rotation, and $(-)$ indicates a counterclockwise, or levorotatory, rotation.

Aldohexoses contain four chiral carbon atoms, and thus there are eight enantiomeric pairs, or 16 isomers. Fortunately for students of biochemistry, only three of the 16 isomers are commonly found in nature, D-$(+)$-glucose, D-$(+)$-mannose, and D-$(+)$-galactose. (All 16 isomers have been prepared synthetically.) There are quite a few ketohexoses, too, but we shall deal with only a single ketohexose here—D-$(-)$-fructose. All of the sugars we discuss in the remainder of this chapter belong to the D family. If no family designation is given, you can assume that the compound is a D sugar.

The penultimate (next to last) carbon has been chosen, by convention, to be the reference carbon atom. It is the chiral carbon atom that is farthest from the aldehyde or ketone group.

D-$(-)$-Ribose $(-)$-2-Deoxy-
                        D-ribose

---

[1] Fischer's arbitrary assignment proved to be correct. In 1951 chemists, with the aid of X-ray crystallography, determined the absolute configurations of the glyceraldehyde enantiomers and found that the D isomer was indeed dextrorotatory.

19.4
HEXOSES

Glucose

D-Glucose is the most abundant sugar found in nature. It is commonly found in fruits, especially in ripe grapes, and for this reason it is often referred to as *grape sugar*. It is also known as *dextrose,* a name that derives from the fact that the predominant natural form of the sugar is dextrorotatory.

Most of the carbohydrates taken in by the body are eventually converted to glucose in a series of metabolic pathways that produce energy for our cells. Since glucose requires no digestion, it can be given intravenously (as a 5% solution) to patients who are unable to take food orally. Glucose is the circulating carbohydrate of animals; hence the name *blood sugar*. The blood contains about 0.08% glucose, and normal urine may contain anywhere from a trace to 0.2% glucose.

Commercially, glucose is made by the hydrolysis of starch. (Corn syrup is mainly glucose.) In the United States, cornstarch is used in the process; in Europe, the starch is obtained from potatoes. Glucose is only 74% as sweet as table sugar (sucrose), but it has the same caloric value.

The structure of D-glucose is illustrated in Figure 19.2. This formula follows our convention of writing the aldehyde group at the top and the primary alcohol group at the bottom. Glucose is a D sugar because the hydroxyl group at the fifth carbon (the one just above the CH_2OH) is on the right. In fact, all the hydroxyl groups except the one at the third carbon are to the right. You should learn to draw this formula for glucose, as well as the analogous formulas for the other three monosaccharides in Figure 19.2.

Mannose

D-Mannose is a component of the polysaccharide mannan, found in some berries. A particularly good source is "vegetable ivory," the endosperm of palm nuts. Buttons were once widely made of this material, with the waste being hydrolyzed to mannose. A structure for mannose is also presented in Figure 19.2. Note that the configuration differs from that of glucose only at the second carbon atom.

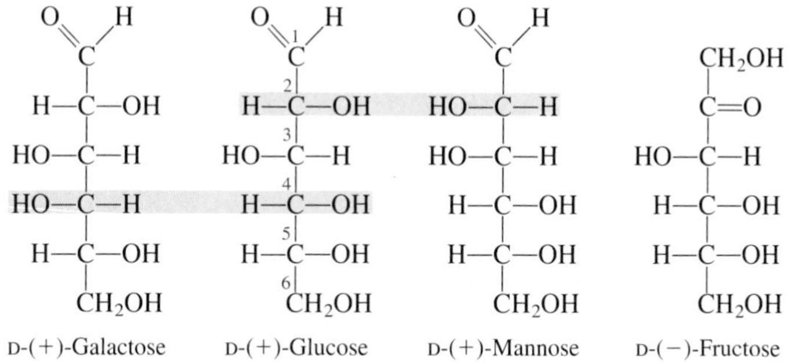

Figure 19.2
Structures of four important hexoses.

monosaccharides

Galactose

D-Galactose is formed by the hydrolysis of lactose, a disaccharide composed of a glucose unit and a galactose unit. It does not occur in nature in the uncombined state. The galactose needed by the human body for the synthesis of lactose (in the mammary glands) is obtained by the conversion of D-glucose into D-galactose. In addition, galactose is an important constituent of the glycolipids (Section 20.8) that occur in the brain and in the myelin sheath of nerve cells. The genetic disease galactosemia results from the absence of an enzyme that converts galactose to glucose (see Section 23.8). Figure 19.2 also shows the structure of galactose. Notice that the configuration differs from that of glucose only at the fourth carbon atom.

Fructose

D-Fructose, whose structure is shown in Figure 19.2, is the only naturally occurring ketohexose. It occurs, along with glucose and sucrose, in honey (which is 40% fructose) and sweet fruits. Fructose (Latin *fructus,* fruit) is also referred to as *levulose* because it has a specific rotation that is strongly levorotatory (-92.4). It is the sweetest sugar (1.7 times sweeter than sucrose). Many nonsugars, however, are several hundred or several thousand times as sweet (Table 19.1). Fructose is the only sugar found in the semen of bulls and men. It is the major energy source for spermatozoa and is formed in the prostate gland. Structurally, fructose is a 2-ketohexose. From the third through the sixth carbon atoms its structure is the same as that of glucose.

Table 19.1
Sweetness of Some Compounds Relative to Sucrose

Compound	Relative Sweetness[a]
Acesulfame K	20,000
Glucose	74
Fructose	173
Lactose	16
Sucrose	100
Maltose	33
P-4000	400,000
Saccharin	50,000
Aspartame	18,000

[a] Sweetness is relative to sucrose at a value of 100.

19.5
MONOSACCHARIDES: CYCLIC STRUCTURES

So far we have represented the monosaccharides as hydroxy aldehydes and ketones. These representations account for many of the properties (Section 19.6) of these simple sugars. However, in Section 15.6 we mentioned that aldehydes and ketones react with alcohols to form hemiacetals and hemiketals.

| Aldehyde | Alcohol | Hemiacetal |

You should not be surprised, then, to find that hydroxyl groups and carbonyl groups conveniently located on the same molecule react with one another and that consequently the monosaccharides exist mainly as hemiacetals.

A sweet taste is one of the four primary taste sensations that can be distinguished by the taste buds on the surface of the tongue. The others are sour, salty, and bitter. Although sweetness is commonly associated with most mono- and disaccharides, it is not a specific property of carbohydrates. Many sugars are sweet to varying degrees, but several organic compounds have been synthesized that are far superior as sweetening agents. These synthetic compounds have no caloric value, and therefore they are useful for those persons (e.g., diabetics) who must minimize their carbohydrate intake.

In 1981, after eight years of extensive testing (and controversy regarding its safety), the FDA approved the use of the low calorie sweetener **aspartame** (L-aspartyl-L-phenylalanine methyl ester—see Chapter 21, Exercise 53). This white crystalline compound is about 180 times sweeter than sucrose and does not leave the bitter aftertaste often associated with saccharin. The FDA approved aspartame for use in more than 70 products, including soft drinks (NutraSweet), cereals, gelatins, and chewing gum, and as tablets to be used as sugar substitutes. It is interesting that the two constituent amino acids are not sweet. L-Aspartic acid has a flat taste, whereas L-phenylalanine is bitter. Aspartame is used as a sweetener for a wide variety of foods because it can blend well with other food flavors. In the body, aspartame is hydrolyzed to aspartic acid, phenylalanine, and methanol. The small amount of methanol does not seem to be a problem, but the release of phenylalanine is a matter of concern to those on low phenylalanine diets (see Section 23.8). A report was released in the early 1980s that pregnant women who consume aspartame may have babies with permanent brain damage, and there was a later report of similar damage to infants who ingested it during the six months following birth. In 1985, the American Medical Association completed its investigations and concluded that aspartame is safe, and only people who are sensitive to phenylalanine need regulate their intake. (It should be noted that the shelf life of aspartame is limited because at high temperatures the molecule breaks down and loses its sweetness.)

Aspartame

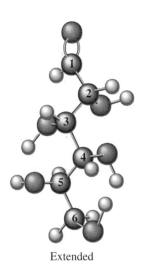

Extended

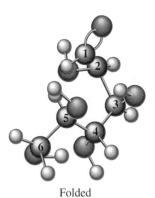

Folded

Figure 19.3

Models of the free-aldehyde form of glucose. Note that in the folded model the oxygen atom on the fifth carbon is near the carbonyl carbon.

"Now wait a minute," you say. "Things are bad enough with all these complications, and now you ignore a hydroxyl group located right next to the carbonyl in order to play around with the hydroxyl at the fifth carbon. And you had to write a long, silly-looking bond to do it!" A reasonable objection—which is why we must introduce a different type of formula. In this new formula, we'll take into account approximately correct bond angles. Figure 19.3 shows a model of the free-aldehyde form of glucose. Notice how it folds around on itself. The center structure in Figure 19.4 is drawn to resemble the model. Note that the hydroxyl on the fifth carbon is quite near the carbonyl carbon, so it's not surprising that this hydroxyl group reacts with the carbonyl. When the reaction occurs, the originally doubly bonded oxygen may be pushed up or down, giving rise to two different hemiacetal forms. The structure on the left, with the hydroxyl on the first carbon projected downward, represents what is called the **alpha** (α) form. That on the right, with the hydroxyl on the first carbon pointed upward, is the **beta** (β) form.

Crystalline glucose may exist in either the alpha or the beta form. The two forms have different properties. The alpha form melts at 146 °C, and the beta form melts at

α-Glucose   Open-chain form   β-Glucose

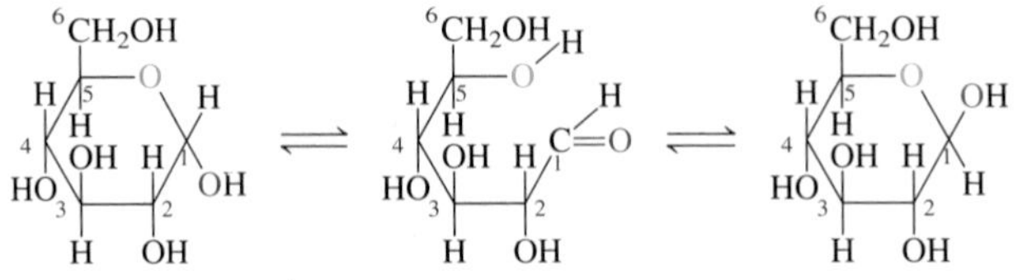

150 °C. In solution, one gets an equilibrium mixture. You can start out with either pure crystalline hemiacetal form, but as soon as it is dissolved in water, the unstable hemiacetal group opens to form the free carbonyl and then closes to either the alpha or beta hemiacetal, reopening and reclosing in succession. This interconversion is referred to as **mutarotation** (Latin *mutare,* to change).[2] At equilibrium, the mixture is about 36% alpha and 64% beta. There is less than 0.02% of the aldehyde form. Nevertheless, that is enough to give most of the characteristic reactions of aldehydes. As the small amount of free aldehyde is used up in a reaction, the hemiacetal forms open up to yield more free aldehyde. Thus, *all* the molecules may eventually react as aldehyde entities, even though very little "free" aldehyde is present at any given time.

In this book, we use the convention first suggested by an English chemist, Sir W. N. Haworth, for representing the formulas of the cyclic forms. The molecules are drawn as planar hexagonal slabs with darkened edges toward the viewer. Ring carbon atoms and the hydrogen atoms directly attached to them are not shown. The disposition of the hydroxyl groups, positioned either above or below the plane of the ring, is sufficient to define the correct configuration of the molecule. Any group of atoms written to the right in the Fischer projection appears below the plane of the ring, and any group written to the left appears above the plane in the cyclic forms.

Intramolecular hemiacetal formation is not unique to glucose. It occurs in galactose, mannose, and the naturally occurring aldopentoses and aldoheptoses. Fructose and other ketoses form intramolecular hemiketals. Figure 19.5 illustrates the equilibrium between the three forms of D-galactose, D-mannose, and D-fructose. Notice that galactose and mannose, like glucose, form a six-member cyclic structure. Fructose can also exist in this form, but it is most commonly found in nature as a five-member ring.

The difference between the alpha and beta forms of the sugars may seem trivial, but keep in mind that such differences are often crucial in biochemical reactions. We shall encounter some examples of this principle later in this chapter.

19.6
PROPERTIES OF SOME MONOSACCHARIDES

Glucose, mannose, galactose, and fructose are crystalline solids at room temperature. With five hydroxyl groups per molecule, these sugars are quite soluble in water.

[2]The two forms of glucose also differ in the way they affect plane-polarized light. α-D-Glucose has a specific rotation of $+112$, and β-D-glucose has a specific rotation of $+18.7$. If either is placed in solution, the observed rotation slowly changes (the substance undergoes mutarotation) to an equilibrium value of $+52.7$.

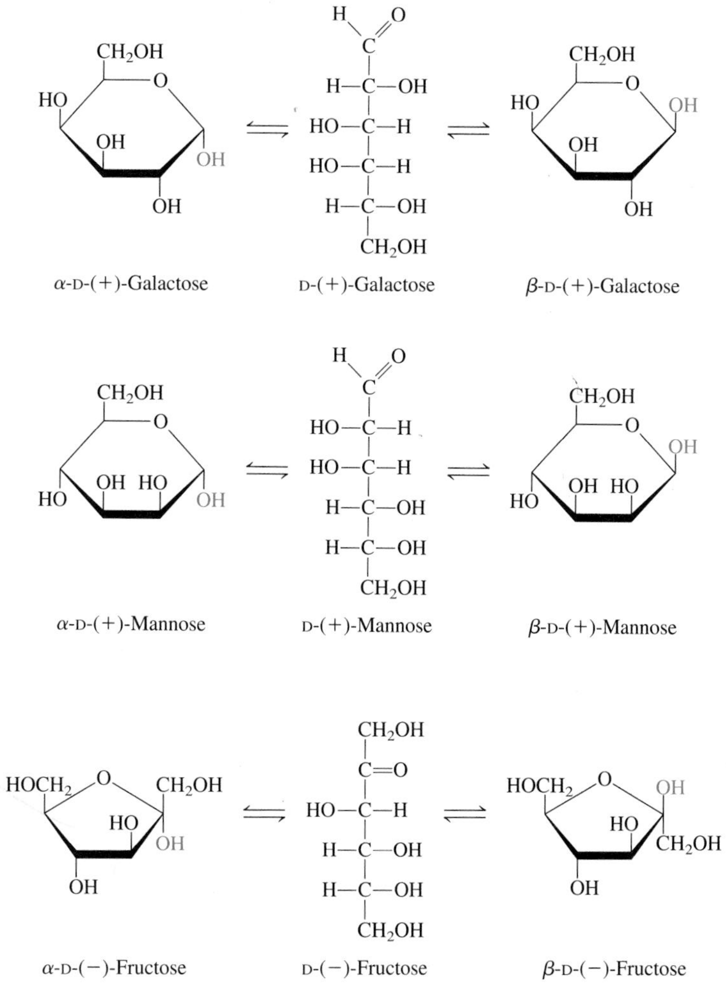

Figure 19.5

Mutarotation of D-galactose, D-mannose, and D-fructose.

Chemically, these monosaccharides undergo the reactions to be expected from their functional groups. The hydroxyl groups react to form esters and ethers. These reactions, though, are more important commercially for the polysaccharide cellulose (Section 19.10) than for the simpler sugars.

One important reaction is the oxidation of the aldehyde group, which is one of the most easily oxidized organic functional groups. This can be accomplished by any mild oxidizing agent. Tollens's, Benedict's, and Fehling's reagents (Section 15.6) are frequently used. The Tollens's test is based on the reduction of silver ions, and both Benedict's and Fehling's tests involve the reduction of copper complexes. Any carbohydrate capable of this reduction without first undergoing hydrolysis is said to be a **reducing sugar.**[3]

---

[3] It should not be surprising that aldoses are reducing sugars, but ketoses also give a positive test. The explanation is based on a reaction we have not discussed (tautomerism—see an organic chemistry text). In alkaline solution, an equilibrium exists between the ketoses and the aldoses. Since the oxidizing reagents commonly used to detect reducing sugars are prepared in a basic solution, all monosaccharides act as reducing sugars.

$$\text{An aldose (CHO)} + Ag(NH_3)_2^+ \longrightarrow \text{Carboxylate anion (COO}^-) + Ag(s)$$

An aldose   Tollens's reagent (clear solution)   Carboxylate anion   Silver mirror

$$\text{CHO} + Cu(citrate)_2^{2-} \longrightarrow \text{COO}^- + Cu_2O(s)$$

Benedict's reagent (blue solution)   Brick-red precipitate

$$\text{CHO} + Cu(tartrate)_2^{2-} \longrightarrow \text{COO}^- + Cu_2O(s)$$

Fehling's reagent (blue solution)   Brick-red precipitate

These reactions have been adopted as simple and rapid diagnostic tests for the presence of glucose in the blood or in the urine. For example, Clinitest tablets, which are used in clinical laboratories to test for sugar in the urine, contain cupric ions and are based on the same principles as Benedict's test. A green color indicates very little sugar, whereas a brick-red color indicates sugar in excess of 2 g/100 mL of urine.

19.7
DISACCHARIDES

Disaccharides ($C_{12}H_{22}O_{11}$) are composed of two monosaccharide units that are joined by acetal linkages (Section 15.6). They differ with respect to their constituent monosaccharides and the type of acetal linkage connecting them. There are three common disaccharides—maltose, lactose, and sucrose. Hydrolysis of 1 mol of disaccharide yields 2 mol of monosaccharide. Using word equations, we can write

$$\text{Maltose} + H_2O \longrightarrow 2 \text{ Glucose}$$
$$\text{Lactose} + H_2O \longrightarrow \text{Glucose} + \text{Galactose}$$
$$\text{Sucrose} + H_2O \longrightarrow \text{Glucose} + \text{Fructose}$$

All three disaccharides are white crystalline solids. Sucrose is quite soluble in water (200 g in 100 mL), and lactose is moderately soluble (20 g in 100 mL). All three molecules are too large to pass through cell membranes. Now let's look at each of these, in turn, in more detail.

Maple syrup is the concentrated sap of the sugar maple tree. It is a solution of sugars—about 65% sucrose with small amounts of glucose and fructose.

Sucrose

Sucrose is known as beet sugar, cane sugar, table sugar, or simply sugar. It is probably the largest selling pure organic compound in the world. As its names imply, sucrose is obtained from sugar canes and sugar beets (whose juices contain 14–20% of the sugar) by evaporation of the water and recrystallization. The dark brown liquid that remains after crystallization of the sugar is sold as molasses.

A molecule of sucrose may be envisioned to result from the combination of one molecule of α-D-glucose and one molecule of β-D-fructose; a molecule of water is eliminated in the process. The unique feature that characterizes the sucrose molecule is its acetal linkage. This linkage involves the hydroxyl group on the carbon in position 1 of α-D-glucose and the hydroxyl group on C-2 of β-D-fructose. By convention, sugars are read from left to right (or top to bottom). This connecting linkage is therefore an **α-1,2-glycosidic linkage.** This bonding bestows certain properties upon sucrose that are quite different from those of the other disaccharides.

Recall that the reaction of an alcohol with a hemiacetal yields an acetal. When two monosaccharides combine, the carbon–oxygen–carbon linkage that joins the components of the acetal is called a *glycosidic linkage.*

Sucrose, unlike the other disaccharides, is incapable of mutarotation. Thus, it exists in only one form both in the solid state and in solution. The presence of the 1,2-glycosidic linkage makes it impossible for sucrose to exist in the alpha or beta configuration or in the open-chain form. This is a direct result of the fact that the acetal carbon of the glucose ring and the ketal carbon of the fructose ring have both been tied up in the formation of the 1,2 (head-to-head) linkage. As long as the sucrose molecule remains intact, it cannot "uncyclize" to form the open-chain structure. Sucrose, therefore, does not undergo reactions that are typical of aldehydes and ketones, and we have now encountered our first nonreducing sugar.

The human body is unable to utilize sucrose or any other disaccharide directly because the molecules are too large to pass through cell membranes. Therefore, the disaccharide must first be broken down by hydrolysis into its two constituent monosaccharide units. In the body, this hydrolysis reaction is catalyzed by enzymes. The same hydrolysis reaction can be carried out in a test tube with dilute acid as a catalyst, but the reaction rate is much slower. The equation for the hydrolysis reaction is the reverse of the one given for the formation of sucrose. The product is an equimolar mixture of glucose and fructose. This 1:1 mixture is called **invert sugar.**[4]

---

[4] Sucrose is dextrorotatory with a specific rotation of 66.5. During hydrolysis, the observed rotation drops off and eventually becomes negative. This is because fructose has a high negative specific rotation (-92.4 at equilibrium) that more than balances the positive rotation of glucose ($+52.7$ at equilibrium). Because the sign of rotation changes during the reaction, the process is known as *inversion* and the products as *invert sugar.*

$$
\begin{array}{c}
\text{(structures of sucrose)} \quad O + HOH \xrightarrow[\text{invertase}]{H^+ \text{ or}} \quad \text{(structures of glucose)} \\
\text{(sucrase)} \quad +
\end{array}
$$

This hydrolysis reaction has several practical applications. Since sucrose can exist in only one molecular configuration, it is one of the most readily crystallizable sugars. Invert sugar has a much greater tendency to remain in solution. In the manufacture of jelly and candy and in the canning of fruit, crystallization of the sugar is undesirable. Therefore, conditions leading to the hydrolysis of sucrose are employed in these processes. Since fructose is sweeter than sucrose, the hydrolysis adds to the sweetening effect. Bees carry out this reaction during their production of honey.

The average American consumes about 100 lb of sucrose every year. Much of it is ingested in soft drinks, presweetened cereals, and other highly processed foods. The widespread use of sucrose has generated much adverse publicity. Various health magazines have reported that excess sugar causes cancer, heart disease, migraine headaches, hyperkinetic children, obesity, and tooth decay. Only the latter two claims have been substantiated. As we shall see in Chapter 26, carbohydrates are converted to fat when the caloric intake exceeds the body's requirements. Sucrose does cause tooth decay by serving as part of the plaque that sticks to a tooth. The bacteria contained within the plaque use sucrose both as an adhesive and as a food source. We shall see in Chapter 24 that bacteria

Chocolate-Covered Cherries

Figure 19.6
The hydrolysis of sucrose is an essential reaction in the production of chocolate-covered cherries.

If you are a devotee of chocolate-covered cherries (Figure 19.6), you may wonder how the manufacturer makes the liquid-filled cherry center liquid without making a hole in the chocolate covering. The secret of liquefying the center depends on a chemical reaction that takes place after the candy is made. The process uses an enzyme, invertase, to convert sucrose into invert sugar.

The cherries are coated with a sugary paste containing invertase before they are dipped in chocolate. The paste hardens, the cherries (coated with the sugary paste) are dipped in chocolate, and then they are stored for one to two weeks. During this storage period, the invertase converts sucrose into a 1:1 ratio of glucose and fructose (invert sugar). In effect, the outer part of the cherry liquefies in its own syrup. It also pleases the sweet tooth of the consumer because one of the sugars formed, fructose, is sweeter than sucrose.

The preparation of the chocolate for the coating is also important. It must be an optimum particle size to please the palate. If the particles are too small, the chocolate will feel slimy when eaten; if too large, it will feel gritty.

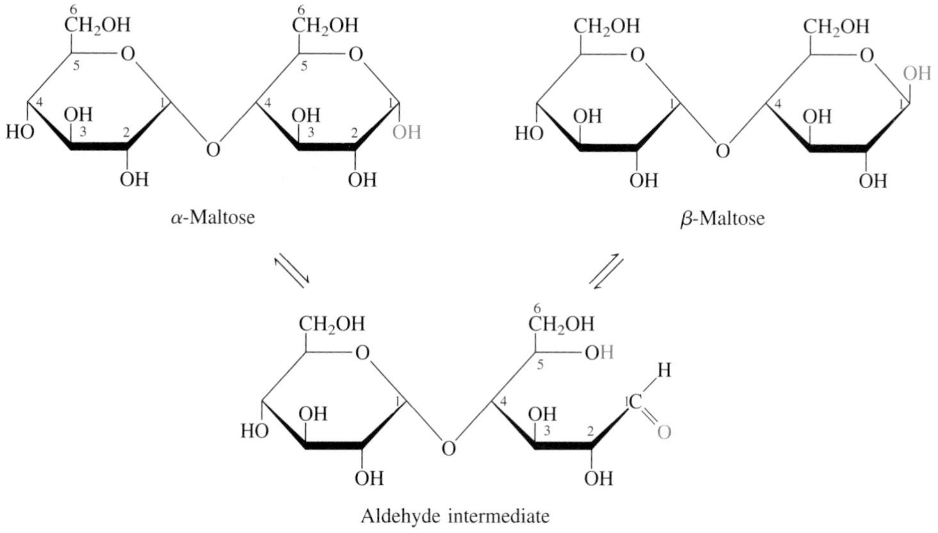

Figure 19.7
Equilibrium mixture of maltose isomers.

α-Maltose

β-Maltose

Aldehyde intermediate

can decompose sugar to lactic acid, and this acid corrodes the teeth and leads to the destruction of the gums. The best way to remove plaque is by daily flossing, and the amount of plaque formed can be decreased by reducing your intake of sucrose.

Some food faddists claim that raw sugar is much better for you than refined sugar. Raw sugar does contain a few trace minerals, but hardly enough to make it a lot more desirable than refined sugar. People in the United States probably consume too much sugar—whether raw or refined.

Maltose

Maltose occurs free, to a limited extent, in sprouting grain. Its major source, however, is the partial hydrolysis of starch. In the manufacture of beer, maltose is liberated by the action of malt (germinating barley) on starch, and for this reason it is often referred to as *malt sugar*. Maltose is about 30% as sweet as sucrose.

Maltose is a reducing sugar, and it exhibits mutarotation. An equilibrium mixture of the maltose isomers is highly dextrorotatory, $[\alpha] = +136$ (Figure 19.7). When maltose is hydrolyzed, either enzymatically or by means of an acid catalyst, two molecules of D-glucose are produced. The formula of maltose must therefore incorporate two glucose molecules in such a way that one free hemiacetal hydroxyl group exists. The glucose units in maltose are joined in a *head-to-tail* fashion through an alpha linkage from C-1 of one glucose molecule to C-4 of the second glucose molecule (that is, an **α-1,4-glycosidic linkage**).

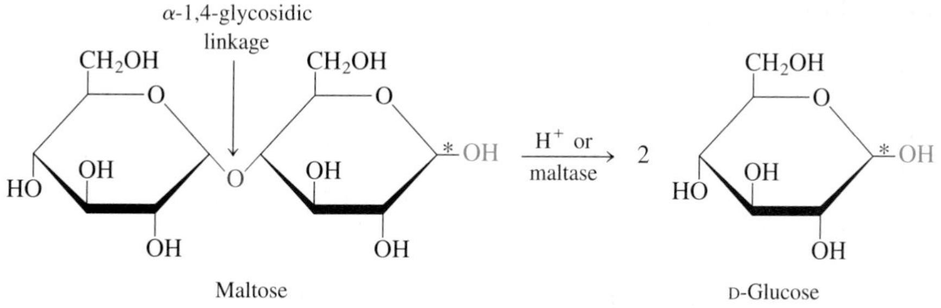

Maltose

D-Glucose

(*We use this convention for writing the hydroxyl group on the hemiacetal carbon when we do not wish to specify either the α or the β isomer.)

Lactose

Lactose is known as *milk sugar* because it occurs in the milk of humans, cows, and other mammals. Human milk contains about 7.5% lactose, whereas cow's milk, which is not as sweet, contains about 4.5% lactose. Unlike most carbohydrates, which are plant products, lactose is one of the few carbohydrates associated exclusively with the animal kingdom. (The biosynthesis of lactose is confined to the mammary tissue.) It is produced commercially from whey, which is obtained as a by-product in the manufacture of cheese. Lactose is one of the lowest ranking sugars in terms of sweetness (about one-sixth as sweet as sucrose). It is a reducing sugar and exhibits mutarotation. An equilibrium mixture of lactose has a specific rotation of +54. The alpha form of the sugar is of commercial importance as an infant food and in the production of penicillin. Drug dealers use lactose to "cut" their heroin and thus increase their profits.

Lactose is composed of one molecule of D-galactose and one molecule of D-glucose joined by a **β-1,4-glycosidic bond.** The two monosaccharides are obtained from lactose by acid hydrolysis or by catalytic action of the enzyme lactase.

Yeasts can metabolize sucrose and maltose but not lactose because they do not contain lactase. Certain bacteria can metabolize lactose, and they form lactic acid as one of the products. As we shall see (Section 21.13), this is responsible for the "souring" of milk.

19.8
POLYSACCHARIDES

The polysaccharides are the most abundant carbohydrates in nature. They serve as reserve food substances and as structural components of plant cells. Polysaccharides are high molecular weight (25,000–15,000,000 g/mol) polymers of monosaccharides joined together by glycosidic linkages. Biochemically, the three most significant polysaccharides are starch, glycogen, and cellulose. They are also referred to as *homopolymers* because each of them yields only one type of monosaccharide (D-glucose) upon complete hydrolysis. (Inulin, another homopolymer, is found in the tubers of the Jerusalem artichoke and is composed entirely of fructose units.) *Heteropolymers* may contain sugar acids, amino sugars, or noncarbohydrate substances. They are very common in nature (gums, pectins, hyaluronic acid) but are not discussed in this text. The polysaccharides are nonreducing carbohydrates, are not sweet tasting (probably because of their limited solubility), and do not undergo mutarotation.

Lactose makes up about 40% of an infant's diet during the first year of life. Infants and small children have one form of lactase within their small intestine. However, adults have a less active form of the enzyme, and about 70% of the world's adult population (especially Africans and Asians) have some lactase deficiency. People who suffer from **lactose intolerance** are unable to digest the sugar found in milk. For some people, the inability to synthesize sufficient enzyme increases with age. Up to 20% of the U.S. population suffers from lactose intolerance. Some of the unhydrolyzed lactose passes into the intestine, and its presence tends to draw water from the interstitial fluid into the intestinal lumen by osmosis. At the same time, intestinal bacteria may act on the lactose to produce organic acids and gases. The intake of water, plus the bacterial decay products, leads to the abdominal distention, cramps, and diarrhea that are symptoms of the condition. Symptoms disappear completely if milk is excluded from the diet. When milk is cooked or fermented, the lactose is at least partially hydrolyzed. People with lactose intolerance may still be able to enjoy cheese, yogurt, or cooked foods containing milk with little or no problem. The most common treatment of lactose intolerance is with lactase preparations (e.g., Lactaid), which are available in liquid and tablet form. These are taken orally with dairy foods.

Starch

Starch is the most important source of carbohydrates in the human diet. (It accounts for more than 50% of our carbohydrate intake.) Starch occurs in plants in the form of granules. Starch granules are particularly abundant in seeds (especially the cereal grains) and tubers, where they serve as a storage form of carbohydrate. The breakdown of starch to glucose nourishes the plant during periods of reduced photosynthetic activity. We often think of potatoes as a ''starchy'' food, yet other plants contain a much greater percentage of starch (e.g., potatoes 15%, wheat 55%, corn 65%, and rice 75%).

Starch is a mixture of two polymers, **amylose** and **amylopectin,** which can be separated from each other by physical and/or chemical methods. Natural starches consist of about 20–30% amylose and 70–80% amylopectin. Amylose is a straight-chain polysaccharide composed entirely of D-glucose units joined by α-1,4-glycosidic linkages, as in maltose. Thus, amylose might be thought of as either polymaltose or polyglucose.[5] There may be 60–300 glucose units per chain.

Amylose   Repeating unit; $n = 30$–150

[5] Experimental evidence indicates that the molecule is not a straight chain of glucose units. Rather, it is coiled like a spring with six glucose monomers per turn (Figure 19.8a). When coiled in this fashion, amylose has just enough room in its core to accommodate an iodine molecule. The characteristic blue-violet color that starch gives when treated with iodine is due to the formation of the amylose–iodine complex. The test is sensitive enough to detect even minute amounts of starch in solution.

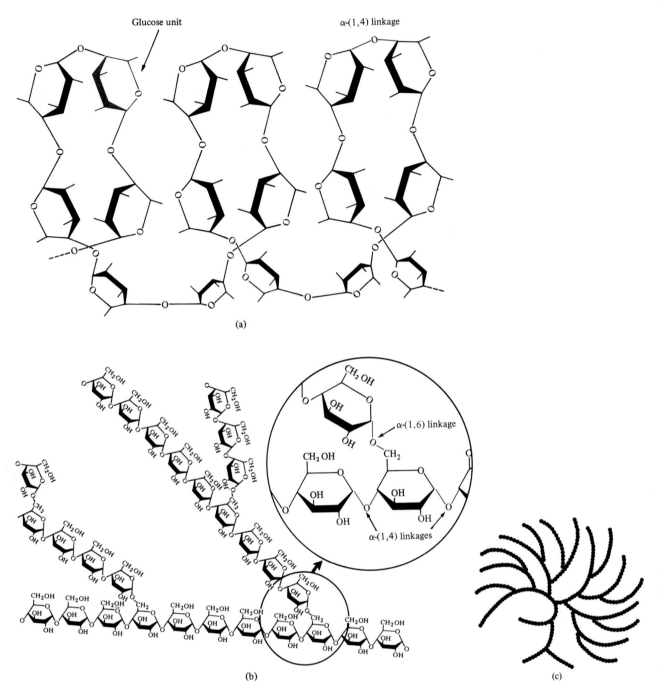

Glucose unit

α-(1,4) linkage

(a)

α-(1,6) linkage

CH₂OH

α-(1,4) linkages

(b)

(c)

Figure 19.8

(a) The conformation of the amylose chain. (b) Branch points of amylopectin. (c) Branched array of glucose units in amylopectin or glycogen.

Amylopectin is a branched-chain polysaccharide composed of glucose units that are linked primarily by α-1,4-glycosidic bonds but have occasional **α-1,6**-glycosidic linkages, which are responsible for the branching. It has been estimated that there may be 300–6000 glucose units in amylopectin and that branching occurs about once every 25–30 units (Figure 19.8b). The helical structure of amylopectin is disrupted by the branching of the chain, so instead of the deep blue-violet color amylose gives with iodine, amylopectin produces a less intense reddish brown.

Commercial starch is a white powder. The complete hydrolysis of starch (amylose and amylopectin) yields, in three successive stages, dextrins, maltose, and glucose. Dextrins are glucose polysaccharides of intermediate size. The shine and stiffness imparted to clothing by starch are due to the presence of dextrins formed when the clothing is ironed. Because of their characteristic stickiness upon wetting, dextrins are used as adhesives on stamps, envelopes, and labels and as pastes and mucilages. Since dextrins are more easily digested than starch, they are extensively used in the commercial preparation of infant foods (Dextrimaltose). A dried mixture of dextrins, maltose, and milk is used in the preparation of malted milk. Starch can be hydrolyzed by heating in the presence of dilute acid. In the human body, it is degraded sequentially by several enzymes known collectively as amylase.

$$\text{Starch} \xrightarrow[\text{amylase}]{H^+, \Delta \text{ or}} \text{Dextrins} \xrightarrow[\text{amylase}]{H^+, \Delta \text{ or}} \text{Maltose} \xrightarrow[\text{maltase}]{H^+, \Delta \text{ or}} \text{Glucose}$$

The symbol Δ is often used to indicate that the reaction requires heat.

Glycogen

Glycogen, often called animal starch, is the reserve carbohydrate of animals. Practically all mammalian cells contain some glycogen for the purpose of storing carbohydrate. However, it is especially abundant in the liver (4–8% per weight of tissue) and in skeletal muscle cells (0.5–1.0%). In liver and muscle tissue, it is arranged in granules (Figure 19.9). These granules are clusters of small particles. When fasting or during periods of starvation, animals draw upon these glycogen reserves to obtain the glucose needed to maintain a state of metabolic balance.

In terms of structure, glycogen is quite similar to amylopectin, but it is more highly branched and its branches are shorter (8–12 glucose units in length). When treated with iodine, glycogen gives a reddish brown color. Glycogen can be broken down into its D-glucose subunits by acid hydrolysis or by means of the same enzymes that attack starch. In animals, the enzyme phosphorylase catalyzes the breakdown of glycogen into phosphate esters of glucose (see Section 24.5).

These percentages may be misleading. Although the percent of glycogen is higher in the liver, the much greater mass of skeletal muscle stores a greater total amount of glycogen. About 70% of the total glycogen in the body is stored in muscle cells.

Cellulose

Cellulose is a fibrous carbohydrate found in all plants; it is the structural component of plant cell walls.[6] Since the Earth is covered with vegetation, cellulose is the most abundant of all carbohydrates, accounting for over 50% of all the carbon found in the vegetable

[6] Different linkages between monosaccharide units result in different three-dimensional forms for cellulose and starch. For example, cellulose in the cell walls of plants is arranged in **fibrils**—bundles of parallel chains. The fibrils, in turn, are arranged parallel to each other in each layer of the cell wall (Figure 19.10). Alternate layers have the fibrils running in opposite directions, imparting great strength to the wall.

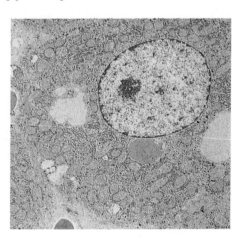

Figure 19.9
Electron micrograph of glycogen granules in a liver cell of a rat.

kingdom. Cotton fibrils and filter paper are almost entirely cellulose (about 95%), wood is about 50% cellulose, and the dry weight of leaves is about 10–20% cellulose. From an industrial and economic standpoint, cellulose is the most important of all the carbohyrates. The largest use of cellulose is in the manufacture of paper and paper products. Although there is increasing use of synthetic fibers, rayon (made from cellulose) and cotton account for over 70% of textile production.

Like amylose, cellulose is a linear polymer of glucose. It differs, however, in that the glucose units (about 2000–3000) are joined by **β-1,4**-glycosidic linkages. The linear nature of the cellulose chains allows a great deal of hydrogen bonding between hydroxyl groups on adjacent chains. As a result, the chains are closely packed into fibers, and there is little interaction with water or with any other solvent. Cotton and wood, for example, are completely insoluble in water and have considerable mechanical strength. Since there is no helical structure, cellulose does not bind to iodine to give a colored product.

$$CH_2OH \qquad Cellulose$$

Cellulose

Cellulose yields D-glucose upon complete acid hydrolysis, yet humans (and all vertebrates) cannot utilize cellulose as a source of glucose. We can eat potatoes, but we can't eat grass. Our digestive juices lack enzymes that can hydrolyze the β-glycosidic linkages found in cellulose. Many microorganisms and herbivorous animals (cows, horses, sheep) can digest cellulose. These higher animals can do so only because there are microorganisms in their digestive tracts whose enzyme (cellulase) catalyzes cellulose hydrolysis. Termites also contain cellulase-secreting microorganisms and thus can subsist on a wood diet. This once again demonstrates the extreme stereospecificity of biochemical processes.

Figure 19.10

Electron micrograph of the cell wall of an alga. The wall consists of successive layers of cellulose fibers in parallel arrangement.

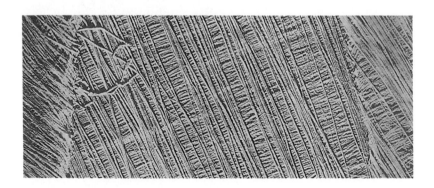

Dietary Fiber

A topic that has aroused considerable interest among nutritionists as well as the general public is the subject of fiber (roughage or bulk) in the diet. Some studies in the 1970s greatly increased public interest in fiber. First, it was noted that people in the developed countries are much more likely to get cancer of the colon than those in the developing nations. Also, people in the developed countries eat diets rich in highly processed, low-bulk foods, while those in more "primitive" areas eat high-fiber diets. High-fiber diets lead to frequent and robust bowel movements. In addition to providing bulk, the fiber absorbs a lot of water, leading to softer stools. Low-bulk diets result in less frequent bowel action, with high retention times for feces in the colon.

Bacteria act upon the materials in the colon. With a high-fiber diet, the materials seldom remain in the colon for more than a day. With a low-bulk diet, the retention time can be as long as 3 days, allowing for prolonged bacterial activity that produces a high level of carcinogenic and/or other harmful chemicals. Some scientists believe that a low-fiber diet (less than 20 g/day) is associated with a higher incidence of colon cancer, diverticulosis, coronary heart disease, atherosclerosis, gallstones, and hemorrhoids. Other scientists dispute some of these claims. They point out that a high-fiber diet (about 100 g/day) is a largely vegetarian diet, and thus individuals consume relatively little meat and saturated fat (see Chapter 20).

A recent proposal is that a high-fiber diet is useful in the treatment of diabetes (see Section 24.4). It has long been known that diabetics ought to avoid ingestion of rapidly digested sugars such as glucose and sucrose and eat only carbohydrates that are slowly digested (e.g., starches high in amylopectin). The presence of fiber in the diet reduces the rate of absorption of glucose, and therefore the peak blood sugar concentration is lowered.

In the past, nutritionists were concerned primarily with the total quantity of dietary fiber present in typical servings of foods. Now, they are becoming more and more concerned about the percentages of the specific components of dietary fiber in foods. Part of the reason for this changing concern is the growing recognition that the various components of dietary fiber act in different ways in the body.

Wheat bran is about 90% cellulose. It and related materials are not dissolved by water or digestive juices. This type of fiber, called *insoluble fiber*, passes unchanged into the colon. In contrast, the fiber in oat bran and rice bran is made up in large part of hemicellulose, a polymer composed of five-carbon sugars. Polymers of hemicellulose tend to be much more soluble in water. Pectins, found in fruits, and guar gum, found in beans, also are made up largely of soluble fibers.

Although it has no nutritive value, cellulose makes up the greater part of dietary fiber. The fibrous portions of plants (stems, peels, and seeds) are rich in fiber. Bran, celery, beans, apples, raspberries, and figs are good sources of dietary fiber.

Since so much of the carbon in the biosphere exists as cellulose, considerable research has gone into converting it to glucose or some other form of food for humans.

There are enzymes that will degrade cellulose, but to do so efficiently requires some pretreatment of the fibers. Perhaps some day, through the ingenuity of research chemists, we will be able to eat grass and straw—after proper conversion, of course. For the moment we must continue to depend on reactions such as

$$\text{Cellulose} + \text{Cow} \longrightarrow \text{Carbohydrates, Proteins, and Fats}$$
$$\text{(grass)} \qquad\qquad\qquad \text{(milk, meat, butter)}$$

EXERCISES

1. Define the following terms.
   a. triose   **b.** aldose
   c. hexose   **d.** disaccharide
   e. polysaccharide   **f.** aldopentose
   g. ketotetrose   **h.** invert sugar
   i. mutarotation   **j.** glycosidic linkage

2. Draw formulas for D-glyceraldehyde and L-glyceraldehyde. What do the prefixes mean?

3. Specify whether each of the following is a D-sugar or an L-sugar.

a. 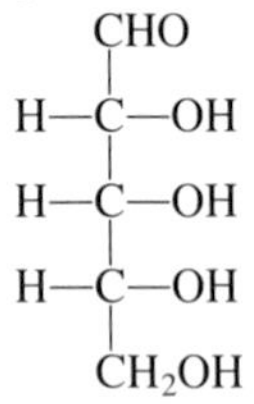
```
    CHO
     |
 H—C—OH
     |
 H—C—OH
     |
 H—C—OH
     |
   CH₂OH
```

b.
```
    CHO
     |
HO—C—H
     |
 H—C—OH
     |
HO—C—H
     |
   CH₂OH
```

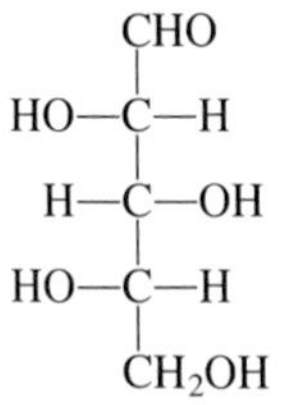

c.
```
    CHO
     |
 H—C—OH
     |
HO—C—H
     |
   CH₂OH
```

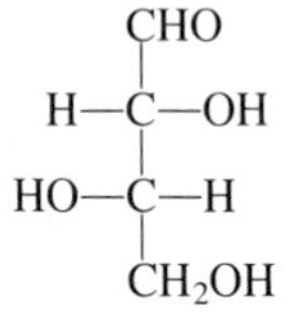

d. 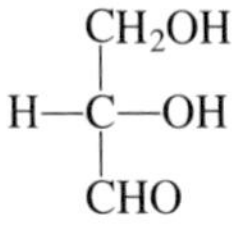
```
   CH₂OH
     |
 H—C—OH
     |
    CHO
```

4. Identify each sugar as an aldose or a ketose.
   a. D-glyceraldehyde   **b.** D-ribose
   c. D-deoxyribose   **d.** D-galactose
   e. D-glucose   **f.** D-fructose
   g. L-fructose   **h.** L-mannose

5. Identify each of the following as a triose, tetrose, pentose, or hexose.
   a. L-glucose   **b.** D-deoxyribose
   c. D-fructose   **d.** L-glyceraldehyde

6. Draw a ketotetrose.

7. Draw an aldoheptose.

8. From memory, draw formulas for the open-chain forms of D-glucose, D-mannose, D-galactose, and D-fructose.

9. Draw cyclic structures for α-D-glucose and β-D-fructose.

10. Knowing that mannose differs from glucose only in the configuration at the second carbon, draw the cyclic structure for α-D-mannose.

11. What is meant by a latent carbonyl group?

12. What is a reducing sugar?

13. For each of these abbreviated sugar formulas, indicate whether the glycosidic linkage is alpha or beta.

a.
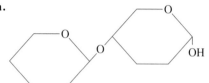

b.

c.

d.
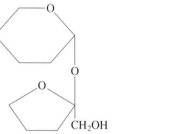

14. What is the orientation of the hydroxyl group at the hemiacetal carbon of structures **a–c** in Exercise 13?

15. Which of the structures shown in Exercise 13 is *not* a reducing sugar?

16. Why are (+)-glucose and (−)-fructose both classified as D sugars?

17. What monosaccharide(s) is(are) obtained from the hydrolysis of each of the following?
   a. starch   **b.** cellulose
   c. maltose   **d.** lactose
   e. sucrose   **f.** glycogen

18. How can it be shown that a solution of α-D-glucose exhibits mutarotation?

19. How does ribose differ from deoxyribose?

20. **a.** What purposes do starch and cellulose serve in plants?
b. What purpose is served by glycogen in animals?

21. What structural characteristics are necessary if a disaccharide is to be a reducing sugar? Draw the structure of a hypothetical nonreducing disaccharide composed of two aldohexoses.

22. Identify these sugars by their proper names.
a. blood sugar **b.** milk sugar
c. dextrose **d.** levulose
e. table sugar **f.** malt sugar

23. Melibiose is a disaccharide that occurs in some plant juices. Its structure is

What monosaccharide units are incorporated in melibiose?

24. What type of linkage (alpha or beta) joins the two rings of melibiose (Exercise 23)?

25. Is melibiose (Exercise 23) a reducing sugar? If so, circle the hemiacetal carbon and indicate whether the hydroxyl group is alpha or beta.

26. Gentiobiose is a disaccharide composed of two glucose units joined by a β-1,6-glycosidic linkage. Draw the structure of gentiobiose.

27. Raffinose is a trisaccharide (found in sugar beets) containing D-galactose, D-glucose, and D-fructose. The enzyme α-galactase catalyzes the hydrolysis of raffinose to galactose and sucrose. Draw the structure of raffinose. (The linkage from galactose to the glucose unit is α-1,6.)

28. The structure of a methyl glycoside of glucose is

a. Is C-1 in the alpha or the beta arrangement?
b. Is the compound a reducing sugar?
c. Will it give a positive test with Benedict's reagent?

29. Which of the following will give a positive Benedict's test?
a. L-galactose **b.** levulose
c. D-mannose **d.** malt sugar
e. cane sugar **f.** invert sugar

g. milk sugar **h.** ribose
i. inulin **j.** starch
k. cellulose **l.** glycogen

30. How do amylose and amylopectin differ? How are they similar?

31. How do amylose and cellulose differ? How are they similar?

32. How do amylopectin and glycogen differ? How are they similar?

33. List the reagents necessary for the following conversions.

34. D-Glucose can be oxidized at C-1 to form D-gluconic acid, at C-6 to yield D-glucuronic acid, and at both C-1 and C-6 to yield D-glucaric acid. Draw structures of these three oxidation products of glucose.

35. Draw the structure for β-D-glucose.

36. By reference to Exercise 35, draw the structure for β-D-galactose.

37. What monosaccharide units make up the disaccharide lactulose?

38. Are all the monosaccharides and disaccharides soluble in water? Explain.

39. The disaccharide cellobiose has two D-glucose units joined by a beta linkage. Draw the alpha form of cellobiose.

40. In the schematic below, Glc represents glucose. What substance is indicated?

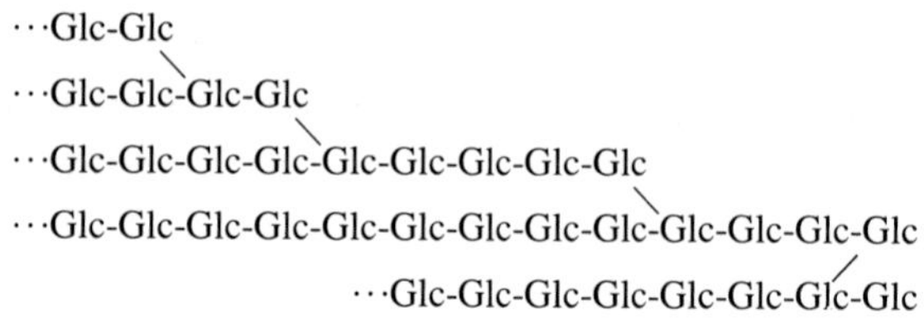

···Glc-Glc
        \
···Glc-Glc-Glc-Glc
              \
···Glc-Glc-Glc-Glc-Glc-Glc-Glc-Glc-Glc
                              \
···Glc-Glc-Glc-Glc-Glc-Glc-Glc-Glc-Glc-Glc-Glc-Glc-Glc
                                            /
         ···Glc-Glc-Glc-Glc-Glc-Glc-Glc-Glc

41. Xylulose is found in the urine of humans with pentosuria. Based on the structure below, classify xylulose as fully as possible.

$$CH_2OH$$
$$|$$
$$C=O$$
$$|$$
$$H-C-OH$$
$$|$$
$$HO-C-H$$
$$|$$
$$CH_2OH$$

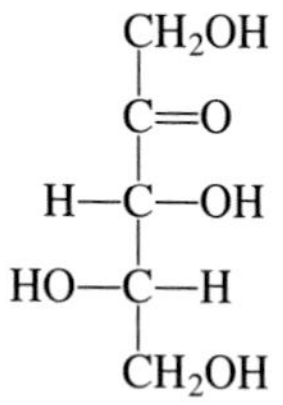

42. Erythrulose can be prepared from D-fructose. Based on the structure below, classify erythrulose as fully as possible.

$$CH_2OH$$
$$|$$
$$C=O$$
$$|$$
$$HO-C-H$$
$$|$$
$$CH_2OH$$

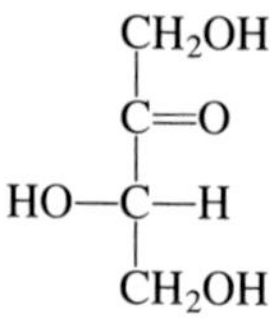

Chapter 20
LIPIDS

A chemist studying a soap film. Soaps are produced in great quantities from lipids.

The food we eat is divided into three primary groups: the carbohydrates (Chapter 19), the proteins (Chapter 21), and the lipids (which we discuss in this chapter). The best-known lipids are the fats. These compounds, in our affluent society, often occupy the lowest estate among the three classes of foods. People dieting to lost weight frequently try to eliminate fats from their diet. Gram for gram, fats pack about twice the caloric content of carbohydrates. While that may be bad news for the dieter, it also says something about the efficiency of nature's designs. The body has a limited capacity for storing carbohydrates. It can tuck away a bit of glycogen in the liver or in muscle tissue, but carbohydrates, primarily in the form of glucose, are meant to serve the body's immediate energy needs. If we intend to store energy reserves, then the more energy we can pack into a given space, the better off we are. The oxidation of fats supplies about 9 kcal/g, whereas the oxidation of carbohydrates supplies only 4 kcal/g (see Chapter 26). The body, an efficient organism, is geared to store fats, and its capacity for doing so is astounding. Most of us store enough energy as fats to last a month or so, but there is a recorded instance of a man weighing 486 kg. If all that energy were stored as carbohydrate, he would have weighed a ton or more.

The body's ability to store fat may elicit from you feelings of disgust or despair rather than awe. But a quick summary of the other functions of lipids in the body may provide a

501

more positive picture of these essential compounds. They play an important role in brain and nervous tissue. Fats serve as protective padding and insulation for vital organs. Without fats in our diets, we'd be deficient in the fat-soluble vitamins, A, D, E, and K. Most important, lipids make up the major part of the membranes of each of the 10 trillion cells in our bodies.

20.1
WHAT IS A LIPID?

Of the three types of foodstuffs, two are classified by functional groups. As Chapter 19 states, carbohydrates are polyhydroxy aldehydes or ketones. The proteins, as we shall soon see, are polyamides. But lipids are not poly anything in particular. They tend to be esters or compounds that can form esters, but that takes in a lot of territory and doesn't even hint at the wide variation in structure found among the lipids.

What makes a lipid a lipid is its solubility. Considering that water is the major solvent in living systems and that reactions of physiological importance tend to take place in aqueous solutions, it is not surprising that insolubility in water should be considered a noteworthy feature when found in important body constituents. Lipids are soluble in relatively nonpolar organic solvents such as carbon tetrachloride, hexane, and diethyl ether (the so-called fat solvents), but they are generally insoluble in water.

Compounds isolated from body tissues are classified as **lipids** if they are more soluble in organic solvents than in water. Included in this category are esters of glycerol and the fatty acids (or phosphoric acid), steroids such as cholesterol, compounds that incorporate sugar units or a complicated amino alcohol called sphingosine, and compounds called prostaglandins, which some regard as potential ''miracle'' drugs. Because of this broad variation in structure, we can't present a general formula for lipids. We shall, instead, consider one subclass at a time and try to point out similarities and differences in structure as we go along. Figure 20.1 indicates a few of the diverse substances that provide dietary lipids.

Figure 20.1
Cream, butter, margarine, cooking oils, and food fried in fat are rich in fats.

Table 20.1

Some Fatty Acids in Natural Fats

Abbreviated Formula[a]	Condensed Structure	Melting Point (°C)	Name	Source
C_3H_7COOH	$CH_3CH_2CH_2COOH$	−8	Butyric acid	Butter
$C_5H_{11}COOH$	$CH_3(CH_2)_4COOH$	−3	Caproic acid	Butter
$C_7H_{15}COOH$	$CH_3(CH_2)_6COOH$	−17	Caprylic acid	Coconut oil
$C_9H_{19}COOH$	$CH_3(CH_2)_8COOH$	31	Capric acid	Coconut oil
$C_{11}H_{23}COOH$	$CH_3(CH_2)_{10}COOH$	44	Lauric acid	Palm kernel oil
$C_{13}H_{27}COOH$	$CH_3(CH_2)_{12}COOH$	54	Myristic acid	Oil of nutmeg
$C_{15}H_{31}COOH$	$CH_3(CH_2)_{14}COOH$	63	Palmitic acid	Palm oil
$C_{17}H_{35}COOH$	$CH_3(CH_2)_{16}COOH$	70	Stearic acid	Beef tallow
$C_{17}H_{33}COOH$	$CH_3(CH_2)_7CH{=}CH(CH_2)_7COOH$	13	Oleic acid	Olive oil
$C_{17}H_{31}COOH$	$CH_3(CH_2)_3(CH_2CH{=}CH)_2(CH_2)_7COOH$	−5	Linoleic acid	Soybean oil
$C_{17}H_{29}COOH$	$CH_3(CH_2CH{=}CH)_3(CH_2)_7COOH$	−11	Linolenic acid	Fish oil
$C_{19}H_{31}COOH$	$CH_3(CH_2)_4(CH{=}CHCH_2)_4CH_2CH_2COOH$	−50	Arachidonic acid	Liver

[a] Saturated fatty acids have the general formula $C_nH_{2n+1}COOH$; unsaturated fatty acids are of the form $C_nH_{2n-1}COOH$, $C_nH_{2n-3}COOH$, $C_nH_{2n-5}COOH$, and so on.

20.2
FATTY ACIDS

Fatty acids are so named because they are structural components of fats. Chemically, fatty acids are generally long-chain carboxylic acids. More than 70 have been identified in nature. Nearly all contain an even number of carbon atoms. Few are branched. Some, the unsaturated fatty acids, contain one or more double bonds. Free fatty acids are rare, occurring in nature in only small amounts. The fats and other lipids, however, provide a reservoir from which the fatty acids can be obtained. Table 20.1 lists some common fatty acids and an important source of each of them.

The normal tetrahedral bond angles of carbon require that the chain of saturated fatty acid molecules assume a zigzag configuration (Figure 20.2a), but the molecule viewed as a whole is relatively straight. (Each angle in these zigzag formulas represents one carbon atom in the fatty acid chain.) Such molecules fit rather nicely into a crystal lattice (Figure 20.2b), a capability that gives these acids and the fats derived from them relatively high

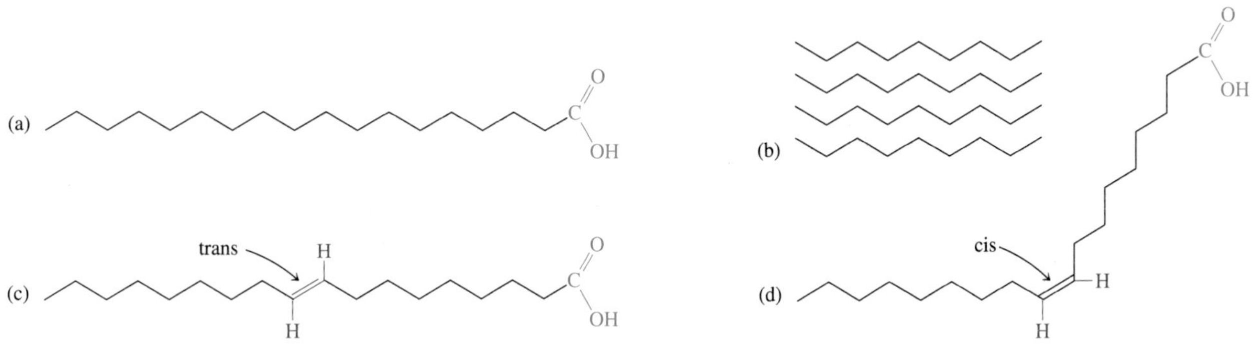

Figure 20.2

(a) A schematic representation of a stearic acid molecule. (b) These saturated acids stack nicely in a crystal lattice. (c) Elaidic acid (rarely found in nature) has a trans double bond and can also stack. (d) Oleic acid, with its cis double bond, will not fit neatly into a crystalline arrangement.

melting points. Unsaturated fatty acids occur almost always in the cis configuration. This results in a severe bend in the molecules (Figure 20.2d). These molecules don't stack neatly, so the van der Waals attractions between molecules are smaller. Consequently, the unsaturated fatty acids and unsaturated fats have lower melting points. Most are liquids at room temperature.

20.3
FATS AND OILS

Do not confuse the term *oil*, used here to refer to a particular group of lipids, with the hydrocarbon petroleum oils.

The systematic name for these esters is *triacylglycerols*, but they are commonly called by the old name, triglycerides.

Fats and oils are the most abundant lipids found in nature. Both types of compounds are called **triglycerides** because they are *esters* composed of *three fatty acids* joined to *glycerol*, a *trihydroxy alcohol*.

$$\text{Acid} \quad + \quad \text{Alcohol} \quad \longrightarrow \quad \text{Ester} \quad + \quad \text{Water}$$

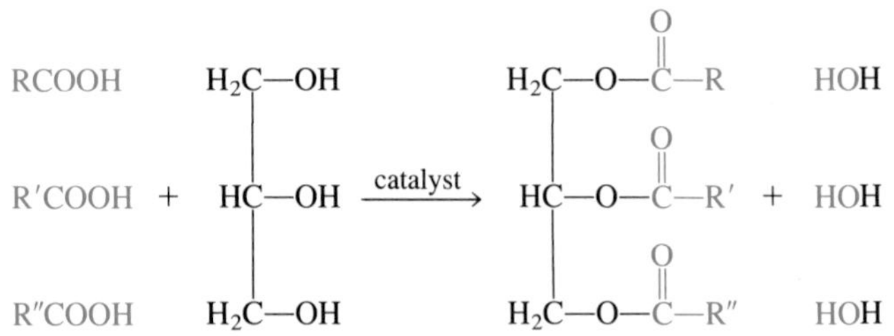

If all three hydroxyl groups of the glycerol molecule are esterified with the same fatty acid, the resulting ester is called a **simple triglyceride.** Although some simple triglycerides have been synthesized in the laboratory, they rarely occur in nature. All of the triglycerides obtained from naturally occurring fats and oils contain two or three different fatty acid components and are thus termed **mixed triglycerides.**

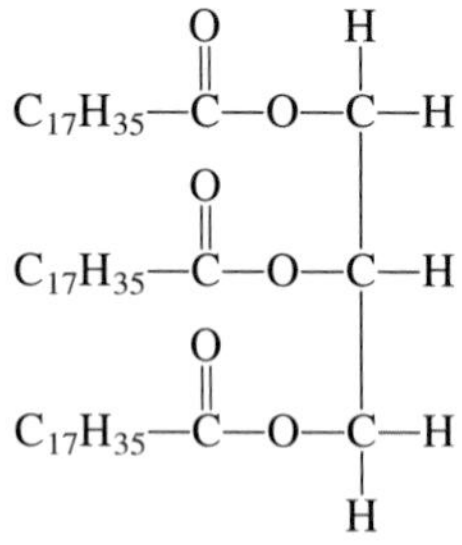

Glyceryl stearate
(Tristearin)
(a simple triglyceride)
mp 71 °C

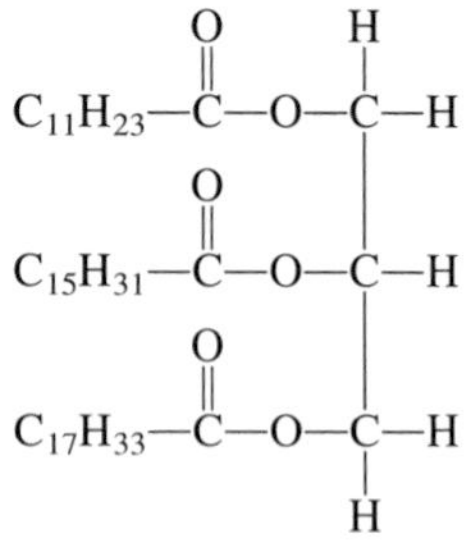

Glyceryl lauropalmitooleate
(a mixed triglyceride)

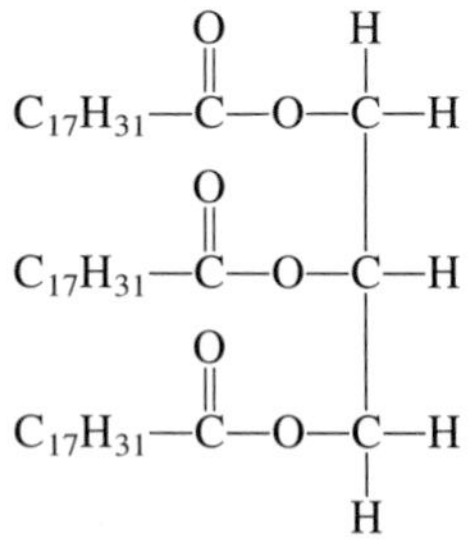

Glyceryl linoleate
(Trilinolein)
(a simple triglyceride)
mp 9 °C

No single formula can be written to represent the naturally occurring fats and oils because they are highly complex mixtures of molecules in which many different fatty acids are represented. Table 20.2 shows the fatty acid composition of some common fats and oils. Notice that there is a fairly wide range of values. The range is wide because the composition of lipids is variable and depends upon the plant or animal species involved as well as

Table 20.2
Fatty Acid Components of
Some Common Fats and
Oils

	Component Acids (%)[a]						
	Lauric (C_{12})	Myristic (C_{14})	Palmitic (C_{16})	Stearic (C_{18})	Oleic (C_{18})	Linoleic (C_{18})	Linolenic (C_{18})
Fats							
Butter	1–4	8–13	25–32	8–13	22–29	2–4	
Tallow (beef)		2–3	24–32	20–25	37–43	2–3	
Lard (hog)		1–2	25–30	12–16	40–50	3–8	
Edible Oils							
Coconut oil[b]	44–50	13–18	7–10	1–4	5–8	1–3	
Palm oil[c]		1–6	32–47	1–6	40–52	2–11	
Olive oil	0–1	0–2	7–20	2–3	53–86	4–22	
Peanut oil		0–1	6–11	3–6	40–65	17–38	
Cottonseed oil		0–3	17–23	1–3	23–44	34–55	
Corn oil		1–2	8–12	2–5	29–49	34–56	
Soybean oil		0–1	6–10	2–5	20–30	50–60	2–10
Safflower oil			6–7	2–3	12–14	75–80	0–2
Nonedible Oil							
Linseed oil		0–1	5–9	4–7	9–29	8–29	45–67

[a]Totals less than 100% indicate the presence of lower or higher acids in small amounts.
[b]Coconut oil is a highly saturated oil. It contains an unusually high percentage (53–70%) of the low-melting C_8, C_{10}, and C_{12} saturated fatty acids. Coconut oil is a liquid in the warmer, tropical climates, but at room temperature in the temperate zone it is a solid.
[c]Palm oil is highly saturated because of the large percentage of palmitic acid.

dietetic and climatic factors. For example, lard from corn-fed hogs is more highly saturated than lard from peanut-fed hogs. Linseed oil obtained from cold climates is more unsaturated than linseed oil from warm climates. Palmitic acid is the most abundant of the saturated fatty acids, and oleic acid is the most abundant unsaturated fatty acid. Unsaturated fatty acids predominate over saturated ones for most plants and animals.

We have seen that one classification of triglycerides is made on the basis of their physical states at room temperature. In general, a lipid is called a **fat** if it is a solid at 25 °C, and an **oil** if it is a liquid at the same temperature. (These differences in melting points reflect differences in the degree of unsaturation of the constituent fatty acids.) Furthermore, lipids obtained from animal sources are usually solids, whereas oils are generally of plant origin. Therefore we commonly speak of *animal fats* and *vegetable oils*. Coconut and palm oils, which are highly saturated, and fish oils, which are relatively unsaturated, are notable exceptions to the general rule.

Fats are often classified according to the degree of unsaturation of the fatty acids they incorporate. *Saturated* fatty acids contain no double bonds, *monounsaturated* fatty acids contain one double bond per molecule, and *polyunsaturated* fatty acids are those that have two or more double bonds. Saturated fats contain a high proportion of saturated fatty acids; the fat molecules have relatively few double bonds. Polyunsaturated fats (oils) incorporate mainly unsaturated fatty acids; these fat molecules have many double bonds.

Saturated fats have been implicated, along with cholesterol, a steroid (Section 20.10), in one type of *arteriosclerosis* (hardening of the arteries). There is a strong correlation between diets rich in saturated fats and incidence of the disease. It is this correlation that has led to a concern over the relative amounts of saturated and unsaturated fats in our diets. Advertisers who recommend that you buy corn oil margarine (prepared from relatively unsaturated vegetable oil) rather than butter (a relatively saturated animal fat) take advantage of this concern.

For reasons that are not fully understood, saturated fats interfere with the removal of cholesterol from the blood and thus tend to raise blood cholesterol levels, whereas unsaturated fats lower blood cholesterol.

20.4
PROPERTIES OF FATS AND OILS

As previously mentioned, the triglycerides may be either liquids or noncrystalline solids at room temperature. Contrary to popular belief, *pure* fats and oils are colorless, odorless, and tasteless. The characteristic colors, odors, and flavors associated with these lipids are imparted to them by foreign substances that have been absorbed by the lipids and are soluble in them. For example, the yellow color of butter is due to the presence of the pigment carotene; the taste of butter is a result of two compounds, diacetyl ($CH_3COCOCH_3$) and 3-hydroxy-2-butanone ($CH_3COCHOHCH_3$), that are produced by bacteria in the ripening of the cream. Fats and oils are lighter than water, having densities of about 0.8 g/cm^3. They are poor conductors of heat and electricity and therefore serve as excellent insulators for the body.

Fats and oils undergo a variety of chemical reactions; the most important is hydrolysis. Triglycerides are esters. They can be hydrolyzed in either acidic or basic media. Acid hydrolysis is of little importance, however, because it is difficult to dissolve fats in acidic media. Basic hydrolysis is of considerable importance in the making of soap and will be discussed in detail in Section 20.5. When we eat fats, they are hydrolyzed by enzymes (lipases). This process is discussed in Chapter 26.

The degree of unsaturation of a fat or an oil is usually measured in terms of the **iodine number.** Recall that halogens such as chlorine and bromine add readily to carbon–carbon double bonds (Section 13.13). Iodine also adds, but less readily.

$$\underset{H}{\overset{H}{>}}C=C\underset{}{\overset{H}{<}} + I_2 \longrightarrow -\overset{\overset{H}{|}}{\underset{\underset{I}{|}}{C}}-\overset{\overset{H}{|}}{\underset{\underset{I}{|}}{C}}-$$

The iodine number of a fat or an oil is the number of grams of iodine that will react with 100 g of the fat or the oil. The more double bonds in a lipid, the more iodine is required for the addition reaction; thus, a high iodine number means a high degree of unsaturation. Representative iodine numbers are given in Table 20.3. Notice the generally lower values for the animal fats (butter, tallow, lard) compared with those for the vegetable oils.

A large-scale commercial industry has been developed for the purpose of transforming vegetable oils into edible fats. The chemistry of this conversion process is essentially identical to the catalytic hydrogenation reaction that was described for the alkenes in Section 13.13.

$$\underset{H}{\overset{H}{>}}C=C\underset{}{\overset{H}{<}} + H_2 \xrightarrow{\text{Ni}} -\overset{\overset{H}{|}}{\underset{\underset{H}{|}}{C}}-\overset{\overset{H}{|}}{\underset{\underset{H}{|}}{C}}-$$

Margarine, a butter substitute, and vegetable shortening, a lard substitute, consist of vegetable oils that have been partially hydrogenated. (If all the bonds were hydrogenated, the product would become hard and brittle like tallow.) If reaction conditions are properly controlled, it is possible to prepare a fat with a desirable physical consistency (soft and pliable). In this manner, inexpensive and abundant vegetable oils (cottonseed, corn, soybean) are converted into oleomargarine and cooking fats (Crisco, for example—Figure 20.3). The consumer would get much greater unsaturation by using the oils directly, but

---

Butter is prepared from milk by churning its cream, which causes the fat globules to coalesce; the liquid that remains is called buttermilk.

Table 20.3
Typical Iodine Numbers for Some Fats and Oils[a]

Fat or Oil	Iodine Number
Coconut oil	8–10
Butter	25–40
Beef tallow	30–45
Palm oil	37–54
Lard	45–70
Olive oil	75–95
Peanut oil	85–100
Cottonseed oil	100–117
Corn oil	115–130
Fish oils	120–180
Soybean oil	125–140
Safflower oil	130–140
Sunflower oil	130–145
Linseed oil	170–205

[a] Most oils are from plant sources. Three fats and one oil, in color, come from animals.

Figure 20.3
Hydrogenation of an oil. A solid vegetable fat like Crisco is made by bubbling hydrogen through a vegetable oil (left).

most people would rather spread margarine than pour oil on their toast. Table 20.4 lists iodine numbers for butter and various kinds of margarines.

In the preparation of margarine, the partially hydrogenated oils are mixed with water, salt, and nonfat dry milk. Flavoring agents, coloring agents, and vitamins A and D are added to approximate butter. (Preservatives and antioxidants are also added.) The peanut oil in peanut butter has been partially hydrogenated to prevent the oil from separating out. Today, because of the possible connection between saturated fats and arterial disease (see Section 20.11), many people are cooking with the vegetable oils (especially safflower seed oil) rather than with the hydrogenated products.

On standing at room temperature in contact with moist air, fats and oils soon turn rancid. This rancidity, characterized by disagreeable odors, results from two reactions. Hydrolysis of the ester bonds releases volatile fatty acids. Butter, for example, yields foul-smelling butyric, caprylic, and capric acids. Microorganisms present in the air furnish the lipases that catalyze the process. Hydrolytic rancidity can easily be prevented by storing butter covered in a refrigerator.

Oxidation of the unsaturated fatty acid components by oxygen in the air also produces a variety of volatile, odorous compounds. The structural unit

$$\sim\sim CH{=}CH{-}CH_2{-}CH{=}CH\sim\sim$$

in linoleic and linolenic acids is readily oxidized. One particularly offensive product, formed by the cleavage of both double bonds, is a compound called malonaldehyde. The

During the hydrogenation of vegetable oils, an isomerization reaction produces some **trans** fatty acids (recall Figure 20.2c). Recent studies showed that these trans acids, like saturated fatty acids, raise cholesterol levels and increase the incidence of coronary heart disease. Consumers are now being told to use soft or liquid margarine and to reduce their total fat consumption.

Malonaldehyde

Food Product	Iodine Number	Comparative Unsaturation[a]
Butter	27	100
Margarines		
Hard type A	68	252
Hard type B	72	267
Hard type C	77	285
Soft type D	84	311
Soft type E	88	326
Liquid type F	90	333
Liquid type G	93	344

[a]Calculated by dividing the iodine number of the substance by the iodine number of butter and multiplying the result by 100.

Table 20.4
Iodine Numbers and Comparative Unsaturation Ratings of Various Margarines and Butter

stale, sweaty odor of the unwashed skin is due in large part to the oxidation of fats and oils excreted by the body.

Rancidity is a major concern of the food industry, and food chemists are continually seeking new and better substances to act as **antioxidants.** Such compounds are added in very small amounts (0.001–0.01%) to suppress rancidity. They have a greater affinity for oxygen than the lipid to which they are added and thus function by preferentially depleting the supply of absorbed oxygen. We shall see in Special Topic I that two of the vitamins (C and E) have antioxidant properties.

20.5
SOAPS

Animal fats are available in large quantities as a by-product of the meat-packing industry. The fatty acids and many other long-chain organic compounds are derived from these fats. The most important derivatives are soaps. The reaction that converts animal fats to soaps is called, logically, **saponification,** which we discussed originally as one of the reactions of esters in general (Section 16.9). Soap-making is one of the oldest organic syntheses known, second only to the production of ethyl alcohol (fermentation). Even though cave people, the Phoenicians (600 B.C.), and the Romans made soap from animal fat and wood ash, the widespread production of soap did not occur until the 1700s.

The old method of soap production consisted of treating molten tallow (the fat of cattle and sheep) with a slight excess of alkali in large open kettles. The mixture was heated, and steam was bubbled through it. After the saponification process was completed, the soap was precipitated by the addition of sodium chloride and then filtered and washed several times with water. It was then dissolved in water and reprecipitated from the aqueous solution by the addition of more sodium chloride. The glycerol was recovered from the aqueous wash solutions.

Today most soaps are prepared by a continuous process wherein triglycerides (frequently tallow and/or coconut oil) are hydrolyzed by water under high pressures and temperatures (700 lb/in.^{2} and 200 °C). Sodium carbonate is used to neutralize the fatty acids.

$$\begin{array}{l} CH_2OOC(CH_2)_nCH_3 \\ | \\ CHOOC(CH_2)_nCH_3 \\ | \\ CH_2OOC(CH_2)_nCH_3 \end{array} \xrightarrow[\substack{\text{heat,} \\ \text{pressure}}]{H_2O} \text{Glycerol} + 3\ CH_3(CH_2)_nCOOH \xrightarrow{Na_2CO_3} 3\ CH_3(CH_2)_nCOO^-\ Na^+$$

Fatty acid   Sodium salt of a fatty acid (a soap)

The crude soap is used as industrial soap without further processing. Pumice or sand may be added to produce scouring soap. Other ingredients, such as dyes, perfumes, and antiseptics, are added to produce colored soaps, fragrant soaps, and deodorizing soaps, respectively. If air is blown through molten soap, a floating soap is produced. Such a soap is not necessarily purer than other soaps; it merely contains more air. Ordinary soap is a mixture of the sodium salts of various fatty acids. Potassium soaps (soft soap) are more expensive but produce a finer lather and are more soluble. They are used in liquid soaps, shampoos, and shaving creams. Tincture of green soap is an alcoholic solution of a potassium soap that is commonly used in hospitals.

Dirt and grime usually adhere to skin, clothing, and other surfaces because they are combined with greases and oils—body oils, cooking fats, lubricating greases, or a variety of similar substances—that act a little like sticky glues. Since oils are not miscible with

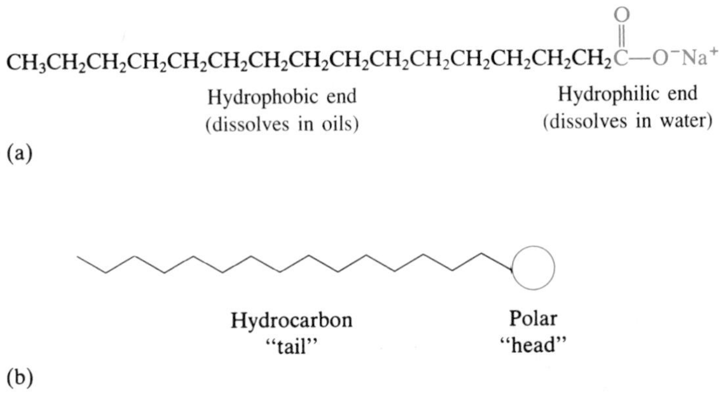

$$CH_3CH_2CH_2CH_2CH_2CH_2CH_2CH_2CH_2CH_2CH_2CH_2CH_2CH_2CH_2\overset{\overset{\displaystyle O}{\|}}{C}-O^-Na^+$$

Hydrophobic end   Hydrophilic end
(dissolves in oils)   (dissolves in water)

(a)

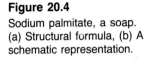

Figure 20.4
Sodium palmitate, a soap.
(a) Structural formula, (b) A
schematic representation.

Hydrocarbon   Polar
"tail"   "head"

(b)

water, washing with water alone does little good. Soap molecules have a dual nature. One end is ionic and dissolves in water. The other end is like a hydrocarbon and dissolves in oils (Figure 20.4). Often, the ionic end is referred to as **hydrophilic** (water-soluble), and the nonpolar end as **hydrophobic** (repelled by water). If we represent the ionic end of the molecule as a circle and the hydrocarbon end as a zigzag line, we can illustrate the cleansing action of soap schematically (Figure 20.5). The hydrocarbon "tails" dissolve in the oil. The ionic "heads" remain in the aqueous phase. In this manner, the oil is broken into tiny droplets and dispersed throughout the solution. The droplets don't coalesce because of the repulsion of the charged groups (the carboxyl anions) on their surfaces. The oil and water form an emulsion, with soap acting as the emulsifying agent. With the oil no longer "gluing" it to the surface, the dirt can be easily removed.

For cleaning clothes and for many other purposes, soap has been largely replaced by synthetic detergents because soaps have two rather serious shortcomings. One of these is that in acidic solutions they are converted to free fatty acids.

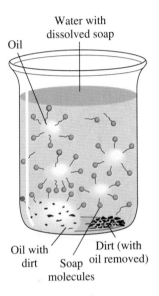

Water with
dissolved soap
Oil

Oil with   Dirt (with
dirt   Soap oil removed)
molecules

$$CH_3(CH_2)_{16}COO^- Na^+ + H^+ \longrightarrow CH_3(CH_2)_{16}COOH + Na^+$$

A soap   A fatty acid

The fatty acids, unlike soap, don't have an ionic end. Lacking the necessary dual nature, they can't emulsify the oil and dirt; that is, they do not exhibit any detergent action. What is more, these fatty acids are insoluble in water and thus precipitate as a greasy scum. To counteract this lack of detergent action in acidic solution, various alkaline substances are added to laundry soap formulations to keep the pH high. These basic compounds include carbonates and silicates.

Figure 20.5
The action of soap in re-
moving dirt.

The second serious disadvantage of soap is that it doesn't work very well in hard water (Figure 20.6). Hard water is just water that contains certain metal ions, particularly magnesium, calcium, and iron ions. The soap anions react with these metal ions to form greasy, insoluble curds.

$$2\ CH_3(CH_2)_{16}COO^- Na^+ + Ca^{2+} \longrightarrow (CH_3(CH_2)_{16}COO^-)_2\ Ca^{2+} + 2\ Na^+$$

Soap (soluble)   Bathtub ring (insoluble)

These deposits make up the familiar bathtub ring. They leave freshly washed hair sticky and are responsible for the "telltale gray" of the family wash.

The term **detergent** is a rather general one meaning any cleansing agent. Soaps would fall under such a broad definition. However, the popular use of the word generally refers to synthetic detergents, also called **syndets.** Syndets have the desirable property of

Figure 20.6
Sudsing quality of hard water versus soft water. From left: detergent in hard water, soap in hard water, detergent in soft water, and soap in soft water. Note that the sudsing of detergent differs little in hard and soft waters. Note also the absence of sudsing and the formation of insoluble material when soap is used in hard water.

not forming precipitates with the ions of hard water. There are close to a thousand synthetic detergents commercially available in the United States, and worldwide production exceeds 25 million tons. Although most synthetic detergents are now made from petroleum, early ones were made from fats. Some, such as sodium lauryl sulfate (sodium dodecyl sulfate), still are. Sodium lauryl sulfate is widely used in specialty products such as toothpastes and shampoos. Sodium dodecylbenzenesulfonate is found in many laundry detergents.

$$CH_3(CH_2)_{11}O-\overset{\displaystyle O}{\underset{\displaystyle O}{\overset{\|}{\underset{\|}{S}}}}-O^-\ Na^+$$

Sodium lauryl sulfate
(in shampoos)

Hydrophobic   Hydrophilic

$$CH_3(CH_2)_{11}-\bigcirc-\overset{\displaystyle O}{\underset{\displaystyle O}{\overset{\|}{\underset{\|}{S}}}}-O^-\ Na^+$$

Sodium dodecylbenzenesulfonate
(in laundry detergents)

Environmental Effects of Synthetic Detergents

Many effective detergents do not foam in water. Experiments have proved that the degree of sudsing has very little to do with the efficiency of a detergent. However, the consumer has come to associate sudsing with cleaning ability, so manufacturers often add sudsing agents.

The large-scale use of synthetic detergents during the 20 years following World War II created a serious disposal problem. Soaps, which contain straight-chain alkyl groups, can be removed from wastewater through degradation by microorganisms in the soil (septic tanks) or in sewage treatment plants. Soaps are therefore said to be **biodegradable.** Many of the synthetic detergents could not be removed in this manner. The metabolism of microorganisms is adapted to the straight-chain alkyl groups found in soaps and natural fats but could not break down the highly branched analogs used in the early syndets. The synthetic detergents continued to foam and make suds, which clogged waste disposal plants, killed fish and wildlife by polluting streams, and even managed to make their way into city drinking water. Since 1966, all U.S. companies have been using straight-chain hydrocarbons in the production of syndets. Although this involves a greater expense, it seems to have alleviated some of the pollution problem.

A second environmental problem caused by detergents has not been solved. In their search for more effective cleansing agents, manufacturers have added "builders" to their detergents. Builders have little detergent effectiveness alone. Their functions are (1) to tie up metal ions, (2) to prevent soil from redepositing on clothes, (3) to maintain a proper level of alkalinity in the wash water, and (4) to add bulk to detergent formulations (i.e., to fill up the detergent box so that consumers believe they are purchasing a full box of detergent). Although a number of inorganic compounds (carbonates, bicarbonates, borates, silicates) have been used as builders, the phosphates are the most effective. When phosphates were first added to detergents, a typical detergent might have contained as much as 50% phosphate. Approximately half of the phosphate content of domestic sewage was contributed by detergents, the remainder being derived from human wastes. (Today, only about 25–30% of the phosphate in sewage comes from detergents.)

$$^{+}Na\ ^{-}O-\overset{\overset{O}{\|}}{\underset{\underset{O^-}{|}}{P}}-O-\overset{\overset{O^-\ Na^+}{|}}{\underset{\underset{O}{\|}}{P}}-O-\overset{\overset{O}{\|}}{\underset{\underset{O^-\ Na^+}{|}}{P}}-O^-\ Na^+$$
$$^{+}Na\ ^{-}O \qquad\qquad\qquad ^{+}Na\ ^{-}O$$

Sodium tripolyphosphate, $Na_5P_3O_{10}$
(most common phosphate builder)

As later chapters will show, phosphate is a nutrient required for plant (and animal) growth. It has thus been implicated as the principal cause of **eutrophication** of lakes and rivers. (Eutrophic derives from the Greek *eu*, well, and *trophos*, pertaining to nourishment.) Eutrophic lakes and rivers contain an overabundance of plants, especially algae (algal blooms), and decreased levels of oxygen (Figure 20.7). Only certain species of fish (carp, crappie, perch, bullhead) can live in such an environment. Moreover, the algae can produce chemicals that cause unpleasant tastes and smells in water and that in some cases are toxic to animals.

As a result of the environmental impact of eutrophication, many states and local communities have banned the sale of laundry detergents containing phosphates. The detergent industry has offered a variety of replacements; the most prominent are sodium carbonate and complex aluminosilicates called *zeolites*. The former acts by precipitating the calcium ions, thus softening the water. (However, the $CaCO_3$ precipitate appears to be harmful to automatic washing machines.)

$$Ca^{2+}\ +\ CO_3^{2-}\ \longrightarrow\ CaCO_3$$

The zeolites are perhaps the most promising of the substitutes. The zeolite anions trap calcium ions by exchanging them for their own sodium ions.

$$Ca^{2+}\ +\ Na_2Al_2Si_2O_7\ \longrightarrow\ 2\ Na^+\ +\ CaAl_2Si_2O_7$$

The calcium ions are held in suspension by the zeolites (rather than being precipitated). Further, zeolite solutions are not strongly basic and are therefore less likely than sodium carbonate and sodium silicate solutions to irritate the skin and eyes.

Figure 20.7
Eutrophication. Algae cover much of the lake surface; their natural growth is enhanced by the presence of phosphates from wastewater and fertilizer runoff.

20.6
WAXES

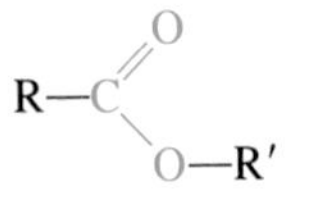

A wax
(a simple ester)

In everyday usage, the word *wax* refers to a substance that is hard when cold yet easily molded when warm. Familiar waxes include the mixture of alkanes called paraffin wax, synthetic polymers such as Carbowax, and carnauba wax, a mixture of esters. In chemistry, however, the term **wax** refers to esters formed from long-chain fatty acids and long-chain monohydroxy alcohols. (Household paraffin wax, which is a mixture of high molecular weight hydrocarbons, has wax-like properties but is not a wax.) The general formula for a wax, then, is the same as that of a simple ester. For a wax, however, R and R' are limited to alkyl groups containing a large number of carbon atoms.

Most natural waxes are mixtures of such esters. Many also contain free alcohols, hydrocarbons, and esters of diprotic acids, hydroxy acids, and diols. All have similar properties; they feel ''waxy,'' are insoluble in water, and melt at temperatures above body temperature (37 °C) and below the boiling point of water (100 °C).

Waxes are not as easily hydrolyzed as the triglycerides and therefore are useful as protective coatings. Plant waxes are found on the surfaces of leaves, stems, flowers, and fruits and serve to protect the plant from dehydration and from invasion by harmful microorganisms. (You can polish an apple to a high luster because of the waxes present in its skin.) Carnauba wax, largely myricyl cerotate ($C_{25}H_{51}COOC_{30}H_{61}$), is obtained from the leaves of certain Brazilian palm trees and is used extensively in floor waxes, automobile waxes, and shoe polishes.

Animal waxes also serve as protective coatings. They are found on the surface of feathers, skin, and hair and help to keep these surfaces pliable and water-repellent. The waxy coating on the feathers of water birds (ducks, gulls) helps them to stay afloat. If this wax is dissolved as a result of the bird swimming in an oil slick, the feathers become wet and heavy; the bird cannot maintain its buoyancy and will drown. Earwax protects the delicate lining of the inner ear.

Beeswax is the material from which bees construct honeycombs. Upon saponification, beeswax yields alcohols and fatty acid salts with even numbers of carbon atoms. The alcohols generally have 24–36 carbon atoms. The fatty acids have up to 36 carbon atoms; about one-fourth are hydroxy acids. About 20% of beeswax is hydrocarbons. These have odd numbers of carbon atoms, from 21 to 33. Figure 20.8 shows some of the typical molecules found in beeswax.

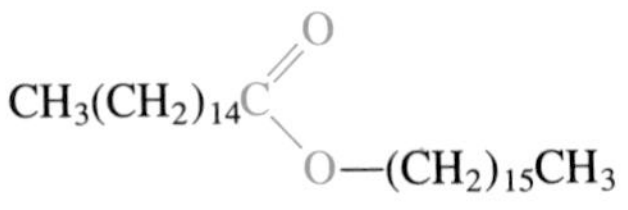

Beeswax is a by-product of the production of honey. Commercial operations can always leave behind enough honey to sustain the colony of bees and ensure future production. Obtaining the wax called spermaceti, however, is a little harder on the creature that produces it. Spermaceti crystallizes when the oil from the head of the sperm whale is cooled. Whales must be killed before the product can be obtained. The principal constituent of spermaceti is cetyl palmitate. Esters of lauric, myristic, and palmitic acids are also

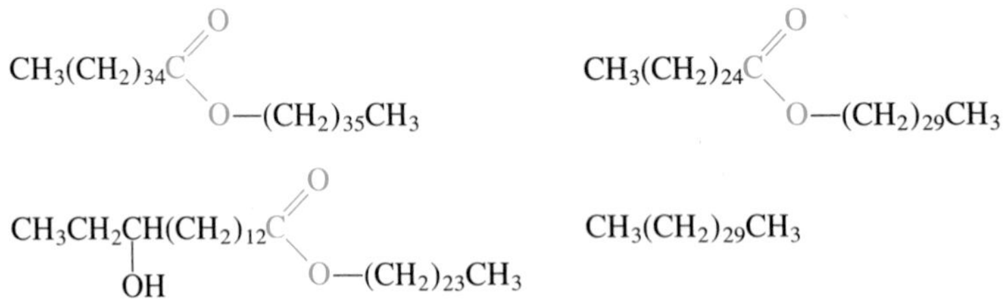

Cetyl palmitate

Figure 20.8
Three esters and an alkane found in beeswax.

present, as are esters of higher alcohols. Whales and other marine species store these waxes as fuels. Spermaceti melts at temperatures from 42 to 50 °C. It was once widely used in ointments, cosmetics, soaps, and candles and for laundering. Demand for spermaceti, sperm oil, and whale meat for human and pet food led to the near extinction of several species of whales.

Lanolin is a wax from sheep's wool. It is a mixture of esters and polyesters of 33 alcohols and 36 fatty acids. Some of the alcohols are steroids (similar to cholesterol). Lanolin is used as a base for ointments and cosmetic lotions.

Many natural waxes have been replaced by synthetic materials, mainly polymers. By careful control of the molecular weight, the properties of natural waxes can be closely duplicated. For example, Carbowax, a polymer of ethylene glycol ($HOCH_2CH_2OH$), is available in a range of average molecular weights. Synthetic waxes are used in cosmetics and ointments and in certain industrial processes.

20.7
PHOSPHOLIPIDS

Glycerol is an alcohol. It forms esters with acids. When the acids are carboxylic acids, the esters formed are fats. But glycerol can also form esters with inorganic acids such as phosphoric acid. The **phosphatides,** a most important class of **phospholipids** (phosphorus-containing lipids), are esters of glycerol in which there are two fatty acid groups and one phosphoric acid unit. Further, the phosphoric acid is also esterified with another alcohol molecule, usually an amino alcohol (Figure 20.9). Since these compounds are getting quite complicated, it is useful to relate their structure to those of other lipids we have studied. Notice that the molecule is identical to the triglycerides (fats and oils) up to the phosphoric acid part. Let's simplify things by representing those parts schematically (as in Figure 20.9b). We will then write out the structure of the amino alcohol to emphasize the parts that are different.

Phospholipids are found in all living organisms. They are particularly abundant in the biological membranes that surround individual cells and in the membranes surrounding

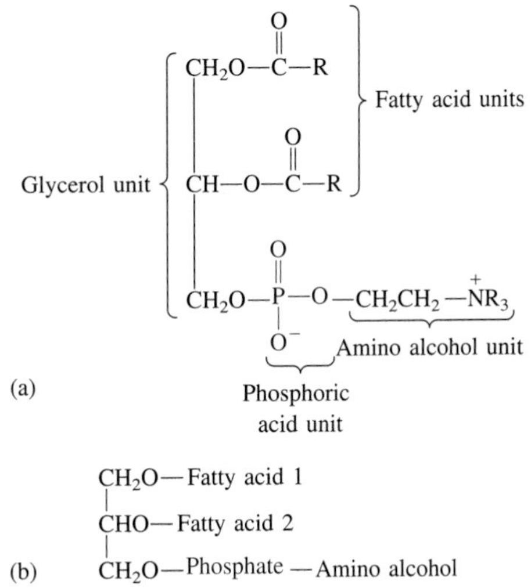

Figure 20.9

Structural formula (a) and schematic representation (b) of a phosphatide.

Figure 20.10

Two common phosphatides.

CH$_2$O—Fatty acid 1
|
CHO—Fatty acid 2
|
CH$_2$O—Phosphate—OCH$_2$CH$_2$NH$_3^+$

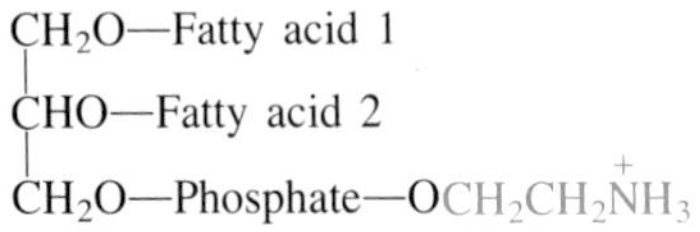

Phosphatidylethanolamine
(a cephalin)

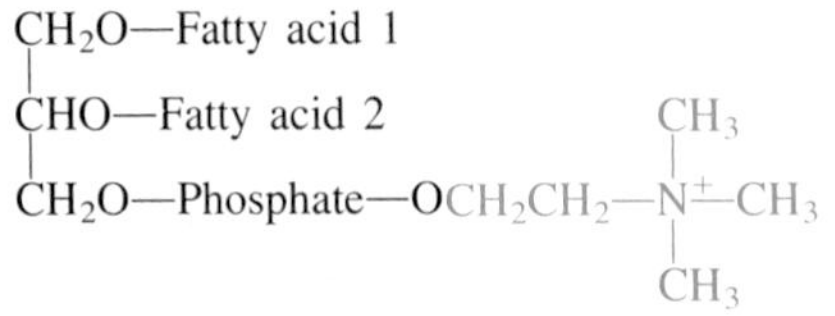

CH$_2$O—Fatty acid 1
|
CHO—Fatty acid 2
| CH$_3$
| |
CH$_2$O—Phosphate—OCH$_2$CH$_2$—N$^+$—CH$_3$
                                        |
                                       CH$_3$

Phosphatidylcholine
(a lecithin)

HOCH$_2$CH$_2$NH$_3^+$

Ethanolamine
(2-Aminoethanol)

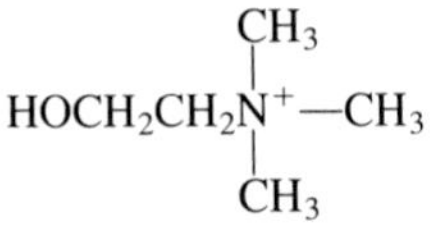

          CH$_3$
           |
HOCH$_2$CH$_2$N$^+$—CH$_3$
           |
          CH$_3$

Choline

CH$_3$(CH$_2$)$_{12}$CH=CHCH—OH
                          |
                       CH—NH$_2$
                          |
                       CH$_2$OH

Sphingosine

certain organelles within the cell. Phospholipids are the most polar of lipids, and they contain both hydrophilic and hydrophobic groups. This characteristic property is important in the structure of membranes (see Section 20.9).

Phospholipids also play other vital roles within the organism. They take part in fat metabolism by promoting the transport of lipids in the bloodstream as lipoprotein complexes (see Section 20.11). They are involved in oxidative phosphorylation, in secretory processes, and in the transport of certain molecules across cell membranes.

Two common phosphatides are shown in Figure 20.10. When the phosphatide contains the *ethanolamine* structural group (we've shown the amine group ionized), the compounds are called **cephalins.** The cephalins are found in brain tissue and nerves. They are also involved in blood clotting.

When *choline* is the amino alcohol unit, the compounds are called **lecithins.** Lecithins occur in all living organisms. They, too, are important constituents of nerve and brain tissue. Egg yolks are especially rich in lecithins. Commercial-grade lecithins are available from soybeans. These lecithins are widely used in foods as emulsifying agents. Many candy bars list lecithin among their ingredients.

Some snake venoms contain an enzyme that can catalyze the hydrolysis of one of the fatty acid ester units of lecithins. The remaining portion of the phospholipid then causes a breakdown of red blood cells.

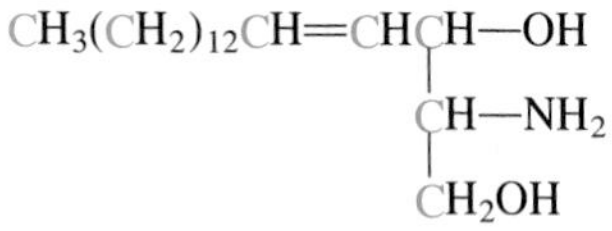

Like the phosphatides, **sphingolipids** contain a phosphoric acid unit. They are classified separately, however, because they are based on the unsaturated amino alcohol sphingosine rather than glycerol. Sphingomyelin is the "simplest" sphingolipid. It contains

Figure 20.11

A sphingomyelin.

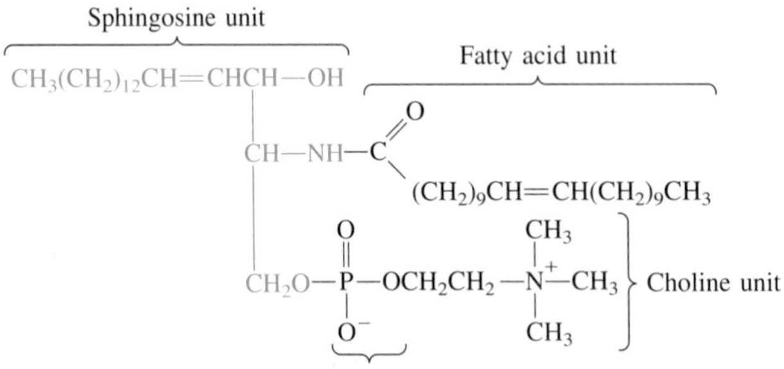

CH$_3$(CH$_2$)$_{12}$CH=CHCH—OH ⎴Sphingosine unit   ⎴Fatty acid unit
                              |
                           CH—NH—C=O
                                    \
                                     (CH$_2$)$_9$CH=CH(CH$_2$)$_9$CH$_3$
                              | CH$_3$
                              | O |
                           CH$_2$O—P—OCH$_2$CH$_2$—N$^+$—CH$_3$ } Choline unit
                                    ‖ |
                                    O$^-$ CH$_3$
                              Phosphoric acid unit

fatty acid, phosphoric acid, sphingosine, and choline units (Figure 20.11). Sphingomyelins are important constituents of the myelin sheath that surrounds the axon of a nerve cell. Multiple sclerosis is one of several diseases related to a fault in the myelin sheath.

20.8
GLYCOLIPIDS

Glycolipids may incorporate glycerol or sphingosine, fatty acids, and, always, a sugar unit. They contain no phosphoric acid. The simplest contain two fatty acid units and one sugar unit, such as a galactose unit, combined with glycerol (Figure 20.12). Again notice the similarity between this molecule and a fat molecule. They are the same except that the glycolipid has a sugar unit where the fat has its third fatty acid unit. These simple glycolipids occur in microorganisms and plants. Most animal cells contain another group of glycolipids called **cerebrosides** (Figure 20.13). Cerebrosides have a galactose unit, a fatty acid unit, and a sphingosine unit. The cerebrosides qualify not only as glycolipids but also as sphingolipids. The structure resembles that of a sphingomyelin, except that a sugar unit is connected where the sphingomyelin has a choline phosphate group. The cerebrosides are important constituents of the membranes of nerve and brain cells. They are believed to play a principal role in the membrane wrapping process that is unique to myelination. Gaucher's disease, a hereditary affliction, results from the substitution of glucose for galactose in the cerebroside. Large amounts of these abnormal cerebrosides accumulate, causing enlargement of the liver and the spleen.

Related compounds, called *gangliosides,* are also found in cell membranes. The structures of these compounds are even more complex than those of the cerebrosides. They usually contain a branched chain of three to eight monosaccharides and/or substituted sugars. Gangliosides are most prevalent in the outer membrane of nerve cells, although they occur in smaller quantities in most cell types. There is considerable variation in their sugar components, and about 130 varieties of glycosphingolipids have been identified. It is the sequence of sugars that most often determines cell-to-cell recognition and communication (e.g., blood group antigens).

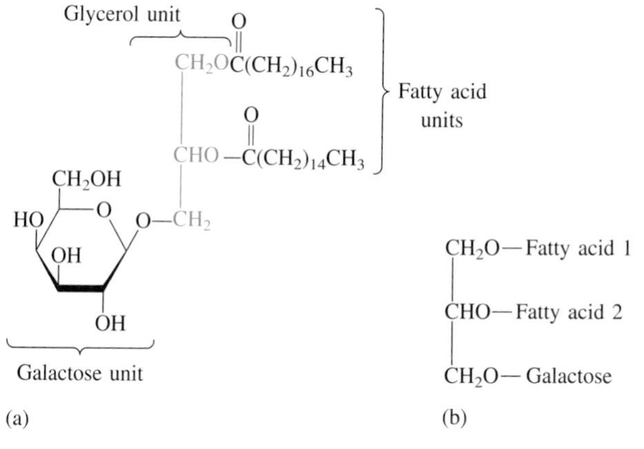

(a)

(b)

Figure 20.12
Structural formula (a) and schematic representation (b) of a simple glycolipid.

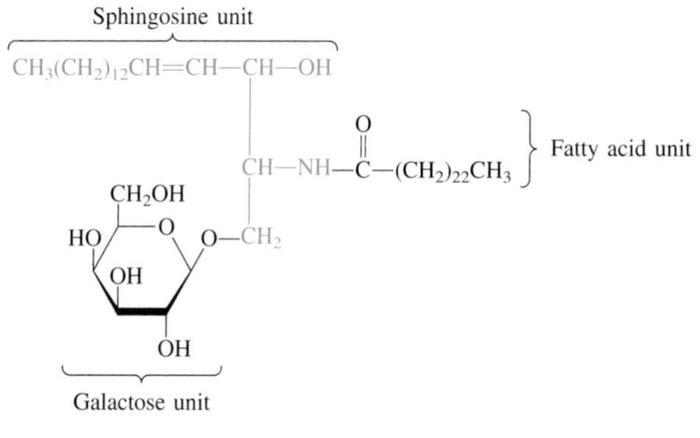

Figure 20.13
Cerebrosides are both glycolipids and sphingolipids.

20.9
CELL MEMBRANES

All the components of a living cell are enclosed within a membrane. Plant cells have rigid walls of cellulose to protect and surround the cell (Figure 20.14), but animal cells have only the cell membrane (Figure 20.15).

Figure 20.14
An idealized "typical" plant cell.

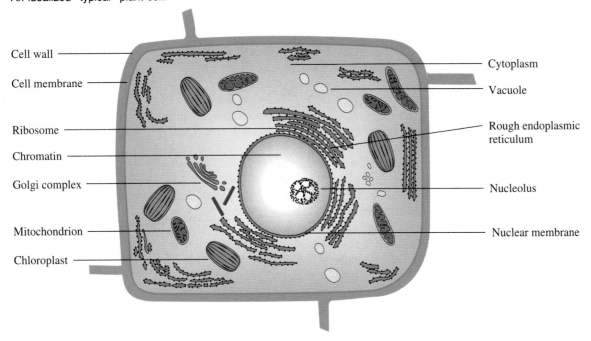

Figure 20.15
An idealized "typical" animal cell.

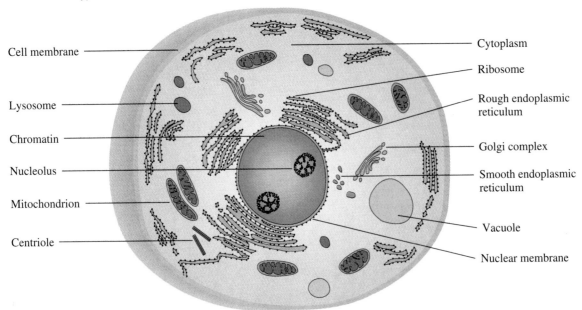

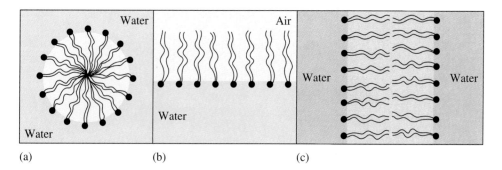

(a)   (b)   (c)

Figure 20.16
(a) Micelle, (b) monolayer, and (c) bilayer formed when polar lipids are added to water.

When polar lipids such as soaps, phospholipids, and glycolipids are placed in water, they disperse and form clusters of molecules called micelles. **Micelles** are aggregations of molecules that contain both polar and nonpolar groups. The hydrocarbon ''tails'' of these lipids are directed inward, away from the polar solvent; the hydrophilic heads are directed outward into the water. Each micelle may contain thousands of the polar lipid molecules (Figure 20.16a).

Polar lipids also form **monolayers,** layers one molecule thick, on the surface of water (Figure 20.16b). The polar heads stick into the water, and the nonpolar tails stick up in the air. In addition, **bilayers** may be formed (Figure 20.16c), with the hydrophobic tails sandwiched between the hydrophilic heads sticking outward into the water. These bilayers have properties quite similar to those of cell membranes.

Biological membranes were once viewed as inert barriers that just served to contain the cell and the organelles (nucleus, mitochondrion, Golgi apparatus, etc.) within the cell. We now recognize them to be intrinsically involved in the transport of materials and as receptors of external stimuli. Most cell membranes are composed of about 50% (by mass) lipid and 50% protein, although large variations from these percentages can exist in certain cells.[1] The membranes are referred to as **semipermeable** because only certain materials are allowed to pass from one side of the membrane to the other.

The three major classes of lipid molecules in the membrane bilayer are phospholipids, glycolipids (about 5% and confined to the outer layer), and cholesterol (see Section 20.10). Roughly equal numbers of phospholipid and cholesterol molecules are present in the membranes of cells. As shown in Figure 20.17, the lipid bilayer consists of two rows of phospholipid molecules arranged tail to tail. The hydrophobic tails (the fatty acid portions) are directed toward each other and can interact by means of hydrophobic and van der Waals forces.[2] At the same time they are isolated from the aqueous environment that exists within and outside the cells. The polar portions of the phospholipids and glycolipids, therefore, project from the inner and outer surfaces of the membrane and interact with water molecules.

---

[1] The percentage of each component of the membrane is related to the function of the particular tissue. For example, the myelin sheath of nerve cells can contain up to 80% lipid because the major function of the membrane is insulation and protection. The inner mitochondrial membrane, on the other hand, is unique in having a large protein component (75–80%). The proteins (enzymes) within the membrane play an integral role in the energy conversion function of the mitochondria (see Section 25.2). It is important to emphasize that these are percents by mass. Lipid molecules are much smaller than proteins; hence there are always more lipid than protein molecules. In an average membrane there are 50 lipid molecules for one protein molecule.

[2] This interaction is relatively weak because of the presence of unsaturated (cis-branching) fatty acids. As a result, the lipid portion is not rigid but is quite fluid and allows movement within the membrane. Cholesterol is also a key moderator of membrane fluidity. It breaks up the hydrophobic and van der Waals interactions of the fatty acid chains by fitting between these chains.

Figure 20.17

Schematic diagram of the fluid mosaic model of membrane structure. The cell membrane is a phospholipid bilayer in which cholesterol and various kinds of proteins are embedded. Attached to the membrane proteins are one or more short polysaccharide chains.

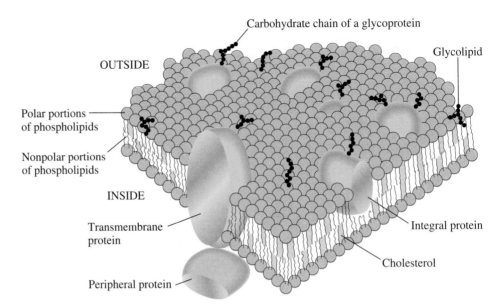

If membranes were composed only of lipids, they would act as a barrier to the passage of ions or polar molecules. The passage of polar species across the membrane is facilitated by proteins that move about in the "sea" of lipids. There are two classes of proteins in the cell membrane. **Integral proteins** extend into the lipid bilayer. Those integral proteins that extend completely across the membrane are called *transmembrane proteins*. **Peripheral proteins,** the other class of proteins, seem to bond only loosely to the lipid bilayer. They appear to be attached to integral proteins by hydrogen bonds and electrostatic forces.

Nonpolar molecules, such as the general anesthetics (pages 365–66), can infiltrate the bilayer and change its permeability. Polar and ionic substances enter by way of channels through the integral proteins. It appears that there are special proteins to facilitate the passage of certain molecules (e.g., hormones and neurotransmitters). A specific interaction occurs between the carrier protein and the molecule. The complex then moves to the other surface of the membrane and discharges its passenger. If this occurs in the direction dictated by the concentration (high to low), it requires no energy and is called **facilitated diffusion.** If the movement takes place *against* the concentration gradient, it requires energy and is called **active transport.** (The source of this energy, cellular ATP, is discussed in Chapter 24.)

20.10
STEROIDS: CHOLESTEROL AND BILE SALTS

All the lipids discussed so far are **saponifiable.** They react with aqueous alkali to yield simpler components such as glycerol, fatty acids, amino alcohols, or sugars. As extracted from cellular material, however, the lipids contain a small but important fraction that does not react with alkali. The most important **nonsaponifiable** lipids belong to a class of compounds called **steroids.** These compounds include the bile salts, cholesterol and related compounds, hormones such as cortisone, and the all-important sex hormones. Those steroids with hormonal activity are discussed in detail in Special Topic H. Very small amounts of steroids or slight variations in structure or in the nature of substituent groups effect profound changes in biological activity.

Over 40 different steroidal compounds have been found in nature. They occur in plant and animal tissues, yeasts, and molds, but not in bacteria, and may exist free or combined with fatty acids or carbohydrates. All steroids have a perhydrocyclopentanophenanthrene ring system, which consists of a completely saturated phenanthrene moiety fused to a cyclopentane ring. The rings are designated by capital letters, and the carbon atoms are numbered as shown.

Cholesterol does not occur in plants, but it is the best known and most abundant (about 240 g) steroid in the human body. About one-half of the total body cholesterol is present in cell membranes interspersed among the phospholipid molecules (recall Figure 20.17). The brain is about 10% cholesterol, but cholesterol's function there is unknown. Much of the cholesterol in the body is converted into cholic acid, which is used in the formation of bile salts. Cholesterol is also an important precursor in the biosynthesis of the sex hormones, adrenal hormones, and vitamin D. Excess cholesterol that is not utilized by the body is released from the liver and transported by the blood to the gallbladder. Normally, it stays in solution and is secreted into the intestine (in the bile) to be eliminated. Sometimes the cholesterol precipitates in the gallbladder, producing gallstones (Figure 20.18). Its name is derived from this source (Greek *chole*, bile; *stereós*, solid).

Perhydrocyclopentanophenanthrene
(the steroid skeleton)

The term *moiety* means part or portion. In chemistry it is used to denote a particular group within a large molecule.

Cholesterol

Most meats and foods derived from animal products such as eggs, butter, cheese, and cream are particularly rich in cholesterol. The average American consumes about 600 mg of cholesterol each day. The liver is the major site of cholesterol biosynthesis, although other tissues (intestines, adrenals, gonads) are also involved. The human body synthesizes about 1 g of cholesterol each day; all 27 carbon atoms are derived from acetyl CoA

One large egg contains 213 mg of cholesterol. Most health authorities recommend a maximum of 300 mg/day of cholesterol in the diet.

Figure 20.18
An assortment of human gallstones. (The dime is present to give an indication of relative sizes.)

molecules (see Section 26.3). The plasma cholesterol level controls the synthesis of cholesterol by the liver. When the cholesterol level in the blood exceeds 150 mg/100 mL, the rate of cholesterol biosynthesis is halved. Hence, if cholesterol is present in the diet, there is a feedback mechanism that suppresses its biosynthesis in the liver. However, this is not a 1:1 ratio. The reduction in biosynthesis does not equal the amount of cholesterol ingested. Fasting also inhibits the biosynthesis of cholesterol because of the limited availability of acetyl CoA. Conversely, diets high in carbohydrate or fat tend to accelerate cholesterol biosynthesis because they increase the amount of acetyl CoA in the liver. The lipids of fish and poultry contain relatively more unsaturated fatty acids than the lipids of beef, lamb, and pork. Fish (fish oils in particular) and poultry are recommended for people who wish to lower their serum cholesterol levels because of the suspected correlation between the cholesterol level in the blood and certain types of heart disease (see Section 20.11). The cholesterol content of blood varies considerably with age, diet, and sex. Young adults average about 170 mg of cholesterol per 100 mL of blood, whereas males at age 55 may have 250 mg/100 mL or higher (because the rate of cholesterol metabolism decreases with age). Females tend to have lower blood cholesterol levels than males.

Bile is a yellowish green liquid with a pH of 7.8–8.6. Bile contains no digestive enzymes, yet it plays a number of vital roles. The composition of typical bile is given in Table 20.5. The bile serves as a route for the excretion of drugs, end products from the breakdown of hemoglobin (Section 28.9), and heavy metal ions.

Bile salts are the most important constituents of bile, and their major function is to aid in the digestion of dietary lipids. The bile salts are sodium salts of amidelike combinations of bile acids and glycine or the rare amino acid taurine ($H_2NCH_2CH_2SO_3^-$). They

Table 20.5
Composition of Bile

Component	Percent
Water	97
Bile salts	0.7
Inorganic salts	0.7
Bile pigments	0.2
Fatty acids	0.15
Lecithin	0.1
Fat	0.1
Cholesterol	0.06

Arteriosclerosis

About 85% of deaths due to cardiovascular disease are directly linked to **arteriosclerosis.** This condition is characterized by the loss of elasticity (hardening) of the arteries, which produces degenerative heart disease, stroke, and other arterial diseases. **Atherosclerosis** (a form of arteriosclerosis) results from the deposition of fatty substances, especially cholesterol, in coronary arteries. When cholesterol and fat leave the blood, they accumulate on the inside of the arterial wall (Figure 20.19). Such deposits are called *plaque.*

Contrary to popular belief, plaque doesn't simply clog arteries as sludge clogs a sewer pipe. Rather, white blood cells migrate into the endothelial cells that line the surface of the artery. The white blood cells attempt to cleanse the endothelial cells by breaking down the plaque. In the process they release growth factors that cause a proliferation of arterial tissue (Figure 20.20). That tissue buildup and the resulting constriction reduce blood flow, diminish oxygen supply, and lead to high blood pressure and the dangers of a blood clot. (Note that high blood pressure is also aggravated by other factors such as lack of exercise, obesity, heredity, stress, and smoking.) Investigators are attempting to establish that reducing the blood cholesterol level by minimizing lipid intake through dietary restrictions, or by suppressing the overactive growth factors, will reduce the incidence of heart disease. Cardiovascular disease accounts for more than half of the deaths each year in the United States (see Section 20.11).

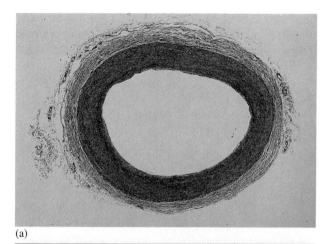

(a)

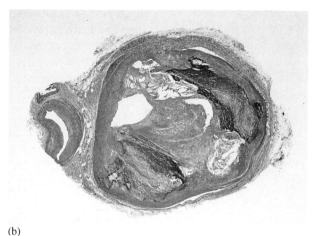

(b)

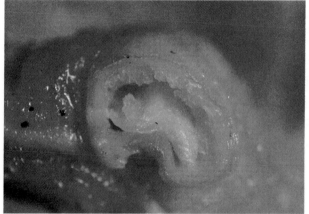

(c)

Figure 20.19

Comparison of normal artery (a) with those containing fatty deposits. (b) Fatty deposits in vessel wall. (c) Plugged artery with fatty deposits and clot.

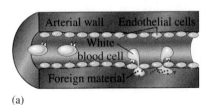

(a)

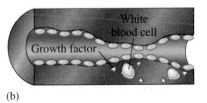

(b)

Figure 20.20

The constriction of arteries. (a) White blood cells enter the arterial tissue to consume foreign material. (b) The cells release growth factors that stimulate tissue production, narrowing the arterial passage.

are synthesized from cholesterol in the liver, stored in the gallbladder, and then secreted in bile into the small intestine. (About 500 mL of bile is secreted each day.) Bile salts are highly effective detergents because they contain both hydrophobic and hydrophilic groups. Thus, they function by acting as emulsifying agents—they break down large fat globules into smaller ones and keep these smaller globules suspended in the aqueous digestive environment (see Section 26.1). The lipases can then hydrolyze the fat molecules more effectively. Bile salts also aid in the absorption of fatty acids, cholesterol, and the fat-soluble vitamins by forming complexes (micelles) that can diffuse into the intestinal epithelial cells.

Cholic acid
(a bile acid)

Sodium glycocholate
(a bile salt)

Emulsions

When violently shaken together with water, fats are broken into tiny (submicroscopic) particles and dispersed through the water. Such a mixture is called an **emulsion.** Unless a third substance has been added, the emulsion breaks down rapidly. The droplets then recombine and float to the surface of the water. Soap, certain types of gum, or protein can stabilize the emulsion by forming protective coatings on the fat droplets that prevent them from coming together.

Many foods are emulsions. Milk is an emulsion of butterfat in water. The stabilizing agent in milk is a protein called casein. Mayonnaise is an emulsion of salad oil in water, stabilized by egg yolk.

20.11
CHOLESTEROL AND CARDIOVASCULAR DISEASE

We get half our total fats, three-fourths of our saturated fats, and *all* our cholesterol from animal products such as meat, milk, cheese, and eggs. Advertising claims that a vegetable oil (for example) contains no cholesterol are simply silly; *no* vegetable product contains cholesterol.

In the past 30 years, few subjects in the nutrition field have attracted as much public attention as cholesterol. Today, everything from margarine and vegetable oils to egg substitutes and meat analogs is advertised on the basis that it contains little or no cholesterol. Cholesterol is believed to be a primary factor in the development of atherosclerosis, coronary heart disease, and stroke. As the leading cause of death in the United States, heart attack and stroke together take about 840,000 lives a year.

The disorder underlying both heart attack and stroke is atherosclerosis, characterized by the buildup of deposits (plaque) on the inner surface of arteries (recall Figures 20.19 and 20.20). If a blood clot forms in such a constricted artery leading to the heart or brain and causes a complete stoppage of blood flow, a heart attack or stroke occurs almost instantly. In atherosclerotic plaques, the lipids in highest concentration are cholesterol and cholesterol esters. Present in lower concentrations are two other types of lipids—phospholipids and triglycerides.

Scientists generally agree that elevated cholesterol levels in the blood, as well as high blood pressure and cigarette smoking, are associated in humans with an increased risk of heart attack. A long-term investigation by NIH showed that, among men aged 30–49, the incidence of coronary heart disease was five times greater for those whose cholesterol level was 260 mg/100 mL of serum or more than for those with cholesterol levels of 200 mg/100 mL or less.

Recent research in animals and humans has demonstrated that a lowered cholesterol level does reduce the risk of coronary heart disease. The gathering of proof to support this contention is exceedingly difficult because large numbers of people must be studied over long periods of time. Useful data can be obtained only by a slow, laborious process, partly because only a small percentage of any group will die of coronary heart disease in any given year.

Several alternative theories have been proposed to explain the cause of atherosclerosis. The most recent (1987) one suggests that defects in the lipid-transporting system are responsible for a buildup of lipids in the blood, which eventually triggers plaque formation. Because lipids such as cholesterol are not soluble in water, they cannot be transported in the blood (an aqueous medium) unless they are complexed with water-soluble proteins as lipoproteins. Lipoproteins generally are classified according to their density and composition. There are four broad categories: *chylomicrons* (density of less than 0.94 g/mL and made in the intestine), *very low density lipoproteins* (density of 0.94–1.006 and made mainly in the liver), *low-density lipoproteins* (1.006–1.063), and *high-density lipoproteins* (1.063–1.21). The density of lipoproteins is determined by the relative content of protein and lipid. Since lipids are less dense than proteins, lipoproteins containing a greater amount of lipid are less dense than those containing a greater proportion of protein. The chylomicrons contain up to 99% by weight of lipids, and the very low density lipoproteins (VLDLs) contain up to about 90% lipids. The low-density lipoproteins (LDLs) contain about 80% lipids, and the high-density lipoproteins (HDLs) only about 50%. The protein component makes up the remainder of the lipoprotein molecule (e.g., HDLs are composed of 50% lipid and 50% protein—Table 20.6).

In research on cholesterol and its role in heart disease, the types of lipoproteins that have received the greatest attention in recent years have been the LDLs and HDLs. The reason is that they almost always contain a higher percentage of cholesterol than do the chylomicrons or the VLDLs, which serve to transport the triglycerides. One of the most fascinating discoveries in this field is that cholesterol that is bound to HDLs reduces a person's risk of developing coronary heart disease. On the other hand, cholesterol that is bound to LDLs increases that risk. Notice from Table 20.6 that LDLs are the major cholesterol-carrying lipoproteins, whereas chylomicrons and VLDLs are the major carriers of triglycerides in the plasma.

Research evidence indicated that atherosclerosis and coronary heart disease are associated with elevated levels of serum LDLs, rather than with serum lipoproteins in general. It was also reported that the level of serum LDLs was a better predictor of coronary heart

Plasma lipoproteins are generally spherical particles with a surface that consists largely of phospholipids, free cholesterol, and protein and a core that contains mostly triglycerides and cholesterol esters.

Lipoprotein Class	Density Range (g/mL)	Composition (wt %)		Cholesterol		Phospholipid
		Protein	Triglyceride	Free	Ester	
Chylomicrons	<0.94	1–2	85–95	1–3	2–4	3–6
VLDL	0.94–1.006	6–10	50–65	4–8	16–22	15–20
LDL	1.006–1.063	18–22	4–8	6–8	45–50	18–24
HDL	1.063–1.21	45–55	2–7	3–5	15–20	26–32

Table 20.6
Composition of Lipoproteins Isolated from Normal Subjects

disease risk than the level of serum cholesterol. (The normal concentration of LDLs in humans is about 120 mg/100 mL of serum). Persons who, because of hereditary factors, have high levels of LDLs in their blood have a higher incidence of coronary heart disease.

Most of the serum cholesterol is transported as an LDL–cholesterol complex, which delivers the cholesterol directly to cells that need it. Low-density lipoproteins contain about 55% cholesterol, whereas high-density lipoproteins contain only about 25% cholesterol. LDLs are believed to promote coronary heart disease by first penetrating the coronary artery wall, where they are broken down enzymatically to cholesterol, cholesterol esters, and protein. The cholesterol and cholesterol esters are then deposited in the artery wall, becoming major parts of the atherosclerotic plaque.

How do HDLs reduce the risk of developing coronary heart disease? No one knows for sure, but two theories have been suggested.

1. HDLs competitively inhibit the uptake of LDLs by cells by occupying, and thus blocking, the LDL receptor sites on the cell membranes.[3]
2. One role of HDLs is to transport excess cholesterol *from* the cells to the liver. Therefore HDLs aid in removing cholesterol from blood and the smooth muscle cells of the arterial wall and transporting it to the liver, where it can be metabolized.

Assuming that HDL helps to protect the body against coronary heart disease, what can be done to increase serum levels of this lipoprotein? One way is to be born a female. Women have a higher average HDL value in the blood than men (55 mg/100 mL versus 45 mg/100 mL), and this may be one reason women (prior to menopause) have a lower rate of heart disease than men. Another way is by sustained exercise (e.g., aerobic training). It was reported that marathon runners had a decidedly higher average HDL level (~65 mg/100 mL) than did a group of sedentary men. Still another way is to lose weight. It was shown that HDL levels can be increased by losing weight. Finally, we should mention a method that has received quite a bit of publicity. It was reported that HDL levels in the blood can be increased (to about 80–100 mg/100 mL) by drinking alcohol *in moderation*. The amount of alcohol used in this study was equivalent to about three 12-oz bottles of beer a day.

To sum up: a reduction of low-density lipoprotein levels is believed by many scientists to be desirable. This may be accomplished by eating a diet low in cholesterol and low in saturated fat. And some physicians recommend that overweight people go on reducing diets to achieve their recommended weight.

There is also statistical evidence that fish oils can prevent heart disease. Researchers at the University of Leiden in the Netherlands found that Greenlanders who eat a lot of fish have a low risk of heart disease, despite a diet that is high in total fat and cholesterol. The probable effective agents are the polyunsaturated fatty acids such as eicosapentaenoic acid.

$$CH_3CH_2(CH=CHCH_2)_5CH_2CH_2COOH$$

Other studies have shown that diets with added fish oil lead to lower cholesterol and triglyceride levels in the blood.

---

[3] LDL receptor proteins on the cell membranes are vital to the uptake of cholesterol from the blood. If these receptors are absent or deficient, the inherited condition called familial hypercholesterolemia results. Cholesterol levels in the blood rise dramatically, and the excess cholesterol that can't enter cells is deposited in certain tissues, particularly in skin and tendons, and in arterial plaques. Most individuals who are homozygous for the disorder have almost no LDL receptors, and they die of coronary artery disease in childhood. Somewhat less than 1% of the U.S. population is heterozygous; they have about half the number of receptors and often develop atherosclerosis by age 50.

Heart disease kills more Americans than all other diseases combined. Every year, approximately 1.5 million people suffer heart attacks. Almost half of them will die. Of those who survive, many are left with permanent damage, and a significant number have subsequent attacks. A heart attack usually does not happen suddenly. The body has an early warning system. The American Heart Association compiled a list of early warning signals.

1. One of the first signs is pressure or pain in the middle of the chest. That's where the heart is, not on the left as many believe.
2. The pain can get worse and spread through the whole chest as well as down the left arm.
3. The pain may also spread to both arms, shoulders, neck, or jaw. A sensation of pressure, fullness, or squeezing may occur in the abdomen and is often mistaken for indigestion.
4. Pain may occur in any one or a combination of these areas at the same time. It could go away and return later. Many times, sweating, nausea, vomiting, or shortness of breath come with the pain.

At the first sign of these symptoms, go to the nearest hospital emergency room at once! Get there as fast as possible. Have someone take you or dial 0 for an operator, say there is a heart attack victim at (location). In the hospital, insist that it is an emergency—you have chest pain (or other symptoms) and may be having a heart attack.

EXERCISES

1. Define, explain, or draw formulas for each of the following.

   a. triglyceride
   b. triolein
   c. palmitic acid
   d. linolenic acid
   e. lipases
   f. iodine number
   g. antioxidant
   h. micelle
   i. semipermeable
   j. biodegradable
   k. builders
   l. eutrophication
   m. phosphatide
   n. steroid
   o. phosphatides and sphingolipids
   p. lecithin and cephalins
   q. arteriosclerosis and atherosclerosis
   r. bile and bile salts

2. Distinguish between the terms in each of the following pairs.

   a. fat and oil
   b. fat and wax
   c. saponifiable liquid and nonsaponifiable lipid
   d. cis double bond and trans double bond
   e. simple triglyceride and mixed triglyceride
   f. butter and margarine
   g. oxidative rancidity and hydrolytic rancidity
   h. hydrophobic and hydrophilic
   i. saponification and emulsification
   j. monolayer and bilayer
   k. hard water and soft water
   l. hard soap and soft soap
   m. soaps and syndets
   n. LDLs and HDLs

3. What functions does fat serve in the body?

4. Contrast the physical properties of fats and fatty acids. In what solvents are fats and oils soluble?

5. Why do unsaturated fatty acids have lower melting points than saturated fatty acids?   → more C=C so more branched chains

6. The melting point of elaidic acid (Figure 20.2c) is 52 °C.

   a. How does this compare with the melting points of stearic acid and oleic acid? Explain.
   b. Would you expect the melting point of *trans*-hexadecenoic acid, $CH_3(CH_2)_5CH=CH(CH_2)_7COOH$, to be lower or higher than that of elaidic acid? Why?

7. Which of these fatty acids are saturated, and which are unsaturated?

   a. caproic acid
   b. oleic acid
   c. stearic acid
   d. palmitic acid
   e. linolenic acid
   f. caprylic acid

8. How many carbon atoms are there in a molecule of each of the compounds listed in Exercise 7?

9. How many mixed triglycerides are possible by combining stearic acid and oleic acid with glycerol?

10. Write structural formulas for the following.
  a. glyceryl palmitate
  b. cetyl (1-hexadecyl) stearate
  c. a triglyceride likely to be found in cottonseed oil
  d. a highly unsaturated oil
  e. sodium oleate
  f. calcium myristate

11. In the determination of the iodine number of a fat, what functional group in the fat molecule reacts with the reagent? *double bonds*

12. Which would have the higher iodine number—tristearin or triolein?

13. Which would you expect to have a higher iodine number—corn oil or beef tallow? Explain your reasoning. *more C=C's*

14. Which would you expect to have the higher iodine number—hard or liquid margarine? Explain your reasoning.

15. What fat would be formed by the complete hydrogenation of triolein? of trilinolein?

16. What products are formed when a fat is **(a)** hydrolyzed? **(b)** saponified?

17. Draw the structural formulas of the products of this saponification reaction.

$$CH_2-O-\overset{\overset{O}{\|}}{C}-CH_2CH_2CH_2CH_2CH_2CH_2CH_3$$

$$CH-O-\overset{\overset{O}{\|}}{C}-CH_2CH_2CH_2CH_2CH_3 \xrightarrow{\text{NaOH}}$$

$$CH_2-O-\overset{\overset{O}{\|}}{C}-CH_2CH_2CH_2CH_2CH_2CH_2CH_2CH_3$$

18. Write the equation for the saponification of glyceryl trilaurate.

19. How would the equation for Exercise 18 differ if the compound being saponified were triolein?

20. a. What disagreeable odor is formed when butter becomes rancid?
  b. What leads to its formation?
  c. How can rancidity be prevented?

21. Can waxes be converted into soaps? Explain.

22. Name four waxes obtained from plants and animals, and give one use for each.

23. Briefly describe how soaps clean.

24. a. What structural features are necessary for a compound to be a good detergent?
  b. What advantages do detergents have over ordinary soap?
  c. What are the disadvantages of detergents?

25. Why are phospholipids referred to as polar lipids?

26. Start with one molecule each of glycerol, palmitic acid, oleic acid, phosphoric acid, and choline, and construct a typical lecithin molecule. Circle all the ester bonds.

27. a. What type of bond joins the fatty acid to sphingosine in cerebrosides?

b. What type of bond joins the sugar unit to sphingosine in cerebrosides?

28. Phosphatidylinositol is another phosphatide that is found in cell membranes. Given the structure of inositol, draw a structural formula for phosphatidylinositol. (Linkage occurs at the colored hydroxyl group.)

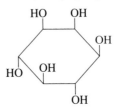

Inositol

29. What general structural feature do phosphatides share with soaps and detergents?

30. Using a circle to represent the ionic group and zigzag lines to represent the two fatty acid units, draw the arrangement of phosphatide molecules in a micelle and in a bilayer.

31. Draw the basic steroid skeleton.

32. Which of the following compounds are classified as steroids?
  a. tristearin   **b.** cholesterol
  c. lecithin   **d.** prostaglandins
  e. cephalin   **f.** cholic acid

33. Which of the following are derived from glycerol, which from sphingosine, and which from neither?
  a. fats   **b.** oils
  c. waxes   **d.** phosphatides
  e. cerebrosides   **f.** steroids

34. Describe the fluid mosaic model of cell membranes.

35. What is an integral protein? What is its role in the function of a cell membrane?

36. What is a peripheral protein? Where is it located on the cell membrane?

37. What is facilitated diffusion?

38. What is active transport?

39. Draw the structure of the cerebroside that has palmitic acid as its fatty acid and glucose as its sugar.

40. A principal wax in spermaceti is an ester of palmitic acid (hexadecanoic acid) and cetyl alcohol (1-hexadecanol). Draw the structure of the ester.

41. Which of the following can diffuse through the lipid portion of a cell membrane?
  a. glucose   **b.** NaCl
  c. $CH_3CH_2OCH_2CH_3$ **d.** CH_3CH_2OH

42. Discuss the roles of cholesterol, saturated fats, and fish oils in arteriosclerosis.

43. The system that transports Na^+ across a cell membrane is called the ''sodium pump.'' Sodium ions are pumped out of a cell in which the concentration of Na^+ is maintained at 0.10 M and into the extracellular fluid where the Na^+ concentration is 0.14 M. Is this an example of facilitated diffusion or of active transport? Explain.

44. A carrier protein complexes a molecule of glucose at the outer surface of a cell membrane. The complex then moves to the other side of the membrane and releases the glucose molecule. The concentration of glucose is higher outside the cell than inside. Is this facilitated diffusion or active transport? Explain.

45. Identify each component of the following phospholipid.

$$CH_2OC(CH_2)_{16}CH_3$$

$$CHOC(CH_2)_7CH=CH(CH_2)_7CH_3$$

$$CH_2OP-OCH_2CH_2\overset{+}{N}-CH_3$$

with O double bonds on the carbonyls and phosphate, and CH_3, CH_3, CH_3 groups on nitrogen.

46. Serine is an amino acid with the structure

$$HOCH_2CHCOOH$$
$$NH_2$$

Give the structure for phosphatidylserine.

47. Examine the labels on margarines and shortenings and list the oils used in the various brands.

48. Examine several different detergents, and note the active ingredient in each. What other additives are present?

Hormones

Like the vitamins (Special Topic I), hormones are organic compounds. Unlike vitamins, hormones can be synthesized in the body. Both vitamins and hormones play critical biochemical roles. The most striking similarity, however, is in their effective amounts. Even in very low concentrations, hormones produce dramatic changes in the body. This special topic, then, deals with a class of compounds that, like the vitamins, illustrate the old adage that good things come in small packages.

H.1
THE ENDOCRINE SYSTEM

The hormones are synthesized in the endocrine glands (Figure H.1) and are then discharged directly into the circulatory system. They serve as "chemical messengers." Hormones released in one part of the body signal profound physiological changes in other parts of the body (heart, liver, muscle, kidney, etc.). They cause reactions to speed up or slow down. In this way they control growth, metabolism, reproduction, and many other functions of body and mind. Hormones can affect (1) the permeability of cell membranes, (2) the rate of enzymatic reactions, or (3) the rate of synthesis of certain proteins.

If we consider hormones as messengers, then the pituitary gland (hypophysis) must be viewed as the central dispatcher or control. Many of the pituitary hormones control the production of hormones by other endocrine glands. The removal of portions of the pituitary gland results in atrophy of other endocrine glands. Shut down the central control, and you ultimately shut down much of the endocrine system. The pituitary itself responds to hormone signals from the hypothalamus. The hypothalamus secretes hormones, called releasing factors, that trigger the production of the pituitary hormones. The hypothalamus is triggered by nerve impulses. The sequence is this: A nerve impulse signals the hypothalamus, the hypothalamus signals the pituitary, the pituitary signals some target endocrine gland, the gland signals some target tissue, and the tissue responds in a specific way (Figure H.2).

Let us take just a moment to consider the extraordinary complexity of this system. For example, in response to neural stimulation, the hypothalamus produces thyrotropin-releasing factor (TRF). The TRF reaches the pituitary and causes that gland to produce thyroid-stimulating hormone (TSH). In the presence of TSH, the thyroid gland releases the hormone thyroxine. Thyroxine signals the cells to increase their metabolic rate. As the level of thyroxine builds up, a feedback mechanism causes the pituitary to slow down its production of thyroid-stimulating hormone. This, in turn, slows the production of thyroxine by the thyroid, which results in a slowing of the metabolic rate, which gets us back to where we started. The cycle includes the synthesis of proteins and peptides (through the complex process described in Chapter 23). It includes

Figure H.1

Approximate locations of endocrine glands in the human body.

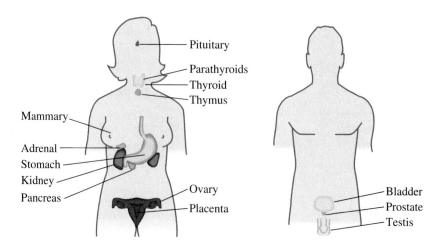

Pituitary
Parathyroids
Thyroid
Thymus
Mammary
Adrenal
Stomach
Kidney
Pancreas
Ovary
Placenta
Bladder
Prostate
Testis

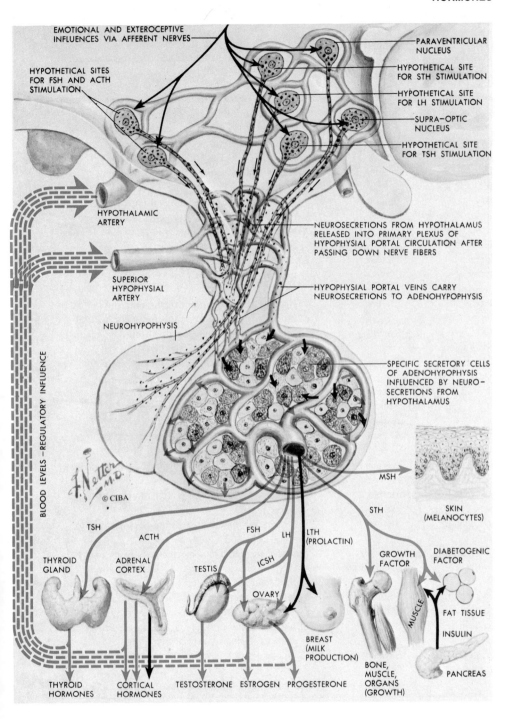

Figure H.2

Schematic diagram illustrating the interrelationship of the hypothalamus, the pituitary, and the organs upon which they work. [Reproduced with permission from the CIBA Collection of Medical Illustration by Frank Netter M.D. Copyright © 1965 by the CIBA Pharmaceutical Co., Division of CIBA-Geigy Corporation. All rights reserved.]

the multistep synthesis of the amino acid thyroxine and, ultimately, the speeding up of that complicated combination of reactions we lump together under the heading *metabolism*.

In a healthy individual all of this happens routinely, in perfect balance, and with no conscious direction. And that's just one of the myriad interrelated biochemical processes that life requires. That it all works, and works so well, is nothing short of miraculous.

Many important human hormones and their physiological effects are listed in Table H.1. We will discuss the hormones vasopressin and oxytocin in Chapter 21. The

Table H.1

Some Human Hormones and Their Physiological Effects

Name	Gland and Tissue	Chemical[a] Nature	Effect
Various releasing and inhibitory factors	Hypothalamus	Peptide	Trigger or inhibit release of pituitary hormones
Human growth hormone (HGH)	Pituitary, anterior lobe	Protein	Controls the general body growth; controls bone growth
Thyroid-stimulating hormone (TSH)	Pituitary, anterior lobe	Protein	Stimulates growth of the thyroid gland and production of thyroxine
Adrenocorticotrophic hormone (ACTH)	Pituitary, anterior lobe	Protein	Stimulates growth of the adrenal cortex and production of cortical hormones
Follicle-stimulating hormone (FSH)	Pituitary, anterior lobe	Protein	Stimulates growth of follicles in ovaries of females, sperm cells in testes of males
Luteinizing hormone (LH)	Pituitary, anterior lobe	Protein	Controls production and release of estrogens and progesterone from ovaries, testosterone from testes
Prolactin	Pituitary, anterior lobe	Protein	Maintains the production of estrogens and progesterone, stimulates the formation of milk
Vasopressin	Pituitary, posterior lobe	Protein	Stimulates contractions of smooth muscle; regulates water uptake by the kidneys
Oxytocin	Pituitary, posterior lobe	Protein	Stimulates contraction of the smooth muscle of the uterus; stimulates secretion of milk
Parathyroid	Parathyroid	Protein	Controls the metabolism of phosphorus and calcium
Thyroxine	Thyroid	Amino acid derivative	Increases rate of cellular metabolism
Insulin	Pancreas, beta cells	Protein	Increases cell usage of glucose; increases glycogen storage
Glucagon	Pancreas, alpha cells	Protein	Stimulates conversion of liver glycogen to glucose
Cortisol	Adrenal gland, cortex	Steroid	Stimulates conversion of proteins to carbohydrates
Aldosterone	Adrenal gland, cortex	Steroid	Regulates salt metabolism; stimulates kidneys to retain Na^+ and excrete K^+
Epinephrine (adrenaline)	Adrenal gland, medulla	Amino acid derivative	Stimulates a variety of mechanisms to prepare the body for emergency action, including the conversion of glycogen to glucose
Norepinephrine (noradrenaline)	Adrenal gland, medulla	Amino acid derivative	Stimulates sympathetic nervous system; constricts blood vessels, stimulates other glands
Estradiol	Ovary, follicle	Steroid	Stimulates female sex characteristics; regulates changes during menstrual cycle
Progesterone	Ovary, corpus luteum	Steroid	Regulates menstrual cycle; maintains pregnancy
Testosterone	Testis	Steroid	Stimulates and maintains male sex characteristics

[a]Protein hormones are discussed in Special Topic F, Chapter 21, and Chapter 24.

pancreatic hormones insulin and glucagon are considered in Chapter 24. In the remaining sections of this special topic, we focus our attention on some steroid hormones and the prostaglandins.

The steroid hormones (e.g., adrenocortical hormones, sex hormones) are lipids. They are soluble in the lipid components of the cell membrane and can easily diffuse into cells. Inside the cell they combine with specific receptor molecules in the cytoplasm. They may influence enzymatic reactions directly, or the steroid–receptor complex can enter the nucleus. In the nucleus, steroids bind to specific sites on DNA, where they increase the rate of synthesis of mRNA (see Section 23.5), thus increasing the rate of biosynthesis of cellular enzymes. Because the primary effect of steroid hormones is on gene expression (the regulation of protein synthesis), their effects are much slower than those of other hormones (i.e., hours rather than minutes).

H.2
ADRENOCORTICAL HORMONES

The outer part, or cortex, of the adrenal[1] glands uses cholesterol to produce a mixture of many steroids that are essential to life. These steroids constitute a family of hormones of which **aldosterone** and **cortisol** are the major representatives. Aldosterone is called a *mineralocorticoid*. This name alludes to its function in regulating the exchange of sodium, potassium, and hydrogen ions. Although aldosterone acts on most cells in the body, it is particularly effective in enhancing the rate of reabsorption of sodium ions in the kidney tubule and increasing the secretion of potassium ions and/or hydrogen ions by the tubule. Since the concentration of sodium ions is the major factor in water retention in the tissues, aldosterone also promotes water retention and reduces urine output. It thus supplements the action of vasopressin (Section 21.6).

Cortisol
(Hydrocortisone)

Aldosterone

Cortisone

Prednisolone

[1] The term *adrenal* comes from the gland's location in the body, *ad*jacent to the *renal* (kidney—Figure H.3).

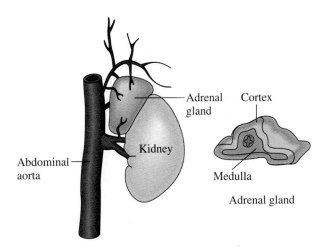

Figure H.3
The anatomical location and composition of the adrenal gland.

Cortisol and its keto derivative, cortisone, are called *glucocorticoids*. These hormones regulate a number of key metabolic reactions (e.g., they increase glucose production and mobilize fatty acids and amino acids). They also inhibit the inflammatory response.[2] Glucocorticoids are used as drugs for immunosuppression after transplant operations and in the treatment of autoimmunity, severe skin allergies, and rheumatoid arthritis. The hormones or their analogs are injected, taken orally, or applied directly to the site of inflammation.

Pharmaceutical companies have been able to obtain large quantities of synthetic cortisone by means of an involved process utilizing cholic acid isolated from the bile juices of cattle. Prolonged use of cortisone can have serious side effects, including high blood pressure, wasting of muscles, and resorption of bone. Cortisone has been supplemented with a synthetic analog, prednisolone, which is effective in much smaller doses, thereby greatly reducing the side effects. Prednisolone has been used in combination with an antibiotic in the treatment of autoimmunity, and for immunosuppression after kidney and liver transplant operations.

[2] **Inflammation** is a tissue response to injury or stress (e.g., chemical irritants, exposure to radiation or to extreme temperature). The blood vessels in the surrounding area become dilated to bring extra blood to the afffected area. (This accounts for the redness associated with inflammation.) Dilated blood vessels are more permeable, and there is a tendency for fluid to leave the blood and enter the damaged tissue, causing swelling. Prostaglandins (Section H.6) contribute to the inflammatory response by (1) promoting vasodilation of the blood vessels, (2) increasing the permeability of the capillaries, and (3) stimulating the pain receptors. It may be that the glucocorticoids are effective anti-inflammatory agents because they inhibit an enzyme (phospholipase) that is necessary for the synthesis of arachidonic acid, a prostaglandin precursor.

H.3
SEX HORMONES AND ANABOLIC STEROIDS

The sex hormones are a class of steroids that are secreted by the gonads (ovaries or testes), the placenta, and the adrenal glands. The primary male sex hormones (called *androgens*) such as *testosterone* and *androstenedione* are produced in the testes (and in lesser amounts by the adrenal cortex and the ovary). They control the primary sexual characteristics of males, that is, the development of the male genital organs and the continued production of sperm. Androgens are also responsible for the development of the secondary male characteristics such as facial hair, deep voice, and muscle strength.

Two sex hormones are of particular importance in females. *Progesterone* prepares the uterus for pregnancy and prevents the further release of eggs from the ovaries during pregnancy (see Section H.4). The *estrogens* are mainly responsible for the development of female secondary sexual characteristics such as breast development and increased deposition of adipose tissue in the breast, buttocks, and thighs. Progesterone is the precursor for the synthesis of the androgens, and the estrogens are synthesized from the androgens. Estrone is derived from androstenedione, whereas estradiol (the major estrogen hormone) is formed from testosterone. Thus, both males and females secrete androgens and estrogens. The difference between the sexes is in the amount of secreted hormones, not the total absence of one or the other group. Notice that the male and female hormones exhibit only very slight structural differences, yet their physiological effects differ enormously.

There has been a great deal of controversy generated by the widespread use of steroids among athletes. The drugs in question are synthetic androgens that stimulate protein synthesis (especially in skeletal muscle cells) without affecting the sex glands. We mentioned that testosterone is responsible for an increase in the amount of muscle cells (and more aggressive behavior), as well as its virilizing activity. However, testosterone is not very active if taken orally because it is metabolized in the liver. The incorporation of a methyl group at C-17 prevents this metabolism. Introduction of a second double bond in ring A produces a compound, methandienone (Dianabol), that has anabolic activity (stimulation of protein synthesis) but little of the virilizing effects of testosterone.

Many male and female athletes take more than 100 mg of steroids each day to increase their muscle bulk and muscle strength. Some reports indicate that under conditions of regular training, anabolic steroids do increase muscle strength, increase aggressiveness, and decrease fatigue. However, the same reports mention the adverse side effects that constitute a serious health hazard with long-term use of steroids. These include kidney, liver, and heart disease, cancer, sterility, impotence, and increased risk of diabetes. In females, there are increased masculinization and termination of menstruation. Chemists can detect the presence of synthetic steroids in the body by monitoring the urine and testing for degradation products. Anabolic steroids are marketed for use in the treatment of senile

Testosterone

Androstenedione

Methandienone
(Dianabol)

Estradiol

Estrone

Estrogens

Progesterone

Mestranol   Ethinyl estradiol

(analogs of the estrogens)

Norethynodrel   Norethindrone

(analogs of the progesterone)

debility, anorexia, and anemia and during convalescence.

Synthetic derivatives of the female sex hormones have attracted widespread attention. When taken regularly, these drugs effectively function to prevent ovulation. The oral contraceptives are usually mixtures of two compounds that are analogs of progesterone and estradiol. For example, Enovid is a combination of norethynodrel and mestranol, whereas Ortho-Novum contains norethindrone and ethinyl estradiol. Most of the combination pills sold in the United States contain 1 mg of the progesterone analog and less than 0.03 mg of the estrogen analog. The prevention of ovulation is also effected by administration of progesterone and estradiol, but these hormones must be injected into the body for maximal results. Only a slight structural difference (incorporation of the acetylenic group at C-17) confers upon the synthetic compounds the ability to be taken orally and to function in the same manner as the steroids produced by the body.

Sex hormones—both natural and synthetic—are sometimes used therapeutically. For example, a woman who has had her ovaries removed may be given female hormones to compensate for those no longer produced by the ovaries. Some of the earliest chemical compounds employed in cancer chemotherapy were sex hormones. The male hormone testosterone was used to treat carcinoma of the breast in females; estrogens, female sex hormones, were given to males to treat carcinoma of the prostate. Sex hormones are also important in sex-change operations. Before corrective surgery, hormones are administered to promote the development of the proper secondary sexual characteristics.

H.4
CONCEPTION AND CONTRACEPTIVES

Control of fertility continues to be of paramount concern throughout the world, even though the population growth rate has begun to decline in some countries. This is due partly to government sanctions (e.g., China), partly to education, and partly to the extensive use of oral contraceptives. Several billion pills are produced annually, and these are consumed by more than 50 million women all over the world. (Oral contraceptives are second only to sterilization as a birth control method in the United States.)

In order to understand how oral contraceptives work, it is first necessary to understand the normal physiology of the female menstrual cycle (Figures H.4 and H.5). This cycle is characterized by rhythmic monthly changes in the rates of secretion of the sex hormones (estrogens and progesterone) and corresponding changes in the sex organs themselves. Menstruation marks the terminal events of the normal ovarian cycle. The cycle begins when the pituitary gland releases *follicle-stimulating hormone* (FSH) into the bloodstream. FSH is primarily responsible for stimulating the growth of one of the eggs in the ovary. Under the influence of FSH, the follicle, or sack surrounding the egg, secretes the estrogen hormones. The estrogens function to maintain secondary sexual traits and to prepare the lining of the uterus to receive the fertilized egg. As the follicle ages, the secretion of estrogens increases, and this

Figure H.4

Changes in the ovary and the uterus during the menstrual cycle. Pregnancy—or the pseudopregnancy caused by the birth control pill—prevents ovulation. [Adapted with permission from Phillip D. Sparks et al., Student Study Guide for the Biological Sciences, 3rd ed., Burgess, Minneapolis, 1973.]

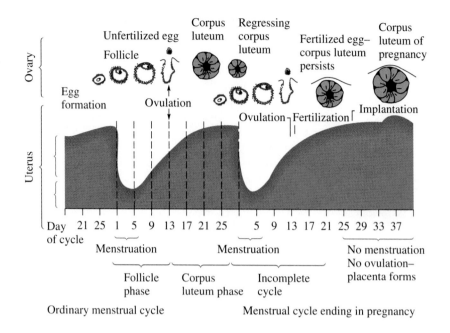

Figure H.5

Hormone interactions in the female reproductive system.

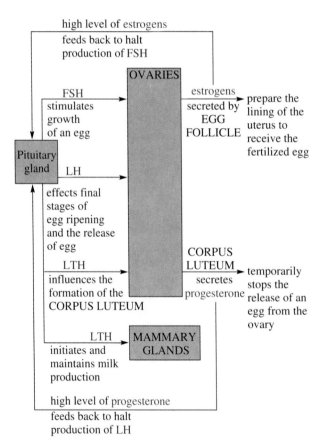

rise in the estrogen level of the body has a negative feedback effect on the pituitary gland. As the amount of estrogens increases, the output of FSH decreases. At this stage another hormone, luteinizing hormone (LH), is secreted by the pituitary gland. LH oversees the final stages of egg maturation and ensures that the egg is ejected from the ovary (ovulation).

Following ovulation, a third hormone, luteotrophic hormone (LTH), is secreted by the pituitary gland. LTH influences the formation of a tissue called the corpus luteum (yellow tissue). As it grows in size, the corpus luteum releases estrogens and progesterone, which temporarily stop the release of another egg from the ovary. Again there is negative feedback control whereby increased amounts of estrogens and progesterone result in a diminution of LH secretion by the pituitary gland. Further egg maturation is thus prevented until the fate of the ovulated egg has been decided.

If the egg is not fertilized, the life of the corpus luteum is short and it disintegrates. When this occurs, many of the small blood vessels that are present in the uterus rupture and bleed. This small amount of blood, along with the fragments of glands and the mucin from these glands, forms the menstrual discharge. Also, if fertilization does not occur, there is a sharp decrease in the amounts of estrogens and progesterone. As a result, the pituitary gland secretes more FSH and LH, and the cycle begins anew.

When an embryo is implanted in the uterus, its cells secrete the hormone HCG[3] (human chorionic gonadotropin). This hormone causes the corpus luteum to be maintained and to continue secreting estrogens and progesterone. Thus, the uterine wall continues to grow and develop and the menstrual cycles are inhibited. In the last 6 months of pregnancy, the secretion of HCG by the implanted embryo declines because the estrogens and progesterone are produced in large quantities by the placenta. The high level of these hormones maintains the uterine wall, stimulates the further development of the mammary glands, and prevents the pituitary from secreting FSH and LH. In this way no additional mature eggs are released during pregnancy.

The two synthetic analogs of the female sex hormones that are in the combination pill deceive the female body into thinking it is pregnant and acting accordingly. Finding an abundant supply of the sex hormones in the blood, the pituitary gland shuts down its supply of FSH and LH.

[3] Pregnancy can usually be diagnosed about 10 days after the first missed menstrual period by a test that detects the presence of HCG in a woman's urine.

The egg follicles are not developed, nor are eggs released into the fallopian tube. The synthetic hormones also enhance the contraceptive effect by increasing the viscosity of the mucus around the cervix, thus setting up a barrier that is difficult for sperm to penetrate.

There have been reports of increased incidence of breast and cervical cancers, thromboembolisms (blood clots), hypertension (high blood pressure), depression and psychiatric disturbances, benign liver tumors, gallbladder disease, urinary tract infections, and other risks among women who have taken the pill. In most cases, the increased risk of adverse side effects has been linked to the estrogen component in the pill. For this reason, the estrogen content in most products was decreased. There are now "minipills" that contain only progesterone analogs (called *progestins*). Minipills have to be taken every day, unlike the combination pill. They are less effective in suppressing ovulation, but they produce fewer side effects. (The FDA now requires a warning about the possible effects of estrogens for all drugs that contain them.)

A report by the Rockefeller Foundation concluded that the pill is a "highly effective and generally safe method of birth control—but not for all women." The major points of the report, based on a review of oral contraceptive research from medical centers around the world, are

1. Oral contraceptives are highly effective and generally safe for most women.
2. The risk of developing serious illness as a consequence of taking the pill is small.
3. The number of deaths associated with the oral contraceptives is of a very low order of magnitude.
4. The pill's long-term side effects must continue to be monitored closely to safeguard users.
5. There is no evidence connecting use of oral contraceptives with human cancers of the breast, uterus, or ovary.
6. Increased risk of venous thrombosis of the legs and of cerebral thrombosis has been established.
7. Smokers, women over 40, and women with a history of blood clots are at a higher risk of experiencing clot formation. Such blood clots can clog a blood vessel and cause death by stroke or coronary heart attack.
8. Women who use oral contraceptives have a greater risk than nonusers of having gallbladder disease requiring surgery. The increased risk may first appear within one year of use and may double after 4 or 5 years.

The death rate associated with use of the pill is about 3 in 100,000. This is only one-tenth that associated with childbirth (which is about 30 in 100,000). For the general population, the pill is probably as safe as aspirin. Women who have blood with an abnormal tendency to clot should not take the pill, nor should those who experience any serious side effects.

Progesterone is essential for the maintenance of pregnancy. If its action is blocked, pregnancy cannot be established or maintained. Rousel-Uclaf, a French subsidiary of Hoechst, has developed a drug that blocks the action of progesterone. Mifepristone (RU-486) is available in Sweden, France, Britain, and China, where it has replaced a substantial number of surgical abortions. A woman who wishes to abort a pregnancy takes three mifepristone tablets, followed in a few days by an injection of a prostaglandin (Section H.6). The lining of the uterine wall and the implanted fertilized egg are sloughed off, and pregnancy is terminated.[4]

Mifepristone

People who believe that human life begins when the ovum is fertilized by the sperm equate the use of mifepristone to abortion. They oppose making the drug available. Other people consider the drug just another method of birth control. They see its use to be safer and more convenient than surgical abortion.

In 1990 the FDA approved the use of an under-the-skin implant for birth control. Levonorgestrel (Norplant), a synthetic analog of progesterone, is incorporated into six thin, tubular plastic capsules, each about 2.5 cm long. These are then implanted under the skin of the upper arm. Release of the drug over time prevents pregnancy for five years with a failure rate of only 0.2%. It has about the same side effects as the progestin-only minipill. (Levonorgestrel is the progestin used in some ordinary birth control pills.)

Why do females have to bear the responsibility for contraception? Why not a pill for males? Recent studies have shown that many men—including a majority of the younger ones—are willing to share the risks and the re-

[4] In 1992, it was reported that RU-486 is an effective "morning after pill." The drug safely prevented pregnancy when taken soon after a woman had sexual intercourse.

sponsibility of contraception. Nevertheless, there are biological reasons for females to bear the burden: women get pregnant when contraception fails, and in females, contraception has only to interfere with one monthly event—ovulation. On the other hand, males produce sperm continuously.

Actually, a good deal of research has gone into male contraceptives. Estrogens would work, but they would bring about the development of female characteristics in men, including a complete loss of interest in sexual relations with women.

Gossypol, a pigment found in cottonseed, has been used in China as a contraceptive for males.

Gossypol

A drug called danazol has also been tested and found safe and effective. Danazol is testosterone enanthate, administered along with testosterone itself. It suppresses sperm production, and the effect is reversed when the drug is withdrawn. Like the pill for women, danazol often causes weight gain in men. The main problem, however, is that the drug is too expensive for widespread use. Perhaps the best bet for a male contraceptive is some sort of drug that doesn't stop sperm formation but does block growth or transport of the sperm.

Testosterone enanthate
(danazol)

The availability of vasectomies—simple surgical procedures that block the emission of sperm—has lessened the demand for a male contraceptive. A major drawback to vasectomy, however, is that it is often irreversible. A safe, cheap, and effective male contraceptive is still a goal of biochemical research.

The control of human reproduction is an area subject to great moral, political, and legal controversy. Much more research is needed to increase our understanding of the biochemistry and physiology of reproduction. Armed with new knowledge, chemists might be able to design drugs acceptable to people of divergent views.

H.5
ESTROGEN REPLACEMENT THERAPY

In most women, menstrual cycles continue to occur into the late forties, at which time they become increasingly irregular and finally cease altogether. This period is called *menopause*. Follicles no longer mature, ovulation does not occur, and the plasma estrogen concentration sharply decreases.

As a result of lower estrogen levels, the female secondary sexual characteristics are modified. In addition, there is no longer the feedback control by estrogens over the secretion of FSH and LH. These hormones are released continuously by the pituitary gland, and they seem to be responsible for some of the unpleasant sensations that menopausal women experience (e.g., hot flashes, irritability, anxiety, and fatigue). These discomforts can be overcome by the administration of daily doses of estrogens or estrogen substitutes. The drugs also seem to reduce the risk of brittle bones (osteoporosis) and may reduce the risk of heart disease and stroke.

Estrogens such as estrone, estradiol, ethinyl estradiol, and mestranol are used in estrogen replacement therapy (ERT), particularly in the menopausal and postmenopausal female. Other synthetic compounds that are structurally similar to the estrogens are also used. One such compound is *diethylstilbestrol* (DES). Other medical applications of DES are suppression of lactation, treatment of breast and prostate carcinoma, treatment of acne, and postcoital contraception. It has also been used to stimulate female fertility. However, DES was found to cause mammary tumors in some strains of mice. Additionally, daughters born to women who took DES during their pregnancy developed rare forms of vaginal cancer (clear cell adenocarcinoma) when they were in their early twenties. The

DES

FDA has ordered that no estrogens may be prescribed during pregnancy. The FDA also removed from the market DES tablets that were used as "morning after pills." The use of DES as a postcoital contraceptive is lawful only following rape or incest. (Notice that the trans form of DES structurally approximates the estrogen steroids.)

Within the past decade, many investigators have suggested that ERT substantially protects women against heart disease. Heart disease is the leading cause of death in women, killing more women each year than all cancers combined. But many women and their physicians have been wary of estrogen because of some reports linking it to a slightly elevated risk of breast and uterine cancers. Of the more than 43 million American women over 50, fewer than one-third take ERT.

In 1991 the *New England Journal of Medicine* published the results of the largest study to date on ERT and women (48,000 menopausal female nurses over a 10-year period). Researchers from Harvard Medical School had found that women who took estrogen had only half the risk of heart attacks and death from coronary heart disease that nonusers did. The report suggested that even healthy women should consider taking estrogen after menopause.

However, the increased risk of contracting cancer is still valid. Usually progesterone is prescribed along with estrogen to counter the increased risk of cancer of the uterine lining that estrogen alone can cause. In considering ERT, women should consult with their doctors to be fully informed of the risks versus the benefits.

H.6
PROSTAGLANDINS

Perhaps no compounds since birth control pills have stimulated as much activity in pharmaceutical companies as have the group of compounds called prostaglandins. Over 6000 papers are published each year on these compounds. **Prostaglandins** are hormonelike substances that were originally isolated from semen found in the prostate gland. In the mature male, the prostate gland secretes about 0.1 mg/day of prostaglandins. However, they are biosynthesized by most mammalian tissue, and they affect almost all organs in the body.

Prostaglandins are a family of unsaturated fatty acids, each containing 20 carbon atoms and having the same basic skeleton as prostanoic acid (Figure H.6a). The major classes are PGA, PGB, PGE, and PGF followed by a subscript that denotes the number of double bonds outside the five-carbon ring. Prostaglandins are not stored as such in cells. Rather they are synthesized on demand from arachidonic acid, a 20-carbon polyunsaturated fatty acid that is released from phospholipids within cell membranes (by

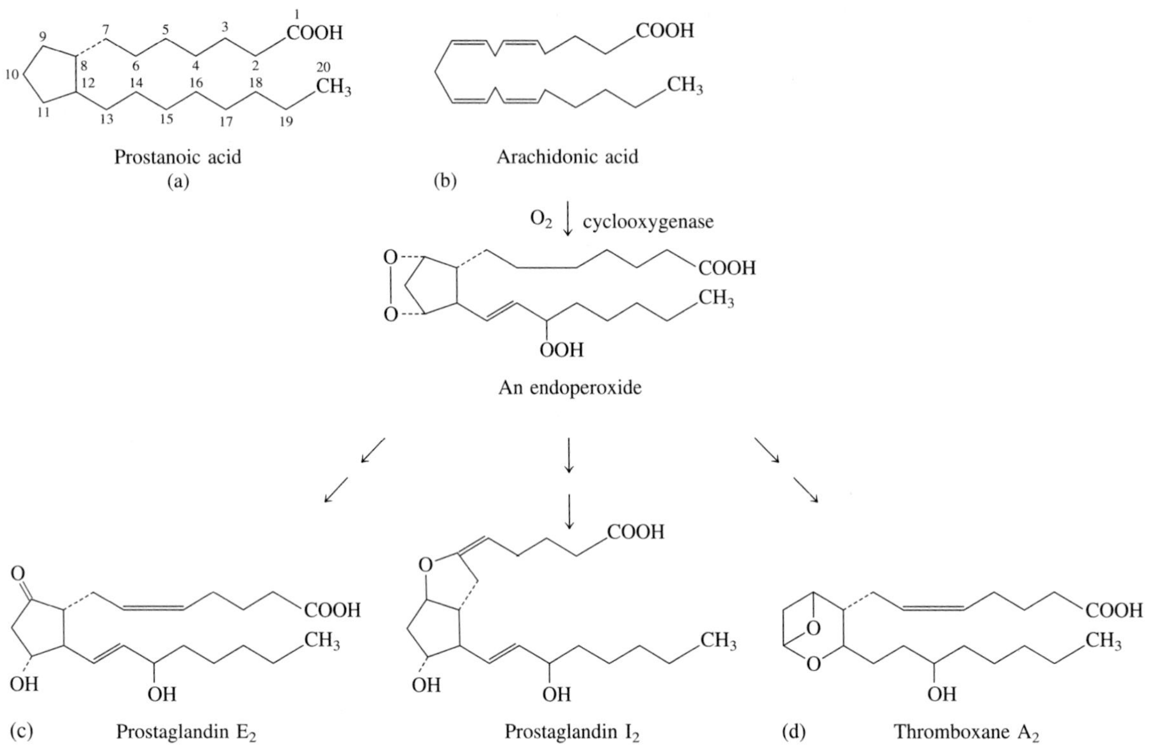

Figure H.6

The arachidonic acid cascade. (a) Prostanoic acid is the parent compound of the prostaglandins. (b) Arachidonic acid is released from membrane phospholipids and is the precursor
for the synthesis of the prostaglandins (c) and the related thromboxanes (d).

the action of the enzyme phospholipase). Arachidonic acid
is then converted into an endoperoxide by an enzyme complex called prostaglandin cyclooxygenase. The endoperoxide intermediate can then be transformed into the prostaglandins or into a related group of compounds, the
thromboxanes. This sequence of reactions is sometimes
referred to as the *arachidonic cascade* (Figure H.6).

The prostaglandins are among the most potent biological substances known. Slight structural changes are responsible for quite distinct biological effects; however, all
prostaglandins exhibit some ability to induce smooth muscle contraction, to lower blood pressure, and to contribute
to the inflammatory response (recall footnote 2, page
531). We mentioned that certain steroid drugs (cortisol
and its synthetic analogs) exert their anti-inflammatory
action by inhibiting the release of arachidonic acid from
membrane phospholipids (i.e., they may inhibit *phospholipase*). On the other hand, aspirin, and the other nonste-
roidal anti-inflammatory agents (e.g., indomethacin—
Indocin; ibuprofen—Motrin, Advil, Nuprin) obstruct the
synthesis of prostaglandins by inhibiting cyclooxygenase,
which converts arachidonic acid to endoperoxides.

The prostaglandins act as mediators of hormone
action, and they are believed to have numerous potential
therapeutic uses. These include regulating menstruation
and fertility, preventing conception, inducing labor, decreasing gastric secretions, controlling blood pressure,
sensitizing the pain receptors, maintaining body temperature, and relieving asthma and nasal congestion. Thromboxane A_2 (made in blood platelets) induces blood clotting
by stimulating blood platelet aggregation, whereas prostaglandin I_2 (made in arterial walls) prevents clotting by
inhibiting platelet aggregation. Recall (Section E.1) that
one of the side effects of aspirin is a prolonged bleeding
time. This is due to the inhibition of platelet aggregation
by blocking biosynthesis of thromboxane A_2.

The major clinical use of prostaglandins and their analogs concerns the induction of abortion. Their mechanism
is uncertain, but it is different from that of the steroids.
The prostaglandins cause regression of the corpus luteum,
uterine contractions, and abortion of the embryo. Because
they induce abortion, prostaglandins would have to be
taken only once a month or only if a menstrual period were
missed. Obviously, a great deal of controversy will be

generated if and when these compounds become commercially available.[5] Practically every major pharmaceutical

[5] As therapeutic agents, prostaglandins must be administered by either intravenous or intrauterine injection. They cannot be taken orally because they are rapidly degraded in the digestive tract. A major goal of researchers is to develop prostaglandin analogs that are orally effective. Several of these drugs will likely be on the market soon, after further testing has demonstrated their relative safety and confirmed their therapeutic value.

company now has active prostaglandin research programs under way to develop new syntheses and to discover new natural sources of prostaglandins. Chemists have developed several completely synthetic sequences for the preparation of prostaglandins. S. K. Bergstrom, B. I. Samuelsson, and J. R. Vane shared the 1982 Nobel prize in physiology and medicine for their work on prostaglandins.

EXERCISES

1. What is the sequence of events that results in the release of hormones from an endocrine gland?
2. Define and give an example of each of the following.
   a. hormone   b. androgen
   c. estrogen   d. progestin
3. Match each hormone to the gland that produces it.
   a. thyroxine   (1) corpus luteum
   b. FSH   (2) pancreas
   c. progesterone   (3) thyroid
   d. insulin   (4) pituitary
   e. epinephrine   (5) adrenal
4. What gland produces releasing factors?
5. What gland is the target of releasing factors?
6. Describe the general sequence of events by which a nerve impulse is translated by the endocrine system into a changed physiological state.
7. Match each hormone with its effect.
   a. FSH   (1) stimulates male sex characteristics
   b. prolactin   (2) stimulates female sex characteristics
   c. thyroxine   (3) regulates cell uptake of glucose
   d. testosterone (4) stimulates milk production
   e. estradiol   (5) increases rate of cellular metabolism
   f. insulin   (6) stimulates growth of follicle in ovary
8. Answer the following questions about the menstrual cycle.
   a. What are the roles of the pituitary hormones FSH and LH?
   b. What is the relationship between the ovarian follicle and the corpus luteum?
   c. Which ovarian hormone affects the lining of the uterus in the early stages of the cycle? Which produces the changes in this lining in the later stages of the cycle?
   d. What triggers the start of a new cycle?
9. What is the general structural classification of the compounds incorporated in birth control pills?
10. How do the birth control pills work?
11. What structural feature renders a synthetic steroid sex hormone effective orally?
12. Which of the two categories of synthetic hormones, estrogens or progestins, is associated with undesirable side effects? Name some of these side effects.

13. The hormone aldosterone causes the kidney to reabsorb sodium, chloride, and bicarbonate ions. Name all the functional groups on aldosterone.

Aldosterone

14. Compare the structural features of the androgen testosterone and the estrogen estradiol. In what ways are the structures similar? In what ways are they different?
15. Androstenedione is an androgen produced in the adrenal cortex. Name its functional groups. How does its structure differ from that of testosterone? How could it be made from testosterone?
16. The birth control pill Ovulen contains mestranol as the estrogen and the following compound as the progestin. Name all the functional groups in the compound.

17. What are the differences in biological function between the mineralocorticoids and the glucocorticoids?
18. What fatty acid is the precursor of the prostaglandins?
19. What are some potential therapeutic uses of prostaglandins?

Chapter 21
PROTEINS

Silkworms and cocoons in an egg box. The worms produce silk, a natural protein used in clothing manufacture.

Carbohydrates, lipids, and proteins—these are the three classes of foods. All are essential to life, but proteins perhaps are closest to the stuff of life itself. No living part of the human body—or of any other organism, for that matter—is completely without protein. There is protein in the blood, in the muscles, in the brain, and even in tooth enamel. The smallest cellular organisms, the bacteria, contain protein. And the viruses, so small that they make the bacteria look like giants, are nothing but large molecules of nucleoproteins. (Nucleoproteins are combinations of proteins with nucleic acids.) Nucleic acids (Chapter 23) cause proteins to be formed, and through these proteins they control the structure and function of cells.

Each type of cell makes its own specific kinds of proteins. Proteins serve as structural material for animals, much as cellulose does for plants. Muscle tissue is largely protein. So are skin and hair. Proteins are made in different forms in different animals: silk, wool, nails, claws, feathers, horns, and hoofs are all proteins (Figure 21.1).

Whereas the carbohydrates and lipids are used primarily as energy sources, the primary function of the proteins is body building and maintenance. Lipids and carbohydrates are stored by the body as energy reserves, but proteins are not stored to any appreciable extent. It is possible for humans to survive for a short period of time on a diet consisting of protein, vitamins, and minerals. We could not survive over the same period of time on a protein-free diet containing lipids, carbohydrates, vitamins, and minerals.

Figure 21.1
Leather, silk, wool, gelatin, and meat tenderizer are all rich in protein.

Proteins are giant polymeric molecules (linear polymers of amino acids) that vary greatly in molecular dimensions. Their molecular weights may range from several thousand to several million daltons. In addition to carbon, hydrogen, and oxygen, all proteins contain nitrogen, and many contain sulfur, phosphorus, and traces of other elements. The composition of most proteins is remarkably constant at about 51% carbon, 7% hydrogen, 23% oxygen, 16% nitrogen, 1–3% sulfur, and less than 1% phosphorus.

In an overcrowded, hungry world, protein is of increasing importance. The rich nations have it—in the form of beefsteak, fish, fowl, soybeans. The poor nations need it but can't afford meat. Many make do with rice or corn and beans. A nation's consumption of sulfuric acid has long been considered an indication of its industrial development. Perhaps its consumption of protein is a better indication of the quality of life of its people, for without protein no nation can have the healthy, vigorous people who are vital to progress.

The **dalton** is a unit of mass used by biologists. It is equivalent to the atomic mass unit: 1/12 the mass of an atom of carbon-12, 1.66×10^{-24} g. A 30,000-dalton protein has a molecular weight of 30,000 g/mol.

21.1
AMINO ACIDS

Proteins may be defined as high molecular weight compounds consisting largely or entirely of chains of amino acids. In this respect, proteins may be considered to be polymers analogous to the polysaccharides. However, 20 *different* structural monomeric units are commonly found in proteins, and these are the amino acids. The proteins in all living species, from bacteria to humans, are constructed from the basic set of 20 amino acids. Several other amino acids (e.g., hydroxyproline), which occur to some extent in certain proteins, are all derivatives of the common amino acids and are modified *after* incorporation into the polypeptide chain. With the exception of proline (which contains an alpha secondary nitrogen atom), the amino acids that are the building blocks of proteins are characterized by a primary amino group bonded to the alpha carbon.

Each amino acid has unique characteristics as a result of the size, shape, solubility, and ionization properties of the different R groups. (The R group is referred to as the *amino acid side chain.*) As we shall see in Section 21.8, the side chains of amino acids exert a profound effect on the conformation and the biological activity of proteins.

Amino acids can be classified in several ways. We choose to group them, according to the nature of their side chains, into four classes: (1) nonpolar, (2) polar and neutral, (3) acidic, and (4) basic. The structures of the common amino acids, their three-letter abbreviations, and certain of their distinctive features are given in Table 21.1.

An α-amino acid

Proline

Hydroxyproline

Table 21.1
Naturally Occurring Amino Acids

Name	Abbreviation	Structural Formula	MW	Distinctive Features
1. Amino Acids with a Nonpolar R— Group				
Alanine	Ala		89	The least hydrophobic member of this class because of its small R group (methyl).
Valine	Val		117	Most animals cannot synthesize branched-chain amino acids. They are therefore essential in the diet.
Leucine	Leu		131	
Isoleucine	Ile		131	
Phenylalanine	Phe		165	
Tryptophan	Trp		204	A heterocyclic amino acid (a derivative of indole).

3. Acidic Amino Acids

Methionine	Met	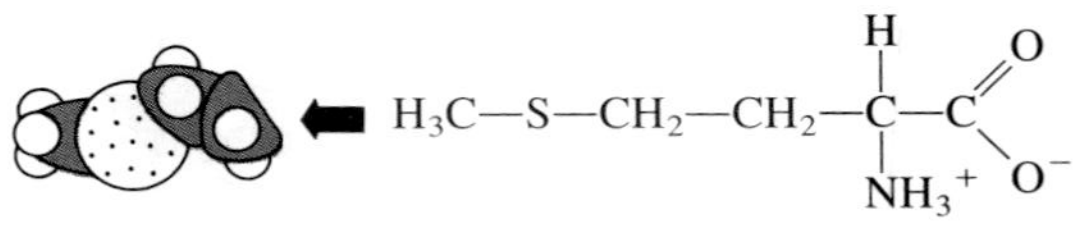 $H_3C-S-CH_2-CH_2-\overset{\overset{\displaystyle H}{\mid}}{\underset{\underset{\displaystyle NH_3^+}{\mid}}{C}}-\overset{\displaystyle O}{\underset{\displaystyle O^-}{C}}$	149	Contains a sulfur atom in the nonpolar side chain and is important as a donor of methyl groups.
Proline	Pro	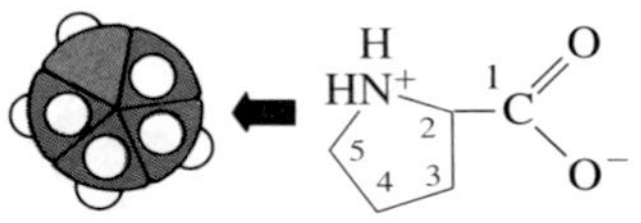	115	Contains a secondary amino group rather than a primary amino group and so is referred to as an *α-imino acid*. A major constituent of the structural protein collagen. Hydroxylation of proline yields 4-hydroxyproline (Hypro), which is also abundant in collagen.

2. Amino Acids with a Polar But Neutral R— Group

Glycine	Gly	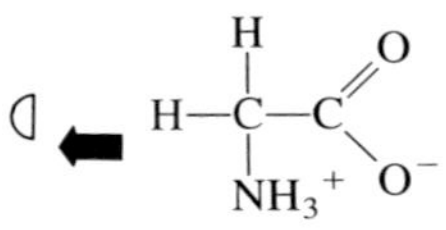	75	The only amino acid lacking a chiral carbon. Sometimes classified as a nonpolar amino acid, but its single hydrogen R group is too small to influence the polarity of the molecule.
Serine	Ser	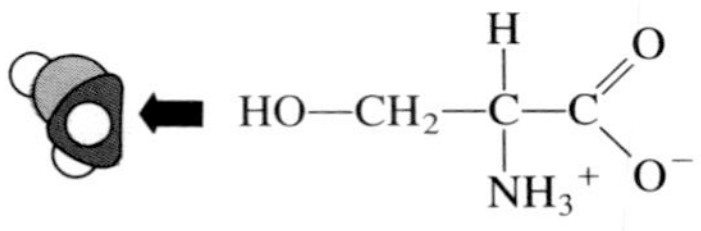	105	Occurs at the active site of many enzymes. The hydroxyl group may take part in the usual alcoholic reactions such as ester formation.
Threonine	Thr	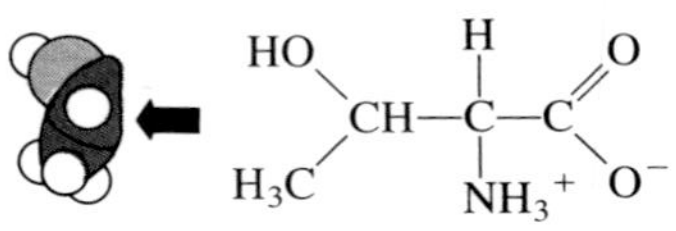	119	Named for its similarity to the sugar threose (contains two chiral carbons).

(Table continues)

21.5
THE SEQUENCE OF AMINO ACIDS

For peptides and proteins to be physiologically active, it is not enough that they incorporate certain amounts of specific amino acids. The order or *sequence* in which the amino acids are connected is also of critical importance. Glycylalanine is different from alanylglycine.

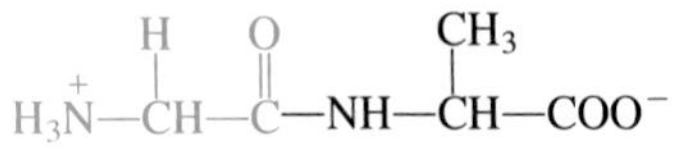

Glycylalanine (Gly-Ala)    Alanylglycine (Ala-Gly)

Although the difference seems minor, the two substances behave differently in the body.

As the length of a peptide chain increases, the possible sequential variations become almost infinite. And this potential for many different arrangements is exactly what one needs in a material that will make up such diverse things as hair and skin and eyeballs and toenails and a thousand different enzymes. To appreciate the enormous complexity of protein molecules, consider that the average protein contains anywhere from 100 to 300 amino acids. It is estimated that a typical animal cell contains about 9000 different proteins, and the human body contains over 100,000 different protein molecules, all of which are characterized by different sequential arrangements of the 20 fundamental building blocks.

Just as we can make millions of different words with our 26-letter English alphabet, we can make millions of different proteins with the 20 or so different amino acids. Just as one can write gibberish with the English alphabet, one can make nonfunctioning proteins by putting together the *wrong sequence* of amino acids. Yet while the correct sequence is ordinarily of utmost importance, it is not always absolutely required. Just as you can sometimes make sense of incorrectly spelled English words, a protein with a small percentage of "incorrect" amino acids may continue to function.[1] It may not function as well, however, as a protein with the correct sequence. And sometimes a seemingly minor change can have a disastrous effect. Some people have hemoglobin with a single incorrect amino acid unit in about 300. That "minor" error is responsible for sickle cell anemia, an inherited condition that ordinarily proves fatal (Section 28.9).

21.6
SOME PEPTIDES OF INTEREST

There are several naturally occurring peptides that possess significant biological activity. In Section E.5, we said that the brain produces a variety of peptides, several of which act like morphine to relieve pain. Rather, we should say that morphine acts like these peptides because the brain was making peptides long before people discovered

[1] An interesting fact about living systems is that different ones contain, in many cases, almost identical protein molecules. Insulin is a striking example of this phenomenon (see Figure 21.7). Not only do a variety of mammalian insulin molecules have similar primary structures, but they also have the same biochemical properties. This is fortunate for some diabetics who develop an allergy to a particular type of insulin—they can often be treated with insulin from another species that does not produce the allergic reaction. In this instance, alteration of some of the amino acids does not cause the protein to have an altered physiological effect. This explanation of the special physiological functions of proteins in terms of their structure will become apparent when we consider the concept of the *active site* in relation to enzymes in Section 22.3. (As we shall see in Section 23.9, human insulin is now being produced by recombinant DNA technology, and this is diminishing the allergy problem.)

Structure or Function Affected	Vasopressin	Oxytocin
Water diuresis	Inhibits	Has no effect on
Blood pressure	Elevates	Slightly lowers
Coronary arteries	Constricts	Slightly dilates
Intestinal contractions	Stimulates	Has questionable effect on
Uterine contractions[a]	Stimulates	Stimulates
Ejection of milk	Slightly stimulates	Stimulates

Table 21.3
A Comparison of the Effects of Oxytocin and Vasopressin

[a]Response varies with species as well as with the stage of the normal and the reproductive cycles.

the painkilling effect of the juice of the opium poppy. As Table H.1 shows, many hormones are peptides or proteins. *Oxytocin* and *vasopressin* are cyclic nonapeptides produced by the pituitary gland. Notice that seven of the nine amino acids are identical in both peptides, yet their physiological effects are markedly different. Oxytocin stimulates lactation and causes the contraction of smooth muscles in the uterine wall. It is often administered at childbirth for the induction of labor.

Vasopressin is called the *antidiuretic*[2] *hormone* (ADH) because it acts on the kidneys to reduce the amount of water excreted. Thus, the major function of vasopressin is to increase water reabsorption in the kidney. After drinking alcohol, many people excrete more urine than can be accounted for by the volume of water in their drinks. It is thought that alcohol inhibits the secretion of vasopressin, and therefore the kidney does not reabsorb as much water. A deficiency of vasopressin, or an inability of the kidney to respond to vasopressin, results in diabetes insipidus, in which too much urine is excreted (>10 L/day). This disease (which is treated by administering vasopressin) should not be confused with diabetes mellitus (see Section 24.4). In addition, vasopressin stimulates the contractions of muscles in the walls of blood vessels and thus increases the blood pressure. It has been used to overcome low blood pressure caused by shock following surgery.

Table 21.3 offers a comparison of the effects of vasopressin and oxytocin. Remember the great similarity in the structure of these compounds as you look at the table.

Bradykinin, a nonapeptide produced in the blood by the cleavage of larger protein molecules, has the amino acid sequence

Arg-Pro-Pro-Gly-Phe-Ser-Pro-Phe-Arg

In terms of activity, bradykinin, a peptide, sounds almost like the prostaglandins (Section H.6), which are lipids. It is a potent biochemical that causes a lowering of blood pressure, stimulation of smooth muscle tissue, an increase in capillary permeability, and pain. The reverse peptide has been synthesized.

Arg-Phe-Pro-Ser-Phe-Gly-Pro-Pro-Arg

It shows none of the activity of bradykinin.

The octapeptide angiotensin II is produced in the kidneys.

Asp-Arg-Val-Tyr-Ile-His-Pro-Phe

This substance is the most powerful vasoconstrictor known. It acts to maintain blood pressure. Some forms of hypertension probably involve overproduction of angiotensin II. Drugs that act to suppress its production are important in the control of hypertension.

```
Phe-Tyr-Cys
 |       |
 |       S
 |       |
 |       S
 |       |
Gln-Asn-Cys-Pro-Arg-Gly
        Vasopressin

Ile-Tyr-Cys
 |       |
 |       S
 |       |
 |       S
 |       |
Gln-Asn-Cys-Pro-Leu-Gly
        Oxytocin
```

[2]**Diuretics** are substances that increase the volume of urine, and therefore any substance that reduces the volume of urine is an antidiuretic. Diuretics are often given to people with high blood pressure (see Section 28.7) to cause the loss of water and sodium ions, both of which contribute to the elevated blood pressure.

Figure 21.8

The sequence of amino acids in proteins varies from one species to another. The variation illustrated here is for a hormone involved in establishing protective coloration in animals.

$\overset{+}{\text{H}_3\text{N}}$-Asp-Ser-Gly-Pro-Tyr-Lys-Met-Glu-His-Phe-Arg-Try-Gly-Ser-Pro-Pro-Lys-Asp-$\overset{\displaystyle O}{\overset{\|}{\text{C}}}$—O$^-$

1   2   3   4   5   6   7   8   9   10   11   12   13   14   15   16   17   18

Ox β-MSH

$\overset{+}{\text{H}_3\text{N}}$-Asp-Glu-Gly-Pro-Tyr-Lys-Met-Glu-His-Phe-Arg-Try-Gly-Ser-Pro-Pro-Lys-Asp-$\overset{\displaystyle O}{\overset{\|}{\text{C}}}$—O$^-$

1   2   3   4   5   6   7   8   9   10   11   12   13   14   15   16   17   18

Porcine β-MSH

$\overset{+}{\text{H}_3\text{N}}$-Asp-Glu-Gly-Pro-Tyr-Lys-Met-Glu-His-Phe-Arg-Try-Gly-Ser-Pro-Arg-Lys-Asp-$\overset{\displaystyle O}{\overset{\|}{\text{C}}}$—O$^-$

1   2   3   4   5   6   7   8   9   10   11   12   13   14   15   16   17   18

Horse β-MSH

$\overset{+}{\text{H}_3\text{N}}$-Asp-Glu-Gly-Pro-Tyr-Arg-Met-Glu-His-Phe-Arg-Try-Gly-Ser-Pro-Pro-Lys-Asp-$\overset{\displaystyle O}{\overset{\|}{\text{C}}}$—O$^-$

1   2   3   4   5   6   7   8   9   10   11   12   13   14   15   16   17   18

Monkey β-MSH

$\overset{+}{\text{H}_3\text{N}}$-Ala-Glu-Lys-Lys-Asp-Glu-Gly-Pro-Tyr-Arg-Met-Glu-His-Phe-Arg-Try-Gly-Ser-Pro-Pro-Lys-Asp-$\overset{\displaystyle O}{\overset{\|}{\text{C}}}$—O$^-$

1   2   3   4   5   6   7   8   9   10   11   12   13   14   15   16   17   18   19   20   21   22

Human β-MSH

As a final example we shall consider a hormone that regulates the coloring in animals, melanophore-stimulating hormone (MSH). Melanophores (melanocytes) are cells that contain the pigments called melanins. MSH is involved in establishing protective coloration in animals. An injection of MSH will cause human skin to darken. There are two MSH groups, one designated alpha and one beta. The α-MSH from all species is identical. The β-MSH differs from one source to another (Figure 21.8). Such variation from species to species is the rule rather than the exception.

There are many other physiologically important peptides. Some of nature's most potent toxins, such as snake venom and bacterial toxins, are peptides. Additional peptides are discussed in Chapter 22. Before leaving the topic, however, we do want to point out that the amino acid sequences are known for over 3000 polypeptides, and partial sequences have been determined for many others (Table 21.4). Some of these polypeptides contain hundreds of amino acid units. Many have been synthesized in laboratories around the world. The first polypeptide hormone synthesized was oxytocin, in 1953. Synthetic peptides have the same physiological properties as the corresponding natural ones.

21.7
CLASSIFICATION OF PROTEINS

Proteins are large polypeptides, those with molecular weights of 10,000 or more. Some have molecular weights ranging into the millions. There are many kinds of proteins. Each has its own characteristic composition, amino acid sequence, and three-dimensional shape.

Table 21.4
Some Proteins Whose
Sequence of Amino Acids
Is Known

Protein	Function	Number of Amino Acid Residues
Enzymes		
Ribonuclease	Hydrolyzes RNA	124
Lysozyme	Cleaves bacterial cell walls	129
Papain	Digests proteins	212
Trypsinogen	Digests proteins	229
Chymotrypsinogen	Digests proteins	245
Carbonic anhydrase (human)	Hydrates CO_2	260
Subtilisin (a bacterial protease)	Digests bacterial proteins	274
Carboxypeptidase A (bovine)	Digests proteins	307
Alcohol dehydrogenase (horse)	Oxidizes alcohol	374
Others		
Nisin	Antibiotic	29
Insulin	Variety of metabolic functions	51
Trypsin inhibitor (bovine pancreas)	Inhibits trypsin	58
Cytochrome c (human, horse, pig, rabbit, chicken)	Electron transport	104
Cytochrome c (yeast)	Electron transport	108
Hemoglobin (human)	Oxygen transport	
alpha chain		141
beta chain		146
Calmodulin (bovine)	Calcium transport	148
Myoglobin	Oxygen storage	153
Tobacco mosaic virus protein subunit	Virus protein	158
Myelin (bovine)	Protects nerve cells	170
Myelin (human)	Protects nerve cells	172
Human growth hormone	Necessary for normal gowth	191
α-Casein (bovine)	Milk protein	199
Human serum albumin	Blood protein	584
Immunoglobulin, IgG	Antibody	1320

Because of their great complexity, protein molecules cannot possibly be classified systematically in the same way as the carbohydrates and lipids are categorized, that is, on the basis of structural similarities. One way to classify proteins is based on solubility. Some proteins, such as those that make up hair, skin, muscles, and connective tissue, are fiberlike. These **fibrous proteins** are insoluble in water. They usually serve structural, connective, and protective functions. Examples of fibrous proteins are keratins, collagens, myosins, and elastins. Hair and the outer layer of skin are composed of *keratin*. Connective tissues contain *collagen*. *Myosins* are muscle proteins and are involved in contraction and extension of muscles. *Elastins* are found in the elastic tissue of artery walls and in ligaments. We will discuss collagen, a typical fibrous protein, in Section 21.11. Myosin is discussed in Section 25.5.

Globular proteins, the other major class, are soluble in aqueous media. The protein chains of globular proteins are folded so that the molecule as a whole is roughly spherical. The mixtures of globular proteins and water are actually colloidal dispersions rather than true solutions. Familiar examples are the *albumins*. Egg albumin is obtained from egg whites. Serum albumin is present in blood, and it plays a major role in maintaining a proper balance of osmotic pressures in the body (Section 28.6). *Globulins* are a second

Figure 21.9
The tertiary structure of a protein. This tube or "sausage" model shows the coiled protein backbone and the approximate volume occupied by the protein.

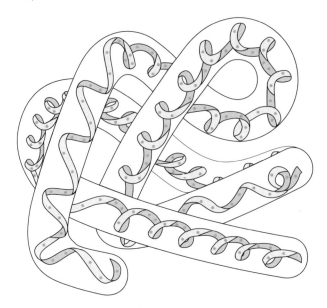

group of globular proteins. Hemoglobin and myoglobin are two examples. Another is the serum globulins, which are a part of our defense against disease. We will discuss one example, myoglobin, in some detail in Section 21.12. Its characteristics are rather typical of globular proteins.

21.8
STRUCTURE OF PROTEINS

The structure of proteins is generally discussed at four organizational levels. **Primary structure** refers to the number and sequence of the amino acids in its polypeptide chain(s). To specify the primary structure, we merely write out the sequence of amino acids in the long-chain molecule. What "holds" the primary structure together are the peptide bonds between the amino acid units.

Molecules of proteins aren't just arranged at random as tangled threads. The chains are held together in unique configurations. The term **secondary structure** refers to the fixed arrangement of the polypeptide backbone. This arrangement may be a pleated sheet, as in silk (Section 21.9), a helix (spiral) as in wool (Section 21.10), or whatever. The term **tertiary structure** refers to the unique three-dimensional shape that results from the precise folding and bending of the protein backbone. An example is the protein chain in globular proteins, which is folded into a compact spherical shape (Figure 21.9). The tertiary structure of a protein is intimately involved with the proper biochemical functioning of that protein, as we shall see in the next chapter.

We can relate these three levels of organization to a more familiar object. Think of the coiled cord on a telephone receiver. The cord starts out as a long, straight wire (Figure 21.10). We'll call that the primary structure. The wire is coiled into a helical arrangement. That's its secondary structure. When the receiver is hung up, the coiled cord folds into a particular pattern. That would be its tertiary structure.

Some proteins contain more than one polypeptide chain. That is, a protein molecule can be an aggregate of subunits. Hemoglobin is the most familiar example. A single hemoglobin molecule contains four polypeptide units. The four units are arranged in a

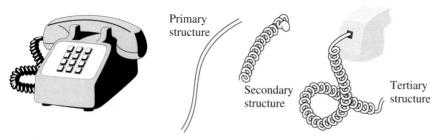

Figure 21.10
Three levels of structure of a telephone cord.

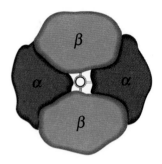

Figure 21.11
The quaternary structure of hemoglobin has the four coiled chains stacked in a nearly tetrahedral arrangement.

specific pattern (Figure 21.11). When we describe the **quaternary structure** of a protein, we describe in detail the way in which the subunits are packed together in the protein molecule. We consider hemoglobin in much greater detail in Chapter 28. The next sections of this chapter will help you gain some insight into secondary, tertiary, and quaternary organization. A schematic representation of the four levels of protein structure is shown in Figure 21.12.

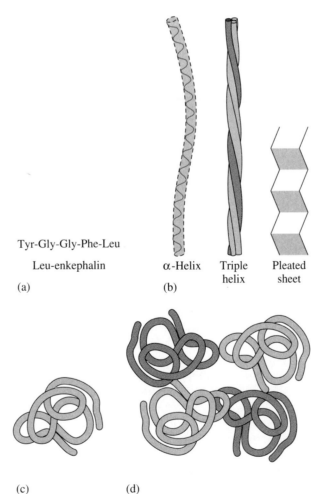

Tyr-Gly-Gly-Phe-Leu

Leu-enkephalin

(a)

α-Helix   Triple helix   Pleated sheet

(b)

(c)   (d)

Figure 21.12
A schematic representation of the four levels of structure in proteins. (a) Primary, (b) secondary, (c) tertiary, (d) quaternary.

Peptide bonds fix the primary structure of proteins. What forces hold proteins in the structural arrangements we have referred to as secondary, tertiary, and quaternary? There are four kinds of forces—hydrogen bonds, ionic forces called salt linkages, disulfide linkages, and hydrophobic interactions.

A number of the amino acids have side chains that can and do participate in *hydrogen bonding* (the hydroxyl group of serine is one example). Nonetheless, the hydrogen bonds of greatest importance are those that stabilize secondary structure. They involve an interaction between the atoms of one peptide bond and those of another. Thus, the carbonyl (C=O) oxygen of one peptide link may form a hydrogen bond to an amide (N—H) hydrogen located some distance away on the same chain or located on an entirely different chain. The secondary structure of wool (Section 21.10) illustrates the former situation; the secondary structure of silk (Section 21.9) illustrates the latter. In both cases you will notice a pattern of such interactions. Because the peptide links are regularly spaced along the chain, there is a regular repetition of hydrogen bonds, both *intramolecular* (in wool) and *intermolecular* (in silk). Such patterns are a fairly common occurrence.

Intramolecular—within the same molecule
Intermolecular—between two molecules

Salt linkages occur when an amino acid with a basic side chain appears opposite one with an acidic side chain. Proton transfer results in opposite charges, which then attract one another. These interactions can occur between relatively distant groups that happen to come into contact with one another because of some folding or coiling of a single chain. They also occur between chains.

Disulfide linkages are formed when two cysteine units (whether on the same chain or two different chains) are oxidized.

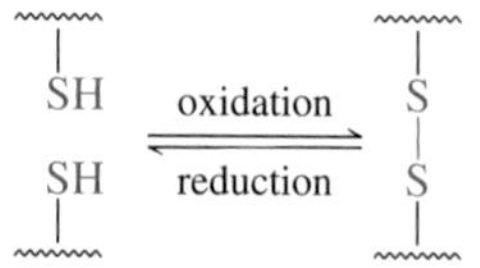

$$
\begin{array}{ccc}
| & & | \\
SH & \xrightleftharpoons[\text{reduction}]{\text{oxidation}} & S \\
| & & | \\
SH & & S \\
| & & |
\end{array}
$$

Intrachain disulfide linkages are frequently found in proteins as a general aid to the stabilization of the tertiary structure. Note, however, that one or more of these bonds may join one portion of a polypeptide chain covalently to another, thus interfering with the helical structure. Interchain disulfide bonds are important forces that link two separate polypeptide chains. Such linkages are clearly indicated in the protein structures given in Figures 21.5–21.7. The disulfide bond is a covalent bond. The strength of this bond is much greater than that of a hydrogen bond. The much larger number of hydrogen bonds does compensate somewhat for their individual weakness.

Still weaker are the *hydrophobic interactions* between nonpolar side chains. These can be important, however, when other types of interactions are missing or are minimized. The hydrophobic interactions are made stronger by the cohesiveness of the water molecules surrounding the protein. Nonpolar side chains minimize their exposure to water by clustering together on the inside folds of the protein in close contact with one another (Figure 21.13). Hydrophobic interactions become fairly significant in structures such as that of silk, in which a high proportion of amino acids in the protein have nonpolar side chains.

The linkages responsible for the tertiary structure of a protein are a function of the nature of the amino acid side chains within the molecule. Globular proteins are extremely compact, almost spherical. Their nonpolar side chains are directed toward the interior of the molecule (the hydrophobic or nonaqueous region), and their polar side chains project outward from the surface of the molecule toward the aqueous environment. The resulting

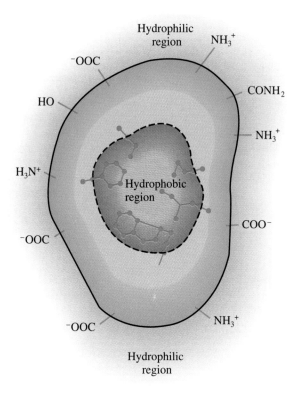

Figure 21.13
A folded protein chain often has a hydrophobic region on the inside (away from water) and a hydrophilic region on the outside with polar groups extending into the water.

picture is very similar to that of a micelle, which was discussed in connection with the properties of cell membranes (Section 20.9). Some of the linkages that contribute to the tertiary structure of proteins are shown in Figure 21.14. Table 21.5 gives an indication of the relative strengths of interactions involving the noncovalent bonds in proteins.

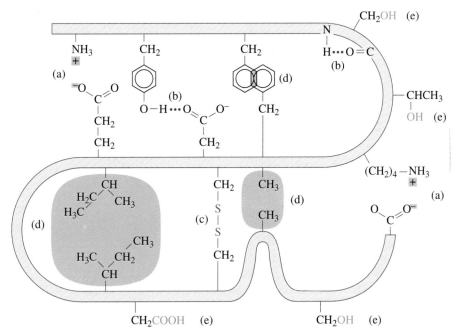

Figure 21.14
Bonds that stabilize the tertiary structure of proteins. (a) Salt linkages, (b) hydrogen bonds, (c) disulfide linkages, (d) hydrophobic interactions, (e) polar group interactions with water.

Table 21.5
Noncovalent Bonds and
Interactions in
Polypeptides

Example	Type of Bond	Approximate Stabilization Energy (kcal/mol)
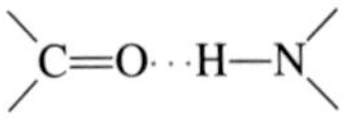 $\diagdown$C=O···H—N$\diagup$	Hydrogen bond between peptides	2–5
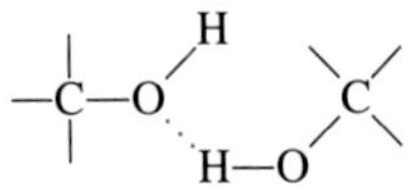 —C—O···H—O···C—	Hydrogen bond between neutral groups	2–5
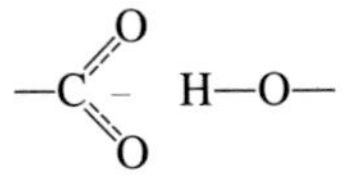 —C(=O)(—O)···H—O—	Hydrogen bond between neutral and charged groups	2–5
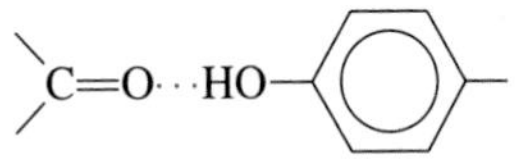 $\diagdown$C=O···HO—⬡—	Hydrogen bond between peptide and R group	2–5
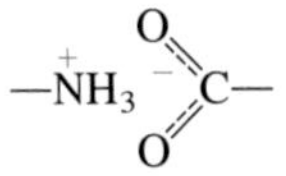 —NH₃⁺ ⁻O=C—	Salt linkage *or* ionic bond between charged groups (strongly dependent on distance)	<10
—CH₃ CH₃—	Hydrophobic interaction	0.3
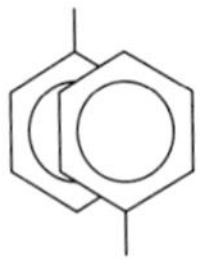 ⬡ (stacked rings)	Hydrophobic interaction—stacking of aromatic rings	1.5
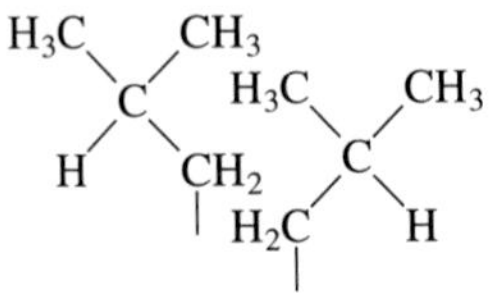 H₃C, CH₃, CH₂ / H₃C, CH₃, H₂C branched chains	Hydrophobic interaction	1.5
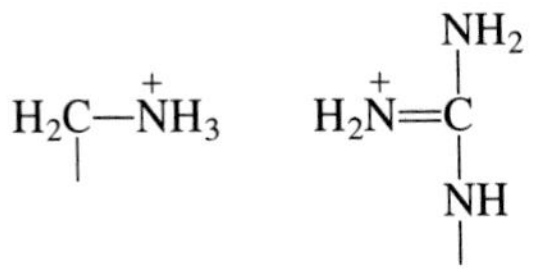 H₂C—NH₃⁺ H₂N⁺=C(NH₂)(NH)	Repulsive interactions between similarly charged groups (strongly dependent on distance)	<−5

21.9
SILK

Silk is a fine, soft, shiny fiber produced by the silkworm as it spins its cocoon. Silk consists of **fibroin,** a fibrous protein whose polypeptide chains exist in an extended zigzag arrangement (Figure 21.15). Unlike most proteins, silk is composed primarily of just three amino acids—glycine (45%), alanine (30%), and serine (12%)—with the remaining 13% being made up of various other amino acids. Indeed, partial degradation shows that much of each chain consists of a repetition of six units.

···Gly-Ser-Gly-Ala-Gly-Ala···

Top view

Intermolecular hydrogen bonds

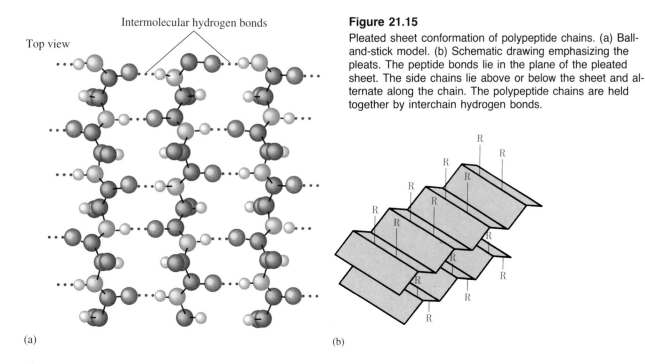

(a)

(b)

Figure 21.15
Pleated sheet conformation of polypeptide chains. (a) Ball-and-stick model. (b) Schematic drawing emphasizing the pleats. The peptide bonds lie in the plane of the pleated sheet. The side chains lie above or below the sheet and alternate along the chain. The polypeptide chains are held together by interchain hydrogen bonds.

In this arrangement, every other amino acid unit is glycine. Most of the remaining side chains, such as the CH_3 in alanine and the CH_2OH in serine, are small. The molecules are stacked in extended arrays, with hydrogen bonds holding adjacent chains together. The appearance of this arrangement has led to its designation as the **pleated sheet conformation.** Silk fibers have these hydrogen-bonded layers arranged one over the other.

The properties of silk—strength, flexibility, and resistance to stretching—are a consequence of its structure. Breaking the fibers involves rupturing thousands of hydrogen bonds or breaking covalent bonds. The interaction of small side chains allows for great flexibility, and because the chains are already fully extended, the fibers cannot be stretched easily.

There is an old saying that "you can't make a silk purse out of a sow's ear." Like many other clichés, this one has its limits. A chemist has proved that you can indeed make silklike fibers from the aural appendages of female hogs. More important commercially, though, is the ability of chemists to make substitutes for silk—such as nylon—out of petroleum, coal, and air.

21.10
WOOL

Wool is another natural protein material, but its composition and properties are quite different from those of silk. Wool has a greater variety of amino acids than silk, including many with large side chains. The structure of proteins such as wool, hair, and muscle was first determined by Linus Pauling and co-workers in 1950. Pauling showed that the amino acid units were arranged in a right-handed helix, or **alpha helix** (Figure 21.16). Each turn of the helix requires 3.6 amino acid units. The NH groups in

Figure 21.16
Three representations of
the alpha-helical conformation of a polypeptide chain.
(a) The skeletal representation best shows the helix.
(b) Hydrogen bonding be-
tween turns of the helix is
shown in the ball-and-stick
model. (c) The space-filling
model shows the actual
shape of a short segment
of the chain.

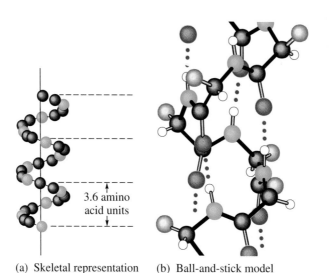

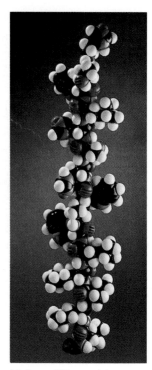

3.6 amino
acid units

(a) Skeletal representation   (b) Ball-and-stick model   (c) Space-filling model

one turn form hydrogen bonds to the carbonyl groups in the next (the dotted lines in the
center model of Figure 21.16).

Unlike silk, wool can be stretched. (Think of how the coiled wire on the telephone
can be stretched.) Such stretching results in an elongation of the hydrogen bonds joining
the turns. When the stretching is discontinued, the helix will return to its original configuration, unless, of course, you have radically disrupted the structure by treating the wool
with very hot water (Section 21.14).

Hair, horn, nails, and muscles also have the alpha-helical structure. This helix extends along the axis of the fiber. Adjacent helices are not precisely parallel but are wound
about one another like the strands of a rope. These ropes, called protofibrils, may have
three or seven alpha helices as strands. Horns, nails, and claws have less flexibility than
wool and hair because of more extensive cross-linking by disulfide bridges.

21.11
COLLAGEN: THE PROTEIN OF CONNECTIVE TISSUES

The principal protein of connective tissues is **collagen** (Figure 21.17). It is the most
abundant of all proteins in higher vertebrates. Most of the organic portion of skin,
bones, tendons, and teeth is collagen. It also occurs in most other parts of the body as
fibrous inclusions. In all, collagen makes up about 25% of all the protein in the human
body.

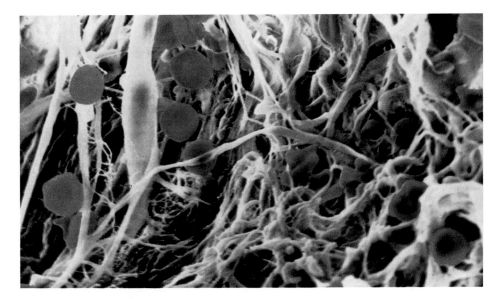

Figure 21.17
Electron micrograph of human connective tissue showing collagen fibers (yellow strands) and red blood cells.

Like other fibrous proteins, collagen is not readily digestible. Treatment with boiling water converts collagen to **gelatin.** Gelatin is not only water-soluble but also digestible. The cooking of meat converts part of the tough connective tissue to gelatin, making the meat more tender.

In composition, collagen is about 33% glycine, with another 20–25% consisting of proline and hydroxyproline. Collagen also contains acidic and basic amino acids. It does not contain enough of the essential amino acids; the gelatin derived from it is a poor-quality protein (see Section 27.3).

Structurally, collagen consists of three protein chains, each wound about its own axis in a left-handed helix (Figure 21.18b). Unlike wool, collagen does not have intrachain hydrogen bonding between coils. Indeed, its structure is much more open than that of the tightly coiled alpha helix. The three chains are wrapped around one another like the three strands of a rope and are cross-linked by interchain hydrogen bonding. In tendons, the collagen fibers are arranged in parallel bundles to yield structures that have nearly the tensile strength of steel wire but little or no capacity to stretch. In bones and teeth, the collagen cables form the matrix upon which the network of calcium salts is built.

Collagen chains are also somewhat cross-linked by covalent bonds. As an animal grows older, the extent of cross-linking increases and the meat gets tougher. Collagen is of considerable commercial importance. The process of tanning increases the degree of cross-linking, converting skin to leather. The soluble gelatin derived from collagen is used in food, film emulsions, and glue and in many other ways.

21.12
MYOGLOBIN

J. C. Kendrew and M. F. Perutz (Nobel prize winners in 1962) were able, through the use of X-ray diffraction studies, to elucidate completely the primary, secondary, and tertiary structures of myoglobin, a globular protein consisting of 153 amino acids

(a)   (b)

Figure 21.18
Schematics of two fibrous proteins. (a) The α-keratins consist of three polypeptide chains in right-handed alpha-helical coils. (b) The protein collagen is a rigid cable made from three left-handed polypeptide chains (each chain contains about 1000 amino acids.) The intertwined chains are maintained by interchain hydrogen bonding.

Figure 21.19

The conformation of myo-globin deduced from X-ray diffraction studies. The disc shape represents the heme group.

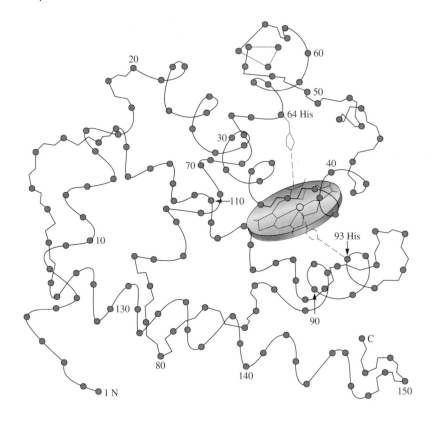

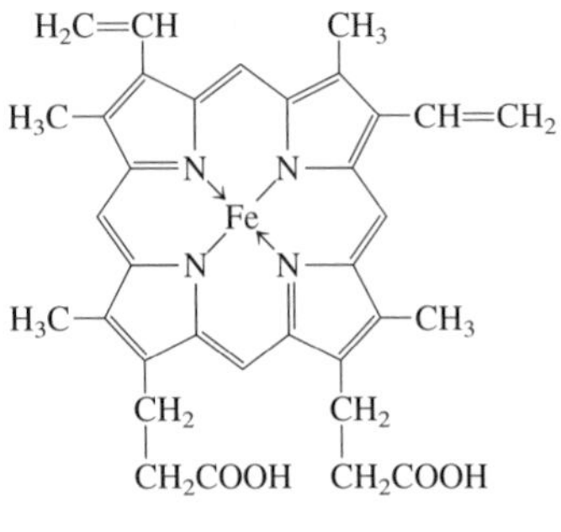

Figure 21.20

Heme is a complex organo-metallic compound that is present in both myoglobin and hemoglobin.

arranged in a single chain (Figure 21.19). Myoglobin can combine reversibly with molecular oxygen, and it functions to store oxygen in muscle cells. It is particularly abundant in marine animals such as whales, seals, and porpoises, enabling them to remain under water for prolonged periods. (In humans, myoglobin is found mainly in heart muscle.) The primary structure of myoglobin results from the formation of the single polypeptide chain. The secondary structure involves the coiling of this chain into an alpha helix (about 70% of the protein strand has a spiral conformation). The tertiary structure results from the nonuniform folding of the chain to form a stable compact structure. The polar side chains are on the outside of the molecule, and almost all of the nonpolar ones are on the inside. The shape of the final molecule includes a hole that nicely accommodates the heme unit (Figure 21.20). The tertiary structure also brings two amino acid side chains into position to anchor the heme unit to the protein portion of the myoglobin molecule.

The structure of the hemoglobin molecule was also deduced by Kendrew and Perutz. It consists of four separate polypeptide chains or subunits—two identical alpha chains (141 amino acids) and two identical beta chains (146 amino acids). Each of the four chains is very similar in structure to the single polypeptide chain of myoglobin (Figure 21.21). Since each subunit contains a heme group, one hemoglobin molecule can bind four molecules of oxygen. The four hemoglobin subunits are held together by noncovalent surface interactions between the polar side chains (and probably hydrophobic interactions as well). As we shall see in Section 28.9, the chief function of hemoglobin is to transport oxygen within the red blood cells.

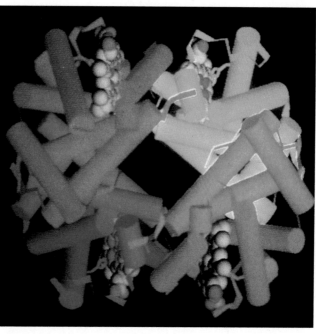

Figure 21.21

The quaternary structure of hemoglobin. The heme molecules are shown within the folds of the four polypeptide chains. Hemoglobin has an almost spherical shape as a result of the fitting together of the four subunits.

Computer graphic representation of the hemoglobin molecule from a human red blood cell (erythrocyte).

21.13
ELECTROCHEMICAL PROPERTIES OF PROTEINS

When amino acids combine to form the polypeptide chain(s) of protein molecules, the majority of their amino and carboxyl groups are tied up in peptide bond formation. However, the side chains of aspartic and glutamic acids and those of lysine, arginine, and histidine all retain their acidic and basic groups. In proteins, just as in the free amino acids, these groups exist in solution as charged species such as $-COO^-$ and $-NH_3^+$. Accordingly, proteins are also amphoteric substances. Since all proteins contain some of the acidic and basic amino acids, positive and negative charges are found throughout the molecule.

At its isoelectric pH, the protein molecule as a whole is electrically neutral. It may contain many ionized groups, but the positively charged side chains are exactly balanced by negatively charged ones. The isoelectric pH is characteristic of a given protein. It is dependent on the number, kind, and arrangement of the acidic and basic groups within the molecule. Proteins that have a high proportion of basic amino acids usually have a relatively high isoelectric pH, and those with a preponderance of acidic amino acids have a relatively low isoelectric pH. Table 21.6 lists the isoelectric pHs of several proteins.

Because of the presence of ionized groups in their structure, proteins behave as either cations or anions, depending upon the pH of the solution. This effect is exploited in the electrophoresis of proteins (recall Figure 21.3). The process of electrophoresis is a very powerful tool that is used to separate and identify specific proteins in a mixture of proteins by subjecting them to an electric field. The protein mixture is applied on a solid support,

Table 21.6

Isoelectric pHs of Various Proteins

Protein	Isoelectric pH
Pepsin	<1.1
Silk fibroin	2.2
Pepsinogen	3.7
Casein	4.6
Egg albumin	4.7
Serum albumin	4.8
Urease	5.0
Insulin	5.3
Fibrinogen	5.5
Catalase	5.6
Hemoglobin	6.8
Myoglobin	7.0
Ribonuclease	9.5
Cytochrome c	10.6
Lysozyme	11.0

such as a strip of cellulose acetate, that is soaked in a buffer solution at a certain pH. A current is then applied, and the proteins migrate toward the oppositely charged electrodes. The proteins with the greatest number of negative (or positive) charges will migrate the most rapidly toward the positive (or negative) electrode. A dye (such as ninhydrin) is used to make the separated protein spots visible. This technique is used on blood samples in hospital laboratories to assess certain diseases by detecting the relative concentrations of the plasma proteins.

The solubility of proteins in water is greatly dependent on pH. As a general rule, a protein is least soluble at its isoelectric pH. The size of many proteins places them in the category of colloids (Section 9.13). At pH values other than the isoelectric pH, the molecules carry a net charge. These charges on the surface of the colloidal proteins repel the other colloidal particles and keep them from coalescing. Thus, they form colloidal dispersions. At the isoelectric pH, however, the colloidal protein molecules are electrically neutral and no longer repel one another. Therefore they come together to form larger aggregates that eventually precipitate from solution.

Casein is the major protein component of milk, and it will precipitate from the milk in the form of white curds at its isoelectric pH of 4.6. The souring of milk results from the production of lactic acid by bacteria. The lactic acid lowers the pH value of milk from its normal value of about 6.6 to about 4.6. Casein is used in the manufacture of cheese. It can be obtained either by adding acid to milk or by bacterial action.

21.14
DENATURATION OF PROTEINS

In many ways, proteins are remarkable compounds. Their highly organized structures are truly masterworks of chemical architecture. But highly organized structures tend to have a certain delicacy, and this is true of many proteins. We shall define **denaturation** as the process in which a protein is rendered incapable of performing its assigned function. If the protein can't do its job, we say it has been **denatured.** (Sometimes denaturation is equated with the precipitation or coagulation of a protein. Our definition is a bit broader.) The process is sometimes reversible, but usually it is not. You have certainly observed the denaturation of egg albumin. The clear egg "white" turns to an opaque white "white" when the egg is boiled or fried. What you have observed is the denaturation and coagulation of the albumin. No one has yet reversed that process!

The primary structure of proteins is quite sturdy. In general, it takes fairly vigorous conditions for peptide bonds to be hydrolyzed (though chemists have devoted much effort to developing gentler methods, and enzymes manage to hydrolyze proteins with remarkable ease). At the secondary and tertiary levels, however, proteins are quite vulnerable to attack (Figure 21.22). A wide variety of reagents and conditions can cause protein denaturation. Some of them are outlined here.

Heat and Ultraviolet Radiation

Heat and ultraviolet radiation supply kinetic energy to protein molecules, causing their atoms to vibrate more rapidly and thus disrupting the relatively weak *hydrogen bonds* and *hydrophobic bonds*. Most proteins are denatured when heated above 50 °C, and this results in coagulation of the protein. Heat and ultraviolet radiation are employed in sterilization techniques because they denature the enzymes in bacteria and, in so doing, destroy them. Denatured proteins are usually easier to chew and easier for enzymes to digest; hence we cook most of our protein-containing food.

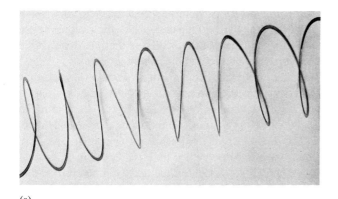

(a)

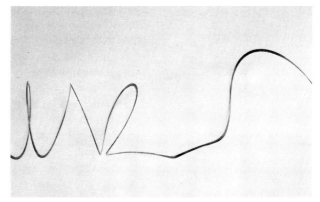

Figure 21.22
Denaturation of a protein. (a) Irreversible denaturation. The coiled spring represents the helical structure of a protein; when the elastic limit of the helix is exceeded, the shape is irreversibly altered. (b) The globular protein is folded into the tertiary conformation necessary for its functioning. The denatured protein can assume various random conformations. It is not active but may, under proper conditions, refold to the active conformation.

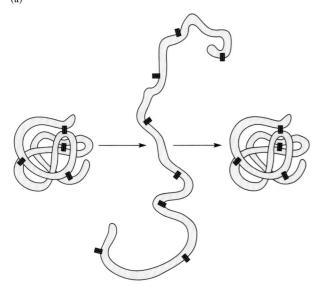

■ = Areas of forces (—S—S—, hydrogen bonding, ionic, etc.) stabilizing conformation

(b)

Treatment with Organic Compounds

Ethyl alcohol, formaldehyde, urea, and rubbing alcohol are capable of forming intermolecular hydrogen bonds with protein molecules, thus disrupting the intramolecular *hydrogen bonding* within the molecule. A 70% alcohol solution is used as a disinfectant in cleansing the skin before an injection. The alcohol functions to denature the protein (enzymes in particular) of any bacteria present in the area of the injection. A 70% alcohol solution effectively penetrates the bacterial cell wall, whereas 100% alcohol coagulates proteins at the surface, forming a crust that prevents the alcohol from entering into the cell (Figure 21.23).

Salts of Heavy Metal Ions

The heavy metal cations Hg^{2+}, Ag^+, and Pb^{2+} form very strong bonds with the carboxylate anions of the acidic amino acids and with the sulfhydryl groups of cysteine. Therefore they disrupt salt linkages and disulfide linkages and cause the protein to precipitate out of

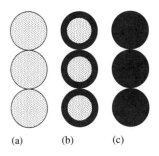

(a)   (b)   (c)

Figure 21.23
Effect of alcohol on bacteria. Dark areas represent coagulated protein. (a) Bacteria before application of alcohol. (b) After application of 100% alcohol. (c) After application of 70% alcohol, which is more effective than 100% alcohol.

solution as insoluble metal–protein salts. This property makes some of the heavy metal salts suitable for use as antiseptics. For example, a 1% solution of silver nitrate (also called lunar caustic), which is used to prevent gonorrhea infections in the eyes of newborn infants, and mercuric chloride, another antiseptic, act to precipitate the proteins in infectious bacteria. Most heavy metal salts are toxic when taken internally because they precipitate the proteins of all the cells with which they come into contact.[3] Substances high in protein, such as egg whites and milk, are used as antidotes for heavy metal poisoning. If a person who has ingested mercury is fed raw eggs immediately, the mercury reacts with egg protein in the stomach rather than with other, more essential proteins. The stomach contents must then be pumped out or vomited to prevent the ultimate digestion of the egg protein and the consequent release of mercury ions within the body. Quite clearly, the technique works only for acute poisonings and not for the far more common chronic mercury poisoning.

Alkaloid Reagents

Picric acid and tannic acid are called alkaloid reagents because they were originally used to study the structure of the alkaloids (morphine, cocaine, quinine). They function in a manner analogous to the heavy metal cations, but the picrate and tannate anions combine with the positively charged amino groups to disrupt the salt linkages. In the manufacture of leather, tannic acid is used to precipitate the proteins in animal hides. This is the process called *tanning*. Tannic and picric acids are sometimes used in the treatment of burns. These acids combine with the protein in the exposed areas to form a crust over the wounds that excludes air and stops the loss of body fluids. The loss of water and salts is the most significant cause of shock and the fatalities that result from burns. In an emergency, tea can serve as a source of tannic acid for the treatment of severe burns.

There are many other ways of denaturing proteins that we have not discussed (introduction of radical changes in pH, for example). The point should be clear, however. The very complexity that makes proteins so versatile also makes them vulnerable. There is a considerable range of vulnerability. The delicately folded globular proteins are much more readily denatured than are the tough, fibrous proteins of hair and skin.

On the other hand, there is increasing evidence that an unfolded protein, given the proper conditions and enough time, will refold and may again exhibit biological activity. Such evidence suggests that, for these molecules, the primary structure determines the secondary and tertiary structures. A given sequence of amino acids seems naturally to adopt its particular three-dimensional arrangement if conditions are right.

We have emphasized structure in this chapter. In other chapters we concentrate on the function of several kinds of proteins, particularly the enzymes.

[3] The danger of lead and/or mercury poisoning has evoked considerable environmental concern. Lead salts are no longer used as pigments in paints, and most of the lead has been removed from gasoline. Mercury compounds still occur in water systems because large quantities of mercury and mercury salts have been dumped by industries into streams and lakes. The mercury is taken in by fish, and then humans eat the fish.

EXERCISES

1. Define, describe, or illustrate each of the following.
   a. peptide bond
   b. tripeptide
   c. disulfide linkage
   d. salt linkage
   e. hydrophobic interaction
   f. primary structure
   g. secondary structure
   h. tertiary structure
   i. quaternary structure
   j. zwitterion
   k. isoelectric pH
   l. diuretic
   m. globular protein
   n. fibrous protein
   o. electrophoresis
   p. denaturation

2. What is the general structure for an α-amino acid?

3. Draw the side chains of the following amino acids.
   a. proline
   b. tryptophan
   c. tyrosine
   d. aspartic acid
   e. cysteine
   f. lysine
4. Write structural formulas for the following.
   a. glycine
   b. alanine
   c. phenylalanine
   d. serine
5. Write structural formulas for the following.
   a. glycylalanine
   b. alanylglycine
6. Write the structural formula for phenylalanylglycylalanine.
7. Write a structural formula for an amino acid with an acidic side chain, ~~and give the name of the compound.~~
8. Write a structural formula for an amino acid with a basic side chain, ~~and give the name for the compound.~~
9. Write structural formulas for the following.
   a. histidine
   b. glutamic acid
10. Write a structural formula for the anion formed when glycine reacts with a base.
11. Write a structural formula for the cation formed when glycine reacts with an acid.
12. To which family of mirror-image isomers do almost all naturally occurring amino acids belong?
13. What is the difference between a polypeptide and a protein?
14. Amino acid units in a protein are connected by peptide bonds. What is another name for the functional group linking the units of proteins?
15. Translate the following abbreviated form into the structural formula for the peptide: Ser-Ala-Gly.
16. Write the abbreviated version of the following structural formula.

$$\underset{H_3\overset{+}{N}CH}{\overset{CH_3}{|}}-\underset{}{\overset{O}{\overset{||}{C}}}-NHCH-\overset{O}{\overset{||}{C}}-NHCH-\overset{O}{\overset{||}{C}}-NHCH-\overset{O}{\overset{||}{C}}-O^-$$

with side chains CH₃, CH₂–OH, CH₂–SH, CH₂–C₆H₅

17. Identify two amino acids that contain more than one chiral carbon atom.
18. Describe the structure of silk and explain how its properties reflect its structure. What name is given to the secondary structure of silk?
19. Describe the structure of wool and relate this to wool's elasticity. What name is given to the secondary structure of wool protein?
20. Describe the structure of collagen.
21. Name the four kinds of interactions that maintain the tertiary structure of proteins.
22. The following sets of amino acids are involved in maintaining the tertiary structure of a peptide. In each case, identify the type of interaction (see Exercise 21) involved.
   a. aspartic acid and lysine
   b. phenylalanine and alanine
   c. serine and lysine
   d. two cysteines

23. Classify these proteins as fibrous or globular.
   a. albumin
   b. myosins
   c. collagen
   d. hemoglobin
   e. keratins
   f. myoglobin
24. Which class of proteins shows greater solubility in aqueous solution—fibrous or globular?
25. Which class of proteins is more easily denatured—fibrous or globular? *more OH's to form H-bonds*
26. Describe some ways of denaturing a protein.
27. What level(s) of structure is(are) ordinarily disrupted in denaturation?
28. Is denaturation of protein usually reversible? *NO*
29. Distinguish between the terms in each pair.
   a. polypeptide and protein
   b. N-terminal amino acid and C-terminal amino acid
   c. oxytocin and vasopressin
   d. alpha-helical and pleated sheet
   e. primary structure and secondary structure
   f. tertiary structure and quaternary structure
   g. interchain and intrachain hydrogen bonds
   h. hemoglobin and myoglobin
30. Identify the amino acid that fits each of the following descriptions.
   a. contains a secondary amino group
   b. contains a heterocyclic ring
   c. contains a benzene ring
   d. not found in proteins
   e. most abundant amino acid in gelatin
   f. abundant in collagen but not found in most other proteins
   g. found in asparagus
   h. most basic of the amino acids
   i. contains a sulfhydryl group
   j. contains a phenolic group
   k. contains a branched chain
31. A direct current was passed through a solution containing alanine, histidine, and aspartic acid at a pH of 6.0. One amino acid migrated to the cathode, one migrated to the anode, and one remained stationary. Match the behavior with the correct amino acid.
32. What would be the products of the acid hydrolysis of the following tetrapeptide? *4 bk bond btw C=O & N-H*

Goal

$$H_3\overset{+}{N}-\underset{\underset{NH_2}{|}}{\underset{(CH_2)_4}{|}}\overset{H}{\underset{|}{C}}-\overset{O}{\overset{||}{C}}-N-\underset{\underset{}{\overset{H}{|}}}{\underset{CH_3}{|}}\overset{}{\underset{|}{C}}-\overset{O}{\overset{||}{C}}-N-\underset{\underset{\hexagon}{|}}{\underset{CH_2}{|}}\overset{H}{\underset{|}{C}}-\overset{O}{\overset{||}{C}}-N-\underset{\underset{COOH}{|}}{\underset{(CH_2)_2}{|}}\overset{H}{\underset{|}{C}}-COO^-$$

33. Glutathione (γ-glutamylcysteinylglycine) is a tripeptide that is found in all the cells of higher animals. It contains a glutamic acid joined in an unusual peptide linkage involving its γ-carboxyl group. Draw the structure of glutathione.

34. What is the difference between the type of hydrogen bonding that occurs in secondary structures of proteins and the type in tertiary structures?

35. Proteins help to maintain the pH of the organism. How can they perform this function?

36. Under what conditions does a protein have **(a)** a net positive charge, **(b)** a net negative charge, and **(c)** a net zero charge?

37. How do the properties of a protein at its isoelectric pH differ from its properties in solutions of other pH values?

38. What occurs when milk becomes sour?

39. One of the neurotransmitters involved in the sensing of pain information is a polypeptide called substance P. Substance P is an undecapeptide that is released by nerve terminals in response to pain. Its primary structure is Arg-Pro-Lys-Pro-Gln-Gln-Phe-Phe-Gly-Leu-Met. What would be the net electric charge (+ or −) of this polypeptide at pH **(a)** 1.0, **(b)** 6.0, and **(c)** 13.0?

40. Briefly describe on the molecular level what occurs during each of the following processes.
a. boiling an egg
b. sterilizing surgical instruments
c. using a dilute solution of silver nitrate as a disinfectant in the eyes of newborn infants

41. Why is a 70% alcohol solution more effective as a disinfectant than a 100% alcohol solution?

42. Discuss the use of egg white as an antidote for heavy metal poisoning.

43. What do we mean when we say that hemoglobin shows species variation?

44. Bacteria synthesize D-alanine and use it in the biosynthesis of cell walls. Draw the structure for D-alanine.

45. A deutrated derivative (one in which ordinary hydrogen, $^1_1$H, has been replaced by deuterium, $^2_1$H or D), 2-deutero-3-fluoro-D-alanine, acts as an antibiotic by inhibiting cell wall synthesis. Draw the structure for this derivative.

46. Two cysteine units joined through a disulfide linkage are sometimes considered a different amino acid called cystine. Draw the structure for cystine.

47. Draw the structure for γ-aminobutyric acid (GABA). Is GABA found in proteins? What is its role in the body?

48. The isoelectric pH of silk fibroin is 2.2. Which of the following amino acids is likely to be present in large amounts?
a. aspartic acid   **b.** histidine
c. lysine

49. Write equations to show how alanine can act as a buffer.

50. Which amino acid is more likely to be in a pleated sheet protein—alanine or phenylalanine?

51. For each of the following amino acids, state whether it is more likely to be on the inside or the outside of a globular protein.
a. phenylalanine   **b.** aspartic acid
c. serine   **d.** lysine
e. leucine   **f.** glutamic acid

52. Carbohydrates are incorporated into *glycoproteins*. How does the incorporation of sugar units affect the solubility of a protein?

53. In 1981, the FDA approved aspartame, L-aspartyl-L-phenylalanine methyl ester, as an artificial sweetener. It is about 180 times sweeter than sucrose, and it has largely replaced saccharin in soft drinks and a wide variety of foods. Aspartame is the active ingredient in the sugar substitute NutraSweet. Draw the structure of aspartame.

Chapter 22
ENZYMES

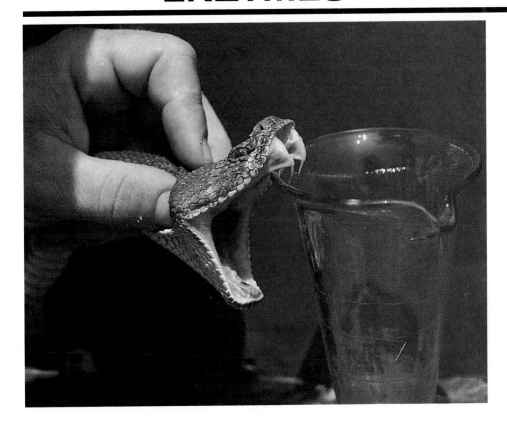

The various reactions that occur in biological systems are in many respects identical to those we have discussed previously (for example, hydration of unsaturated bonds, interconversions of alcohols, aldehydes, and acids, hydrolysis of esters and amides). In fact, scientists have been able to duplicate in the test tube (in vitro conditions) many of the reactions commonly carried out in living organisms (in vivo conditions). There is, however, one significant difference—the rate at which the two types of reactions occur. The in vivo reactions take place about 100 to 1 million times faster than the corresponding in vitro reactions. Some enzymatic reactions occur in milliseconds (1 ms = 10^{-3} s); others are more rapid, and rates are measured in microseconds (1 μs = 10^{-6} s). For example, it has been estimated that the reactions involved in the transmittal of a nerve impulse take between 1 and 3 millionths of a second (1–3 μs).

If the in vitro reactions are to take place at an appreciable rate, drastic reaction conditions must be employed. Such conditions, which include high temperatures and the use of potent oxidizing or reducing agents and/or strong acids or bases, are all lethal to living systems. The in vivo reactions are carried out at body temperature (~37 °C) and in the physiological pH range (pH ~7). The agents employed by the cell to bring about reactions under these conditions are the highly efficient, highly specific catalysts called enzymes. Life as we know it would be impossible without enzymes because nearly all functions of the cell depend directly or indirectly upon them.

571

Recall that a *catalyst* is any substance that increases the rate of a chemical reaction without being consumed in the reaction. The catalyst does *not* change the position of equilibrium in a reversible reaction; it only increases the rate of attainment of equilibrium. An **enzyme** can be defined as a complex organic catalyst produced by living cells. Most enzymes operate within the cell that produces them and are thus termed *intracellular*. A typical human cell contains about 2000 enzymes, and these enzymes catalyze over 100 reactions each minute. If the enzyme's usual site of catalytic activity is outside the cell that produces it (as in the case of the digestive enzymes), the enzyme is designated as *extracellular*.

The hydrolysis of sucrose affords a good example by which to distinguish enzyme action from the classic concept of a catalyst. If we were to exclude bacteria and molds from it, a solution of sucrose in water could be kept indefinitely without undergoing hydrolysis to any appreciable extent. If we added hydrochloric acid and heated the reaction mixture, hydrolysis would take place, producing glucose and fructose. If we added the enzyme invertase (sucrase) instead of the acid, the reaction would take place at a greater rate and the solution would not have to be heated at all. Furthermore, hydrochloric acid catalyzes the hydrolysis of lactose and maltose as well as that of sucrose. Invertase is specific for sucrose alone and will not catalyze the hydrolysis of any other disaccharide.

Several thousand enzymes are known. Many are available commercially in varying grades of purity. Some have even been isolated as highly purified crystals.

22.1
CLASSIFICATION AND NAMING OF ENZYMES

The first enzymes to be discovered were named according to their source or method of discovery. The enzyme pepsin, which aids in the hydrolysis of proteins, is found in the digestive juices of the stomach (Greek *pepsis,* digestion). Ptyalin, found in saliva (Greek *ptyalon,* spittle), starts the hydrolysis of starches.

As more enzymes were discovered, a more systematic nomenclature developed. The substance upon which an enzyme acts is known as its **substrate.** Enzymes are most commonly named by adding the suffix *-ase* to the root of the name of the substrate. Thus, for example, urease is the enzyme that catalyzes the hydrolysis of urea, and sucrase is the enzyme that catalyzes the hydrolysis of sucrose. A lipase is an enzyme that catalyzes the cleavage of lipids. A dipeptidase aids in the splitting of dipeptides to amino acids. Sometimes an enzyme is named after the products that are formed as a result of its catalytic activity, as in the case of invertase, another name for sucrase.

To avoid the continued haphazard naming of enzymes, the International Union of Biochemistry, in 1961, recommended that enzymes be systematically classified according to the general type of reaction they catalyze. There are six major types.

1. **Oxidoreductases** catalyze all the reactions in which one compound is oxidized and another is reduced. *Oxidases* operate in physiological oxidation reactions. They are essential in cells that use oxygen to produce energy. The oxidoreductases also include the *dehydrogenases,* enzymes that oxidize by the removal of hydrogen.

2. **Transferases** *(mutases)* facilitate the transfer of groups such as methyl, amino, and acetyl from one molecule to another. *Transaminases* catalyze the transfer of an amino group from one molecule to another. This reaction is involved in the removal of the

amino group during the metabolism of amino acids. *Kinases* are involved in the transfer of phosphate groups.

3. **Hydrolases** are enzymes that catalyze hydrolysis reactions. These include *lipases,* which act on fat and other lipids; *carbohydrases,* which speed up the hydrolysis of carbohydrates to monosaccharides; and *proteases* and *peptidases,* which catalyze the hydrolysis of proteins and peptides.

4. **Lyases** aid in the removal of certain groups without hydrolysis. Examples are the *decarboxylases,* which catalyze the removal of carboxyl groups.

5. **Isomerases,** as their name implies, catalyze the conversion of a compound into another that is isomeric with it.

6. **Ligases** *(synthetases)* are involved in the formation of new bonds from carbon to a nitrogen, an oxygen, a sulfur, or another carbon atom.

We encounter additional examples, particularly of the first four classes, in other chapters.

22.2
CHARACTERISTICS OF ENZYMES

Enzymes are globular proteins. Some enzymes, such as pepsin, trypsin, and ribonuclease, are simple proteins—they consist entirely of amino acid chains. Others contain some other chemical component necessary to the proper functioning of the enzyme. Such a component is called a **cofactor.** The cofactor may be a metal ion such as zinc (Zn^{2+}), manganese (Mn^{2+}), magnesium (Mg^{2+}), iron(II) (Fe^{2+}), or copper(II) (Cu^{2+}). For example, alcohol dehydrogenase requires zinc ion as a cofactor, and arginase requires manganese ion. If the cofactor is an organic molecule it is called a **coenzyme.** Coenzymes are nonprotein. The polypeptide segment of an enzyme is called an **apoenzyme.** Neither the coenzyme nor the apoenzyme has enzymatic activity by itself. The catalytically active enzyme–coenzyme complex is sometimes called the *holoenzyme.*

$$\text{Coenzyme} \; + \; \text{Apoenzyme} \; \rightleftharpoons \; \text{Holoenzyme}$$

Nonprotein	Protein	(active)
(inactive)	(inactive)	

Many coenzymes are vitamins or are derived from vitamin molecules. These important chemical substances and their relationships to vitamins are discussed in detail in Special Topic I.

Some enzymes, such as the protein-digesting enzymes (peptidases), are secreted in larger, inactive forms known as **zymogens** or **proenzymes.** This is a protective feature that prevents the active form of these enzymes from digesting the proteins in the walls of the digestive tract. Pepsinogen and trypsinogen are two such compounds that have been carefully studied.

Pepsinogen is secreted from the cells of the stomach. The acid in the stomach converts pepsinogen into pepsin, an active protease. The reaction is also **autocatalytic:** it is catalyzed by the product pepsin. That is, once some pepsin is formed, it speeds the reaction, which produces more pepsin. Activation involves the removal of 42 amino acid residues, as small peptides, from the pepsinogen molecule.

$$\text{Pepsinogen} \; + \; H_3O^+ \; \xrightarrow{\text{pepsin}} \; \text{Pepsin} \; + \; \text{Small peptides}$$

For over 50 years, one of the central tenets of biochemistry was that *all* enzymes were proteins. In the 1980s, Thomas Cech and Sidney Altman identified some ribonucleic acids (RNA—see Chapter 23) that could catalyze cellular reactions. In 1989, they were awarded the Nobel prize in chemistry for their discovery of "ribozymes."

Trypsinogen is synthesized in the pancreas and is secreted into the small intestine, where it is activated in response to the presence of food in the intestine. A hexapeptide is cleaved from the molecule, enabling the new protein to attain the conformation essential to its catalytic activity (see Figure 27.1).

Activation of proenzymes is quite important. We wouldn't want an enzyme breaking down the proteins of our pancreas. Better to have a proenzyme made there that can be activated in the small intestine, where it can hydrolyze the proteins in our food. We shall see that zymogens play an important role in the mechanism of blood clotting (Section 28.8).

22.3
MODE OF ENZYME ACTION

In 1888 the Swedish chemist Svante Arrhenius proposed a scheme to account for catalytic activity. He suggested that a catalyst combines with a reactant to form an intermediate compound that is more reactive than the initial uncombined species. The formation of an intermediate compound therefore affords a lower energy pathway than that of the uncatalyzed reaction (Figure 22.1). This, in effect, lowers the activation energy of the reaction, accounting for the increased rate of reaction. Enzymes reduce activation energies more effectively than other catalysts, thus enabling biochemical reactions to proceed at relatively low temperatures. Note that the amount of energy absorbed or released in the reaction is not altered by the enzyme.

This scheme applies to all catalytic reactions, whether inorganic, organic, or biochemical. It is generally believed that enzymatic reactions occur in at least two steps. In the first step a molecule of the enzyme (E) and a molecule of the substrate (S) collide and react to form an intermediate compound, which is called the **enzyme–substrate** complex (E—S). (This step is reversible because the complex can break apart, yielding the original substrate and the free enzyme.) The enzyme–substrate complex may or may not react with additional substances (water, oxidizing or reducing agents, ATP, etc.) to form products (P), which are then released from the surface of the enzyme.

Figure 22.1

Energy diagram for the progress of a chemical reaction and the effect of an enzyme.

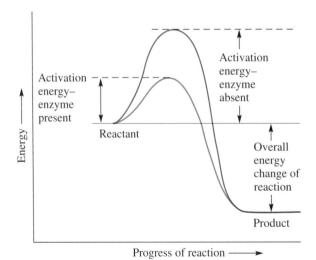

General Reaction

$$S + E \rightleftharpoons E{-}S$$
$$E{-}S \longrightarrow P + E$$

Specific Example

Sucrose + Sucrase $\rightleftharpoons$ Sucrose–sucrase complex

Sucrose–sucrase + H_2O $\longrightarrow$ Glucose + Fructose + Sucrase

The existence of an enzyme–substrate complex has been verified by spectroscopic and kinetic experiments. The bonds that hold the enzyme and the substrate together are the same forces that help to maintain protein structure—electrostatic interactions, covalent bonds, hydrogen bonds, etc. In addition, it has been demonstrated that the structural features or functional groups essential to the formation of the enzyme–substrate complex occur at a specific location on the surface of the enzyme. This section of the enzyme, which combines with the substrate and at which the substrate is transformed to products, is called the **active site** of the enzyme. The active site is often a cleft or a crevice in the exterior of the molecule. It possesses a unique conformation (and correctly positioned bonding groups) that is complementary to the structure of the substrate. Thus the two molecules are able to fit together in much the same manner as a key fits into a lock. The **lock-and-key theory** of enzyme action is illustrated in Figure 22.2. This theory portrays an enzyme as conformationally rigid and able to bond only to substrates that are structurally suitable. This would explain the high degree of specificity of enzymes for their substrates.

The use of X-ray crystallography in determining the precise three-dimensional structure of enzymes was a major advance in understanding their catalytic activity. It was observed that the binding of some substrates leads to a large conformational change in the enzyme. In 1963 D. E. Koshland, Jr., augmented the older lock-and-key theory by suggesting that the binding site of an enzyme is not a rigid structure. Instead, some enzymes undergo a change in conformation when they react with a substrate molecule to form an activated complex. The active site has a shape complementary to that of the substrate only *after* the substrate is bound. After catalysis, the enzyme resumes its original structure.

> To explain the enzyme's ability to discriminate between similar compounds, a ''fit'' between the substrate and a portion of the enzyme surface seems essential. However, it appears possible that this fit is not a static one in which a rigid ''positive'' substrate fits on a rigid ''negative'' template, but rather, is a dynamic interaction in which the substrate induces a structural change in the enzyme molecule, as a hand changes the shape of a glove.[1]

[1] D. E. Koshland, Jr., *Science*, **142**, 1533 (1963).

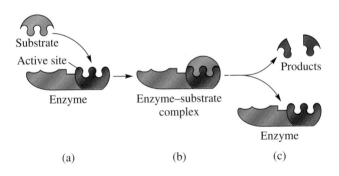

(a)   (b)   (c)

Figure 22.2

A schematic representation of the interaction of enzyme and substrate. (a) The active site on the enzyme and the substrate have complementary structures (and complementary bonding groups) and hence fit together as a key fits a lock. (b) While they are bonded together in the enzyme–substrate complex, the catalytic reaction occurs. (c) The products of the reaction leave the surface of the enzyme, freeing the enzyme to combine with another molecule of substrate.

Figure 22.3

A schematic representation of an active site. Specificity is determined by the fit of "contact" amino acid R-groups with the substrate. Catalytic R-groups act on the substrate bond, which is indicated by the zigzag line. The R-groups that interact with each other maintain the three-dimensional structure of the enzyme.

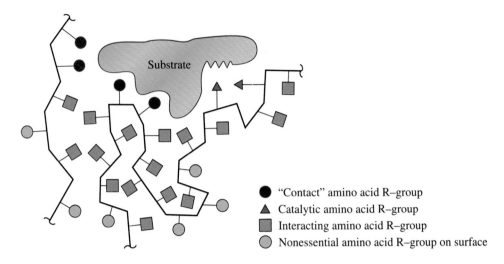

● "Contact" amino acid R–group
▲ Catalytic amino acid R–group
■ Interacting amino acid R–group
○ Nonessential amino acid R–group on surface

This **induced-fit theory** is an attractive proposal because it explains several experimental findings that are incompatible with the older theory. Koshland cites examples of compounds that bind to enzymes without undergoing further reaction as well as other compounds that are sterically not suited for the active site but nevertheless react catalytically. According to Koshland, the active site of an enzyme consists of two components. One (a *contact group*) is responsible for substrate specificity, and the other (a *catalytic group*) is responsible for catalysis. The active site is a flexible region that can be induced to fit several structurally similar compounds. However, only the proper substrate is capable of correct alignment with the catalytic groups as well (Figures 22.3 and 22.4).

The manner in which an enzyme transforms a substrate into product(s) has been extensively studied. We know that the reactants are brought into proximity as they bind to the enzyme, and this increases the frequency of collisions. Since the enzyme properly aligns each reactant, the effectiveness of each collision will also be increased. As yet, however, the detailed mechanism by which enzymes increase the rate of reactions more efficiently than other catalysts is incompletely understood. We still do not know, for example, whether the full catalytic activity of an enzyme resides in its protein structure as a whole or in a small region associated with the active site. Small peptides have been cleaved from some enzymes (such as ribonuclease) without appreciable loss of catalytic activity.

Specific amino acid side chains (such as the hydroxyl group of serine, the sulfhydryl group of cysteine, the imidazole group of histidine) are part of the active sites of various

Figure 22.4

Schematic representation of a flexible active site.
(a) Substrate binding induces proper alignment of catalytic groups A and B so that reaction ensues.
(b) Compounds that are either too large or too small are bound but fail to produce proper alignment of catalytic groups; hence there is no reaction.

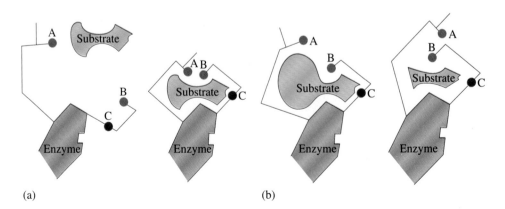

(a)   (b)

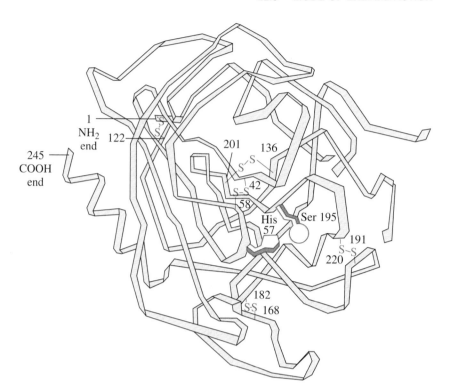

Figure 22.5
Model of α-chymotrypsin. The enzyme consists of 245 amino acids in three polypeptide chains, which are interconnected by five disulfide bonds. The active site is in the region of histidine-57 and serine-195. Chymotrypsin is a digestive enzyme that catalyzes the hydrolysis of proteins within the small intestine.

enzymes. The active site probably consists of amino acids from several different positions along the protein chain. These amino acids can be brought next to one another as a result of the folding and bending of the polypeptide chain. The fact that enzymes are inactivated by denaturation points to the importance of the secondary and tertiary structures in maintaining the active site in a precise three-dimensional arrangement. It is known that two amino acids, histidine-57 and serine-195, are involved in the active site of chymotrypsin (Figure 22.5). The polar groups at this site no doubt interact strongly with the peptide linkage that is to be hydrolyzed. The remaining amino acids presumably function in maintaining the active site in the correct geometrical configuration to provide for maximum catalytic activity. They also impart specificity to the molecule.

Table 22.1 lists some important enzymes and gives an indication of how rapidly some of them work. The **turnover number** is the number of substrate molecules converted to product in 1 min by one enzyme molecule. Typical turnover numbers are about 1000/min. Some, like carbonic anhydrase, are much faster. Others, such as lysozyme, are much slower.

Table 22.1
Some Important Enzymes

Enzyme	Reaction Catalyzed	Turnover Number (per minute)
Carbonic anhydrase	$H_2O + CO_2 \rightleftharpoons H_2CO_3$	36,000,000
Catalase	$2\ H_2O_2 \rightleftharpoons 2\ H_2O + O_2$	5,600,000
Fumarase	Fumaric acid $\rightleftharpoons$ Malic acid	1,200,000
Lactate dehydrogenase	Pyruvic acid $\rightleftharpoons$ Lactic acid	60,000
Succinate dehydrogenase	Succinic acid $\rightleftharpoons$ Fumaric acid	1,150
DNA polymerase I	Addition of nucleotides to DNA chain	900
Lysozyme	Hydrolysis of specific polysaccharide bonds	30

22.4
SPECIFICITY OF ENZYMES

One characteristic that distinguishes an enzyme from all other types of catalysts is its *substrate specificity*. Recall, for example, that hydrogen ions catalyze the hydrolysis of disaccharides, polysaccharides, lipids, and proteins with complete impartiality, whereas different enzymes are required in all four cases. Enzyme specificity is a result of the uniqueness of the active site of each enzyme. This uniqueness is a function of the chemical nature, electric charge, and spatial arrangements of the groups located there. Enzyme specificity is crucial in chemical reactions in the cell. It ensures, for the most part, that the proper reactions occur in the proper place at the proper time. A wide range of enzyme specificities exist, and they are arbitrarily grouped as follows.

Absolute Specificity

Enzymes that have **absolute specificity** catalyze a particular reaction for one particular substrate only and have no catalytic effect on substrates that are closely related. Urease, for example, catalyzes the hydrolysis of urea but not of methylurea, thiourea, or biuret. Such absolute specificity, however, is rather rare among enzymes characterized to date.

$$H_2N-\overset{\overset{\displaystyle O}{\|}}{C}-NH_2 \ + \ H_2O \ \underset{\text{urease}}{\rightleftharpoons} \ CO_2 \ + \ 2\,NH_3$$

Urea

$$H_2N-\overset{\overset{\displaystyle O}{\|}}{C}-NH-CH_3 \qquad H_2N-\overset{\overset{\displaystyle S}{\|}}{C}-NH_2 \qquad H_2N-\overset{\overset{\displaystyle O}{\|}}{C}-NH-\overset{\overset{\displaystyle O}{\|}}{C}-NH_2$$

Methylurea   Thiourea   Biuret

Stereochemical Specificity

Because enzymes are chiral molecules, they show a markedly high degree of **stereochemical specificity**—they have specificity for one stereoisomeric form of the substrate. This is analogous to the binding of monosodium glutamate to the taste buds (see Figure 18.18). L-Lactic acid dehydrogenase catalyzes the oxidation of the L-lactic acid in muscle cells. D-Lactic acid, found in certain microorganisms, does not bind to the enzyme. Fumarase adds water to fumaric acid but not to its cis isomer, maleic acid.

Group Specificity

Enzymes that have **group specificity** are less selective in that they act upon structurally similar molecules that have the same functional groups. Many of the peptidases fall into this category. Pepsin hydrolyzes all peptides that have adjacent aromatic amino acids. Carboxypeptidase attacks peptides from the carboxyl end of the chain, cleaving the amino acids one at a time (see Section 27.1).

Linkage Specificity

Enzymes that have **linkage specificity** are the least specific of all because they attack a particular kind of chemical bond, irrespective of the structural features in the vicinity of the linkage. The lipases, which catalyze the hydrolysis of ester linkages in lipids, are an example of this type of enzyme.

22.5
FACTORS THAT INFLUENCE ENZYME ACTIVITY

The single most important property of an enzyme is its catalytic activity. Since enzymes are protein catalysts, they are affected by those factors that act upon proteins and upon catalysts in general. The activity of an enzyme can be measured by monitoring the reaction that it catalyzes at fixed time intervals. The rate of the reaction is determined by observing either the rate of disappearance of the substrate or the rate of formation of the product(s). In such experiments, the rate is the only variable; all other experimental conditions are held constant.

Concentration of Substrate

The rate of an enzymatic reaction increases as the substrate concentration increases until a limiting rate is reached. At this point, further increase in the substrate concentration produces no significant change in the reaction rate. At excess substrate concentrations, practically all the enzyme molecules are saturated with the substrate at any given instant. Extra substrate molecules must wait until the enzyme–substrate complexes have dissociated to yield products and the free enzymes before they can undergo reaction.

It is as if you had 10 taxis (enzymes) waiting to take people (substrates) to a certain destination. If there were only three people at the taxi stand and the trip took 5 min, the rate at which they arrived at their destination would be three people every 5 min. If the concentration of people at the stand were increased to five, the rate would increase to five arrivals in 5 min. With 10 people, you would have 10 arrivals every 5 min. With 20 people at the stand, the rate would still be 10 arrivals in 5 min. The taxis have been saturated (in our analogy, each taxi can carry only one passenger). If the taxis could carry two or three passengers each, the same principle would apply. The rate would simply be higher (20 or 30 people in 5 min) before it leveled off.

This relationship is illustrated in Figure 22.6 and can be summarized in terms of the two equations given in Section 22.3. At low substrate concentrations the formation of the E—S complex is the rate-determining step, whereas at high substrate concentrations the

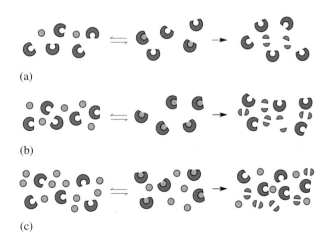

(a)

(b)

(c)

Figure 22.6

A schematic representation of relative concentrations of enzyme and substrate.
(a) Low substrate concentration. (b) Adequate substrate concentration.
(c) Excess substrate concentration.

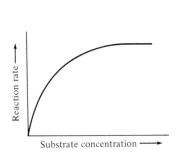

Figure 22.7

Effect of substrate concentration on the rate of a reaction that is catalyzed by a fixed amount of enzyme.

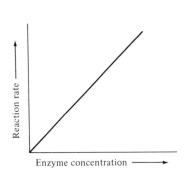

Figure 22.8

Effect of enzyme concentration on the rate of a reaction.

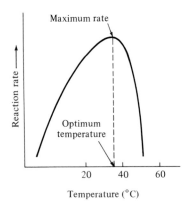

Figure 22.9

Effect of temperature on the rate of an enzymatic reaction.

slowest step is the dissociation of the E—S complex. Figure 22.7 is a characteristic plot of an enzyme-catalyzed reaction, and it is taken as further evidence of the existence of the enzyme–substrate intermediate.

Concentration of Enzyme

Since in essentially all practical cases the concentration of enzyme is much lower than the concentration of substrate (we always have more people than taxis), the rate of an enzyme-catalyzed reaction is directly dependent upon the enzyme concentration (Figure 22.8). This is not a new concept. The reaction rate of any catalytic reaction increases as the concentration of the catalyst is increased (the more taxis, the more people can be transported). At any given time the concentration of enzyme in a cell is determined by its rate of synthesis and its rate of degradation. The concentration of enzymes present in a cell can be increased (enzyme induction) or decreased (enzyme suppression) according to the needs of the organism.

Temperature

A rule of thumb for most chemical reactions is that a 10 °C rise in temperature approximately doubles or triples the reaction rate. (This is due to an increase in the number of molecules that possess sufficient kinetic energy to exceed the activation energy.) To some extent, this is true of all enzymatic reactions. After a certain point, however, an increase in temperature causes a decrease in the rate of reaction, as indicated in Figure 22.9. The temperature that affords maximum activity is known as the **optimum temperature** for the enzyme in question. Most enzymes of warm-blooded animals have optimum temperatures of about 37 °C (98 °F). The decrease in rate is a direct consequence of the fact that enzymes are proteins and thus are denatured by heat. Heating disrupts the secondary and tertiary structures of the enzymes, causing a disorientation of the active site. This disorientation renders the active site inaccessible to the substrate (Figure 22.10).

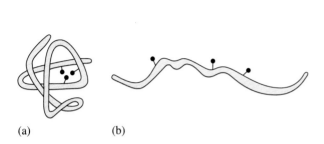

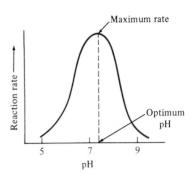

(a)   (b)

Figure 22.10

(a) Representation of an active site in an enzyme. (b) Heating denatures the enzyme, and the groups of the active site are no longer in close proximity.

Figure 22.11

Effect of pH on the rate of an enzymatic reaction.

At temperatures of 0 °C and 100 °C the rate of enzyme-catalyzed reactions is nearly zero. This fact has several practical applications. We sterilize objects by placing them in boiling water so as to denature the enzymes of any bacteria that may be in or on them. We refrigerate and freeze our food to slow down enzyme activity and preserve the food. Animals go into hibernation because of a decrease in their body temperature (in the winter), and as a result the rates of their metabolic processes decrease. The food required to maintain this lowered metabolic rate is provided by reserves stored in their tissues.

Hydrogen Ion Concentration

Again, as a consequence of their protein nature, enzymes are sensitive to changes in the pH of their environments. Extreme values of pH (whether high or low) can cause denaturation of the protein. However, any change in the hydrogen ion concentration alters the degree of ionization of acidic and basic groups both on the enzyme and on the substrate. If any ionizable groups are located at the active site, and if a certain charge is necessary in order for the enzyme to bind its substrate, then an enzyme molecule that has even one of these charges neutralized will lose its catalytic activity. An enzyme will exhibit maximum activity over a narrow pH range in which the molecule exists in its proper charged form. The median value of this pH range is known as the **optimum pH** of the enzyme (Figure 22.11). With the notable exception of gastric juice, most body fluids have pH values between 6 and 8. This is essential because most enzymes exhibit optimal activity in the physiological pH range of 7.0–7.5. However, each enzyme has a characteristic optimum pH, and in a few cases this value is outside the usual physiological range. For example, the optimal pH for pepsin is 2.0 in the stomach, and that for trypsin is 8.0 in the small intestine.

The enzyme lysozyme destroys bacteria by hydrolyzing certain polysaccharide bonds that occur in their cell walls. The optimal pH for lysozyme is 5.0. Activity declines rapidly at both higher and lower pH. Lysozyme is active only when an aspartic acid side chain is ionized (as COO⁻) and a glutamic acid side chain is not (as COOH). At low pH, both are protonated. At high pH, both are ionized. In either case, the enzyme doesn't function.

Table 22.2

Poisons as Enzyme Inhibitors

Poison	Formula	Enzyme Inhibited	Action
Cyanide	CN^-	Cytochrome oxidase, catalase	Binds Fe^{3+} cofactor
Fluoride	F^-	Enolase	Binds Mg^{2+} cofactor
Sulfide	S^{2-}	Phenolase	Binds Cu^{2+} cofactor
Arsenate	$AsO_4{}^{3-}$	Glyceraldehyde 3-phosphate dehydrogenase	Substitutes for phosphate
Iodoacetate	ICH_2COO^-	Triose phosphate dehydrogenase	Binds to cysteine sulfhydryl group
Nerve gas	$\begin{array}{c} O \\ \parallel \\ F-P-OCH(CH_3)_2 \\ \mid \\ OCH(CH_3)_2 \end{array}$	Acetylcholinesterase	Binds to serine hydroxyl group

22.6
ENZYME INHIBITION: POISONS

In the preceding section, we noted that enzymes are inactivated by increased temperatures and by changes in pH. In a sense, then, temperature and hydrogen ion concentration can be considered to be factors that inhibit enzyme activity. In fact, any physical change or chemical reagent that denatures protein will adversely affect the rate of an enzymatic reaction. This type of enzyme inhibition is referred to as *nonspecific* inhibition because it affects all enzymes in the same manner. In contrast, a *specific* inhibitor exerts its effect upon a single enzyme or a group of related enzymes. The inhibition of enzyme activity is one of the most important control mechanisms in living organisms. Many poisons act to inhibit specific enzymes (Table 22.2).

Competitive Inhibition

A **competitive inhibitor** is any compound that bears a close structural resemblance to a particular substrate and competes with that substrate for binding at the same active site on the enzyme. The inhibitor is not acted upon by the enzyme and so remains bound to the enzyme, preventing the substrate from approaching the active site. The degree of competitive inhibition depends upon the relative concentrations of substrate and inhibitor. If the inhibitor is present in relatively large quantities, it blocks the active sites on all the enzyme molecules, and complete inhibition results. However, formation of the inhibitor–enzyme complex is reversible. Increased substrate concentration permits displacement of the inhibitor from the active site. Competitive inhibition can be completely reversed by addition of large excesses of substrate. The reversible nature of competitive inhibition has provided much information about the enzyme–substrate complex and about the specific groups involved at the active sites of various enzymes. Pharmaceutical companies have synthesized drugs that can competitively inhibit metabolic processes in bacteria (Section 22.9) and in cancer cells (see Section J.6).

    A classic example of competitive inhibition is the effect of malonic acid on the enzyme activity of succinic acid dehydrogenase. Malonic acid is a homolog of the enzyme's normal substrate, succinic acid. It will bind to the active site because the spacing of its carboxyl groups is not greatly different from that of succinic acid. No catalytic

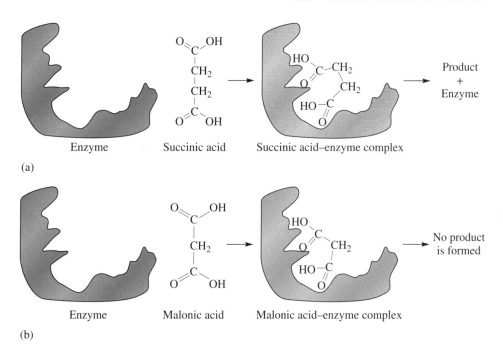

(a)

(b)

Figure 22.12
(a) Succinic acid binds to the enzyme succinic acid dehydrogenase. A dehydrogenation reaction occurs, and the product (fumaric acid) is released from the enzyme. (b) Malonic acid can also bind to the active site of succinic acid dehydrogenase. However, in this case no subsequent reaction occurs, and malonic acid remains bound to the enzyme.

reaction occurs, and malonic acid remains bonded to the enzyme (Figure 22.12). We discuss this reaction again in connection with carbohydrate metabolism (Section 25.1).

Noncompetitive Inhibition

A **noncompetitive inhibitor** is a substance that can combine with either the free enzyme or the enzyme–substrate complex. The noncompetitive inhibitor binds to the enzyme at a position relatively remote from the active site and in so doing alters the three-dimensional conformation of the enzyme. This effects a change in the configuration of the active site, so that either the E—S complex does not form at its normal rate or, once formed, it does not decompose at the normal rate to yield products. Since the inhibitor does not structurally resemble the substrate, the addition of excess substrate does *not* reverse the inhibitory effects.

Many enzymes contain reactive groups, such as —COO⁻, —NH₃⁺, —SH, or —OH, that are essential for maintaining the proper three-dimensional conformation of the enzyme. Any chemical reagent that is capable of combining with one or more of these groups will inhibit the enzyme. The heavy metal ions Ag^+, Hg^{2+}, and Pb^{2+} have strong affinities for carboxylate and sulfhydryl groups. We have already discussed their toxic effects with regard to protein denaturation (Section 21.14). Similarly, these ions react with the sulfhydryl groups of enzymes, rendering them inactive. This can occur at a position removed from the active site (Figure 22.13). Poisoning by lead or mercury ions is an example of noncompetitive inhibition.

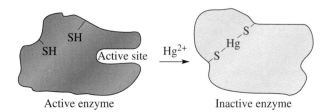

Figure 22.13
Mercury poisoning is an example of noncompetitive inhibition of an enzyme. Mercury ions react with sulfhydryl groups to change the conformation of the enzyme and destroy the active site.

Irreversible Inhibition

An **irreversible inhibitor** inactivates enzymes by forming strong bonds to a particular group at the active site. The inhibitor does not resemble the substrate, and the inhibitor–enzyme bond is so strong that the inhibition cannot be reversed by adding excess substrate. Irreversible inhibition was formerly considered to be another type of noncompetitive inhibition, but it is now recognized as a distinct type of inhibition.

The nerve gases, especially diisopropylfluorophosphate (DFP), irreversibly inhibit biological systems by forming an enzyme–inhibitor complex with a specific hydroxyl group (of serine) situated at the active site of certain enzymes. The peptidases trypsin and chymotrypsin contain such serine groups and are inhibited by DFP.

$$\boxed{\text{Enzyme}}\!-\!\text{OH} \;+\; \text{F}-\overset{\displaystyle\overset{\text{O}}{\|}}{\text{P}}-(\text{OR})_2 \;\longrightarrow\; \boxed{\text{Enzyme}}\!-\!\text{O}-\overset{\displaystyle\overset{\text{O}}{\|}}{\text{P}}(\text{OR})_2 \;+\; \text{HF}$$

Metalloenzymes (enzymes that require the presence of a metal ion for activation) are irreversibly inhibited by substances that form strong complexes with the metal. Traces of hydrogen cyanide inactivate iron-containing enzymes such as catalase and cytochrome oxidase. Oxalic and citric acids inhibit blood clotting by forming complexes with calcium ions necessary for the activation of the enzyme thromboplastin (see Section 28.8).

Inhibition of Nerve Transmission

The action of organophosphorus compounds is another example of irreversible inhibition. These compounds act on an enzyme that is an essential part of the process of nerve transmission. Nerve cells (neurons) interact with other nerve cells, and with muscles and glands, at junctions called *synapses* (refer to Figure F.3). Nerve impulses are transported across synapses by small molecules known as **neurotransmitters.** Neurons use a number of different molecules as neurotransmitters. The two major neurotransmitters are acetylcholine and norepinephrine (see Section F.2).

Acetylcholine is synthesized from acetyl coenzyme A (Section 25.1), and choline and is stored in special vesicles at the axon ends of neurons. The arrival of a nerve impulse leads to the release of acetylcholine into the synapse.[2] The acetylcholine molecules then diffuse across the synapse, where they combine with specific receptor protein molecules embedded in the postsynaptic membrane of the adjacent neuron (or muscle, or gland). Binding of acetylcholine to the receptors causes a change in membrane permeability of the receiving neuron (or muscle, or gland). Sodium ions move into the cell, potassium ions move out, and this ion flux causes the signal (the action potential) to be sent along the entire neuron until it reaches another synapse.

Once the impulse has been passed on, the acetylcholine must be immediately deactivated so that the receptor molecules can receive the next stimulus. The deactivation occurs by the hydrolysis of acetylcholine to choline and acetic acid, through the catalytic activity of the enzyme acetylcholinesterase (which is located in the synaptic cleft). This enzyme is characterized by an extremely high turnover number. It is estimated that this

[2] Botulism toxin, one of the most poisonous substances known, and certain snake venoms act by *inhibiting the release of acetylcholine* from the axon. Thus these toxic substances effectively block nerve transmissions that use acetylcholine as their neurotransmitter.

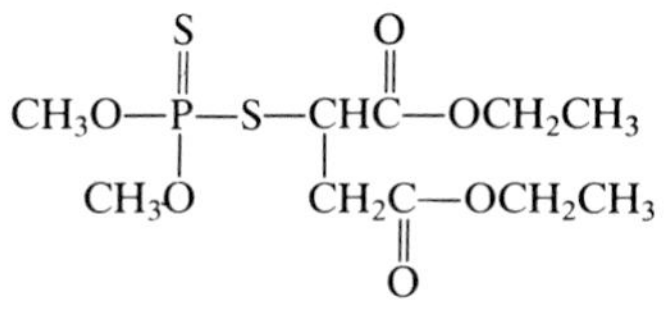

Malathion

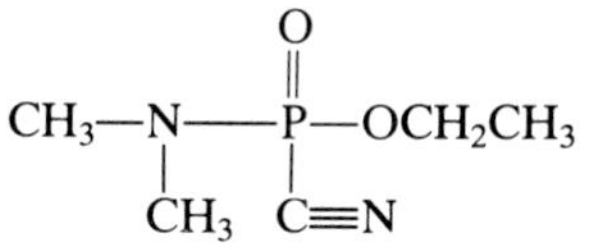

Tabun
(agent GA)

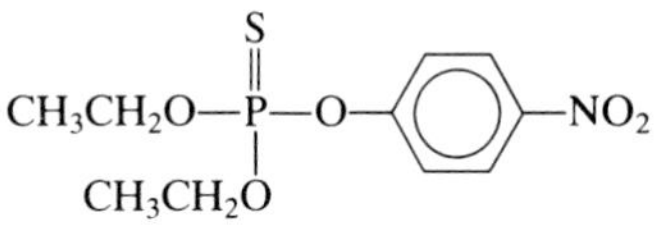

Parathion

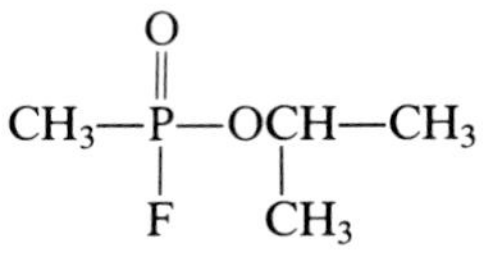

Sarin
(agent GB)

Figure 22.14
Some organophosphorus compounds. Malathion and parathion are insecticides. Tabun and sarin are nerve gases for use in chemical warfare.

enzyme-catalyzed hydrolysis reaction occurs in 40 μs (40×10^{-6} s). This speed is essential because nerve fibers can transmit 1000 impulses per second as long as the postsynaptic membrane is continually available to receive new acetylcholine molecules.[3]

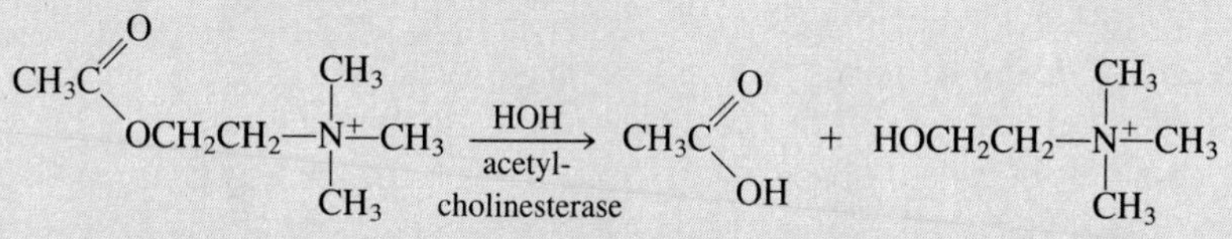

Acetylcholine Acetic acid Choline

The organophosphorus nerve poisons (Figure 22.14) affect all biological systems in a similar manner. The polar phosphorus–oxygen bond attaches tightly to acetylcholinesterase, preventing the enzyme from performing its normal function (Figure 22.15). If the breakdown of acetylcholine is blocked, then this messenger compound builds up, causing the receptor nerves to "fire" repeatedly, to be continuously "on." This overstimulates the muscles, glands, and organs. The heart beats wildly and irregularly. The victim goes into convulsions and dies quickly.

Although the organophosphorus compounds bind tightly to acetylcholinesterase, some can be displaced. Pralidoxime (2-PAM) is one antidote for organophosphorus poisons.

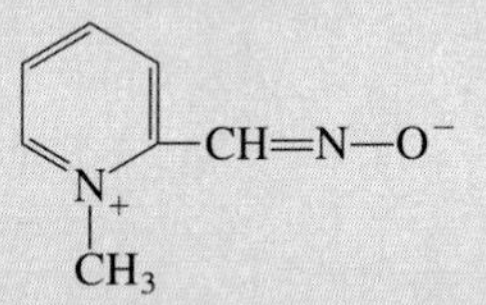

It is thought that the positive charge on nitrogen enables 2-PAM to bind to the site normally occupied by the quaternary nitrogen of acetylcholine, thus displacing the organophosphorus compound.

[3]Curare, used for centuries on the arrows of South American Indians, causes skeletal muscle paralysis. Curare exerts its effects by competing with acetylcholine for the receptor sites on the postsynaptic muscle cell membranes. Thus it *blocks the receptor sites* and prevents the message from being transmitted from nerve to muscle.

Figure 22.15
(a) Acetylcholinesterase catalyzes the hydrolysis of acetylcholine to acetic acid and choline. (b) An organophosphate ties up acetylcholinesterase, preventing it from breaking down acetylcholine. (c) Pralidoxime (2-PAM) displaces the organophosphorus compound.

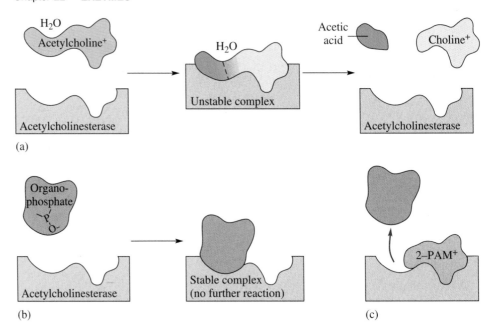

(a)

(b)   (c)

The nerve cells, muscle cells, or other receptor cells that are sensitive to acetylcholine are said to be *cholinergic*. Pharmacologists have developed drugs that affect cholinergic nerves by either enhancing or blocking the action of acetylcholine on the receptor cells. Succinylcholine, an inhibitor of acetylcholine, is used to produce muscular relaxation in surgical procedures. On the other hand, an insufficient secretion of acetylcholine by motor neurons results in the condition called myasthenia gravis. The person suffers from muscular weakness and may have trouble contracting the muscles associated with breathing, chewing, eye movements, and speaking. This condition can be treated by administering neostigmine, which inhibits acetylcholinesterase. If the enzyme is inhibited, enough acetylcholine may accumulate in the neuromuscular synapse to stimulate muscle contraction.

$$(CH_3)_3\overset{+}{N}CH_2CH_2OCCH_2CH_2COCH_2CH_2\overset{+}{N}(CH_3)_3$$
Succinylcholine

Neostigmine

22.7
ENZYME REGULATION AND ALLOSTERISM

Most biochemical processes take place in several steps, each catalyzed by a different enzyme. The product of each step becomes the substrate for the next enzyme. A reaction product of one enzyme may control the activity of another enzyme. Consider the system

$$A \xrightarrow{E_1} B \xrightarrow{E_2} C \xrightarrow{E_3} D \xrightarrow{E_4} F$$

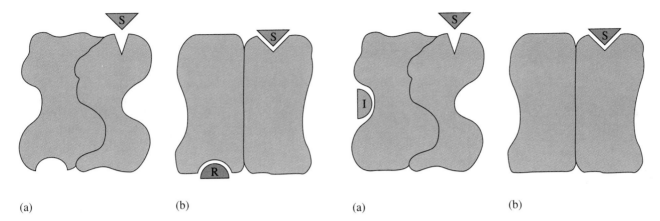

Figure 22.16

Negative modulation of an allosteric enzyme. With an inhibitor attached at a regulatory site (a), the substrate cannot bind effectively, and the reaction slows. Conformations are different without the inhibitor present (b)—the substrate binds readily, and the reaction proceeds rapidly.

Figure 22.17

Positive modulation of an allosteric enzyme. When the regulator is absent (a), the active site is relatively inaccessible. When the regulator is bound (b), both protein subunits change conformation, and the substrate binds readily.

where the E's are various enzymes controlling the steps of a four-stage reaction. A through D represent substrates of the different enzymes. The last product in the chain, F, may inhibit the activity of enzyme E_1. (This inhibition is noncompetitive.) When the concentration of F is low, all the reactions proceed rapidly. As the concentration of F increases, though, the action of E_1 slows and eventually stops. The buildup of the concentration of F signals E_1 to quit because the cell has enough F for a while. When E_1 shuts down, the whole chain of reactions stops. This mechanism for regulating enzyme activity is called **feedback control.**

Sometimes enzyme regulation takes place by a mechanism involving a site other than the active site. An enzyme so regulated is called an **allosteric enzyme,** and the process is called **allosterism** (Greek *allo,* other, and *stereos,* space or site). The substance binding at the regulatory site is called a **modulator** or a *regulator*. It may inhibit enzyme activity *(negative modulation)* or enhance it *(positive modulation)*. Allosteric enzymes are often large and contain two or more subunits. Often the regulatory site is on one protein chain and the active site on another.

Figure 22.16 shows a simple model for negative modulation of an allosteric enzyme with two subunits. When an inhibitor binds to the regulatory site, the conformations of the proteins change, and the active site is less accessible to the substrate. The reaction rate drops. In the absence of the inhibitor, the substrate binds readily, and reaction proceeds rapidly.

A simple model for positive modulation is shown in Figure 22.17. When a regulator molecule binds to a site on one protein subunit, the conformations of that subunit and a neighboring subunit both change. The neighbor can then bind substrate more effectively, and the reaction rate is enhanced.

Allosteric regulation also occurs in nonenzyme proteins. Positive modulation explains how hemoglobin's ability to transport oxygen is enhanced by the binding of an oxygen molecule to one subunit (see Section 28.9).

22.8
CHEMOTHERAPY

Chemotherapy is the use of chemicals (drugs) to destroy infectious microorganisms (and cancer cells) without damaging the cells of the host. From bacteria to humans, the metabolic pathways of all living organisms are quite similar, and so the discovery of safe and effective chemotherapeutic agents is a formidable task. It is now well established that many drugs function through their inhibitory effect on a critical enzyme in the cells of the invading organism.

Chemotherapy is widely used in the treatment of cancer patients. **Antineoplastic** drugs (substances that inhibit the growth of cancer cells) such as 5-fluorouracil and 6-mercaptopurine interfere with the production of DNA and RNA in tumor cells by substituting for the pyrimidine and purine bases (see Section 23.1).

Antimetabolites

In Section J.6, we examine some antimetabolites that are anticancer agents.

An **antimetabolite** is a substance that possesses a structure closely related to the normal substrate (the *metabolite*) of an enzyme and *competitively* inhibits a significant metabolic reaction. One of the earliest (1935) and best understood antimetabolites is the synthetic antibacterial agent sulfanilamide. Its effectiveness rests on its structural similarity to *p*-aminobenzoic acid, a compound vital to the growth of many pathogenic bacteria. The vitamin folic acid (Section I.6) serves as a coenzyme for several important biochemical processes. We obtain folic acid in our diets and from bacteria in our digestive tracts. Bacteria can synthesize folic acid *if* they have access to *p*-aminobenzoic acid. When bacteria encounter sulfanilamide, a bacterial enzyme readily incorporates the drug into a false folic acid. This altered folic acid not only cannot function as a proper coenzyme, it also serves as a competitive inhibitor of the enzyme. The bacteria are unable to make compounds such as certain amino acids and nucleotides, so they die.

Sulfanilamide is not harmful to humans (or to other mammals) because we cannot synthesize folic acid but must obtain it, preformed, from our diets. After the drug was recognized as an antibacterial agent, many other sulfanilamide derivatives (sulfa drugs) were synthesized and found to be even more effective in this capacity. Many lives were saved during World War II as a result of these popularly named ''wonder drugs.'' Soldiers carried packages of powdered sulfa drugs to sprinkle on open wounds to prevent infection. Unfortunately, prolonged use of sulfa drugs causes a number of side effects, particularly kidney damage, so they have been largely replaced by the penicillins and other antibiotics. However, they are still prescribed for some specific infections against which they are highly effective, such as infections of the bladder and urinary tract. Some newer sulfa drugs are used in the treatment of tuberculosis and leprosy, and they are widely used in veterinary medicine. Structures of sulfa drugs used as urinary antiseptics are given in Figure 22.18, along with the structure of the metabolite *p*-aminobenzoic acid.[4]

---

[4]It is interesting to note that *p*-aminobenzoic acid (PABA) has become widely known as an ingredient in suntan lotions. PABA acts as a sun filter by absorbing the short-wavelength ultraviolet rays that are responsible for causing sunburn. The most effective sunscreen products are those that contain 5% PABA as the active ingredient. The ethyl ester of PABA is Benzocaine, a local anesthetic found in a wide variety of over-the-counter products including first aid and sunburn sprays, foot powders, cough medicines, and appetite control products.

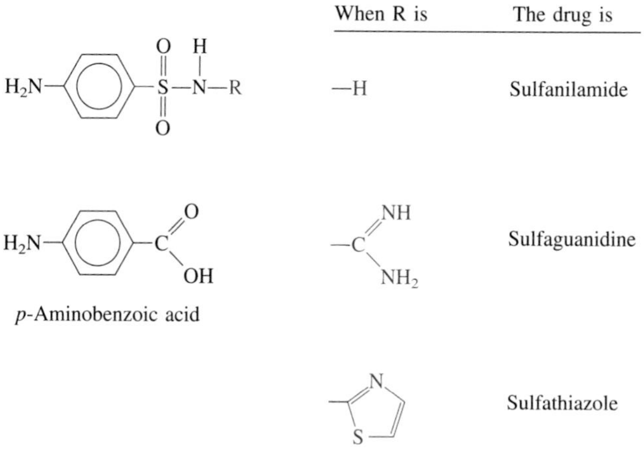

	When R is	The drug is

H_2N—⟨benzene ring⟩—$\overset{\overset{O}{\|}}{\underset{\overset{\|}{O}}{S}}$—$\overset{H}{N}$—R

—H   Sulfanilamide

H_2N—⟨benzene ring⟩—$\overset{\overset{O}{\|}}{C}$—OH

p-Aminobenzoic acid

$-\overset{\overset{NH}{\|}}{C}-NH_2$ Sulfaguanidine

(thiazole ring)   Sulfathiazole

Figure 22.18
The sulfa drugs interfere with the normal metabolism of *p*-aminobenzoic acid.

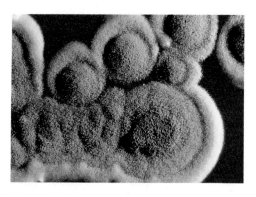

Figure 22.19
Penicillin molds. These symmetrical colonies of mold are *Penicillium chrysogenum*, a mutant form of which now produces almost all of the world's commercial penicillin.

Antibiotics

Although some antibiotics are believed to function as antimetabolites, the terms are not identical. An **antibiotic** is a compound produced by one microorganism (bacterium, mold, yeast) that is toxic to another microorganism (Figure 22.19). Antibiotics, many of which can now be synthesized in the laboratory, constitute no well-defined class of chemically related substances but instead possess the common property of effectively inhibiting a variety of enzymes essential to bacterial growth.

Penicillin, one of the most widely used antibiotics in the world, was fortuitously discovered by Alexander Fleming in 1928 (Figure 22.20). In 1938, Ernst Chain and Howard Florey isolated it in pure form and proved its effectiveness as an antibiotic. (The three scientists received the Nobel prize for physiology or medicine in 1945.) Penicillin was first introduced into medical practice in 1941. It functions by interfering with the synthesis of cell walls of reproducing bacteria. Penicillin inhibits an enzyme (transpeptidase) that catalyzes the last step in bacterial cell wall biosynthesis. This step involves the joining of long polysaccharide chains by short peptide chains. The new cell walls are defective, and subsequently the bacterial cells burst. Since human cells have cell membranes and not cell walls, they are not affected. Several naturally occurring penicillins have been isolated. All have the empirical formula $C_9H_{11}O_4SN_2R$ and contain a four-member ring fused to a five-member ring (Figure 22.21). The various R groups are obtained by the addition of the appropriate organic compounds to the culture medium.

The penicillins are effective against gram-positive bacteria (bacteria that are stained by Gram's dye) and a few gram-negative bacteria (including *E. coli*). They have proved effective in the treatment of diphtheria, gonorrhea, pneumonia, syphilis, many pus infections, and certain types of boils. Penicillin G was the earliest penicillin to be used on a wide scale. However, it cannot be administered orally because it is quite unstable and the acid pH of the stomach causes a rearrangement to an inactive derivative. Penicillin V,

The term *antibiotic* is considered to be synonymous with antibacterial, although many antiviral and anticancer drugs are also antibiotics.

Figure 22.20
Sir Alexander Fleming, discoverer of penicillin, shown here on a 1981 Hungarian stamp.

Figure 22.21

The penicillins differ only in the identity of their R groups. Notice that the amino acids valine and cysteine are incorporated into the penicillin structure.

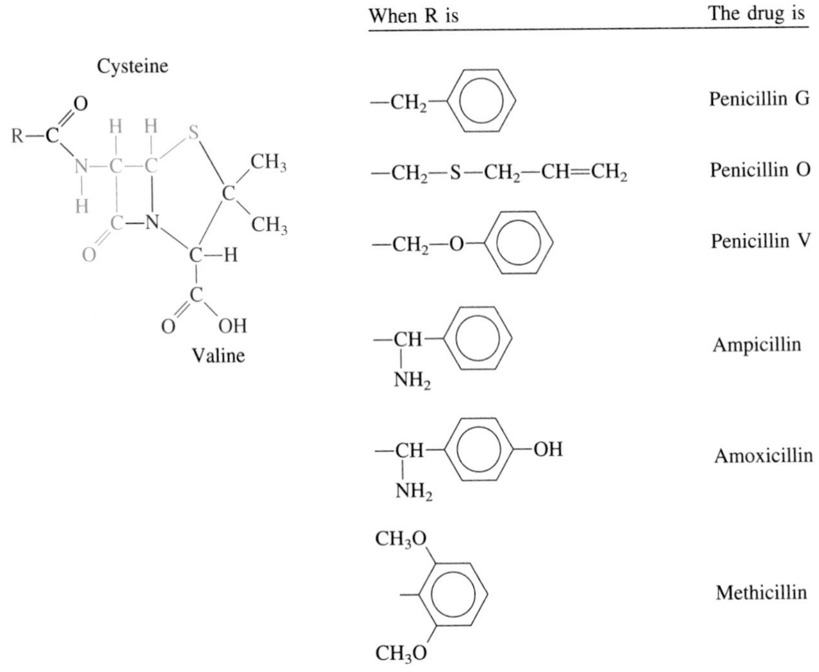

ampicillin, and amoxicillin, on the other hand, are acid-stable, and they are the major oral penicillins. Amoxicillin was the most widely prescribed drug in the United States in 1988.

Some strains of bacteria become resistant to penicillin by a mutation that allows them to synthesize an enzyme, penicillinase, that breaks down the antibiotic (by cleavage of the amide linkage in the four-member ring). To combat these strains, scientists have been able to synthesize penicillin analogs (such as methicillin) that are not inactivated by penicillinase. Another way is to use an inhibitor such as clavulinic acid, which forms a covalent bond to the active site of penicillinase, irreversibly inhibiting the enzyme. When given along with penicillin, clavulinic acid deactivates the enzyme that would otherwise deactivate the penicillin. The penicillin can then destroy the bacteria in the usual way.

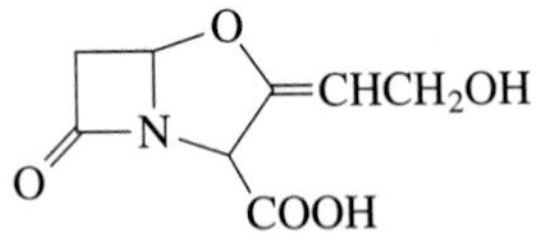

Clavulinic acid

Some people are allergic to penicillin and therefore must be treated with other antibiotics. (The allergic reaction is so severe that a fatal coma may occur if penicillin is inadvertently administered to a sensitive individual.) Fortunately, a number of antibiotics have been discovered (Figure 22.22). Most are the complete products of microbial synthesis (e.g., aureomycin, streptomycin). Others are made by chemical modifications of antibiotics (e.g., semisynthetic penicillins, tetracyclines), and some are manufactured entirely by chemical synthesis (e.g., chloramphenicol). They have proved to be as effective as penicillin in destroying infectious microorganisms. Many of these antibiotics exert their effects by blocking protein synthesis in microorganisms. (See Table J.1 for a listing of some of the major bacterial diseases.)

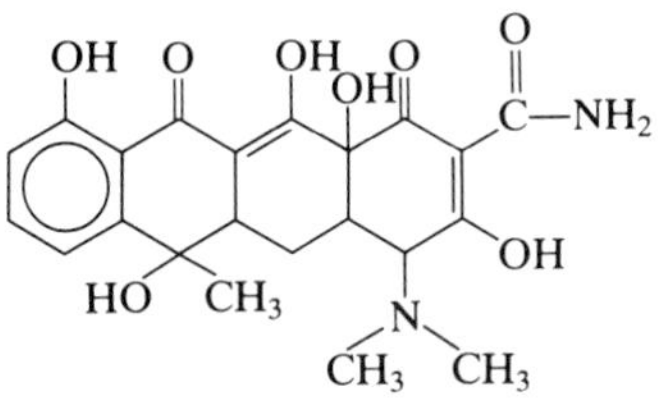

Tetracycline

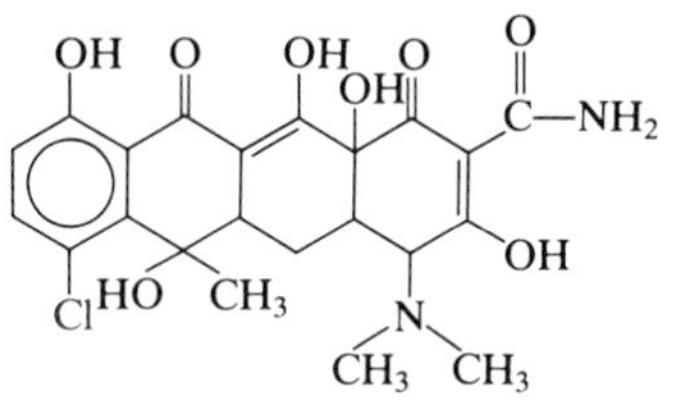

Aureomycin
(Chlorotetracycline)

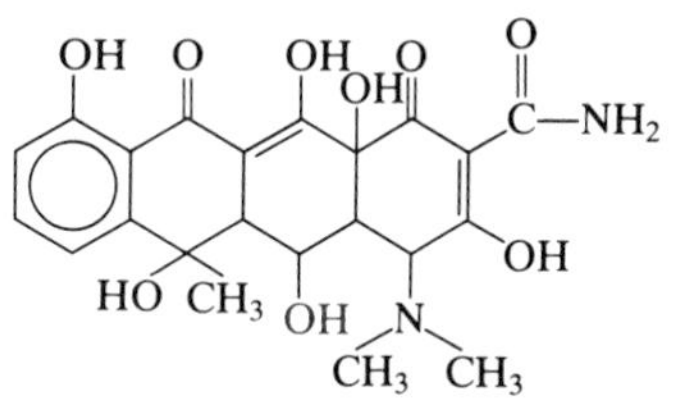

Terramycin
(Oxytetracycline)

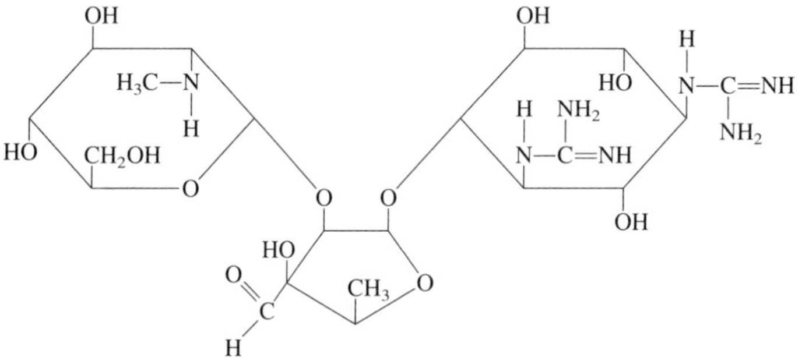

Streptomycin

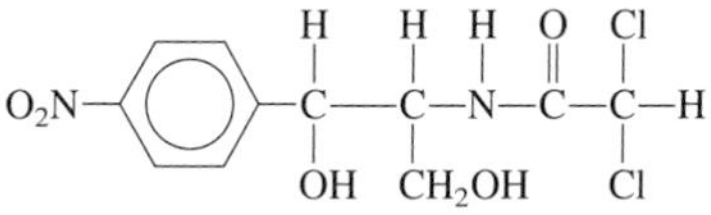

Chloramphenicol
(Chloromycetin)

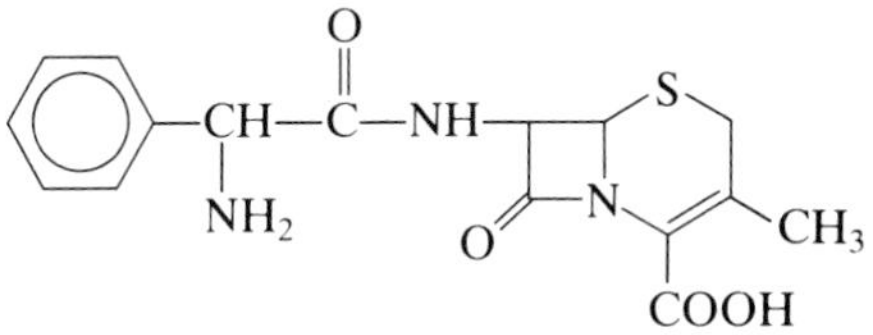

Cephalexin
(Keflex)

Figure 22.22
Structures of some common antibiotics. Notice the structural similarities among the tetracyclines. Streptomycin is a glycoside containing an amino derivative of glucose, and chloramphenicol bears a resemblance to epinephrine. Cephalexin is a member of the cephalosporins, antibiotics that are related to the penicillins.

Diagnostic Applications of Enzymes

The measurement of enzyme activity in such body fluids as plasma or serum has become a valuable tool in medical diagnosis. Certain enzymes that function in the plasma, such as the enzymes involved in blood clotting, are continually secreted into the blood by the liver. Most other enzymes, however, are normally present in plasma in very low concentrations. They are derived from the routine destruction of erythrocytes, leukocytes, and other cells. When cells die, their soluble enzymes leak out of the cells and enter the bloodstream. Since not all cells contain the same complement of enzymes, those that are specific to a particular organ can be important in aiding diagnosis. Therefore, an abnormally high level of a particular enzyme in the blood often indicates specific tissue damage, as in hepatitis and myocardial infarction (*myo,* muscle; *cardi,* heart; an *infarct* is an area of dead tissue). For example, elevated blood levels of creatine kinase (CK; see Section 24.8) and glutamic-oxaloacetic transaminase (GOT; see Section 27.5) accompany some forms of severe heart disease. A blood analy-

LAB NO.____________

☐ ROUTINE
☐ PRE-OP
☐ ASAP
☐ STAT

SPECIMEN:
☐ BLOOD
☐ OTHER
DATE____________
TIME____________
TECH____________

IF IMPRINT PLATE NOT USED PRINT HERE

PATIENT NO. ROOM NO.
PATIENT NAME AGE
ADDRESS
DOCTOR

✓	TEST	NORM	RESULT	✓	TEST	NORM	RESULT	✓	TEST	NORM	RESULT	✓	TEST	NORM	RESULT
	LDH				Amylase				Albumin				Calcium		
	LDH-1				Acid Phos.				Globulin				Phosphorus		
	CPK				Cholesterol				BUN				Magnesium		
	CPK-MB				Triglyceride				Creat				Ethanol		
	Alk. Phos.				HDL-Chol.				Total Bili				Hgb AIC		
	SGOT (AST)				Iron				Direct Bili				Ammonia		
	SGPT (ALT)				TIBC				Indirect Bili				Lithium		
	GGPT				Total Prot.				Uric Acid			TECH:		DATE:	

BRIGGS, DES MOINES, IA 50306 CHEMISTRY I PRINTED IN U.S.A.

Figure 22.23
A hospital form for clinical analysis of a blood sample.

sis that shows high levels of CK may indicate that the heart muscle has suffered serious damage. On the other hand, many forms of strenuous (and healthful) physical activity will also result in elevated CK levels. The enzyme mediates the reaction that serves as one source of energy for muscle contraction. Indeed, it is even possible for the CK level to rise simply because someone who hates needles has tensed up while waiting for the blood sample to be taken. Nonetheless, as Figure 22.23 suggests, analysis for specific enzymes is considered one valuable source of data on which to base a medical diagnosis. Table 22.3 lists the commonly assayed enzymes that are used in clinical diagnosis.

Table 22.3
Some Important Enzymes for Clinical Diagnoses

Enzyme Assayed	Organ or Tissue Affected
α-Amylase	Pancreas
Alkaline phosphatase	Bone, liver
Acid phosphatase	Prostate
Creatine kinase (CK)	Muscle, heart
Glutamic-oxaloacetic transaminase (GOT)	Heart, liver
Glutamic-pyruvic transaminase (GPT)	Liver
Lactic dehydrogenase (LDH)	Heart, liver
Alanine aminotransferase	Liver
Aspartate aminotransferase	Heart, liver

Modern medical practices have automated and computerized the assay procedures for most of these serum enzymes. It is important to note that the precise patterns of enzyme changes in certain tissue diseases are characteristic. For example, in a myocardial infarction, the GOT/GPT ratio is usually high; the reverse is true in liver disease. A summary of changes in serum enzyme levels following a heart attack is illustrated in Figure 22.24.

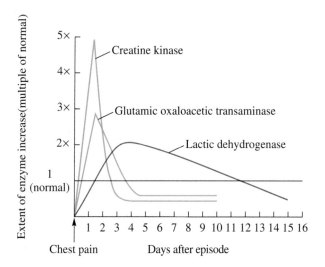

Figure 22.24
Typical changes in serum enzyme levels following a heart attack.

11-23

EXERCISES

1. Define and, where appropriate, give an example for each of the following.
   a. enzyme
   b. substrate
   c. holoenzyme
   d. proenzyme
   e. active site
   f. chemotherapy
   g. cofactor
   h. enzyme specificity
   i. irreversible inhibition
   j. feedback control
   k. regulatory site
   l. sulfa drug

2. Distinguish between the terms in each of the following pairs.
   a. in vitro and in vivo
   b. intracellular enzyme and extracellular enzyme
   c. sucrose and sucrase
   d. trypsin and trypsinogen
   e. contact group and catalytic group
   f. absolute specificity and stereochemical specificity
   g. group specificity and linkage specificity
   h. optimum temperature and optimum pH
   i. competitive inhibitor and noncompetitive inhibitor
   j. antimetabolite and antibiotic.

3. What is the substrate for each of the following enzymes?
   a. maltase
   b. cellulase
   c. peptidase
   d. lipase

4. In what ways are enzymes similar to ordinary chemical catalysts? How do they differ? What is the effect of an enzyme on an equilibrium system?

5. Why are enzymes more specific than inorganic catalysts?

6. Animals can digest starch but not cellulose. Explain.

7. Identify three intracellular enzymes and three extracellular enzymes.

8. Which enzyme is more specific—urease or carboxypeptidase?

9. To which of the six major types of enzymes does each of the following belong?
   a. decarboxylase
   b. peptidase
   c. transaminase
   d. dehydrogenase
   e. kinase
   f. lipase

10. Why does the body synthesize trypsin as the proenzyme trypsinogen rather than make it directly?

11. How does the body activate trypsinogen, that is, how is it converted to trypsin?

12. Alcohol dehydrogenase is an enzyme that catalyzes the conversion of ethanol to acetaldehyde. The active enzyme consists of a protein molecule and a zinc ion. Identify the following components of this reaction system.
    a. substrate
    b. cofactor
    c. apoenzyme

13. Can the zinc ion mentioned in Exercise 12 be called a coenzyme? Explain your answer.

14. Succinate dehydrogenase is an enzyme that is only active in combination with a nonprotein organic molecule called flavin adenine dinucleotide (FAD). Is FAD a cofactor? Is it a coenzyme?

15. The concentration of substrate X is low. What happens to the rate of the enzyme-catalyzed reaction if the concentration of X is doubled?

16. An enzyme has an optimum pH of 7.4. What happens to the activity of the enzyme (a) if the pH drops to 6.8? (b) if the pH rises to 8.0?

17. A bacterial enzyme has an optimum temperature of 35 °C. Will the enzyme be more or less active at normal body temperature? Will it be more or less active if the patient has a fever of 40 °C?

18. What is an enzyme inhibitor?

19. Why it is that only a relatively few enzyme molecules can catalyze the conversion of many molecules of substrate?

20. Compare the lock-and-key theory with the induced-fit theory of Koshland.

21. How is it possible that two amino acids that are relatively far apart in the primary structure of the protein chain can be in proximity at the active site?

22. What is allosterism?

23. Why should enzymes consist of 100 or more amino acid units when only a few amino acid units are involved in the active site?

24. What is the turnover number for an enzyme?

25. What is negative modulation? Give an example.

26. What is positive modulation? Give an example.

27. How do mercury ions act as poisons? Is their action reversible or irreversible?

28. What is acetylcholine? Describe the acetylcholine cycle.

29. How do organophosphorus compounds act as poisons?

30. How does 2-PAM act as an antidote for poisoning by organophosphorus compounds?

31. How do sulfa drugs kill bacteria?

32. How does penicillin kill bacteria?

33. How do resistant bacteria deactivate penicillin?

34. Describe two ways to get around the problem of bacterial resistance.

35. What is the active ingredient in meat tenderizer? How does it work?

36. Which has the longer polypeptide chain—prothrombin or thrombin?

37. What enzymes are monitored in the blood of a patient who is suspected of having a heart attack? Why? Why is the test not always reliable?

38. L-Aspartic acid is readily converted to fumaric acid in the presence of the enzyme aspartase, but D-aspartic acid does not react. Explain.

39. What enzyme is involved in the conversion of lactose to galactose and glucose? To what class of enzymes does it belong?

40. In a commercial process, glucose (in corn syrup) is converted enzymatically to fructose (to form high-fructose corn syrup). What type of enzyme is involved? (Recall that both glucose and fructose have the formula $C_6H_{12}O_6$.)

41. Both wool (protein) and nylon (synthetic polymer) are large molecules in which smaller molecules have been joined by amide linkages. Moths can digest wool, but they can't digest nylon. Explain.

42. Would you expect an apoenzyme to bind to more than one type of coenzyme? Explain.

43. Explain why enzymes become inactive above and below (a) the optimum temperature and (b) the optimum pH.

44. How does increased enzyme concentration affect the rate of a reaction?

45. Acetylcholinesterase has aspartic acid, histidine, and serine residues at the active site. In acid solution, the enzyme is inactive, but activity increases as the pH rises. Explain.

46. Which type of R-group interaction (i.e., hydrogen bonding, hydrophobic, etc.) at the active site of an enzyme would bind to the following groups on the substrate?

   a. —COOH   b. —COO⁻
   c. —NH₂   d. —NH₃⁺
   e. —OH   f. —SH
   g. —CH(CH₃)₂   h. phenyl

47. Sketch the general shape of a curve showing how the rate of a reaction varies with the concentration of substrate (if the concentration of enzyme is kept constant).

48. Alcohol dehydrogenase is involved in the oxidation of both methanol and ethanol. How might ethanol work as an antidote for methanol poisoning?

49. In Section 21.14 we discussed the denaturation of proteins. What is the effect of denaturing an enzyme? Be specific.

50. Experimentally, how could you distinguish a competitive inhibitor from a noncompetitive inhibitor?

51. Is oxaloacetic acid ($HOOCCH_2COCOOH$) an inhibitor of succinic acid dehydrogenase? What type of inhibitor is it?

52. Explain why antimetabolites and antibiotics can both be classified as antiseptics.

53. Derivatives of PABA are ubiquitous. Ethyl p-aminobenzoate (benzocaine) is an antiseptic and a local anesthetic. Ointments containing 5–10% benzocaine are used to treat sunburn and minor scrapes. Draw the structure of benzocaine.

54. List five antibiotics.

Vitamins

Carbohydrates, fats, and proteins are the three major classes of foods. To remain healthy we must take in relatively large amounts of each of these types of substances. They are not, however, the only nutrients we require. Some of our needs are satisfied only by vitamins and minerals. The minerals, inorganic ions of critical importance to our health, were discussed as a group in Section 12.9. Vitamins are considered in this special topic.

We normally eat three meals a day to satisfy our need for carbohydrates, fats, and proteins, but all the necessary vitamins can be packed into a single small pill. So small are the required amounts that not even a vitamin pill is necessary. We can get all we need simply by eating a balanced diet. Nature has thoughtfully incorporated minute but adequate amounts of vitamins into our various foodstuffs.

I.1
WHAT ARE VITAMINS?

Vitamins are organic compounds that cannot be synthesized by an organism but nevertheless are essential for the maintenance of normal metabolism and therefore must be included in the diet. The absence or shortage of a vitamin results in a vitamin deficiency disease.

One such disease, called scurvy, had plagued seamen since early times. In 1747, a British naval captain, James Lind, showed that scurvy could be prevented by the inclusion of fresh fruit or vegetables in the diet. A convenient fresh fruit to carry on long voyages was the lime. British ships put to sea with barrels of limes aboard. The sailors ate a lime or two every day and remained free of scurvy. British sailors came to be known as "lime eaters" or simply "limeys."

In 1897, the Dutch scientist Christiaan Eijkman showed that polished rice lacked something found in the hulls of whole-grain rice. Lack of that "something" caused the disease beriberi, which was then quite a problem in the Dutch East Indies. Within the next two decades a British scientist, F. G. Hopkins, fed a synthetic diet of carbohydrates, fats, proteins, and minerals to a group of rats. The rats were unable to sustain healthy growth. Again, something was missing. Eijkman and Hopkins shared the 1929 Nobel prize in medicine or physiology for their important discoveries.

In 1912, Casimir Funk, a Polish biochemist, coined the word *vitamine* (from the Latin *vita,* life) for these missing factors. Funk thought that they all contained the amino group. The final *e* was dropped after it was found that not all the factors were amines. The generic term became *vitamin*.

Since organisms differ in their synthetic abilities, a substance that is a vitamin for one species may not be so for another. Over the years, scientists have isolated 13 vitamins needed by humans. Researchers today are in general agreement that no more vitamins are likely to be found. One reason is that about 50 years has elapsed since the last one (vitamin B_{12}) was discovered, despite an intensive search for others. Another reason is that many people have lived for years on intravenous solutions containing only the known vitamins and other nutrients without exhibiting signs of vitamin deficiency.

The vitamins are divided into two broad categories, the **fat-soluble** group, including A, D, E, and K, and the **water-soluble** group, made up of the B complex and vitamin C. All the fat-soluble vitamins incorporate a high proportion of hydrocarbon structural elements. There are one or two oxygen atoms present, but the compounds as a whole are nonpolar. In contrast, a water-soluble vitamin contains a high proportion of the electronegative atoms oxygen and nitrogen, which can form hydrogen bonds to water; therefore, the molecule as a whole is soluble in water.

Most water-soluble vitamins act as coenzymes or are required for the synthesis of coenzymes (see Section I.6). The fat-soluble vitamins have more varied functions. In general, the fat-soluble vitamins are obtained from fish, liver, dairy products, green vegetables, and vegetable oils. Fat-soluble vitamins, if taken in high doses, can accumulate in hazardously large amounts (hypervitaminosis) because they are stored in body fat. People who consume too much vitamin D, for example, can develop bone pain, bonelike deposits in the kidneys, and mental retardation. Water-soluble vitamins, on the other hand, generally are rapidly excreted in the urine. Thus, even when taken in relatively large amounts, they usually are not toxic.

Minimum daily requirements (MDRs) of the vitamins

Figure I.1

Animals are able to convert β-carotene to vitamin A.

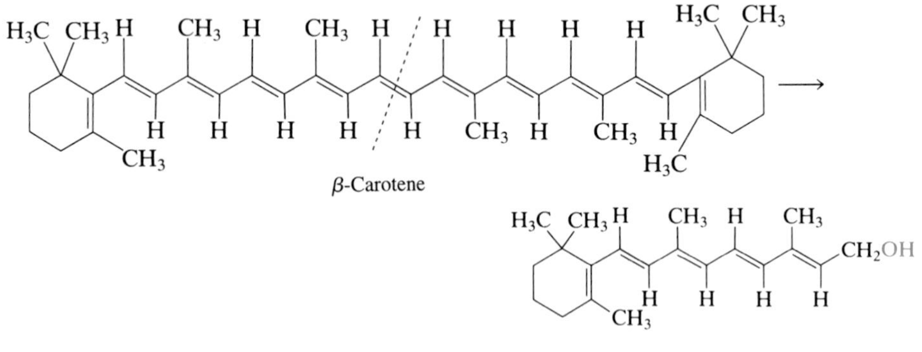

β-Carotene

Vitamin A
(Retinol)

have been set by examining the levels below which deficiency diseases occur.[1] (General warning signs of vitamin deficiency are slow healing of wounds, tiredness, and frequent illness.) There is no agreement, however, on the optimum levels of dietary vitamins. Vitamins have been the subject of more fads and more misrepresentations than any other group of nutrients. Claims have been made that, among other things, vitamins cure cancer, arthritis, and mental illness; increase sexual potency; prevent colds; and overcome muscular weakness. It is small wonder that the public is baffled by such claims, which most registered dietitians reject.

I.2

VITAMIN A

Vitamin A is a compound that occurs only in the animal kingdom. It was first isolated from halibut oil and is also present in cod liver oil and in butter. However, the plant pigment β-carotene is a precursor substance (a provitamin) that can be converted to vitamin A by animals, and thus most green and yellow vegetables (carrots, lettuce, spinach, yams) are a good source of the vitamin (Figure I.1).

Some foods are fortified by the addition of vitamin A obtained by extraction from fish liver oils or made synthetically. The recommended daily allowance of vitamin A is 0.7 mg. If vitamin A is present in excess, it is stored in the

[1] The RDA (recommended daily allowance) is a government estimate of the amount of vitamins that the average healthy person should eat daily to maintain good nutrition and health. A committee of the National Academy of Sciences is presently revising the RDAs, but the new figures have not yet (1992) been released. The RDA should not be confused with the MDR. The MDR is the smallest amount of a nutrient that if ingested will prevent a nutritional deficiency. The RDA significantly exceeds the MDR for each nutrient.

liver. Adult livers can store enough vitamin A to last for several months. On the other hand, the livers of infants and children do not store much of the vitamin. Consequently, infants and children are more likely to develop vitamin A deficiencies if their diets are inadequate.

In Special Topic G we discussed the well-known role of vitamin A in vision. We also know that a deficiency of this vitamin affects most of the body's organs. What we do not know is the detailed biochemistry of vitamin A as it relates to all these other organs. We can, however, describe some of the effects of vitamin A deficiency.

Vitamin A is required for normal growth. Young animals fed a diet lacking in the vitamin simply fail to grow. One of the earliest manifestations of vitamin A deficiency is a loss of night vision (a function of the rod cells in the retina). Mucous membranes may harden, dry, and crack. In cases of severe deprivation, victims may exhibit xerophthalmia, a condition characterized by inflammation of the eyes and eyelids, leading ultimately to infection and blindness. Vitamin A also stimulates fluid secretion by the epithelial cells of the eye. Thus, if the dietary supply of vitamin A is inadequate, the cornea of the eye becomes dried, or keratinized. The cornea then becomes extremely vulnerable, and even the slightest nick or scratch may cause it to perforate, which leads to blindness. Blindness in this case is brought about by a vitamin A deficiency disease called keratomalacia. This disease is the major cause of blindness in young children in most of the developing countries, and it is estimated to affect hundreds of thousands of children throughout the world. Health workers in those countries often carry injectible solutions of the vitamin for emergency treatment of such cases.

In addition, vitamin A is important to the growth and maintenance of epithelial tissue. In the past, some physicians used large doses of vitamin A for treating acne. Not only were the doses potentially harmful (painful joints,

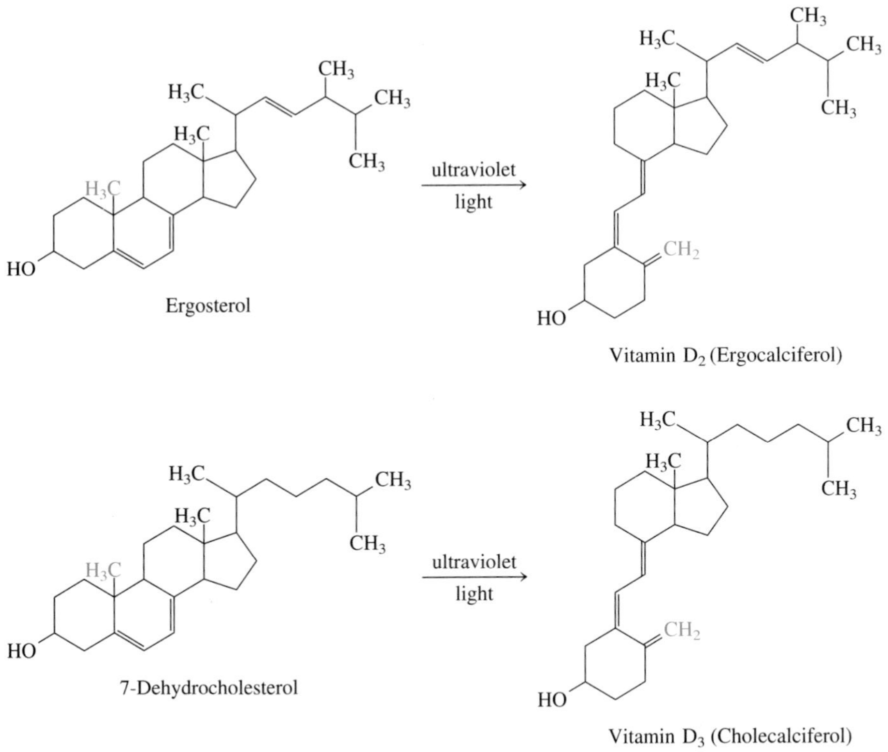

Figure I.2

Formation of two forms of vitamin D by action of ultraviolet light on corresponding provitamins.

Ergosterol

Vitamin D$_2$ (Ergocalciferol)

7-Dehydrocholesterol

Vitamin D$_3$ (Cholecalciferol)

loss of hair), but they were not really effective. However, it has been found that a synthetic derivative of vitamin A (13-*cis*-retinoic acid) appears promising as a drug for treating severe acne.

Population studies suggest that vitamin A may be protective against cancers of the lungs, bladder, mouth, esophagus, larynx, breast, and cervix. A 1984 study of lung cancer patients and healthy volunteers at the Cancer Center of Hawaii found that men who consumed a diet with low levels of vitamin A had twice the lung cancer rates of those with high levels of the vitamin. Other studies have shown a statistical correlation between diets rich in carotenes and a decreased incidence of lung cancer (see Section J.7). The effect was noted even in smokers. More research into this phenomenon is currently under way.

Vitamin A is found in high concentration in fish liver oils. Liver, eggs, fish, butter, and cheese are also good sources. It is interesting to note that polar bear liver has so much vitamin A that it is toxic. These large animals eat seals that eat fish. Fat-soluble vitamin A is concentrated in each step of the food chain. Eskimos who kill a polar bear know better than to eat its liver. Large excesses cause irritability, dry skin, and a feeling of pressure inside the head. Massive doses of vitamin A administered to pregnant rats result in malformed offspring.

I.3 VITAMIN D

Several chemical compounds have vitamin D activity. Only two commonly occur in foods or are used in drugs and food supplements. Each is formed from a precursor by the action of ultraviolet light (Figure I.2). Vitamin D$_2$ (ergocalciferol) is synthesized by irradiation of ergosterol, a compound found in yeast and other molds. Vitamin D$_3$ (cholecalciferol) is formed in the skin of animals by the action of sunlight on 7-dehydrocholesterol. The two vitamins differ only in the structure of their side chains; D$_2$ contains an extra carbon and a double bond. (There is no vitamin D$_1$. The material that was originally given this designation proved to be a mixture of vitamins D$_2$ and D$_3$.)

Vitamin D increases the utilization of calcium and phosphorus by the body. Deficiency in infants and growing children results in abnormal bone formation, a condition known as **rickets** (Figure I.3). The condition is characterized by bowed legs, knobby bone growths where the ribs join the breastbone (called a "rachitic rosary"), pigeon breast, and poor tooth development. In adults, with completed bone growth, rickets does not develop. Women

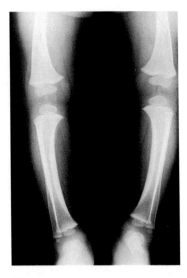

Figure I.3
This X-ray shows the typical distortion of leg bones in rickets, a disease caused by a deficiency of vitamin D.

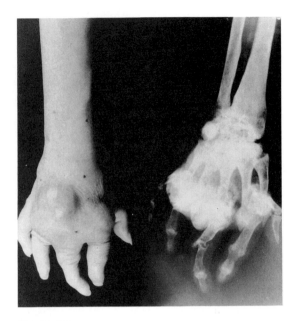

Figure I.4
Too much vitamin D causes the deposit of bonelike material in joints.

who are deficient in vitamin D may develop osteomalacia, a condition characterized by fragile bone structure. This condition is rare but may occur after several pregnancies.

Most foods contain little or no vitamin D. The best natural sources of the vitamin are fish liver oils and egg yolks. Irradiated ergosterol (from yeast) is added to milk (10 μg per quart) and margarine as a supplemental source of vitamin D.

Vitamin D is the ''sunshine vitamin.'' Individuals with a reasonable proportion of their skin exposed to sunlight rarely suffer from a vitamin D deficiency. It is recommended that children receive about 20 μg of vitamin D in their daily diet. For adults, exposure to sunlight for 30 minutes satisfies the daily requirements of the vitamin.

Vitamin D, like vitamin A, is fat-soluble. Amounts taken in excess are stored in body fat. The effects of large overdoses are even more severe than with vitamin A. Too much vitamin D can cause pain in the bones, nausea, diarrhea, and weight loss. Bonelike material may be deposited in kidney tubules, in blood vessels, in heart, stomach, and lung tissue, and in joints (Figure I.4).

I.4
VITAMIN E

As was true for vitamin D, there are several compounds with vitamin E activity. The compounds are called tocopherols; the most potent of these is α-tocopherol.

$$H_3C \overset{CH_3}{\underset{HO}{\bigcirc}} \overset{CH_3}{O} (CH_2)_3CH(CH_2)_3CH(CH_2)_3CHCH_3$$

α-Tocopherol

Rats deprived of vitamin E become sterile. Because of this, vitamin E is called the antisterility vitamin. Some food faddists promote the ingestion of large amounts of vitamin E to increase sexual prowess, combat wrinkled skin, and prevent heart attacks. There is no unambiguous evidence for any of these claims. Although vitamin E, in small amounts, prevents sterility in rats, we cannot assume that massive dosages will enhance sexual activity in humans.

Vitamin E is available in wheat germ oil, green vegetables, vegetable oil, egg yolks, and meat. Most nutritionists contend that it would be nearly impossible to eat a diet deficient in vitamin E. Vitamin E is fat-soluble, and some is stored in the body. Large doses may waste money, but they do not seem to have harmful effects. A belief in the efficacy of vitamin E, however, might lead some people to postpone needed medical treatment.

A vitamin E deficiency can lead to a vitamin A deficiency. Vitamin A can be oxidized to an inactive form when vitamin E is no longer present to act as an antioxidant. Indeed, the loss of vitamin E's antioxidant effect is believed to be responsible for all the symptoms of vitamin E deficiency. Polyunsaturated fatty acids, in cellular membranes, are oxidized at increased rates. Muscular dystrophy, sterility, and other symptoms manifested by animals deficient in vitamin E are believed to result from this "simple" change in body chemistry.

I.5
VITAMIN K

Chemically, vitamin K has a fused ring system related to the structure of naphthalene (Section 13.17). One of the rings contains two carbonyl groups, an arrangement that has the special name *quinone*. Attached to the quinone ring are alkyl groups. One of these is usually methyl. The other has 20 or more carbon atoms. Many compounds have vitamin K activity. The structure of vitamin K_1 is given here.

Vitamin K_1 (Phylloquinone)

Vitamin K, like vitamins A, D, and E, is insoluble in water but soluble in fats and fat solvents. Vitamin K is necessary for the formation of prothrombin (Section 28.8), one of the enzyme precursors involved in blood clotting. (The vitamin got its name from the Danish word *koagulation*.) A vitamin K deficiency will increase the time required for clotting of the blood. Symptoms are bleeding under the skin and in muscles, leading to ugly "bruises" from what would otherwise be minor blows. Infants lacking in vitamin K may die from hemorrhaging in the brain. Increased vitamin K intake by pregnant women has lowered the incidence of this disease in newborn infants. Good sources of vitamin K are spinach and other green leafy vegetables. Synthetic compounds with vitamin K activity are readily available. Vitamin K deficiencies in humans are rare because intestinal bacteria synthesize the body's requirements. Prolonged treatment with antibiotics has the adverse effect of killing these bacteria, and the body's supply of vitamin K is temporarily reduced.

I.6
THE B COMPLEX

In Chapter 22 we discussed how some enzymes require cofactors, nonprotein components, for proper function. Organic cofactors are called coenzymes, and many coenzymes are vitamin B derivatives.

There is really no vitamin B. What was once called vitamin B has long since been recognized to be a complicated mixture of factors. The term **B complex** is now used to designate a group of water-soluble vitamins found together in many food sources. Table I.1 lists the members of the B complex, their structures and sources, and the coenzymes derived from the vitamins.

The B complex vitamins are water-soluble. The body has a limited capacity to store water-soluble vitamins. It will excrete anything over the amount that it can immediately use. Thus, water-soluble vitamins must be taken in at frequent intervals, whereas a single large dose of a fat-soluble vitamin can be used by the body over a period of several weeks or longer.

Thiamine (Vitamin B_1)

Thiamine is necessary for the normal metabolism of carbohydrates. It is converted in the body to the pyrophosphate, which is a coenzyme in the decarboxylation of pyruvic acid to acetyl CoA and α-ketoglutaric acid to succinyl CoA (see Section 25.1). The coenzyme is also involved in the synthesis of ribose, which in turn is used by the body to produce nucleotides and nucleic acids. A deficiency of thiamine in the diet leads to the *beriberi* syndrome characterized by deterioration of the cardiovascular and nervous systems. This disease is a serious health problem in the Far East because rice, the major food in the region, has a relatively low content of thiamine. Alcoholism is the most common cause of thiamine deficiency in the United States because alcohol is the major caloric contributor to an alcoholic's diet, and therefore there is a low vitamin intake. A severe form of beriberi can also occur in infants of nursing mothers whose diets are deficient in thiamine. Synthetic vitamin B_1 is added to enrich the vitamin content of bread and flour. Thiamine is not stored in the body to any significant degree; excesses are excreted in the urine. The vitamin is destroyed in foods that are cooked for prolonged periods at temperatures over 100 °C.

Riboflavin (Vitamin B_2)

Riboflavin is essential for mammalian cells. A lack in the human diet causes well-defined symptoms, among them

Table I.1
Vitamins and Coenzymes

Vitamin and RDA[a]	Coenzyme	Source of vitamin
Thiamine (B₁) RDA = 1.5 mg	Thiamine pyrophosphate (TPP)	Germ of cereal grains, legumes, nuts, milk, and brewers yeast
Riboflavin (B₂) RDA = 1.7 mg	Flavin adenine dinucleotide (FAD)[b]	Milk, red meat, liver, egg white, green vegetables, whole wheat flour (or fortified white flour), and fish
Pyridoxine (B₆) RDA = 2.0 mg	Pyridoxal phosphate	Eggs, liver, yeast, peas, beans, and milk

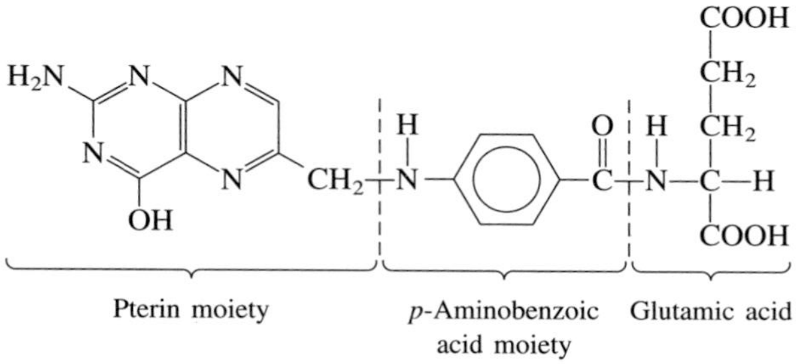

Liver, kidney, mushrooms, yeast, and green leafy vegetables

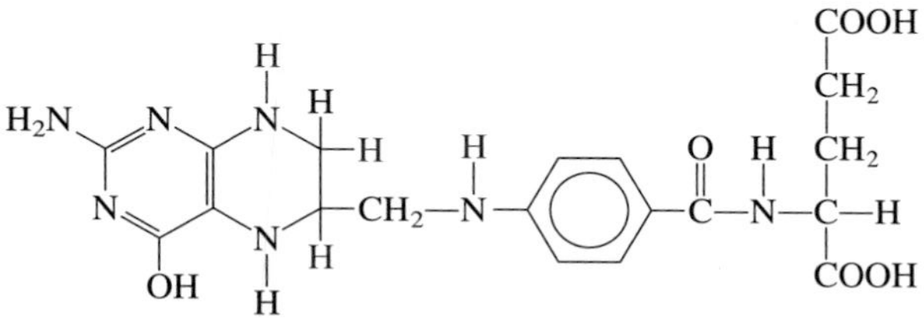

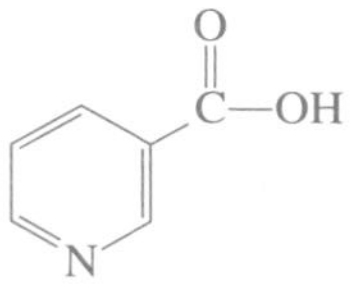

Tetrahydrofolic acid (FH$_4$)

Pterin moiety

p-Aminobenzoic acid moiety

Glutamic acid

Folic acid (F)

RDA = 0.2 mg

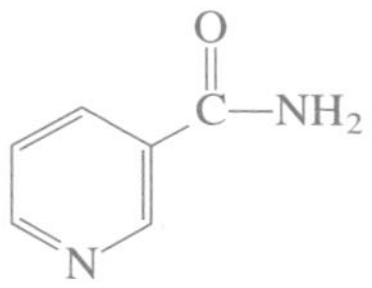

Red meat, liver, collards, turnip greens, yeast, and tomato juice

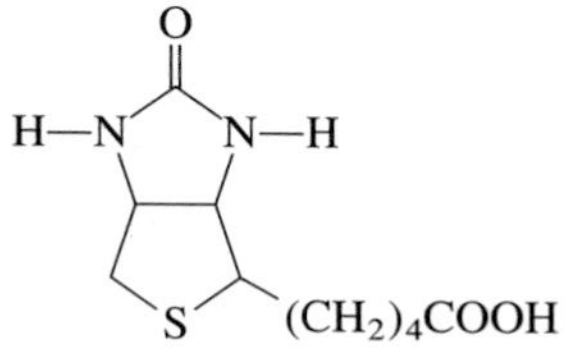

Ribose

Adenine

Ribose

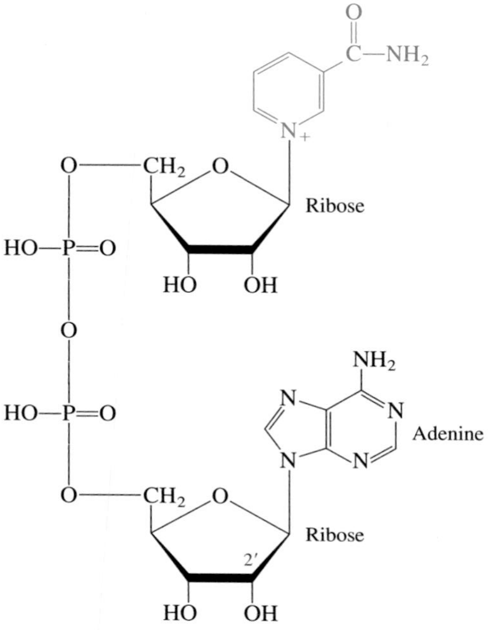

Nicotinamide adenine dinucleotide (NAD$^+$)[c]

Biotin is both a vitamin and a coenzyme

Nicotinic acid (Niacin)

Nicotinamide

RDA = 20 mg

Beef liver, yeast, peanuts, chocolate, and eggs.

Biotin

RDA = 0.3 mg

(continued)

Table I.1
(continued)

Vitamin and RDA[a]	Coenzyme	Source of vitamin

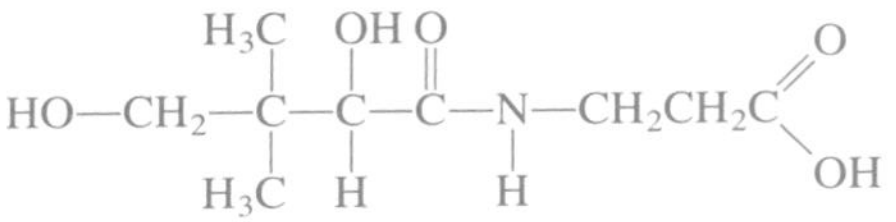

Pantothenic acid

RDA = 10 mg

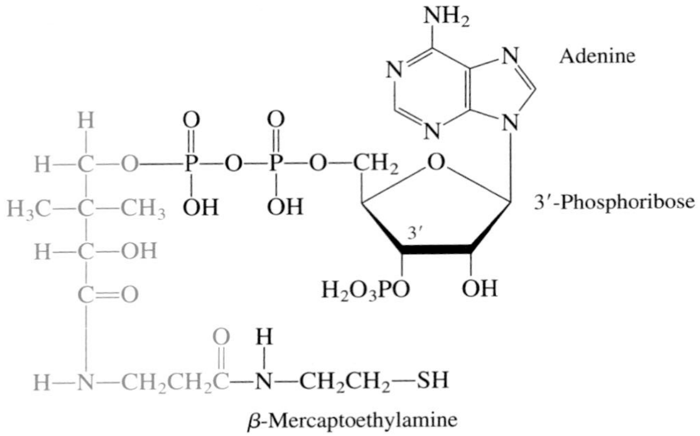

Adenine

3′-Phosphoribose

H_2O_3PO

β-Mercaptoethylamine

Coenzyme A (CoA-SH)

Liver, eggs, yeast, and milk

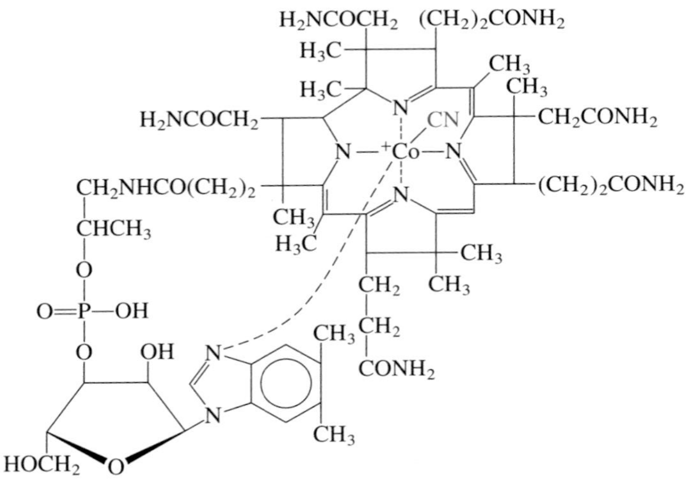

Cyanocobalamin (B_{12})

RDA = 2 μg

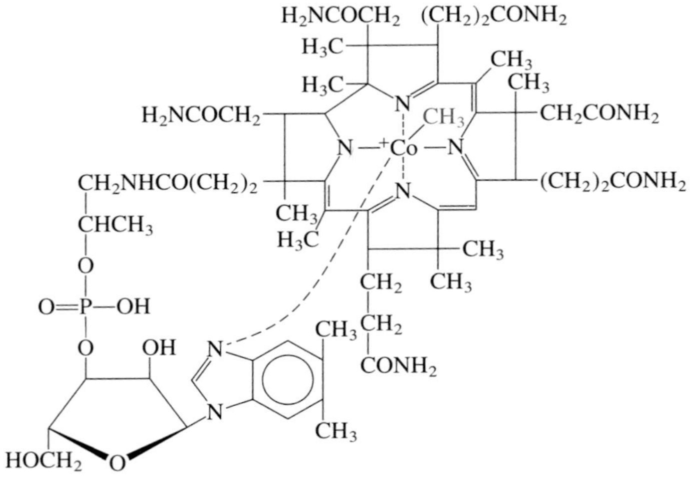

Methylcobalamin

Liver, meat, eggs, and fish (not found in plants)

---

[a] The Recommended Daily Allowance (RDA) is based on a 70-kg adult consuming 3000 cal/day. The body's requirement for most of the B vitamins increases during pregnancy and lactation. See also footnote 1, Section I.1.

[b] When there is *only a phosphate group* bonded to the terminal carbon of riboflavin, the coenzyme is named flavin mononucleotide (FMN).

[c] When there is an *additional phosphate group* on the 2′-hydroxyl group of the ribose moiety, the coenzyme is named nicotinamide adenine dinucleotide phosphate (NADP⁺).

602

dermatitis (skin inflammation), glossitis (tongue inflammation), and anemia. Riboflavin is converted in the body to the coenzymes FAD and FMN, and these function in oxidation–reduction reactions in the metabolism of carbohydrates and lipids. Riboflavin is destroyed by light and thus does not have a long stability in food products. It is stable at ordinary cooking temperatures.

Pyridoxine (Vitamin B$_6$)

Pyridoxine occurs in the tissues and body fluids of virtually all living organisms. The coenzyme pyridoxal phosphate is required for a wide variety of metabolic transformations of amino acids (Section 27.5). Clinical symptoms of vitamin B$_6$ deficiency include lesions of the skin and mucosa, anemia, irritability, apathy, and neuronal dysfunction including convulsions. Vitamin B$_6$ enhances the decarboxylation of L-dopa, so the vitamin should be avoided by patients receiving L-dopa to treat Parkinson's disease (see Section 27.5). Vitamin B$_6$ is stable at normal cooking temperatures but is sensitive to light.

Folic Acid

Folic acid is reduced to its coenzyme, tetrahydrofolic acid, which acts as a carrier of one-carbon units (e.g., as formyl or methyl groups) in the formation of compounds such as choline, heme, and nucleic acids. Deficiency of folic acid affects purine biosynthesis, and clinical symptoms include anemia and gastrointestinal disturbances. Folic acid is synthesized by intestinal microorganisms, and it can be absorbed into the general circulation. Folic acid is readily destroyed by cooking. As we saw in Section 22.8, the sulfa drugs act as antimetabolites by interfering with the bacterial biosynthesis of folic acid. The anticancer drug methotrexate inhibits the conversion of folic acid to tetrahydrofolic acid. Without the coenzyme, cells cannot grow because they cannot replicate their DNA.

Nicotinic Acid (Niacin) and Nicotinamide

Nicotinic acid and its amide are equally effective in supplying human needs. The vitamin is best known for its ability to prevent *pellagra* in humans. In the early 1900s pellagra was particularly prevalent in the southern United States and was directly associated with low-grade starchy (corn) diets. Pellagra is characterized by loss of appetite and weakness, followed by diarrhea, dermatitis, mental disorders, and death in severe cases. Nicotinamide serves

as a component of coenzymes (NAD^+ and $NADP^+$) for a wide variety of enzymes that catalyze oxidation–reduction reactions. Some nicotinic acid (about 10%) is synthesized from tryptophan. The vitamin is not destroyed by cooking, although some is lost to dissolution in the cooking water.

Biotin

Biotin is widely distributed as a cell constituent of animal and human tissue. Biochemically, biotin functions as a coenzyme in carboxylation reactions. It is a carbon carrier in both carbohydrate and lipid metabolism. Because biotin is synthesized by intestinal microorganisms in large quantities, biotin deficiency seldom occurs in humans. Biotin deficiency can be produced by antibiotics that inhibit the growth of intestinal bacteria. Also, raw egg white contains a protein, avidin, that binds biotin and prevents its absorption from the intestinal tract. An artificially produced deficiency of biotin in humans causes dermatitis, anorexia, nausea, muscle pains, and depression. Biotin is stable at normal cooking temperatures.

Pantothenic Acid

Pantothenic acid is a precursor for the biosynthesis of coenzyme A, which is important as a carrier of acyl groups. (The name coenzyme A resulted from its involvement in enzymatic acetylation reactions.) Pantothenic (from the Greek word meaning *from everywhere*) acid has a widespread distribution in foods, and deficiency in humans is practically unknown. Symptoms produced by experimental feeding of an antagonist include nausea, fatigue, and burning cramps in the limbs. This vitamin is stable at moderate cooking temperatures but is destroyed at high temperatures.

Cyanocobalamin (Vitamin B$_{12}$)

Vitamin B$_{12}$ is a complex cobalt-containing structure that has similarities to the heme group of hemoglobin. The vitamin is converted into two coenzymes, one of which is methylcobalamin (the —CN group is replaced by —CH$_3$). Methylcobalamin acts as a methyl group donor, and it is essential for cell growth and replication and for the maintenance of neural function (maintaining the myelin sheath). Vitamin B$_{12}$ is formed only by certain bacteria that live in a symbiotic relationship with their hosts. Vitamin B$_{12}$ is stored in various tissues, particularly the liver.

Vitamin B$_{12}$ is associated with the disease *pernicious anemia*. This dietary disease is characterized by the presence of abnormally large, immature, and fragile red blood cells. It is accompanied by gastrointestinal disturbances and lesions of the spinal cord with loss of muscular coordination (ataxia).

Pernicious anemia is usually caused by poor absorption of the vitamin from the intestinal tract rather than by a lack of vitamin B_{12}. Cells of the stomach lining synthesize a glycoprotein, called the *intrinsic factor,* that specifically binds vitamin B_{12} and transports the vitamin into intestinal cells for its subsequent transfer to the blood. Pernicious anemia patients lack or have a deficiency of the intrinsic factor and cannot absorb the ingested vitamin B_{12}. Elderly people often have a decreased synthesis of intrinsic factor and must receive vitamin B_{12} by injection directly into the bloodstream in order to avoid anemia. Since plants do not contain vitamin B_{12}, pernicious anemia symptoms are sometimes observed among strict vegetarians. Vitamin B_{12} is stable during most cooking procedures.

I.7
VITAMIN C

Chemically, vitamin C is ascorbic acid. It is a white, crystalline solid that is quite soluble in water. Like vitamin E, the role of vitamin C in nutrition is still a subject of controversy. Whereas vitamin E protects the lipid portion of cells, vitamin C (a highly polar compound) serves as an antioxidant in the aqueous regions. Vitamin C participates in several biological oxidation reactions, such as the hydroxylation of proline and lysine groups in collagen. It reacts with oxygen and/or oxidizing agents to form dehydroascorbic acid.

$$HOH_2C-\overset{\overset{\displaystyle H}{|}}{\underset{\underset{\displaystyle HO}{|}}{C}} \quad \text{Ascorbic acid (Vitamin C)} \quad \xrightarrow{O_2} \quad HOH_2C-\overset{\overset{\displaystyle H}{|}}{\underset{\underset{\displaystyle HO}{|}}{C}} \quad \text{Dehydroascorbic acid}$$

In 1747, James Lind discovered that citrus fruit was effective in treating sailors suffering from scurvy, a weakening of the collagenous tissues (Figure I.5). The symptoms are swollen gums; loose teeth; sore joints; thin, porous bones; bleeding under the skin; and slow healing of wounds. It was not until 1932, however, that the vitamin was isolated from citrus fruit. Vitamin C was the first dietary component to be recognized as essential for pre-

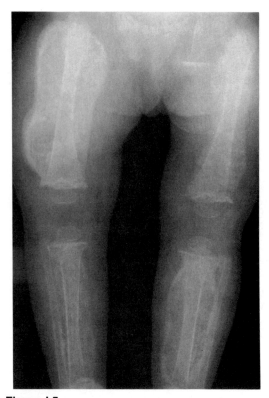

Figure I.5
An X-ray photograph of the legs of a patient with infantile scurvy, a result of vitamin C deficiency.

venting a human disease. It is widely distributed in plants[2] and animals. However, humans, other primates, guinea pigs, and some bats, birds, and fish lack an enzyme for its biosynthesis.

The controversy surrounding vitamin C received its impetus in 1970, when Linus C. Pauling (Figure I.6) published his best-selling book *Vitamin C and the Common Cold.* He stated that vitamin C in doses ranging from 1 to 5 g a day could prevent colds and that as much as 15 g a day could cure a cold. Scientists investigating Pauling's claims have obtained conflicting results. (The RDA of vitamin C for adults is 60 mg.)

Some scientists report that taking 1.5 g of vitamin C every half hour over a period of 2 hours can cure a cold.

---

[2] As is well known, vitamin C is particularly abundant in vegetables and citrus fruits, but for the vitamin to be useful the food must be reasonably fresh because ascorbic acid is slowly oxidized by air. Vitamin C is one of the least stable vitamins. It can be destroyed by heat, light, and alkali as well as by oxidizing agents. Therefore, cooking vegetables destroys an appreciable amount of vitamin C activity. It even deteriorates when kept for long periods in the refrigerator in well-capped bottles.

Figure I.6
Linus Pauling, winner of two Nobel prizes, advocates massive doses of vitamin C to prevent the common cold.

large-scale controlled studies have been quite unimpressive.''

There have also been claims that vitamin C can prevent cancer and that it is useful in the therapy and management of cancer patients. Its role in preventing cancer and colds is related to its supposed stimulation of the immune system and its activity as an antioxidant in suppressing the damage of free radicals and in blocking the formation of nitrosoamines. There is as yet no concrete evidence to either substantiate or invalidate these claims.

Because vitamin C is water-soluble, excessive amounts are excreted rather than stored in the body. However, some scientists point to the possible dangers of taking massive doses of this vitamin. Such doses, for example, may raise the uric acid level in body fluids and thus cause gout in people predisposed to this disease (see Section 27.7). Also, it is known that the ingestion of 5 g of ascorbic acid by the normal adult human will cause diarrhea.

Others concede that this may be true for a small group of the population. However, as one researcher put it: ''For the remainder of the population, vitamin C is relatively or completely ineffective in curing the common cold, since

EXERCISES

1. Define and give an example of each of the following.
   a. vitamin   **b.** provitamin   **c.** coenzyme
2. Compare and contrast vitamins and minerals (Section 12.9) with regard to the following.
   a. inorganic or organic
   b. essentiality in the diet
   c. amounts needed by the body
3. Match each compound with its designation as a vitamin.

Compound	Designation
Ascorbic acid	Vitamin A
Ergocalciferol	Vitamin B$_{12}$
Cyanocobalamin	Vitamin C
Retinol	Vitamin D
Tocopherol	Vitamin E

4. Which of the following are B vitamins?
   a. folic acid   **b.** insulin
   c. niacin   **d.** riboflavin
   e. thiamine   **f.** biotin
5. Identify the vitamin deficiency associated with each of the following diseases.
   a. scurvy   **b.** rickets
   c. night blindness
6. In each case, identify the deficiency disease associated with a diet lacking in the indicated vitamin.
   a. vitamin B$_1$ (thiamine)
   b. niacin
   c. vitamin B$_{12}$ (cyanocobalamin)
7. What is the structural difference between water-soluble and fat-soluble vitamins?

8. Identify each of the following vitamins as water-soluble or fat-soluble.
   a. vitamin A   **b.** vitamin B$_6$
   c. vitamin B$_{12}$ **d.** vitamin C
   e. vitamin K
9. Identify each of the following vitamins as water-soluble or fat-soluble.
   a. ergocalciferol   **b.** niacin
   c. riboflavin   **d.** tocopherol
10. Identify each of the following vitamins as water-soluble or fat-soluble.

    a. $$HOCH_2\overset{\overset{\displaystyle CH_3}{|}}{C}-\overset{\overset{\displaystyle OH}{|}}{\underset{\underset{\displaystyle H}{|}}{C}}-\overset{\overset{\displaystyle O}{\|}}{C}-NHCH_2CH_2COOH$$

    b. $$H_3C\quad CH_3$$
    $$CH=CH\overset{CH_3}{C}=CHCH=CH\overset{CH_3}{C}=CHCH_2OH$$
    (ring with CH_3 substituents)

11. Which is likely to be the more dangerous—an excess of a water-soluble vitamin or an excess of a fat-soluble vitamin? Why?
12. Could a one-a-month vitamin pill satisfy all human requirements? Explain your answer.
13. Name one function and one deficiency disease associated with each of the following.
    a. vitamin A   **b.** vitamin D
    c. vitamin E   **d.** vitamin C

14. How are vitamins related to coenzymes?
15. Does synthetic vitamin C differ from natural vitamin C?
16. If boiled vegetables are served as part of a meal, would the water-soluble or fat-soluble vitamins originally present be lost? Why?
17. What foods, in general, are good sources of the B vitamins?
18. Why is vitamin D called the "sunshine vitamin"?
19. What biochemicals are protected by vitamin E's antioxidant effect?
20. What is meant by the term *B complex?*
21. Which vitamin is a part of the coenzyme NAD^+?
22. Which vitamin is a part of the coenzyme FAD?
23. Which vitamin is a part of coenzyme A?

24. Vitamin B_{12} has the formula $C_{63}H_{88}CoN_{14}O_{14}P$ and a molecular weight of 1355 g/mol, yet it is soluble in water. Explain.
25. The phrases below refer to the structural changes that occur to vitamins to convert them to coenzymes. Consult Table I.1 and identify the vitamin(s).
    a. are converted into coenzymes that catalyze oxidation–reduction reactions
    b. undergoes no further change (i.e., it is also a coenzyme)
    c. pyrophosphate group is added
    d. is oxidized and phosphorylated
    e. is reduced
    f. methyl replaces cyanide

Chapter 23
NUCLEIC ACIDS AND
PROTEIN SYNTHESIS

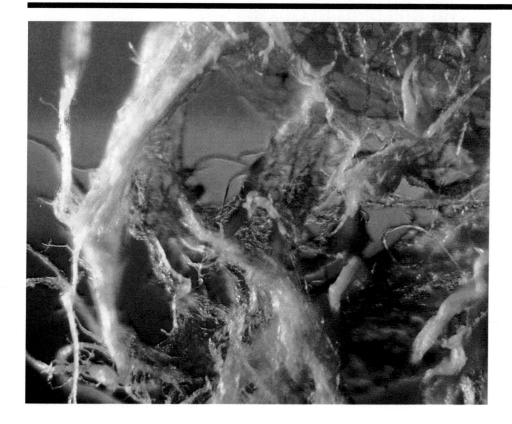

Artificial DNA synthesized from animal DNA. DNA is termed the "blueprint" of life because it contains the information for the synthesis of all the body proteins.

In contrast to what many people believe, the complexity of the sciences increases as one proceeds from physics to chemistry to biology. Because the language of physics is mathematics, most people regard physics as the most difficult of the sciences. Yet physical phenomena can be described with mathematical precision because the relationships involved are comparatively simple. We can write an equation that accurately describes the behavior of gases or of subatomic particles. A functioning cell defies such analysis.

Nonetheless, the cell is slowly yielding its secrets. One of these secrets, perhaps the most important one, is the method by which the cell stores and transmits information on how to reproduce itself. Nucleic acids are the molecules that store the patterns of life. It is through nucleic acids that these patterns are passed from one generation to the next. Nucleic acids also control the synthesis of proteins, including the enzymes that mediate those biochemical reactions that make an organism what it is.

With understanding comes control. The biochemists and molecular biologists who are unraveling these mechanisms are also learning how to manipulate the structure of living matter. The repair of defective genes, the design of precise molecular medicines,

607

and control—for better or worse—of our heredity may lie in the future. The twentieth century may be remembered as the nuclear age, but the twenty-first century could well become the age of molecular biology.

23.1
THE BUILDING BLOCKS: SUGARS, PHOSPHATES, AND BASES

Nucleoproteins are found in every living cell. They are exceedingly complex, as might be expected from the role they play: they are the information and control centers of the cell. More about that later in the chapter. Let's look first at what nucleoproteins are made of. One way to find out is to take them apart.

Working carefully, chemists can separate nucleoproteins into a nucleic acid portion and a protein portion.

$$\text{Nucleoprotein} \longrightarrow \text{Nucleic acid} + \text{Protein}$$

The protein is highly basic. Hydrolysis reveals that it contains many units of the amino acids lysine and arginine.

$$\text{Protein} \xrightarrow{\text{H}_2\text{O}} \text{Lysine} + \text{Arginine} + \text{Other amino acids}$$

The nucleic acids can also be hydrolyzed. Controlled hydrolysis gives units called nucleotides. These can be further hydrolyzed to phosphoric acid and compounds called nucleosides. Nucleosides can be hydrolyzed to the ultimate molecular constituents that are purine and pyrimidine bases (Chapter 17) and a pentose sugar (Chapter 19).

$$\text{Nucleic acids} \xrightarrow{\text{H}_2\text{O}} \text{Nucleotides} \xrightarrow{\text{H}_2\text{O}} \begin{cases} \text{Nucleosides} \\ + \\ \text{H}_3\text{PO}_4 \end{cases} \xrightarrow{\text{H}_2\text{O}} \begin{cases} \text{Two purine bases} \\ + \\ \text{Two pyrimidine bases} \\ + \\ \text{A pentose sugar} \end{cases}$$

There are actually two kinds of nucleic acids. Each is a gigantic polymer with nucleotides as the repeating units. Deoxyribonucleic acid (DNA) occurs in the cell nucleus. Ribonucleic acid (RNA) is found in all parts of the cell. The two nucleic acids differ only slightly in composition. Complete hydrolysis of DNA ultimately gives two purine bases, adenine and guanine, and two pyrimidine bases, cytosine and thymine. Ribonucleic acid (RNA) differs from DNA only in that one of the pyrimidine bases, thymine, is replaced by another, uracil, and the sugar deoxyribose is replaced by ribose. These differences in composition are summarized in Table 23.1.

Table 23.1
Ultimate Hydrolysis Products of DNA and RNA

	DNA	RNA
Purine bases	⎰ Adenine ⎱ Guanine	⎰ Adenine ⎱ Guanine
Pyrimidine bases	⎰ Cytosine ⎱ Thymine	⎰ Cytosine ⎱ Uracil
Pentose sugar	2-Deoxyribose	Ribose
Inorganic acid	Phosphoric acid	Phosphoric acid

Pyrimidine Bases

The pyrimidine bases that occur in nucleic acids are substituted derivatives of the parent compound, pyrimidine. Pyrimidine is a heterocyclic six-member ring compound containing two ring-nitrogen atoms. It does not occur free in nature, but its derivatives *uracil, thymine,* and *cytosine* occur in nucleic acids.

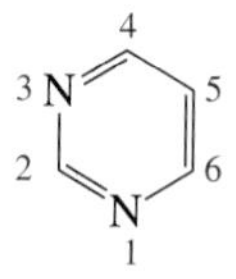

Pyrimidine

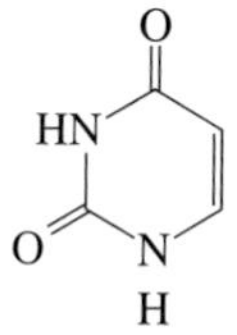

Uracil
(2,4-Dioxypyrimidine)

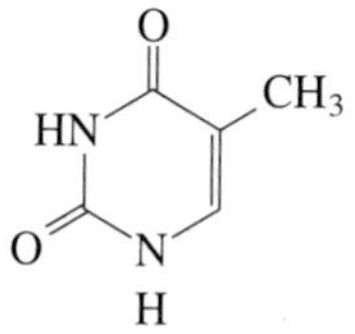

Thymine
(5-Methyl-2,4-dioxypyrimidine)

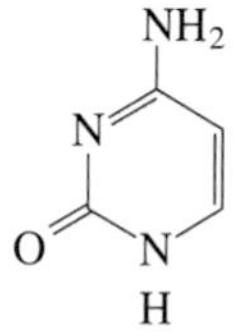

Cytosine
(4-Amino-2-oxypyrimidine)

Several other pyrimidine derivatives (called modified or minor bases) also are found in various nucleic acids. Among these are 5-methylcytosine and 5-hydroxymethyl-cytosine.

Purine Bases

The naturally occurring purine bases are derivatives of the parent compound purine, a heterocyclic amine consisting of a pyrimidine ring fused to an imidazole ring. Adenine and guanine are the major purine constituents of nucleic acids.

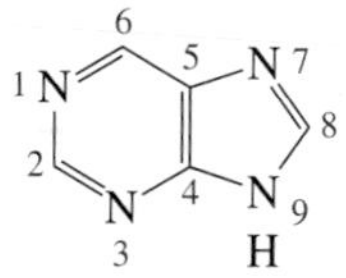

Purine

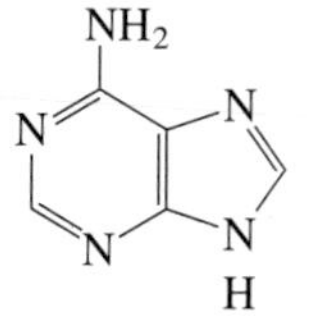

Adenine
(6-Aminopurine)

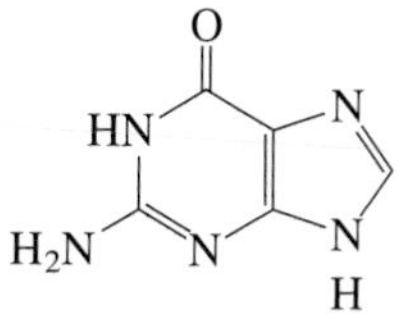

Guanine
(2-Amino-6-oxypurine)

Methylation is the most common form of purine modification, and methylated purines occur in varying amounts in nucleic acids. 6-Methyladenine and 2-methylguanine are two of the minor purine bases that occur in certain nucleic acids.

Nucleosides

When a purine or pyrimidine base is combined with one of the pentose sugars, a compound called a **nucleoside** is formed. If the sugar is ribose, the compound is a **ribonucleoside.** If 2-deoxyribose is the sugar involved, the product is a **deoxyribonucleoside.**

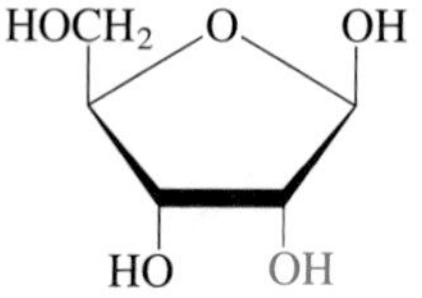

β-Ribose

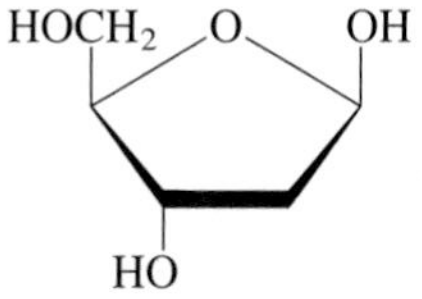

β-2-Deoxyribose

The bond joining the pentose to the nitrogen base is termed an *N-glycosyl linkage,* and it is always beta in naturally occurring nucleosides. The N-glycosyl linkage is formed between C-1′ of the sugar and N-1 of the pyrimidine base or N-9 of the purine base. A molecule of water is eliminated in the process. The following equations are given only to help the student visualize the joining of the sugar to the base; nucleosides are *not* synthesized in this fashion in the cell.

Illustrative Equations

The convention used in numbering is that atoms of the pentose ring are designated by primed numbers, whereas atoms of the purine or pyrimidine ring are designated by unprimed numbers.

Ribose   Adenine   Adenosine

Deoxyribose   Thymine   Deoxythymidine

The common names of the ribonucleosides are derived from the names of the nitrogenous bases. The suffix *-osine* denotes purine nucleosides, and the suffix *-idine* is used for pyrimidine nucleosides. The prefix *deoxy-* is used if the base is combined with deoxyribose (deoxynucleosides)—deoxyadenosine, deoxyguanosine, deoxycytidine, and deoxythymidine. Structures and names of the major ribonucleosides and one of the deoxyribonucleosides are given in Table 23.2.

Some Pharmacological Nucleosides

Adenosine may serve as a chemical regulator throughout the body. Receptor sites for it have been identified. It appears that adenosine may regulate the function of neurons in the brain, dilate blood vessels in the heart, constrict bronchial tubes, and inhibit the aggregation of platelets. Caffeine may act as a stimulant by blocking adenosine receptors.

Several nucleoside derivatives have been used in medicine. One, puromycin (Figure 23.1a), is derived from adenosine. It is an antibiotic. First obtained from cultures of the fungus *Streptomyces alboniger,* puromycin is effective against protozoa and has shown some antitumor activity. Vidarabine (Figure 23.1b) is an antiviral drug. The base in vidarabine is adenine, but the sugar is arabinose, an isomer of ribose. 5-Fluorodeoxyuridine (see Section J.6) is an anticancer drug and azidothymidine (see Section J.2) is a nucleoside used in the treatment of AIDS.

PYRIMIDINE NUCLEOSIDES

Cytidine

Uridine

Deoxythymidine

PURINE NUCLEOSIDES

Adenosine

Guanosine

Table 23.2
The Major Pyrimidine and Purine Nucleosides

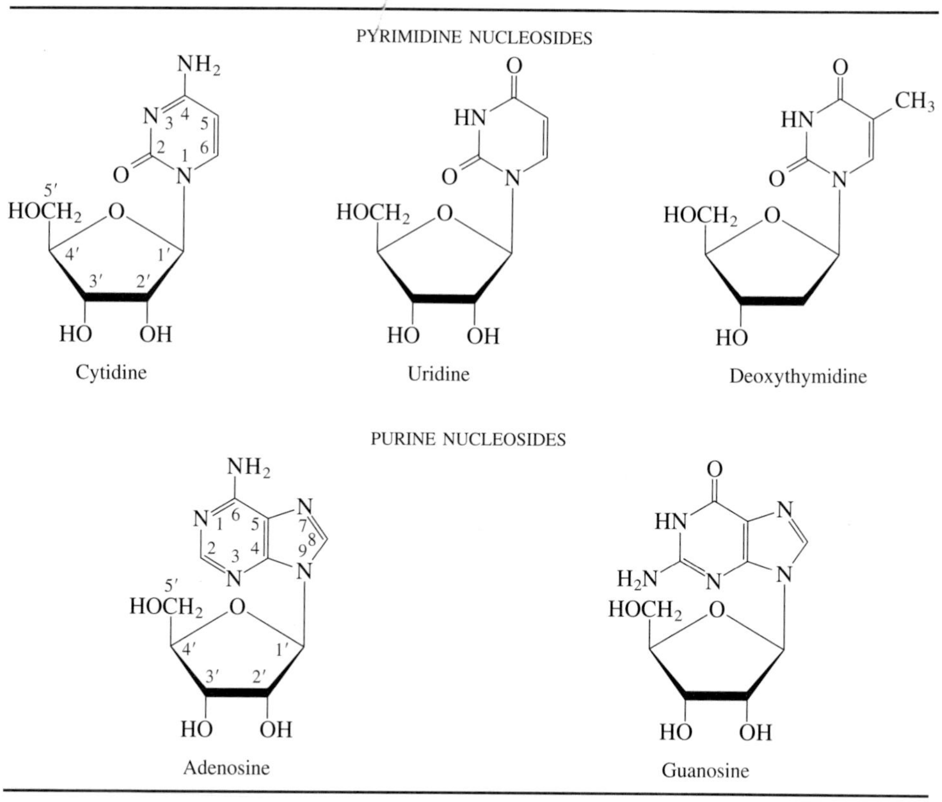

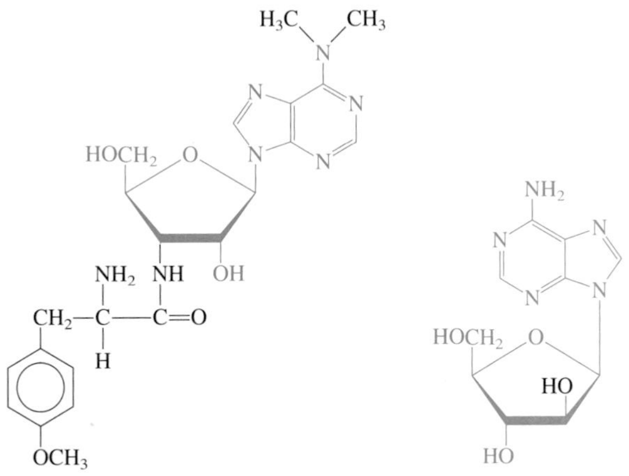

(a) Puromycin

(b) Vidarabine

Figure 23.1
Puromycin (a) and vidarabine (b), two nucleoside derivatives used in medicine. See Figure J.3 for other antiviral drugs that are derived from nucleosides.

Nucleotides

The **nucleotides** are phosphate esters of the nucleosides and may be envisioned to result from the esterification of phosphoric acid with one of the free pentose hydroxyl groups. Note that again we are illustrating the combination of phosphate and a nucleoside. Nucleotides are *not* formed in this manner in the cell.

Illustrative Equation

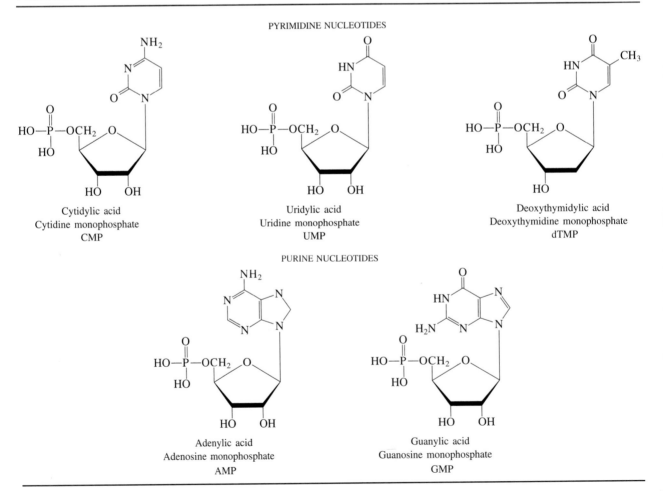

The nucleotides are named in two ways. In one system you drop the ending from the name of the corresponding nucleoside (either *-ine* or *-osine*) and add the ending *-ylic acid*. Thus, the nucleoside uridine, upon esterification with phosphoric acid, becomes uridylic acid. Similarly, guanosine becomes guanylic acid. In the other system, which is simpler and more frequently used, you use the nucleoside name as is and add the word *monophosphate*. Thus, adenosine, upon esterification, becomes adenosine monophosphate, often abbreviated AMP. The prefix *deoxy-* indicates that the sugar involved in deoxyribose rather than ribose. The names and structures of some nucleotides are given in Table 23.3.

Table 23.3
The Pyrimidine and
Purine Nucleotides

PYRIMIDINE NUCLEOTIDES

Cytidylic acid
Cytidine monophosphate
CMP

Uridylic acid
Uridine monophosphate
UMP

Deoxythymidylic acid
Deoxythymidine monophosphate
dTMP

PURINE NUCLEOTIDES

Adenylic acid
Adenosine monophosphate
AMP

Guanylic acid
Guanosine monophosphate
GMP

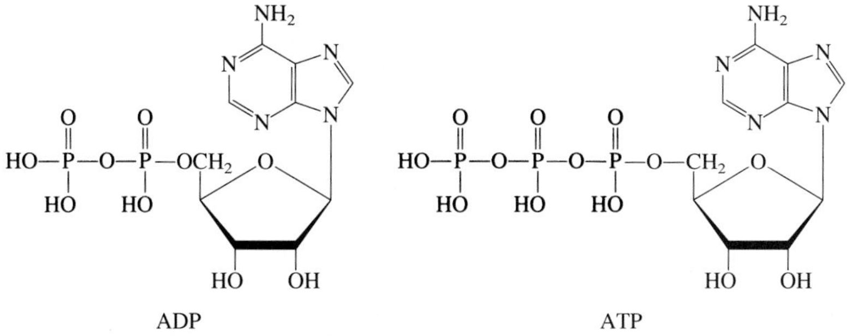

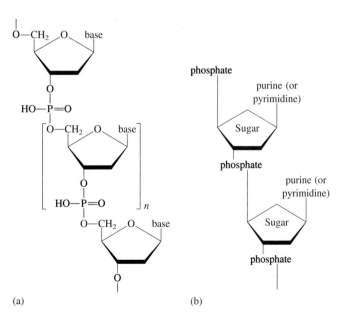

Figure 23.2
Structures of two important nucleotide derivatives.

ADP   ATP

Nucleotides are the monomers from which DNA and RNA are synthesized. In addition, the nucleotides and some of their derivatives perform a variety of other functions in the cell. Adenosine diphosphate (ADP) and adenosine triphosphate (ATP) are involved in many metabolic and biosynthetic processes. We encounter them often in other chapters. Structures of these nucleotide derivatives are shown in Figure 23.2. In addition, a cyclic 3′,5′-phosphate of adenosine occurs in which the phosphate group is bonded to two of the ribose carbons. The compound is adenosine 3′,5′-monophosphate (cyclic AMP) and, as we shall see in Section 24.4, it plays a crucial role in metabolism. Certain other nucleotides are structural components of a number of important coenzymes. Refer to Table I.1 and notice that FAD, NAD^+, and coenzyme A are all adenine nucleotides.

23.2
THE BASE SEQUENCE: PRIMARY STRUCTURE OF NUCLEIC ACIDS

The nucleic acids are polymers of nucleotides. The backbone of the polymer is a polyester chain. This backbone involves the phosphoric acid and pentose sugar portions of the nucleotides. The connection between one nucleotide unit and the next is made when a phosphate hydroxyl group in one unit reacts with a sugar hydroxyl on the next (Figure 23.3).

Figure 23.3
(a) The polymeric backbone of a nucleic acid, shown here for deoxyribonucleic acid. (b) A schematic representation.

(a)   (b)

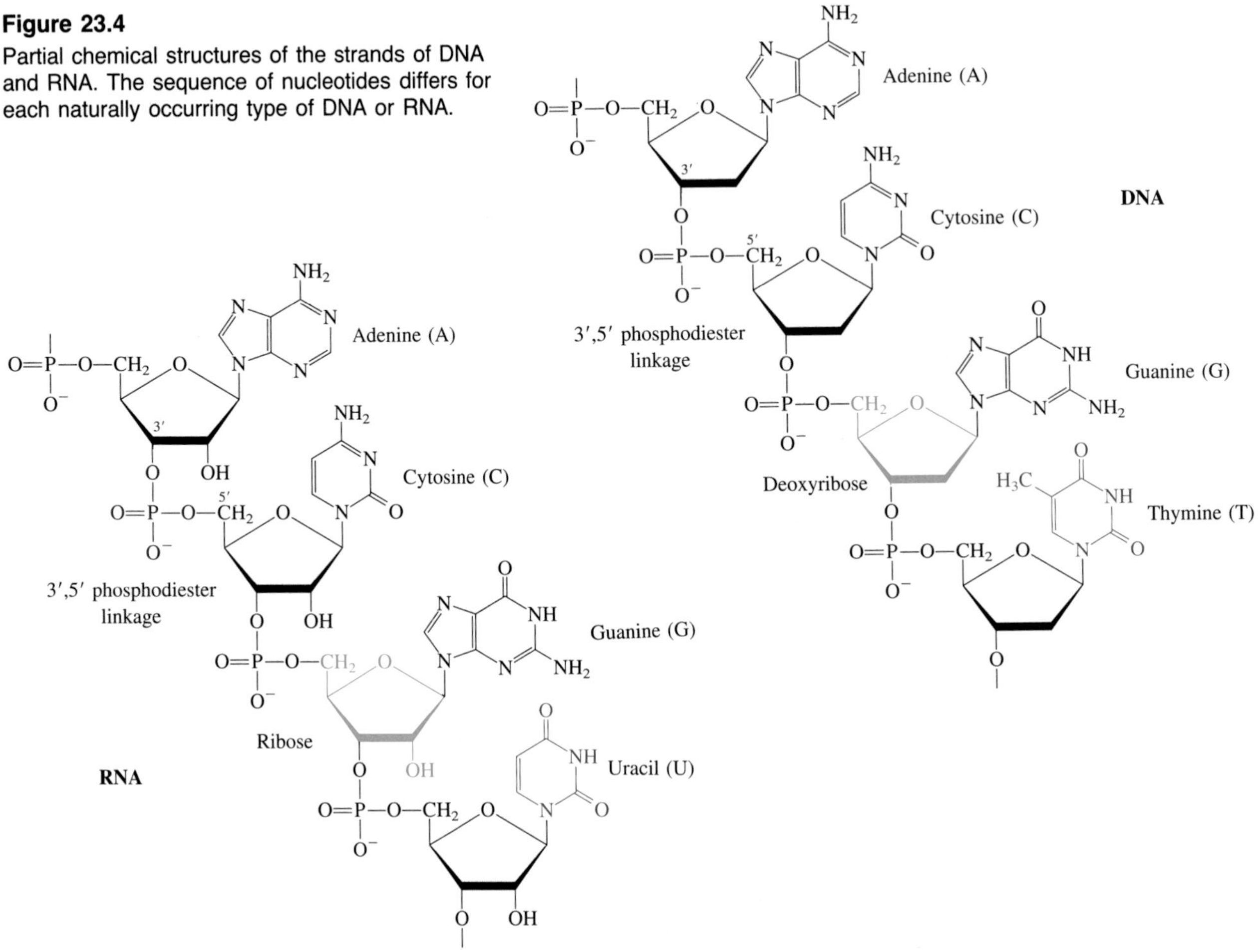

Figure 23.4

Partial chemical structures of the strands of DNA and RNA. The sequence of nucleotides differs for each naturally occurring type of DNA or RNA.

As we have seen, the sugar in DNA is 2-deoxyribose, and the one in RNA is ribose. The bases in DNA are adenine, guanine, cytosine, and thymine. Those in RNA are adenine, guanine, cytosine, and uracil. Note the one ionizable hydrogen on each phosphate unit. That is what makes these compounds nucleic *acids*. In solution or combined with basic proteins as nucleoproteins, the acid is ionized. Partial structures for DNA and RNA are shown in Figure 23.4.

Nucleic acids resemble proteins in one respect. To completely specify the primary structure of a nucleic acid, one must specify the sequence of bases. Unlike the proteins which have 20 different amino acids, there are only four different bases in a nucleic acid. However, the molecular weight of a nucleic acid is often much greater than that of a protein, ranging into the billions for mammalian DNA. In the 1960s, sequencing was extremely laborious. For example, the primary structure of the nucleic acid called alanine transfer RNA, a molecule with 77 nucleotide units, was determined in that decade. The work, done by Robert W. Holley and co-workers at Cornell University, took 7 years. Holley was rewarded with a share of the Nobel prize in 1968 for his part in the project.

The sequence of bases in short strands of nucleic acids is determined by using gel electrophoresis. Enzymes are used to cleave the nucleic acids at specific base sequences. The primary structure of each fragment is determined. Then overlapping parts are matched to give the base sequence of the whole strand. Now the work is largely automated, and sequences of several thousand base units per molecule have been determined.

23.3
SECONDARY STRUCTURE OF DNA: BASE PAIRING AND THE DOUBLE HELIX

The shape of the giant DNA molecules was long a mystery. Early studies revealed no more than the fact that the structure exhibited a periodic pattern. A real breakthrough occurred in 1950, when Erwin Chargoff, of Columbia University, showed that the molar amount of adenine (A) in DNA was always equal to that of thymine (T). Similarly, the molar amount of guanine (G) was the same as that of cytosine (C). The bases must be paired, A to T and G to C. But how? The race was on, with an almost certain Nobel prize for the winner. Many illustrious scientists, including Linus Pauling, were working on the problem. However, at Cambridge University in 1953, two relative unknowns in the world of science announced that they had worked out the structure of DNA. Using data from the X-ray studies of Rosalind Franklin and Maurice Wilkins, which involved quite sophisticated chemistry, physics, and mathematics, and working with models not unlike a child's construction set, James D. Watson and Francis Crick determined that DNA must be composed of two helices wound about one another to form a **double helix.** The phosphate and sugar groups (the backbone of the nuclei acid polymer) form the outside of the structure, which is rather like a spiral staircase. The purine and pyrimidine bases are paired on the inside—with guanine always opposite cytosine and adenine always opposite thymine. These specific base pairs are referred to as **complementary** bases. In our staircase analogy, these base pairs are the stairsteps (Figure 23.5). This structure can explain how cells are able to divide and go on functioning, how genetic data are passed on to new generations, and even how proteins are built to required specifications. It all depends on the base pairing. Figure 23.6 shows a space-filling model of the DNA double helix.

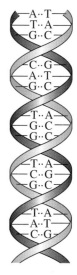

Figure 23.5
The DNA double helix as portrayed by Watson and Crick.

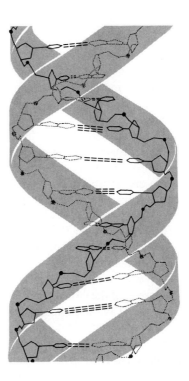

Figure 23.6
Two models of the DNA double helix.

Figure 23.7

Pairing of the complementary bases thymine and adenine (a) and of cytosine and guanine (b) by means of hydrogen bonding as in DNA.

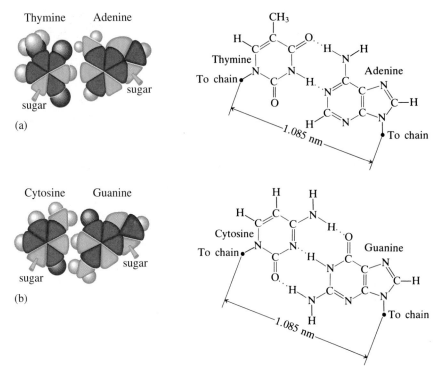

Figure 23.8

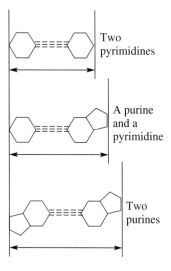

Difference in widths of possible base pairs.

Which brings up the still unanswered question "Why do the bases pair in that precise pattern, always A to T and T to A, always G to C and C to G?" The answer is hydrogen bonding and a truly elegant molecular design. Figure 23.7 shows the two sets of base pairs. Notice two things. First, a pyrimidine is paired with a purine in each case, and the long dimensions of both pairs are identical (1.085 nm). If two pyrimidines were paired or two purines were paired, the two pyrimidines would take up less space than a purine and a pyrimidine, and the two purines would take up more space, as is illustrated in Figure 23.8. If this were the situation, the structure of DNA would be like a staircase made with stairs of different widths. In order for the two strands of the double helix to fit neatly, a pyrimidine must always be paired with a purine.

The other thing you should notice from Figure 23.7 is that when guanine is paired with cytosine, three hydrogen bonds can be drawn between them, and between adenine and thymine there are two hydrogen bonds. It is the additive contribution of all these hydrogen bonds that imparts great stability to the DNA molecule.

There are about 10 base pairs per turn of the double helix. The acidic phosphate units are on the outside. In nucleoproteins, the highly basic proteins probably wrap around the double helix. Proton transfers from the phosphate units of DNA to the lysine and arginine side chains of the protein result in ionic charges. The protein is held to the nucleic acid, at least in part, by these salt linkages. It has been calculated that the total amount of DNA in a typical mammalian cell contains about 5.5×10^9 nucleotides. If all this DNA were stretched out end to end, it would extend more than 2 meters.

Watson and Crick received the Nobel prize in 1962 for discovering, as Crick put it, "the secret of life" (Figure 23.9). It was not long after the development of the models that DNA was synthesized in the laboratory. In 1967, Arthur Kornberg of Stanford University carried out a test-tube synthesis of a single strand of DNA that was able to reproduce itself. Kornberg, of course, had to add the appropriate precursors, and he added the enzymes and cofactors essential to the process. Synthesis of life in a test tube? Hardly. A strand of DNA is still a long way from even the simplest functioning cell.

Figure 23.9

Francis Crick and James D. Watson, who proposed the double-helix model of DNA.

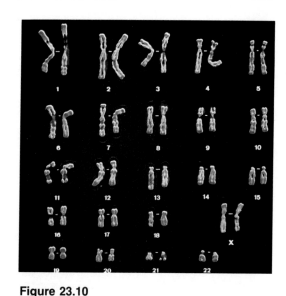

Figure 23.10

Female chromosomes arranged in numbered homologous pairs. Male and female sets differ only in the sex pair (bottom right); a male would have an X and a Y instead of two Xs. The nucleus of each human cell contains a total of 46 chromosomes, 23 of maternal origin and 23 of paternal origin.

23.4
DNA: SELF-REPLICATION

Cats have kittens that grow up to be cats. Bears have cubs that grow up to be bears. Why is it always so that each species reproduces after its own kind? How does a fertilized egg "know" that it should develop as a kangaroo and not as a koala?

The physical basis of heredity has been known for a long time. Most higher organisms reproduce sexually. A sperm cell from the male unites with an egg cell from the female. The fertilized egg so formed must carry all the information needed to make all the various cells, tissues, and organs necessary for the functioning of the new individual. For humans, that single cell must carry the information for the making of legs, liver, lungs, heart, head, hair, and hands—in short, all the instructions ever needed for growth and maintenance of the individual. In addition, if the species is to survive, information must be set aside in germ cells—either sperm or eggs—for the production of new individuals.

The basic hereditary material is found in the nuclei of all cells, concentrated in elongated, threadlike bodies called *chromosomes* (Figure 23.10). The number of chromosomes varies with the species. Human body cells have 46. The basic units of heredity, called *genes,* are arranged along the chromosomes in a linear fashion. During cell division, each chromosome produces an exact duplicate of itself. Sperm and egg cells carry only half the chromosomes of the body cells. Thus, in sexual reproduction, the entire complement of chromosomes is achieved only when the egg and sperm combine; a new individual receives half its hereditary material from each parent.

Calling the unit of heredity a gene merely gives it a name. What are genes? What are they made of? The material of genes is nothing other than a distinct segment of a long DNA strand. (Some viruses carry genetic information in RNA; see Section J.1.) Each gene codes for a specific polypeptide. Transmission of genetic information involves the **self-replication** (copying or duplication) of the DNA strand.

Chromosomes are fibers consisting of complex structures of DNA and proteins. Human chromosomes are composed of about 25% DNA and 75% protein. The DNA contains the genetic information; the proteins are a major factor in the regulation of gene expression.

Humans have about 100,000 genes. Scientists have embarked on an ambitious endeavor, called the *human genome project,* to map the location of each of these genes on the chromosomes. (The *genome* of an organism is its complete set of genes.)

Figure 23.11

A schematic view of the formation of a complementary polymer upon a template surface.

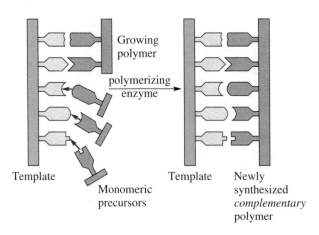

Figure 23.11 diagram labels: Growing polymer, polymerizing enzyme, Template, Monomeric precursors, Template, Newly synthesized *complementary* polymer

The Watson–Crick double helix provides a ready model for genetic replication. If the two chains of the double helix are pulled apart and the hydrogen bonds holding the base pairs together are broken, then each chain can act as a **template,** or pattern, and direct the synthesis of a new DNA chain (Figure 23.11). In the cellular fluid surrounding the DNA are all the necessary nucleotides (monomers). It is simply a matter of a nucleotide with the proper base combining with its complementary base on the DNA strand. Keep in mind that adenine can pair only with thymine and guanine only with cytosine. Each base unit in the separated strand can pick up only a unit identical to the one that it had before. Each of the separating chains serves as a template for the formation of a new complementary chain.

As the nucleotides become aligned, they react with one another to form the sugar–phosphate backbone of the new chain. In this way, each strand of the original DNA molecule forms a duplicate of its former partner. Whatever information was encoded in the original DNA double helix is now contained in each of the replicates. When the cell divides, each daughter cell gets one of the DNA molecules and all of the information that was available to the parent cell (Figure 23.12).

Figure 23.12

A schematic diagram of DNA replication. Replication is assumed to occur by sequential "unzipping" of the double helix. The new nucleotides are positioned (by an enzyme called DNA polymerase), and phosphate bridges are formed, thus restoring the original double-helix configuration. Each newly formed double helix consists of one old strand and one new strand, a process referred to as semiconservative replication. (This representation is simplified. The nucleotides are actually triphosphate derivatives—ATP, TTP, GTP, and CTP.)

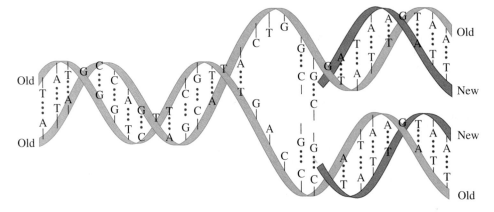

In 1985, the British biologist Alec Jeffreys invented a technique and coined the term "DNA fingerprinting." Like fingerprints, a person's DNA is unique to that individual. Any cells—skin, blood, semen, etc.—can supply the necessary DNA sample. The technique is intricate and requires several chemical steps. In the final step, a "print" is represented as a series of horizontal bars resembling the bar codes imprinted on packaged goods sold in supermarkets.

DNA fingerprinting has been hailed as a major advance in criminal investigation. Several hundred criminal cases have been solved with this technology. Also, because children inherit half their DNA from each parent, DNA fingerprinting has been used to establish the parentage of a child of contested origin. The odds in favor of being right in such cases are said to be excellent—at least 100,000:1.

We keep saying that there is information encoded in the DNA molecule. DNA is often compared to a set of directions for putting together a model airplane or for knitting a sweater. Knitting directions store information as words on paper. Letters of the alphabet are arranged in a certain way (e.g., "knit one, purl two"), and these words direct the knitter to perform a certain operation with needles and yarn. If all of the directions are correctly followed, the ball of yarn becomes a sweater.

How is information stored in DNA? It is the particular arrangement of bases along the DNA chain that encodes the directions for building an organism. Just as *saw* means one thing in English and *was* means another, the sequence of bases CGT means something, and TGC means something else. Although there are only four "letters"—the four bases—in the genetic code of DNA, their sequence along the long strands can vary so widely that there is an essentially unlimited information storage system. Even a tiny bacterium, 2 μm long and 1 μm in diameter, has 3 million base pairs. The genetic material of a human cell consists of 5 billion base pairs, and these can specify 20 billion bits of information—enough information to fill 1000 books of 2000 pages each. Thus each cell can carry all the information it needs to determine all the hereditary characteristics of even the most complex organism. We shall see how this information is conveyed to the cell in Section 23.6.

23.5
RNA: THE DIFFERENT RIBONUCLEIC ACIDS

The molecules of RNA consist of single strands of the nucleic acid. Some internal (intramolecular) base pairing may occur in sections where the molecule folds back on itself. Because of this, portions of the molecule may exist in a double-helix form. The importance of base pairing to the proper functioning of the various types of RNA will be shown in later sections.

All of the cellular RNAs appear to be synthesized from a part of the DNA molecule by a template mechanism that is analogous to DNA replication in many respects. To initiate RNA biosynthesis, the two strands of the DNA molecule begin to uncoil. This occurs at specific sites, called *promoters,* on the DNA template. The nucleotides are

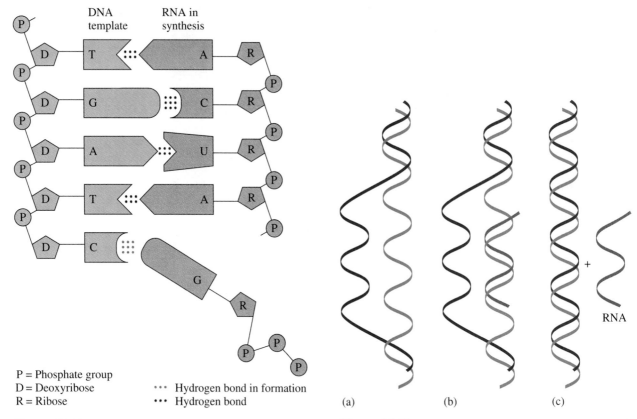

P = Phosphate group
D = Deoxyribose
R = Ribose

••• Hydrogen bond in formation
••• Hydrogen bond

(a)   (b)   (c)

RNA

Figure 23.13

Transcription-synthesis of RNA upon a DNA template. One strand of the DNA double helix is used as a template for alignment of RNA bases. Once a nucleotide is in position, a phosphate bridge is formed, resulting in a single strand of RNA. The nucleotides are joined by the enzyme RNA polymerase.

Figure 23.14

The DNA double helix is partly unwound (a). An RNA is formed on the separated portion (b) and then released (c).

attracted to the uncoiling region of the DNA molecule according to the rules of base pairing. Thymine in DNA calls for adenine in RNA, cytosine specifies guanine, guanine calls for cytosine, and adenine requires uracil. Recall from Section 23.1 that, in RNA molecules, uracil replaced DNA's thymine. Notice the similarity between the structures of these two bases.

DNA Base	Complementary RNA Base
Adenine	Uracil
Thymine	Adenine
Cytosine	Guanine
Guanine	Cytosine

Because the RNA is a complementary copy of the information contained in the DNA, the process of RNA biosynthesis is referred to as **transcription.** We see later (Section 23.6) why this process is vital to all growth and development. Figures 23.13 and 23.14 depict the process schematically.

RNA Type	Relative Amount (%)	Average MW	Approx. Number of Nucleotides
Messenger RNA (mRNA)	5	4×10^5	1200
Transfer RNA (tRNA)	15	3×10^4	70–90
Ribosomal RNA (rRNA)	80	6×10^5	1800

Table 23.4
RNA Molecules in *E. coli*

Transcription differs from DNA replication in several ways. During transcription, in a given DNA region only one of the DNA strands serves as a template and is copied, but in another region of the DNA the other strand may serve as a template for the biosynthesis of a different RNA. Thus, RNA molecules are much shorter than DNA molecules. Also, RNA molecules do not remain hydrogen-bonded to DNA for any length of time. As soon as transcription is completed, the RNA is released and the DNA helix re-forms.

Three basic types of cellular RNAs are known to exist. The distinctions among them are made primarily on the basis of biochemical function. However, they also differ in molecular weight and in secondary structure (Table 23.4).

Messenger RNA (mRNA)

Messenger RNA (mRNA) makes up only a few percent of the total amount of RNA within the cell. It has been shown to be complementary to a given segment of the DNA of the organism from which it is isolated. A molecule of mRNA exists for a relatively short time. Like proteins, it is continuously being degraded and resynthesized. The rate of mRNA degradation differs from species to species and also from one type of cell to another. In bacteria, one-half of the total mRNA is degraded every 2 min, whereas in rat liver the half-life is several days.

The molecular dimensions of the mRNA molecule vary according to the amount of genetic information that the molecule is meant to encode. It is known, however, that there is very little intramolecular hydrogen bonding in this type of RNA and that the molecule exists in a fairly random coil. After transcription, the mRNA passes into the cytoplasm, carrying the genetic message from DNA to the ribosomes, the sites of protein synthesis. In Section 23.6 we shall see how mRNA directly governs that synthesis.

Ribosomal RNA (rRNA)

Ribosomal RNA (rRNA) makes up 80% of the total cellular complement of ribonucleic acid. The **ribosome** is a cellular substructure that serves as the site for protein synthesis. Its composition is about 65% rRNA and 35% protein. The ribonucleic acids and the proteins are bonded together by a large number of noncovalent forces such as hydrogen bonds and hydrophobic interactions. Structurally, a ribosome is composed of two spherical particles of unequal size. The smaller of them has a distinct affinity for mRNA; the larger has an attraction for tRNA. In terms of cellular structure, ribosomes are extremely small particles visible only with the aid of an electron microscope. More often than not, they are seen as clusters known as *polyribosomes,* or *polysomes,* bound to the endoplasmic reticulum of animal and plant cells or to the cell membrane of microorganisms. When ribosomes occur in such aggregates, they are held together by strands of mRNA. On the average, five to eight ribosomes are simultaneously synthesizing the same polypeptide from the information in one mRNA strand (large proteins require long strands of mRNA, and as many as 100 individual ribosomes may be attached). The time required for the

Figure 23.15
Cloverleaf diagram of a tRNA molecule. The molecule is a single chain, but in folding back on itself hydrogen bonds are formed and large regions of the molecule are characterized by base pairing.

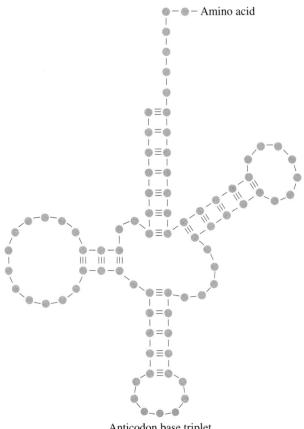

Anticodon base triplet

synthesis of an average size polypeptide (~300 amino acids) is about 15 seconds in a bacterial cell and 2 or 3 minutes in a mammalian cell.

Transfer RNA (tRNA)

Transfer RNA (tRNA) is a relatively low molecular weight nucleic acid, soluble in solvents commonly used to isolate the higher molecular weight nucleic acids. It functions by attaching itself (with the aid of a specific enzyme) to a particular amino acid and carrying that amino acid to the site of protein synthesis at the precise moment specified by the genetic code. Each of the 20 amino acids found in proteins has at least one corresponding tRNA, and most amino acids have more than one (see Table 23.5). For example, there are two different tRNAs specific for the transfer of lysine, three for isoleucine, four for glycine, and six for serine. The existence of several tRNAs for the same amino acid is termed **multiplicity.**

A tRNA molecule has a "cloverleaf" structure (Figure 23.15). On one of the loops is a unique sequence of three bases that is different in the tRNAs for different amino acids. This triplet is called the **anticodon** (see Figure 23.16). A specific amino acid becomes attached to the other end of the tRNA. In the discussion of polypeptide synthesis (Section 23.6), we shall use the figure ⬍ to represent the tRNA molecule.

23.6
PROTEIN SYNTHESIS

Recall (Section 23.4), we said that a **gene** is the segment of a DNA molecule that codes for the biosynthesis of one complete polypeptide chain. (If a given protein contains two or more polypeptide chains, each chain is coded by a different gene.) Each gene in a section of DNA contains about 1000–2000 nucleotides. A human cell contains about 100,000 genes (although there is sufficient DNA to code for 80–100 times this number of genes).

Even if we accept the fact that DNA carries a message, there is still the problem of how this message is read. How is a particular sequence of bases along the DNA chain translated into a complex organism? The answer is that *DNA directs the synthesis of proteins.* Proteins serve as building materials and, most important, as enzymes. The mechanism by which the DNA blueprint is transformed into protein molecules involves the intermediacy of RNA molecules.

How can a molecule with just four different monomeric units specify the sequence of the 20 different amino acids that occur in proteins? If each different nucleotide coded for one different amino acid, then obviously the nucleic acids could code for only four of the 20 amino acids. Suppose we consider the nucleotides in groups of two. There are 4^2 or 16 different combinations of pairs of the four distinct nucleotides. Such a code is more extensive but still inadequate. If, however, the nucleotides are considered in groups of three, there are 4^3 or *64 different combinations.* Here we have a code that is extensive enough to govern the primary structure of the protein molecule because it contains more than enough coding units to designate all 20 amino acids. Now we shall see how this code directs protein synthesis.

If the sequence of bases along the DNA strand determines the sequence of amino acids along the polypeptide chain, then the information contained in the DNA must be conveyed from the nucleus to the site of protein synthesis. This is accomplished by the orderly interactions of the nucleic acids with over 100 different enzymes. Recall that mRNA is made from a DNA template and so contains a base sequence that is complementary to that of the DNA upon whose surface it was synthesized. Once it is formed, the mRNA is transported across the nuclear membrane into the cytoplasm (and hence to the ribosomes), carrying with it the genetic instructions. *Each group of three bases along the mRNA strand now specifies a particular amino acid, and the sequence of these triplet groups dictates the sequence of the amino acids in the protein.* Because the code involves three bases per coding unit, it is referred to as a **triplet code.** The coding unit is called a **codon.**

Now the cell faces the problem of lining up the amino acids according to the sequence called for by the mRNA and joining them together by means of peptide linkages. Because this process involves the transfer of the information encoded in the mRNA to the ultimate structure of the protein molecule, it is often referred to as **translation.**

Before the amino acids can be incorporated into a polypeptide chain they must first be activated. Activation occurs before the amino acid reacts with its particular tRNA carrier molecule. This crucial process requires certain "activation" enzymes and the participation of an ATP molecule. *Both the enzymes (aminoacyl tRNA synthetases) and the tRNAs are each highly specific for a particular amino acid.* The high degree of specificity of the synthetase enzymes is vital to the correct incorporation of amino acids into proteins. After the amino acid molecules have been activated and have undergone reaction with the tRNA carriers, protein synthesis can take place.

It has been estimated that about 5% of the DNA of higher organisms conveys the information for the synthesis of proteins. The remainder is involved in the regulation of protein synthesis, in the synthesis of tRNA and rRNA, and in maintaining the correct conformation of the DNA. See also "Genetic Regulation" (page 627).

When an amino acid is bound to its tRNA "carrier," the combination is referred to as an aminoacyl tRNA complex.

Figure 23.16 depicts a schematic stepwise representation of this all-important process. When a certain codon of the mRNA strand has been *read* by a given ribosome, another ribosome may attach itself to the strand and begin to read it as the first ribosome moves on to read the next codon.

(a) Protein synthesis is already in progress at the ribosome. The growing polypeptide chain is bound to the peptidyl (P) site. At this point the aminoacyl (A) site is vacant. The codon UUU is lined up above the A site. An activated tRNA molecule whose anticodon is AAA approaches the ribosome. (The tetracyclines—Figure 22.22—block the binding of the aminoacyl tRNA to the A site of bacterial ribosomes. This inhibits bacterial protein synthesis and blocks bacterial growth. The tetracyclines do not bind to mammalian ribosomes and thus do not affect protein synthesis in host cells.)

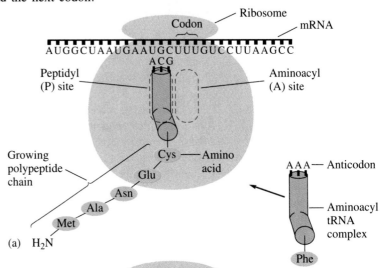

(b) The activated tRNA molecule has become bound to the ribosome at the A site. It is also bound to the mRNA molecule by means of base pairing between codon and anticodon. Amino acid Phe is about to be incorporated into the polypeptide chain. The peptide linkage will be formed between the carboxyl group of amino acid Cys and the amino group of amino acid Phe. This is catalyzed by the enzyme peptidyl transferase, which is a component of the ribosome. (Chloramphenicol blocks the peptidyl transferase reaction.)

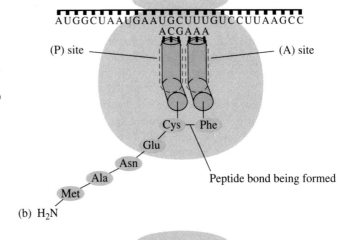

(c) The peptide linkage has been formed, and the growing polypeptide chain is now attached to the A site. The tRNA molecule has dissociated from the P site and is about to move into the cytosol to pick up another amino acid.

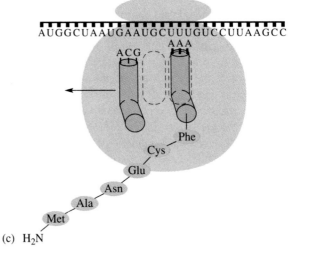

Figure 23.16
The elongation steps in protein synthesis.

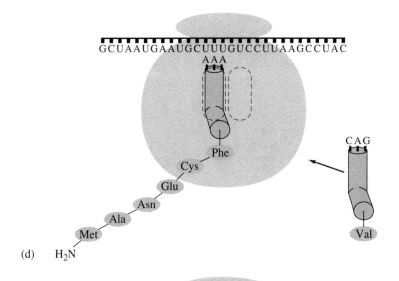

(d)   H₂N

(d) The ribosome moves (translocates) from left to right along the mRNA strand. (The energy for translocation is supplied by GTP.) The polypeptide chain, along with the tRNA molecule to which it is bound, is simultaneously shifted from the A site to the P site. This brings the next codon, GUC, into line over the A site. Notice that an activated tRNA molecule (containing the next amino acid to be incorporated into the chain) is moving into position on the surface of the ribosome. Its anticodon is CAG. (Erythromycin blocks the translocation reaction in bacteria.)

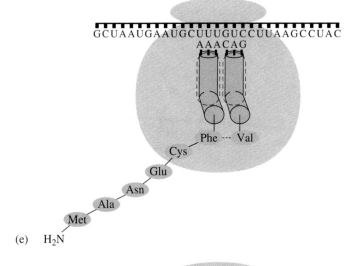

(e)   H₂N

(e) The activated tRNA molecule carrying amino acid Val is now in place on the ribosome. The peptide linkage between the carboxyl group of Phe and the amino group of Val is about to be made.

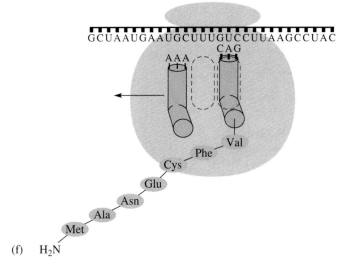

(f)   H₂N

(f) The peptide linkage has been formed, and the growing chain is attached, through a tRNA molecule, to the A site. The polypeptide chain is now seven amino acid units in length. The ribosome will translocate again, and the tRNA–polypeptide complex will be in position at the P site. This process will continue until the polypeptide chain is complete (i.e., when one of the three termination codons appears at the A site). When the ribosome reaches the end of the message, both it and the polypeptide are released from the mRNA molecule.

Thus in cells active in protein synthesis we find clusters of ribosomes connected by a single strand of mRNA (Figure 23.17). The amount of any particular protein in a cell depends on the balance between the rate at which it is synthesized (which is largely controlled by the rate at which its mRNA is synthesized in the nucleus) and the rate at which it is degraded. These events are summarized in Figure 23.18.

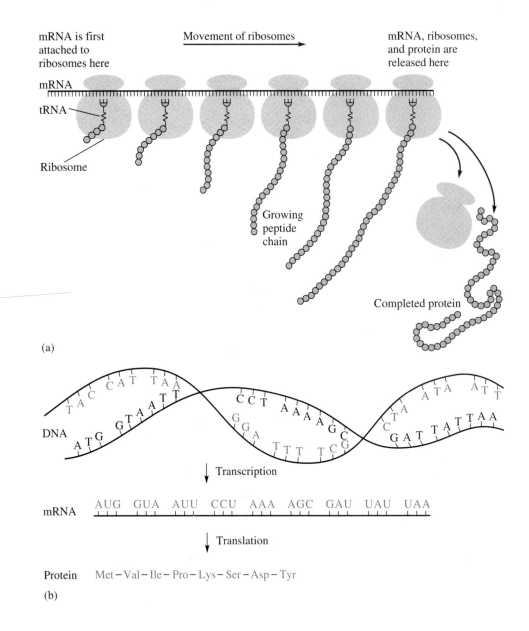

(a)

(b)

Figure 23.17

(a) Diagrammatic representation of a polysome. Several ribosomes are attached to mRNA. The ribosomes have progressed varying distances along the mRNA during translation, and each one is associated with a progressively longer protein chain. At the end of the message, the mRNA and ribosome separate and the complete protein is released. (b) The relationship between transcription and translation.

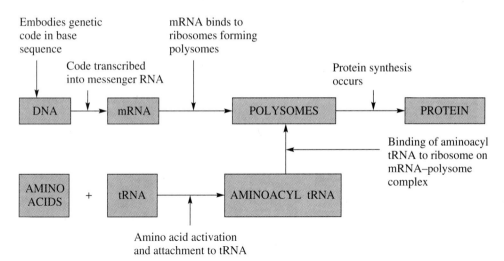

Embodies genetic code in base sequence

Code transcribed into messenger RNA

mRNA binds to ribosomes forming polysomes

Protein synthesis occurs

DNA → mRNA → POLYSOMES → PROTEIN

Binding of aminoacyl tRNA to ribosome on mRNA–polysome complex

AMINO ACIDS + tRNA → AMINOACYL tRNA

Amino acid activation and attachment to tRNA

Figure 23.18
Outline of events in protein synthesis from DNA transcription and amino acid activation to completed protein.

It is essential at this point to note that the preceding discussions of DNA replication, RNA synthesis, and protein synthesis are correct, but they are an oversimplification. Several molecular components (including RNA primers) are required to initiate DNA replication. Additional proteins are required for the elongation and termination processes. Furthermore, ultraviolet radiation and certain chemicals are known to damage DNA by disrupting the sugar–phosphate backbone and/or by altering the purine and pyrimidine bases. DNA repair enzymes exist that can mend the backbone and correct the base sequence.

Also, the majority of plant and animal genes occur in pieces, spread out along the DNA. Parts of the gene are expressed (exons); the intervening regions (introns) are not expressed. Therefore, the RNAs transcribed from these genes have been found to be spliced. That is, the entire length of DNA is copied in a longer transcript that is cut once or several times and hooked back together to give a shorter, functional messenger. In addition, protein synthesis is critically controlled at the level of transcription. We now know that there are structural genes, operator genes, promoter genes, regulatory genes, and repressor molecules (which are proteins), plus an involvement of GTP and cyclic AMP. Postribosomal modification of proteins occurs after synthesis on the mRNA–ribosome complex and involves many types of modifications (e.g., methylation or hydroxylation of amino acid side chains, breakage of peptide bonds to activate a zymogen or a hormone). An advanced biochemistry text will provide you with a detailed explanation of the fascinating studies in molecular genetics.

Genetic Regulation

23.7
THE GENETIC CODE

As we have stated, the sequence of bases on the mRNA directs the precise sequence of the amino acids for each protein. We have indicated that the codon, the unit that codes for a particular amino acid, consists of a group of three adjacent nucleotides on

Table 23.5
The Genetic Code

First Base	U	C	A	G	Third Base
U	UUU ⎱ Phe UUC ⎰ UUA ⎱ Leu UUG ⎰	UCU ⎱ UCC ⎱ Ser UCA ⎰ UCG ⎰	UAU ⎱ Tyr UAC ⎰ UAA Termination UAG Termination	UGU ⎱ Cys UGC ⎰ UGA Termination UGG Trp	U C A G
C	CUU ⎱ CUC ⎱ Leu CUA ⎰ CUG ⎰	CCU ⎱ CCC ⎱ Pro CCA ⎰ CCG ⎰	CAU ⎱ His CAC ⎰ CAA ⎱ Gln CAG ⎰	CGU ⎱ CGC ⎱ Arg CGA ⎰ CGG ⎰	U C G G
A	AUU ⎱ AUC ⎱ Ile AUA ⎰ AUG Met	ACU ⎱ ACC ⎱ Thr ACA ⎰ ACG ⎰	AAU ⎱ Asn AAC ⎰ AAA ⎱ Lys AAG ⎰	AGU ⎱ Ser AGC ⎰ AGA ⎱ Arg AGG ⎰	U C A G
G	GUU ⎱ GUC ⎱ Val GUA ⎰ GUG ⎰	GCU ⎱ GCC ⎱ Ala GCA ⎰ GCG ⎰	GAU ⎱ Asp GAC ⎰ GAA ⎱ Glu GAG ⎰	GGU ⎱ GGC ⎱ Gly GGA ⎰ GGG ⎰	U C A G

Second Base: U, C, A, G

the mRNA. There are 64 possible triplet codons. Early experimenters were faced with the task of determining which codon (or perhaps codons) stood for each of the 20 amino acids. The cracking of the genetic code was the joint accomplishment of several well-known geneticists, notably H. Khorana, M. Nirenberg, P. Leder, and S. Ochoa (1961–1964). A genetic dictionary has been compiled and is given in Table 23.5. Of the 64 possible codons, 61 code for amino acids and three serve as signals for the termination of polypeptide synthesis (that is, as periods at the end of a sentence). Notice that only methionine (AUG) and tryptophan (UGG) have a single codon. All other amino acids have two or more codons.

Further experimentation by Nirenberg threw much light on the nature of the genetic code. It now appears that

1. The code is essentially universal—animal, plant, and bacterial cells use the same codons to specify each amino acid. (A few exceptions have been discovered.)
2. The code is degenerate—in all but two cases (methionine and tryptophan), more than one triplet codes for a given amino acid.
3. The first two bases of each codon are most significant; the third base often varies. This suggests that a change in the third base by a mutation may still permit the correct incorporation of a given amino acid into a protein (see Section 23.8). The third base is sometimes called the "wobble" base.
4. In general, codons with C or U as the second base specify the nonpolar amino acids, whereas codons with A or G as the second base specify the polar amino acids (see Table 21.1).
5. The code is continuous and nonoverlapping—there are no special signals, and adjacent codons do not overlap (except in the case of a few viruses that do have overlapping genes).
6. There are three codons that do not code for any amino acid. These are the termination codons; they are read by special proteins (called release factors) and signal the end of the translation process.

7. The codon AUG codes for methionine and is also the initiation codon. Thus methionine is the first amino acid in each newly synthesized polypeptide. This first amino acid is usually removed enzymatically before the polypeptide chain is completed; the vast majority of polypeptides do not begin with methionine.

23.8
MUTATIONS AND GENETIC DISEASES

We have seen that DNA directs the synthesis of proteins through the intermediary mRNA and that the sequence of bases in the DNA is critical and specific for the proper sequence of amino acids in proteins. On rare occasions, however, the base sequence in DNA may be modified either spontaneously (about 1 in 10 billion) or by exposure to heat, radiation, or certain chemicals. Any chemical or physical change that alters the sequence of bases in the DNA molecule is termed a **mutation.** The most common types of mutations are *substitution* (a different base is substituted), *insertion* (addition of a new base), and *deletion* (loss of a base). These changes within the DNA are called **point mutations** because the change occurs at a single nucleotide position (Figure 23.19).

Each step in the replication–transcription–translation process is subject to error. In replication alone, each time a human cell divides, a copy is made of 4 billion bases to make a new strand of DNA. There are perhaps 2000 errors each time replication occurs. Most such errors are unimportant, but some have terrible consequences—genetic disease or even death.

Figure 23.19
The three types of point mutations.

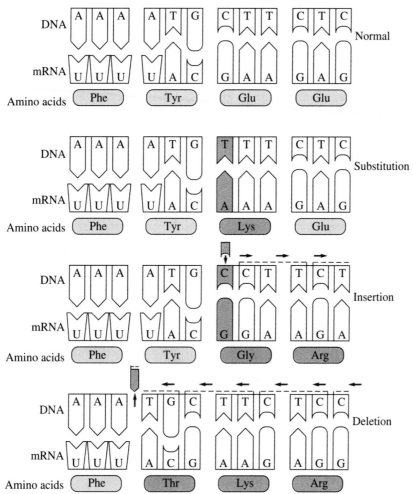

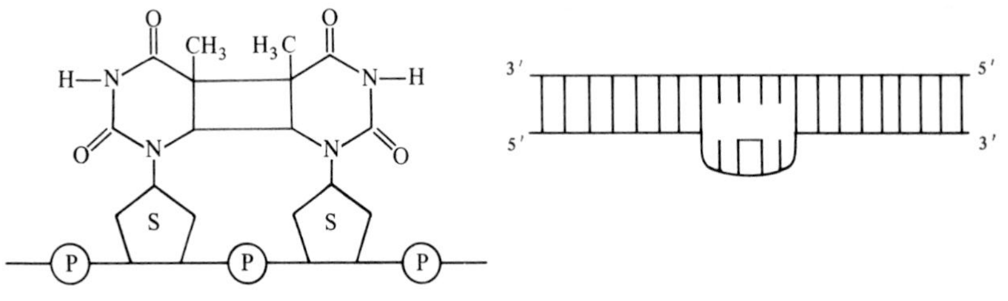

(a)   (b)

Figure 23.20

(a) Structure of a DNA intrastrand thymine dimer. (b) Defect in the double strand produced by the thymine dimer. The dimerization is caused by ultraviolet light. This temporarily stops DNA replication, but the dimer can be enzymatically excised and the strand can be repaired.

The chemical and/or physical agents that cause mutations are termed **mutagens.** Examples of physical mutagens are ultraviolet and gamma radiation. They exert their mutagenic effects either directly or via free radicals induced by the radiation. Radiation and free radicals are known to cause covalent modification (often cross-linkage) of bases already incorporated into DNA. For example, upon exposure to UV light, two adjacent thymines on a DNA strand can become covalently linked, producing a thymine dimer (Figure 23.20). If not repaired, the dimer prevents formation of the double helix at the point at which it occurs. The genetic disease xeroderma pigmentosum is transmitted as an autosomal recessive trait—an example of a defective mechanism for the repair of pyrimidine dimers in DNA. (The enzyme that cuts out the damaged thymine dimers is not synthesized by the cells.) Individuals affected by this condition are abnormally sensitive to light and are more prone to skin cancer than normal individuals. During replication, an abnormal DNA is produced, which apparently has no stop signal and results in the proliferation of cancer cells (see Special Topic J).

Among the chemical mutagens are two base analogs, 5-bromouracil and 2-aminopurine. They can be incorporated into the new DNA strand, but they exhibit faulty base pairing. 5-Bromouracil is incorporated into DNA in place of thymine, but it can base-pair with guanine (instead of adenine). 2-Aminopurine substitutes for adenine, yet it sometimes base-pairs with cytosine (instead of thymine). Hydroxylamine and nitrous acid are other chemical mutagens. Hydroxylamine (NH_2OH) deaminates cytosine, yielding a product that pairs with adenine instead of guanine. Nitrous acid (HNO_2) can convert cytosine to uracil, which will also bond to adenine instead of guanine. That these compounds are both carcinogenic and mutagenic strongly indicates that disruption of DNA is fundamental to both processes.

Many mutations are perpetuated by replication and are thus inherited. Although it is possible for a gene mutation to be beneficial (thus permitting evolutionary advances through natural selection), most mutations are detrimental. If a point mutation occurs at a crucial position, the defective protein will lack biological activity and may result in the death of the cell. In such cases the altered DNA sequence is lost and will not be copied into daughter cells. Nonlethal mutations often lead to metabolic abnormalities or to hereditary diseases. Such diseases are called **inborn errors of metabolism** or **genetic diseases.** A partial listing of genetic diseases is presented in Table 23.6, and a few specific conditions are discussed here. In most cases the defective gene results in the failure to synthesize a particular enzyme.

Table 23.6
A Partial Listing of
Genetic Diseases in
Humans and the
Malfunctional or Deficient
Protein or Enzyme

Disease	Responsible Protein or Enzyme
Acatalasia	Catalase (red blood cells)
Albinism	Tyrosinase
Alkaptonuria	Homogentistic acid oxidase
Cystathioninuria	Cystathionase
Fabry's disease	α-Galactosidase
Galactosemia	Galactose 1-phosphate uridyl transferase
Gaucher's disease	Glucocerebrosidase
Glycogen storage disease	Various types:
	$\quad$ α-Amylase
	$\quad$ Debranching enzyme
	$\quad$ Glucose 1-phosphatase
	$\quad$ Liver phosphorylase
	$\quad$ Muscle phosphofructokinase
	$\quad$ Muscle phosphorylase
Goiter	Iodotyrosine dehalogenase
Gout and Lesch–Nyhan syndrome	Hypoxanthine–guanine phosphoribosyl transferase
Hemolytic anemias	Various types:
	$\quad$ Glucose 6-phosphate dehydrogenase
	$\quad$ Glutathione reductase
	$\quad$ Phosphoglucoisomerase
	$\quad$ Pyruvate kinase
	$\quad$ Triose phosphate isomerase
Hemophilia	Antihemophilic factor (factor VIII)
Histidinemia	Histidase
Homocystinuria	Cystathionine synthetase
Hyperammonemia	Ornithine transcarbamylase
Hypophosphatasia	Alkaline phosphatase
Isovaleric acidemia	Isovaleryl-SCoA dehydrogenase
Maple syrup urine disease	α-Keto acid decarboxylase
McArdle's syndrome	Muscle phosphorylase
Metachromatic leukodystrophy	Sphingolipid sulfatase
Methemoglobinemia	NADPH–methemoglobin reductase and
	$\quad$ NADH–methemoglobin reductase
Niemann–Pick disease	Sphingomyelinase
Phenylketonuria	Phenylalanine hydroxylase
Pulmonary emphysema	α-Globulin of blood
Sickle cell anemia	Hemoglobin
Tay–Sachs disease	Hexosaminidase A
Tyrosinemia	Hydroxyphenylpyruvate oxidase
Von Gierke's disease	Glucose 6-phosphatase
Wilson's disease	Ceruloplasmin (blood protein)

Phenylketonuria (PKU) results when the enzyme phenylalanine hydroxylase is absent. A person with PKU cannot convert phenylalanine to tyrosine, which is the precursor of the neurotransmitters dopamine and norepinephrine as well as the skin pigment melanin (see Figure F.5).

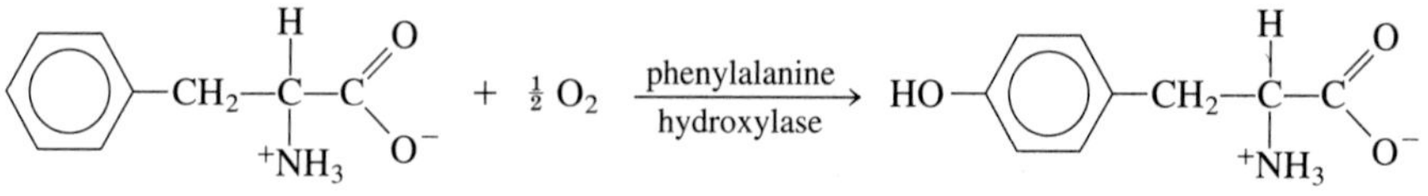

Phenylalanine $\quad\quad\quad\quad\quad\quad\quad\quad\quad\quad\quad\quad\quad\quad$ Tyrosine

In the absence of this step, phenylalanine accumulates, and the transamination (see Section 27.5) of phenylalanine to phenylpyruvate, normally a very minor process, becomes important.

Phenylalanine   Phenylpyruvate

Excessive amounts of phenylpyruvate impair normal brain development, causing severe mental retardation (a mean IQ of 20)—about 1% of patients in mental institutions are phenylketonuric. The life span of untreated PKU individuals is significantly shorter than normal (75% are dead before age 30). The disease acquired its name from the high levels of this phenyl ketone in the urine. PKU may be diagnosed by assaying a sample of blood or urine for phenylalanine or one of its metabolites, and federal law requires that all newborns be tested (usually within the first two weeks). If the condition is detected, mental retardation can be prevented by giving the afflicted infant a diet containing little or no phenylalanine. Because phenylalanine is so prevalent in natural foods, the low-phenylalanine diet is composed of a synthetic protein substitute plus very small measured amounts of natural foods. The diet is maintained until the child is at least 3 years old, by which time brain development is completed. The incidence of PKU in newborns is about 1 in 15,000.

Another inborn error in the metabolism of tyrosine leads to **albinism.** Tyrosine serves as a precursor for the melanins, the pigments that color the skin, hair, and eyes. The absence of the enzyme tyrosinase prevents the occurrence of one of the reactions necessary for this conversion. The lack of pigmentation characteristic of albinism is the result.

Galactosemia results from the lack of the enzyme that catalyzes the formation of glucose from galactose. The blood galactose level is markedly elevated, and galactose is found in the urine. The baby experiences a lack of appetite, weight loss, diarrhea, and jaundice. The disease may result in impaired liver function, cataracts, mental retardation, and even death. If recognized in early infancy, the effects of galactosemia can be eliminated by removing milk and all other sources of galactose from the diet. As the children grow older, they normally develop an alternate pathway for metabolizing galactose, and thus the need to restrict milk is not permanent. The incidence of galactosemia in the United States is 1 in every 65,000 newborn babies.

There are several genetic diseases that are collectively categorized as lipid storage diseases. As we shall learn in Chapter 26, lipids are constantly being synthesized and broken down. If the enzymes that catalyze lipid decomposition are missing, the lipids tend to accumulate and cause a variety of medical problems. The enzymes that are responsible for lipid storage diseases are known. The juncture at which the metabolic pathways go awry can be pinpointed. Unfortunately, however, no cure for these diseases has yet been developed. At present, genetic counseling of prospective parents who carry the defective gene is the only approach to control of the diseases.

In **Niemann–Pick disease,** a disease of infancy or early childhood, sphingomyelins accumulate in the brain, liver, and spleen because the enzyme sphingomyelinase is lacking. The accumulation of the sphingomyelins causes mental retardation and early death.

In **Gaucher's disease,** cerebrosides accumulate in the brain and cause severe mental retardation and death. Juvenile and adult forms of this disease are characterized by enlarged spleen and kidneys, hemorrhaging, mild anemia, and fragile bones. This disease is caused by the lack of a specific enzyme called glucocerebrosidase, which cleaves glucocerebrosides into glucose and sphingosine.

In the absence of another particular enzyme, hexosaminidase A, gangliosides accumulate in brain tissue. The ganglion cells of the brain become greatly enlarged and nonfunctional. This effect, called **Tay–Sachs disease,** results in retardation of development, dementia, paralysis, and blindness. Death usually occurs before the age of 3. Tay–Sachs disease can be diagnosed by assaying the amniotic fluid (amniocentesis) for the enzyme. The absence of hexosaminidase A in the amniotic fluid allows for a recommendation for a therapeutic abortion because the disease is incurable. Genetic screening can identify Tay–Sachs carriers because they produce only half the normal amount of hexosaminidase A (although they do not exhibit symptoms of the disease). Tay–Sachs is most common in persons of Eastern European Jewish ancestry.

23.9
GENETIC ENGINEERING: BIOTECHNOLOGY

Over 3000 human diseases have a genetic component. Over the last decade or so, researchers have linked specific genes to specific diseases. Now the ability to use this information to diagnose and cure genetic diseases appears to be within our grasp. By determining the location of a gene on the DNA molecule, scientists have been able to identify and isolate genes with specific functions.

A gene is a rather elusive substance. There are approximately 10,000 genes on each human chromosome, and isolating the one defective gene that causes a particular genetic disease is a monumental task. One approach to gene location is to treat the DNA with enzymes called restriction endonucleases. This method yields segments of genetic material called restriction fragment length polymorphisms (RFLPs; pronounced ''rif lips''). These segments are much easier to work with because they contain fewer genes. The RFLP pattern obtained upon enzyme treatment is an inherited one. RFLP patterns characteristic of certain families can be isolated. If the pattern of a relative matches that of a person with a genetic disease, that relative will probably develop the disease. Thus it is possible to identify and even predict the occurrence of a genetic disease.

A further hope of genetic engineering is that we will be able to introduce a functioning gene into a person's cells (Figure 23.21), thus correcting the action of a defective gene.

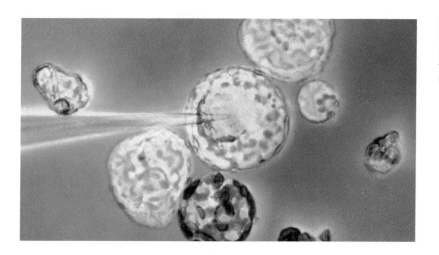

Figure 23.21
Under a powerful microscope, a tiny pipet is used to introduce foreign genes into a cell nucleus.

All living organisms (except some viruses) have DNA as their hereditary material. It should be possible, then, to place a gene from one organism into the genetic material of another. **Recombinant DNA technology** does just that.

By working backward from the amino acid sequence of the protein, scientists can work out the base sequence of the gene that codes for the protein. When isolated by the RFLP process, the gene is spliced into a special kind of bacterial DNA called a **plasmid.** The recombined plasmid is then inserted into the host organism (usually the bacterium *E. coli*). Figure 23.22 illustrates the production of recombinant DNA. The steps are as follows.

Figure 23.22
Recombinant DNA in *E. coli*.

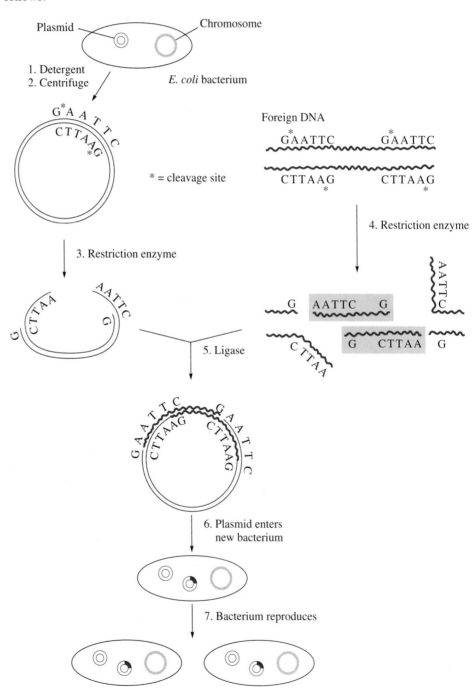

1. *Escherichia coli* bacteria are placed in a detergent solution to break open the cells.
2. The plasmids are separated from the chromosomal DNA by differential centrifugation.
3. Restriction enzymes (endonucleases) are used to cleave the plasmid at a specific short sequence in a way that creates overlapping, cohesive (''sticky'') ends. Each restriction enzyme can recognize a specific sequence of four to six nucleotides in DNA. For example, the endonuclease designated as *Eco*RI cuts at the sequence GAATTC, and *Sma*I cuts at CCCGGG. More than 100 restriction enzymes, purified from different species of bacteria, are now commercially available.
4. In vitro combination of the same restriction enzyme with DNA from another organism (foreign DNA) or with synthetic DNA produces segments of DNA with cohesive ends that are complementary to those of the plasmid. [Since different restriction enzymes have different cleavage sites, a given strand of DNA can be separated into many different segments of varying lengths. It is therefore possible to insert almost any foreign gene(s) into *E. coli.*]
5. The enzyme DNA ligase seals the foreign DNA segment into place in the plasmid.
6. The resealed plasmid is placed in a solution of calcium chloride containing *E. coli.* When the solution is heated, the bacterium cell membrane becomes permeable, allowing the plasmid to enter.
7. *Escherichia coli* reproduces by dividing (and thus doubling its population) at a rate of about once every 20–30 min. The new *E. coli* bacteria that are created have characteristics dictated by their own genes as well as those that have been transplanted from a different species.

The modified plasmids do all the things DNA does. They replicate, and they control the synthesis of proteins. Microorganisms multiply rapidly. Soon vast vats of these modified creatures are producing a protein specified by a ''foreign'' gene.

Many valuable materials, difficult to obtain in any other way, are now made using recombinant DNA technology. People with diabetes formerly had to use insulin from pigs or cattle. Now human insulin, a protein coded by human DNA, is being produced by the cell machinery of bacteria (Figure 23.23). All newly diagnosed insulin-dependent diabetics in the United States are now treated with human insulin produced by recombinant DNA technology. The hope for the future is that a functioning gene for insulin can be incorporated into the cells of insulin-dependent diabetics.

Many viruses readily penetrate human cells where they use the cell machinery to replicate themselves (see Section J.1). A human gene can be cloned into a modified virus that will then transport that gene into the human cells where it produces multiple copies of the human gene. The first human gene transplant was carried out in 1989. It was a demonstration project that proved the technique works, although it was not used to treat disease. The first human gene therapy was attempted in 1990. Modified human white blood cells were infused into a young girl with a severe immune deficiency.

Figure 23.23
A drop of the first human insulin produced through recombinant DNA technology.

Human growth hormone, used to treat children who fail to grow properly, was formerly available only in tiny amounts obtained from cadavers. Now it is readily available through recombinant DNA technology. This technology also yields interferon, a promising anticancer agent. The gene for epidermal growth factor, which stimulates the growth of skin cells, has been cloned. It has been used to speed the healing of burns and other skin wounds. Scientists have even designed bacteria that will "eat" the oil released in an oil spill, although their success in actual spills has been minimal.

Proponents of recombinant DNA research are excited about its great potential benefits. Recombinant techniques are an enormous aid to scientists in mapping and sequencing genes and in determining the functions of different segments of an organism's DNA. The complete DNA sequences of more than 100 mammalian genes have been determined using recombinant DNA technology. An understanding of gene function and gene regulation is a primary goal of scientists working to cure cancer. Human gene therapy has begun. Skin cancer patients have received injections of genetically altered white blood cells. The cells circulate in the blood seeking out malignant tissue. When they contact a cancer cell, their foreign gene produces a toxic enzyme called tumor necrosis factor. Cancer cells will get lethal doses, and the rest of the body will be spared.

It is conceivable that recombinant DNA could lead to cures for genetic diseases. When appropriate genes are successfully inserted in *E. coli,* the bacteria can become miniature pharmaceutical factories, producing great quantities of insulin, clotting factor for hemophiliacs, missing enzymes, hormones, vitamins, antibodies, vaccines, and so on. The production of DNA-recombinant molecules containing synthetic genes for the production of tissue plasminogen activator (TPA—a clot-dissolving enzyme that can rescue heart attack victims) has been accomplished in *E. coli.* Vaccines against hepatitis B (humans) and hoof and mouth disease (cattle) have been produced. In addition it may be possible to breed human intestinal bacteria that can digest cellulose and to create new plants that can obtain their nitrogen directly from the air rather than from costly petroleum-based fertilizers.

Besides *E. coli,* scientists have used other bacteria as well as yeast and fungi in gene-splicing experiments. The enzyme rennin has been produced in fungi. Rennin is used commercially to coagulate milk into curds in the production of cheese. A bacterial plasmid has been used as a vector by plant molecular biologists. The bacterium is *Agrobacterium tumefaciens,* which can cause tumors in many plants. Scientists have succeeded in introducing genes for several foreign proteins (including animal protein) into plants by means of *A. tumefaciens,* at the same time eliminating its tumor-causing ability. One practical application would be to transfer the gene necessary for the synthesis of a deficient amino acid into a particular plant (e.g., the gene for methionine synthesis into soybeans).

Nor have the young people of the world been left out. Bacteria have had genes inserted that enable them to produce indigo, the dye used for blue jeans. Designer genes for your designer jeans! In 1980 the U.S. Supreme Court, in a landmark decision, decreed that "man-made life forms" (i.e., the mutant bacteria, yeast, fungi, etc.) are patentable. That decision was expected to increase the vigor with which companies pursue their quest for marketable proteins via genetic manipulation. (To date only a dozen or so medicines have reached the market, but more than 100 are in development.)

Concern over the potential for disaster in this type of research has lessened somewhat in recent years. Initially, scientists worried about the possibility of producing a deadly "artificial" organism. What if a gene that causes cancer were spliced into the DNA of a bacterium that normally inhabits our intestine? We would have no natural immunity against such an artificial organism. To protect against such a development, strict guidelines for recombinant DNA research have been instituted.

Another point of contention is that this research can be misused for political and social purposes. These techniques might be exploited for genetically engineered control of human behavior, even enhancing people's IQs. There are some who believe that we should not meddle with evolution by creating new forms of life different from any that exist on Earth.

The new molecular genetics has already resulted in some impressive achievements. Its possibilities are mind-boggling—elimination of the genetic defects, a cure for cancer, a race of geniuses, and who knows what else? Knowledge gives power. It does not necessarily give wisdom. Who will decide what sort of creatures the human species should be? The greatest problem we are likely to face in our use of bioengineering is that of choosing who is to play God with the new "secret of life."

EXERCISES

1. Explain what is meant by each of the following terms.
   - **a.** ribosome
   - **b.** template replication
   - **c.** multiplicity
   - **d.** complementary bases
   - **e.** genetic code
   - **f.** translocation
   - **g.** code degeneracy
   - **h.** mutagens
   - **i.** genetic disease
   - **j.** PKU
   - **k.** amniocentesis
   - **l.** recombinant DNA
   - **m.** plasmid
   - **n.** restriction enzyme

2. Distinguish between the terms in each of the following pairs.
   - **a.** purine base and pyrimidine base
   - **b.** major base and minor base
   - **c.** ribose and deoxyribose
   - **d.** nucleoside and nucleotide
   - **e.** nucleotide and nucleic acid
   - **f.** DNA polymerase and DNA ligase
   - **g.** codon and anticodon
   - **h.** chromosomes and genes
   - **i.** transcription and translation
   - **j.** mutation and point mutation

3. **a.** Name the two kinds of nucleic acids.
   b. Which of these is concentrated in the nucleus of the cell?

4. **a.** What sugar is incorporated in the RNA polymer?
   b. What is the sugar unit in DNA?

5. Compare DNA and RNA with respect to the major bases present in each type of nucleic acid.

6. How do DNA and RNA differ in secondary structure?

7. For each of the following, indicate whether the compound is a nucleoside, a nucleotide, or neither.

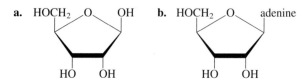

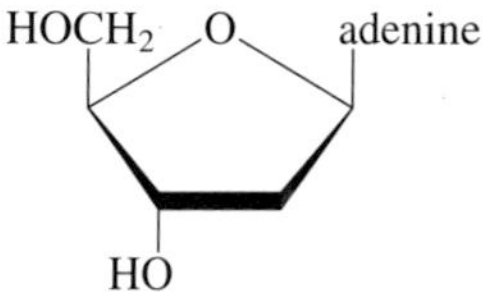

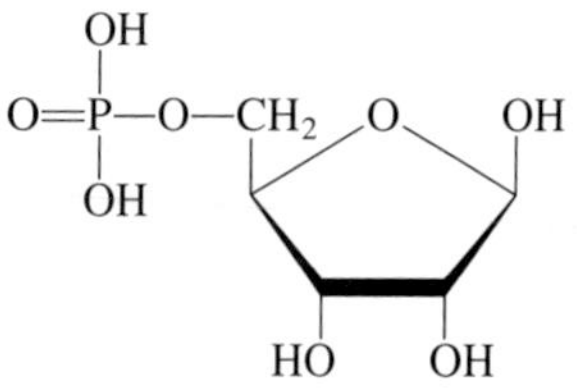

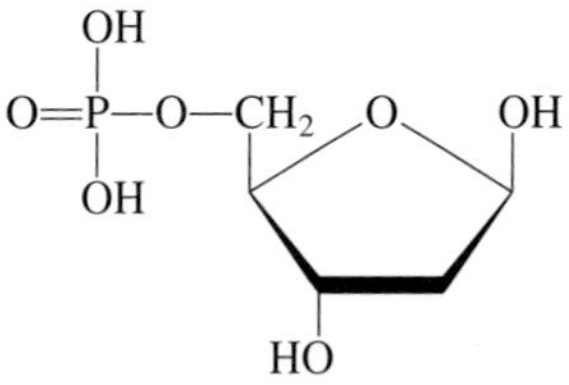

8. For each of the structures shown in Exercise 7, indicate whether the sugar unit is ribose or deoxyribose.

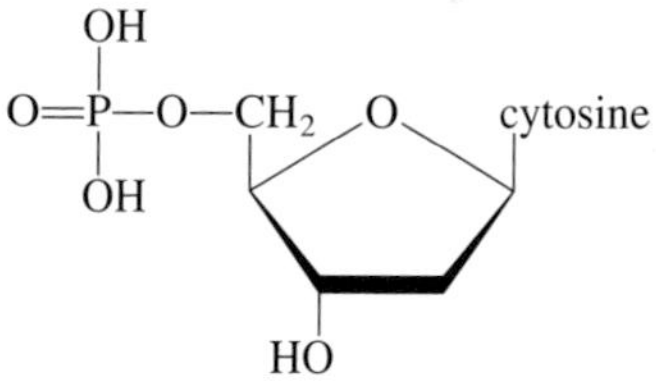

9. Give the names of all the compounds that can be obtained from the complete hydrolysis of **(a)** DNA and **(b)** RNA.

10. Draw structural formulas for the following minor pyrimidine and purine bases.
    - **a.** 5-methylcytosine
    - **b.** 5-hydroxymethylcytosine
    - **c.** 6-methyladenine
    - **d.** 2-methyladenine
    - **e.** 6-oxypurine
    - **f.** 4-thiouracil
    - **g.** 5,6-dihydrouracil
    - **h.** 1-methylguanine

11. The possibilities of hydrogen bond formation are similar for uracil and thymine. Explain.

12. DNA and RNA are termed nucleic acids. What makes them acidic?

13. What constitutes the backbone of a DNA chain?

14. What is the difference between AMP, ADP, and ATP?

15. Using a schematic representation, show a length of nucleic acid polymer chain, indicating the positions of the sugar, phosphoric acid, and base units.

16. The primary structure of a protein is defined by the sequence of amino acids. What defines the primary structure of nucleic acids? *base sequence*

17. With the same sort of schematic representation you used for Exercise 15, show the overall design of the double helix.

18. Why is it structurally important in the DNA double helix that a purine base always pair with a pyrimidine base?

19. What kind of intermolecular force is involved in base pairing? *H-Bonding*

20. In DNA, which base would be paired with each of the following bases?
    a. cytosine   b. adenine
    c. guanine   d. thymine

21. In an RNA molecule, which base would pair with each of the following bases?
    a. adenine   b. guanine
    c. uracil   d. cytosine

22. Describe the process of DNA replication.

23. In DNA replication, a parent DNA molecule produces two daughter molecules. What is the fate of each strand of the parent DNA double helix?

24. We say that DNA controls protein synthesis, yet most DNA resides within the cell nucleus whereas protein synthesis occurs outside of the nucleus. How does DNA exercise its control? *via RNA*

25. Explain the role of mRNA in protein synthesis.

26. Explain the role of tRNA in protein synthesis.

27. Which nucleic acid(s) is(are) involved in the process referred to as transcription?

28. Which nucleic acid(s) is(are) involved in the process referred to as translation?

29. a. Which nucleic acid contains the codon?
    b. Which nucleic acid contains the anticodon?

30. a. How many nucleotide units are present in a codon?
    b. Why is it that a triplet of bases codes for each amino acid?

31. The base sequence along one strand of DNA is ATTCG. What would be the sequence of the complementary strand of DNA?

32. What sequence of bases would appear in the mRNA molecule copied from the original DNA strand shown in Exercise 31?

33. If the sequence of bases along a mRNA strand is UCCGAU, what was the sequence along the DNA template?

34. What are the complementary triplets on tRNA for the following triplets on mRNA?
    a. UUU   b. CAU
    c. AGC   d. CCG

35. What are the complementary triplets on mRNA for the following triplets on tRNA molecules?
    a. UUG   b. GAA
    c. UCC   d. CAC

36. Using Table 23.5, identify the amino acids carried by the tRNA molecules in Exercise 34.

37. Using Table 23.5, identify the amino acids carried by the tRNA molecules in Exercise 35. Remember that Table 23.5 lists the mRNA codons.

38. Refer to Table 23.5. What amino acid sequence would result for each of the following base sequences on mRNA?
    a. UUACCUCGA
    b. GCGUCAUAA
    c. CCCCCCCCC

39. If the DNA base sequence TTACTCTCA acted as a template for mRNA formation, what amino acid sequence would eventually be produced from the mRNA?

40. What is the relationship between the cell parts called chromosomes, the units of heredity called genes, and the nucleic acid DNA?

41. Show the replication of the following DNA segment.

42. Write the RNA base sequence that would be obtained upon transcription of the lower chain of the following DNA segment.

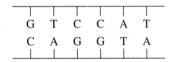

43. What are the two most important sites on tRNA molecules?

44. A hypothetical protein has a molecular weight of 60,000. Assume that the average molecular weight of an amino acid is 120.
    a. How many amino acids are present in the protein?
    b. How many codons occur in the mRNA that codes for this protein?
    c. How many nucleotide bases are found in the mRNA?

45. If the sequence of bases in a section of mRNA is

    AUGUACCACGGUACGCGGGUAUUGCUA-
    GCCGAUGGGUAA

    what would be the amino acid sequence in the peptide produced from this mRNA? (See Table 23.5.)

46. If the base sequence of a gene is

TACGAATCTAGAATACTTCCAAAAGTA-
TTTTGATACATC

what would be the amino acid sequence of the polypeptide that is synthesized?

47. The hormone somatostatin, produced in the pancreas, is composed of 14 amino acids. Somatostatin inhibits the release of a variety of hormones including glucagon, insulin, and human growth hormone. It is believed to keep the pancreas's output of insulin and glucagon in a proper balance (see Section 24.4).

a. Given the structure of somatostatin, postulate a possible base sequence of the mRNA that would direct the synthesis of somatostatin. Include an initiation codon and a termination codon.

```
        Ala-Gly-Cys-Lys-Asn-Phe-Phe
                 |               |
                 S              Trp
                 |               |
                 S              Lys
                 |               |
        Cys-Ser-Thr-Phe-Thr
```

b. What is the base sequence of the DNA that codes for this mRNA?

48. What is the basic process in recombinant DNA technology?

49. Discuss some applications of genetic engineering.

50. Chemical bonds can be broken by ultraviolet radiation, gamma rays, certain chemical substances, and other means. What are the biological implications of these facts? (*Hint:* What effect might the breaking of chemical bonds have on DNA?)

51. Certain genes are implicated in cancer (see Section J.3). Some of these *oncogenes* can be activated by single point mutations. Ordinarily the DNA triplet GGT codes for the amino acid proline. What amino acid is formed when the DNA triplet is changed by mutation to (**a**) TGT? (**b**) to GTT? (**c**) to CGT?

52. Give two examples of physical mutagens and four examples of chemical mutagens.

53. Name three genetic diseases, and indicate which enzyme is lacking for each.

54. We shall see in Section 28.9 that sickle cell hemoglobin differs from normal hemoglobin as a result of the substitution of valine for glutamic acid as the sixth amino acid from the N-terminal end of the polypeptide chain.

a. What alteration in the base sequence of the DNA could have caused this substitution?

b. What would be the resulting base sequence in the mRNA?

55. The following table contains just a few of the 200 or so known hemoglobin mutants. For each mutation, give the codon for the normal amino acid, and then give the codon for the mutant amino acid.

Type	Chain	Position	Normal	Mutant
J	α	5	Ala	Asp
I	α	16	Lys	Glu
M	α	58	His	Tyr
D	β	121	Glu	Gln
K	β	136	Gly	Asp

56. Consider the following segment of DNA

ACGTTAGCCCCAGCT

a. Write the sequence of bases in the corresponding mRNA.

b. What would be the amino acid sequence formed by translation?

c. What amino acid sequence would result from (i) replacement of the colored guanine by adenine, (ii) insertion of thymine immediately after the colored guanine, and (iii) deletion of the colored guanine?

57. Assume that a segment of a gene that coded for a particular enzyme has the following base sequence.

TACGACGTAACAAGC

a. What effect would result from a point mutation in which an adenine substituted for the colored guanine?

b. What effect would result from a point mutation in which a thymine substituted for the colored adenine?

58. Following are the results of two different point mutations. Which is likely to be more serious?

a. Valine is substituted for leucine.

b. Glutamic acid is substituted for leucine.

Viruses and Cancer

A discussion of viruses seems particularly appropriate here because viruses are composed almost entirely of proteins and nucleic acids. The field of virology is a rapidly expanding one, and recent research efforts to establish connections between viruses and some cancers have yielded much important and exciting information.

J.1
THE NATURE OF VIRUSES

Viruses are a unique group of infectious agents composed of a tightly packed central core of nucleic acids that is enclosed in one or more protein coats (Figure J.1). They are divided into two main classes on the basis of the nucleic acid content of the central core. Viruses contain either DNA or RNA *but never both*. (Recall that the cells of higher organisms, from bacteria to humans, contain both kinds of nucleic acids.) Viruses differ from one another in both size[1] and shape. The influenza virus, for example, is about ten times bigger than the polio virus. Viruses may be spherical, rod-shaped, or threadlike.

A *DNA virus* enters a host cell where its DNA is replicated and directs the host cell to produce viral proteins.

---

[1] Viruses are exceedingly small, much smaller than bacteria. The tobacco mosaic virus is approximately 300 nm long by 18 nm wide (i.e. 300×10^{-9} m by 18×10^{-9} m) and has a molecular weight of about 40 million. In general, animal viruses are larger than bacterial viruses. Most viruses are visible only under the electron microscope. The size of an average bacterial cell is 1500 nm long by 750 nm wide.

Table J.1
Human Infectious Diseases

Diseases of Bacterial Origin	Diseases of Viral Origin	
Cholera	AIDS	Measles
Diphtheria	Chicken pox	Meningitis
Dysentery	Cold sores	Mumps
Gonorrhea	Common cold	Pneumonia
Plague	Encephalitis	Polio
Syphilis	Gastroenteritis	Rabies
Tetanus	Genital herpes	Shingles
Tuberculosis	German measles	Smallpox
Typhoid fever	Hepatitis	Warts
Whooping cough	Influenza	Yellow fever

The viral proteins and viral DNA assemble into new viruses, which are released by the host cell. These new viruses can then invade other cells and continue the process. Figure J.2 depicts the life cycle of a bacteriophage (a virus that infects bacteria). Cell death and the production of new viruses account for the symptoms of viral infections. As Table J.1 indicates, a greater number of human diseases are of viral origin than of bacterial origin. Infectious diseases of viral origin (especially the common cold, influenza, and AIDS) are among the most significant health problems in our society.

Most *RNA viruses* use their nucleic acids in much the same way as the DNA viruses. The virus penetrates a host cell where the RNA strands are replicated and induce the synthesis of viral proteins. The new RNA strands and viral proteins are then assembled into new viruses. Some RNA viruses, called **retroviruses,** synthesize DNA in the host cell. This process is the opposite of the transcription of a DNA code into RNA that normally occurs in cells. The synthesis of DNA from an RNA template is catalyzed by the enzyme reverse transcriptase. The human immunodeficiency virus (HIV) that causes AIDS is perhaps the best known of the retroviruses. The HIV invades and eventually destroys T cells, a group of white blood cells that normally help protect the body from infections. With the T cells destroyed, the AIDS victim succumbs to infectious diseases such as pneumonia *(Pneumocystis carnii)* and Kaposi's sarcoma (a rare cancer).

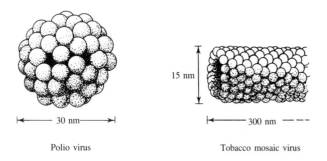

15 nm

|← 30 nm →|

Polio virus

|← 300 nm — − − |

Tobacco mosaic virus

Figure J.1
Schematic diagrams of the polio virus and the tobacco mosaic virus.

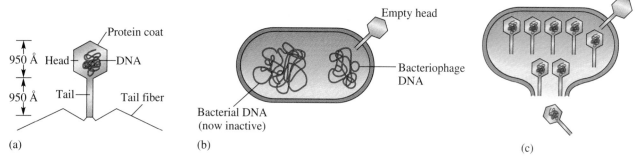

Figure J.2

Life cycle of a bacteriophage. (a) Drawing of T2 bacterio-phage. The bacteriophage DNA is located in the head of the particle, surrounded and protected by the protein coat. The end of the tail contains lysozyme, which can hydrolyze the polysaccharide of the bacterial cell wall. (b) When the bac-teriophage infects a bacterium, the phage DNA is injected into the bacterial cell. Synthesis of bacterial nucleic acids and protein stops, and the synthetic machinery of the bacte-rial cells begins to produce phage DNA and protein. (c) As soon as all the phage constituents are synthesized, they begin to form new phage particles. When the bacterial cell becomes filled with phage particles, lysis occurs, and the progeny are released. A total of 150–300 progeny viruses are produced in about 30–60 min from the infection of one bacterial cell by a single virus.

J.2
ANTIVIRAL DRUGS

Scientists had to learn the normal biochemistry of cells before they could develop drugs that treat abnormal conditions caused by the invasion of viruses. Gertrude Elion and George Hitchings of Burroughs Wellcome Research Laboratories in North Carolina and James Black of Kings College in London did the basic biochemistry that led to the development of antiviral drugs and many of the anticancer drugs (Section J.6) and shared the 1988 Nobel prize for physiology or medicine. They determined the shapes of cell membrane receptors, and they learned how normal cells work. Then they and other scientists were able to design drugs to block receptors in infected cells. Today scientists use powerful computers to design mole-cules to fit receptors. Drug design often was hit or miss in its early decades. It is rapidly becoming a precise science.

Scientists have developed drugs that are effective against some viruses (Figure J.3). Amantadine helps pre-vent some influenza A infections. Acyclovir (Zovirax) controls flareups of the herpes viruses that cause genital sores, chicken pox, shingles, mononucleosis, and cold sores. Azidothymidine (Zidovudine—AZT) slows the onslaught of the AIDS virus, but the disease is still relent-lessly fatal to those infected by the virus. AZT apparently acts by substituting for thymine. The reverse transcriptase

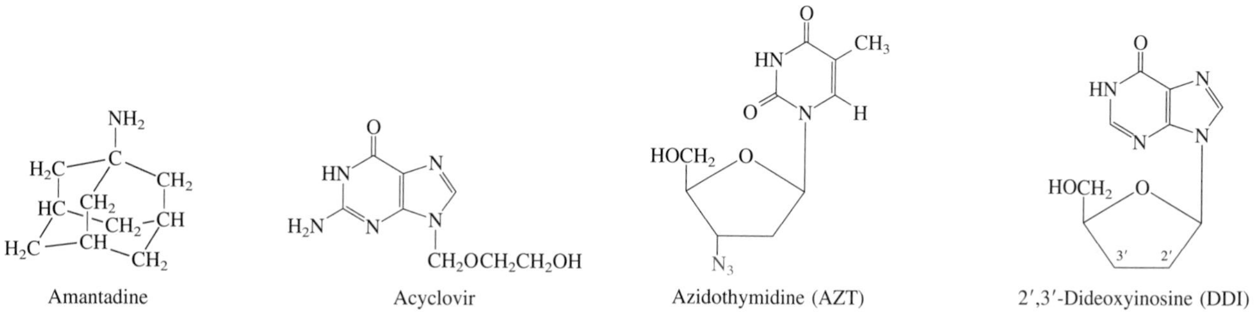

Amantadine

Acyclovir

Azidothymidine (AZT)

2',3'-Dideoxyinosine (DDI)

Figure J.3

Some antiviral drugs. Amantadine is totally synthetic. Acyclovir is derived from the purine base guanine, azidothymidine is derived from a thymine base attached to a deoxyribose sugar, and 2',3'-dideoxyinosine is derived from the purine base inosine and dideoxyribose.

incorporates AZT into the DNA chain, blocking the synthesis of the DNA. Unfortunately, the toxicity of AZT prevents its use in quantities sufficient to stop HIV replication completely. Other drugs similar to AZT are undergoing testing.

In 1991, the FDA approved the use of a second drug, $2',3'$-dideoxyinosine (DDI), to treat AIDS. The approval was for those patients who could not tolerate AZT treatment or for whom AZT was no longer effective. Like AZT, DDI blocks the replication of the HIV virus. The long-term effects of DDI are not yet known; the FDA will be closely monitoring the results of its use.

J.3
CANCER AND ITS CAUSES

Carcinogens were mentioned often in earlier chapters. These substances cause the growth of tumors. A tumor is an abnormal growth of new tissue. Tumors may be either benign or malignant. **Benign tumors** are characterized by slow growth, they often regress spontaneously, and they do not invade neighboring tissues. **Malignant tumors** may grow slowly or rapidly, but their growth is generally irreversible. They often are called **cancers.** Malignant growths invade and destroy neighboring tissues. Actually, cancer is not a single disease. It is a catchall term for over a hundred different afflictions. Many are not even closely related to each other.

The World Health Organization estimates that 80–90% of cancer cases are caused by environmental factors, 10–20% by genetic factors and (perhaps) viruses. Included most prominently among those "environmental" causes are cigarette smoking (40%), dietary factors (25–30%), and occupational exposure (10%).[2] That leaves 10–15% that may be caused by environmental *pollutants*. If you read the newspapers or watch television, you may get the idea that "everything" causes cancer. Even among those chemicals that have been suspect, however, many cannot be shown to be carcinogenic. Only about 30 chemical compounds have been identified as human carcinogens. Another 300 or so have been shown to cause cancer in laboratory animals. Some of the 300 are widely used, though.

Not all carcinogens are synthetic chemicals. Some, such as safrole in sassafras and the aflatoxins produced by

molds on foods, occur naturally. Some researchers estimate that 99.99% of all carcinogens that we ingest are natural ones. Plants produce compounds to protect themselves from fungi, insects, and higher animals, including humans. Carcinogenic compounds are found in mushrooms, basil, celery, figs, mustard, pepper, fennel, parsnips, and citrus oils—almost everywhere a curious chemist looks. Carcinogens are also produced during cooking and as products of normal metabolism. We must have some way of protecting ourselves from carcinogens.

How do chemicals cause cancer? Their mechanisms of action are probably as varied as their chemical structures. Some carcinogens chemically modify DNA, thus scrambling the code for replication and for the synthesis of proteins. For example, aflatoxin B is known to bind to guanine residues in DNA. Just how this initiates cancer, however, is not known for sure.

In the 1970s, scientists began identifying genes that seemed to trigger or sustain the processes that convert normal cells to cancerous ones. So far about 20 or 30 of these genes have been identified. They have been named **oncogenes** (Greek *onkos,* mass). (Oncology is the branch of medicine that treats tumors.) Oncogenes arise from ordinary genes that regulate cell growth and cell division. They can be activated by chemical carcinogens, radiation, or perhaps some viruses. It seems that more than one oncogene must be turned on, perhaps at different stages of the process, before a cancer develops. We also have suppressor genes that ordinarily prevent the development of cancers. These genes must be inactivated before a cancer develops. Suppressor gene inactivation can occur through mutation, alteration, or loss. In all, 10 or 15 mutations may be required in a cell before it turns cancerous. There is hope that someday suppressor genes can be produced through genetic engineering and used in therapy.

J.4
CHEMICAL CARCINOGENS

A variety of widely different chemical compounds are carcinogenic. We could not attempt to cover all the types of carcinogens here, so we will concentrate on a few major classes.

Some of the more notorious carcinogens are the polycyclic aromatic hydrocarbons (Section 13.17), of which 3,4-benzpyrene is perhaps the best known. Carcinogenic hydrocarbons are formed during the incomplete burning of nearly any organic material. They have been found in charcoal-grilled meats, cigarette smoke, automobile ex-

---

[2] There are 400,000 cancer deaths each year in the United States. Of these 150,000 are related to cigarette smoking. Another 150,000 are related to our diet.

hausts, coffee, burnt sugar, and many other materials. Not all polycyclic aromatic hydrocarbons are carcinogenic. There are strong correlations between carcinogenicity and certain molecular sizes and shapes. The mechanism of their action is currently under intense investigation. It already appears rather certain that the actual carcinogens are not the hydrocarbons themselves but the oxidation products formed in the liver.

Another important class of carcinogens is the aromatic amines. Two prominent ones are β-naphthylamine and benzidine.

β-Naphthylamine   Benzidine

These compounds once were used widely in the dye industry. They were responsible for a high incidence of bladder cancer among the workers whose jobs brought them into prolonged contact with them.

Several aminoazo dyes have been shown to be carcinogenic. An interesting example is 4-dimethylaminoazobenzene. This compound is also known as "butter yellow." It was used widely as a coloring for butter and oleomargarine before its carcinogenicity became known.

4-Dimethylaminoazobenzene

Not all carcinogens are aromatic. Prominent among the aliphatic ones are dimethylnitrosoamine (Section 17.5) and vinyl chloride (Table 13.11). Others include three- and four-member heterocyclic rings containing nitrogen or oxygen. The epoxides and derivatives of ethyleneimine are examples. Others are cyclic esters called lactones.

Bis(epoxy)butane   N-Laurylethyleneimine

β-Propiolactone

Keep in mind that this list is not all-inclusive. Rather, its purpose is to give you an idea of the kinds of compounds that have tumor-inducing properties—and the list grows almost daily as the results of research are released.

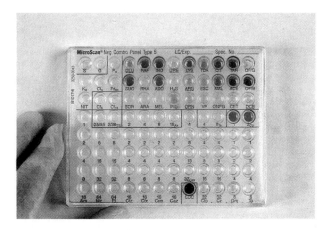

Figure J.4
Ames test using *Salmonella* isolate. Color changes represent *Salmonella* activity.

J.5
TESTING FOR CARCINOGENS

How do we know that a chemical causes cancer? Obviously, we can't experiment on humans to see what happens. That leaves us with no way to prove beyond doubt that a chemical does or does not cause cancer in humans. There are three ways, however, to gain evidence against a compound: bacterial screening for mutagenesis, animal tests, and epidemiological studies.

The cheapest way to gain evidence that a substance may be carcinogenic is by use of a screening test (Figure J.4) developed by Bruce N. Ames of the University of California, Berkeley. The Ames procedures use a variant of the bacterium *Salmonella typhimurium* to test for mutations. The test is simple, but it is not completely accurate. Some chemicals that are mutagens are not carcinogens. About 90% of the chemicals that fail the Ames test, though, are found to be carcinogens when tested by the other methods. One shortcoming of the Ames test is that some chemicals, not carcinogenic themselves, are converted to carcinogens by metabolism. These might well pass the Ames test and still cause cancer. A modification of the test uses metabolites from urine or feces in the screening process to overcome this problem.

Chemicals suspected of being carcinogens can be tested on animals.[3] Tests using low dosages on millions of

[3] Animal studies cost about $1 million each and take 2 years.

rats would cost too much, so tests usually are done by using large doses on 30 or so rats. An equal number of rats serve as controls. The control group is exposed to the same diet and environment as the experimental group, except that the control group does not get the suspected carcinogen. A higher incidence of cancer in the experimental animals than in the controls indicates that the compound is carcinogenic.

Animal tests are not conclusive. Humans usually are not exposed to comparable doses, and there may be a threshold below which a compound is not carcinogenic. Further, human metabolism is different from the metabolisms of the test animals. The carcinogen might be active in the rat but not in humans (or vice versa!).[4]

The best evidence that a substance causes cancer in humans comes from the epidemiological studies. A population that has a higher than normal rate for a particular kind of cancer is studied for common factors in their background. It was this sort of study, for example, that showed that cigarette smoking causes lung cancer, that vinyl chloride causes a rare form of liver cancer, and that asbestos causes cancer of the lining of the pleural cavity (the body cavity that contains the lungs). These studies require sophisticated mathematical analyses. There is always the chance that some other (unknown) factor is involved in the carcinogenesis.

J.6
CHEMICALS AGAINST CANCER

Chemists have designed molecules to relieve headache, cure infectious diseases, and prevent conception. Why can't they do something about cancer? They have done a lot, but much remains to be done. Treatment with drugs, radiation, and surgery has led to high rates of cures for some forms of cancer (for example, one form of skin cancer). For other forms, such as lung cancer, the rate of cure is still quite low. Over 30 different chemical substances are used widely in the treatment of cancer. That number will no doubt increase rapidly as our understanding of basic cell chemistry increases. We examine a few representative anticancer drugs here.

In cancer chemotherapy (Section 22.8), *antimetabolites* usually are compounds that inhibit the synthesis of

nucleic acids. Rapidly dividing cells, characteristic of cancer, require large quantities of DNA. The anticancer metabolites block DNA synthesis and therefore block the increase of the number of cancer cells. Because cancer cells are undergoing rapid growth and cell division, they generally are affected to a greater extent than normal cells.[5]

The most widely used anticancer drug is cisplatin, a platinum-containing compound. Cisplatin binds to DNA and blocks its replication. Transplatin, an isomer of cisplatin, is ineffective. The *shape* of the molecule is all-important.

Cisplatin   Transplatin

Two other prominent antimetabolites are 5-fluorouracil and its deoxyribose nucleoside, 5-fluorodeoxyuridine. In the body, both of these can be incorporated into a *nucleotide* (Section 23.1). The fluorine-containing derivatives inhibit the formation of thymine-containing nucleotides required for DNA synthesis. Thus, both compounds slow the division of cancer cells. These compounds have been employed against a variety of cancers, especially those of the breast and the digestive tract.

5-Fluorouracil   Uracil   5-Fluorodeoxyuridine

Another common antimetabolite is 6-mercaptopurine. This compound can substitute for adenine in a nucleotide.

6-Mercaptopurine   Adenine

---

[4] There is only a 70% correlation between the carcinogenesis of a chemical in rats and that in mice. The correlation between carcinogenesis in either rodent and that in humans probably is less.

[5] Cancer chemotherapy also affects body cells that undergo rapid replacement. These include those cells that line the digestive tract and those that produce hair. Side effects of the therapy include nausea and loss of hair. Eventually the normal cells are affected to such a degree that treatment must be discontinued.

The pseudonucleotide then inhibits the synthesis of nucleotides that incorporate adenine and guanine. Hence, DNA synthesis and cell division are slowed. 6-Mercaptopurine has been used in the treatment of leukemia.

Methotrexate

Folic acid

Another antimetabolite, methotrexate, acts in a somewhat different manner. Note the similarity between its structure and that of folic acid. Like the pseudofolic acid formed from sulfanilamide, methotrexate competes successfully with folic acid for an enzyme but cannot perform the growth-enhancing function of folic acid. Again, cell division is slowed, and cancer growth is retarded. Methotrexate is used frequently against leukemia.

J.7
MISCELLANEOUS ANTICANCER AGENTS

There is a bewildering variety of anticancer agents that defy ready classification. Alkaloids from vinca plants have been shown to be effective against leukemia and Hodgkin's disease. Actinomycin, a mixture of complex compounds obtained from the molds *Streptomyces antibioticus* and *Streptomyces parvus,* is used against Hodgkin's disease and other types of cancer. It is quite effective but extremely toxic. Actinomycin acts by binding to the double helix of DNA, thus blocking the replication of RNA on the DNA template. Protein synthesis is inhibited.

Sex hormones can be used against cancers of the reproductive system. For example, the female hormones estradiol (a natural hormone) and DES (a synthetic hormone) can be used against cancer of the prostate gland. Conversely, male hormones such as testosterone can be used against breast cancer. Such treatment often brings about a temporary cessation—or even a regression—in the growth of cancer cells.

The food additive butylated hydroxytoluene (BHT) has been shown to be anticarcinogenic in tests involving laboratory animals. Some people involved in cancer research speculate that the use of this additive as a preservative in foods may account for the declining rate of stomach cancer in the United States.

Similarly, there is some evidence that vitamin A may confer a resistance to some cancers. For example, persons suffering from vitamin A deficiencies exhibit a higher incidence of lung cancer. Vitamin C also may have an anticancer function. It has been shown to inhibit the formation of nitrosoamines (Section 17.5) under conditions similar to those in the human stomach. Fiber is thought to protect against colon cancer.

Perhaps most notable is the protective value of a diet high in cruciferous vegetables (cabbage, brussels sprouts, broccoli, cauliflower, and kale). This kind of diet has been shown to reduce the incidence of cancer both in animal studies and in studies of human population groups. The exact chemical substances in these vegetables that act as anticarcinogens are not known. It seems quite likely, though, that by eating a balanced diet, including fresh fruit and vegetables, we also balance our carcinogens and anticarcinogens.

Chemotherapy is only a part of the treatment of cancer. Surgical removal of tumors and radiation treatment remain major weapons in the war on cancer. Modern management of cancers can involve surgery, radiation, and one or more anticancer drugs. Indeed, a combination of drugs is often considerably more effective than any one alone. It is unlikely that a single agent will be found to cure all cancers. Steady progress is being made, however. Rates of cure should improve as research progresses. Perhaps a greater hope lies in the prevention of cancer. Much active research is under way on the mechanisms of carcinogenesis. Once we find the causes, rapid progress toward cures should follow.

EXERCISES

1. What is the composition of viruses?
2. Distinguish a DNA virus from an RNA virus.
3. What are the shapes of viruses?
4. Compare the sizes of viruses and bacteria.
5. How does a DNA virus invade and destroy a cell?
6. How does an RNA virus invade and destroy a cell?
7. Name five diseases of viral origin and five diseases of bacterial origin.
8. List three antiviral drugs. Which, if any, cure viral diseases?
9. What is a tumor?
10. How are benign and malignant tumors different?
11. What is the single leading cause of cancer?
12. Name several natural carcinogens.
13. What are oncogenes? How are they involved in the development of cancer?
14. What are suppressor genes? How are they involved in the development of cancer?
15. List some conditions under which carcinogenic hydrocarbons are formed.
16. Name two aromatic amines that are carcinogens.
17. What is butter yellow? How was it used before it was found to be a carcinogen?
18. Name two aliphatic carcinogens.
19. List some of the limitations involved in testing compounds for carcinogenicity by using laboratory animals.
20. What is an epidemiological study? Can such a study prove absolutely that a compound causes cancer?
21. What is a mutagen?
22. Describe the Ames test for mutagenicity. What are its limitations as a screening test for carcinogens?
23. List two major classes of anticancer drugs.
24. How does cisplatin work as an anticancer agent? Why is transplatin not effective as an anticancer agent?
25. What natural substance does 6-mercaptopurine resemble? How does 6-mercaptopurine act against cancer?
26. How does 5-fluorouracil act against cancer?
27. What natural substance does methotrexate resemble? How does methotrexate act against cancer?
28. What is actinomycin? How does it act against cancer?

Chapter 24
CARBOHYDRATE METABOLISM I

Weight lifting is an anaerobic process. Energy is required for short bursts of vigorous activity.

Life requires energy. Living cells are inherently unstable and avoid falling apart only because of a continued input of energy. Living organisms are restricted to using certain forms of energy. Supplying a plant with heat energy by holding it in a flame will do little to prolong its life. On the other hand, a green plant is uniquely able to tap the richest source of energy on Earth, sunlight.

In our earliest discussion of carbohydrates, it was mentioned that human existence on this planet is directly dependent upon the plant kingdom. The animal world derives its foodstuffs, and hence its energy, from plant life (Figure 24.1). Plants obtain water, inorganic salts, and nitrogenous compounds from the soil, and carbon dioxide and oxygen from the atmosphere. With these raw materials, they are able to synthesize carbohydrates, lipids, and proteins. The energy required for these synthetic reactions is obtained from the sun—radiant energy (as sunlight) is the ultimate source of biological activity. The overall process by which glucose is formed from carbon dioxide and water at the expense of solar energy is termed **photosynthesis.** The general photosynthetic equation can be written as follows.

$$6\ CO_2\ +\ 6\ H_2O\ +\ 686,000\ cal\ \xrightarrow{\text{sunlight}}\ C_6H_{12}O_6\ +\ 6\ O_2$$
$$\text{Glucose}$$

647

Figure 24.1
Some energy transforma-
tions in living systems.

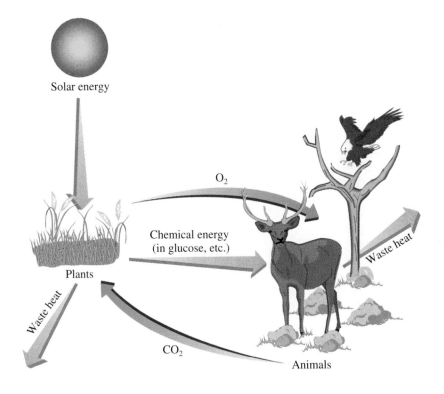

This synthesis is a distinguishing characteristic of green plants, and from the stand-point of human survival it represents the most important series of reactions (>100 en-zyme-catalyzed steps) that occur on the surface of the Earth. Notice that the formation of 1 mol of glucose requires the expenditure of 686,000 cal. Compared to the energy requirement of most other endothermic chemical reactions, this is a very large sum. (It is roughly equivalent to the heat required to raise the temperature of 7 quarts of water from 0 to 100 °C.) On the solar scale, however, it is a drop in the bucket.[1]

Animals cannot directly use the energy of sunlight. They must eat plants, or other animals that eat plants, in order to get carbohydrates, fats, and proteins with their stored chemical energy. Once digested and transported to the cell, a food molecule can be used in either of two ways. It can be used as a building block to make new cell parts or to repair old ones, or it can be "burned" for energy.

The entire series of coordinated chemical reactions that keep cells alive is called **metabolism.** In general, metabolic reactions are divided into two classes. The breaking down of molecules to provide energy is called **catabolism.** The process of building up the molecules of living systems is called **anabolism.**

Carbohydrates are the primary metabolites of the animal kingdom. Recall that over half of the food that we ourselves consume is composed of carbohydrates. Carbohydrates serve as the chief fuel of biological systems, supplying living cells with usable energy. Like all fuels, carbohydrates must be burned or oxidized if energy is to be released. The combustion (oxidation) process ultimately results in the conversion of the carbohydrate

Any chemical compound that is involved in a metabolic reaction is referred to as a **metabolite.**

---

[1] It has been estimated that on an average sunny day each square centimeter of Earth receives about 1 cal of solar radiation every minute. Yet the energy intercepted by the Earth is but a tiny part ($2 \times 10^{-7}\%$) of the total energy given off by the sun, and only a small fraction of this intercepted energy ($\sim 1\%$) is utilized in photosynthesis. The most important photosynthesizing organisms are the phytoplankton living in the world's oceans. Phytoplankton are the chief source of energy for aquatic organisms.

into carbon dioxide and water. The energy stored in the carbohydrate molecule during photosynthesis is released in the reaction.

$$C_6H_{12}O_6 \ + \ 6\,O_2 \ \longrightarrow \ 6\,CO_2 \ + \ 6\,H_2O \ + \ 686{,}000 \ cal$$

This equation summarizes the biological combustion of foodstuff molecules by the cell (respiration). The term **respiration** is often used in a broader sense to include all metabolic processes by which gaseous oxygen is used to oxidize organic matter to carbon dioxide, water, and energy.

Both respiration and the combustion of the common fuels (wood, coal, gasoline) use oxygen from the air to break down complex organic substances to carbon dioxide and water. The energy released in the burning of wood is manifested entirely in the form of heat, but excess heat energy is useless and even injurious to the living cell. Organisms conserve almost half of the 686,000 cal by a series of stepwise reactions that liberate small amounts of usable energy, which is transmitted to the phosphate bond of ATP. The remainder of the energy is used to heat the body and thus to maintain proper body temperature. Before continuing, let us more closely examine the compound ATP.

Plant and animal cells exist in a *symbiotic* cycle—each requires the products of the other for life. Although H_2O and O_2 are abundant in the atmosphere, CO_2 is present only to the extent of 0.02–0.05%. It has been estimated that if animals were removed from the Earth, all the atmospheric CO_2 would be consumed in 1–2 years.

24.1
ATP: UNIVERSAL ENERGY CURRENCY

In Section 16.10, we discussed phosphate esters in general. In Chapter 23, we learned of the nucleotide adenosine monophosphate (AMP). Probably the most important phosphate compound is **adenosine triphosphate (ATP).** Adenosine triphosphate was first isolated from skeletal muscle tissue and has since been shown to occur in all types of plant and animal cells. The concentration of ATP in the cell varies from 0.5 to 2.5 mg/mL of cell fluid. ATP is a nucleoside triphosphate composed of adenine, ribose, and three phosphate groups (Figure 24.2).

The most significant feature of the ATP molecule is the presence of the phosphoric acid anhydride, or pyrophosphate, linkage.

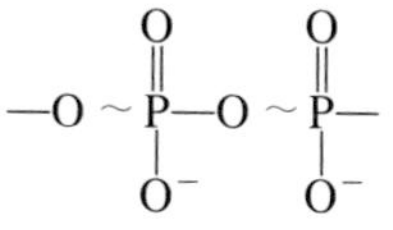

$$-O \sim \overset{\overset{\textstyle O}{\|}}{\underset{\underset{\textstyle O^-}{|}}{P}} -O \sim \overset{\overset{\textstyle O}{\|}}{\underset{\underset{\textstyle O^-}{|}}{P}} -$$

ATP is often called an *energy-rich compound*. Its pyrophosphate bonds are referred to as high-energy bonds and are sometimes symbolized by a squiggle bond ($\sim$). A **high-energy bond** is one that releases a relatively large amount of energy (>7000 cal/mol) when it is

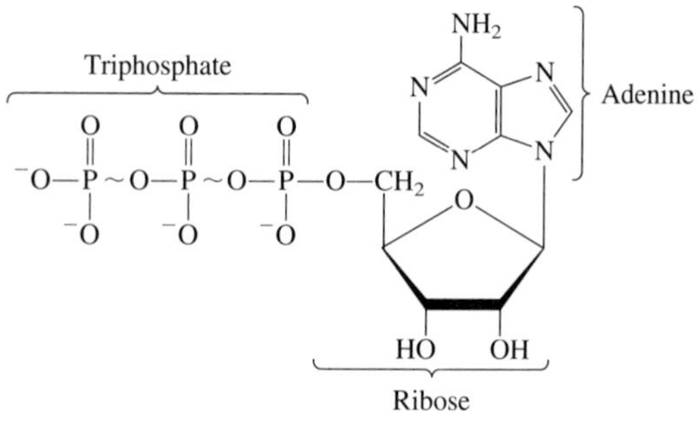

Figure 24.2

The chemical structure of adenosine triphosphate (ATP).

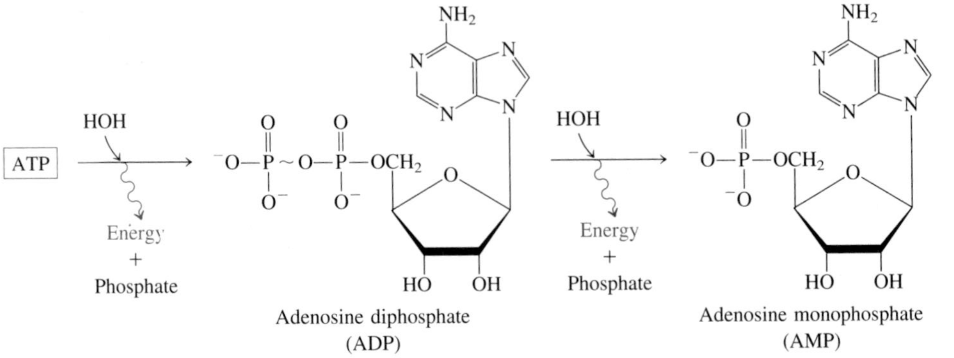

Figure 24.3
The relationships of ATP, ADP, and AMP.

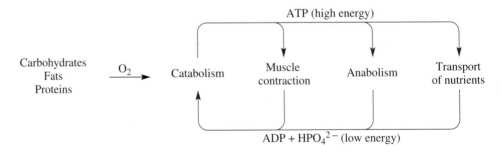

Adenosine diphosphate
(ADP)

Adenosine monophosphate
(AMP)

P_i is the symbol for the inorganic phosphate anions $H_2PO_4^-$ and HPO_4^{2-} that are present in the intra- and extracellular fluids. About 1.2 g of phosphate is needed in the daily diet to replace the amount excreted in the urine.

hydrolyzed. In this case, one of the driving forces for the reaction is to relieve the electron–electron repulsions associated with the negatively charged phosphate groups. It should be noted, however, that there is nothing special about the bonds themselves. The symbol $\sim$ is just a device to focus attention on a portion of the molecule that undergoes reaction. Energy-rich compounds, then, are substances with particular structural features that yield high energies of hydrolysis, and for this reason they are able to supply energy for energy-requiring biochemical processes (Figure 24.3).[2]

$$\text{ATP} \xrightleftharpoons{H_2O} \text{ADP} + P_i + 7300 \text{ cal/mol}$$

The important feature of this biochemical reaction is its reversibility. The hydrolysis of ATP releases energy; its synthesis requires energy. In a typical cell, an ATP molecule is consumed ("turned over") within 1 min after its formation. Thus ATP is produced by those processes that supply energy to an organism (absorption of radiant energy of the sun in green plants and breakdown of foodstuffs in animals), and ATP is hydrolyzed by those processes that require energy (syntheses of carbohydrates, lipids, proteins; transmission of nerve impulses; muscle contraction; etc.). This coupling of the synthesis of ATP to processes that release energy, and of the hydrolysis of ATP to processes that require energy, is one of the striking characteristics of living matter (Figure 24.4).

Because ATP is the principal medium of energy exchange in biological systems, it is referred to as the energy currency of the cell. However, it is not the only high-energy

[2]The values in the literature for the energy released when ATP is hydrolyzed to ADP vary somewhat. This situation is due in part to the fact that reaction conditions (concentration, temperature, pH) have not always been the same in different laboratories and in part to the difficulties in obtaining exact values for the equilibrium constants. Most texts and research articles report values between -7000 and -8000 cal/mol for the free energy of hydrolysis of ATP at 25 °C and a pH of 7.0. We shall use a value of -7300 cal/mol throughout this text. However, this is only an approximate value because the concentrations of cellular reactants are not molar, as required for energy calculations.

Figure 24.4
ATP is called the energy currency of the cell. Energy in ATP, from catabolic reactions, can be used for mechanical work such as muscle contraction, for chemical synthesis (anabolism), and to transport nutrients.

ATP (high energy)

Carbohydrates
Fats
Proteins

$\xrightarrow{O_2}$ Catabolism   Muscle contraction   Anabolism   Transport of nutrients

$\text{ADP} + HPO_4^{2-}$ (low energy)

Type	Example	$\Delta G°$ (cal/mol)
Acyl phosphate	1,3-Diphosphoglyceric acid	−11,800
	Acetyl phosphate	−10,300
Guanidine phosphate	Creatine phosphate	−10,300
	Arginine phosphate	−7,700
Pyrophosphate	ATP ⟶ AMP + PP$_i$	−7,700
	ATP ⟶ ADP + P$_i$	−7,300
	ADP ⟶ AMP + P$_i$	−7,300
	PP$_i$ ⟶ 2 P$_i$	−6,500
Sugar phosphate	Glucose 1-phosphate	−5,000
	Fructose 6-phosphate	−3,800
	AMP ⟶ Adenosine + P$_i$	−3,400
	Glucose 6-phosphate	−3,300
	Glycerol 3-phosphate	−2,200

compound. There are several other phosphate esters that provide energy for certain energy-requiring reactions. Table 24.1 lists a number of phosphate compounds. The use of these compounds will be illustrated in subsequent pages. Notice that the free energy of hydrolysis of ATP is approximately midway between those of the high-energy and the low-energy phosphate compounds.

24.2
DIGESTION AND ABSORPTION
OF CARBOHYDRATES

Digestion can be defined as a hydrolytic process whereby food molecules are broken down into simpler chemical units that can be absorbed by the body. In humans, digestion takes place in the digestive tract (Figure 24.5), and absorption occurs primarily in the small intestine. The digestive tract is a tunnel that runs *through* the body. Food in the digestive tract is in the tunnel, not in the body. The alternative name for the digestive tract, the alimentary canal, perhaps conveys this image more clearly.

Figure 24.5

The human digestive tract.

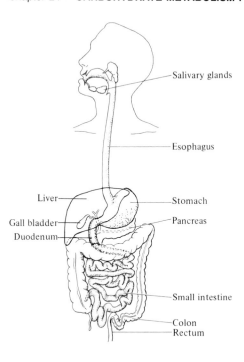

- Salivary glands
- Esophagus
- Liver
- Stomach
- Gall bladder
- Pancreas
- Duodenum
- Small intestine
- Colon
- Rectum

The average American consumes approximately 325 g of carbohydrates each day—about 160 g of starch, 120 g of sucrose, 30 g of lactose, 10 g of glucose, 5 g of fructose, and traces of maltose. A typical diet supplies about 50% of total body energy in the form of carbohydrate.

Food is sluiced through the canal by a flow of digestive juices and by physical pushes imparted by sections of the canal. Compounds that are changed during this journey into suitable forms are absorbed through the walls of the canal into the circulatory systems of the body. Those materials that can't be absorbed make their way through the entire length of the canal and out again. Without the process of digestion, very little of the food we eat would nourish us.

Starch is the principal carbohydrate ingested by humans. Its digestion begins in the mouth, where the food encounters **saliva,** a digestive fluid secreted by the salivary glands. Saliva is over 99% water. It contains several inorganic ions and a variety of organic molecules typical of other body fluids. In the mouth, food is chewed, that is, torn or crushed to a finer consistency. Reduction of food to smaller particles facilitates digestion by providing greater surface area for enzymes to work on. Chewing also coats the food particles with mucin, a glycoprotein constituent of saliva. The mucin lubricates the food and makes it easier to swallow.

The secretion of saliva can be triggered by the sight, taste, smell, or even the thought of food. An average person produces about 1.5 L of saliva a day. Excessive flow may be caused by certain pathological conditions, such as mercury poisoning.

The principal digestive enzyme in the mouth is **α-amylase,** sometimes called ptyalin. This enzyme attacks the α-glycosidic linkages in starch more or less at random. The pH of saliva is about 6.8, the optimum pH for α-amylase. Cleavage of the glycosidic linkages produces a mixture of dextrins, maltose, and glucose (Figure 24.6). Digestion of starch in the mouth is probably not necessary, except perhaps as a means of removing particles of starchy food lodged between the teeth.

α-Amylase continues to function as food passes through the esophagus, but it is quickly inactivated when it comes into contact with the acidic environment of the stomach. Very little carbohydrate digestion occurs in the stomach—acid-catalyzed hydrolysis proceeds too slowly at body temperature to be effective. The primary site of carbohydrate digestion is the small intestine, where another amylase, *amylopsin* (secreted from the

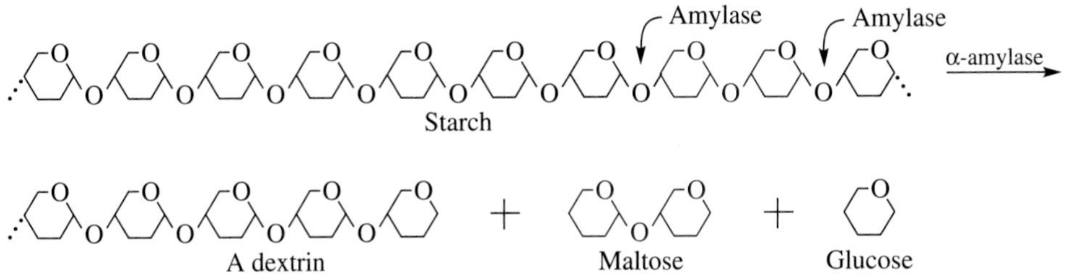

Figure 24.6

A schematic representation of the hydrolysis of starch to dextrins, maltose, and glucose. See Chapter 19 for details of structure.

pancreas), converts the remaining starch molecules, along with the dextrins, to maltose. Maltose is then cleaved into two glucose molecules by the enzyme maltase. Disaccharides such as sucrose and lactose are not digested until they reach the small intestine, where they are acted upon by sucrase and lactase. The enzymes that catalyze the hydrolysis of disaccharides are termed *disaccharidases* and are located on the membranes of cells that line the inner surface of the small intestine. Ultimately, the complete hydrolysis of disaccharides and polysaccharides produces three monosaccharide units—glucose, fructose, and galactose. These monosaccharides are then absorbed through the wall of the small intestine into the bloodstream.

Absorption of most digested food takes place in the small intestine through the fingerlike projections, called *villi,* that line the inner surface (Figure 24.7). Each villus is richly supplied with a fine network of blood vessels and a central lymph vessel (Figure 24.8). Monosaccharides are absorbed through the semipermeable membranous wall of each villus into the blood capillaries. However, the absorption does not occur by means of a simple process of osmosis or diffusion through an inert membrane (i.e., *passive transport*). All cell membranes are selective in their action, a fact that implies that they play an active role in the absorption process. The passage of monosaccharides across the intestinal wall is an energy-requiring process, and so the term *active transport* is used to describe this type of absorption.

Any polysaccharides or disaccharides that escape hydrolysis by intestinal enzymes cannot be absorbed. Intestinal bacteria metabolize these carbohydrates into lactose, short-chain carboxylic acids, and gases (CO_2, CH_4, and H_2), causing fluid secretion, increased intestinal mobility, and cramps.

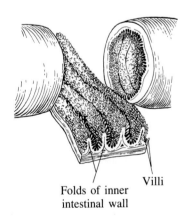

Figure 24.7

A section of the small intestine opened to reveal the folds of the inner wall and the lining covered with villi.

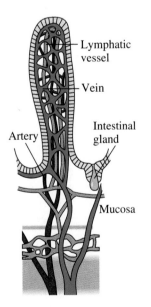

Figure 24.8

Diagram of one of the intestinal villi. Each villus contains arteries, veins, and lymphatic vessels.

Following absorption, the monosaccharides are carried by the portal vein to the liver. Here galactose and fructose are enzymatically converted to glucose or into phosphorylated intermediates to be metabolized by the liver. Some of the glucose passes into the general circulatory system to be transported to other tissues. The glucose in the tissues may be oxidized to CO_2 and H_2O, or it may be converted to muscle glycogen, thus serving as a source of readily available energy for the muscle. The majority of the absorbed glucose remains in the liver and is either stored as glycogen (as a reservoir for the maintenance of the normal level of blood glucose) or converted to fat and exported to adipose tissue as lipoprotein complexes (the VLDLs—see Section 20.11). The average person has a sufficient amount of stored glycogen (in liver and muscle cells) to supply energy for about 18 hr.

24.3
BLOOD GLUCOSE

The concentration of glucose in the blood is referred to as the **blood sugar level.** Under normal circumstances the blood sugar level remains remarkably constant at about 80 mg/100 mL of blood. However, since individuals differ in their chemical makeup, a normal concentration of glucose may range from 70 to 100 mg/100 mL. (This amounts to a total of about 5–6 g of glucose, or 1 teaspoonful, in the entire body.) Soon after eating a meal, the blood sugar level may rise to 120 mg/100 mL of blood, but it returns to the normal level within 2 hr.

As we shall see in Section 24.4, regulation of the blood sugar level is vital to the well-being of an individual. When the control mechanisms function improperly, there may be dire consequences. The condition of high blood sugar is called **hyperglycemia.** After severe starvation or vigorous exercise, the blood sugar concentration may fall below normal, leading to the condition called **hypoglycemia.** Neither condition, if temporary, is necessarily pathological, because the body has several methods of regulating the level of glucose in the blood.

The kidneys may excrete excess glucose. The **renal threshold** value is fairly high, however, ranging from 150 to 170 mg of glucose per 100 mL of blood. In general, the kidneys are designed to conserve the glucose in the blood. Only when the blood glucose level goes well above ''normal'' will the kidneys shunt some of the glucose into the urine. The liver also helps to regulate the blood glucose level by converting excess sugar to glycogen. Excess glucose may also be converted to fats for storage. And, of course, the glucose may be oxidized in the cells, producing energy.

Extreme hypoglycemia[3] can cause unconsciousness and lowered blood pressure and may result in death. Loss of consciousness is most likely due to the lack of glucose in the brain tissue, which has no capacity for glycogen storage and thus is dependent upon a continuous supply of glucose for its energy requirements. The brain uses about 125 g/day of glucose. At rest the total glucose requirement of all other tissues of the body (heart, liver, kidney, muscle, etc.) is about 200 g/day. The total daily glucose requirements are normally met from the dietary intake of carbohydrates.

The major cause of hyperglycemia is diabetes mellitus (see Section 24.4). Over 10

If the concentration of a substance in the blood exceeds the renal threshold, the substance will appear in the urine (see Section 28.12).

---

[3] There is a great deal of controversy regarding hypoglycemia. There are some clear-cut cases in which easily recognized symptoms are evident. These include general weakness, trembling, and rapid heartbeat. Severe cases can lead to delirium, coma, and even death. The treatment for hypoglycemia is simple. Give the patient some sugar. If the person is unconscious, glucose solution can be administered intravenously. A common cause of hypoglycemia is an overdose of insulin. Diabetics generally carry candy to counteract excess insulin.

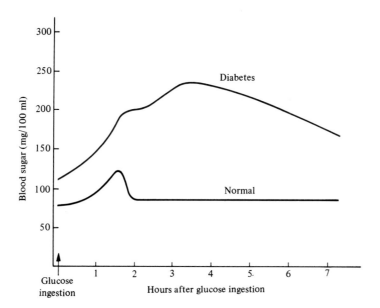

Figure 24.9
Glucose tolerance test for a normal patient and for a diabetic person.

million Americans have diabetes (either mildly or severely), and this disease alone is the third leading cause of death (either outright or through side effects) in the United States. Diabetes is characterized by an abnormal metabolism of carbohydrates as well as of proteins and lipids (see Section 26.6). Because a diabetic is unable to use glucose properly, excessive quantities accumulate in the blood and urine. Characteristic symptoms of diabetes are constant hunger, weight loss, extreme thirst, and frequent urination because the kidneys excrete large amounts of water in an attempt to remove excess sugar from the blood. High levels of glucose in the blood can cause various types of body damage. Among the many complications of diabetes are blindness, cardiovascular disease, gangrene, and kidney disease.

The **glucose tolerance test** is the most important diagnostic test for diabetes mellitus. A patient's blood sugar level is determined after an overnight fast. Then a known amount (~75 g) of glucose is dissolved in about 400 mL of water and administered orally over a period of 5–10 min. Blood is drawn from the patient's fingertip at 30-min intervals after ingestion, and the blood sugar concentration is determined. After an initial rise, the blood sugar level falls rapidly in a normal individual. In a diabetic person, on the other hand, the increase in the blood sugar level is greater than normal, and the level will remain elevated for several hours (Figure 24.9).

24.4
HORMONAL REGULATION OF BLOOD SUGAR LEVEL

The most important contributing process in the maintenance of a constant blood glucose concentration is the synthesis and breakdown of glycogen in the liver. The liver is responsible for removing glucose from the blood when the concentration is too high (after a meal) and releasing it when the blood sugar level is too low (e.g., between meals). The activity of the liver, in this regard, is controlled by several hormones; among these are insulin, epinephrine (adrenaline), and glucagon.

Insulin, which is produced by the beta cells of the islets of Langerhans in the pancreas, *is the most important regulator of metabolism in the body.* It is released from the beta cells in response to a high blood sugar level. In general, insulin promotes anabolic reactions and inhibits catabolic reactions in the liver, muscle, and adipose tissue. (It has no effect on carbohydrate metabolism in the brain or kidneys.) Specifically, insulin performs the following functions.

1. It enhances **glycogenesis** (the formation of glycogen from glucose) in both the liver and muscle.
2. It promotes the entry of glucose into muscle, liver, and adipose tissue. (Note, however, that red blood cells and cells in the brain, kidney, and intestinal tract do not require insulin for glucose uptake.)
3. It accelerates the conversion of glucose into fatty acids and hence the synthesis and storage of triglycerides in adipose tissue (see Section 26.2).
4. It inhibits the breakdown of glycogen and stored fat.
5. It promotes the transport of amino acids into cells and stimulates protein synthesis. It inhibits the intracellular degradation of proteins.
6. It suppresses **gluconeogenesis,** the synthesis of glucose from noncarbohydrate precursors (e.g., amino acids, glycerol, lactic acid). Gluconeogenesis occurs primarily in the liver and is particularly important in maintaining a constant blood sugar level during periods of starvation or strenuous exercise to supply glucose for brain tissue.

Hence, the principal role of insulin is to remove glucose rapidly from the blood, thus lowering the blood sugar level. The mechanism by which insulin accomplishes its prodigious tasks is largely unknown. Insulin does bind to specific receptor proteins in the membranes of target cells. Just how the hormone–receptor interaction is coupled to the physiologic responses is being investigated in many laboratories throughout the world.

If the pancreas does not secrete enough insulin and/or if there are insufficient (or defective) insulin receptors on the cell membranes, **diabetes mellitus** develops. There are two types of this disease. Insulin-dependent diabetics do not produce sufficient amounts of insulin to regulate their blood sugar level. This type of diabetes develops early in life and is termed *Type I diabetes.* It is rapidly reversed by the administration of insulin (usually injected subcutaneously), and Type I diabetics can lead active lives provided they receive insulin as needed. Since insulin is a protein, it cannot be taken orally because it would be digested. Therefore, the Type I diabetic must be treated with daily injections of insulin. Limited success has been achieved by implanting insulin-producing cells from cadavers into adults with Type I diabetes. In some patients, insulin production lasted from 3 to 12 weeks and significantly decreased the patient's needs for injected insulin.

A current theory is that in Type I diabetics the islet cells of the pancreas are destroyed by the body's own immune system. These cells may have been altered by a viral infection (and thus appear as "foreign" to the immune system). Researchers have developed a simple blood test capable of predicting who will develop Type I diabetes several years before the disease becomes apparent. The blood test searches for antibodies that destroy the body's insulin-producing cells. The antibodies require several years to destroy enough islet cells to cause diabetes.

Type II (non-insulin-dependent) *diabetes* is by far the more common (about 90% of diabetic cases—about 10 million Americans), and it occurs late in life. Type II diabetics produce sufficient amounts of insulin, but the beta cells are not secreting enough of it, or it is not utilized properly (because there is a lack of insulin-receptor proteins on the target cells, or the insulin-receptor proteins are defective). For these people the disease can usually be controlled with a combination of diet and exercise alone, without insulin

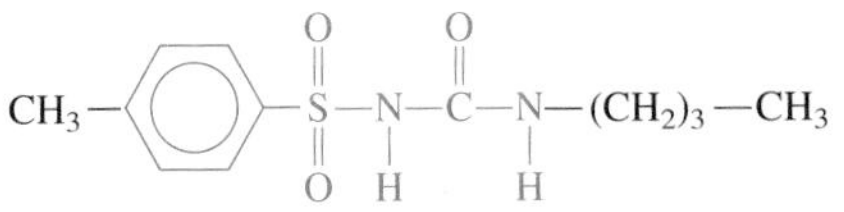

Tolbutamide
(Orinase)

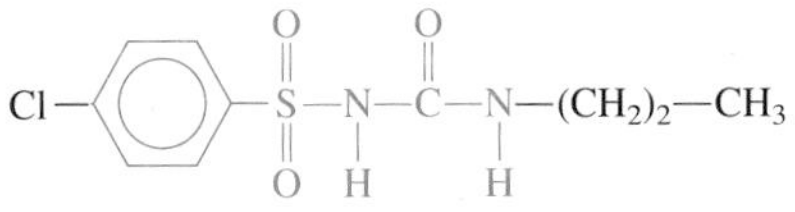

Chlorpropamide
(Diabinese)

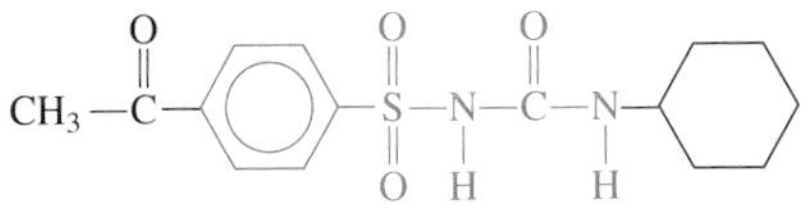

Acetohexamide
(Dymelor)

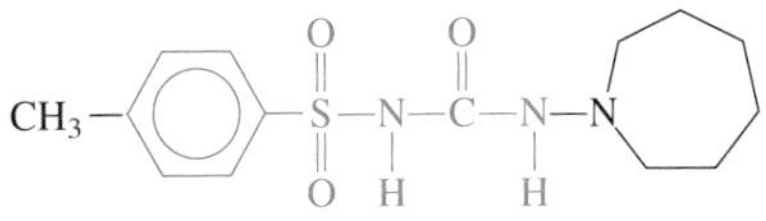

Tolazamide
(Tolinase)

Figure 24.10
The antidiabetic drugs are structurally similar to the sulfa drugs (see Figure 22.18). The replacement of the *p*-amino group on the benzene ring with other substituents accounts for the loss of antibacterial properties.

injections. Alternatively, there are oral antidiabetic drugs that stimulate the islet cells to secrete insulin (Figure 24.10). These compounds are only effective for persons who manufacture their own insulin but who fail to release the insulin in response to increased blood sugar level. The drugs have been shown to increase the risk of heart disease. Their use has become rather controversial. Second-generation drugs, such as glyburide, are now available. They stimulate the release of insulin, as do the first-generation drugs, but also increase the sensitivity of the cell receptors to the insulin. Like the earlier drugs, they must carry warnings about the increased risk of cardiovascular disease.

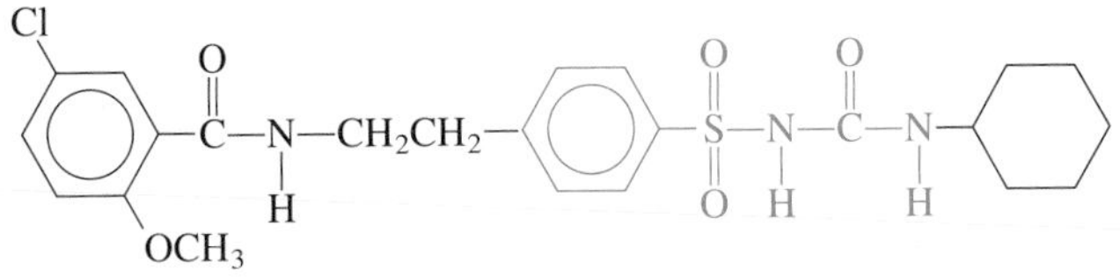

Glyburide

All other hormones that affect glucose metabolism act to raise the blood sugar level. Both epinephrine and glucagon exert their effects by binding to a receptor protein on the outside of the cell membrane (epinephrine and glucagon each bind to a different specific receptor). These receptor proteins are linked to the enzyme *adenyl cyclase,* which is bound to the inner membrane. When the outer receptor protein is unbound, the enzyme is not active. In a manner similar to allosteric activation (Section 22.7), the binding of epinephrine or glucagon to their receptors causes conformational changes in other membrane proteins such that adenyl cyclase becomes active. The function of the enzyme is to catalyze the conversion of ATP to adenosine 3′,5′-monophosphate (cyclic AMP or cAMP). Adenyl cyclase is extremely efficient, and many cAMP molecules can be synthesized by a single activated enzyme.

$$\text{ATP} \xrightarrow[\text{adenyl cyclase}]{Mg^{2+}} \text{cAMP} + \text{PP}_i$$

Cyclic AMP is often referred to as a *second messenger* because it transmits messages (delivered via the blood by the extracellular hormones—the primary messengers) from the cell membrane to enzymes within the cell (Figure 24.11). Cyclic AMP performs its function by binding to and activating certain inactive enzyme precursors, eventually resulting in a cascade of cellular events that results in the stimulation of a wide range of catabolic processes and the inhibition of several anabolic reactions. (See an advanced biochemistry text for a discussion of the cAMP cascade.) It should be noted that cAMP

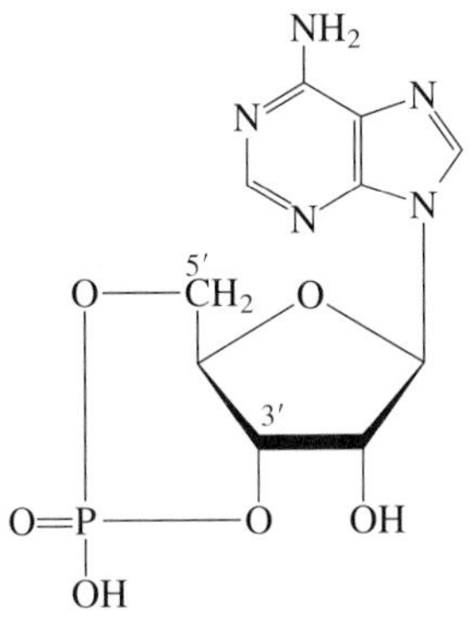

Adenosine 3′,5′-monophosphate
(cAMP)

Figure 24.11
Binding of an extracellular hormone to a receptor protein activates adenyl cyclase, which catalyzes the synthesis of cAMP.

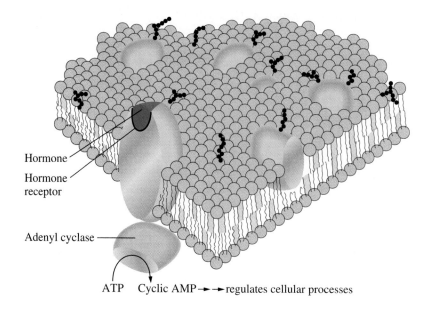

Hormone

Hormone receptor

Adenyl cyclase

ATP   Cyclic AMP ──►──► regulates cellular processes

serves as the second messenger for some neurotransmitters and several other hormones in addition to epinephrine and glucagon (Table 24.2).

Epinephrine (Section F.2) is secreted by the adrenal medulla in response to a low blood sugar level. It is also released during exercise or during periods of emotional stress, such as anger or fright, to provide the organism with additional energy. Epinephrine binds to its receptors (called *adrenergic receptors* after its other name, adrenaline), primarily on the membranes of muscle cells and to a lesser extent on the membranes of liver cells. It markedly stimulates **glycogenolysis,** the breakdown of glycogen to form glucose. Epinephrine also promotes gluconeogenesis in the liver. Both of these processes increase the amount of glucose in the blood, making that additional glucose available to the body tissues. Epinephrine increases the heart rate and raises the blood pressure.

Glucagon, like insulin, is a polypeptide (29 amino acids) hormone that is produced by the pancreas (the alpha cells of the islets of Langerhans), but glucagon is an antagonist of insulin. Like epinephrine, it is secreted in response to a low blood sugar level, and its

Table 24.2
Hormones That Increase cAMP Levels

Tissue	Hormone	Principal Response
Bone	Parathyroid hormone	Calcium resorption
Muscle	Epinephrine	Glycogenolysis
Adipose	Epinephrine	Lipolysis
	Adrenocorticotrophic hormone	Lipolysis
	Glucagon	Lipolysis
Brain	Norepinephrine	Discharge of Purkinje cells
Thyroid	Thyroid-stimulating hormone	Thyroxine secretion
Heart	Epinephrine	Increased contractility
Liver	Epinephrine	Glycogenolysis
Kidney	Parathyroid hormone	Phosphate excretion
	Vasopressin	Water reabsorption
Adrenal	Adrenocorticotrophic hormone	Hydrocortisone secretion
Ovary	Luteinizing hormone	Progesterone secretion

task is to increase the glucose concentration. Glucagon acts primarily on the liver (and adipose tissue) and not on skeletal muscle because there are no receptors for glucagon on muscle cell membranes. Glucagon stimulates gluconeogenesis and glycogenolysis in the liver, to restore the blood glucose to its normal level. The concentration of glucagon in the blood of a diabetic is above normal, and this may be a significant contributor to the problems associated with diabetes.

Glycogen metabolism is strongly influenced by the ratio of insulin to glucagon in the blood. Higher amounts of insulin lead to glycogen storage after a meal, whereas higher amounts of glucagon favor the breakdown of liver glycogen to add more glucose to the blood. It is the insulin/glucagon ratio that determines the outcome of carbohydrate metabolism. A high ratio leads to carbohydrate anabolism and storage, whereas a low ratio results in carbohydrate catabolism and utilization. The secretion of these hormones is directly governed by the blood sugar level.

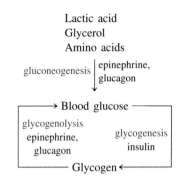

24.5
EMBDEN–MEYERHOF PATHWAY

To fully appreciate the role of carbohydrates in living systems, it is necessary to understand the biochemical nature of their metabolic pathways. A **metabolic pathway** is represented by a type of flow diagram that enables us to explain how an organism converts a certain reactant into a desired end product. In the 1930s, the German biochemists G. Embden and O. Meyerhof elucidated the sequence of reactions by which glycogen and glucose are degraded in the absence of oxygen (**anaerobic** conditions) to pyruvic acid. It was discovered that the two apparently dissimilar processes of alcoholic fermentation in yeast and muscle contraction in animals proceed by the same pathway as far as pyruvic acid. In the absence of oxygen, muscle cells (and certain other animal tissues) convert pyruvic acid into lactic acid, but under similar conditions the enzymes in yeast convert the pyruvic acid to ethyl alcohol and carbon dioxide. The former process is known as **glycolysis** (glucose splitting), and the latter is **fermentation.**

$$C_6H_{12}O_6 \longrightarrow \longrightarrow \longrightarrow 2\ CH_3-\overset{O}{\underset{}{C}}-C\overset{O}{\underset{OH}{}}$$

Glucose   Pyruvic acid

in yeast / fermentation → $2\ C_2H_5OH\ +\ 2\ CO_2\ +\ $ Energy

Ethyl alcohol   Carbon dioxide

without oxygen

in muscle / glycolysis → $2\ CH_3CHOHCOOH\ +\ $ Energy

Lactic acid

It is important to notice that the conversion of glucose to pyruvic acid represents an oxidation reaction (that is, $C_6H_{12}O_6 \rightarrow 2\ CH_3COCOOH + 4\ H^+ + 4\ e^-$), and yet oxygen is not required for the process to occur. We shall see that the outstanding characteristic of the Embden–Meyerhof pathway is the utilization of the coenzyme NAD^+ as the electron acceptor (oxidizing agent). Thus glycolysis and fermentation differ only in the eventual fate of the pyruvic acid, and hence in the means employed for the regeneration of the NAD^+.

A summary of the reactions and the metabolites involved in the Embden–Meyerhof pathway is given in Figure 24.12. This sequence of reactions is probably the best understood of all the metabolic pathways. Over a dozen different, specific enzyme molecules act in such a manner that the product of the first enzyme-catalyzed reaction becomes the

Figure 24.12
Embden–Meyerhof pathway.

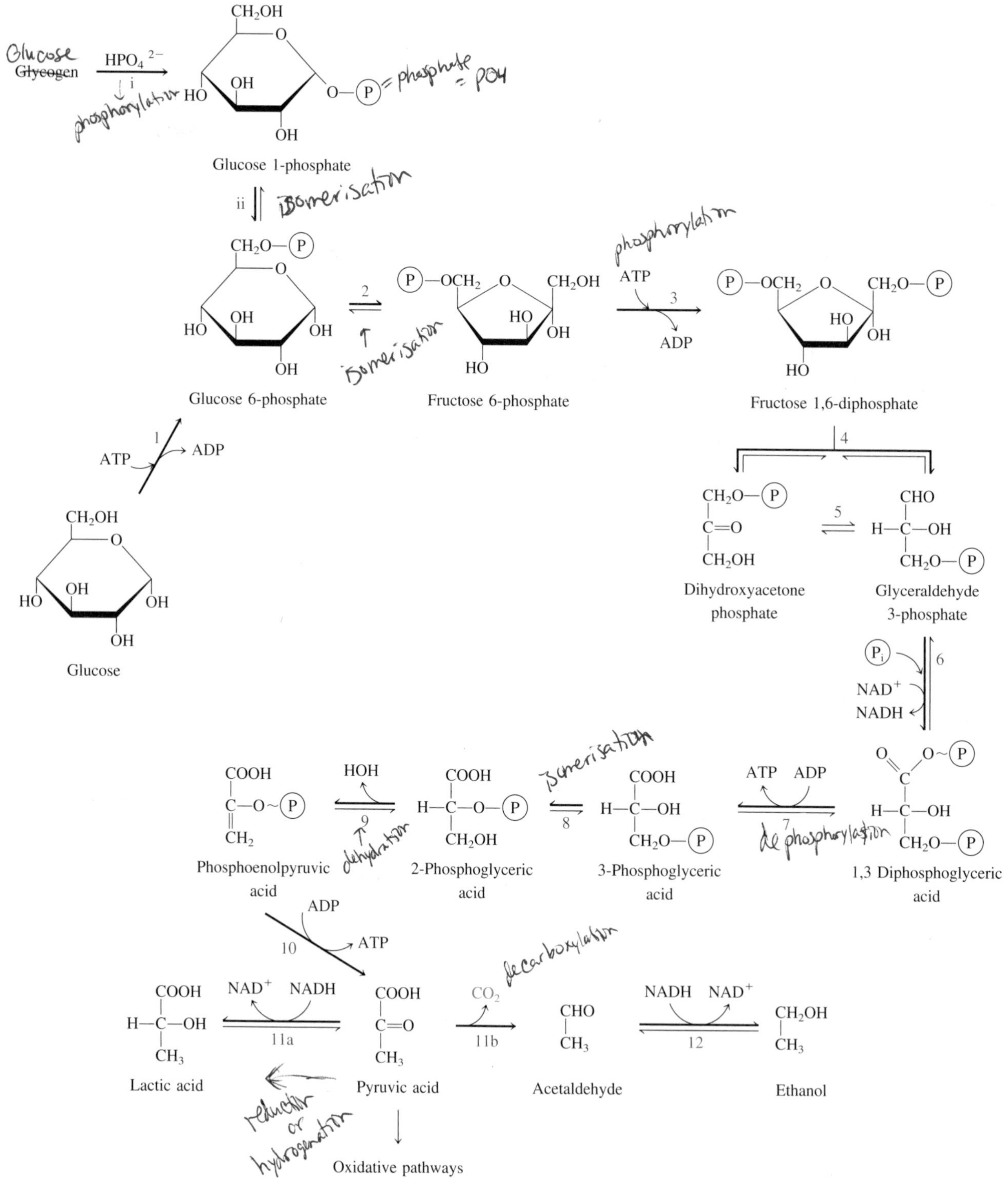
Glycogen
HPO4 2−
i
Glucose 1-phosphate
ii
Glucose 6-phosphate
2
Fructose 6-phosphate
ATP
3
ADP
Fructose 1,6-diphosphate
1
ATP
ADP
Glucose
4
5
Dihydroxyacetone phosphate
Glyceraldehyde 3-phosphate
6
NAD+
NADH
HOH
9
8
ATP ADP
7
Phosphoenolpyruvic acid
2-Phosphoglyceric acid
3-Phosphoglyceric acid
1,3 Diphosphoglyceric acid
ADP
10
ATP
NAD+ NADH
11a
CO2
11b
NADH NAD+
12
Lactic acid
Pyruvic acid
Acetaldehyde
Ethanol
Oxidative pathways

substrate of the next. All the reactions of the pathway occur in the cytoplasm, where the enzymes involved are in solution. The transfer of intermediates from one enzyme to the next occurs by diffusion. Since these enzymes are found in a soluble form in the cell, they were relatively easy to isolate and characterize. Each of the reactions given in Figure 24.12 has been assigned a number that corresponds to the discussion that follows. There is often the tendency to become lost in the complexity of this process and to lose the overall perspective. We are interested in seeing how the principles of organic chemistry apply to these biochemical reactions, but it must always be kept in mind that the central theme of metabolism is the extraction of chemical energy from foodstuff and not merely the degradation of these molecules into simpler substances. In the discussion that follows, the full names of the pertinent enzymes are given in the body of the text. Only the general names are written above or below the arrows of the biochemical equations.

Step i The liver contains the only store of glycogen that can be converted to glucose and released into the blood for transport to all other tissues. Muscle glycogen, on the other hand, is used solely within the muscle. During stress and exercise, skeletal muscle cells use glycogen as the primary source of energy production.

Liver and muscle cells contain enzymes called *phosphorylases* that catalyze the phosphorolytic cleavage of the α-1,4-glycosidic bonds at the nonreducing end of the glycogen chain to produce glucose 1-phosphate. The reaction is analogous to a hydrolytic cleavage, except that here inorganic phosphate takes the place of a water molecule. (Two other enzymes, a *transferase* and a *debranching enzyme*, are required to hydrolytically split off glucose molecules at the branch points of glycogen.) Although this reaction is reversible, different enzymes and different reaction sequences are employed for the biosynthesis of glycogen.

McArdle's disease is a genetic disease in which there is a deficiency in, or an absence of, phosphorylase in muscle cells. Thus, stored glycogen cannot be used to provide energy. The patients have a limited capacity for exercise, and they suffer from painful muscle cramps.

Glycogen

Glucose 1-phosphate

Step ii Glucose 1-phosphate and glucose 6-phosphate are readily interconvertible in the presence of Mg^{2+} ions and the enzyme *phosphoglucomutase*. (In general, a *mutase* is an enzyme that catalyzes the intramolecular transfer of a chemical group.) For simplicity, the isomerization reaction may be thought of as an intramolecular transfer of the phosphate group from C-1 to C-6 and, although this process is reversible, glucose 6-phosphate formation is favored.

Glucose 1-phosphate   Glucose 6-phosphate

The symbol (P) is a shorthand notation for the phosphite group, PO_3^{2-}.

Step 1   The initial step, upon entry of glucose into yeast cells and most animal cells, is phosphorylation to glucose 6-phosphate. (With the exception of liver cells, there is very little free glucose inside most cells.) The phosphate donor in this reaction is ATP, and the enzyme, which requires Mg^{2+} ions for its activity, is *hexokinase*.

Glucose   Glucose 6-phosphate

The reaction is accompanied by an expenditure of energy because a molecule of ATP is being used rather than synthesized. However, this step is necessary for the activation of the glucose molecule. (We shall see that a second molecule of ATP is expended in step 3. These two initiating molecules of ATP are recovered at a later stage of the pathway.) Reaction 1 is essentially irreversible in the cell. ATP is not formed to an appreciable extent by the reaction of ADP with a simple phosphate ester.

Step 2   Glucose 6-phosphate is isomerized to fructose 6-phosphate by the action of the enzyme *phosphoglucoisomerase*. (In general, an *isomerase* catalyzes the interconversion of one isomeric form of a sugar to another isomeric form.) Phosphoglucoisomerase is highly specific for glucose 6-phosphate.

Glucose 6-phosphate   Fructose 6-phosphate

A mechanism for this aldose–ketose transformation involves the formation of an enediol intermediate and is best understood if the open-chain structures of the sugars are considered.

The fate of glucose 6-phosphate is not the same in muscle cells as in the liver. Because the major role of glycogen in muscle is to provide energy for muscle contraction, the glucose 6-phosphate is immediately converted to fructose 6-phosphate. However, in the liver the chief function of glycogen is to serve as a reservoir of glucose molecules. The liver stores and releases glucose to meet the needs of other tissues. (The liver relies mainly on the oxidation of fatty acids for its own energy needs—see Chapter 26.)

When the blood sugar level decreases, glucose 6-phosphate in the liver is converted into glucose via a hydrolytic reaction catalyzed by glucose 6-phosphatase. (Muscle cells lack this enzyme and therefore cannot export glucose.)

$$\text{Glucose 6-phosphate} + \text{HOH} \xrightarrow[\text{(only in the liver)}]{\text{phosphatase}} \text{Glucose} + P_i$$

The glucose molecules then exit the liver and are transported by the blood for use by other tissues (primarily brain and skeletal muscle). Another genetic disease, von Gierke's disease, is caused by a lack of the enzyme glucose 6-phosphatase in liver cells. Hypoglycemia occurs because glucose cannot be formed from glucose 6-phosphate.

| Open-chain form of glucose 6-phosphate | Enediol intermediate | Open-chain form of fructose 6-phosphate |

Step 3 Next follows another phosphorylation reaction, again involving the utilization of ATP as the phosphate group donor. The enzyme *phosphofructokinase*[4] is specific for fructose 6-phosphate and, like hexokinase, requires Mg^{2+} ions for activity. The reaction is irreversible and necessitates the expenditure of energy from a second molecule of ATP.

Fructose 6-phosphate   Fructose 1,6-diphosphate

[4] Phosphofructokinase is an allosteric enzyme that is primarily responsible for controlling the rate of glycolysis. This enzyme is strictly regulated by the intracellular concentration of ATP. When there is a high concentration of ATP in the cell, ATP binds to the regulatory site, inhibiting the enzyme, and the pace of glycolysis decreases. When ATP is being used to provide energy, the regulatory site is vacant, the enzyme is active, and the glycolytic rate increases.

Step 4   Fructose 1,6-diphosphate is then enzymatically cleaved by the enzyme *aldolase* into two molecules of triose phosphate. Again, a better understanding of the reaction is achieved if the fructose 1,6-diphosphate is written in its open-chain form.

Fructose 1,6-diphosphate   Dihydroxyacetone   Glyceraldehyde
                                  phosphate   3-phosphate

Step 5   The next step is concerned with the interconversion of the triose phosphates. This is essential because only glyceraldehyde 3-phosphate can be further metabolized by the body. If cells were unable to convert dihydroxyacetone phosphate to glyceraldehyde 3-phosphate, then half of the energy stored in the original glucose molecule would be lost, and would accumulate in the cell as the nonmetabolizable ketotriose phosphate. The enzyme for the isomerization reaction is *triose phosphate isomerase,* and the mechanism involves another enediol intermediate.

Dihydroxyacetone   Enediol   Glyceraldehyde
phosphate   intermediate   3-phosphate

A summation of steps 4 and 5 indicates that the enzymes aldolase and triose phosphate isomerase have effectively accomplished the conversion of one molecule of fructose 1,6-diphosphate into *two* molecules of glyceraldehyde 3-phosphate. All the remaining steps of glycolysis involve three-carbon compounds. Thus far, the glycolytic pathway has required an energy input in the form of two molecules of ATP but has not yet released any of the energy stored in the glucose.

Step 6   We now come to the first oxidation–reduction reaction in the glycolytic sequence. The enzyme *glyceraldehyde 3-phosphate dehydrogenase* contains the coenzyme NAD^+. In the process of oxidizing the aldehyde to a carboxylic acid, the coenzyme is reduced to NADH. Notice that the same enzyme also catalyzes a phosphorylation reaction; inorganic phosphate is the phosphate donor. The energy released by the oxidation reaction is used for the subsequent phosphorylation reaction. The poisonous effect of iodoacetic acid (recall Table 22.2) in blocking glycolysis occurs at this step. The dehydrogenase enzyme contains free sulfhydryl groups at its active site. Iodoacetate noncompetitively inhibits the enzyme by covalently bonding to these catalytic groups.

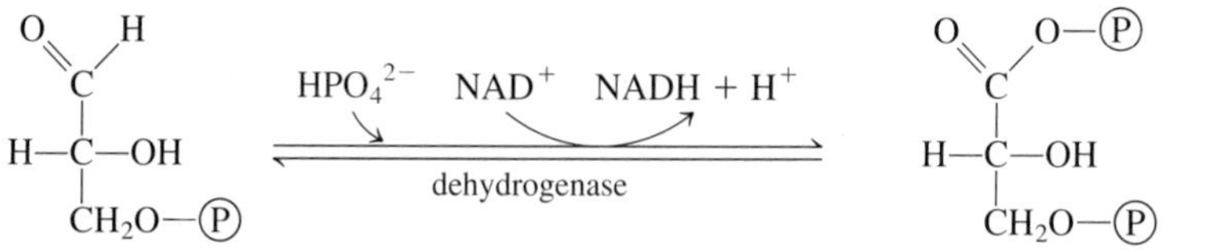

Glyceraldehyde
3-phosphate

1,3-Diphosphoglyceric acid

The reaction is more easily understood if it is considered to occur in two steps: (a) oxidation and (b) phosphorylation. Note, however, that this presentation is an oversimplification that in no way implies the actual enzyme-catalyzed sequence of events.

(a) Oxidation

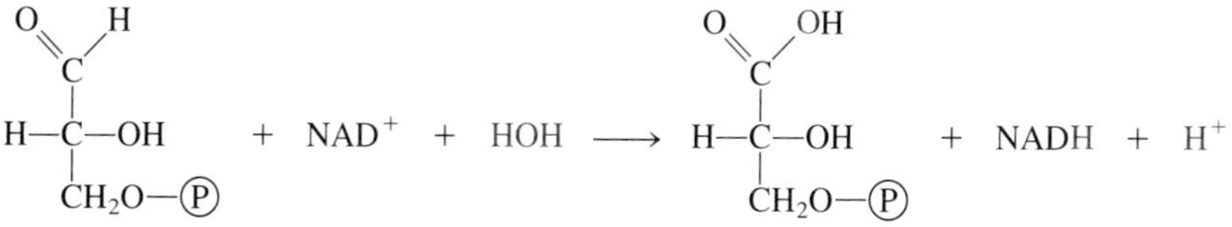

(b) Phosphorylation

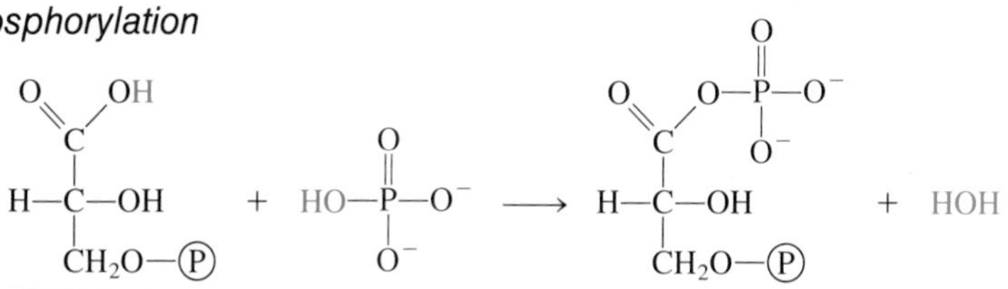

Step 7 1,3-Diphosphoglyceric acid, the product of the reaction in step 6, contains a high-energy acyl phosphate bond (recall Table 24.1). It can transfer the phosphate group on C-1 directly to a molecule of ADP, thus forming a molecule of ATP. The enzyme that catalyzes the reaction is *phosphoglycerokinase,* which, like all kinases, requires Mg^{2+} ions for activity. It is in this reaction that ATP is first produced in the pathway. Since the ATP is formed by a direct transfer of a phosphate group from a metabolite to ADP, the process is referred to as **substrate-level phosphorylation** to distinguish it from another ATP-synthesizing process we shall consider in Chapter 25.

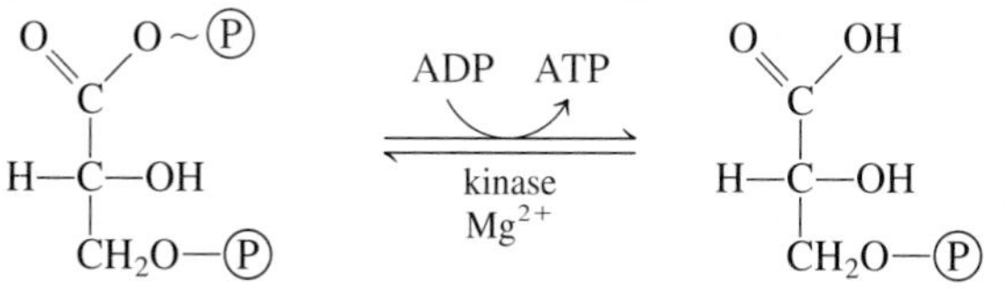

1,3-Diphosphoglyceric acid

3-Phosphoglyceric acid

Step 8 This step in the pathway is very similar to the intramolecular transfer of phosphate as seen in step ii. The enzyme *phosphoglyceromutase* catalyzes the exchange of a phosphate group from the hydroxyl group of C-3 to the hydroxyl group of C-2.

3-Phosphoglyceric acid

2-Phosphoglyceric acid

Step 9 The enzyme *enolase,* which also requires Mg^{2+} ions for activity, catalyzes an alcohol dehydration reaction to produce phosphoenolpyruvic acid (PEP), a compound with a high-energy enol phosphate group (recall Table 24.1). Fluoride is an effective poison of the glycolytic pathway because of its inhibition of enolase (Table 22.2). Fluoride ions bind to the activator Mg^{2+} ions to form a magnesium fluorophosphate complex.

2-Phosphoglyceric acid   Phosphoenolpyruvic acid
                                            (PEP)

Step 10 This irreversible step provides a second example of substrate-level phosphorylation. The phosphate group of PEP is transferred to ADP; one molecule of ATP is produced per molecule of PEP. It is likely that the reaction proceeds via the enol form of pyruvic acid, which is unstable with respect to rearrangement to pyruvic acid. The enzyme *pyruvate kinase* requires both Mg^{2+} and K^+ ions for activity.

PEP   Enol form   Pyruvic acid
                  of pyruvic acid

Steps 1 through 10 are identical for glycolysis and fermentation. Pyruvic acid, as we shall see, is the *crossroads* compound—its metabolic fate depends on the availability of oxygen, the organism under consideration, and the tissue involved.

24.6
GLYCOLYSIS

Under usual conditions and during moderate exercise, muscle cells respire **aerobically** (in the presence of oxygen—see Chapter 25). However, during strenuous exercise, the energy demand on the muscles is enormous, and the respiratory and circulatory systems are unable to deliver oxygen to these cells in sufficient amounts to meet this demand. (This condition, called **oxygen debt,** occurs when not enough oxygen is available for cellular activities.) As a result the muscle cells must obtain energy via the anaerobic pathway.

Step 11a In the presence of NADH the enzyme *lactic acid dehydrogenase*[5] catalyzes the reduction of pyruvic acid to lactic acid (ketone to secondary alcohol).

---

[5]This enzyme derives its name from its catalytic role in reversing this reaction. Enzymes catalyze both the forward and reverse reactions of an equilibrium mixture and may be named accordingly. An equally descriptive name for this enzyme is *pyruvic acid reductase* or *hydrogenase.*

Step 12 T
dehydrogen
NAD$^+$. (W

Pyruvic acid   Lactic acid

This reaction is essential because it regenerates the NAD$^+$ that is needed in step 6. If NAD$^+$ were not replenished, the cell's supply of this coenzyme could be swiftly depleted, the Embden–Meyerhof pathway would cease, and glyceraldehyde 3-phosphate would accumulate in the cell.

Lactic acid, then, is the end product of glycolysis.[6] If there were not some mechanism for its removal, it would accumulate in the muscle cells, raising the level of acidity in these cells. The generation of energy via the anaerobic pathway is self-limiting. As lactic *acid* builds up in muscle tissue, the pH drops (from a resting value of 7.0 to about 6.5) and deactivates the enzymes required for glycolysis. The muscle's response to stimuli becomes weaker, and in extreme cases there may be no response at all. In this state the muscle is described as fatigued. Two processes act to maintain a proper level of lactic acid, and both require oxygen.

24.8
REV
AND

It has b
pathway
cate tha
with the fact
can be conve
Therefore, o
sal of steps
common phe
the metabolis
the same inte
cal. Similarly
degradative a
without affec
energy and p
reactions are

1. Most (70–80%) of the lactic acid diffuses out of the muscle and is transported to the liver. There it may be oxidized to pyruvic acid and hence to carbon dioxide and water (via the Krebs cycle—see Section 25.1) or converted back into glucose (gluconeogenesis). The anaerobic catabolism of glucose to lactic acid in muscle cells, the transport of the lactic acid via the blood to the liver, and the reconversion of lactic acid to glucose comprise the **Cori cycle.**

The Cori cycle is named in honor of Gerty and Carl Cori who first described it. The Coris won a Nobel prize in 1947, the third husband-wife team to achieve this distinction (Marie and Pierre Curie were the first; Irène and Frédéric Joliot-Curie, the second).

24.9
BIOE
GLYC

The net
obtained
lated fro
One mol

2. The 20–30% of the lactic acid that remains within the muscle cells can be reoxidized to pyruvic acid, which then enters the Krebs cycle and is further oxidized to carbon dioxide and water during those periods when the muscle cells receive an ample supply of oxygen.

We have seen that when muscles use anaerobic pathways, they incur an oxygen debt. It is as if the body regards oxidation as the only proper source of energy for muscular activity and uses anaerobic metabolism as a temporary expedient. As soon as it can, the body oxidizes some of the resulting lactic acid back to pyruvic acid and ultimately to carbon dioxide and water. The energy released in the process is used to convert the rest of the lactic acid back to glucose and hence to glycogen to be stored for future use.

When is this oxygen debt repaid? Just as soon as the very high level of muscular activity ceases. Sprinters, while running a 100-m dash, breathe in oxygen but still obtain only a fraction of the energy required for their intense muscular activity through aerobic

A second mo

[6]The average individual, expending normal energy, produces about 120 g of lactic acid each day. (Vigorous exercise produces significantly more lactic acid.) It is estimated that one-third of this daily lactic acid is produced by tissues that are strictly anaerobic (i.e., erythrocytes, retina, renal medulla, epidermis)—tissues that do not have mitochondria.

Figure 2

Immediat
the runne
quantities
the oxyge

About 3 g/day of ethyl alcohol is produced continuously by microbial fermentation in the large intestine. For many people this amount is insignificant compared with the amount ingested from alcoholic beverages. In humans the principal route of metabolism of ingested alcohol is believed to be oxidation in the liver, first to acetaldehyde and subsequently to acetic acid. Most of the acetic acid is released by the liver and transported to other tissues, where it is converted to acetyl CoA. (As we shall see, the acetyl CoA is metabolized to carbon dioxide and water, or it is used in the biosynthesis of fat.)

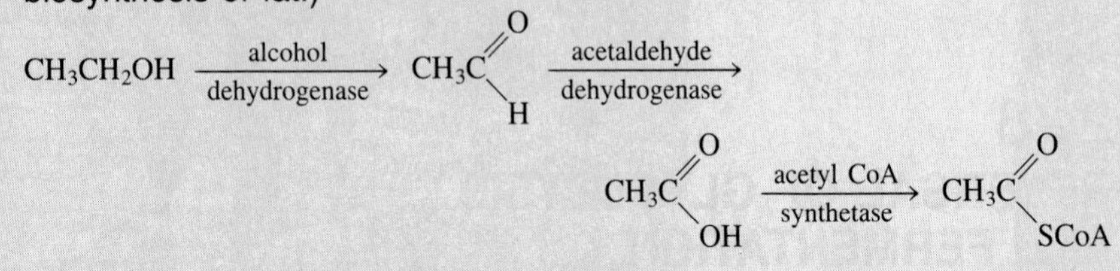

$$CH_3CH_2OH \xrightarrow[\text{dehydrogenase}]{\text{alcohol}} CH_3C\!\!\begin{array}{c}O\\\\H\end{array} \xrightarrow[\text{dehydrogenase}]{\text{acetaldehyde}}$$

$$CH_3C\!\!\begin{array}{c}O\\\\OH\end{array} \xrightarrow[\text{synthetase}]{\text{acetyl CoA}} CH_3C\!\!\begin{array}{c}O\\\\SCoA\end{array}$$

Alcohol readily crosses the placental membrane and builds up in the fetus because the liver of the fetus does not contain the detoxifying enzymes. Therefore, children of alcoholic mothers have a high risk of *fetal alcohol syndrome* (FAS). FAS consists of facial deformities, growth deficiency, and mental retardation. FAS is believed to be the third most common cause of mental deficiency. (IQs average 35–40 points below normal.) Some researchers say that FAS can occur even if a pregnant woman drinks only moderately (one drink daily). In the United States, more than 10,000 children are born each year suffering from FAS.

After heart disease and cancer, alcohol addiction is the third largest health problem in the United States. More than 200,000 people die of it each year (mainly as a result of cirrhosis of the liver, but also of cardiovascular disease and cancers of the mouth and larynx), and it is a factor in one of every 10 deaths. People who drive while intoxicated are responsible for about 50% of U.S. traffic fatalities. Alcohol-impaired driving is the leading cause of death and injury among those under 25 years of age.

One proven method in the treatment of chronic alcoholism is to administer a drug that inhibits the second step of alcohol metabolism. The drug disulfiram (Antabuse) successfully competes with acetaldehyde for the active site on the enzyme acetaldehyde dehydrogenase. As a result, when alcohol is ingested by an individual previously treated with disulfiram, the blood acetaldehyde concentrations rise five to ten times higher than in an untreated individual. This effect is accompanied by severe discomfort. Characteristic physiological responses are an increase of the heartbeat and a reduction of the blood pressure. The individual experiences respiratory difficulties, nausea, sweating, vomiting, chest pains, and blurred vision. Once elicited, the effect lasts between 30 min and several hours. To avoid such discomfort, the patients must abstain from alcohol (and the treatments are continued until they can abstain voluntarily). Disulfiram should be administered only by a physician, and therapy is usually begun in a hospital.

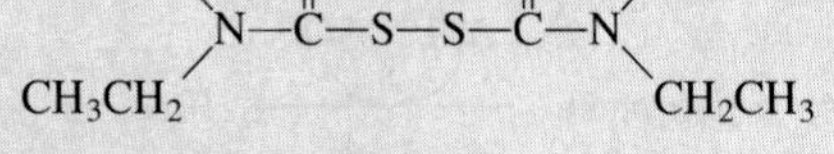

$$\begin{array}{c}CH_3CH_2\\\\CH_3CH_2\end{array}\!\!N-\!\!\overset{S}{\underset{}{C}}\!\!-S-S-\!\!\overset{S}{\underset{}{C}}\!\!-N\!\!\begin{array}{c}CH_2CH_3\\\\CH_2CH_3\end{array}$$

Disulfiram

Steps 4 and 5 are very significant, for it is in these reactions that each mole of six-carbon sugar is transformed into 2 mol of the triose phosphate glyceraldehyde 3-phosphate. In step 7, each mole of 1,3-diphosphoglyceric acid is converted into 1 mol of 3-phosphoglyceric acid; 1 mol of ATP is produced per mole of triose phosphate, or *2 mol of ATP per mole of glucose.*

$$\text{1,3-Diphosphoglyceric acid} \xrightarrow{\underset{\overset{\text{ADP}}{\diagdown}\overset{\text{ATP}}{\diagup}}{}} \text{3-Phosphoglyceric acid}$$

In step 10, 1 mol of ATP is generated for each mole of pyruvic acid that is formed from phosphoenolpyruvic acid; again 2 mol of ATP is produced for each mole of glucose that entered the pathway.

$$\text{Phosphoenolpyruvic acid} \xrightarrow{\underset{\overset{\text{ADP}}{\diagdown}\overset{\text{ATP}}{\diagup}}{}} \text{Pyruvic acid}$$

A summation of all these steps reveals that for every mole of glucose degraded, 2 mol of ATP is initially consumed and 4 mol of ATP is ultimately produced. The net production of ATP is thus 2 mol per mole of glucose converted to lactic acid or to ethanol. If, however, glycogen is used as the source of glucose, steps i and ii are operative instead of step 1. It would then be necessary to expend only 1 mol of ATP (in step 3) to produce fructose 1,6-diphosphate, and the net yield of ATP would be 3 mol per mole of glucose 1-phosphate. Thus, the utilization of glycogen for anaerobic energy production represents a 50% increase in efficiency compared to the utilization of glucose.

In Yeast

$$C_6H_{12}O_6 + 2\,ADP + 2\,P_i \longrightarrow 2\,C_2H_5OH + 2\,CO_2 + 2\,ATP$$

In Muscle—Starting with Glycogen

$$(C_6H_{12}O_6)_n + 3\,ADP + 3\,P_i \longrightarrow (C_6H_{12}O_6)_{n-1} + 2\,CH_3CHOHCOOH + 3\,ATP$$

Recall that about 7300 cal of free energy is conserved per mole of ATP produced (Section 24.1). Recall also that the total amount of energy that can theoretically be obtained from the complete oxidation of 1 mol of glucose is 686,000 cal. The energy conserved by anaerobic metabolism, then, is only a minute amount of the total energy available, that is, either 2.1% (14,600/686,000) or 3.2% (21,900/686,000). Thus, anaerobic cells extract only a small fraction of the total energy of the glucose molecule, yet this amount is sufficient for their survival. Since they are not nearly as efficient as aerobic cells, it is necessary that they use more glucose per unit of time to accomplish the same amount of cellular work.

24.10
STORAGE OF CHEMICAL ENERGY

We now consider how muscle cells store the energy that they extract from carbohydrates (and other compounds). In 1927, a compound called *creatine phosphate* was isolated from mammalian muscle. It was subsequently demonstrated to be the storage form of energy in the muscles of vertebrates and in nerve tissue. *Arginine phosphate* serves the same purpose in the muscles of invertebrates. These high-energy phosphate

$$\begin{array}{c} \overset{\displaystyle O}{\underset{\displaystyle \parallel}{}} \\ H{-}N \sim P{-}O^- \\ | \\ O^- \\ | \\ C{=}NH \\ | \\ N{-}CH_3 \\ | \\ CH_2COO^- \end{array}$$

Creatine phosphate
(Phosphocreatine)

$$\begin{array}{c} \overset{\displaystyle O}{\underset{\displaystyle \parallel}{}} \\ H{-}N \sim P{-}O^- \\ | \\ O^- \\ | \\ C{=}NH \\ | \\ N{-}H \\ | \\ (CH_2)_3 \\ | \\ H{-}C{-}NH_3{}^+ \\ | \\ COO^- \end{array}$$

Arginine phosphate
(Phosphoarginine)

compounds, which serve as reservoirs of phosphate bond energy, are often called *phosphagens*. The high-energy bond of these particular phosphagens occurs between phosphorus and nitrogen rather than between phosphorus and oxygen.

The turnover rate of ATP is very high, and this precludes its use as a storage form of energy. A 70-kg man will hydrolyze and resynthesize about 70 kg of ATP per day. Therefore, the concentration of ATP in muscle is relatively low and cannot meet the demands of muscular exertion for more than a fraction of a second. At rest, mammalian muscle contains four to six times as much creatine phosphate as ATP. As ATP is utilized, creatine phosphate, in the presence of creatine kinase, reacts with ADP to produce more ATP and creatine.

$$\text{Creatine phosphate} \;+\; \text{ADP} \underset{\text{Mg}^{2+}}{\overset{\text{kinase}}{\rightleftharpoons}} \text{Creatine} + \text{ATP}$$

The reaction is readily reversible. When muscular activity is required, the reaction proceeds to the right. When there is an abundant amount of ATP (formed from the catabolism of foodstuffs), the reaction proceeds to the left, and creatine phosphate is stored in the muscle cells. Since the concentration of creatine phosphate in the muscle is limited, it is only useful in generating a quick source of utilizable energy. It has been estimated that the creatine phosphate can provide energy for about 20 s of strenuous exercise (e.g., long enough to sprint 200 m). Recall we said earlier that energy for prolonged activity is obtained from the anaerobic breakdown of glycogen that is synthesized during long periods of muscular inactivity. The glycogen stored in muscle cells is more readily available for energy production than is glucose from the blood. In the days just preceding a marathon, competitors will "load up" on carbohydrates in order to maximize the amount of glycogen stored in muscle cells.[7]

The human body can store roughly 450 g of glucose as glycogen (about 350 g of glycogen in all of the muscle cells and about 100 g of glycogen in the liver). The liver glycogen reservoir allows us to eat intermittently because it provides the glucose to maintain the blood sugar level. Stored glycogen, however, does not last long. After a 24-hr fast, very little glycogen remains in the liver. We shall see in Chapter 26 that when carbohydrate intake exceeds the amount required for energy and for storage as glycogen, the excess is converted to fat to be stored in adipose tissue. When the glycogen reserves are depleted, fatty acids provide the bulk of the energy for muscle cells.

---

[7] In carbohydrate loading, glycogen stores are first depleted by limiting intake of carbohydrates and training vigorously. Then, a few days before competition, the athlete cuts back on training and eats a diet high in carbohydrates. The presumption is that, under this regimen, the body will store more glycogen than usual. The benefits of carbohydrate loading, even for top athletes, are questionable. For casual athletes, the technique is probably of very little value.

EXERCISES

1. Define or give an explanation for each of the following terms.

   a. metabolite
   b. villi
   c. blood sugar level
   d. renal threshold
   e. gluconeogenesis
   f. glucose tolerance test
   g. cyclic AMP
   h. adenyl cyclase
   i. Ⓟ
   j. metabolic pathway
   k. oxygen debt
   l. kinase
   m. FAS
   n. Cori cycle
   o. carbohydrate loading
   p. Antabuse

2. Distinguish between the terms in each of the following pairs.

   a. anabolism and catabolism
   b. photosynthesis and respiration
   c. metabolism and digestion
   d. active transport and passive transport
   e. hypoglycemia and hyperglycemia
   f. glycogenesis and glycogenolysis
   g. anaerobic and aerobic
   h. glycolysis and fermentation

i. a mutase and an isomerase

j. a phosphatase and a phosphorylase

k. facultative anaerobes and strict anaerobes

3. What is the structural difference among ATP, ADP, and AMP?

4. Why is ATP referred to as the energy currency of the cell?

5. Referring to Table 24.1, indicate which of these compounds are classified as high-energy phosphates.

    a. ATP   **b.** AMP

    c. ADP   **d.** creatine phosphate

    e. glucose 1-phosphate   **f.** glucose 6-phosphate

6. Give the location of action and the function of each digestive enzyme.

    a. salivary amylase (ptyalin)

    b. pancreatic amylase (amylopsin)

    c. sucrase

    d. lactase

    e. maltase

7. What is the general type of reaction used in digestion?

8. What is mucin? What is its function in saliva?

9. What are the end products of carbohydrate digestion?

10. In what section of the digestive tract does most of the digestion of carbohydrates take place?

11. If a cracker, which is rich in starch, is chewed for a long time, it begins to develop a sweet, sugary taste. Why?

12. What is the major factor in maintaining a constant blood sugar level?

13. What is the role of insulin in regulating the blood sugar level?

14. How do epinephrine and glucagon act to raise the blood sugar level?

15. Draw the structure of cyclic AMP.

16. When a hormone causes an increase in cAMP levels within the cells, where is the binding site for the hormone located?

17. Explain how the binding of a hormone at its receptor site can release cAMP within a target cell.

18. Glucose appears in the urine of a diabetic patient. Does glucose in the urine always indicate diabetes mellitus? Explain.

19. Why can insulin not be taken orally?

20. Briefly describe the two types of diabetes mellitus.

21. In structure and purpose, how do the oral drugs such as Orinase differ from insulin?

22. What is meant when glyburide is called a second-generation drug?

23. What is the storage form of carbohydrate in the body?

24. In what tissues or organs are carbohydrates stored?

25. What happens to the monosaccharide galactose after it reaches the liver?

26. How many molecules of pyruvic acid are produced from one molecule of glucose in glycolysis?

27. How do muscle cells and liver cells differ in their metabolism of glucose 6-phosphate and glycogen?

28. When aldolase catalyzes the cleavage of ketose mono- and diphosphates, dihydroxyacetone phosphate is always one of the products. Complete the following catalytic reactions.

    a.

$$\underset{\substack{\text{Sedoheptulose}\\\text{1,7-diphosphate}}}{\begin{array}{c}\text{CH}_2\text{O—}\textcircled{P}\\|\\\text{C}=\text{O}\\|\\\text{HO—C—H}\\|\\\text{H—C—OH}\\|\\\text{H—C—OH}\\|\\\text{H—C—OH}\\|\\\text{CH}_2\text{O—}\textcircled{P}\end{array}}\quad\underset{\text{aldolase}}{\rightleftharpoons}$$

    b. fructose 1-phosphate $\underset{\text{aldolase}}{\rightleftharpoons}$

29. Iodoacetic acid and fluoride ions are poisonous because they inhibit enzymes of the Embden–Meyerhof pathway. What enzymes are inhibited?

30. What is the fate of the lactic acid formed by muscular activity?

31. Draw a molecule of glucose, and label the carbons in the 3 and 4 positions.

    a. Which of these two carbon atoms will appear in lactic acid?

    b. Which of these two carbon atoms will appear in ethanol or in carbon dioxide?

32. **a.** Which is the oxidative step in the Embden–Meyerhof pathway?

    b. What is the oxidizing agent?

    c. How is the reduced coenzyme reoxidized in yeast cells?

    d. How is the reduced coenzyme reoxidized in muscle cells?

33. The enzyme lactic acid dehydrogenase is not specific for pyruvic acid but will catalyze the reduction of other keto acids. Write an equation for the enzymatic reduction of phenylpyruvic acid. Does this surprise you in light of what was said about the stereochemical specificity of this enzyme in Section 22.4? Comment.

34. Write a balanced half-reaction for each of the following conversions.

    a. Glucose $\longrightarrow$ Pyruvic acid

    b. Pyruvic acid $\longrightarrow$ Lactic acid

    c. Pyruvic acid $\longrightarrow$ Ethanol + Carbon dioxide

35. List four phosphorylated and four nonphosphorylated metabolites of glycolysis and fermentation.

36. What critical role is played by both 1,3-diphosphoglyceric acid and phosphoenolpyruvic acid (PEP) in the Embden–Meyerhof pathway?

37. Assign names to the question marks in the following word equations.

    a. ? $\underset{\text{dehydrogenase}}{\overset{\text{alcohol}}{\longrightarrow}}$ Ethanol

b. Fructose 1,6-diphosphate $\xrightarrow{\text{aldolase}}$? + ?

c. 2-Phosphoglyceric acid $\xrightarrow{?}$ Phosphoenolpyruvic acid

d. ? $\xrightarrow{\text{glucose 6-phosphatase}}$ Glucose

e. Glycogen $\xrightarrow{\text{phosphorylase}}$?

f. Dihydroxyacetone phosphate $\xrightarrow{?}$ Glyceraldehyde 3-phosphate

g. Glucose $\xrightarrow{\text{hexokinase}}$?

h. ? $\xrightarrow{\text{lactic acid dehydrogenase}}$ Lactic acid

i. Fructose 6-phosphate $\xrightarrow{?}$ Fructose 1,6-diphosphate

j. ? $\xrightarrow{\text{phosphoglucoisomerase}}$ Fructose 6-phosphate

k. Glyceraldehyde 3-phosphate $\xrightarrow{\substack{\text{glyceraldehyde} \\ \text{3-phosphate} \\ \text{dehydrogenase}}}$?

l. 1,3-Diphosphoglyceric acid $\xrightarrow{?}$ 3-Phosphoglyeric acid

m. 3-Phosphoglyeric acid $\xrightarrow{\text{phosphoglyceromutase}}$?

n. Pyruvic acid $\xrightarrow{?}$ Acetaldehyde

o. ? $\xrightarrow{\text{pyruvate kinase}}$ Pyruvic acid

p. Glucose 1-phosphate $\xrightarrow{\text{phosphoglucomutase}}$?

38. Refer to Exercise 37, and select the equations (by letter) in which the following processes occur.
  a. the expenditure of phosphate bond energy
  b. the formation of high-energy phosphate bonds
  c. isomerization reactions
  d. an oxidation reaction
  e. reduction reactions
  f. a dehydration reaction
  g. a decarboxylation reaction

39. The average adult consumes about 65 g of fructose daily (either as the free sugar or as part of sucrose). In the liver, fructose is first phosphorylated to fructose 1-phosphate, which is then split into dihydroxyacetone phosphate and glyceraldehyde. The latter compound is phosphorylated to glyceraldehyde 3-phosphate. Write out equations (using formulas) for these three steps, and give the specific names of the enzymes. Indicate which steps utilize ATP.

40. The alcohol we drink is detoxified by enzymes in the liver. Write out the sequence of reactions for the metabolism of alcohol.
  a. Which enzyme is inhibited by disulfiram?
  b. Does alcohol supply energy (calories) for the body? Explain.

41. Why is it more efficient for anerobic cells to utilize glycogen as a source of energy than for them to use glucose?

42. Write the structural formula of creatine phosphate. What is its role in muscle contraction?

Chapter 25
CARBOHYDRATE METABOLISM II

Prolonged exercise is fueled primarily by aerobic oxidation. This type of training is preferable for cardiovascular fitness.

In Chapter 24 we mentioned that pyruvic acid is a "crossroads" compound. All living cells metabolize glucose to pyruvic acid via the Embden–Meyerhof pathway. Cells that function anaerobically convert the pyruvic acid into lactic acid or ethanol. Recall (Section 24.9) that anaerobic metabolism accounts for only a small fraction of the total available energy of the glucose molecule. Cells that function aerobically oxidize the pyruvic acid in a number of discrete enzymatic reactions to carbon dioxide and water.

$$C_6H_{12}O_6 \xrightarrow{\text{glycolysis}} 2\ CH_3CHOHCOOH\ +\ {\sim}47{,}000\ cal$$

$$2\ CH_3COCOOH\ +\ 5\ O_2 \longrightarrow 6\ CO_2\ +\ 4\ H_2O\ +\ {\sim}639{,}000\ cal$$

From the standpoint of energy production, the oxidation of pyruvic acid is of considerable significance because it liberates most of the energy (93%) stored in the glucose molecule. Much of this energy is then conserved in a chemical form in the high-energy phosphate bonds of ATP.

675

25.1
THE KREBS CYCLE

The Krebs cycle is also known by two other names. It is called the *citric acid cycle* because citric acid is a most important intermediate. It is also known as the *tricarboxylic acid cycle* because four of the intermediate compounds have three carboxyl groups each.

A scheme for the complex series of reactions that brings about the oxidation of pyruvic acid to carbon dioxide and water was first proposed by Hans Krebs in 1937 (Nobel prize in physiology or medicine, 1953). Pyruvic acid is transported from the cytoplasm into the inner compartment of the mitochondrion (see Figure 25.2). Within the mitochondrion, pyruvic acid is first decarboxylated to a two-carbon compound, which then enters a cyclic sequence of reactions known collectively as the **Krebs cycle.** The Krebs cycle functions to produce ATP and to provide metabolic intermediates for the biosynthesis of needed compounds. As we shall see in later chapters, the Krebs cycle is not restricted to the metabolism of carbohydrates. It also plays a vital role in the metabolism of lipids and proteins and thus occurs in almost all cells of higher animals.

Taken as a whole, the Krebs cycle may seem rather complex. All the reactions, though, are familiar types in organic chemistry: condensations, dehydrations, hydrations, oxidations, decarboxylations, and hydrolyses. The difference here is one of *conditions*. These reactions take place at constant temperature (37 °C in humans) and constant pH. Each is catalyzed by an enzyme. Every enzyme that participates in the cycle has been identified, and the operation of this metabolic pathway in vivo has been completely verified by the use of isotopic tracers. All of the enzymes involved in the Krebs cycle are located within the mitochondria. Aerobic metabolism is thus separated from the reactions of glycolysis, which occur in the cytoplasm.

Figure 25.1 is a schematic outline of the Krebs cycle. Each reaction is numbered, and the individual steps of the sequence are now considered in detail.

Step 1   The pyruvic acid formed in the Embden–Meyerhof pathway is not an intermediate in the Krebs cycle. It must first be enzymatically decarboxylated and oxidized (*oxidative decarboxylation*) to yield the extremely important intermediate **acetyl coenzyme A.** The formation of acetyl CoA from pyruvic acid requires the sequential action of three different enzymes and the participation of five coenzymes: coenzyme A, thiamine pyrophosphate (TPP), lipoic acid, NAD^+, and FAD. (See Table I.1 for structures of the coenzymes.)

$$CH_3-\overset{\overset{\displaystyle O}{\|}}{C}-\overset{\overset{\displaystyle O}{\|}}{C}\diagdown_{OH} \ + \ CoASH \ + \ NAD^+ \ \xrightarrow[\text{TPP, FAD}]{\text{lipoic acid}} \ CH_3-\overset{\overset{\displaystyle O}{\|}}{C}\diagdown_{SCoA} \ + \ NADH \ + \ CO_2 \ + \ H^+$$

Pyruvic acid   Acetyl CoA

The name given to this multienzyme system is *pyruvic acid dehydrogenase complex.* The mechanism of the reaction is believed to involve four distinct steps—decarboxylation, reductive acetylation, acetyl transfer, and electron transport. The initial decarboxylation step is irreversible. Therefore, the overall conversion of pyruvic acid to acetyl CoA is irreversible.

Step 2   Acetyl CoA is likewise not a true intermediate in the Krebs cycle. It enters the cycle by condensing with the four-carbon dicarboxylic acid oxaloacetic acid, yielding citric acid. The reaction is highly exothermic because of the energy released by the subsequent hydrolysis of the thioester bond of acetyl CoA (7500 cal/mol). The two car-

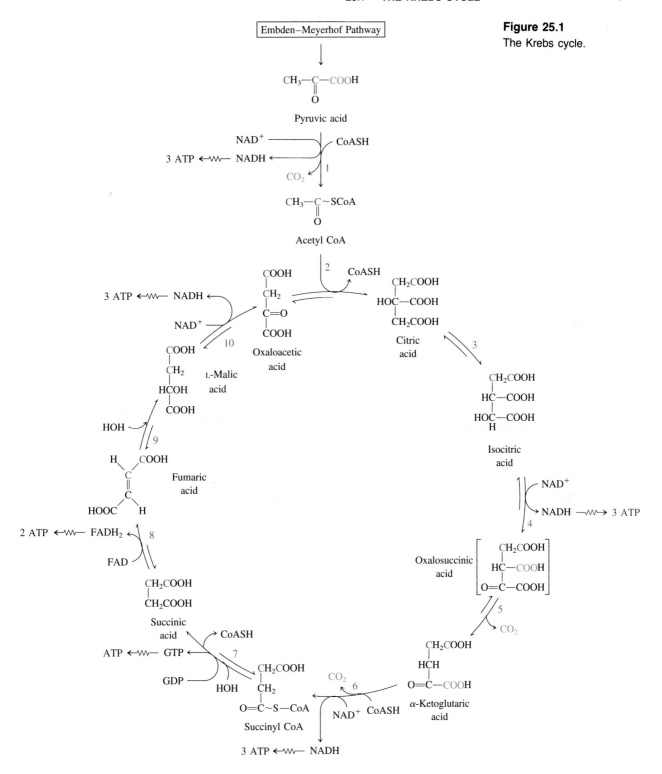

Figure 25.1
The Krebs cycle.

bon atoms that originate from acetyl CoA are shown in red here and in subsequent reactions. Note that this step regenerates coenzyme A.

The Krebs cycle is regulated in part at this step. The reaction is catalyzed by *citric acid synthetase,* an enzyme that is inhibited by ATP and NADH. When the cell has sufficient ATP for its immediate needs, ATP molecules interact with the enzyme to reduce its affinity for acetyl CoA. The latter is then shunted to the biosynthesis of fatty acids.

Step 3   The third step is sometimes considered to be two separate reactions. The single enzyme *aconitase* catalyzes successive dehydration and hydration reactions. The net result is the isomerization of citric acid to its less symmetrical isomer, isocitric acid. The intermediate is *cis*-aconitic acid, and it normally remains bound to the enzyme.

Steps 4 and 5   The next two steps are usually discussed together because experimental evidence indicates that the intermediate, oxalosuccinic acid, does not exist free but is firmly bound to the surface of the enzyme. The enzyme *isocitric acid dehydrogenase* catalyzes the oxidative decarboxylation of isocitric acid to α-ketoglutaric acid.

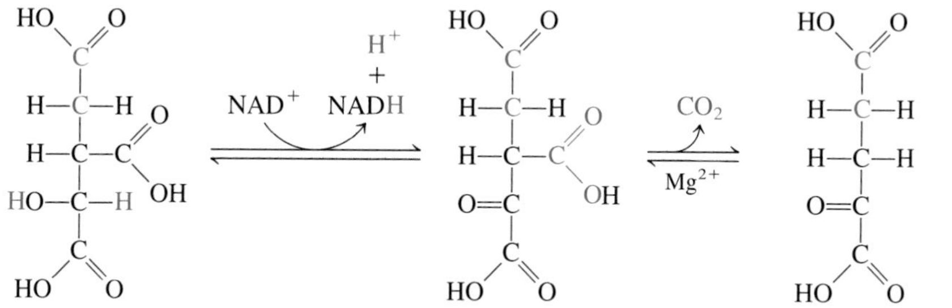

Isocitric acid

Oxalosuccinic acid
(enzyme-bound)

α-Ketoglutaric acid

Step 6   This step is analogous to step 1; it is catalyzed by another multienzyme system known as *α-ketoglutaric acid dehydrogenase complex,* which requires the same five coenzymes as pyruvic acid dehydrogenase. This is the only irreversible reaction in the Krebs cycle. As such, it prevents the cycle from operating in the reverse direction.

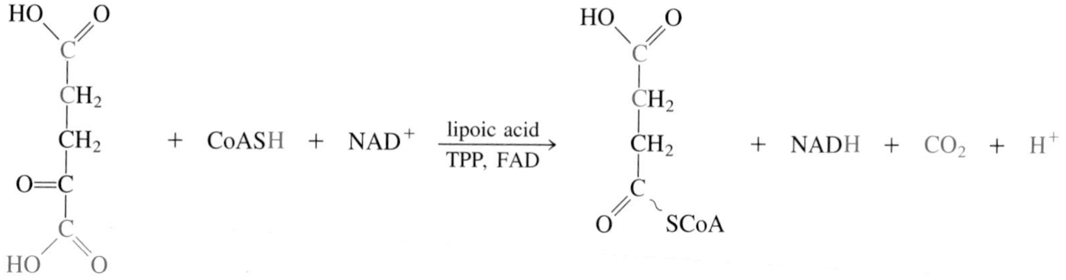

α-Ketoglutaric acid

Succinyl CoA

Step 7   In this reaction the energy released by the hydrolysis of the high-energy thioester bond of succinyl CoA ($\sim$8000 cal/mol) is used to form guanosine triphosphate (GTP) from guanosine diphosphate (GDP) and inorganic phosphate. This reaction is significant because GTP has a higher free energy of hydrolysis than ATP and can readily transfer its terminal phosphate group to ADP to generate ATP in the presence of the enzyme *nucleoside diphosphokinase.* Here we have another example of substrate-level phosphorylation. It is the only reaction in the Krebs cycle that directly involves a high-energy phosphate bond.

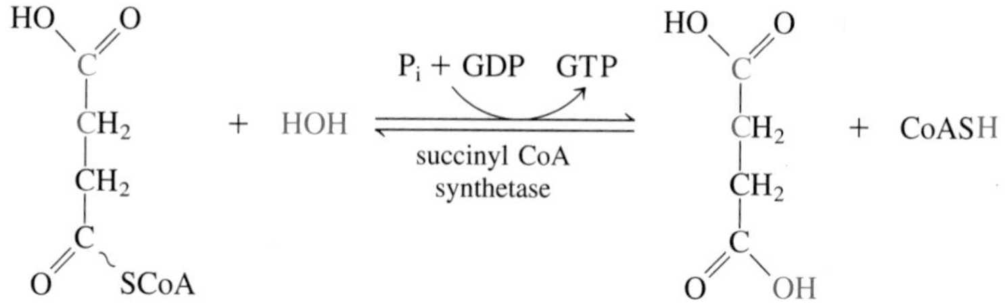

Succinyl CoA

Succinic acid

$$GTP \; + \; ADP \; \underset{}{\overset{kinase}{\rightleftharpoons}} \; GDP \; + \; ATP$$

Step 8   The enzyme *succinic acid dehydrogenase* catalyzes the removal of two hydrogen atoms from succinic acid, thus forming fumaric acid. You will recall that this enzyme is competitively inhibited by malonic acid (Section 22.6). This dehydrogenation reaction is the only one in the cycle that uses the coenzyme FAD rather than NAD^+. Succinic acid dehydrogenase is the only enzyme of the Krebs cycle that is located within the inner mitochondrial membrane. This permits electrons to be transferred directly into the electron transport chain when the $FADH_2$ is reoxidized (see footnote on page 684).

Succinic acid   Fumaric acid

Step 9   The addition of a molecule of water across the double bond of fumaric acid to form L-malic acid is catalyzed by the enzyme *fumarase*. This enzyme is highly stereospecific; trans addition occurs, so only the L isomer of malic acid is produced.

Fumaric acid   L-Malic acid

Step 10   One revolution of the cycle is completed with the oxidation of L-malic acid to oxaloacetic acid, brought about by the enzyme *malic acid dehydrogenase*. This is the fourth oxidation–reduction reaction that uses NAD^+ as the oxidizing agent.

L-Malic acid   Oxaloacetic acid

Oxaloacetic acid can accept an acetyl group from acetyl CoA, and the cycle is ready for another spin. Every time we go around the cycle, two carbons are fed into the system as acetyl CoA, and two carbons are kicked out as carbon dioxide molecules.

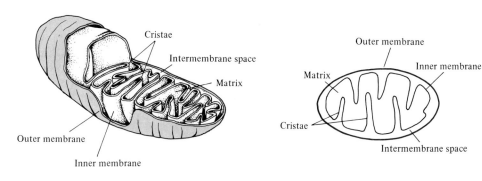

(a)   (b)

Figure 25.2
Schematic three-dimen-
sional (a) and two-dimen-
sional (b) representations of
a liver mitochondrion.

25.2
RESPIRATORY CHAIN: ELECTRON-TRANSPORT CHAIN

We have stated that aerobic metabolism occurs only in the presence of molecular oxygen. It has also been mentioned that the major portion of solar energy stored in carbohydrates is conserved in the process. Yet nowhere in the Krebs cycle has the utilization of oxygen or the conservation of energy been indicated. None of the intermediates of the cycle has been shown to be linked to phosphate groups, and no direct synthesis of ATP has taken place from ADP and inorganic phosphate. The Krebs cycle deals primarily with the fate of the carbon skeleton of pyruvic acid, describing the metabolites involved in its conversion to carbon dioxide. Two carbon atoms enter the cycle as acetyl CoA (step 2), and two different carbon atoms exit the cycle as carbon dioxide (steps 5 and 6). The coenzymes NAD^+ and FAD are reduced to NADH and $FADH_2$; no mechanism has yet been indicated for their regeneration. The reduced coenzymes must be reoxidized if the aerobic phase of carbohydrate metabolism is to continue.

All the enzymes and coenzymes that are necessary for the Krebs cycle and for the conservation of energy are localized in the mitochondria, small organelles that are often referred to as the "power plants" of the cell (Figure 25.2). **Mitochondria** are oval, dual-membrane structures. They may be randomly distributed throughout the cytoplasm, or they may be organized in regular rows or clusters. A cell may contain 100–1000 mitochondria, depending on its function, and the mitochondria can reproduce themselves if the energy requirements of the cell increase. A mitochondrion has two lipid/protein membranes—an *outer membrane* and an *inner membrane* that is extensively folded into a series of internal ridges called *cristae*. Thus there are two compartments in mitochondria—the *intermembrane space* and the *matrix*, which is surrounded by the inner membrane. The outer membrane is permeable, whereas the inner membrane is impermeable to most molecules and ions. (Water, oxygen, and carbon dioxide can freely penetrate both membranes.) The mitochondrial matrix contains all the enzymes of the Krebs cycle, with the exception of succinic acid dehydrogenase, which is embedded in the inner membrane. The enzymes that operate to provide energy for the cell are also contained within the inner mitochondrial membrane and are positioned in geometrically specific arrays in such a way that they are capable of functioning as extremely efficient assembly lines. This sequence of highly organized oxidation–reduction enzymes is known as the **respiratory chain** or the **electron transport chain.** As we shall see, the operation of the chain is analogous to that of a bucket brigade.

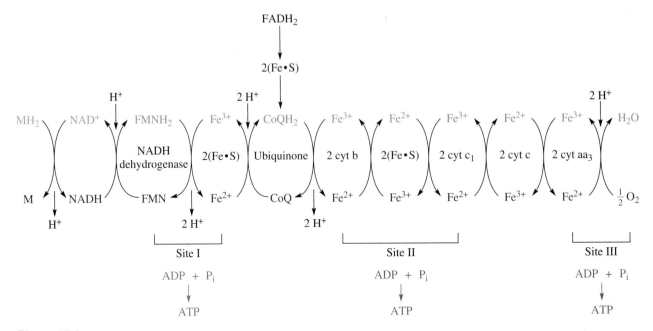

Figure 25.3
A schematic diagram of the respiratory chain depicting the oxidized and reduced forms of the carriers, the input and output of protons from the mitochondrial matrix, and the sites along the chain where oxidative phosphorylation occurs.

Figure 25.3 illustrates the respiratory chain in the mitochondria. The sequence in which the electron carriers of the chain operate is determined by their respective reduction potentials. The **reduction potential** is a measure of the tendency of a substance to gain electrons compared to the standard hydrogen electrode (25 °C, 1 atm H_2, and 1 M H^+), which is arbitrarily assigned a value of 0.0 V. Since in biological systems we are more interested in neutral solutions, a correction is made for changes in pH. At pH 7 the hydrogen electrode has a potential difference of -0.42 V when measured against the standard hydrogen electrode. The reduction potentials of some of the intermediates of the respiratory chain are listed in Table 25.1.

For each molecule of pyruvic acid that is converted to carbon dioxide and water via the Krebs cycle, five dehydrogenation reactions occur. NAD^+ serves as the electron acceptor in four of these (steps 1, 4, 6, and 10), and FAD is the oxidizing agent in the fifth (step 8). In Figure 25.3, MH_2 symbolizes the reduced metabolites (pyruvic acid, isocitric

Table 25.1
Standard Reduction Potentials for Some Respiratory Chain Components

System (Oxidant/Reductant)	E (at pH 7) (V)
NAD^+/NADH	-0.32
FMN/$FMNH_2$	-0.30
CoQ/$CoQH_2$	$+0.04$
Cyt b-Fe(III)/cyt b-Fe(II)	$+0.07$
Cyt c_1-Fe(III)/cyt c_1-Fe(II)	$+0.23$
Cyt c-Fe(III)/cyt c-Fe(II)	$+0.25$
Cyt aa_3-Fe(III)/cyt aa_3-Fe(II)	$+0.29$
$\frac{1}{2} O_2$/H_2O	$+0.82$

acid, α-ketoglutaric acid, and malic acid) and M signifies the oxidized metabolites (acetyl CoA, oxalosuccinic acid, succinyl CoA, and oxaloacetic acid).

We now examine one of these conversions by writing a balanced half-reaction.

$$\text{HOOC}-\text{CH}_2-\overset{\overset{\displaystyle OH}{|}}{\underset{\underset{\displaystyle H}{|}}{C}}-\text{COOH} \longrightarrow \text{HOOC}-\text{CH}_2-\overset{\overset{\displaystyle O}{\|}}{C}-\text{COOH} + 2\,H^+ + \boxed{2\,e^-}$$

L-Malic acid   Oxaloacetic acid

It has been stated that the function of the coenzyme NAD^+ is to accept this pair of electrons. We can also write a balanced half-reaction for the reduction of the coenzyme; R— represents the remaining portion of the NAD^+ molecule (see Table I.1).

NAD$^+$ NADH

Two hydrogen ions and two electrons are removed from the substrate. The NAD^+ molecule accepts both electrons and one hydrogen ion. The other hydrogen ion is transported from the matrix, across the inner mitochondrial membrane, to the intermembrane space.

In the next step of the respiratory chain, both electrons are passed on to an enzyme called *NADH dehydrogenase* whose coenzyme is FMN (see Table I.1). By passing the electrons along, NADH is reoxidized back to NAD^+ and FMN is reduced to $FMNH_2$. It is this step that accounts for the regeneration of NAD^+. Again we depict only the relevant portion of the FMN molecule.

$$\text{NADH} + H^+ \longrightarrow NAD^+ + 2\,H^+ + \boxed{2\,e^-}$$

FMN   FMNH$_2$

The electrons are then transferred from $FMNH_2$ to a series of iron–sulfur complexes (abbreviated as Fe · S). A number of iron–sulfur proteins have been identified in the respiratory chain, complexed with other electron carriers. The iron atom in these proteins is in the Fe(III) (ferric) form; by accepting an electron, it is reduced to the Fe(II) (ferrous) form. (Since each iron–sulfur complex can transfer only one electron, two molecules are needed to transport the two electrons and regenerate FMN.)

$$\text{FMNH}_2 \longrightarrow \text{FMN} + 2\,H^+ + \boxed{2\,e^-}$$
$$2\,\text{Fe(III)} \cdot \text{S} + \boxed{2\,e^-} \longrightarrow 2\,\text{Fe(II)} \cdot \text{S}$$

The two electrons are next transferred from Fe(II) · S to **coenzyme Q** (CoQ). This coenzyme is a quinone derivative with a long isoprenoid side chain. Coenzyme Q is also called *ubiquinone* because it is ubiquitous in living systems. Several ubiquinones are known, differing only in the number of isoprene units in their side chain. The most common ubiquinone found in the mitochondria of animal tissues contains 10 isoprene units and has been designated as CoQ_{10}. The quinone ring is reversibly reducible to a hydroquinone, and so CoQ serves as the electron carrier between the flavin coenzymes and the cytochromes.[1] It is in this step that the oxidized form of the iron–sulfur complex is regenerated.

$$2\ Fe(II)\cdot S\ \longrightarrow\ 2\ Fe(III)\cdot S\ +\ \boxed{2\ e^-}$$

Coenzyme Q_{10} $\qquad\qquad\qquad$ Coenzyme $Q_{10}H_2$

The combination of cyanide with the ferric ions of cytochrome aa_3 completely inhibits their reduction to the ferrous state, thus blocking the transfer of electrons to molecular oxygen. This accounts for the extreme toxicity of cyanide to living organisms—just 50 mg of HCN constitutes a lethal dose for a human.

A series of compounds called **cytochromes** were probably the first entities to be associated with electron-transferring reactions. A number of these substances exist, and we have included four of them in Figure 25.3. (Notice that another Fe · S protein is located between two of the cytochromes.) The cytochromes are conjugated protein enzymes. Their prosthetic groups are iron porphyrins, which resemble heme, the pigment in hemoglobin and myoglobin (see Figure 21.20). The various cytochromes differ with respect to (1) their protein constituents, (2) the manner in which the porphyrin is bound to the protein, and (3) the substituents on the periphery of the porphyrin ring. Such slight differences in structure bestow differences in reduction potential upon the different cytochromes. Like the iron in the Fe · S complexes, the characteristic feature of the cytochromes is the ability of their iron atoms to exist in either ferrous or ferric form. Thus each cytochrome in its oxidized form, Fe(III), can accept one electron and be reduced to the Fe(II)-containing form. This change in oxidation state is reversible, and the reduced form can donate its electron to the next cytochrome, and so on. Only the last cytochrome, cytochrome aa_3 (called *cytochrome oxidase*), has the ability to transfer electrons to molecular oxygen. Since the Fe(III)/Fe(II) system is only a one-electron exchange, two cytochrome b molecules are necessary to complete the oxidation of coenzyme $Q_{10}H_2$.

$$CoQ_{10}H_2\ \longrightarrow\ CoQ_{10}\ +\ 2\ H^+\ +\ \boxed{2\ e^-}$$
$$2\ Cyt\ b\text{-}Fe(III)\ +\ \boxed{2\ e^-}\ \longrightarrow\ 2\ Cyt\ b\text{-}Fe(II)$$

Then

$$Cyt\ b\ \longrightarrow\ Fe\cdot S\ \longrightarrow\ Cyt\ c_1\ \longrightarrow\ Cyt\ c\ \longrightarrow\ Cyt\ aa_3$$

In the final step, two molecules of the terminal electron carrier, cytochrome aa_3, pass their electrons to molecular oxygen, the ultimate electron acceptor. It has been estimated that about 95% of the oxygen used by cells reacts in this single process.

$$2\ Cyt\ aa_3\text{-}Fe(II)\ \longrightarrow\ 2\ Cyt\ aa_3\text{-}Fe(III)\ +\ \boxed{2\ e^-}$$
$$\tfrac{1}{2}\ O_2\ +\ 2\ H^+\ +\ \boxed{2\ e^-}\ \longrightarrow\ H_2O$$

[1] Recall that we said the enzyme succinic acid dehydrogenase (which catalyzes the oxidation of succinic acid to fumaric acid—step 8) contains the coenzyme FAD, which is reduced to $FADH_2$. The oxidized form of the coenzyme must be regenerated, and this is accomplished via the respiratory chain. The electrons from $FADH_2$ require a two-step transfer to enter the chain. First they are passed to an iron–sulfur protein,

$$FADH_2\ \longrightarrow\ FAD\ +\ 2\ H^+\ +\ \boxed{2\ e^-}$$
$$2\ Fe(III)\cdot S\ +\ \boxed{2\ e^-}\ \longrightarrow\ 2\ Fe(II)\cdot S$$

which in turn passes them on to CoQ_{10} (see Figure 25.3).

$$2\ Fe(II)\cdot S\ \longrightarrow\ 2\ Fe(III)\cdot S\ +\ \boxed{2\ e^-}$$
$$CoQ_{10}\ +\ 2\ H^+\ +\ \boxed{2\ e^-}\ \longrightarrow\ CoQ_{10}H_2$$

Each intermediate compound in the respiratory chain is reduced by the addition of electrons in one reaction and is subsequently restored to its original form when it delivers the electrons to the next compound. Thus each pair of electrons that is removed from the substrates of the Krebs cycle ultimately reduces one atom of oxygen.

25.3
OXIDATIVE PHOSPHORYLATION

The process whereby ATP synthesis is linked to the consumption of oxygen in the respiratory chain is referred to as **oxidative phosphorylation.** A model (the chemiosmotic hypothesis) that links the formation of ATP to the operation of the respiratory chain has been proposed by Peter Mitchell (Nobel prize, 1978—see an advanced biochemistry text for details). The energy required for the production of ATP results from the passage of a pair of electrons from NADH or $FADH_2$ to oxygen through a series of electron carriers. The electron transport chain can be thought to function as a biochemical battery; that is, energy is obtained from oxidation–reduction reactions.

Electron transport is tightly coupled to oxidative phosphorylation. The reduced forms of the coenzymes, NADH and $FADH_2$, are oxidized by the respiratory chain *only* if ADP is simultaneously phosphorylated to ATP. Within energy-utilizing cells, the turnover of ATP is very high. These cells, then, contain high levels of ADP, and they must consume large quantities of oxygen to continuously phosphorylate the ADP back to ATP. For example, at rest about 20% of an adult's oxygen consumption occurs within brain tissue to supply energy to brain cells. (Much of this energy is used to maintain the concentration gradients across the membranes of the trillions of brain cells.) Resting skeletal muscles utilize about 30%, whereas during strenuous exercise these muscles account for almost 90% of the total oxygen consumption of the organism. The enzymes of oxidative phosphorylation are embedded in the inner mitochondrial membrane in association with the enzymes of the respiratory chain. The sites on the respiratory chain at which the oxidative phosphorylations are believed to occur are shown in Figure 25.3. From the data contained in Table 25.1, we can calculate the maximum amount of energy (E) that is made available when a single pair of electrons travels from NADH to oxygen along the chain.

$$
\begin{array}{lll}
& & E \\
NADH \longrightarrow NAD^+ + H^+ + 2\,e^- & & +0.32 \\
\tfrac{1}{2}\,O_2 + 2\,H^+ + 2\,e^- \longrightarrow H_2O & & +0.82 \\
\hline
NADH + \tfrac{1}{2}\,O_2 + H^+ \longrightarrow NAD^+ + H_2O & & +1.14\ V
\end{array}
$$

The energy change for the reaction can be obtained by use of the following equation.

$$\text{Energy change} = -nF\Delta E$$

where n = number of electrons transferred
F = Faraday's constant (23,062 cal/V equivalent)

$$\text{Energy change} = -(2)(23{,}062)(1.14)$$
$$= -52{,}600\ \text{cal}$$

This value of 52,600 cal represents a considerable amount of energy. If it were released all at once, much of it would be dissipated as heat and it might prove damaging to the cell. Therefore, the respiratory chain serves as a device for delivering this energy in

Since the actual concentrations of the various compounds are unknown, the use of reduction potentials can yield only a rough estimate of the actual energy change for the reaction.

small increments to be used to phosphorylate ADP. It has been experimentally observed that three molecules of ATP are formed for every molecule of NADH that is oxidized in the chain (but only two ATPs are formed if the primary acceptor is FAD, as is the case when succinic acid serves as a substrate). The net equation for the respiratory chain is

$$NADH + \tfrac{1}{2}O_2 + H^+ + 3\,ADP + 3\,P_i \longrightarrow NAD^+ + H_2O + 3\,ATP$$

Recall that 7300 cal is required for the conversion of 1 mol of ADP to ATP. It can be determined that almost half of the energy released in the electron transport chain is conserved in the formation of high-energy phosphate bonds.

$$\text{Energy conserved by respiratory chain} = \frac{\text{energy conserved}}{\text{energy available}}(100\%)$$

$$= \frac{(3)(7300)}{52{,}600}(100\%) = 42\%$$

25.4
ENERGY YIELD OF CARBOHYDRATE METABOLISM

It is now possible to summarize the energy conserved (in ATP production) from the complete oxidation of one molecule of glucose. It must be recalled that, under aerobic conditions, the glycolytic sequence terminates with the formation of two molecules of pyruvic acid. Two molecules of ATP are obtained from substrate-level phosphorylation. In addition, since the pyruvic acid is not reduced to lactic acid, there are two molecules of NADH (from step 6) that remain in the cytoplasm. We know that NAD^+ must be regenerated from NADH for glycolysis to continue. The problem, however, is that NADH cannot pass across the mitochondrial membrane to be oxidized by the respiratory chain. A solution is achieved whereby electrons from NADH are transported across the membrane via a shuttle process involving glycerol 3-phosphate and dihydroxyacetone phosphate. These compounds can penetrate the outer mitochondrial membrane.

The first step in this shuttle (Figure 25.4) is the reduction of dihydroxyacetone phosphate by NADH to form glycerol 3-phosphate and regenerate the NAD^+. This reaction occurs in the cytoplasm. Glycerol 3-phosphate then enters the mitochondrion, where

Figure 25.4

The glycerol phosphate shuttle.

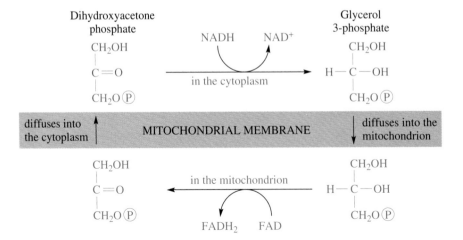

Reaction	Type of Phosphorylation	Number of ATP Molecules Formed
Glucose $\longrightarrow$ 2 Pyruvic acid	Substrate level	2
Glyceraldehyde 3-phosphate $\underset{\text{FAD} \quad \text{FADH}_2}{\overset{\text{NAD}^+ \quad \text{NADH}}{\longrightarrow}}$ 1,3-Diphosphoglyceric acid	Oxidative (via glycerol phosphate shuttle)	2×2
Pyruvic acid $\overset{\text{NAD}^+ \quad \text{NADH}}{\longrightarrow}$ Acetyl CoA	Oxidative	2×3
Isocitric acid $\overset{\text{NAD}^+ \quad \text{NADH}}{\longrightarrow}$ Oxalosuccinic acid	Oxidative	2×3
α-Ketoglutaric acid $\overset{\text{NAD}^+ \quad \text{NADH}}{\longrightarrow}$ Succinyl CoA	Oxidative	2×3
Succinyl CoA $\overset{\text{GDP} \quad \text{GTP}}{\longrightarrow}$ Succinic acid	Substrate level	2×1
Succinic acid $\overset{\text{FAD} \quad \text{FADH}_2}{\longrightarrow}$ Fumaric acid	Oxidative	2×2
Malic acid $\overset{\text{NAD}^+ \quad \text{NADH}}{\longrightarrow}$ Oxaloacetic acid	Oxidative	$\underline{2 \times 3}$
	Sum	36

Table 25.2

Production of ATP During the Oxidation of One Molecule of Glucose to Carbon Dioxide and Water

it is reoxidized to dihydroxyacetone phosphate, this time by a dehydrogenase that contains the coenzyme FAD instead of NAD^+. The dihydroxyacetone phosphate diffuses out of the mitochondrion and returns to the cytoplasm to complete the shuttle. The reduced form of the flavin coenzyme, $FADH_2$, is reoxidized by passing its electrons to the respiratory chain at the level of coenzyme Q_{10}. As a result, two molecules of ATP (instead of three) are formed every time a cytoplasm NADH molecule is reoxidized via the glycerol phosphate shuttle and the respiratory chain.[2]

The aerobic continuation of glycolysis yields a total of 15 molecules of ATP from each molecule of pyruvic acid, or 30 molecules of ATP per molecule of glucose. Table 25.2 lists the various reactions that result in ATP synthesis.

$$C_6H_{12}O_6 \; + \; 6\,O_2 \; + \; 36\,ADP \; + \; 36\,P_i \; \longrightarrow \; 6\,CO_2 \; + \; 42\,H_2O \; + \; 36\,ATP$$

The above equation summarizes the oxidation of 1 mol of glucose in certain aerobic cells (e.g., skeletal muscle, nerve). The energy released (686,000 cal) is coupled to the synthesis of 36 mol of ATP from ADP and inorganic phosphate. Assuming that 7300 cal is required for the synthesis of each mole of ATP, then $36 \times 7300 = 263,000$ cal is conserved by the cell. The efficiency of conservation therefore is

$$\frac{263,000}{686,000}(100\%) = 38\%$$

[2]The glycerol phosphate shuttle operates in most cells (e.g., skeletal muscle and nerve cells). In certain other tissues, particularly the liver and heart, another type of shuttle exists. This is the malic acid–aspartic acid shuttle. By a quite complex mechanism, it produces three molecules of ATP from the reoxidation of a cytoplasmic NADH molecule.

This recovery of energy compares favorably with the efficiency of any machine, and it represents a remarkable achievement on the part of the living organism. In comparison, automobiles are only about 5% efficient in utilizing the energy released in the combustion of gasoline. (Exercise 26 asks you to determine the number of ATPs and the energy efficiency in heart and liver cells.)

What happens to the 62% of the energy that is not trapped? It is released as heat to the surroundings, that is, to the cell. It is this heat that maintains body temperature. If we are exercising strenuously and our metabolism speeds up to provide the necessary energy for muscle contraction, then more heat is also produced. We begin to sweat to dissipate some of that heat. As the sweat evaporates, the excess heat is carried away from the body by the departing water vapor.

25.5
MUSCLE POWER

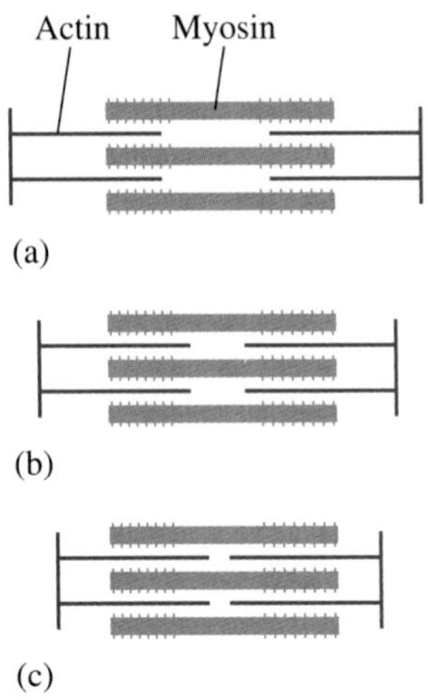

Actin   Myosin

(a)

(b)

(c)

Figure 25.5
Diagram of actomyosin complex in muscle. (a) Extended muscle. (b) Resting muscle. (c) Partially contracted muscle.

The stimulation of muscle causes it to contract. That contraction is work and requires energy. The immediate source of energy for muscle contraction is the splitting of ATP to ADP and inorganic phosphate. It is the energy stored in the ATP molecule that is used to bring about the physical movement of muscle tissue. Two proteins, **actin** and **myosin**, play important roles in this process. Together actin and myosin form a loose complex called **actomyosin,** the contractile protein of which muscles are made (Figure 25.5). When ATP is added to isolated actomyosin, the protein fibers contract. It seems likely that the same process occurs in vivo, that is, in muscle in living animals.

In the resting person, muscle activity (including that of the heart muscle) accounts for only about 15–30% of the energy requirements of the body. Other activities, such as cell repair or the transmission of nerve impulses or even the maintenance of body temperature, account for the remaining energy needs. During intense physical activity, the energy requirements of muscle may be more than 200 times the resting level.

Fats are the major source of energy for sustained, low- or moderate-intensity activity (see Section 26.3). Prolonged exercise is fueled primarily by aerobic oxidation of both fats and carbohydrates. For example, long-distance runners derive only a small percentage of their energy needs from glycolysis. Aerobic oxidation (the Krebs cycle coupled to oxidative phosphorylation) provides most of the required ATP. Some glycogenolysis (hydrolysis of glycogen) and glycolysis (anaerobic metabolism of glucose) does occur, and after very long periods of this moderate muscle activity, lactic acid does build up and muscles do become fatigued. But since fats and some carbohydrate are supplying most of the energy through aerobic oxidation, the buildup of lactic acid takes much longer than during anaerobic activity.

Muscle tissue seems to have been designed to provide for both short, intense bursts of activity and sustained, moderate levels of activity. Muscle fibers are divided into two categories, described as *fast twitch* and *slow twitch*. Table 25.3 lists some characteristics of these different types of muscle fibers. The Type I (slow twitch) fibers described in Table 25.3 are called on during activity of light or moderate intensity. The respiratory capacity of these fibers is high, which means they can provide much energy via aerobic pathways, or, to put it another way, they are geared to oxidative phosphorylation. Notice that for Type I fibers, myoglobin levels are also high. Myoglobin (Section 21.12) is the heme-containing protein in muscle that transports oxygen (as hemoglobin does in the blood). Aerobic oxidation requires oxygen, and this muscle tissue is geared to supply high

	Type I	Type IIB[a]
Category	Slow twitch	Fast twitch
Color	Red	White
Respiratory capacity	High	Low
Myoglobin level	High	Low
Catalytic activity of actomyosin	Low	High
Capacity for glycogenolysis	Low	High
Number of mitochondria	High	Low

[a]There is a Type IIA fiber that resembles Type I in some respects and Type IIB in others. We will discuss only the two types described in the table.

Table 25.3
A Comparison of Types of Muscle Fibers

The relative percentages of each type of muscle fiber within an individual seem to be inherited.

levels of oxygen. The capacity of Type I muscle fibers for glycogenolysis is low. This tissue is not geared to anaerobic generation of energy and does not require the hydrolysis of glycogen. The number of mitochondria in Type I muscle tissue is high, as we would expect, because oxidative phosphorylation takes place in the mitochondria. The catalytic activity of the actomyosin complex is low. Actomyosin is not only the structural unit in muscle that actually undergoes contraction—it is also responsible for catalyzing the hydrolysis of ATP to provide energy for the contraction. Low catalytic activity means that the energy is parceled out more slowly. That is not good if you want to lift 200 kg, but it is perfect for a run of 15 km (Figure 25.6).

The Type IIB (fast twitch) fibers described in Table 25.3 have characteristics just the opposite of those of Type I fibers. Low respiratory capacity, low myoglobin levels, and fewer mitochondria all argue against aerobic oxidation. A high capacity for glycogenolysis and high catalytic activity of actomyosin allow this tissue to generate ATP rapidly via glycolysis and also to hydrolyze that ATP rapidly in intense muscle activity. Thus, this type of muscle tissue gives you the capacity to do short bursts of vigorous work. We say *bursts* because this type of muscle fatigues relatively quickly. A period of recovery in which lactic acid is cleared from the muscle is required between brief periods of activity.

Figure 25.6
The New York Marathon, a test of the respiratory capacity of muscle.

Figure 25.7

Strength exercises build muscle mass but do not increase the respiratory capacity of muscles.

Fast twitch fibers increase in size and strength with repeated anaerobic exercise. Weight training does not increase respiratory capacity.

The fields of sports medicine and exercise physiology have done much to increase our understanding of muscle action. Endurance exercise training (e.g., running for long distances) increases the size and number of mitochondria in skeletal muscles. There is an increase in the level of enzymes required for the transport and oxidation of fatty acids (Section 26.3), for the Krebs cycle, and for oxidative phosphorylation. The increase in the mitochondrial enzymes is much greater for Type I fibers (used in prolonged, moderate-intensity activity) than for Type IIB fibers (brief, intense activity). Endurance training also increases myoglobin levels in skeletal muscles, providing for faster oxygen transport, and stimulates additional capillaries to grow within the muscles, bringing additional oxygen-carrying blood to the muscles.

These changes can be observed shortly after training begins, that is, within a week or two. Muscle changes resulting from endurance training do not necessarily include a significant increase in the size of the muscle,[3] in contrast to the effect of strength exercises such as weight lifting. Weight lifting develops fast twitch muscles but does not result in the mitochondrial changes we have just described (Figure 25.7). The mitochondria of heart muscle, which is working constantly anyway, also undergo no change during endurance training.

Athletes usually emphasize one type of training (anaerobic or aerobic) over the other. For example, an athlete training for a 60-m dash will do mainly anaerobic work, but one planning to run a 10-km race will do mainly aerobic training. Sprinting and weight lifting are largely anaerobic activities; a marathon run is largely aerobic. Research shows that world-class marathon runners may possess up to 80–90% slow twitch fibers, compared with championship sprinters, who may have up to 70% fast twitch muscle fiber. Some exceptions have been noted, however.

We hope we have made clear the centrol role ATP plays in many life-sustaining metabolic reactions. Our cells can make ATP, using the energy stored in foods. But only green plants can make the food. They also replenish the oxygen we breathe. Have you thanked a green plant lately?

---

[3]Conversely, a muscle that is not used to any great extent tends to decrease in size and strength. There is a reduction in the number of mitochondria within each muscle cell and a reduction of the number of capillaries within the muscle fibers. This condition is called *atrophy* and is particularly noticeable in an individual upon removal of a cast after several weeks of limb immobilization.

EXERCISES

1. Define or give an explanation for the following terms.
   a. Krebs cycle
   b. oxidative decarboxylation
   c. mitochondria
   d. respiratory chain
   e. cytochromes
   f. coenzyme Q_{10}
2. What is the main function of the Krebs cycle?
3. Two carbon atoms are fed into the Krebs cycle as acetyl coenzyme A. In what form are two carbon atoms removed from the cycle?
4. What are the oxidizing agents most immediately involved in the Krebs cycle?
5. How many molecules of ATP are formed by the oxidation of NADH in the respiratory chain?
6. How many molecules of ATP are formed by the oxidation of $FADH_2$ in the respiratory chain?
7. What is GTP?
8. What is substrate-level phosphorylation?
9. What is oxidative phosphorylation?
10. List the component parts of the NAD^+ molecule. Which part is a B vitamin?
11. List the component parts of the FAD molecule. Which two parts combined make up a B vitamin? What is the name of that vitamin?
12. When NAD^+ and FAD act as oxidizing agents, which portion of each is reduced?
13. If the carbon atom of the methyl group of pyruvic acid is labeled, where would the label appear after the oxidation of pyruvic acid by one turn of the Krebs cycle?
14. Assign names to the question marks in the following word equations.
    a. $? \xrightarrow{\text{aconitase}}$ Isocitric acid
    b. $? \xrightarrow[\text{synthetase}]{\text{citric acid}}$ Citric acid
    c. Fumaric acid $\xrightarrow{\text{fumarase}}$?
    d. Isocitric acid $\xrightarrow{?}$ α-Ketoglutaric acid
    e. α-Ketoglutaric acid $\xrightarrow{?}$ Succinyl CoA
    f. Malic acid $\xrightarrow[\text{dehydrogenase}]{\text{malic acid}}$?
    g. $? + ? \xrightarrow[\text{diphosphokinase}]{\text{nucleoside}}$ GTP + ADP
    h. Pyruvic acid $\xrightarrow{?}$ Acetyl CoA
    i. Succinyl CoA $\xrightarrow[\text{synthetase}]{\text{succinyl CoA}}$?
    j. Succinic acid $\xrightarrow[\text{dehydrogenase}]{\text{succinic acid}}$?
15. Refer to Exercise 14 and select the equations (by letter) in which the following processes occur.
    a. isomerization
    b. hydration
    c. dehydration
    d. oxidation
    e. decarboxylation
    f. phosphorylation
16. Fluoroacetic acid (CH_2FCOOH) is the toxic component in a South African plant known as gifblaar (*Dichapetalum*

cymosum). Fluoroacetic acid is highly toxic to humans because it can easily enter cells and pass across the mitochondrial membrane. Within the mitochondrion, fluoroacetic acid is converted by acetyl CoA synthetase to fluoroacetyl CoA. That compound then condenses with oxaloacetic acid to form fluorocitric acid, which is an inhibitor of the enzyme aconitase. (Fluorocitric acid is a lethal poison and is used as a rodenticide.) Write equations showing the conversion of fluoroacetic acid to fluorocitric acid. What type of inhibitor (see Section 22.6) is fluorocitric acid?

17. Sketch and label the pertinent features of a mitochondrion.
18. Outline, in detail, the sequence of reactions that occur in the respiratory chain.
19. Write balanced half-reactions for all the oxidation–reduction reactions that occur in the Krebs cycle and in the respiratory chain.
20. How is the energy of carbohydrate metabolism made available to the body cells?
21. The complete oxidation of 1 mol of acetic acid in a calorimeter yields about 200 kcal.
    a. How much energy is stored as ATP when 1 mol of acetic acid is converted to acetyl CoA and metabolized via the Krebs cycle?
    b. What is the percent energy efficiency?
22. How many moles of ATP can be formed from each mole of lactic acid that is oxidized to CO_2 and H_2O?
23. What are the names of the two shuttle systems that transport NADH from the cytoplasm into a mitochondrion?
24. Only two molecules of ATP are produced by the aerobic conversion of (a) succinic acid to fumaric acid and (b) glyceraldehyde 3-phosphate to 1,3-diphosphoglyceric acid. Explain.
25. Write word equations for all the substrate-level phosphorylation reactions that occur during the metabolism of glucose.
26. a. How much ATP (in moles) is produced from 1 mol of glucose in a typical liver or heart cell? (Recall footnote 2.)
    b. What is the efficiency associated with ATP production in a liver or heart cell?
27. Write the net equation for the complete oxidation of 1 mol of glucose in a skeletal muscle cell that is respiring under aerobic conditions.
28. a. What purpose do carbohydrates fulfill in metabolism?
    b. In what respect does carbohydrate metabolism resemble the burning of table sugar in a pan?
    c. In what way does it differ?
29. What is the electron acceptor at the end of the respiratory chain? To what product is this compound reduced?
30. What is the function of the cytochromes in the respiratory chain?
31. Why is cyanide ion so toxic?
32. What is the function of the mitochondria?

33. What are the two functions of actomyosin?
34. Which energy reserves (that is, carbohydrates or fats) are tapped in intense bursts of vigorous activity?
35. Which energy reserves (carbohydrates or fats) are mobilized to fuel prolonged low levels of activity?
36. Which type of metabolism (aerobic or anaerobic) is primarily responsible for providing energy for intense bursts of vigorous activity?
37. Which type of metabolism (aerobic or anaerobic) is primarily responsible for providing energy for prolonged low levels of activity?
38. Identify Type I and Type IIB muscle fibers as
    a. fast twitch or slow twitch
    b. suited to aerobic oxidation or to anaerobic glycolysis
39. Explain why high levels of myoglobin and mitochondria are appropriate for muscle tissue that is geared to aerobic oxidation.
40. Why does the high catalytic activity of actomyosin in Type IIB fibers suggest that these are the muscle fibers engaged in brief, intense physical activity?
41. Why can the muscle tissue that utilizes anaerobic glycolysis for its primary source of energy be called on only for *brief* periods of intense activity?
42. Which type of muscle fiber is most affected by endurance training exercises? What changes occur in the muscle tissue?
43. Birds use large, well-developed breast muscles for flying. Pheasants can fly 80 km/hr, but only for short distances. Great blue herons can fly only about 35 km/hr, but they can cruise great distances. What kind of fibers would each have in its breast muscles?

Chapter 26
LIPID METABOLISM

Lipids serve as the principal storage form of energy as well as to insulate the organism against drastic changes in temperature.

Nearly all the energy required by the animal organism is generated by the oxidation of carbohydrates and lipids. Of all major nutrients, triglycerides are the richest energy source. The oxidation of 1 g of a typical fat or oil liberates about 9500 cal; the oxidation of an equal mass of carbohydrate liberates only about 4200 cal. A lipid molecule has a high proportion of carbon–hydrogen bonds ($—CH_2—$) compared to a carbohydrate molecule ($—CHOH—$). Therefore, lipids have a greater capacity to combine with oxygen and consequently have a higher heat content.

Carbohydrates provide a readily available source of energy. Lipids function as the principal energy reserve. Men and women differ in their capacity to store lipids. The average male stores about 12% of his body weight as fat, whereas about 20% of the weight of an average female is due to body fat. (In contrast, marathon runners average about 5–6% body fat.) Americans consume about 100–125 g of lipids each day. This represents 38% of the daily calorie requirements. The National Cancer Institute, the American Heart Association, and most other health authorities recommend that Americans reduce their lipid intake and that not more than 30% of total calories be provided by lipids. Normally, triglycerides constitute about 98% of the total dietary lipids. However, as we shall see, significant quantities can be synthesized from carbohydrates when carbohydrate intake is high and lipid intake is low.

693

Fats as Fuels

A large percentage of the lipid molecule is like a saturated hydrocarbon (i.e., highly reduced). Thus fats may be thought to be analogous to combustible petroleum products, whereas carbohydrates may be seen as analogous to alcohols (more oxidized), which are not nearly as effective as fuels.

The fat reserves in the average individual provide sufficient energy to survive starvation for 30 days (given sufficient water). In comparison, glycogen in the liver (about 100 g) is depleted within 1 day. From the standpoint of efficiency of fuel storage, fats are far superior. Glycogen, because of its many hydroxyl groups, is extremely hydrated. About 2 g of water is bound to every gram of stored glycogen. Thus in comparing stored mass of each fuel in the body, 1 g of fat contains more than six times the energy content of 1 g of hydrated glycogen. The ability to store greater amounts of the more energy-efficient lipids is especially important for migrating birds and terrestrial animals. The camel's hump, for example, is almost all adipose tissue (Section 26.2), and migratory birds rely on stored fats to supply the energy for long, sustained flight.

EXAMPLE 26.1

One of McDonald's Big Mac sandwiches furnishes a total of 541 kcal, of which 279 kcal comes from fat. What percent of total calories is from fat?

SOLUTION
Simply divide the calories from fat by the total calories. Then multiply by 100 to get percent (parts per 100).

$$\% \text{ calories from fat} = \frac{279 \text{ kcal}}{541 \text{ kcal}} \times 100 = 51.6\%$$

Practice Exercise
One of Burger King's Whopper sandwiches furnishes a total of 606 kcal, of which 288 kcal comes from fat. What percent of total calories is from fat?

The nutritional aspects of lipids are still not completely understood. Lipids are not dietary necessities. An organism can survive on a lipid-free diet if carbohydrates and proteins are supplied as a source of metabolic energy. Certain lipids, however, are required for normal growth and development. These lipids supply certain fatty acids that the organism cannot synthesize. The unsaturated acids containing more than one double bond—linoleic and linolenic acids—are the **essential fatty acids.** Linoleic acid is used by the body to synthesize many of the other unsaturated fatty acids such as arachidonic acid. (Recall from Section H.6 that arachidonic acid is required as a precursor for the biosynthesis of prostaglandins.) In addition, the essential fatty acids are incorporated into the structures of the membrane lipids (Section 20.9), and they are necessary for the efficient transport and metabolism of cholesterol. The average daily diet should contain about 2–3 g of linoleic acid. Infants lacking essential fatty acids in their diet will lose weight and develop eczema. **Eczema** is an inflammatory skin disease characterized by scales and crusts on the skin.

In this chapter, we look at how our bodies store and metabolize fats. We'll also consider some of the problems associated with lipid storage and metabolism. Lipids offer one of the best illustrations of an old saying. Too much of a good thing can be bad!

26.1
DIGESTION AND ABSORPTION OF LIPIDS

Lipids are not digested by the body until they reach the upper portion of the small intestine (duodenum). In this region, a hormone is secreted that stimulates the gall-bladder to discharge bile into the duodenum. In the context of lipid digestion, the principal constituents of the bile are the bile salts (Section 20.10). The bile salts act as emulsifiers and serve to disrupt some of the hydrophobic bonds that tend to hold the lipid molecules together. This emulsification process is essential to lipid digestion. Bile salts act much like soap molecules. They break down large, water-insoluble lipids into smaller globules (micelles) and keep the smaller globules suspended in the aqueous digestive medium. The greatly increased surface area of the lipid particles and the opportunity afforded for more intimate contact with the lipases results in a much more rapid digestion of the fats. Another hormone then promotes the secretion of the pancreatic juice. This juice contains the lipases that catalyze the digestion of triglycerides into 2-monoglycerides and fatty acids. Phospholipids and cholesterol esters are also hydrolyzed into their component molecules.

Biochemists do not agree upon the extent to which a lipid must be hydrolyzed in order to be absorbed. It is probable that after emulsification some lipids are absorbed directly through the intestinal membrane before being hydrolyzed. (Any lipid that is not absorbed is eliminated in the feces.) Once the monoglycerides, fatty acids, and free cholesterol pass into the cells of the intestinal epithelium, they are immediately resynthesized into triglycerides, phospholipids, or cholesterol esters. It has been shown experimentally that the glycerol needed to form these compounds is supplied by the metabolic pool within the intestinal cells and is not the glycerol obtained by lipid hydrolysis.

Lipids are, by definition, relatively insoluble in water.[1] Since blood is an aqueous solution, lipids as such are not soluble in blood. But lipids can be efficiently transported by the blood if they are first complexed with water-soluble proteins in the plasma. Such

---

[1] There is one advantage to the low solubility of fats. These compounds can be separated from other components of the blood by extraction with fat solvents such as toluene or methylene chloride. Such an extraction is always one of the first steps in any clinical procedure aimed at determining lipid levels.

Table 26.1
Range of Normal Levels
of Lipids in Blood Plasma
of a Fasting Person

Constituent	Concentration (mg/100 mL)
Free cholesterol	30–60
Cholesterol esters	75–150
Total cholesterol	120–250
Triglycerides	25–260
Lecithin	100–225
Sphingomyelin	10–47
Total phospholipids	150–250
Total lipids	400–700

complexes are called **lipoproteins.** The triglycerides that are formed within the intestinal cells become associated with proteins and form the chylomicrons (recall Section 20.11).

The chylomicrons are transported by the lymphatic system into the bloodstream.[2] Some of the triglycerides are carried to the liver, where they are modified and/or used to provide energy for liver functions. The remaining lipids (and the triglycerides that are synthesized in the liver) are transported to specialized fat-storage cells (via VLDLs) or are circulated through the blood as lipoprotein complexes (LDLs and HDLs). The circulating lipids are distributed to the various tissues to be incorporated into membrane lipids or oxidized for the generation of energy.

The concentration of lipids in the blood changes constantly. As lipids are absorbed from the digestive tract after a meal, the level in the blood increases. As blood lipids are removed to storage or are oxidized in certain tissues, the blood lipid level falls. On demand, the body can synthesize lipids from other foods or remove already-formed lipids from storage. Both activities would result in an increase in the blood lipid level. Finally, the body excretes some lipids as a normal component of feces. These excreted lipids may be in the form of fats or soaps or fatty acids. Their removal via the intestine would tend to lower the blood lipid level.

To standardize the conditions under which lipid concentrations are determined, lipids in the plasma are measured after fasting. This is the same procedure followed in the measurement of blood sugar levels. Typical *normal fasting levels* of lipids are given in Table 26.1. Abnormally high levels of triglycerides and cholesterol are thought by many medical practitioners to be involved in hardening of the arteries, a condition that may lead to rupture or blockage of vessels in the brain (a stroke) or in the heart (a heart attack, or coronary—Section 20.11).

26.2
FAT DEPOTS

Fats are stored throughout the body. Principally, though, they are deposited in a special kind of connective tissue called **adipose tissue.** Storage places are called fat depots. **Fat depots,** such as those directly beneath the skin (subcutaneous) and in the abdominal region, are specialized tissue in which a relatively large percentage of cytoplasm is replaced by large droplets of triglycerides (approximately 90% of the mass of the

---

[2] Lymph is tissue fluid (water and dissolved substances) that has entered a lymphatic capillary. Lymph vessels transport excess fluid from interstitial spaces and return it to the bloodstream. The blood system and the lymphatic system are separate, but there is a crossover point via the thoracic duct. After a high-fat meal, the lymph changes from a transparent yellow to a milky white because of the emulsified lipid.

fat cell—Figure 26.1). Adults have approximately 30–40 billion fat cells that swell or shrink like a sponge depending on the amount of fat inside them. Adipose tissue is the only tissue in which free triglycerides occur in appreciable amounts. Elsewhere, in cells or in the blood plasma, lipids are bound to proteins as lipoprotein complexes.

Considerable fat is stored around vital organs such as the heart, liver, kidneys, and spleen. There it serves as a protective cushion, helping prevent injury to the organs. Fat is also stored under the skin, where it helps insulate against sudden temperature changes. The fat acts just like the insulation in the walls of a house, trapping body heat and preventing it from escaping to the surroundings. In some ways, the fatty tissue also acts like the furnace of a house. When the outside temperature drops, metabolic activity in the cells generates heat to compensate for the heat lost to the environment.

When food is taken into the body in excess of immediate needs, it is shunted to the storage areas. Fats are tucked away in the adipose tissue. Carbohydrates are converted to glycogen. When no more glycogen storage area is available, carbohydrates are converted to fats and stored in the depots. There is a continuous, dynamic change as molecules come and go from storage. If your weight is constant, there is an ''equilibrium'' in which the number of molecules arriving for storage is equal to the number being removed from the depot. We will discuss two situations in which there is an imbalance—obesity and starvation—in subsequent sections. The fate of lipids in the animal body is outlined in Figure 26.2.

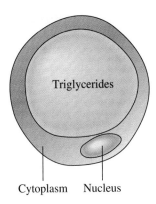

Figure 26.1

A typical fat cell found in adipose tissue. Fat cells function to produce and store lipids; they are among the largest cells in the body. Notice that the nucleus and cytoplasm are displaced to the cell periphery by the large lipid droplet.

26.3
FATTY ACID OXIDATION

In order for triglycerides to be used for the production of energy, they must first be cleaved to fatty acids and glycerol by the lipases contained within the adipose tissue.

Fat metabolism, like carbohydrate metabolism, is regulated by hormones. Again, it is the same two energy-invoking hormones, epinephrine and glucagon, along with the

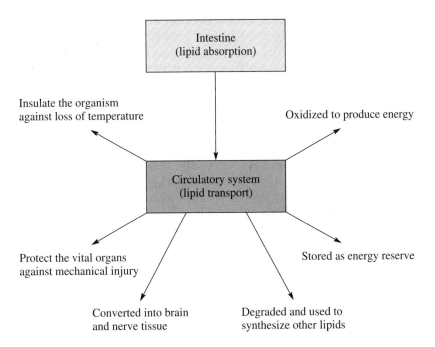

Figure 26.2

Metabolic fate of lipids.

Brown Fat

> In addition to the normal white adipose tissue, some organisms contain another type of fatty tissue called **brown fat.** Its brown color is due to the presence of large numbers of mitochondria (stocked with red-brown cytochromes). In these specialized mitochondria, the oxidation of fat is used to generate heat instead of being converted into ATP. Brown fat is particularly abundant in hibernating animals and in the neck and upper back of newborn infants (as well as other mammals that are born hairless). These cells serve as heat factories that revive hibernating animals and maintain the body temperature of young animals. Adults have few brown fat cells because metabolic reactions usually generate more than enough heat so that body temperature can be regulated by the dissipation of this heat.

neurotransmitter norepinephrine that bind to receptor proteins in the cell membranes of adipose tissue. They activate the enzyme *adenyl cyclase* (Section 24.4) to form cAMP from ATP. The cAMP then stimulates the hydrolysis *(lipolysis)* of triglycerides (by the lipases) and the release *(mobilization)* of fatty acids from adipose tissue. Insulin, on the other hand, is concerned with the storage of metabolic fuels. It enhances the synthesis of triglycerides and the storage of fats in adipose tissue. Insulin will be present in the blood when fatty acids are not needed for energy, that is, when there is a plentiful supply of blood glucose.

The glycerol obtained from fat hydrolysis is transported to the liver, where it is converted to a carbohydrate derivative (see Section 26.5) and is metabolized via the Embden–Meyerhof pathway. The fatty acids, bound to serum albumins, are transported by the blood to other tissues for oxidation. It should be emphasized that in a normal individual, certain organs use both glucose and fatty acids (in varying amounts) for their energy needs. The utilization rate of a specific fuel depends on the physiological state of the individual. For example, glucose (or glycogen) is the chief fuel for the immediate energy needs of active skeletal muscle (as in a sprint), whereas fatty acids are the major fuel for resting skeletal muscle. Fatty acids are the major fuel for cardiac muscle, although ketone bodies (see Section 26.6), glucose, and lactic acid are also used by the heart. During fasting and prolonged exercise, fatty acids are the preferred fuel for all types of muscles. The kidneys and adipose tissue use both glucose and fatty acids, whereas fatty acids are the preferred fuel of the liver. The brain uses glucose and ketone bodies (Section 26.6). It does not use fatty acids because they are bound to albumins and cannot diffuse across the blood–brain barrier. Red blood cells can only use glucose because they lack mitochondria and hence cannot obtain energy via oxidative phosphorylation.

The fatty acids, which contain the bulk of the energy of the lipids, are broken down in a series of sequential reactions accompanied by the gradual release of utilizable energy. Some of these reactions are oxidative and require the same coenzymes (NAD^+, FAD) as those that take part in the oxidation of carbohydrates. The enzymes involved in fatty acid catabolism are localized in the mitochondria along with the enzymes of the Krebs cycle, respiratory chain, and oxidative phosphorylation. This factor (the localization within the mitochondria) is of the utmost importance because it provides for the efficient utilization of the energy stored in the fatty acid molecules.

In the beginning of the twentieth century, the German biochemist Franz Knoop showed that the breakdown of fatty acids occurs in a stepwise fashion with the removal of two carbon atoms at a time. Subsequent investigations have resulted in the separation and purification of the enzymes involved. The details of the degradation sequence are now fully understood. The reaction scheme is shown in Figure 26.3.

Step 1 The first phase of fatty acid metabolism occurs on the outer mitochondrial membrane. Fatty acids, like carbohydrates and amino acids, are relatively inert and must first be activated by conversion to an energy-rich fatty acid derivative of coenzyme A (called *fatty acyl CoA*). This activation process is a two-step reaction that is catalyzed by the enzyme *acyl CoA synthetase*. For each molecule of fatty acid that is activated, one molecule of coenzyme A and one molecule of ATP are used. First the fatty acid reacts with ATP to form a fatty acyl adenylate. Then the sulfhydryl group of coenzyme A attacks the acyl adenylate to yield fatty acyl CoA and AMP. Finally, the pyrophosphate formed in the

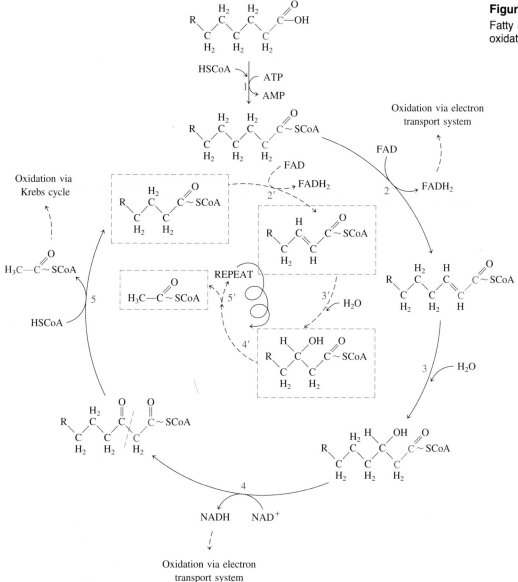

Figure 26.3

Fatty acid oxidation (beta-oxidation).

initial reaction is hydrolyzed to phosphate ions by the enzyme *pyrophosphatase*. The net effect is the utilization of two high-energy bonds of ATP to activate each molecule of a fatty acid.

$$
\text{R—CH}_2\text{CH}_2\text{CH}_2\text{CH}_2\text{CH}_2\text{C}\overset{\displaystyle O}{\underset{OH}{\diagup}} \;+\; \text{ATP} \;\xrightleftharpoons[\;]{\text{synthetase}}\; \text{R—CH}_2\text{CH}_2\text{CH}_2\text{CH}_2\text{CH}_2\text{C}\overset{\displaystyle O}{\underset{AMP}{\diagup}} \;+\; \text{PP}_i
$$

Fatty acid   Fatty acyl adenylate

$$
\text{R—CH}_2\text{CH}_2\text{CH}_2\text{CH}_2\text{CH}_2\text{C}\overset{\displaystyle O}{\underset{AMP}{\diagup}} \;+\; \text{HSCoA} \;\xrightleftharpoons[\;]{\text{synthetase}}\; \text{R—CH}_2\text{CH}_2\text{CH}_2\text{CH}_2\text{CH}_2\text{C}\overset{\displaystyle O}{\underset{SCoA}{\diagup}} \;+\; \text{AMP}
$$

Fatty acyl adenylate   Coenzyme A   Fatty acyl CoA

$$
\text{PP}_i \;+\; \text{HOH} \;\xrightarrow{\text{pyrophosphatase}}\; 2\,\text{P}_i
$$

The fatty acyl CoA combines with a carrier molecule (carnitine—See Exercise 23) and is transported into the mitochondrial matrix. Further metabolism of the fatty acyl CoA occurs entirely within the mitochondrial matrix via a sequence of reactions known as **β oxidation** (because the β carbon undergoes successive oxidations) or the **fatty acid spiral** (because it involves the progressive removal of two-carbon units from the carboxyl end of the fatty acyl CoA).

The reactions involved in steps 2–4 (dehydrogenation, hydration, and dehydrogenation) are analogous to those involved in the conversion of succinic acid to oxaloacetic acid in the Krebs cycle (succinic → fumaric → malic → oxaloacetic). The functional group at the β carbon can be visualized as undergoing the following changes: alkane → alkene → secondary alcohol → ketone. The net effect is the introduction of a keto group beta to a carboxyl group.

Step 2   This first oxidation reaction is catalyzed by the enzyme *acyl CoA dehydrogenase*. The coenzyme FAD accepts two hydrogen atoms from adjacent carbons, one from the α carbon and one from the β carbon. The enzyme is stereospecific in that only the *trans*-alkene is obtained.

$$
\text{R—CH}_2\text{CH}_2\text{CH}_2\overset{\displaystyle H}{\underset{\displaystyle H}{-\overset{|}{\underset{|}{C}}}} \overset{\displaystyle H}{\underset{\displaystyle H}{-\overset{|}{\underset{|}{C}}}} -\text{C}\overset{\displaystyle O}{\underset{SCoA}{\diagup}} \;+\; \text{FAD} \;\xrightarrow{\text{dehydrogenase}}\; \text{R—CH}_2\text{CH}_2\text{CH}_2\overset{\displaystyle H}{\underset{\;}{-\overset{|}{C}}}{=}\overset{\;}{\underset{\displaystyle H}{\overset{|}{C}}}-\text{C}\overset{\displaystyle O}{\underset{SCoA}{\diagup}} \;+\; \text{FADH}_2
$$

Fatty acyl CoA   Enoyl CoA

Each molecule of $FADH_2$ that is reoxidized via the electron transport chain supplies energy to form two molecules of ATP.

Step 3   Again, as in the Krebs cycle, only the ʟ isomer is formed when the stereospecific enzyme *enoyl CoA hydratase* adds water across the trans double bond.

$$
\text{R—CH}_2\text{CH}_2\text{CH}_2\overset{\displaystyle H}{\underset{\displaystyle H}{-\overset{|}{C}}}{=}\overset{\;}{\underset{\;}{\overset{|}{C}}}-\text{C}\overset{\displaystyle O}{\underset{SCoA}{\diagup}} \;+\; \text{HOH} \;\xrightleftharpoons[\;]{\text{hydratase}}\; \text{R—CH}_2\text{CH}_2\text{CH}_2\overset{\displaystyle HO}{\underset{\displaystyle H}{-\overset{|}{C}}}\overset{\displaystyle H}{\underset{\displaystyle H}{-\overset{|}{C}}}-\text{C}\overset{\displaystyle O}{\underset{SCoA}{\diagup}}
$$

Enoyl CoA   ʟ-β-Hydroxyacyl CoA

Step 4   The second oxidation reaction is catalyzed by *β-hydroxyacyl CoA dehydrogenase,* which exhibits an absolute stereospecificity for the ʟ isomer. Here the coenzyme NAD^+ is the hydrogen acceptor.

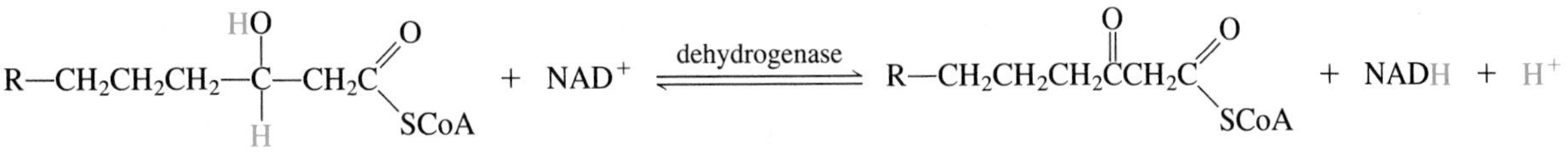

R—CH$_2$CH$_2$CH$_2$—$\overset{\text{HO}}{\underset{\text{H}}{\text{C}}}$—CH$_2C\overset{\text{O}}{\underset{\text{SCoA}}{}}$ + NAD$^+$ $\xrightleftharpoons{\text{dehydrogenase}}$ R—CH$_2$CH$_2$CH$_2$$\overset{\text{O}}{\text{C}}CH_2C\overset{\text{O}}{\underset{\text{SCoA}}{}}$ + NADH + H$^+$

L-β-Hydroxyacyl CoA   β-Ketoacyl CoA

The reoxidation of NADH and the transport of hydrogen ions and electrons through the respiratory chain furnishes three molecules of ATP.

Step 5 The final reaction is a cleavage of the β-ketoacyl CoA by a molecule of coenzyme A. The products of the reaction are acetyl CoA and a fatty acyl CoA whose chain length is shortened by two carbon atoms. The enzyme is β-ketothiolase *(thiolase)*.

R—CH$_2$CH$_2$CH$_2$$\overset{\text{O}}{\text{C}}CH_2C\overset{\text{O}}{\underset{\text{SCoA}}{}}$ + HSCoA $\xrightarrow{\text{thiolase}}$ R—CH$_2$CH$_2$CH$_2$C$\overset{\text{O}}{\underset{\text{SCoA}}{}}$ + H—$\overset{\text{H}}{\underset{\text{H}}{\text{C}}}$—C$\overset{\text{O}}{\underset{\text{SCoA}}{}}$

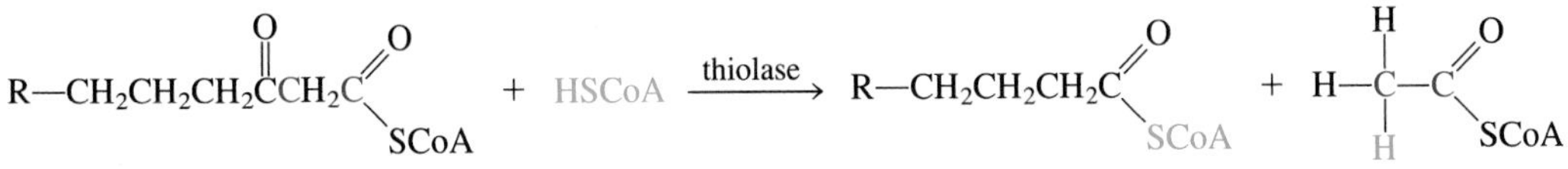

β-Ketoacyl CoA   Shortened fatty acyl CoA   Acetyl CoA

The newly formed fatty acyl CoA can then be degraded further by repetition of steps 2–5; a molecule of acetyl CoA is liberated at each turn of the spiral. Normally, the spiral is repeated as many times as is necessary to break down a fatty acid containing an even number of carbon atoms, n, into $n/2$ molecules of acetyl CoA. It should be noted that no further addition of ATP is necessary because the shortened fatty acids already contain the thiol ester. One molecule of ATP is sufficient to activate any fatty acid regardless of the number of carbon atoms in its hydrocarbon chain. The unsaturated fatty acids (which comprise about 50% of the fatty acids found in humans) are also incorporated into the β oxidation sequence. Two ancillary enzymes are required to convert them into normal substrates of the fatty acid spiral. The overall equation for the β oxidation of palmitoyl CoA (16 carbons) is as follows.

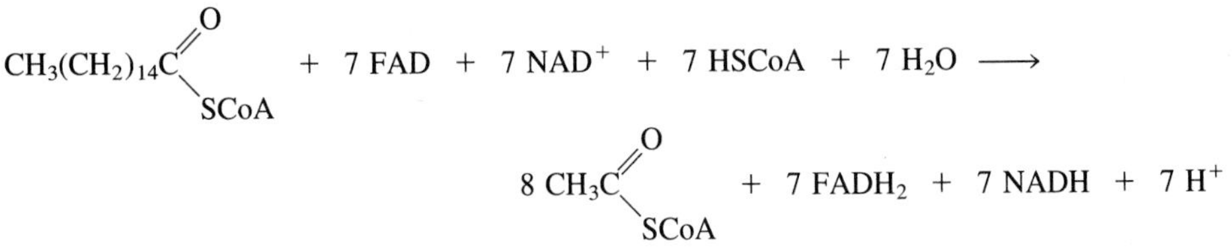

CH$_3$(CH$_2$)$_{14}$C$\overset{\text{O}}{\underset{\text{SCoA}}{}}$ + 7 FAD + 7 NAD$^+$ + 7 HSCoA + 7 H$_2$O $\longrightarrow$

8 CH$_3$C$\overset{\text{O}}{\underset{\text{SCoA}}{}}$ + 7 FADH$_2$ + 7 NADH + 7 H$^+$

The acetyl CoA formed in the fatty acid spiral (plus the acetyl CoA obtained from glucose metabolism) is involved in a myriad of biochemical pathways. It may enter the Krebs cycle to be oxidized to produce energy, or it may be used as the starting material for biosynthesis of lipids (triglycerides, phospholipids, cholesterol, and other steroids). Acetyl CoA is also used in the formation of the ketone bodies (Section 26.5). The various routes available to acetyl CoA are summarized in Figure 26.4.

When the body ingests more carbohydrate than it needs for energy and for glycogen synthesis, the excess is converted into fatty acids via the common intermediate, acetyl

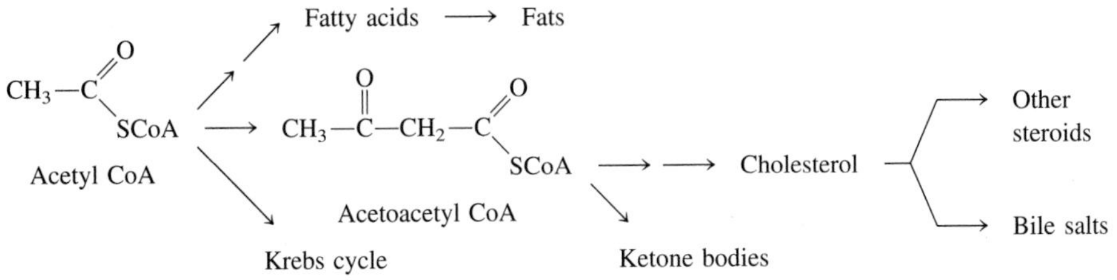

Figure 26.4

Acetyl coenzyme A plays a variety of roles in cellular chemistry.

CoA. (In normal adults, roughly one-third of the total ingested carbohydrates may be converted into fatty acids and stored as fat.) *Fatty acids are built up from two-carbon units that come from acetyl CoA.* Although we will not discuss the biosynthetic pathway in this text, it should be pointed out that fatty acid synthesis is not simply the reverse of fatty acid breakdown. In humans, the liver is the major site of fatty acid synthesis. Synthesis occurs in the cytoplasm of the liver, whereas fatty acid oxidation occurs in the mitochondria of various tissues. A completely different set of enzymes are involved. The newly synthesized fatty acids are used to provide energy, or they are esterified with glycerol and stored as triglycerides in adipose tissue.

Finally, it is important to recall that the conversion of pyruvic acid to acetyl CoA is irreversible (Section 25.1) in mammals. Therefore the acetyl CoA molecules derived from fatty acids cannot be used to synthesize glucose. *Mammals can convert carbohydrates to lipids (via acetyl CoA). They cannot convert lipids to carbohydrates.*

26.4
BIOENERGETICS OF FATTY ACID OXIDATION

Note that the oxidation of fatty acids also produces large quantities of water, which sustains migratory birds and animals (e.g., the camel) for long periods of time.

The combustion of 1 mol of palmitic acid releases a considerable amount of energy, and the precise amount can be determined by conducting the experiment in a calorimeter.

$$C_{16}H_{32}O_2 + 23\ O_2 \longrightarrow 16\ CO_2 + 16\ H_2O + 2340\ kcal$$

The amount of energy that is made available to the cell and conserved in the form of ATP is readily calculable.

The breakdown by an organism of 1 mol of palmitic acid requires 1 mol of ATP, and 8 mol of acetyl CoA is formed. You will recall (Figure 25.1) that each mole of acetyl CoA metabolized by the Krebs cycle yields 12 mol of ATP. Each turn of the fatty acid spiral produces 1 mol of NADH and 1 mol of $FADH_2$. Reoxidation of these compounds by the respiratory chain (within skeletal muscle cells) yields 3 and 2 mol of ATP, respectively. The complete degradation of 1 mol of palmitic acid requires seven turns of the spiral, and a total of 7 mol of $FADH_2$ and NADH is therefore formed. The energy calculations can be summarized as follows.

	Yield of ATP
1 mol of ATP is split to AMP and 2 P_i	−2
8 mol of acetyl CoA formed (8 × 12)	96
7 mol of $FADH_2$ formed (7 × 2)	14
7 mol of NADH formed (7 × 3)	21
Total moles of ATP	129

The percentage of available energy that can theoretically be conserved by the cell in the form of ATP is calculated as follows.

$$\frac{\text{Energy conserved}}{\text{Total energy available}}(100\%) = \frac{(129)(7300)}{2,340,000}(100\%) = 40\%$$

The efficiency with which fatty acids are metabolized is seen to be comparable to that of the carbohydrate metabolism system. Recall we said that carbohydrates are used for the normal energy requirements of skeletal muscle, whereas lipids provide energy for prolonged activity. It has been estimated that after 4 hr of sustained exercise, fatty acid oxidation supplies more than 60% of the muscle's energy demands.

26.5
GLYCEROL METABOLISM

One molecule of glycerol is obtained from each molecule of triglyceride or phospholipid that is hydrolyzed. Glycerol is transported to the liver, where it is readily incorporated into the scheme of carbohydrate metabolism by conversion to dihydroxyacetone phosphate. This two-step process includes the phosphorylation of a primary hydroxyl group followed by the oxidation of a secondary hydroxyl group to a ketone. Dihydroxyacetone phosphate can be used to provide energy by conversion to pyruvic acid (via the glycolytic pathway), or it can be transformed into glucose. Since glycerol is the starting compound for glucose synthesis, this is another important example of gluconeogenesis.

$$
\begin{array}{ccc}
\text{CH}_2\text{OH} & \text{CH}_2\text{OH} & \text{CH}_2\text{OH} \\
| & \xrightarrow[\text{kinase}]{\text{ATP ADP}} \quad | & \underset{\text{dehydrogenase}}{\overset{\text{NAD}^+ \quad \text{NADH} + \text{H}^+}{\rightleftharpoons}} \quad | \\
\text{CHOH} & \text{CHOH} & \text{C}=\text{O} \\
| & | & | \\
\text{CH}_2\text{OH} & \text{CH}_2\text{O}\,\text{\textcircled{P}} & \text{CH}_2\text{O}\,\text{\textcircled{P}} \\
\text{Glycerol} & \text{Glycerol} & \text{Dihydroxyacetone} \\
& \text{phosphate} & \text{phosphate}
\end{array}
$$

A summary of the steps involved in the catabolism of a triglyceride molecule is illustrated in Figure 26.5.

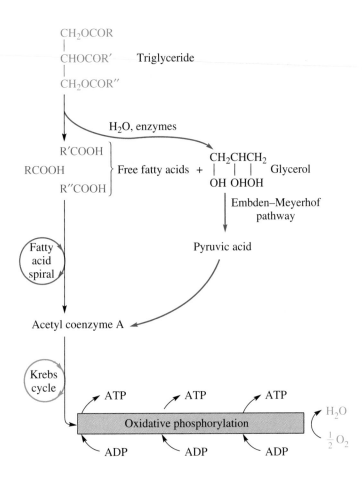

Figure 26.5

An outline showing how fat reserves (triglycerides) are used for the production of energy.

Chapter 27
PROTEIN METABOLISM

Meats and vegetables supply the amino acids necessary to build body proteins.

Consider green plants. If proper amounts of nutrients—nitrates, phosphates, sulfur compounds, and so on—are available, green plants can make all the amino acids they need. From these the plants put together all their necessary protein. We poor animals are not quite so versatile. We too can put together all the proteins we require, but only if we eat properly.

In Chapter 21 we indicated the essential nature of protein molecules and their vital functions in all living organisms. In Chapter 23 the biosynthesis of proteins was discussed in connection with the study of nucleic acids. In this chapter we deal primarily with the catabolic aspects of protein metabolism, with particular reference to the metabolic role of the amino acids.

27.1
DIGESTION AND ABSORPTION OF PROTEINS

Like polysaccharides and lipids, intact proteins cannot normally be absorbed across intestinal membranes. They must first be hydrolyzed into their constituent amino acids. Protein digestion begins in the stomach, where the action of the gastric juice effects the hydrolysis of about 10% of the peptide bonds. Gastric juice is a mixture of secretions of the stomach. The chief components are water (more than 99%), mucin, inorganic ions, hydrochloric acid, and some enzymes. A protein hormone called *gastrin,* produced in the stomach, starts the flow of gastric juices. Flow can also be started by histamine. Indeed, it may well be that gastrin acts by releasing histamine (see Section 27.5), which in turn stimulates the secretion of gastric juices.

Hydrochloric acid is secreted by certain glands in the lining of the stomach. The pH of freshly secreted gastric juice is about 1.0, but the contents of the stomach may react with it, raising the pH to between 1.5 and 2.5. (The pain of a gastric ulcer is at least partially due to the irritation of the ulcerated tissue by the acidic gastric fluid.) Hydrochloric acid is involved in the denaturation of food protein—it opens up the folds to expose the chains to more efficient enzyme action.

The principal digestive component of gastric juice is *pepsinogen,* a zymogen that is produced in secretory cells located in the stomach wall. Pepsinogen is catalytically converted by hydrogen ions into its active form, *pepsin.* Once some pepsin is formed, it also assists in the activation of the remaining pepsinogen. This phenomenon is known as **autocatalysis,** the catalysis of a reaction by one of the products of that reaction. Pepsin is an endopeptidase that catalyzes the hydrolysis of peptide linkages within the protein molecule. It has a fairly broad specificity but acts preferentially on linkages involving the aromatic amino acids tryptophan, tyrosine, and phenylalanine as well as methionine and leucine. The gastric juice of infants contains *rennin,* an enzyme having a specificity very similar to that of pepsin.

Food stays in the stomach for 2–5 hr. It is broken down by pepsin and by the mechanical churning of the stomach into a thin, watery liquid called *chyme.* This material then passes, in small portions, into the duodenum, the first 30 cm of the small intestine. Much coiled, the small intestine is about 7 m long in adult humans. Most of the enzymes of the intestinal juice are secreted from the duodenum.

Protein digestion is completed in the small intestine by the action of the digestive juices of the pancreas and of the intestinal mucosal cells. The pancreas is a large organ lying just below the stomach (see Figure 24.5). It produces two kinds of secretions. One kind includes insulin, a hormone of protein nature that is emptied into the blood stream, where it influences carbohydrate metabolism. The other, called pancreatic juice, passes through the pancreatic duct into the duodenum. This juice is sufficiently alkaline (pH ~7.5–8.5) to neutralize the acidic material passed on from the stomach. The pancreatic juice contains the zymogens trypsinogen and chymotrypsinogen.

The intestinal mucosal cells secrete the proteolytic enzyme enteropeptidase, which converts trypsinogen to trypsin (Figure 27.1). Trypsin then activates chymotrypsinogen to chymotrypsin. Both of these active enzymes are endopeptidases. Chymotrypsin preferentially attacks peptide bonds involving the carboxyl groups of the aromatic amino acids (phenylalanine, tryptophan, and tyrosine). Trypsin attacks peptide bonds involving the carboxyl groups of the basic amino acids (lysine and arginine). Pancreatic juice also contains the zymogen procarboxypeptidase, which is cleaved by trypsin to carboxypeptidase. The latter is an exopeptidase—it catalyzes the hydrolysis of the peptide linkages at the free carboxyl end of the peptide chain, resulting in the stepwise liberation of free amino acids from the carboxyl end of the polypeptide.

Two types of peptidases are secreted in the intestinal juice: (1) an exopeptidase, *aminopeptidase,* which acts upon the peptide linkages of terminal amino acids possessing a free amino group, and (2) *dipeptidase* and *tripeptidase,* which cleave dipeptides and tripeptides. Figure 27.2 illustrates the specificity of protein-digesting enzymes.

The combined action of the proteolytic enzymes of the gastrointestinal tract results in the hydrolysis not only of the dietary (exogenous) proteins but also of endogenous proteins (the digestive enzymes, other secreted proteins, and dead epithelial cells). (The

All enzymes that digest proteins (peptidases) are secreted in the form of their inactive precursors, probably to protect tissues from digestion by their own enzymes. A mucous lining protects the digestive tract from hydrolysis by the proteolytic enzymes that function within the tract.

Occasionally, small polypeptides may be absorbed into the bloodstream. These foreign polypeptides act as *antigens*—they stimulate the formation of specific *antibodies* (see Section 28.5) and are probably responsible for the allergic reactions that some individuals develop for certain foods.

Figure 27.1

Schematic representation of the structural changes involved in the activation of trypsinogen. Rupture of the lysyl–isoleucine bond in the N-terminal region (dashed arrow) leads to the liberation of the activation peptide and causes the newly formed N-terminal region of the polypeptide chain to assume a more nearly helical configuration. This in turn permits a histidine and a serine side chain to come into juxtaposition so that these catalytic groups are properly aligned. The specificity site of the protein (X) is believed to be preexistent in the zymogen molecule.

Figure 27.2

Specificity of peptidase hydrolysis.

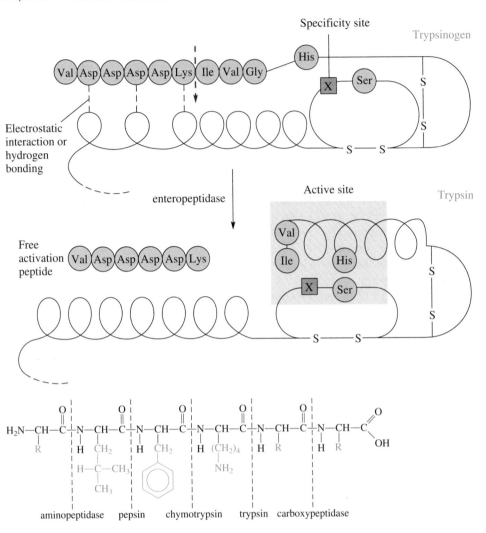

quantity of endogenous proteins hydrolyzed to amino acids is about 50–70 g each day.) The amino acids are actively transported (an energy-requiring process) across the intestinal wall into the portal vein and are carried to the liver. The liver is the principal organ responsible for the degradation and synthesis of amino acids. Amino acids have a high turnover rate (half-lives of a few days). The liver is also the site of synthesis of most blood proteins, such as albumins, globulins, fibrinogen, and prothrombin (see Chapter 28). The amino acids synthesized in the liver and the amino acids obtained from protein digestion, along with the amino acids derived from the turnover of tissue proteins, are transported by the blood to all the cells of the body.

27.2
NITROGEN BALANCE

The cellular proteins are in a constant state of flux. Our bodies continually take in proteins and break them down into their constituent amino acids. The liver and other body tissues take these amino acids and incorporate them into new proteins. Structural proteins, enzymes, and proteinlike hormones are continually being degraded and rebuilt. Our proteins are in a state of dynamic equilibrium as "old" ones are hydrolyzed and "new" ones are synthesized.

The amount of protein synthesized or degraded per unit time gives the turnover rate. The protein turnover rate in a normal adult is about 300 g/day. The rate is different for proteins in different kinds of body tissues. Turnover rates represent the average residence time for a protein molecule in the tissues. Turnover rates are usually expressed as half-lives. A half-life is the time interval in which one-half of the protein molecules in a given tissue have been replaced. Liver proteins and those in the blood plasma have rapid turnovers, with half-lives of 2–10 days. Muscle proteins have a more stable employment, with half-lives of about 6 months. Some collagen molecules may hang around for 3 years. The half-lives of enzymes vary widely, depending on their metabolic importance and the cell in which they function. Half-lives of enzymes may range from 10 min to 6 hr. Hair has no half-life because only synthesis (and not degradation) of hair protein occurs within the ectodermal (skin) cells.

A dietary intake of nitrogen is required to provide for the biosynthesis of the various nitrogenous compounds. Nitrogen is lost in the continuous degradation of tissue protein and in the excretion of certain nitrogen-containing waste materials. Under normal conditions, an individual's intake of dietary nitrogen is equal to the amount of nitrogen lost in the feces, urine, and sweat. Such a condition is referred to as **nitrogen balance** or **nitrogen equilibrium.** Organisms are said to be in *positive* nitrogen balance (intake exceeds excretion) whenever tissue is being synthesized—for example, during periods of growth, pregnancy, and convalescence from disease. *Negative* nitrogen balance results from (1) an inadequate intake of protein (for example, fasting); (2) fever, infection, surgery, or a wasting disease; or (3) a diet that lacks, or is deficient in, any one of the essential amino acids. These factors cause an accelerated breakdown of tissue protein (in an attempt to supply the missing amino acids), and nitrogen excretion exceeds intake.

27.3
THE ESSENTIAL AMINO ACIDS

We have said that higher plants and certain microorganisms are capable of synthesizing all of their amino acids from carbon dioxide, water, and inorganic salts. They obtain their required nitrogen either from soil nitrates or from atmospheric nitrogen (via nitrogen-fixing bacteria). Thus these organisms can grow on a medium that does not contain any preformed amino acids. Animals, however, can synthesize only about half of their naturally occurring amino acids. The remainder must be supplied in the diet. All of the amino acids required for the synthesis of a particular protein must be available to the cell at the time of protein synthesis. If just one amino acid is absent, or is present in insufficient quantity, the protein will not be synthesized. For example, if a given protein requires four units of phenylalanine per molecule and 40 are available, only 10 protein molecules can be made, even though there may be enough of all the other amino acids to make a million protein molecules. Within these limits, proteins are synthesized as needed by our bodies.

An **essential** or *indispensable* amino acid is one that cannot be synthesized by an organism, from the substrates ordinarily present in its diet, at a rate rapid enough to supply the normal requirements of protein biosynthesis. A list of essential amino acids is given in Table 27.1. Notice that, as a rule, these compounds contain carbon chains or aromatic rings that are not present as intermediates of carbohydrate or lipid metabolism. The inability to synthesize these amino acids does not arise from a lack of the necessary nitrogen but

Table 27.1
Essential and Nonessential Amino Acids for Humans

Essential	Nonessential
Lysine	Glycine
Leucine	Alanine
Isoleucine	Serine
Methionine	Tyrosine[a]
Threonine	Cysteine
Tryptophan	Aspartic acid
Valine	Asparagine
Phenylalanine	Glutamic acid
Histidine	Glutamine
Arginine[b]	Proline

[a] In the presence of adequate amounts of phenylalanine.
[b] Essential for growing children, not for adults.

Does Protein Build Muscles?

The pregame steak dinner consumed by football players in the past was based on a myth that protein builds muscles. If athletes want more muscle, says the myth, they should eat more protein. This is just not true. Although athletes do need the RDA for protein (based on grams per kilogram of body weight), they do not need an excess. Protein consumed in amounts greater than the amount needed for synthesis and repair of tissue will only make the athlete fatter (due to excessive calorie intake), not more muscular.

Some extreme endurance athletes such as triathletes and ultramarathoners may need a bit more than the RDA for protein. Since nearly all Americans eat 50% more protein than they need, even these special athletes seldom need protein supplements.

Muscles are built through exercise, not through eating excess protein. When a muscle contracts, against a resistance, creatine (Section 24.10) is released. Creatine stimulates the production of the protein myosin (Section 25.5), thus building more muscle tissue. If the exercise stops, the muscle begins to shrink after about 2 days. After about 2 months without exercise, muscle built through the exercise program is almost completely gone. (The muscle does *not* turn to fat, as some athletes believe. Former athletes often get fat, though, because they continue to take in the same number of calories and expend fewer.)

results rather from the animal's inability to manufacture the correct carbon skeleton. If, for example, an animal is supplied with phenylpyruvic acid, it can readily synthesize the amino analog phenylalanine. Lysine appears to be an exception because the entire preformed amino acid must be supplied. The general effect of a deficiency of one or more essential amino acids is to restrict growth and protein synthesis and produce a negative nitrogen balance.

A **complete protein** source supplies all the essential amino acids in the quantities needed for growth and repair of body tissues. Essential amino acids are best provided by animal protein. (Casein, the protein from milk, is especially beneficial because it is well balanced in its amino acid distribution.) Proteins that lack an adequate amount of the essential amino acids are termed **incomplete proteins.** Gelatin is an example of an incomplete animal protein because it is deficient in tryptophan. Most plant proteins are deficient in lysine and/or one other essential amino acid. Zein, the protein in corn, is deficient in lysine and tryptophan. People whose diet consists chiefly of corn may suffer from malnutrition even though the amount of calories supplied by the food is adequate. Protein from rice is short of lysine, methionine, and threonine. Wheat protein is lacking in lysine. Even the legumes, good nonanimal protein sources, are lacking in the essential amino acids methionine and tryptophan. People on vegetarian diets should consume a larger total quantity of protein than would be required from animal protein diets and should include diverse vegetable protein sources to provide the proper amino acid requirements. The average individual requires about 50–60 g (about 2 oz) of protein in the daily diet. This is equivalent to the protein supplied by one quarter-pound hamburger or one regular cheeseburger. (In the industrialized countries, the average daily intake of proteins is about 100 g.)

Plants, such as grasses or grains, trap a small fraction of the solar energy that falls upon them. They use this energy to convert carbon dioxide, water, and mineral nutrients

From a health standpoint, "liquid protein" reducing diets are harmful to the individual because the protein is denatured and partially hydrolyzed gelatin. If this is the chief source of protein, the diet lacks the essential amino acid tryptophan.

**Amino Acid
Supplements**

Amino acids have been touted in articles and advertisements in newspapers and magazines as a cure-all for a variety of ailments including cold sores, depression, fatigue, insomnia, obesity, and pain. For example, there are claims that tryptophan and tyrosine cure depression and insomnia and that leucine and phenylalanine are effective as pain relievers. However, the FDA states that it is dangerous for consumers to ingest large amounts of any one amino acid. Animal studies have shown that amino acid imbalances can be created in the body by abnormal intakes of specific individual amino acids.

In 1989, products containing the *manufactured* amino acid L-tryptophan were banned in most states. Pills and capsules containing doses of L-tryptophan were taken by some people for insomnia, depression, premenstrual syndrome, or weight control. At the time, over 600 cases of a rare blood disorder (eosinophilia-myalgia syndrome) were reported nationwide for users of the supplement. Over half the people were hospitalized, many in intensive care, and at least one death was directly linked to the product.

After months of intensive investigation, it was learned that all of the L-tryptophan was manufactured in Japan and then distributed worldwide. A contaminant in the manufacturing process, and not L-tryptophan itself, was found to be the cause of the blood disorder.

such as nitrates, phosphates, and sulfates into protein. Cattle eat the plant protein, digest it, and convert a small portion of it into animal protein. People eat this animal protein, digest it, and reassemble some of the amino acids into human protein. Some of the energy originally trapped by the green plants is lost at every step. If people ate the plant protein directly, one highly inefficient step would be skipped. Extreme vegetarianism is dangerous, however. One would have to eat a wide variety of plant materials to be sure of getting enough of all the essential amino acids. Even then, an all-vegetable diet is likely to be lacking in vitamin B_{12} because this nutrient is not found in plants. Other nutrients scarce in all-plant diets include calcium, iron, riboflavin, and (for children not exposed to sunlight) vitamin D. A modified vegetarian diet that includes milk, eggs, cheese, and fish can provide excellent nutrition even with red meat totally excluded.

27.4
THE CHEMISTRY OF STARVATION

When the human body is totally deprived of food, whether voluntarily or involuntarily, the condition is known as **starvation.** During total fasting, the body's glycogen stores are depleted rapidly and the body calls on its fat reserves. Fat is first obtained from around the kidneys and heart. Then it is removed from other locations. Ultimately even the bone marrow (which is also a fat storage depot) is depleted, and it becomes red and jellylike rather than white and firm.

In the early stages of a total fast, body protein is also metabolized at a relatively rapid rate. The preferred energy source for brain cells is glucose. If none is available in the diet,

Figure 27.3

A lack of proteins and vitamins causes the deficiency disease kwashiorkor. The symptoms include retarded growth, discoloration of skin and hair, bloating, swollen belly, and mental apathy.

the cells will make it from amino acids.[1] A starving human may lose as much as 6% of muscle mass per day. The loss of plasma proteins, especially albumins, occurs to an even greater extent. The nitrogenous metabolites from protein catabolism must be excreted through the kidneys. This requires large volumes of water. Starving people often die of dehydration.

After several weeks, the rate of protein breakdown slows considerably as the brain adjusts to using fatty acid metabolites for its energy source. When no more fat reserves remain, the body must again draw heavily on structural protein for its energy requirements. The emaciated appearance of a starving individual is due to depleted muscle protein.

Starvation is seldom the sole cause of death. Weakened by starvation, a person succumbs to disease. Even usually minor diseases such as chicken pox and measles become life-threatening. Barring disease, starvation alone will lead eventually to death from circulatory failure when the heart muscle becomes too weak to pump blood.

A similar situation occurs in the case of the protein deficiency disease kwashiorkor (Figure 27.3). Kwashiorkor results in extreme emaciation, bloated abdomen, mental apathy, diarrhea, lack of pigmentation of the skin and hair, and eventual death. This disease is prevalent in Latin America, Asia, and Africa where corn or rice is the major food. It is said to be the most severe and widespread nutritional disorder among young children. Kwashiorkor can best be treated by the administration of adequate amounts of well-balanced protein. The problem, of course, is that animal protein is a scarce commodity in many of these areas.

[1] During fasting or starvation, the liver glycogen is depleted very rapidly and blood glucose must be manufactured from other sources to supply energy for the brain cells. The major source of this blood glucose is the amino acids obtained from the degradation of tissue protein (e.g., proteins from skeletal muscles and from membranes that line the digestive tract). The minimum requirement for carbohydrates is only about 75 g/day because glucose can be synthesized from amino acids.

27.5
AMINO ACID METABOLISM

Carbohydrates can be stored in the liver and muscles as glycogen. Fats can be placed on reserve in the fat depots. For proteins, however, there are no comparable storage facilities. Proteins are digested in the small intestine, and the resulting amino acids are absorbed directly into the bloodstream. Here they join a circulating pool of amino acids.

Members of the **amino acid pool** are there only on temporary assignment. New members are constantly added to the pool, and old ones are regularly withdrawn. The body can supply the pool by synthesizing amino acids from scratch or by breaking down tissue protein to obtain the constituent amino acids. It drains the pool to obtain raw materials for new protein synthesis or for the synthesis of other nitrogen-containing compounds such as heme. It can also use the amino acids as a source of energy. The nitrogen stored in the pool cannot be recycled endlessly within the body. Each day some of the amino acids are catabolized, and the nitrogen is eliminated from the body as urea. This represents a net drain of material from the pool. It is to compensate for this drain that we require proteins in our diet.

The human body contains about 100g of *free* (surplus) amino acids. Two amino acids, glutamic acid and glutamine, account for half of this total, whereas the essential amino acids constitute about 10g. Most of the amino acids in the metabolic pool are utilized for the synthesis of any of the myriad of proteins necessary to the living organism. This is the major function of amino acids in the body. About 75% of the metabolized amino acids in the normal adult are used for the protein synthesis made necessary by the constant destruction of body proteins by wear and tear. However, amino acids also play an essential role in the metabolism of all nitrogenous compounds. Although the liver is the principal organ for amino acid metabolism, other tissues such as the kidney, intestine, muscle, and adipose tissue are also involved. Generally, the first step in the breakdown of amino acids is the separation of the amino group from the carbon skeleton. The amino groups are then incorporated into almost all the other nonprotein nitrogen-containing compounds, such as neurotransmitters, hormones, nucleic acids, porphyrins, and creatine (Figure 27.4). A discussion of these specific metabolic pathways is beyond the scope of this text.

The carbon skeletons resulting from the deaminated amino acids are used to form either glucose (via gluconeogenesis) or fats (via acetyl CoA), or they are converted to a metabolic intermediate that can be oxidized by the Krebs cycle. Amino acid catabolism is particularly prevalent during hypoglycemia and fasting (also starvation). Since fatty acids cannot be converted to glucose, proteins must be hydrolyzed to amino acids that can provide the proper carbon skeleton to form the glucose needed by the brain. By the end of the overnight fast, gluconeogenesis is providing blood glucose from amino acids.

Transamination

Transamination is an exchange of functional groups between amino compounds and keto compounds. The α-amino group of any amino acid except lysine, threonine, proline, and hydroxyproline can be removed by transamination. The amino group is transferred to one of three α-keto compounds—pyruvic acid, α-ketoglutaric acid, or oxaloacetic acid. It is

Figure 27.4

Scheme of general paths of amino acids in metabolism (average human).

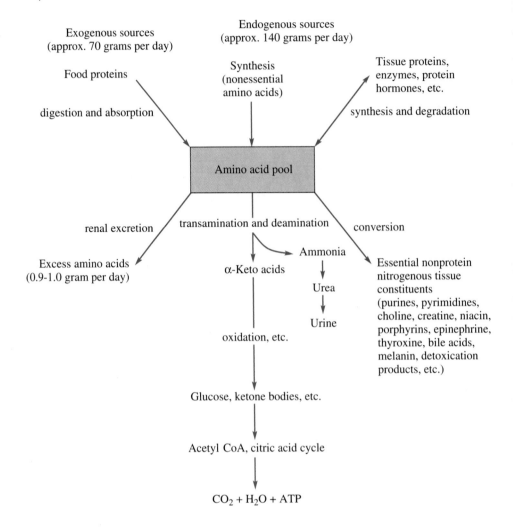

probable that there are a large number of transaminases (also called aminotransferases), each one catalyzing a reaction between a specific α-amino acid and some α-keto acid.

The three possible receptor compounds are converted to alanine, glutamic acid, and aspartic acid, respectively. These reactions are catalyzed by three transaminases. The enzymes, named for the product amino acid, are alanine transaminase, glutamate transaminase, and aspartate transaminase.

Two of the product amino acids, alanine and aspartic acid, are ultimately converted to the third, glutamic acid, through transamination reactions (Figure 27.5). These two particular transamination reactions are of special clinical interest. Normally, the blood contains a low concentration of transaminases. However, when extensive tissue destruction occurs, it is accompanied by rapid and striking increases in the blood transaminase levels. Glutamic-pyruvic transaminase (GPT) has a particularly high activity in the cyto-

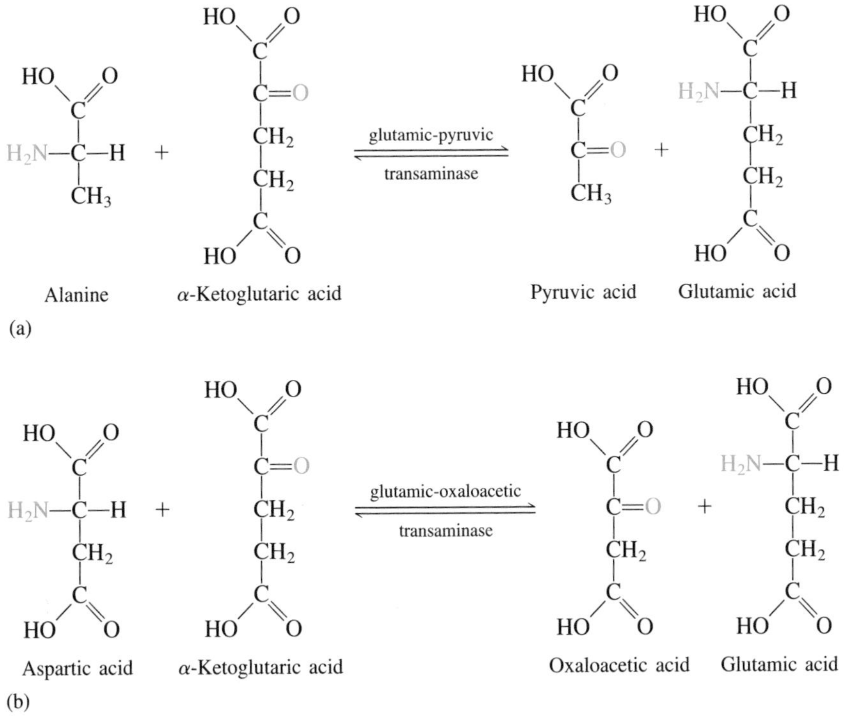

Figure 27.5
Transamination reactions catalyzed by GPT (a) and GOT (b). Note that whatever the pathway, the final receptor for the amino group from several different amino acids is α-ketoglutaric acid, and the final product amino acid is glutamic acid. The reverse of these reactions can occur. In this case, glutamic acid acts as the donor of its amino group to α-keto acids.

plasm of the liver, and an elevated serum level of this enzyme is indicative of liver damage. Glutamic-oxaloacetic transaminase (GOT) is abundant in heart muscle (recall Section 22.9), and a sharp rise in the concentration of GOT in the blood is an indication of myocardial infarction.

Oxidative Deamination

We saw that in transamination reactions the α-amino group of many amino acids is transferred to α-ketoglutaric acid to form glutamic acid. The latter compound then loses the amino group as ammonia and is oxidized back to α-ketoglutaric acid. This process is termed **oxidative deamination,** and it occurs in the liver mitochondria. Glutamic acid dehydrogenase is unusual in that it uses either NAD^+ or $NADP^+$ as a coenzyme. (The reduced forms of the coenzymes, NADH or NADPH, are ultimately oxidized by the respiratory chain.) The ammonia formed from glutamic acid by oxidative deamination is converted into urea, which is excreted in the urine (see Section 27.7).

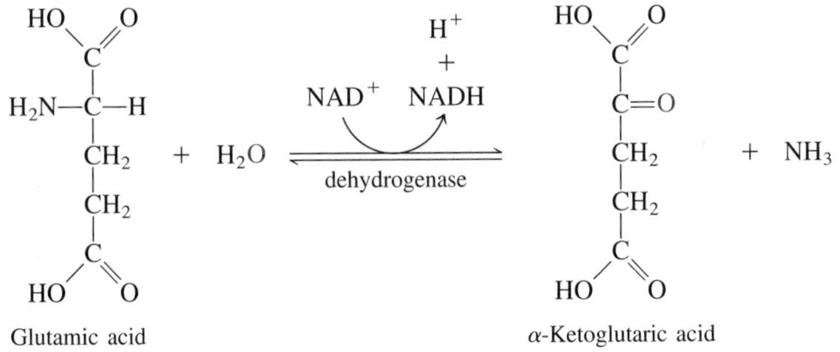

Table 27.2
Glucogenic and Ketogenic
Amino Acids

Glucogenic		Ketogenic	Glucogenic and Ketogenic
Alanine	Glycine	Leucine	Isoleucine
Arginine	Histidine		Lysine
Asparagine	Methionine		Phenylalanine
Aspartic acid	Proline		Tyrosine
Cysteine	Serine		Tryptophan
Glutamic acid	Threonine		
Glutamine	Valine		

This equilibrium reaction is of central importance in linking protein metabolism and carbohydrate metabolism. The reverse reaction is significant in nitrogen metabolism because it is one of the few reactions in animals that can convert inorganic nitrogen (NH_3) into organic nitrogen (amino acids). The amino group in the glutamic acid can be passed on through transamination reactions, producing all the other cellular amino acids, provided the appropriate α-keto acids are available.

All of the amino acids can make their way from the amino acid pool to the Krebs cycle. Each amino acid has a unique pathway. Most of the amino acids lose their amino groups through transamination. Each α-keto acid thus formed then follows its own special pathway. Some of these pathways are quite devious. Phenylalanine undergoes a six-step reaction before it splits into fumaric acid and acetoacetic acid. The fumaric acid is an intermediate in the Krebs cycle, but acetoacetic acid must be converted to acetoacetyl CoA and then to acetyl CoA before it enters the cycle. Eventually the 20 or more individual pathways leading from the amino acid pool converge into five routes that join the Krebs cycle.

Those amino acids that can form any of the metabolites of carbohydrate metabolism can be converted to glucose (via gluconeogenesis) or oxidized to carbon dioxide, water, and energy; they are referred to as **glucogenic.** Amino acids that give rise to acetoacetyl CoA or acetyl CoA are precursors for fatty acid synthesis; they are said to be **ketogenic** (because they yield ketone bodies). Certain amino acids fall into both categories. Leucine is the only amino acid that is exclusively ketogenic. Table 27.2 classifies the amino acids with regard to the metabolic fate of their carbon skeletons. The products and metabolic routes of the 20 amino acids are summarized in Figure 27.6.

Decarboxylation

Several amino acid decarboxylases eliminate carbon dioxide from amino acids to form primary amines. Pyridoxal phosphate is the necessary coenzyme (Table I.1).

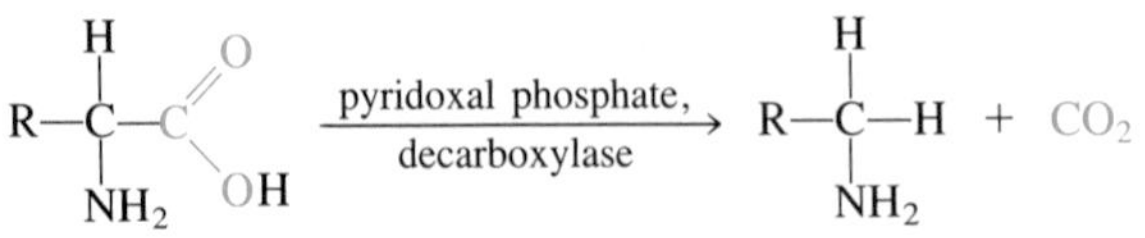

These enzymes are found primarily in microorganisms, but they are also found in some animal tissues. Intestinal bacteria are responsible for amino acid decarboxylation, and the foul smell of feces is due in part to the resulting amines. (Cadaverine and putrescine are formed upon decarboxylation of lysine and ornithine—see Figure 17.3).

Figure 27.6
Fates of the carbon skeletons of amino acids.

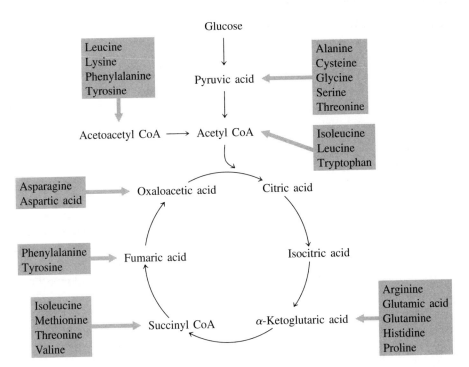

Some of the amines produced as a result of decarboxylation have important physiological effects. Many people are allergic to pollen, dust, insect stings, and so on. By a
mechanism not completely understood, the body responds to these foreign substances by
decarboxylating histidine to *histamine*. (Allergic individuals synthesize abnormally high
amounts of histidine.)

Histamine dilates blood vessels and thus initiates the inflammatory response (see Section
H.2). The discomfort associated with hay fever and other allergies is due to the inflammation of the eyes, nose, and throat. Furthermore, the expansion of the blood capillaries
causes a decrease in the blood pressure that, if severe enough, may induce shock. (Recall
that prostaglandins are also released, and they increase the sensitivity to pain associated
with inflammation—Section H.6.)

Besides histamine, several other neurotransmitters are synthesized by the decarboxylation of specific amino acids. Tyrosine is converted to tyramine by bacterial action.
Tyramine has a physiological action similar to, but weaker than, that of norepinephrine,
which it resembles structurally.

Histamine is sometimes used
in medicine. As little as 1 μg by
injection causes a pronounced
drop in blood pressure by dilat-
ing the blood vessels.

Antihistamines

Antihistamines are compounds that are structurally similar to histamine. They can occupy the receptor sites normally occupied by histamine (e.g., on smooth muscles that surround capillaries), thereby preventing the physiological changes produced by histamine (i.e., they are antagonists). Figure 27.7 illustrates some of the common antihistamines found in nasal decongestants, combination pain relievers, and hay fever preparations. The most common side effect of antihistamines is sedation. This can impair one's ability to operate machinery or drive a motor vehicle. Terfenadine (Seldane), an antihistamine approved for prescription use in 1985, does not cause drowsiness.

There are two types of receptors for histamine, designated as H_1 and H_2. H_1 receptors are found in the walls of capillaries and in the smooth muscle of the respiratory tract. They affect the vascular (dilation and increased permeability of capillaries) and muscular (bronchoconstriction) changes associated with hay fever and asthma. H_1 receptors are blocked by the classical antihistamines (such as Benadryl), which relieve these symptoms. H_2 receptors occur mainly in the wall of the stomach, and their activation causes an increased secretion of hydrochloric acid. Cimetidine blocks H_2 receptors and thus, by reducing acid secretion, is an effective drug for people with ulcers.

Figure 27.7

The antihistamines act as competitive antagonists to histamine. Notice that the structural similarity to histamine is the substituted ethylamine moiety (in blue).

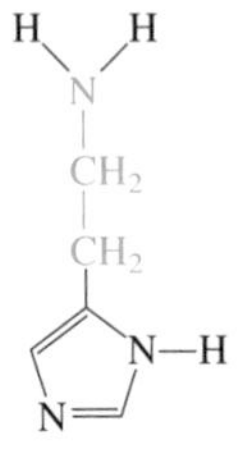

Histamine

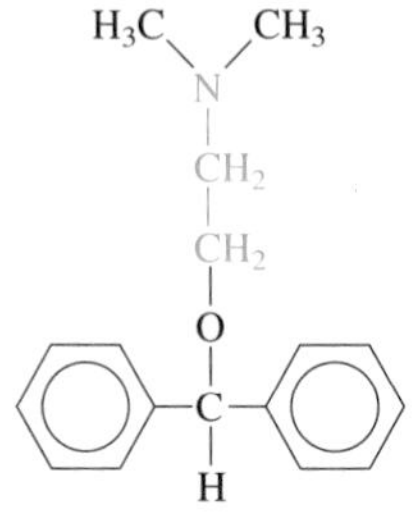

Benadryl
(Diphenhydramine)

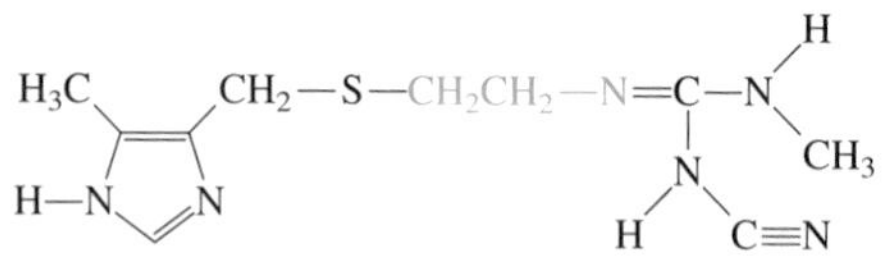

Cimetidine
(Tagamet)

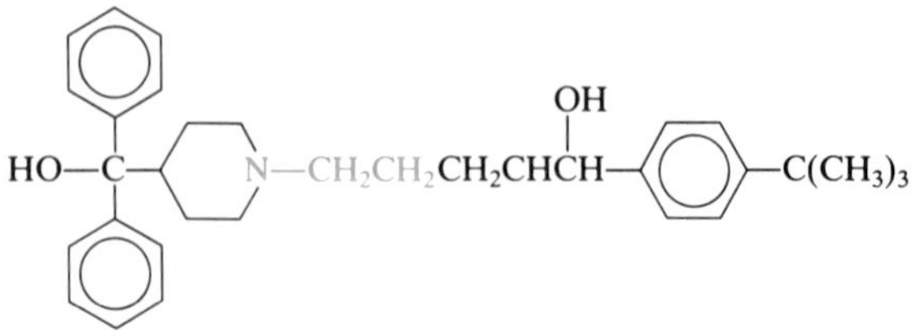

Terfenadine
(Seldane)

Serotonin (5-hydroxytryptamine—Section F.2) is formed by the action of a specific decarboxylase on 5-hydroxytryptophan. The cell bodies of serotonin-containing neurons are located almost exclusively in the upper brainstem, from which axons project to other areas of the central nervous system. Serotonin constricts blood vessels, stimulates smooth muscle, and has a potent inhibitory effect on its postsynaptic neurons.

5-Hydroxytryptophan   Serotonin
                                                      (5-Hydroxytryptamine)

An important inhibitory neurotransmitter, GABA (γ-aminobutyric acid), is formed in the brain and spinal cord from the decarboxylation of glutamic acid. GABA is thought to inhibit dopamine neurons in particular, as well as other neurons throughout the central nervous system.

Glutamic acid γ-Aminobutyric acid
                                                    (GABA)

Another decarboxylase catalyzes the decarboxylation of 3,4-dihydroxyphenylalanine (L-dopa) to form dopamine. A deficiency of dopamine in the brain cells is a primary cause of Parkinson's disease, a disorder of the central nervous system that involves a progressive paralytic rigidity, tremors of the extremities, and unresponsiveness to external stimuli. Dopamine itself cannot be administered because it does not pass across the blood–brain barrier. A major breakthrough in the treatment of Parkinson's disease has been the use of L-dopa. (The L enantiomer is more effective and less toxic than the D form of the drug.) Large doses of L-dopa are administered orally; the drug is able to pass from the digestive system into the blood and then cross the blood–brain barrier. L-Dopa is decarboxylated in the brain into the necessary neurotransmitter, dopamine. (Incidentally, numerous studies have linked schizophrenia to an *overabundance* of dopamine in brain cells.)

3,4-Dihydroxyphenylalanine   3,4-Dihydroxyphenylethylamine
        (L-Dopa)   (Dopamine)

27.6
STORAGE OF NITROGEN

In contrast to carbohydrates and lipids, proteins are not stored to an appreciable extent in living organisms. Small amounts of amino acids are excreted by humans, but this is essentially a wasteful process because amino acids contain utilizable energy. (One gram of protein liberates about 4300 cal.) Accordingly, most excess amino acids are deaminated. The resulting carbon skeleton is oxidized to produce energy or is stored as glycogen or fat. Most of the ammonia that is liberated is converted to urea. A relatively small proportion combines with glutamic acid in the presence of glutamine synthetase and ATP to form glutamine. This is another example of the conversion of inorganic nitrogen to organic nitrogen.

Glutamic acid   Glutamine

Glutamine is present in many tissues and in the blood and serves as a temporary storage and transport form of nitrogen. The formation of glutamine is the major method of disposing of ammonia from the brain.

In the liver and in the kidneys, the amide group of glutamine can be donated in specific reactions to appropriate acceptor molecules to effect the biosynthesis of many nitrogen-containing compounds.

$$\text{Glutamine} + H_2O \xrightarrow{\text{glutaminase}} \text{Glutamic acid} + NH_3$$

27.7
EXCRETION OF NITROGEN

Whenever amino acids are used for energy production or for the synthesis of glucose or fat, an amino group is liberated in the form of ammonia. Living organisms must have some mechanism for removing this ammonia from the cell environment because even low concentrations of ammonia are poisonous. Levels of only 5 mg of ammonia per 100 mL of blood (hyperammonemia) are toxic to humans. The normal concentration of this compound is about $1-3$ $\mu g/100$ mL. Organisms differ biochemically in the manner in which they excrete excess nitrogen. Most vertebrates and most adult amphibia excrete nitrogen as urea in the urine. Birds, reptiles, and insects convert nitrogen to uric acid. All marine organisms, from unicellular organisms to fish, excrete free ammonia. The ammonia is very soluble in water and is rapidly diluted in the aqueous environment.

In humans, 95% of nitrogenous waste products results from amino acid catabolism; the remaining 5% comes from the metabolism of other nitrogen-containing compounds (e.g., creatine, neurotransmitters, pyrimidines). The breakdown of purine bases in the human body results in the production of uric acid, and very small concentrations of this acid are found in the urine and in the body fluids.

Under certain pathological conditions (impairment of purine metabolism), large quantities of uric acid are produced, and the plasma concentration of uric acid becomes abnormally high (>7 mg/100 mL). At physiological pH, the excess uric acid precipitates as the sparingly soluble monosodium salt. When such deposits occur in the joints of the body's digital regions, the surrounding tissue can become inflamed, causing the painful arthritic characteristics of **gout.** Crystals of monosodium urate may also accumulate in the kidneys (kidney stones) and cause extensive impairment of renal function. Kidney failure is another serious medical consequence of gout. (For some unknown reason, men suffer from gout to a much greater extent than women.)

Another serious impairment of purine metabolism results from a genetic defect. In the *Lesch–Nyhan* syndrome, an enzyme necessary for the normal utilization of purines is absent. Affected children are mentally defective and exhibit a compulsive, aggressive behavior toward others (extreme hostility), as well as toward themselves (self-mutilation by chewing their tongue, lips, and fingers). The relationship between the absence of the enzyme and the aberrant behavior is still unknown.

In mammals, the liver is the principal organ concerned with the formation of urea (by a cyclic series of reactions known as the **urea cycle**—see an advanced biochemistry text). Extensive liver damage (e.g., cirrhosis) or a genetic defect of any of the urea cycle enzymes results in hyperammonemia and causes tremor, slurred speech, and blurred vision. A continual rise in ammonia concentrations can lead to coma and death. Once produced, urea is transported by the bloodstream to the kidneys and is eliminated in the urine. The kidneys filter about 100 L of blood each day. The normal individual excretes about 1.5 L of water, containing approximately 30 g of urea, daily. This value varies greatly from day to day and can rise dramatically (to about 100 g of urea) upon ingestion of a high-protein diet. A long-term decrease in urea excretion is indicative of liver and/or kidney disease. Loss of nitrogen also occurs through the skin, mainly as sweat (about 2 g of urea) and via the feces (smaller amounts).

27.8
RELATIONSHIPS AMONG THE METABOLIC PATHWAYS

A great variety of organic compounds can be derived from carbohydrates, lipids, and proteins. All organisms utilize the three major foodstuffs to form acetyl CoA or the metabolites of the Krebs cycle. These in turn supply energy upon subsequent oxidation by the cycle. A brief summary of the interrelationships of the metabolic pathways is given in Figure 27.8. We should mention once again that in mammals the conversion of

Figure 27.8
Interrelationships of metabolic pathways.

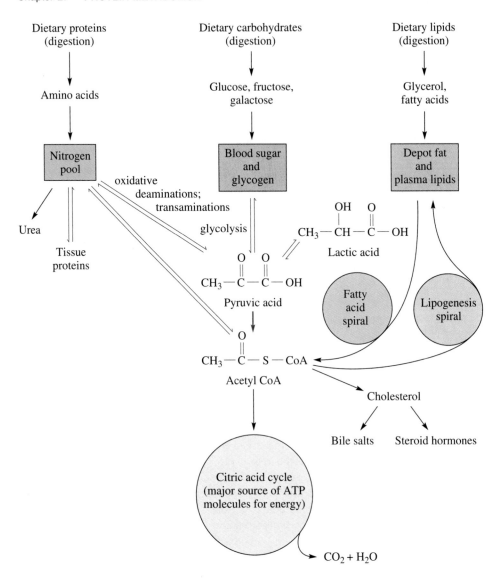

pyruvic acid to acetyl CoA is irreversible. Thus, an acetyl CoA molecule derived from fatty acid degradation cannot be transformed directly into pyruvic acid. Consequently, mammals are unable to synthesize carbohydrates from lipids.

EXERCISES

1. Define or give an explanation for each of the following terms.

   a. peptidase
   b. autocatalysis
   c. protein turnover
   d. nitrogen balance
   e. kwashiorkor
   f. starvation
   g. GABA
   h. Parkinson's disease
   i. L-dopa
   j. hyperammonemia
   k. gout
   l. urea

2. Distinguish between the terms in each of the following pairs.

   a. pepsin and pepsinogen
   b. chymotrypsin and trypsin
   c. endopeptidase and exopeptidase
   d. positive and negative nitrogen balance
   e. essential and nonessential amino acid
   f. complete and incomplete protein

g. transamination and deamination

h. GPT and GOT

i. glucogenic and ketogenic amino acid

j. histamine and antihistamine

3. Give the location of action and the function of each of the following enzymes.

   a. pepsin   **b.** trypsin

   c. chymotrypsin   **d.** carboxypeptidase

   e. dipeptidase   **f.** enteropeptidase

4. What is the defense mechanism employed by the body to protect its tissues from the digestive action of the proteolytic enzymes?

5. In the treatment of diabetes, insulin is administered by injection rather than orally. Explain.

6. Indicate the expected products from the enzyme action of pepsin, chymotrypsin, and trypsin on each of the following tripeptides.

   a. Ala-Phe-Tyr   **b.** Ile-Tyr-Ser

   c. Phe-Arg-Leu   **d.** Thr-Glu-Lys

7. Indicate the location in a polypeptide chain that is cleaved by (a) an aminopeptidase and (b) a carboxypeptidase.

8. What are the products of protein digestion?

9. What is meant by the term "metabolic pool of amino acids"?

10. Name three processes that add amino acids to the amino acid pool.

11. For what purposes are amino acids removed from the amino acid pool?

12. What happens to the nitrogen balance when the diet is lacking in one of the essential amino acids?

13. What happens to the nitrogen balance during fasting?

14. What happens to the nitrogen balance during recovery from a wasting illness?

15. The RDA for protein is about 0.8 g per kilogram of body weight. How much protein is required each day by a 125-kg football player?

16. Compare the turnover rates of enzymes and muscle proteins.

17. Vegetable protein is converted in the animal body into animal protein. Explain.

18. **a.** Why are vegetarians instructed to eat a wide variety of vegetables?

   b. What is the best source of essential amino acids?

19. **a.** What is a unique structural feature of most of the essential amino acids?

   b. What is the effect of a deficiency of one or more of the essential amino acids?

20. What are three functions of amino acids in the body?

21. What is the significance of the transamination reaction in the metabolism of amino acids?

22. The following compound can supply the body with one of the essential amino acids. Show, with an equation, how this is possible.

$$CH_3CH_2CH(CH_3)\overset{\displaystyle O}{\overset{\|}{C}}-\overset{\displaystyle O}{\overset{\|}{C}}_{OH}$$

23. Write the equation for the transamination reaction that occurs between phenylalanine and pyruvic acid.

24. What reaction is catalyzed by the enzyme GOT?

25. What reaction is catalyzed by the enzyme GPT?

26. What keto acid serves as a reactant in a transamination that produces alanine as a product?

27. If the essential amino acid leucine (2-amino-4-methyl-pentanoic acid) is lacking in the diet, a keto acid can substitute for it. Give the structure of the keto acid.

28. Rationalize the substitution described in Exercise 27.

29. What is oxidative deamination?

30. What product is formed by the oxidative deamination of glutamic acid?

31. Write equations for the formation of each of the following compounds from an amino acid.

   a. pyruvic acid   **b.** histamine

   c. α-ketoglutaric acid   **d.** glutamine

   e. cadaverine   **f.** oxaloacetic acid

   g. tyramine   **h.** dopamine

32. If 1 mol of alanine is converted into pyruvic acid, and the pyruvic acid is metabolized via the Krebs cycle, how many moles of ATP could be produced?

33. **a.** What toxic compound is formed in the oxidative deamination reaction?

   b. How does the body get rid of this compound?

34. **a.** How does Benadryl exert its effect?

   b. What is the difference between an H_1 receptor and an H_2 receptor?

35. Name the compound that is described.

   a. the amino acid that is exclusively ketogenic

   b. the principal amino group acceptor in transamination reactions

   c. the α-keto acid that is usually formed in oxidative deamination reactions

   d. the compound that is decarboxylated to serotonin

   e. the compound that is decarboxylated to GABA

   f. the compound that serves as the temporary storage form of nitrogen in the body

36. What is the end product of purine metabolism?

37. **a.** What organ is responsible for the synthesis of urea?

   b. What organ is responsible for the excretion of urea?

Chapter 28
BODY FLUIDS

The testing of blood is one of the best methods to diagnose a variety of abnormal conditions.

The processes of life occur for the most part in solution. The solvent for life processes is water, and the solutes are many—simple ions, small molecules, large molecules, and colloidal aggregates. These are found in a variety of body fluids, and each in its own way is essential to life.

A 70-kg person has about 5 L of blood and about 10 L of interstitial fluid, the fluid that fills the space between cells. Fluid within the cells amounts to about 35 L. Each day the individual excretes an average of 1.5 L of fluid urine, 0.2 L of water in the feces, and 1 L of water through the skin and lungs. All of these losses are balanced by water taken into or synthesized during metabolism within the body.

You are indeed a solution—or, rather, several solutions. But the solutions of which you are composed are not static—they constantly renew themselves, so their composition is quite dependent on the state of your health. Analysis of various body fluids represents one of the most powerful diagnostic techniques available to medical personnel.

28.1
BLOOD: AN INTRODUCTION

Blood is the principal transport medium of the human body. It moves through a 100,000-km-long network of blood vessels. Some of these are so small that blood cells have to line up to pass through. Blood carries (1) oxygen from the lungs to the tissues, (2) carbon dioxide from the tissues to the lungs, (3) nutrients from the intestines to the tissues, (4) metabolic wastes from the tissues to the excretory organs, (5) hormones from the endocrine glands to their target tissues, and (6) three major kinds of blood cells. In addition, blood helps maintain a fairly constant body temperature, an acid–base balance, an electrolyte balance, and a water balance. A rather remarkable substance, blood. And we've only mentioned a few highlights.

Circulatory blood consists of a straw-colored fluid portion (the plasma) and the formed elements (blood cells). It accounts for about one-twelfth of the body weight of the average individual. Moderate amounts lost through bleeding or blood donation are readily replaced. The volume of blood varies with body size. A 150-lb (68-kg) human has about 5–6 L of blood, roughly 55–60% of which is plasma. **Plasma** is an extremely complex solution, containing all of the biochemically significant compounds (carbohydrates, amino acids, proteins, lipoproteins, enzymes, hormones, vitamins) and inorganic ions. Although these compounds and ions are continuously entering and leaving the circulatory system, the overall composition of the plasma remains remarkably constant (a state of dynamic equilibrium exists). Approximately 90–92% of plasma is water; the remaining 8–10% consists of dissolved solids, the greatest component of which is the plasma proteins. Dispersed throughout the plasma are the **formed elements,** which account for about 40–45% of the total blood volume. The formed elements of the blood are the **erythrocytes** (red blood cells), **leukocytes** (white blood cells), and **thrombocytes** (platelets). The chemical analysis of blood samples is of great clinical significance. Increased or decreased amounts of certain substances may indicate a particular disease or condition.

28.2
ELECTROLYTES IN PLASMA AND ERYTHROCYTES

A variety of ions are found in solution in both plasma and blood cells. These electrolytes play important roles in several life processes.

Sodium ions are found mainly in the plasma, and potassium ions are found chiefly in the erythrocytes. These positive ions do not migrate readily across cell membranes. Calcium and magnesium ions are the main dipositive ions in blood. (Calcium ions are necessary for blood clotting; if they are removed, the blood will not clot—see Section 28.8.) Calcium ions are not found in erythrocytes, but magnesium ions are. The metabolism of calcium and the metabolism of phosphorus are closely related, and levels of the two usually vary in a reciprocal manner. Magnesium levels, on the other hand, sometimes parallel the variations in calcium levels and at other times parallel the variations of phosphorus levels.

The principal negative ions present in the blood, in addition to inorganic phosphorus (as HPO_4^{2-}), are chloride, bicarbonate, and sulfate. Bicarbonate and hydrogen phosphate are involved in the acid–base balance of the blood (see Section 28.10). Table 28.1 lists some conditions that influence plasma levels of certain electrolytes.

Table 28.1
Normal Composition of Blood

Constituent	Normal Range (mg/100 mL)[a]	Clinical Significance	
		Increased in	Decreased in
Calcium	8.5–10.3 (4.5–5.3 mEq/L)[b]	Hyperparathyroidism, Addison's disease, malignant bone tumor, hypervitaminosis D	Hypoparathyroidism, rickets, malnutrition, diarrhea, chronic kidney disease, celiac disease
Cholesterol, total	150–265	Diabetes mellitus, obstructive jaundice, hypothyroidism, pregnancy	Pernicious anemia, hemolytic jaundice, hyperthyroidism, tuberculosis
Uric acid	Male, 3–9 Female, 2.5–7.5	Gout, leukemia, pneumonia, liver and kidney disease	
Urea nitrogen	8–25	Mercury poisoning, acute glomerulonephritis, kidney disease	Pregnancy, low-protein diet, severe hepatic failure
Nonprotein nitrogen	15–35	Kidney disease, pregnancy, intestinal obstruction, congestive heart failure	Low-protein diet
Creatine	3–7	Nephritis, renal destruction, biliary obstruction, pregnancy	
Creatinine	0.7–1.5	Nephritis, chronic renal disease	
Glucose	70–100	Diabetes mellitus, hyperthyroidism, infections, pregnancy, emotional stress, after meals	Starvation, hyperinsulinism, Addison's disease, hypothyroidism, extensive hepatic damage
Chlorides	96–106 mEq/L	Nephritis, anemia, urinary obstruction	Diabetes, diarrhea, pneumonia, vomiting, burns
Phosphate, inorganic	3–4.5	Hypoparathyroidism, Addison's disease, chronic nephritis	Hyperparathyroidism, diabetes mellitus
Sodium	136–145 mEq/L	Kidney disease, heart disease, pyloric obstruction	Vomiting, diarrhea, Addison's disease, myxedema, pneumonia, diabetes mellitus
Potassium	3.5–5 mEq/L	Addison's disease, oliguria, anuria, tissue breakdown	Vomiting, diarrhea
Carbon dioxide	Adults, 24–29 mEq/L Infants, 20–26 mEq/L	Tetany, vomiting, intestinal obstruction, respiratory disease	Acidosis, diarrhea, anesthesia, nephritis
Hemoglobin	Male, 14–18 g/100 mL Female, 12–16 g/100 mL	Polycythemia	Anemia

[a] Milligrams per 100 mL is also called milligram percent.   [b] mEq/L = milliequivalents per liter.

28.3
PROTEINS IN THE PLASMA

The protein content of the blood is about the same as that of the muscle and other tissues. More than 100 proteins have been identified in the blood; most of them are synthesized in the liver. The normal concentration range of blood proteins is 7.0–8.0 g/100 mL of plasma. These proteins remain within the circlatory system and ordinarily are not used as sources of energy. At times of protein deprivation, the plasma protein concentration is maintained at the expense of tissue protein. Plasma proteins are grouped into three main classes on the basis of their solubility properties and methods of isolation.

Albumins

Albumins are the most abundant proteins (~55% by mass) in the plasma. Their major function is to maintain osmotic pressure (see Section 28.6) by controlling the water balance. They are also important for their buffering capacity and in the transport of fatty acids, certain metal ions, and many drugs (particularly the nonpolar ones). Normal concentrations of albumins range from 3.5 to 4.5 g per 100 mL of plasma.

Globulins

Globulins account for about 40% by mass of the total plasma proteins. They have higher molecular weights than the albumins (150,000 as compared to 70,000). Three subclasses of globulins are recognized, and they differ from one another with respect to their rates of movement in an electric field; α-globulins (0.7–1.5 g/100 mL of plasma), β-globulins (0.6–1.1 g/100 mL), and γ-globulins (0.7–1.5 g/100 mL). α-Globulins and β-globulins are synthesized in the liver. They form complexes (i.e., VLDLs, LDLs, and HDLs) with the water-insoluble lipids and thus serve to transport these compounds through the aqueous media. Unlike the other plasma proteins the γ-globulins are synthesized by white blood cells in the lymph nodes. They combat certain infectious diseases (e.g., diphtheria, influenza, measles, mumps, typhoid), and they are referred to as *antibodies* or *immunoglobulins* (see Section 28.5). Based on molecular weights and chemical properties, five classes of immunoglobulins are recognized (IgG, IgM, IgA, IgD, and IgE). IgG is the most abundant (~80%) immunoglobulin in plasma. IgA accounts for about 13% of the antibodies, IgM about 6%, and IgD and IgE the remaining 1%. IgA is the principal antibody in external secretions (e.g., mucus, saliva, tears). Thus it represents the initial defense against bacteria and viruses.

Fibrinogen

Fibrinogen is a large protein (MW 340,000) consisting of six polypeptide chains. It is synthesized in the liver and constitutes about 5% (0.3–0.4 g/100 mL of plasma) of the total blood protein. It functions in blood coagulation (see Section 28.8). The fibrinogen content of plasma increases when inflammatory or infectious conditions exist and during menstruation and pregnancy.

28.4
THE FORMED ELEMENTS

Apart from the respiratory function of the hemoglobin molecules in the erythrocytes (Section 28.9), the blood cells have specific roles that are not directly concerned with the general metabolic processes. This is in contrast to the plasma, which serves as the metabolic transport medium of the organism.

Erythrocytes

The red blood cells (erythrocytes) are formed in the red bone marrow, and they are the most numerous of the formed elements. The blood of the average adult female contains about 5 million of these cells, and that of the average male about 5.5 million, in every cubic millimeter (mm^3) of blood. (The volume of one drop of blood is about 100 mm^3, so there are about 500 million erythrocytes in each drop of blood.) Any condition that tends to lower the oxygen content of the blood causes an increase in the number of erythrocytes. Persons who live at high altitudes generally have a higher erythrocyte count than those who live at sea level. Conversely, increased barometric pressure results in a decrease in the erythrocyte count. The term *hematocrit value* is applied to the volume (in percent) of packed red blood cells in a sample (usually 10 mL) that has been centrifuged under standard conditions. The cells are spun to the bottom of a centrifuge tube and the supernatant liquid (the plasma) is drawn off the top. Usually about 45% of a blood sample is red cells, so 45 is the normal hematocrit value. Variations from the normal value indicate the existence of certain pathological conditions. When **anemia** occurs, for instance, *the percentage of erythrocytes (and/or the percentage of hemoglobin) is abnormally low*. Anemia may result from (1) a decreased rate of production of erythrocytes (aplastic anemia), (2) an increased destruction of erythrocytes (hemolytic anemia), or (3) an increased loss of erythrocytes (as in hemorrhaging). *Polycythemia* is the condition arising from an abnormally high percentage of red blood cells.

Red blood cells are disc-shaped with slight depressions at the center, like a solid doughnut (Figure 28.1). Unlike most other cells, erythrocytes contain neither mitochondria nor a nucleus. (The nucleus is lost during the development and maturation of the erythrocyte.) They cannot reproduce, have no aerobic metabolism, and are unable to synthesize carbohydrates, lipids, or proteins. They obtain all of their energy from the pentose phosphate shunt (see an advanced text) and from substrate-level phosphorylation in the Embden–Meyerhof pathway. The most significant components of the red blood cell are the hemoglobin molecules. An individual's blood type is determined by specific short-chain polysaccharides that are bound to certain proteins (glycoproteins) on the membranes of the red blood cells.

The major function of the erythrocyte is to transport oxygen from the lungs to the cells. Also, it assists in the transport of carbon dioxide from the tissues to the lungs. Human red blood cells have a life span of about 4 months. (During this time interval, each red blood cell makes about 120,000 trips around the body.) During this period, there is no degradation or resynthesis of hemoglobin molecules in the erythrocyte. To maintain a constant level of erythrocytes, new cells are formed in the bone marrow at the same rate that old cells are eliminated by special tissues in the liver and the spleen. It has been estimated that of the approximately 30 trillion red blood cells in an average adult male, about 3 million are destroyed each second. Assuming that there are 300 million hemoglo-

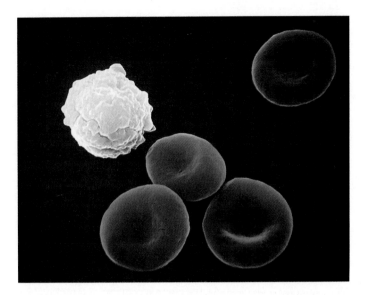

Figure 28.1
Red blood cells are doughnut-shaped discs, whereas white blood cells are spherical. The photo was obtained through a scanning electron microscope at 5200× magnification.

bin molecules in each erythrocyte, then 900 trillion molecules of hemoglobin must be synthesized every second (by cells in the bone marrow) in order to maintain a constant supply.

Leukocytes

The composition of white blood cells resembles that of other tissue cells. They are nucleated, and they contain glucose, lipids, proteins, and other soluble organic substances and inorganic salts. White blood cells constitute the body's primary defenders against foreign organisms (e.g., viruses, bacteria), and the blood is the vehicle that transports them to sites of infection.

The different varieties of leukocytes have specialized functions. *Lymphocytes* are involved in the synthesis and storage of antibodies (see Section 28.5). *Phagocytes* (*macrophages*) can leave the blood by squeezing between the endothelial cells that line the capillary wall. They are attracted toward sites of inflammation by chemicals released from injured tissue. Phagocytes contain lysosomal enzymes, and their function is to engulf and digest the invading organisms.

Billions of leukocytes are produced each day in the bone marrow to replace the ones that die. On the average, there are about 7000 leukocytes per cubic millimeter of blood, but this value is subject to considerable variation. A higher than normal leukocyte count occurs during acute infections such as *appendicitis* (16,000–20,000 per mm^3). High numbers of leukocytes may also appear during emotional disturbances and following vigorous exercise and/or excessive loss of body fluids. Viral diseases such as chicken pox, influenza, measles, mumps, and polio are accompanied by an abnormally low count (<5000 per mm^3), because in fighting the viruses the leukocytes are killed faster than they can be produced. **Leukemia** is a cancer that is characterized by the uncontrolled production of leukocytes that fail to mature. Thus, despite their numbers, these cells are unable to destroy invading pathogens, and the person has a lowered resistance to infections. Invariably some of the cancer cells *metastasize* (spread out) from the bone marrow or lymph nodes to other parts of the body. In these other tissues, the leukemic cells eventually crowd out the normal cells.

Figure 28.2
Relative sizes of the
formed elements.

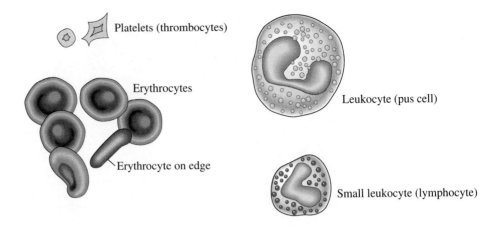

Platelets

There are about 250,000 platelets (thrombocytes) in every cubic millimeter of blood. These small, nonnucleated cells contain proteins and relatively large amounts of phospholipids, mostly cephalin. The blood platelets liberate species that are instrumental in the mechanism of blood clotting (see Section 28.8). An abnormally low platelet count ($<$100,000 per mm^3) is related to a tendency to bleed.

Figure 28.2 shows the relative sizes of the various types of formed elements.

28.5
THE IMMUNE RESPONSE

Some of the plasma proteins, the **gamma globulins,** are associated with a sophisticated defense mechanism in the body termed the **immune response.** When the body is invaded by foreign macromolecules (such as proteins), it responds by synthesizing its own specialized proteins called **antibodies** (or *immunoglobulins*). Since the foreign substance triggers the synthesis of antibodies, the invader is called an **antigen** (it causes antibody generation). Each distinctive antigen causes a unique set of antibodies to be produced. The antibodies are designed to lock onto the particular antigen. They bind to it and (to oversimplify a bit) incapacitate it. In some instances the incapacitation involves the formation of a precipitate, with the antigen being tied up as part of an antibody–antigen clot.

When the body is invaded by pathogenic antigens (e.g., disease-causing bacteria or viruses), its ability to produce specific antibodies represents an effective defense mechanism. It is even possible to prime this mechanism. A **vaccine** containing a weakened form of the antigen (e.g., dead bacteria) will cause an individual to build up a certain level of antibodies for this particular antigen. If the same individual is then subjected to attack by a more virulent form of the antigen, the body responds much more rapidly and effectively to its presence. Having practiced on the dead ''bug,'' the system can easily handle the live one—it has gained an immunity to that particular organism (Figure 28.3).

Unfortunately, the body can't tell ''good'' antigens from ''bad'' antigens. We want it to respond to a viral infection, but we wish it wouldn't attack purposely transplanted tissue. Nonetheless, the host body will respond to a donor skin graft, a kidney transplant, or a heart transplant as it does to a virus, by generating antibodies to attack it. When the body responds to foreign tissues in this way, the process is called *rejection*. This is a

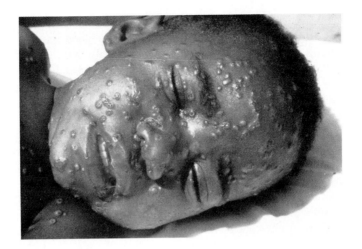

Figure 28.3
Smallpox, a disease that once killed and disfigured millions, has been eradicated by the use of vaccines.

major problem associated with organ transplants. There are drugs that suppress the immune response, but when these are used on a transplant recipient to protect the transplanted organ, the patient also becomes more susceptible to infection. If you turn off the immune response, you turn off both the good and bad aspects of it.

There are a variety of diseases associated with the immune system. An overactive or misguided immune system can attack its own body organs, causing an **autoimmune** disease. It is thought that multiple sclerosis, some forms of arthritis, and some cases of diabetes are autoimmune diseases. The system senses something as foreign even though it isn't. It attacks and destroys the myelin sheath of nerves (multiple sclerosis), the connective tissues in the joints (arthritis), or the insulin-producing cells of the pancreas (diabetes).

At the opposite end of the scale is *acquired immunodeficiency syndrome* (AIDS). In this deadly affliction, the immune system is destroyed by the HIV virus (Section J.1). The victims then succumb to a variety of infections or to a rare form of cancer.

Scientists have sought to enhance certain immune responses as weapons against diseases. By genetic engineering, researchers have cloned the genes for specific antibodies and placed them in cells that then produce *monoclonal antibodies*. In theory, monoclonal antibodies should attack and overwhelm specific antigens. In practice, they have been a bit disappointing, but they still hold considerable promise.

28.6
OSMOTIC PRESSURE

How does material get from the blood to the cells? Diffusion of water, electrolytes, glucose, amino acids, and other materials occurs at a rapid rate through pores in the capillary walls. It occurs in both directions—from the blood into the interstitial fluid outside the capillary and from the interstitial fluid back into the blood. Diffusion through capillary walls occurs at a fantastic rate, estimated as equivalent to 1500 L of water per minute for someone weighing 70 kg. Ultimately material makes its way from the interstitial space through cell walls and into (or out of) the cells themselves.

The blood doesn't just sit in the capillaries while the diffusion takes place. It circulates—it continues to move through the vascular system. What keeps the blood moving is the heart, a pump that pushes the blood around the system. The pressure imparted to the blood by this pumping action falls off as the blood travels from the heart through the arteries, the capillaries, and the veins and finally back to the heart, where it gets a fresh push. Blood moves from the high-pressure arterial end of the capillaries to the low-pressure venous end. The diffusion process described above occurs as the blood circulates through the capillaries. The pressure that keeps the blood moving around the system also has a tendency to push the blood plasma out of the porous capillaries into the interstitial space. In the absence of any countereffect, the rapid diffusion of material back and forth between blood and interstitial fluid would be accompanied by a slow but steady net loss of fluid from the blood to the interstitial space. Blood volume would drop, and tissue would swell with the extra fluid. There is a countereffect, though—osmotic pressure.

When a living cell is placed in distilled water, water gradually flows through the semipermeable membrane into the cell. The environment outside the cell is 100% water, while the percentage of water in the internal environment is considerably lower owing to the presence of dissolved substances. Since the semipermeable membrane prevents these substances from leaving the cell, equalization of concentration can only be attained by the passage of the small water molecules into the cell. The diffusion of water from a dilute solution (high concentration of water) through a semipermeable membrane into a more concentrated solution (low concentration of water) is known as the process of *osmosis* (recall Section 9.12). **Osmotic pressure** is defined as the pressure required to prevent the occurrence of osmosis when two solutions of unequal concentrations are separated by a semipermeable membrane. The osmotic pressure depends solely upon the concentration of solute particles (either ions or molecules) in the solutions involved.

The concentration of protein in the plasma far exceeds the concentration of protein in the interstitial fluid outside the blood vessels. This concentration gradient results in an osmotic pressure of about 25 mm Hg. Therefore, if no external force were applied, fluids would be expected to diffuse from the interstitial fluid into the bloodstream. However, the pumping action of the heart creates the so-called blood pressure (see Section 28.7), and this pressure is greater at the arterial end of a capillary (~32 mm Hg) than at the venous end (~17 mm Hg). Since the blood pressure at the arterial end is higher than the osmotic pressure, the natural tendency is reversed, and there is a net flow *from* the capillary *into* the interstitial fluid. The fluid that leaves the capillary contains the dissolved nutrients, oxygen, hormones, and vitamins needed by the tissue cells. As the blood moves along the capillary branches, the blood pressure decreases until at the venous end the osmotic pressure is greater than the blood pressure and there is a net flow of fluid *from* the interstitial fluid *into* the capillary. The incoming fluid contains the metabolic waste products such as carbon dioxide and excess water (Figure 28.4). If we are in good health,

Figure 28.4

Oxygen and nutrients leave at the arteriole end of the capillaries. Carbon dioxide and other cellular waste products enter the venule end of the capillaries.

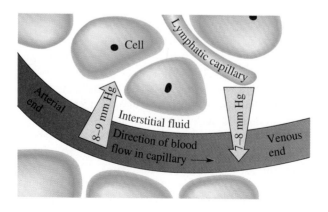

everything balances rather nicely. To be sure, nutrients have diffused from blood to interstitial fluid and wastes have diffused from interstitial fluid to blood, but the volume of fluid in the two systems has remained constant.

Osmotic pressure, and hence the delicate balance of fluid exchange, is directly related to the concentration of albumins in the plasma. The normal half-life of albumins in the plasma is 20 days. If the albumin level is low, as might be the case from (1) malnutrition (low protein intake), (2) abnormal protein synthesis (liver disease), or (3) the loss of protein in the urine as a result of kidney disease (nephrosis), the osmotic pressure decreases. This results in a net efflux of fluids from the capillaries into the interstitial and cellular regions. This abnormal accumulation of fluids within the interstitial spaces produces noticeable swelling, particularly in the lower extremities. The condition so characterized is known as *edema*. Recall that children who suffer from the protein deficiency disease kwashiorkor (Section 27.4) characteristically have bloated abdomens. This swelling is caused by the accumulation of water, which leaves the blood because there are insufficient albumins to maintain the osmotic pressure of the blood.

The medical condition called **shock** also results from a loss of fluid from the vascular system. Capillary permeability increases, and albumins are lost to the interstitial fluid. The resulting outflow of fluid from the vascular system reduces blood volume enough to cause a dramatic decrease in blood pressure. There is a consequent decrease in oxygen-transporting capability with potentially fatal results. When someone goes into shock following a traumatic injury, it is because the delicately balanced transport system in the body has gone awry and could fail completely. Shock is a physiological, not a psychological, state. Treatment of the condition involves bringing the blood volume back up to normal levels.

> Edema can also result from the increased capillary permeability that accompanies an inflammatory reaction.

28.7
BLOOD PRESSURE

Blood pressure is the force exerted by the blood against the inner walls of the arteries. The **systolic pressure** is the maximum pressure achieved during contraction of the heart ventricles. When the ventricles relax, the blood pressure drops, and the lowest pressure that remains in the arteries before the next ventricular contraction is called the **diastolic pressure.** The alternate expansion and contraction of the arterial walls can be felt as a *pulse* (when the artery is near the surface of the skin). Blood pressure measurements are reported as a *ratio of systolic pressure* (in mm Hg) to *diastolic pressure* (in mm Hg), for example, 120/80.

Blood pressure depends on several factors including total volume of blood, heart action, and the smooth muscles that surround the arteries. It is directly proportional to the volume of blood. Thus, if there is a loss of blood due to hemorrhaging, the blood pressure drops, but it returns to normal when the lost blood is replaced by a blood transfusion. Contraction of the smooth muscles in the walls of the arteries constricts the vessels (vasoconstriction), causing an increase in blood pressure. Conversely, dilation of the smooth muscles causes vasodilation and a decrease in blood pressure.

One of the major health problems in the United States is high blood pressure or **hypertension.** Over 20% of the world's population are afflicted, and in the United States it is estimated that hypertension occurs in about 25 million people. To treat this disorder, it is necessary to attempt to counteract those factors that cause the increased pressure. Blood pressure can be reduced by any of the following procedures.

1. Administer diuretics (Section 21.6) and/or reduce the sodium ion intake. Either of these steps will increase the excretion of urine and so decrease the volume of blood.

2. Negate the stimulating effects of epinephrine on the cardiac muscle with drugs that specifically bind to epinephrine binding sites (so-called beta blockers) and/or reduce stress and therefore reduce the levels of epinephrine.

3. Administer vasodilators, drugs that cause relaxation of the smooth muscles in the arterial walls and/or use drugs that block the action of smooth muscle vasoconstrictors.

If hypertension is not brought under control, the left ventricle must contract with a greater force against the increased arterial pressure. As a result of this increased workload, the heart muscle cells require additional oxygen, which cannot be delivered fast enough and in sufficient quantities. Heart cells die, and a myocardial infarction results.

28.8
CLOTTING OF BLOOD

The phenomenon of blood clotting or coagulation is of the utmost importance to the organism. If such a mechanism did not exist, loss of blood would occur whenever a blood vessel was injured. The details of the coagulation process, which is quite complex, have been clarified considerably by research in recent years. The theory accounting for the clotting of blood has evolved from a two-step mechanism to the present model, which depicts a multistep cascade of activation of protein factors. We will restrict our discussion to a general description of the final two steps. (See an advanced biochemistry text for a more detailed explanation.)

When blood clots, the soluble plasma protein fibrinogen is converted to fibrin. Fibrin monomers undergo a polymerization reaction that results in the formation of insoluble needlelike threads. These threads enmesh the blood cells and effectively seal off the area where the blood vessel has been damaged (Figure 28.5). Once the fibrinogen and the formed elements have been removed from the plasma (by a centrifuge), the fluid that remains is called the **serum.** The only distinction, therefore, between blood plasma and blood serum is the presence of fibrinogen in the plasma. *Because blood serum lacks fibrinogen, it is unable to clot.*

Figure 28.5

Scanning electron micrograph of erythrocytes enmeshed in fibrin fibrils—a part of a typical blood clot.

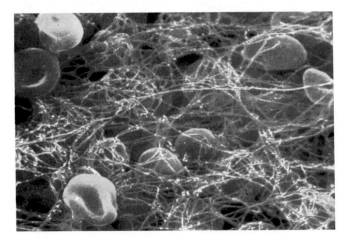

The clotting mechanism becomes operative only when a tissue is cut or injured. Blood platelets and damaged tissue cells are somehow activated and release a group of compounds collectively referred to as *thromboplastin.* In the presence of calcium ions and other cofactors, thromboplastin converts prothrombin to thrombin. Prothrombin is a zymogen (Section 22.2), and its activation is analogous to the activation of the various digestive enzymes. *Thrombin* is the actual clotting enzyme. It brings about the conversion of fibrinogen to fibrin by hydrolysis of two peptide fragments from the former. Hemophilia is a sex-linked hereditary disease in which one of the protein factors involved in the formation of thrombin is lacking or inactive.[1]

Normally, blood takes 5–8 min to form a clot. After the tissue is repaired by the body, the fibrin clot is digested by the enzyme *plasmin* (which circulates in the blood as the zymogen called *plasminogen*). The process involved in blood coagulation can be summarized as follows (recall Section 22.9).

$$\text{Prothrombin} \xrightarrow[\substack{\text{Ca}^{2+},\ \text{phospholipids,} \\ \text{other factors}}]{\text{thromboplastin}} \text{Thrombin}$$

$$\text{Fibrinogen} \xrightarrow{\text{thrombin}} \text{Fibrin}$$

A number of substances, the *anticoagulants,* inhibit the clotting of blood by interfering with one or another of the reactions. Heparin is one of the principal anticoagulating agents. It is a polysaccharide rich in sulfate ester groups and is believed to block the catalytic activity of both *thromboplastin* and *thrombin.* Low concentrations of heparin are normally secreted into the circulatory system to prevent **thrombosis,** the formation of a clot within a blood vessel. (If the clot, or a fragment of the clot, breaks loose and is carried away by the blood to be lodged in small blood vessels elsewhere, e.g., the lungs, then it is called an *embolism.*) Certain sodium salts are employed as anticoagulants when blood is collected for clinical purposes. The anions of these salts—citrate, oxalate, and fluoride— form strong complexes with calcium ions, thus preventing them from existing in the free ionic form. Without calcium ions in the plasma, blood will not clot.

Two other compounds, vitamin K and dicumarol, are known to affect the clotting process. Vitamin K is a nutritional factor necessary for normal blood clotting (Section I.5). Animal blood deficient in vitamin K has a prolonged coagulation time because of a lack of prothrombin in the plasma. Vitamin K is a coenzyme in the oxygen-dependent carboxylation of the glutamic acid side chains of prothrombin to yield γ-carboxyglutamate residues. These carboxyglutamate groups must be present in prothrombin to enable it to bind to calcium ions during its conversion to thrombin. Dicumarol is believed to act as a metabolic antagonist of vitamin K. It prevents blood clotting either by repressing prothrombin formation or by inhibiting the enzyme for which vitamin K is a coenzyme. Dicumarol is frequently administered to patients who have suffered heart attacks caused by thrombosis as a preventive measure against further clotting in the blood vessels. Chemists have succeeded in synthesizing new anticoagulants that have a greater potency than dicumarol. One of these is warfarin sodium (Coumadin),[2] which is unique in that it can be administered orally, intravenously, intramuscularly, or rectally. These drugs are usually administered before an operation or after heart attacks to minimize thrombosis.

If one wishes to store blood plasma, an anticoagulant must be added to the freshly drawn blood to prevent clotting.

Dicumarol

Warfarin sodium

[1] Traditionally, hemophilia has been treated by plasma transfusion. This transfused blood contains the missing factor that can initiate clotting. A more recent development is the use of commercially available plasma concentrate that contains a high percentage of the antihemolytic factor (called factor VIII).

[2] Warfarin is also employed as a rat poison. It is safe for use as a rodenticide because regular ingestion of massive doses is fatal to rodents (producing internal hemorrhaging), whereas a single, accidental ingestion by a child or pet is harmless.

**Aspirin Against
Thrombosis**

Some doctors recommend that elderly persons and patients with a history of heart attacks and strokes should take aspirin regularly (one or two tablets daily). Recall (Section H.6) that among its other physiological effects, aspirin inhibits the synthesis of prostaglandins and thromboxanes, thereby interfering with the aggregation of blood platelets and diminishing the rate at which the blood platelets release the blood-clotting factors (thromboplastin). Hence, aspirin prevents heart attacks by inhibiting thrombosis. The greatest danger from thrombosis is that the clot may become detached and travel through the blood to some vital organ such as the heart or the brain. If the clot becomes lodged in and obstructs the blood vessels to these organs, their tissue cells are starved for oxygen and the cells die. If tissue death occurs in the brain, the condition is termed a *stroke;* if heart muscle tissue is destroyed, the condition is called a *coronary thrombosis* or a *myocardial infarction.*

28.9
HEMOGLOBIN: OXYGEN TRANSPORT

One of the main functions of blood is the transport of oxygen. The oxygen-carrying component is a conjugated protein called hemoglobin. The hemoglobin molecule has a protein portion (globin) and a nonprotein portion, or prosthetic group. The prosthetic group is heme (refer to Figures 21.20 and 21.21). A hemoglobin molecule contains four heme units, each with a central iron atom. This iron atom in each heme unit is attached to the nitrogen atoms of four pyrrole rings. Iron is in the $+2$ oxidation state and is capable of forming two additional bonds.

The protein portion of the hemoglobin molecule has four polypeptide chains, which are not covalently bonded to one another. There are two identical α chains, each with 141 amino acid units, and two β chains, each with 146 amino acid units. The protein chains play an active role in the transport of oxygen. The fifth bonding site on the iron atoms is occupied by a nitrogen atom of a histidine that is the 87th amino acid unit in the α chain or the 92nd in the β chain (Figure 28.6).

Each hemoglobin molecule is capable of transporting four molecules of oxygen. Each oxygen molecule occupies the sixth bonding site of an iron in one of the heme units. Binding of one oxygen molecule facilitates the attachment of another because oxygenation changes the conformation of the chain. This seems also to induce changes in a neighboring heme–peptide unit, giving it a greater affinity for oxygen. Conversely, release of oxygen from one site facilitates release from a neighboring site. These factors enable hemoglobin to load and unload oxygen quite rapidly.

Normally, there is about 15 g of hemoglobin per 100 mL of blood. This amount of hemoglobin can combine with about 20 mL of gaseous oxygen (at STP). Without the hemoglobin, only 0.3 mL of gaseous oxygen could physically dissolve in 100 mL of plasma. Hemoglobin makes up about 90% of the total protein of a red blood cell. The characteristic red color of blood is due entirely to the presence of hemoglobin or, more precisely, to the presence of the heme groups, which absorb strongly in the blue region of the spectrum ($\sim$400 mm).

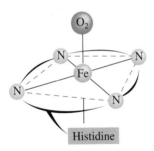

Figure 28.6
Schematic illustration of oxyhemoglobin.

The human body requires an enormous amount of oxygen to satisfy the demands of the energy-yielding oxidative phosphorylation reactions. The hemoglobin molecule is well suited to meet these demands because of its affinity for oxygen and because the attachment of oxygen to heme is readily reversible. In the alveoli of the lungs, hemoglobin comes into direct contact with a rich supply of oxygen (90–100 mm Hg) and is converted to oxyhemoglobin. The oxyhemoglobin is carried by the arterial circulation to the cells in which there is a low oxygen concentration (25–40 mm Hg) and a relatively high concentration of carbon dioxide (~60 mm Hg). The oxygen is given up to the cells. The resulting hemoglobin carries some of the carbon dioxide back to the lungs to be expelled, and more oxyhemoglobin is formed (see Figure 7.18).

Oxyhemoglobin does not transfer all of its oxygen to the tissue cells. Normally, every 100 mL of arterial blood combines with about 20 mL of oxygen. In the resting individual, the venous blood carries about 13 mL of oxygen per 100 mL of blood. Therefore, about 65% of the hemoglobin in venous blood is still combined with oxygen. When people are engaged in strenuous exercise, their oxygen demand is high and the percentage of oxyhemoglobin in the venous blood may fall as low as 25%. Arterial blood is crimson; venous blood is a darker red, but it is not purple or blue.

Various chemical substances act as poisons by interfering with the transport of oxygen by hemoglobin. Oxidizing agents such as potassium ferricyanide can oxidize the iron of hemoglobin to the ferric state. The same result is achieved in vivo by the action of nitrites and certain organic compounds (e.g., acetanilide, nitrobenzene, the sulfa drugs). The resulting compound, which contains iron in the +3 oxidation state, is called *methemoglobin* and is incapable of oxygen transport. Small amounts of methemoglobin are normally present (about 0.3 g/100 mL blood) in the erythrocytes, but appreciable amounts of this substance result in the pathological condition *methemoglobinemia*.[3] The brown color of stale meat and dried blood results from the air oxidation of hemoglobin to methemoglobin.

Carbon monoxide hemoglobin (CO-hemoglobin) is formed by the combination of carbon monoxide with hemoglobin. The ferrous ions in hemoglobin have a much greater affinity for carbon monoxide than they do for oxygen (by a factor of 200). Thus they will preferentially combine with any carbon monoxide that is in the blood. CO-hemoglobin will not transport oxygen, because all of the iron-binding sites are tied up by the carbon monoxide molecules. If sufficiently large numbers of hemoglobin molecules become saturated with carbon monoxide (about 60%), death occurs as a result of a failure of the blood to supply the brain with oxygen. Carbon monoxide poisoning is treated by greatly increasing the concentration of oxygen in the blood, either by artificial respiration in fresh air or by breathing in pure oxygen from an oxygen tank. The blood of cigarette smokers contains a relatively high percentage of CO-hemoglobin. The strain of pumping a greater volume of blood to compensate for insufficient oxygen is likely one of the major contributors to heart disease in smokers.

When red blood cells are destroyed, their hemoglobin molecules are completely catabolized. The porphyrin ring is first cleaved; then the globin and the iron are removed. The globin is digested, and its amino acids join the others in the metabolic pool. The iron

[3] Salami and other preserved meat products contain nitrite salts as preservatives. (Recall that nitrites have been implicated in nitrosoamine formation—Section 17.5). People who consume relatively large quantities of these foods have a tendency to develop methemoglobinemia. Also, nitrate ions are reduced to nitrite ions by microorganisms in the digestive tract. Concern exists over the high level of nitrates from fertilizer in the groundwater in some areas. Babies are particularly sensitive. Nitrites or nitrates cause methemoglobinemia and result in the blue baby syndrome.

An average diet supplies about 12–15 mg of iron per day, only about 10% being absorbed. To maintain sufficient iron for hemoglobin synthesis, the body must retain the 20–25 mg of iron that is released each day.

is set free and incorporated into the iron-storage protein *ferritin*. This iron will be reused for the synthesis of new hemoglobin in the bone marrow. The porphyrin skeleton is of no further use to the body. It undergoes a series of degradation reactions that lead to the production of the bile pigments, chiefly *biliverdin* and *bilirubin*. These are colored substances that give the bile its yellow color. The degraded pigments are stored in the gallbladder and released into the small intestine. As they travel down the intestinal tract, they undergo additional transformations that result in a darkening of their color, thus accounting for the characteristic colors of feces and urine. The sometimes spectacular color changes observed in a bruise have a related cause. A bruise consists of blood released and trapped beneath the skin. As the blood is gradually broken down, a series of colored products are formed.

An excess of bilirubin in the blood is responsible for the yellow color of the skin in jaundice (French *jaune,* yellow). Jaundice caused by an excess of bilirubin can arise from (1) infectious hepatitis, a condition during which the liver malfunctions and cannot remove sufficient bilirubin, (2) the obstruction of bile ducts by gallstones, and (3) an acceleration of erythrocyte destruction in the spleen (hemolytic jaundice). Jaundice occurs in a large percentage of newborn infants because of insufficient synthesis of the liver enzymes that decompose bilirubin. A common treatment of neonatal jaundice is to shine a special fluorescent light onto the baby's skin. The energy of the light is able to decompose some of the bilirubin just beneath the surface of the skin.

There are a variety of abnormal hemoglobins, each responsible for a particular disease. Perhaps the most notorious of the abnormal hemoglobins is the one responsible for sickle cell anemia. This hemoglobin, called sickle cell hemoglobin (HbS), differs from ordinary hemoglobin in only one amino acid—at the sixth position of the β chains, HbS has valine rather than glutamic acid.

Normal Hemoglobin

$$\underset{1\quad2\quad3\quad4\quad5\quad6\quad7\quad8}{\text{Val-His-Leu-Thr-Pro-Glu-Glu-Lys}}\ldots$$

Sickle Cell Hemoglobin

$$\text{Val-His-Leu-Thr-Pro-Val-Glu-Lys}\ldots$$

28.10
BLOOD BUFFERS

Recall that the maintenance of pH within narrow limits is vital to the well-being of an organism. Any slight change in hydrogen ion concentration (± 0.2–0.4 pH unit) inhibits oxygen transport and alters the rates of the metabolic processes by decreasing the catalytic efficiency of enzymes. The blood plasma and the erythrocytes contain four buffering systems that maintain the pH of the blood between 7.35 and 7.45. These are (1) the bicarbonate pair, (2) the phosphate pair, (3) the plasma proteins, and (4) the hemoglobin of red blood cells. In Section 11.6 we mentioned the buffering actions of the bicarbonate pair H_2CO_3/HCO_3^- and the phosphate pair $H_2PO_4^-/HPO_4^{2-}$. In Section 21.3 we indicated that the amino acids, in their zwitterionic forms, can neutralize small concentrations of either acids or bases. Since the plasma proteins and the globin of hemoglobin contain both acidic and basic amino acids, they tend to minimize changes in pH by combining with, or liberating, hydrogen ions. Thus they serve as excellent buffering agents over a wide range of pH values.

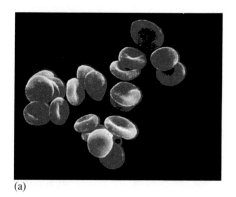 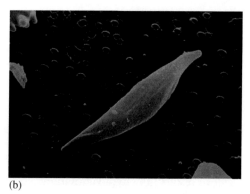

(a)   (b)

Figure 28.7
Scanning electron micrographs of normal (a) and sickled (b) red blood cells.

Sickle Cell Anemia

Sickle cell anemia is an inherited disease that, if left untreated, may be fatal (due to infection, blood clots, or cardiac or renal failure). The change from a polar amino acid (Glu) to a nonpolar one (Val) reduces the overall charge on the hemoglobin molecule. If the altered hemoglobin molecule is fully oxygenated, there is no problem and it remains in solution. However, if the level of oxygenation decreases (e.g., at high altitudes or during vigorous physical exercise), the less soluble, deoxygenated hemoglobin molecules clump together, forming long insoluble fibers, and force the red blood cell to change from a round to a crescent or sickled shape (Figure 28.7). The abnormal red blood cells become trapped in the capillaries and impair circulation. The resultant blockage of blood further decreases the oxygen supply to the affected areas of the body and increases the sickling of additional red blood cells. The abnormal cells are subsequently destroyed by the spleen, and this leads to anemia (and subsequent tiredness and other permanent damage). In the United States about 10% (more than 2 million) of the black population are genetic carriers of the disease (heterozygotes), and 0.25% have the symptoms of sickle cell anemia (homozygotes). Genetic screening tests can determine whether prospective parents carry the sickle cell trait. Sickle cell anemia is the most prevalent of the genetic diseases (Section 23.8). The genetic code has one error in its message, calling for valine instead of glutamic acid. What a difference one "minor mistake" in the code can make.

The principal *intracellular* buffer of the body cells is the phosphate pair, whereas hemoglobin molecules are the most important buffers within the red blood cells. The major *extracellular* buffer in the blood and interstitial fluid is the bicarbonate pair, owing to its intimate connection with the respiration processes. Under normal conditions the primary metabolic factor that tends to lower the pH of an organism is the continuous production of carbon dioxide and the acidic metabolites (acetoacetic acid, pyruvic acid, lactic acid, α-ketoglutaric acid, etc.). All of these compounds vary in concentration according to metabolic circumstances. When acids enter the blood, they are neutralized by bicarbonate ions, and the slightly dissociable acid, carbonic acid, is formed.

$$H^+ + HCO_3^- \rightleftharpoons H_2CO_3$$

It would seem that this reaction would alter the buffer ratio by decreasing the bicarbonate ion concentration and increasing the concentration of carbonic acid. The excess carbonic acid, however, is readily decomposed to water and carbon dioxide by the enzyme carbonic anhydrase. The respiration rate is increased, and the carbon dioxide is eliminated at the lungs, thus preserving the proper buffer ratio.

$$H_2CO_3 \xrightleftharpoons[\text{}]{\text{carbonic anhydrase}} H_2O + CO_2$$

The respiration rate is another factor that influences the pH of the blood. *Respiratory acidosis* results from **hypoventilation,** a condition that arises when the rate of breathing is too slow. Hypoventilation is brought on by an obstruction to respiration (e.g., asthma, pneumonia, or pulmonary emphysema), by coronary attack, or by drugs that depress the brain's respiratory center (e.g., morphine, barbiturates). When the respiration rate is very low, carbon dioxide is not expelled from the lungs fast enough, and the preceding equilibrium is shifted to the left. This increases the concentration of carbonic acid and results in a higher than normal H_2CO_3/HCO_3^- ratio and a subsequent decrease in the blood pH.

Hyperventilation, when the rate of breathing is too rapid, causes *respiratory alkalosis.* Hyperventilation arises during strenuous exercise, anxiety, crying, and hysteria. At high altitudes, hyperventilation may occur in response to low oxygen pressure. An increased rate of respiration accelerates the removal of carbon dioxide from the lungs, and the equilibrium shifts to the right. The concentration of carbonic acid in the blood decreases, and the H_2CO_3/HCO_3^- ratio becomes lower than normal with a subsequent increase in the blood pH.

28.11
LYMPH: A SECONDARY TRANSPORT SYSTEM

The lymphatic system also plays a role in the transport of materials from one part of the body to another. It serves to return components of the interstitial fluid to the bloodstream. The lymphatic system (Figure 28.8) is composed of veins and capillaries but no arteries. Lymph capillaries are closed at one end. Lymph flow is quite slow compared with blood circulation. Interstitial fluid is absorbed into the lymph capillaries, in which it is called **lymph.** This lymph flows into larger and larger lymph veins. Eventually, two large lymph veins empty into veins of the blood circulatory system.

Lymph serves other functions. Fat absorbed in the intestine is picked up by lymph capillaries rather than blood capillaries. **Lymph nodes** serve as filters and as factories for the production of some forms of white blood cells. The white blood cells located there remove dead cells, bacteria, and other foreign elements from the lymph. It is in the nodes that antibodies are synthesized. These nodes are lumpy enlargements in the lymph veins. Sometimes the nodes are so effective at filtering out bacteria from an infected area that they become swollen. Someone suffering from a sore throat may also exhibit swollen and tender lymph nodes in the neck area.

The role of lymph, though secondary to that of blood, is nonetheless an important one.

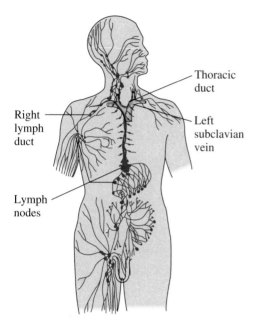

Figure 28.8
The lymphatic system.

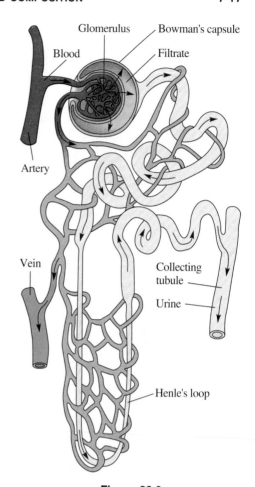

Figure 28.9
Nephrons are the functional units of kidneys, where metabolic wastes are removed from the blood. The arrows indicate the path of fluids through the unit.

28.12
URINE: FORMATION AND COMPOSITION

The kidneys operate to remove metabolic waste products from the blood. The functional units of the kidneys are called **nephrons** (Figure 28.9). Each nephron has a bulbous *Bowman's capsule,* which tails into a long, highly convoluted urinary tubule. The capsule surrounds a network of arterial capillaries called the *glomerulus.* These capillaries rejoin to form a small artery (arteriole) and then divide into another network of capillaries that surrounds the tubule of the nephron. Finally these capillaries join again and form a small vein. Blood flows from the artery into the glomerulus (capillaries), then through the arteriole and the second capillary network surrounding the tubule, and finally out the vein.

Most of the constituents of the blood, except the formed elements and large protein molecules, are filtered out through the capillary walls of the glomerulus. The glomerulus seems to act as a simple filter, with particle size being the main factor that decides what goes and what stays (Figure 28.10). The fluid that enters Bowman's capsule and eventu-

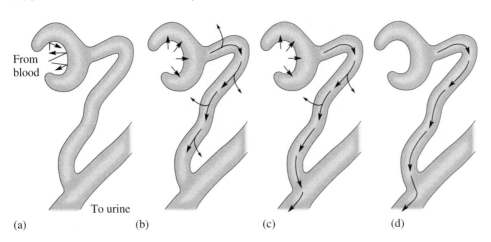

Figure 28.10

The fates of various substances in the kidney tubule. (a) Particles of colloidal size (proteins) or larger (blood cells) never filter into Bowman's capsule and therefore do not ordinarily appear in the urine. (b) Substances with a high threshold (glucose, etc.) filter into the tubules but are then reabsorbed by the blood. They do not ordinarily appear in the urine either. (c) Substances with intermediate thresholds (inorganic ions such as Na^+ or Cl^-) filter into Bowman's capsule and are then partially reabsorbed by the blood. Dietary levels determine how much finally remains in the urine. (d) Substances with low thresholds filter into the tubule and are not reabsorbed. Urine is rich in urea.

ally collects in the tubule contains many valuable components as well as wastes. Consider water alone. About 170 L per day is filtered into the tubules. Most of this water—and valuable constituents such as glucose, amino acids, and salt—is reabsorbed by the blood through the walls of the tubules. Waste products such as urea, uric acid, and excess salts are passed on into collecting tubules and eventually excreted as **urine.** Knowing that healthy adults pass 1.1–1.5 L of urine each day, you can calculate that over 99% of the water that is filtered through the glomerulus is reabsorbed through the tubules. We'd have quite a drinking problem if all that water weren't conserved.

The amount and composition of the urine are quite variable. Its volume depends on liquid intake, amount of perspiration, presence of fever or diarrhea, and other factors. The composition varies with diet and state of health. Most substances have a renal threshold level. If the concentration of a substance in the blood exceeds this threshold value, the excess is not reabsorbed through the tubule but appears in the urine. The threshold level for glucose (blood sugar), for example, is quite high (recall Section 24.3). Normally, nearly all glucose is reabsorbed. If the glucose level in the blood is too high, however (as happens in cases of uncontrolled diabetes), glucose shows up in the urine (Figure 28.11).

Some components of urine have a relatively low solubility. When some condition (such as an increase in concentration or a change in pH) leads to the precipitation of these materials in the kidney, a stone is formed. *Kidney stones* (or renal calculi) usually consist of calcium phosphate [$Ca_3(PO_4)_2$], magnesium ammonium phosphate ($MgNH_4PO_4$), calcium carbonate ($CaCO_3$), calcium oxalate (CaC_2O_4), or a mixture of these. Stone formation may accompany increased concentration of calcium ion caused by disease or increased ingestion of calcium ion. People who eat foods rich in oxalates (such as spinach) have a relatively high incidence of oxalate kidney stones. (What a great excuse for kids who don't want to eat their spinach!)

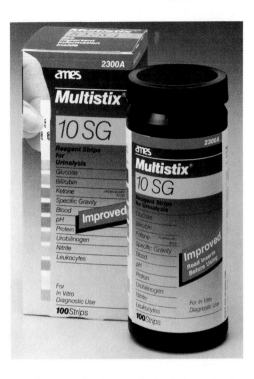

Figure 28.11
Urinalysis test strips. The reagent strip is dipped into the urine sample and then compared to a color chart for each constituent.

The kidney does far more than just get rid of wastes. It plays a vital role in maintaining the balance of water, electrolytes, acids and bases, and other components of the body fluid. When the kidneys malfunction, the body is in trouble. Analysis of the urine can often give a good indication of the health of the individual.

Wilhelm Kolff invented the artificial kidney at the Cleveland Clinic in the 1950s. Before that time, kidney failure meant death. There are now 60,000 people on dialysis at an annual cost of $1.8 billion. This constitutes 3.6% of all costs of Medicare and Medicaid, which pay for dialysis. The number of dialysis patients and the costs are rising rapidly.

28.13
SWEAT

The skin is also an organ of excretion. Through it we lose water, electrolytes, nitrogenous wastes, and lipids. Water is lost directly through the skin and through the respiratory tract at a rate of about 700 mL per day. This water loss is called *insensible perspiration*. Perspiration from the 2.5 million sweat glands is called *sensible perspiration*. It is activated by a rise in blood temperature.

Sweat is about 99% water. It contains sodium ions, chloride ions, calcium ions, and smaller amounts of other minerals. A person working in a hot environment might lose 12 L of sweat per day, including 70 g of salt. Organic constituents of sweat include urea, lipids (body oils), creatinine, lactic acid, and pyruvic acid. Drugs such as morphine, nicotine, and alcohol will also appear in sweat. The normal American diet probably contains too much sodium chloride. Thus, it makes the most sense to replace the water component of lost sweat with pure water itself (Figure 28.12). Commercial ''thirst quenchers'' are quite popular with both serious and weekend athletes. These drinks, designed to replace the salts and water your body loses during long, sweat-provoking

It is interesting to note that female mosquitoes, seeking their meal of blood, find us by following a warm stream of air laden with carbon dioxide and lactic acid. We could foil them by not sweating and not breathing!

Figure 28.12

The best replacement for fluids lost during exercise is plain water. Electrolyte replacement fluids help little, if any, and salt tablets will likely do more harm than good.

periods of exercise, may be too concentrated—a hazard that could lead to diarrhea. At best, the thirst quenchers are of marginal value.

If the air around us is not too humid, the water in sweat evaporates. Perspiration carries off not only wastes but also heat. It helps us keep cool on hot days. Each gram of water that evaporates absorbs 540 cal of heat from the body (Section 8.7).

28.14
TEARS: THE CHEMISTRY OF CRYING

If ever you should cry over your chemistry grade, take consolation from the fact that this lacrimal fluid is responsible for maintaining the health of your eyes. Tears keep the eyes moist. The eyelids sweep the secretions of the lacrimal glands over the surface of the eye at regular intervals.

Tears are actually three-layered. There is an inner layer of mucus, then a layer of lacrimal secretions, and finally an outer layer of oily film that retards evaporation from the watery middle layer. Total normal secretion is about 1 g per day.

Chemically, tears have about the same composition as other body fluids. They have approximately the same salt content as blood plasma. Tears contain *lysozyme,* an enzyme that ruptures bacterial cell walls. This bactericidal action helps prevent eye infections.

Copious flows of tears may be caused by irritants, such as pepper, acid fumes, or a variety of chemical lachrymators. Tear gas, usually α-chloroacetophenone, is a specially designed eye irritant. The flow of tears can also be triggered by emotional upsets. Such crying is undoubtedly controlled by hormones. And if you feel like crying, go right ahead. Most psychologists say it's good for you. Indeed, many people have long believed that crying is beneficial. Richard Crashaw, an English poet of the seventeenth century, called tears "the ease of woe."

α-Chloroacetophenone

Mammal	Fat	Protein	Lactose
Cow	4.4	3.8	4.5
Human	3.8	1.6	7.5
Goat	4.1	3.7	4.2
Reindeer	22.5	10.3	2.5
Porpoise	49.0	11.0	1.3

Table 28.2
Composition of Milks
(g/100 g)

28.15
THE CHEMISTRY OF MOTHER'S MILK

Newborn mammals are nourished by a secretion of the mammary glands of their mothers. Indeed, the presence of such glands in females characterizes the animal class Mammalia. The composition of milk varies from one species to another (Table 28.2). Amounts of fat and protein tend to be greater in marine mammals and those that live in cold climates.

Colloidal proteins and emulsified fats give milk its characteristic milky appearance. Casein is the precipitated protein from cow's milk. Humans beyond infancy often include some milk or milk products in their diet. This food is an excellent source of most nutrients. It includes all the essential amino acids. It also contains most vitamins and minerals. Cow's milk, though, is a poor source of iron, copper, and vitamin C. It is also short of vitamin D as it comes from the cow, so this vitamin is often added as a supplement to cow's milk intended for human consumption.

Cow's milk is sometimes given to infant humans as a substitute for their mother's milk. Evidence now indicates that this may not always be a wise substitution. Human milk not only more closely matches the nutritional needs of human infants, but it may also add to the infant's immunological defenses against disease. Newborn infants are unable to make antibodies. They have *temporary passive immunity* because their blood contains antibodies made by the mother and passed across the placenta. Mother's milk produced within the first few days after childbirth is a rich source of gamma globulins. There is clear evidence that these immunoglobulins protect some newborn animals (ungulates such as cows, horses, and sheep). A similar function for the immunoglobins in mother's milk in human infants is presumed.

And there you have it! Chemistry from molecules to mother's milk. Carbohydrate, lipid, and protein metabolism are all tied together. Indeed, in a living organism, everything is connected to everything else. Life is one huge, complicated set of chemical reactions—and, of course, a lot more. The whole of life is certainly much more than the sum of a set of chemical reactions.

We hope that we have enriched your life by helping you to learn something of the basis of chemistry and of the many ways chemistry touches your life every day. And we wish for you the proper reward for the many hours you have spent studying chemistry: the joy of success in your chosen profession.

EXERCISES

1. Define or give an explanation for the following terms.
   a. lymph
   b. formed elements
   c. hematocrit value
   d. leukemia
   e. vaccine
   f. osmosis
   g. edema
   h. hypertension
   i. hemophilia
   j. anticoagulant
   k. thrombosis
   l. embolism
   m. methemoglobinemia
   n. jaundice

2. Distinguish between the terms in each of the following pairs.
   a. aplastic anemia and hemolytic anemia
   b. anemia and polycythemia
   c. lymphocytes and phagocytes
   d. antibodies and antigens
   e. osmotic pressure and blood pressure
   f. systolic pressure and diastolic pressure
   g. vasodilation and vasoconstriction
   h. plasma and serum
   i. stroke and coronary thrombosis
   j. hypoventilation and hyperventilation

3. What are the three circulating fluids in the body?

4. List five functions of the blood.

5. The blood accounts for about 8% of the total body weight. Calculate the volume of blood that your body contains (density of blood = 1.06 g/mL).

6. What are some of the nonprotein constituents of the blood?

7. Give the formulas for five of the inorganic ions present in the blood.

8. What are the functions of (a) albumins, (b) α- and β-globulins, and (c) fibrinogen? Where are they synthesized?

9. a. What are the functions of the γ-globulins?
   b. Where are γ-globulins synthesized?
   c. What are the designations of the five classes of γ-globulins?
   d. Which class is most abundant?

10. Name the three formed elements of the blood, and give a function for each.

11. a. In what way are red blood cells similar to other cells?
    b. How do they differ?
    c. Where are erythrocytes formed?
    d. Where are they broken down?

12. Briefly describe the immune response.

13. Explain the processes involved when fluid is exchanged between the plasma and the cells (a) at the arterial end and (b) at the venous end of a capillary.

14. What determines the osmotic pressure within the blood vessels?

15. How are blood pressure measurements reported?

16. Cite three ways of treating hypertension.

17. How does each of the following function in the clotting of blood?
    a. prothrombin
    b. thrombin
    c. fibrinogen
    d. thromboplastin
    e. calcium ions
    f. fibrin

18. a. What is the role of vitamin K in blood coagulation? How do (b) heparin and (c) dicumarol act to prevent thrombosis?
    d. How do anions such as oxalates, fluorides, and citrates prevent blood coagulation?
    e. Write a structural formula for γ-carboxyglutamic acid.

19. Describe the chemical structure of hemoglobin.

20. Contrast the structural features of the following pairs of compounds.
    a. heme and hemoglobin
    b. hemoglobin and myoglobin
    c. hemoglobin and oxyhemoglobin
    d. oxyhemoglobin and CO-hemoglobin
    e. oxyhemoglobin and methemoglobin
    f. hemoglobin and sickle cell hemoglobia

21. What is the oxidation state of iron in (a) hemoglobin, (b) oxyhemoglobin, (c) methemoglobin, and (d) CO-hemoglobin?

22. a. Why is carbon monoxide such a deadly poison?
    b. How is carbon monoxide poisoning treated?

23. Name three heme-containing compounds.

24. Name the two bile pigments that are formed from the porphyrin skeleton of hemoglobin.

25. What are the causes of jaundice?

26. a. Contrast the contents of arterial blood and venous blood.
    b. What are the colors of arterial blood and venous blood?

27. What is the normal pH range of the blood?

28. What is the relationship between acidosis and oxygen transport?

29. Illustrate, with equations, how the blood buffers prevent the change in pH of the blood when small amounts of acid or base are produced during metabolic reactions.

30. Name the two blood buffer pairs. Why is the bicarbonate buffer pair effective in spite of the fact that the ratio of acid to anion is $1:10$?

31. Make use of the following equilibrium to explain what occurs to cause (a) respiratory acidosis and (b) respiratory alkalosis.

$$H^+ + HCO_3^- \rightleftharpoons H_2O + CO_2$$

32. What is interstitial fluid?

33. What is shock?

34. Which of the body fluids functions to cool the body? By what mechanism does it work?

35. List three functions of the lymphatic system.

36. How is urine formed in the kidney?

37. List four factors that affect the volume of urine excreted.
38. List three organic components of urine.
39. What is the difference between sensible and insensible perspiration?
40. List three inorganic and three organic constituents of sweat.
41. What are the three layers of tears?
42. What is the function of lysozyme in tears?
43. Is cow's milk a "perfect" food? Does it contain a complete protein?
44. What advantage does human milk possess over cow's milk as a food for a human infant?
45. Which process results in the transfer of more solutes among fluids of the body—filtration or diffusion?

46. If the amount of colloidal protein in the interstitial compartment matched the amount in the blood plasma, would more or less fluid be likely to filter from the capillaries to the interstitial space?
47. When pneumonia blocks respiratory passageways and limits the transfer of carbon dioxide from blood to the atmosphere, blood CO_2 levels build up. Is the pH of the blood higher or lower than normal in this situation?
48. A common cause of death in young children is acidosis resulting from severe diarrhea. In severe diarrhea, large amounts of bicarbonate ion are excreted from the body. Why should this lead to acidosis?

The International System of Measurement

Measurement is discussed in Chapter 1. Conversions within a system of measurement and between systems are discussed in Special Topic A. Further discussion and additional tables are provided here.

The standard unit of length in the International System of Measurement is the **meter.** This distance was once meant to be 0.0000001 of the Earth's quadrant, that is, of the distance from the North Pole to the equator measured along a meridian. The quadrant was difficult to measure accurately. Consequently, for many years the meter was defined as the distance between two etched lines on a metal bar (made of a platinum–iridium alloy) kept in the International Bureau of Weights and Measures at Sèvres, France. Today, the meter is defined even more precisely as being the distance that light travels in a vacuum during 1/299,792,458 of a second.

The primary unit of mass is the **kilogram** (1 kg = 1000 g). It is based on a standard platinum–iridium bar kept at the International Bureau of Weights and Measures. The **gram** is a more convenient unit for many chemical operations.

The basic SI unit of volume is the cubic meter. The unit more frequently employed in chemistry, however, is the cubic decimeter, which is often called a liter.

$$1 \text{ dm}^3 = 1 \text{ L} = 0.001 \text{ m}^3$$

All other SI units of length, mass, and volume are derived from these basic units.

Prefix	Abbreviation	Connotation
Pico-	p	$0.000000000001 \times$ (or $10^{-12} \times$)
Nano-	n	$0.000000001 \times$ (or $10^{-9} \times$)
Micro-	μ	$0.000001 \times$ (or $10^{-6} \times$)
Milli-	m	$0.001 \times$ (or $10^{-3} \times$)
Centi-	c	$0.01 \times$ (or $10^{-2} \times$)
Deci-	d	$0.1 \times$ (or $10^{-1} \times$)
Deka-	da	$10 \times$ (or $10^{1} \times$)
Hecto-	h	$100 \times$ (or $10^{2} \times$)
Kilo-	k	$1,000 \times$ (or $10^{3} \times$)
Mega-	M	$1,000,000 \times$ (or $10^{6} \times$)
Giga-	G	$1,000,000,000 \times$ (or $10^{9} \times$)
Tera-	T	$1,000,000,000,000 \times$ (or $10^{12} \times$)

Table I.1
Some SI Prefixes and Their Relationship to the Basic Units

1 kilometer (km) = 1000 meters (m)
1 meter (m) = 100 centimeters (cm)
1 centimeter (cm) = 10 millimeters (mm)
1 millimeter (mm) = 1000 micrometers (μm)

Table I.2
Some Metric Units of Length

1 kilogram (kg) = 1000 grams (g)
1 gram (g) = 1000 milligrams (mg)
1 milligram (mg) = 1000 micrograms (μg)

Table I.3
Some Metric Units of Mass

Table I.4
Some Metric Units
of Volume

1 liter (L) = 1000 milliliters (mL)
1 milliliter (mL) = 1000 microliters (μL)
1 milliliter (mL) = 1 cubic centimeter (cm^3)

Table I.5
Some Common Metric Conversions

Length	Mass	Volume
1 mile (mi) = 1.61 kilometers (km)	1 pound (lb) = 454 grams (g)	1 U.S. quart (qt) = 0.946 liter (L)
1 yard (yd) = 0.914 meter (m)	1 ounce (oz) = 28.4 grams (g)	1 U.S. pint (pt) = 0.473 liter (L)
1 inch (in.) = 2.54 centimeters (cm)	1 pound (lb) = 0.454 kilogram (kg)	1 fluid ounce (fl oz) = 29.6 milliliters (mL)
	1 grain (gr) = 0.0648 gram (g)	1 gallon (gal) = 3.78 liters (L)
	1 carat (car) = 200 milligrams (mg)	

Table I.6
Some Conversion Units
for Pressure

1 millimeter of mercury (mm Hg) = 1 torr
1 atmosphere (atm) = 760 millimeters of mercury (mm Hg)
= 760 torr
1 atmosphere (atm) = 29.9 inches of mercury (in. Hg)
= 14.7 pounds per square inch (lb/in.^{2})
= 101 kPa

Table I.7
Some Temperature
Equivalents[a]

Phenomenon	Fahrenheit	Celsius
Absolute zero	−459.69 °F	−273.16 °C
Nitrogen boils/liquefies	−320.4 °F	−195.8 °C
Carbon dioxide solidifies/sublimes	−109.3 °F	−78.5 °C
Bitter cold night, northern Minnesota	−40 °F	−40 °C
Cold night, Indiana	0 °F	−18 °C
Water freezes/ice melts	32 °F	0 °C
Pleasant room temperature	72 °F	22 °C
Body temperature	98.6 °F	37.0 °C
Very hot day	100 °F	38 °C
Water boils/steam condenses	212 °F	100 °C
Temperature for baking biscuits	450 °F	232 °C

[a]Keep these equations in mind.

$$°F = \frac{9}{5}(°C) + 32$$

$$°C = (°F - 32)\frac{5}{9}$$

Table I.8
Some Conversion Units
for Energy

1 calorie (cal) = 4.184 joules (J)
1 British thermal unit (Btu) = 1053 joules (J)
= 252 calories (cal)
1 food "Calorie" = 1 kilocalorie (kcal)
= 1000 calories (cal)
= 4184 joules (J)

APPENDIX II

Exponential Notation

Scientists often use numbers that are so large—or so small—that they boggle the mind. For example, light travels at 30,000,000,000 cm/s. There are 602,300,000,000,000,000,000,000 carbon atoms in 12 g of carbon. On the small side, the diameter of an atom is about 0.0000000001 m. The diameter of an atomic nucleus is about 0.000000000000001 m.

It is obviously difficult to keep track of the zeros in such quantities. Scientists find it convenient to express such numbers as **powers of ten.** Tables II.1 and II.2 contain partial lists of such numbers.

The speed of light is usually expressed as 3×10^{10} (i.e., $3 \times 10 \times 10 \times 10 \times 10 \times 10 \times 10 \times 10 \times 10 \times 10$) cm/s. The mass of an atom of cesium (Cs) is expressed as 2.21×10^{-22} g, which is

$$2.21 \times \frac{1}{10,000,000,000,000,000,000,000} \text{ g}$$

Table II.1
Positive Powers of Ten

$10^0 = 1$
$10^1 = 10$
$10^2 = 10 \times 10 = 100$
$10^3 = 10 \times 10 \times 10 = 1000$
$10^4 = 10 \times 10 \times 10 \times 10 = 10,000$
$10^5 = 10 \times 10 \times 10 \times 10 \times 10 = 100,000$
$10^6 = 10 \times 10 \times 10 \times 10 \times 10 \times 10 = 1,000,000$
$\vdots$
$10^{23} = 100,000,000,000,000,000,000,000$

Table II.2
Negative Powers of Ten

$10^{-1} = 1/10 = 0.1$
$10^{-2} = 1/100 = 0.01$
$10^{-3} = 1/1000 = 0.001$
$10^{-4} = 1/10,000 = 0.0001$
$10^{-5} = 1/100,000 = 0.00001$
$10^{-6} = 1/1,000,000 = 0.000001$
$\vdots$
$10^{-13} = 1/10,000,000,000,000 = 0.0000000000001$

Numbers such as 10^6 are called exponential numbers, where 10 is the **base** and 6 is the **exponent.** Numbers in the form 6.02×10^{23} are said to be written in **scientific notation.**

Exponential numbers are often used in calculations. The most common operations are multiplication and division. Two rules must be followed: (1) to **multiply** exponentials, **add** the exponents, and (2) to **divide** exponentials, **subtract** the exponents. These rules can be stated algebraically as

$$(x^a)(x^b) = x^{a+b}$$
$$\frac{x^a}{x^b} = x^{a-b}$$

A-3

Some examples follow.

$$(10^6)(10^4) = 10^{6+4} = 10^{10}$$
$$(10^6)(10^{-4}) = 10^{6+(-4)} = 10^{6-4} = 10^2$$
$$(10^{-5})(10^2) = 10^{(-5)+2} = 10^{-5+2} = 10^{-3}$$
$$(10^{-7})(10^{-3}) = 10^{(-7)+(-3)} = 10^{-7-3} = 10^{-10}$$
$$\frac{10^{14}}{10^6} = 10^{14-6} = 10^8$$
$$\frac{10^6}{10^{23}} = 10^{6-23} = 10^{-17}$$
$$\frac{10^{-10}}{10^{-6}} = 10^{(-10)-(-6)} = 10^{-10+6} = 10^{-4}$$
$$\frac{10^3}{10^{-2}} = 10^{3-(-2)} = 10^{3+2} = 10^5$$
$$\frac{10^{-8}}{10^4} = 10^{(-8)-4} = 10^{-12}$$
$$\frac{10^7}{10^7} = 10^{7-7} = 10^0 = 1$$

Problems involving both a coefficient (a numerical part) and an exponential are solved by multiplying (or dividing) coefficients and exponentials separately.

EXAMPLE II.1

To what is the following expression equivalent?

$$(1.2 \times 10^5)(2.0 \times 10^9)$$

SOLUTION
First, multiply the coefficients.

$$1.2 \times 2.0 = 2.4$$

Then multiply the exponentials.

$$10^5 \times 10^9 = 10^{5+9} = 10^{14}$$

The complete answer is

$$2.4 \times 10^{14}$$

EXAMPLE II.2

To what is the following expression equivalent?

$$\frac{(8.0 \times 10^{11})}{(1.6 \times 10^4)}$$

SOLUTION

First, divide the coefficients.

$$\frac{8.0}{1.6} = 5.0$$

Then divide the exponentials.

$$\frac{10^{11}}{10^4} = 10^{11-4} = 10^7$$

The answer is

$$5.0 \times 10^7$$

EXAMPLE II.3

Give an equivalent for the following expression.

$$\frac{(1.2 \times 10^{14})}{(4.0 \times 10^6)}$$

SOLUTION

It is convenient, before carrying out the division, to rewrite the dividend (the numerator) so that the coefficient is larger than that of the divisor (the denominator).

$$1.2 - 10^{14} = 12 \times 10^{13}$$

Note that the coefficient was made larger by a factor of 10 and the exponential was made smaller by a factor of 10. The quantity as a whole is unchanged. Now divide.

$$\frac{12 \times 10^{13}}{4.0 \times 10^6} = 3.0 \times 10^7$$

EXAMPLE II.4

Give an equivalent for the following expression.

$$\frac{(3 \times 10^7)(8 \times 10^{-3})}{(6 \times 10^2)(2 \times 10^{-1})}$$

SOLUTION

In problems such as this, you can carry out the multiplications specified in the numerator and in the denominator separately and then divide the resulting numbers.

$$(3 \times 10^7)(8 \times 10^{-3}) = 24 \times 10^4$$
$$(6 \times 10^2)(2 \times 10^{-1}) = 12 \times 10^1$$
$$\frac{24 \times 10^4}{12 \times 10^1} = 2 \times 10^3$$

The multiplications and divisions in problems like this can be carried out in any convenient order.

There is only one other mathematical function involving exponentials that is of importance to us. What happens when you raise an exponential to a power? You just multiply the exponent by the power. To illustrate,

$$(10^3)^3 = 10^9$$
$$(10^{-2})^4 = 10^{-8}$$
$$(10^{-5})^{-3} = 10^{15}$$

If the exponential is combined with a coefficient, the two parts of the number are dealt with separately, as in the following example.

$$(2 \times 10^3)^2 = 2^2 \times (10^3)^2 = 4 \times 10^6$$

For a further discussion of—and more practice with —exponential numbers, see the following reference.

Goldish, Dorothy M., *Basic Mathematics for Beginning Chemistry*, 4th ed., Macmillan, New York, 1990. Chapter 3 covers exponential notation.

EXERCISES

1. Express each of the following in scientific notation:
   - **a.** 0.00001
   - **b.** 10,000,000
   - **c.** 0.0034
   - **d.** 0.0000107
   - **e.** 4,500,000,000
   - **f.** 406,000
   - **g.** 0.02
   - **h.** 124×10^3

2. Carry out the following operations. Express the answers in scientific notation.
   - **a.** $(4.5 \times 10^{13})(1.9 \times 10^{-5})$
   - **b.** $(6.2 \times 10^{-5})(4.1 \times 10^{-12})$
   - **c.** $(2.1 \times 10^{-6})^2$
   - **d.** $\dfrac{(4.6 \times 10^{-12})}{(2.1 \times 10^3)}$
   - **e.** $\dfrac{(9.3 \times 10^9)}{(3.7 \times 10^{-7})}$
   - **f.** $\dfrac{(2.1 \times 10^5)}{(9.8 \times 10^7)}$
   - **g.** $\dfrac{(4.3 \times 10^{-7})}{(7.6 \times 10^{22})}$

Significant Figures

Unlike counting, measurement is never exact. You can *count* exactly 10 people in a room. If you asked each of those 10 people to *measure* the length of the room to the nearest 0.01 m, however, the values they determine are likely to differ slightly. Table III.1 presents such a set of measurements.

Note that all 10 students agree on the first 3 digits of the measurement; differences occur in the fourth digit. Which values are correct? Actually, all are accurate within the accepted range of uncertainty for this physical measurement. The accuracy of measurement depends on the type of measuring instrument and the skill and care of the person making the measurement. Measured values are usually recorded with the last digit regarded as uncertain. The data in Table III.1 allow us to state that the length of the room is between 14.1 m and 14.2 m, but we are not sure of the fourth digit. The measurements in the table have four *significant figures,* which means that the first three are known with confidence and the fourth conveys an approximate value. **Significant figures** include all digits known with certainty plus one uncertain digit.

In any properly reported measurement, all nonzero digits are significant. The zero presents problems, however, because it can be used in two ways: to position the decimal point or to indicate a measured value. For zeros, follow these rules.

1. A zero between two other digits is always significant.
   Examples: The number 1107 contains four significant figures.
   The number 50.002 contains five significant figures.
2. Zeros to the left of *all* nonzero digits are not significant.
   Examples: The number 0.000163 has three significant figures.
   The number 0.06801 has four significant figures.
3. Zeros that are *both* to the right of the decimal point *and* to the right of nonzero digits are significant.
   Examples: The number 0.2000 has four significant figures.
   The number 0.050120 has five significant figures.
   The number 802.760 has six significant figures.
4. Zeros in numbers such as 40,000 (that is, zeros to the right of *all* nonzero digits in a number that is written without a decimal point) may or may not be significant. Without more information, we simply do not know whether 40,000 was measured to the nearest unit or ten or hundred or thousand or ten-thousand. To avoid this confusion, scientists use exponential notation (Appendix II) for writing numbers. In exponential notation, 40,000 would be recorded as 4×10^4 or 4.0×10^4 or 4.0000×10^4 to indicate one, two, and five significant figures, respectively.

Table III.1
A Set of Measurements of the Length of a Room

Student	Length (m)
1	14.14
2	14.15
3	14.17
4	14.14
5	14.16
6	14.14
7	14.17
8	14.17
9	14.16
10	14.17

III.1
ADDITION OR SUBTRACTION

In addition or subtraction, the result should contain no more digits to the right of the decimal point than the quantity that has the least digits to the right of the decimal point. Align the quantities to be added or subtracted on the decimal point, then perform the operation, assuming blank spaces are zeros. Determine the correct number of digits after the decimal point in the answer and round off to this number. In rounding off, you should increase the last significant figure by one if the following digit is 5 through 9.

EXAMPLE III.1

Add the following numbers: 49.146, 72.13, 5.9432. Align the numbers on the decimal point and carry out the addition.

$$
\begin{array}{r}
49.146 \\
72.13 \\
5.9432 \\
\hline
127.2192
\end{array}
$$

The quantity with the least digits after the decimal point is 72.13. The answer should have only two digits after the decimal point. Since the third digit after the decimal point is 9, the second digit after the decimal point should be rounded up to 2. Correct answer: 127.22

EXAMPLE III.2

Perform the following addition.

$$
\begin{array}{r}
744 \\
2.6 \\
14.812 \\
\hline
761.412
\end{array} \longrightarrow 761
$$

The first quantity has no digits to the right of the decimal point (which is understood to be at the right of the digits). The answer must therefore be rounded so that it, too, contains no digits to the right of the decimal point.

EXAMPLE III.3

Perform the indicated operation.

$$
\begin{array}{r}
71.12496 \\
-\ 9.143 \\
\hline
61.98196
\end{array} \longrightarrow 61.982
$$

Since the second quantity has only three digits to the right of the decimal point, so must the answer.

III.2
MULTIPLICATION AND DIVISION

In multiplication and division, our answers can have no more significant figures than the factor that has the least number of significant figures. In these operations, the *position* of the decimal point makes no difference.

EXAMPLE III.4

Multiply 10.4 by 3.1416.

$$10.4 \times 3.1416 = 32.67264 \longrightarrow 32.7$$

The answer has only three significant figures because the first term has only three.

EXAMPLE III.5

Divide 5.973 by 3.0.

$$\frac{5.973}{3.0} = 1.991 \longrightarrow 2.0$$

The answer has only two significant figures because the divisor has only two.

III.3
EXACT VALUES

Some quantities are not measured but are defined. A kilometer is defined as 1000 meters: 1 km = 1000 m. Similarly, 1 foot can be defined as 12 inches: 1 ft = 12 in. The ''1 km'' should not be regarded as containing one significant figure; nor should ''12 in.'' be considered to have two significant figures. In fact, these values can be considered to have an infinite number of significant figures (1.0000000000000000 . . .) or, more correctly, to be *exact*. Such defined values are frequently used as conversion factors in problems (Chapter 1). When you are determining the number of significant figures for the answer to a problem, you should ignore such exact values. Use only the measured quantities in the problem to determine the number of significant figures in the answer.

EXERCISES

Perform the indicated operations and give answers with the proper number of significant figures.

1. **a.** 48.2 m + 3.82 m + 48.4394 m
   b. 151 g + 2.39 g + 0.0124 g
   c. 15.436 mL + 9.1 mL + 105 mL
2. **a.** 100.53 cm − 46.1 cm
   b. 451 g − 15.46 g
   c. 19.71 L − 10.4 L

3. **a.** 0.137 cm × 1.43 cm   **b.** 3.146 cm × 5.4 cm
   c. 73 m × 1.340 m × 0.41 m
4. **a.** $\dfrac{5.179 \text{ g}}{4.6 \text{ mL}}$ **b.** $\dfrac{4561 \text{ g}}{3.1 \text{ mol}}$ **c.** $\dfrac{40.00 \text{ g}}{3.2 \text{ mL}}$
5. $\dfrac{1.426 \text{ mL} \times 373 \text{ K}}{204 \text{ K}}$

Glossary

absolute alcohol   100% alcohol.

absolute scale   A temperature scale in which the zero point is the coldest temperature possible, or absolute zero.

acetal   Product of hemiacetal–alcohol reaction; a 1,1-diether.

acid   A proton donor.

acid anhydride   A substance that reacts with water to produce an acid; a nonmetal oxide.

acid rain   Rain having a pH less than 5.6.

acidosis   Condition that results when the pH of the blood falls below 7 and oxygen transport is hindered.

actin   A major protein in muscle, active in relaxation and contraction of the muscle.

activation energy   Minimum energy necessary to initiate a reaction.

active site   Portion of the enzyme into which substrate fits.

active transport   Movement of substances against a concentration gradient; requires energy.

activity series   A listing of metals with the most reactive metals at the top and the least reactive at the bottom.

actomyosin   The contractile protein of which muscles are made; contains actin and myosin.

adipose tissue   Connective tissue where fat is stored.

adrenaline   Epinephrine, a secondary amine causing increase in blood pressure.

aerobic process   A process that occurs in the presence of oxygen.

agonist   A molecule that fits and activates a specific receptor.

air   A mixture of gases, primarily nitrogen (78%) and oxygen (21%).

alcohols   Organic compounds containing the —OH group.

aldehyde   A compound that has a hydrogen attached to a carbonyl group.

aldose   Sugar with an aldehyde group.

alkali metal   An element in Group IA of the periodic table.

alkaline earth metal   An element in Group IIA of the periodic table.

alkaloid   A nitrogen-containing organic compound obtained from plants.

alkalosis   A physiological condition in which the pH of the blood rises above 7.5.

alkane   A hydrocarbon with only single bonds; a saturated hydrocarbon.

alkene   A hydrocarbon containing one or more double bonds.

alkyl group   The group of atoms that results when a hydrogen atom is removed from an alkane.

alkyl halide   A compound resulting from the replacement of one (or more) hydrogen atom(s) of an alkane with halogen atom(s).

alkyne   A hydrocarbon containing one or more triple bonds.

allergen   A substance that triggers an allergic reaction.

allotropes   Different forms of the same element in the same state.

alpha ray   A radioactive beam consisting of helium nuclei.

amide   An organic compound having the functional group CON in which the carbon is doubly bonded to the oxygen and singly bonded to the nitrogen.

amide linkage   The carbonyl-carbon-to-nitrogen bond.

amine   A compound that contains the elements carbon, hydrogen, and nitrogen; can be viewed as derived from ammonia by replacement of one, two, or three of the hydrogens by alkyl groups.

amino acid   An organic compound that contains both an amino group and a carboxylic acid group; amino acids combine to produce proteins.

amino group   An NH_2 unit.

amphetamines   Synthetic amines used as stimulants.

amylase   An enzyme that catalyzes the hydrolysis of starches.

amylopectin   A starch with branched chains of glucose units.

amylose   A starch with the glucose units joined in a continuous chain.

anabolic steroid   A drug that aids in the building (anabolism) of body proteins and thus of muscle tissue.

anabolism   The process of building up the molecules of living systems.

anaerobic process   A process that occurs in the absence of oxygen.

analgesic   A pain reliever.

androgen   A male sex hormone.

anemia   Abnormally low percentage of red blood cells.

anesthetic, general   A substance that produces unconsciousness as well as insensitivity to pain.

anesthetic, local   A substance that produces insensitivity to pain yet leaves the patient conscious.

anhydrous   Without water.

aniline   A benzene ring with an NH_2 group attached.

anion   A negatively charged ion.

anode   The electrode at which oxidation occurs.

antagonist   A drug that blocks the action of an agonist by blocking the receptor(s).

antibiotic   A soluble substance produced by a mold or bacterium that inhibits growth of other microorganisms.

antibody   Proteins that bind to and destroy foreign substances.

anticoagulant   A substance that inhibits the clotting of blood.

anticodon   The sequence of three adjacent nucleotides in a tRNA molecule that is complementary to a codon on mRNA.

antidiuretic   A water-conserving substance.

antihistamine   A substance that relieves the symptoms of allergies: sneezing, itchy eyes, and runny nose.

anti-inflammatory   Pertaining to a substance that inhibits inflammation.

antimetabolite A compound that inhibits the synthesis of nucleic acids.

antioxidant A substance that combines with oxygen; used to suppress rancidity.

antipyretic A fever-reducing substance.

antiseptic A compound applied to living tissue to kill or prevent the growth of microorganisms.

apoenzyme Protein part of an enzyme.

aqueous Water-containing or in water solution.

aromatic compound Any organic compound that contains a benzene ring.

arteriosclerosis Hardening of the arteries.

atmosphere The gaseous envelope surrounding the Earth (or other planet). A pressure equal to 760 mm Hg.

atom Smallest characteristic particle of an element.

atomic mass The sum of the number of protons and neutrons in the nucleus of an atom.

atomic number The number of protons in the nucleus of an atom of an element.

atomic weight A term that is commonly used to designate the atomic mass.

autocatalysis The catalysis of a reaction by one of the products of that reaction.

Avogadro's number 6.02×10^{23}.

azeotrope A constant-boiling mixture of two liquids.

background radiation Ever-present radiation from cosmic rays and from natural radioisotopes in air, water, soil, and rocks.

barbiturate A depressant anticonvulsant drug.

barometer An instrument to measure atmospheric pressure.

base A proton acceptor.

basic anhydride A substance that reacts with water to produce a basic solution; a metal oxide.

battery A series of electrochemical cells.

bends A condition resulting from dissolved nitrogen in the blood.

beriberi The disease caused by a deficiency of thiamine.

beta ray A beam of electrons emitted from atomic nuclei.

bile Yellowish green liquid with pH of 7.8–8.6.

binary compound A compound containing two elements.

biodegradable Able to be broken down by microorganisms.

bleach A compound used to remove unwanted color from fabrics, hair, or other materials.

blood pressure The force exerted by the blood against the inner walls of the arteries.

blood sugar Glucose, a simple sugar circulated in the bloodstream.

boiling point The temperature at which the vapor pressure of a liquid becomes equal to atmospheric pressure.

bond Force that holds atoms together to form a molecule or crystal.

Boyle's law For a given mass of a gas at constant temperature, the volume varies inversely with the pressure.

broad-spectrum antibiotic An antibiotic effective against a wide variety of microorganisms.

broad-spectrum insecticide An insecticide that kills many kinds of insects.

buffer A compound that consumes either acid or base to keep the pH of a solution essentially constant.

Calorie 1000 calories (or 1 kcal); used to measure the energy content of foods.

calorie The amount of heat required to raise the temperature of 1 g of water 1 °C.

carbohydrate A compound consisting of carbon, hydrogen, and oxygen; a starch or sugar.

carbonyl functional group A carbon atom doubly bonded to an oxygen atom.

carboxyl group —COOH, the functional group of the organic acids.

carboxylic acid An organic compound that contains the —COOH functional group.

carcinogens Compounds that cause cancer.

catabolism The process of breaking down molecules to provide energy in living systems.

catalyst A substance that increases the rate of a chemical reaction without itself being used up.

catenation A process in which carbon atoms join together to form long chains of hundreds or even thousands of atoms.

cathode The electrode at which reduction occurs.

cathode ray A stream of high-speed electrons.

cation A positively charged ion.

cell nucleus A membrane-enclosed structure within plant and animal cells that contains the genetic material.

Celsius A temperature scale that defines the freezing point of water as 0 °C and the boiling point of water as 100 °C.

cephalin A phosphatide lipid.

chain reaction A self-sustaining change in which one or more products of one event cause one or more new events.

Charles's law For a given mass of gas at constant pressure, the volume varies directly with the temperature (on the absolute scale).

chemical bond The force of attraction that holds atoms together in compounds.

chemical equation A before-and-after description in which chemical formulas and coefficients represent a chemical reaction.

chemical properties Descriptions of the ways in which a substance reacts with other substances to change its composition.

chemistry The study of matter and the changes it undergoes.

chemotherapy The use of chemicals to control or cure diseases.

chiral carbon A carbon atom that is attached to four different groups.

chlorination Addition or substitution of chlorine.

chlorofluorocarbon A carbon compound that contains chlorine and fluorine.

chlorophyll A green plant pigment that absorbs energy from sunlight for use in the synthesis of carbohydrates from carbon dioxide and water.

chromosomes　Threadlike bodies in cells that contain the hereditary materials.

cloning　Reproducing in identical form.

codon　A sequence of three adjacent nucleotides in mRNA that codes for one amino acid.

coenzyme　Organic cofactor.

cofactor　Enzyme activator.

collagen　Proteins of connective tissue.

colligative properties　Properties that depend on the number of particles in solution.

colloid　Solute particles intermediate in size between solution solutes and suspended matter.

combustion　Oxidation via exothermic reaction; burning.

complementary bases　The base pairs adenine–thymine and guanine–cytosine.

complete protein　A protein that supplies all the essential amino acids in the quantities needed for the growth and repair of body tissues.

compound　A pure substance made up of two or more elements combined in fixed proportions.

condensation　The reverse of vaporization; the change from the gaseous state to the liquid state.

condensed structural formula　An organic chemical formula that shows the atoms of hydrogen right next to the carbons to which they are attached.

continuous spectrum　A spectrum in which there is a continuous variation from one color to another.

cosmic rays　Radiation that comes from outer space and consists of high-energy atomic nuclei.

covalent bond　A bond formed by a shared pair of electrons.

crenation　Shriveling of a cell.

critical mass　The mass of an isotope above which a self-sustaining nuclear chain reaction can occur.

crystal　A solid with plane surfaces at definite angles.

curie　A measure of the rate of disintegration of a radioactive material.

cyclic　Ring-containing.

cytochrome　Iron-containing globular protein.

Dalton's law　The total pressure of a mixture of gases is equal to the sum of partial pressures exerted by the gases.

dehydration　Removal of water.

denature　Alter properties by adding other chemicals or by subjecting to heat or radiation.

density　The amount of mass (or weight) per unit volume.

deoxyribonucleic acid (DNA)　The type of nucleic acid found primarily in the nuclei of cells.

depressant　A drug that slows down both physical and mental activity.

detergent　Any cleansing agent.

deuterium　An isotope of hydrogen with a proton and a neutron in the nucleus (mass of 2 amu).

dextrorotatory　Capable of rotating the plane of polarized light to the right.

diabetes　A disease characterized by an abnormally high level of sugar in the blood.

dialysis　Separation of small molecules and ions from large ones.

diastereomers　Stereoisomers that are not enantiomers.

diffusion　The movement of one substance into another by molecular motion (from regions of higher concentration to regions of lower concentration).

digestion　The hydrolytic process whereby food molecules are broken down into simpler chemical units.

dimer　A two-molecule unit.

dipeptide　A compound composed of two bonded amino acids.

dipole　A molecule that has a positive end and a negative end.

dipole interactions　The attractive intermolecular forces that exist among polar covalent substances.

diprotic acid　An acid that can donate two protons per molecule.

disaccharide　A carbohydrate that can be hydrolyzed to two monosaccharides.

dispersion forces　The momentary, usually weak, attractive forces between molecules.

distillation　The boiling off of the volatile compounds such as alcohol or water, leaving behind the solids and high-boiling compounds.

disulfide unit　A covalent linkage through two sulfur atoms.

diuretic　A substance that increases the body's output of urine.

double bond　The sharing of two pairs of electrons between two atoms.

drug　Any chemical substance that affects an individual in such a way as to bring about physiological, emotional or behavioral change.

drug abuse　Use of a drug for its intoxicating effect or for a purpose other than the intended use.

edema　Retention of fluids that causes swelling.

electrodes　The substances that are connected to a source of electricity and inserted into solutions under study.

electrolysis　The splitting of compounds by means of electricity. The use of electricity to bring about chemical change.

electrolyte　A compound that, when melted or taken into solution, conducts an electric current.

electron　The unit of negative charge.

electron configuration　The arrangement of electrons about an atomic nucleus.

electron-dot formula　A representation in which the outer electrons of an atom are indicated by dots.

electronegativity　The ability of an atom to attract shared electrons toward itself when joined to another atom by a chemical bond.

electrophoresis　The process of separating a mixture by application of a buffer solution and an electric current.

element　A fundamental substance in which all atoms have the same number of protons.

emollient　An oil or grease used as a skin softener.

emulsion　A suspension of submicroscopic (colloidal) particles of fat or oil in water.

enantiomer　A molecule that is nonsuperimposable on its mirror image.

endorphin A naturally occurring peptide that bonds to the same receptor site as an opiate drug.

endothermic reaction A chemical reaction to which energy must be supplied as heat.

endpoint The point at which an indicator dye changes color during a titration.

energy The capacity for doing work.

energy of activation Minimum energy needed to initiate a reaction.

energy level Region of electron density around the nucleus of an atom.

enkephalin A compound composed of a peptide chain of five amino acid units; morphinelike substance produced by the body.

entropy A measure of the randomness of a system.

enzyme A biological catalyst.

equilibrium The condition under which the rates of the forward and reverse reactions are equal.

erythrocytes Red blood cells.

essential amino acid One of eight amino acids not produced in the body that must be included in the human diet.

essential fatty acid One of two fatty acids not produced in the body that must be included in the human diet.

ester A compound derived from carboxylic acids and alcohols. The —OH of the acid is replaced by an —OR group.

estrogen A female sex hormone.

ether A compound having two organic groups attached to an oxygen atom.

eutrophication Overabundance of nutrients and consequently of plant life in a body of water.

excited state The state in which an atom is supplied energy and an electron is moved from a lower to a higher energy level.

exothermic reaction A chemical reaction that releases heat.

Fahrenheit A temperature scale that defines the freezing point of water as 32 °F and the boiling point of water as 212 °F.

fast-twitch fibers Muscle fibers that are stronger and larger and most suited for anaerobic activity.

fat A compound formed by the reaction of glycerol with three fatty acid units; solid at room temperature.

fat-soluble vitamins Nonpolar vitamins with a high proportion of hydrocarbon structural groups that dissolve in the fatty tissue of the body and are stored for future use.

fatty acid A carboxylic acid that contains from 4 to 20 or more carbon atoms in a chain.

fermentation The process by which yeast produces alcohol.

first law of thermodynamics Energy is neither created nor destroyed.

fission The splitting of an atomic nucleus into two large fragments.

force Energy required to overcome inertia.

formalin 40% formaldehyde solution.

formula weight The sum of the atomic weights of all the atoms represented in the chemical formula.

free radical A reactive neutral chemical species that contains an unpaired electron.

freezing The reverse of melting; the change from liquid to solid state.

Freon A carbon compound containing fluorine as well as chlorine; Du Pont trade name for chlorofluorocarbon.

fructose A simple sugar found in fruits and honey or made by isomerization of glucose.

fuel cell A device in which chemical reactions are used to produce electricity directly from fuels and oxygen.

fuels Substances that burn readily with the release of significant amounts of energy.

functional group A group of atoms that confers characteristic properties on a family of organic compounds.

fundamental particles Basic units from which more complicated structures can be fashioned; protons, neutrons, and electrons.

fusion Combination of atomic nuclei to form a larger one.

gamma rays Rays similar to X-rays that are emitted from radioactive substances; have higher energy and are more penetrating than X-rays.

gas A state of matter in which the substance maintains neither shape nor volume.

gene The segment of a DNA molecule that codes for the biosynthesis of one polypeptide chain.

geometric isomers Compounds that have different configurations because of the presence of a rigid structure in the molecule.

globulin A globular protein.

gluconeogenesis The synthesis of glucose from noncarbohydrate precursors.

glucose The simple sugar that is circulated in the bloodstream; also called dextrose or blood sugar.

glycogen An animal starch composed of branched chains of glucose units.

glycogenesis The formation of glycogen from glucose.

glycogenolysis The breakdown of glycogen to form glucose.

glycol Dihydroxy hydrocarbon.

glycolysis Splitting of sugar.

glycosidic linkage The bond that joins monosaccharide units.

greenhouse effect The retention of the sun's heat energy by the Earth as a result of excess carbon dioxide or other substances in the atmosphere; causes an increase in the Earth's atmospheric and surface temperature.

ground state The state of an atom in which all electrons are in their lowest possible energy levels.

group A vertical column of the periodic table; a family of elements.

half-life The time during which half of a given radioactive sample disintegrates.

hallucinogen A drug that produces visions and sensations that are not part of reality.

halogen An element in Group VIIA of the periodic table.

hard water Water containing ions of calcium, magnesium, and/or iron.

heat A measure of the quantity of energy that a substance contains.

heat of vaporization The amount of heat involved in the evaporation or condensation of 1 g of a material.

hemiacetal Product of aldehyde–alcohol reaction.

hemiketal Product of ketone–alcohol reaction.

hemolysis The rupture of red blood cells.

herbicide A chemical used to kill weeds.

heterocyclic compound A cyclic compound in which one or more atoms in the ring are not carbon.

heterogeneous mixture A mixture with definite separate component phases.

high-energy bond A bond that releases a relatively large amount of energy when it is hydrolyzed.

homogeneous mixture A solution with its solute dissolved and mixed evenly throughout.

homologs Compounds that differ by one carbon atom and two hydrogen atoms.

hormone A chemical messenger that is secreted into the blood by an endocrine gland.

humidity The measure of the amount of water vapor in the air.

hydration The process in which water molecules surround solute ions.

hydrocarbon An organic compound that contains only carbon and hydrogen.

hydrogen bond The dipole interaction between a hydrogen atom bonded to F, O, or N in a donor molecule and an F, O, or N atom in a receptor molecule.

hydrolysis The reaction of a substance with water; literally, a splitting by water.

hydrometer A device used to measure the specific gravity of a liquid.

hydronium ion (H_3O^+) A water molecule to which a hydrogen ion (H^+) has been added; the characteristic ion of an aqueous acid.

hydrophilic Water-soluble portion; literally, water loving.

hydrophobic Water-insoluble portion; literally, water hating.

hydrosphere The oceans, seas, rivers, and lakes of the Earth.

hydroxide ion (OH^-) Responsible for the properties of base in water.

hydroxyl group The —OH group.

hygroscopic Readily absorbing water from the air.

hyperacidity An excess of acid in the stomach.

hyperglycemia Blood sugar concentration above normal.

hypertonic solution Solution of relatively high osmolarity.

hypnotic A substance that induces sleep.

hypoglycemia Blood sugar concentration below normal.

hypotonic solution Solution of relatively low osmolarity.

immiscible Cannot be mixed; pertaining to liquids that are insoluble in each other.

immune response Synthesis of antibodies in response to invading bacteria or viruses.

indicator A dye the color of which depends on the acidity of the solution.

inflammation A tissue response to injury or stress.

inorganic chemistry The study of the compounds of all elements other than carbon.

insulin A hormone affecting blood sugar level.

interionic forces The electrostatic forces between ions.

intermolecular forces The forces of attraction between molecules.

intramolecular forces The forces of attraction within a molecule.

invert sugar An equimolar mixture of glucose and fructose.

iodine number The number of grams of iodine that will be consumed by 100 g of fat or oil; an indication of the degree of unsaturation.

ion A charged atom or group of atoms.

ionic bond The chemical bond that results when electrons are transferred from one atom to another; the electrostatic interaction of oppositely charged ions.

ionization energy The energy needed to remove an electron from an atom.

ionizing radiation Radiation that causes the formation of ions from neutral particles.

isoelectric pH The pH value at which an amino acid exists as a zwitterion.

isomerization The conversion of a compound from one isomeric form to another.

isomers Compounds that have the same molecular formula but different structural formulas and properties.

isotonic solutions Solutions of the same osmotic pressure.

isotopes Atoms that have the same number of protons but different numbers of neutrons.

joule The SI unit for energy.

kelvin The SI unit for temperature.

keratin The tough fibrous protein that comprises most of the outermost layer of the epidermis and is the main component of hair and nails.

ketal Product of hemiketal–alcohol reaction.

ketohexose A six-carbon monosaccharide with a ketone group.

ketone Compound with two organic groups attached to a carbonyl group.

ketone bodies Acetoacetic acid, β-hydroxybutyric acid, and acetone.

ketose Sugar with a ketone group.

ketosis Elevated levels of ketone bodies in the blood.

kilocalorie Measurement of energy content; equal to 1000 calories.

kilogram The SI unit of mass; a quantity slightly greater than 2 lb.

kinetic energy The energy of motion.

kinetic molecular theory A model that uses the motion of molecules to explain the behavior of the three states of matter.

lactose The sugar found in milk; composed of two simpler sugars, glucose and galactose.

lactose intolerance The inability of some individuals to break down the sugar lactose; caused by the absence of a necessary enzyme.

law of conservation of energy The amount of energy within the universe is constant; energy cannot be created or destroyed, only transferred.

law of conservation of mass   Matter is neither created nor destroyed during a chemical change.

law of constant composition   A compound always contains elements in certain definite proportions. Also called law of definite proportions.

LD$_{50}$ The dosage that would be lethal to 50% of the population of test animals.

Le Châtelier's principle   If a stress is applied to a system at equilibrium, the equilibrium shifts in the direction that will relieve the stress.

lecithin   Cephalin that contains choline.

leukemia   A cancer that is characterized by the uncontrolled production of leukocytes that fail to mature.

leukocytes   White blood cells.

levorotatory   Capable of rotating the plane of polarized light to the left.

limiting reagent   The reactant that is used up first in a reaction, after which the reaction ceases no matter how much remains of other reactants.

lipases   Enzymes that catalyze the hydrolysis of fats.

liquid   A state of matter in which the substance assumes the shape of its container, flows readily, and maintains a fairly constant volume.

liter   A measurement of volume equal to a cubic decimeter.

lymph   Interstitial fluid absorbed into the lymph capillaries.

macromolecule   A molecule with a very high molecular mass; a polymer.

marijuana   A preparation made from the leaves, flowers, seeds, and small stems of the *Cannabis* plant.

mass   A measure of the quantity of matter.

matter   Stuff of which all materials are made; has mass and occupies space.

mechanism   A series of individual steps in a chemical reaction that gives the net overall change.

melting   The process in which a substance changes from the solid to the liquid state.

melting point   The temperature at which a solid changes to a liquid.

meso compound   A stereoisomer that contains at least two chiral centers but has an internal symmetry plane.

messenger RNA (mRNA)   The type of RNA that contains the codons for a protein; mRNA travels from the nucleus of the cell to a ribosome.

metabolism   The set of all chemical reactions in living systems that break down large molecules for energy and component parts and build large molecules from component parts.

metals   The group of elements to the left of the heavy stepped line on the periodic table.

metastable   Pertaining to an isotope that can release some energy and form a more stable isotope.

meter   The SI unit of length, slightly longer than 3 feet.

methemoglobin   Hemoglobin in which the iron ion has a 3+ charge.

micelle   A cluster of molecules that contain both polar and nonpolar groups.

micronutrients   Substances needed by the body only in tiny amounts.

mineral acids   Acids derived from inorganic materials.

miscible   Pertaining to liquids that are completely soluble in each other.

mitochondria   Cell organelles. The "power plants" of the cell.

mixture   Matter with a variable composition.

molar volume   The volume occupied by a mole of most gases at STP (about 22.4 L).

molarity   Moles of solute per liter of solution.

mole   A quantity of a chemical substance that contains 6.02×10^{23} units of the substance.

molecular weight   A term that is commonly used to designate the molar mass.

molecule   Discrete group of atoms held together by shared pairs of electrons.

monomer   A molecule of relatively low molecular mass. Monomers are combined to make polymers.

monosaccharide   A carbohydrate that cannot be hydrolyzed into simpler sugars.

monounsatured fatty acid   Fatty acid that contains one double bond per molecule.

multiplicity   The existence of several tRNAs for the same amino acid.

mutarotation   Interconversion of α and β forms of sugars.

mutation   Any chemical or physical change that alters the sequence of bases in the DNA molecule.

myosin   The main protein in muscle; provides the energy for muscle contraction by hydrolyzing ATP.

narcotic   A drug that produces both narcosis and relief of pain.

neuron   Nerve cell.

neurotransmitter   A chemical that carries an impulse across the synapse from one nerve cell to the next.

neutralization   The reaction of an acid and a base to produce a salt and water.

neutron   A fundamental particle with a mass of approximately 1 amu and no electric charge.

nitrogen balance   An individual's intake of dietary nitrogen is equal to the amount of nitrogen lost.

noble gases   Generally unreactive gases that appear on the far right of the periodic table.

nonbonding pair   Pair of electrons not involved in a bond.

nonmetals   The group of elements to the right of the heavy stepped line on the periodic table.

nonpolar covalent bond   A covalent bond in which there is an equal sharing of electrons.

nucleic acid   The molecule of heredity, found in every living cell; built from nucleotides.

nucleoprotein   A combination of a protein with a nucleic acid.

nucleoside   A sugar–base combination.

nucleotide   A combination of a heterocyclic amine, a pentose sugar, and phosphoric acid; the monomer unit of nucleic acid.

nucleus   Concentrated, positively charged matter at the center of an atom; composed of protons and neutrons. *See also* cell nucleus

octet rule Atoms seek an arrangement that will surround them with eight electrons in the outermost energy level.

oil Substance formed from glycerol and unsaturated fatty acids; liquid at room temperature.

optically active Able to rotate a beam of polarized light.

optimum pH The pH at which an enzyme exhibits maximum activity.

optimum temperature The temperature that affords maximum enzyme activity.

orbital Space occupied by directional location of electrons.

organic chemistry The study of the compounds of carbon.

osmosis The net diffusion of water through a semipermeable membrane.

osmotic pressure The pressure needed to prevent the net flow of solvent from dilute to concentrated solutions.

oxidation A substance is oxidized when it loses electrons.

oxide A binary compound of oxygen and another element.

oxidizing agent A substance that causes oxidation and is itself reduced.

oxygen debt An oxygen imbalance resulting from an anaerobic pathway.

paraffin An alkane; a saturated hydrocarbon.

partial pressure The pressure of an individual gas in a mixture of gases.

pathogenic Disease-causing.

penultimate carbon The chiral carbon that is farthest from the carbonyl carbon.

peptide bond The amide linkage that bonds amino acids in chains of proteins, polypeptides, and peptides.

perfluorocarbon A compound in which all hydrogen atoms have been replaced by fluorine atoms.

periodic table A systematic arrangement of the elements in columns and rows; elements in a given column have similar properties.

periods The horizontal rows of the periodic table.

pH scale An exponential scale of acidity; below 7, acidic; 7, neutral; above 7, basic.

phenol A benzene ring with a hydroxyl group attached.

phenyl Benzene ring as substituent.

pheromones Chemicals that are used for communication between members of the same species of insects.

phospholipid Phosphorus-containing lipid.

photoscan A permanent visual record showing the differential uptake of a radioisotope by various tissues.

photosynthesis The process used by green plants to form glucose from carbon dioxide and water at the expense of solar energy.

physical dependence Acute withdrawal symptoms occur if drug is not available.

physical properties Properties that describe equalities that can be demonstrated without changing the composition of the substance.

placebo A substance that looks and tastes like a real drug but has no active ingredients.

plasmid A circular piece of DNA that occurs outside the nucleus in bacteria.

plasmolysis The rupture of a cell.

polar covalent bond A covalent bond in which the electrons are unequally shared.

polarimeter An instrument that is used to detect optically active compounds.

polarity Charge separation occurring in a covalent bond where electrons are shared unequally.

polarized light Light that vibrates in a single plane.

polyatomic ion A charged particle containing two or more covalently bonded atoms.

polymer A molecule with a large molecular mass; a chain formed of repeating smaller units.

polypeptide A polymer of amino acids; usually of lower molecular mass than protein.

polysaccharide A carbohydrate that can be hydrolyzed into many monosaccharide units.

polyunsaturated Containing two or more double bonds.

polycyclic Pertaining to organic molecule containing several fused rings.

positron A positively charged particle with the mass of an electron.

potential energy Energy by virtue of position or composition.

precipitate An insoluble substance.

pressure The force exerted by a substance upon a surface.

product A substance produced by a chemical reaction and whose formula follows the arrow in a chemical equation.

progestin A compound that mimics the action of progesterone.

proof Twice the percentage of alcohol by volume.

prostaglandins Hormonelike compounds derived from arachidonic acid that are involved in increased blood pressure, the contraction of smooth muscle, and other physiological processes.

prosthetic group Organic or inorganic component introduced in conjugated protein.

protein Amino acid polymer.

proton The unit of positive charge in the nucleus of an atom.

proton donor An acid; a substance that gives up an H^+ (proton).

psychedelic drug Drug that induces colorful visions.

psychological dependence An uncontrollable desire for a drug.

psychotomimetic drug Drug that induces psychosis.

psychotropic drug Drug that affects the mind.

pure substance Matter with a definite, or fixed, composition.

racemic mixture A mixture of enantiomers that is optically inactive because it contains equal amounts of molecules with opposite rotatory power.

radioactivity Spontaneous emission of alpha, beta, and/or gamma rays by the disintegration of the nuclei of atoms.

radioisotope Radioactive isotope.

reactant A starting material or original substance in a chemical change; one whose formula precedes the arrow in a chemical equation.

recombinant DNA DNA in an organism that contains genetic material from another organism.

Recommended Daily Allowance (RDA) The recommended level of a nutrient necessary for a balanced diet.

redox reaction A reaction in which oxidation and reduction occur; an oxidation-reduction reaction.

reducing agent A substance that causes reduction and is itself oxidized.

reducing sugar Any carbohydrate that is capable of acting as a reducing agent.

reduction A substance is reduced when it gains electrons.

replication Copying or duplication; the process by which DNA reproduces itself.

resonance A term used to describe the phenomenon in which no single classical Lewis structure adequately accounts for the properties of a molecule.

respiration "Burning" of glucose by living cells; oxygen is absorbed and carbon dioxide is given off.

respiratory chain The sequence of highly organized oxidation reduction enzymes.

restriction endonuclease Enzyme that cleaves DNA molecule at specific location of specific base sequence.

ribonucleic acid (RNA) The form of nucleic acid found mainly in the cytoplasm but present in all parts of the cell.

ribosome Cellular substructure that serves as the site for protein synthesis.

salt linkage An interaction between an acidic side chain on one amino acid residue and a basic side chain on another; the resulting charges serve as an ionic bond between peptide chains or between two parts of the same chain.

saponification Soap making: alkaline hydrolysis of fat.

saturated fat A fat composed of a large proportion of saturated fatty acids.

saturated fatty acid A fatty acid that contains no double bonds.

saturated hydrocarbon A compound of carbon and hydrogen with only single bonds.

saturated solution A solution that contains all the solute that it can at equilibrium.

scientific law A summary of experimental data; often expressed in the form of a mathematical equation.

semipermeable Pertaining to a membrane permeable to some solute but not to others.

single bond The sharing of one pair of electrons.

slow-twitch fiber Muscle fiber that is best suited for aerobic work.

smog A contraction of the words smoke and fog.

solid A state of matter in which the substance maintains its shape and volume.

solute The substance that is dissolved in another substance to form a solution; usually present in a smaller amount than the solvent.

solution A homogeneous mixture of two or more substances.

solvent The substance that dissolves another substance to form a solution; usually present in a larger amount than the solute.

specific gravity The ratio of the density of a substance to that of water.

specific heat The amount of heat required to raise the temperature of 1 g of a substance by 1 °C.

starvation The withholding of nutrition from the body whether voluntary or involuntary.

state of matter A condition or stage in the physical being of matter; a solid, liquid, or gas.

stereoisomers Isomers that have the same molecular formulas but differ in the arrangement of atoms in three-dimensional space.

steroid A lipid substance with four fused rings.

stimulant A drug that increases alertness, speeds up mental processes, and generally elevates the mood.

STP Standard temperature (273 K) and pressure (1 atm).

strong acid An acid that reacts completely, is 100% ionized in water; a powerful proton donor.

strong base A powerful proton acceptor; dissociates 100% in water.

structural formula A chemical formula that shows how the atoms of a molecule are arranged, to which other atom(s) they are bonded, and the kinds of bonds.

substituent Any atom or group of atoms that substitutes for a hydrogen atom in an organic molecule.

substitution Replacement of one atom for another in a molecule.

substrate Substance that reacts with enzymes to create a new reaction product.

sucrose Common table sugar derived principally from sugarcane or sugar beets; a disaccharide that hydrolyzes to glucose and fructose.

supersaturated solution A solution that contains solute in excess of that in a solution at equilibrium.

surface tension The special force that exists at the surface of a liquid.

suspension Temporary dispersion of solute in solution.

synapse The gap between nerve fibers.

synergistic effect An effect much greater than just the sum of the expected effects.

temperature A measure of heat intensity, or how energetic the particles of a sample are.

thrombosis The formation of a clot within a blood vessel.

titration A technique used to detect the amount of acid or base present in a sample.

tolerance Increasing dosages of a drug are required to produce the same degree of narcosis or analgesia.

tracer Radioactive isotope used to trace movement or locate the site of radioactivity in physical, chemical, and biological systems.

transamination Formation of glutamic acid from other amino acids.

transcription The process by which DNA directs the synthesis of RNA molecules during protein synthesis.

transfer RNA (tRNA)   A small RNA molecule that contains the anticodon nucleotides; the RNA molecule that bonds to an amino acid.

transition elements   Metallic elements situated in the center portion of the periodic table in the B groups.

translation   The process by which the information contained in the codon of a mRNA molecule is converted to a protein structure.

transmutation   Change of one element into another.

triglyceride   Three long-chain fatty acids bonded to glycerol.

triose   A three-carbon sugar.

tripeptide   A compound composed of three bonded amino acids.

triple bond   The sharing of three pairs of electrons between two atoms.

triprotic acid   An acid that can donate three protons per molecule.

tritium   A rare radioactive isotope of hydrogen with two neutrons and one proton in the nucleus (a mass of 3 amu).

turnover number   The number of substrate molecules converted to product in 1 min by one enzyme molecule.

Tyndall effect   The scattering of a beam of light as it passes through a colloid.

unsaturated hydrocarbon   A hydrocarbon containing a double or a triple bond.

unsaturated solution   A solution that contains less solute than a saturated solution.

valence   The number of covalent bonds that an atom can form.

valence shell   Outermost shell of electrons of an atom.

valence-shell electron-pair repulsion theory   A theory of chemical bonding; that valence-shell electrons locate themselves as far apart as possible.

vapor pressure   The partial pressure exerted by the molecules of a substance that are in the gas phase above the liquid phase of the substance.

vaporization   The process in which a substance changes from the liquid to the gaseous (vapor) state.

viscosity   Resistance to flow.

vital capacity   The maximum amount of air that can be forced from the lungs.

vitamins   Organic compounds that the body can not produce in the amounts required for good body health.

wax   Esters formed from long-chained fatty acids and long-chained monohydroxy alcohols.

weak acid   An acid that reacts only slightly; a poor proton donor with a low degree of ionization in solution.

weight   A measure of the force of attraction of the Earth on an object.

X-ray   Radiation similar to visible light but of much higher energy and much more penetrating.

zwitterion   A compound that contains both a positive and a negative charge; a dipolar ion.

Answers to Practice Exercises
and Selected Exercises

CHAPTER 1

EXERCISES

3. a, b, d, e
5. yes
7. no
9. a. physical change   **b.** chemical change
11. a. physical change   **b.** chemical change
13. steam
15. a pure substance
19. a. helium   **b.** nitrogen   **c.** fluorine
   d. magnesium   **e.** silicon   **f.** sulfur
   g. potassium   **h.** iron   **i.** copper   **j.** bromine
   k. tin   **l.** barium   **m.** gold   **n.** potassium
   o. plutonium
21. elements: a, b, d, f; compounds: c, e, g, h, i
23. sprinter
25. automobile
27. on 10-m platform
29. at top of hill
31. length: meter; volume: liter; mass: kilogram; in laboratory: centimeter, milliliter, and gram
33. a. cm   **b.** kg   **c.** dL
35. a. m   **b.** lb   **c.** gal
37. a. 50,000 m   **b.** 0.25 m
39. a. 10 dL   **b.** 0.2 dL
41. a. 15 g   **b.** 86 mg
43. a. 1500 mL   **b.** 0.018 L
45. 0 K, 0 °F, 0 °C
47. 100 °C
49. a. 2750 cal   **b.** 740 cal
51. 1000 cal (only 2 significant figures)
53. 30 g
55. 1.1 g/cm^3
57. 1.6 g/mL
59. 680 g
61. 1.05 g/mL
63. 39.5 g
65. 7.3 g/mL
67. 1.1044

SPECIAL TOPIC A

1. a. 3400 m   **b.** 5.70 m   **c.** 0.121 m   **d.** 91.4 m
3. a. 1290 g   **b.** 1.575 g   **c.** 421 mg   **d.** 2.55 m
   e. 18.3 mm   **f.** 4220 mL
5. 141 lb

7. 317 lb
11. 0.325 g, 0.0114 oz
13. 7.5 mi/min
15. 72 km/hr
17. 299,000,000 m/s
19. $2.09 per gal
21. 750 lb
23. 23 mi/gal
25. 633 g

CHAPTER 2

PRACTICE EXERCISES

2.1 $^{90}_{42}Mo$
2.2 38 protons, 52 neutrons, 38 electrons
2.3 $^{90}_{37}X$ and $^{88}_{37}X$; $^{90}_{38}X$ and $^{93}_{38}X$
2.4 32
2.7 $1s^2 2s^2 2p^5$
2.8 $1s^2 2s^2 2p^6 3s^2 3p^5$
2.9 Rb: $5s^1$; Se: $4s^2 4p^4$

EXERCISES

5. no
9. attract
11. protons and neutrons
17. A, 21 amu; B, 21 amu; C, 22 amu; D, 20 amu
19. a. 2   **b.** 11   **c.** 17   **d.** 8   **e.** 12   **f.** 16
21. $^{10}_5B$
23. $^{125}_{53}I$
25. a. 30 p, 32 n   **b.** 94 p, 147 n   **c.** 43 p, 56 n
   d. 36 p, 45 n
27. lithium-7
31. electrons
33. 12
35. Electron has moved to a lower energy level.
39. a. 6   **b.** dumbbell   **c.** 3
43. a. 4   **b.** 7   **c.** 13
45. O: 4 $2p$ electrons; F: 5 $2p$ electrons
47. argon
49. no
53. a. IA   **b.** IB–VIIIB   **c.** VIIA   **d.** IIA
55. b
57. b, d, e
59. b, c
61. 2
63. a. As   **b.** Cl   **c.** Be
65. a. Cs   **b.** Sb   **c.** Ac

CHAPTER 3

PRACTICE EXERCISES

3.1 $^{246}_{98}Cf$

3.2 $^{85}_{35}Br$

3.3 0.3 mg

3.4 22,920 years

3.5 $^{1}_{0}n$

3.6 $^{188}_{79}Au \rightarrow {}^{0}_{+1}e + {}^{188}_{78}Pt$

EXERCISES

3. a. $^{4}_{2}He$ **b.** $^{0}_{-1}e$ **c.** $^{1}_{0}n$ **d.** $^{0}_{+1}e$

5. $^{209}_{82}Pb \rightarrow {}^{0}_{-1}e + {}^{209}_{83}B$

7. $^{225}_{90}Th \rightarrow {}^{221}_{88}Ra + {}^{4}_{2}He$

9. $^{186m}_{79}Au \rightarrow {}^{186}_{79}Au + \gamma$

11. $^{82}_{36}Kr$

13. 26 s

15. $^{31}_{16}S \rightarrow {}^{0}_{+1}e + {}^{31}_{15}P$

17. $^{87}_{35}Br \rightarrow {}^{1}_{0}n + {}^{86}_{35}Br$

19. $^{21}_{12}Mg \rightarrow {}^{1}_{1}H + {}^{20}_{11}Na$

21. a. $^{7}_{3}Li$ **b.** $^{1}_{1}H$ **c.** $^{1}_{1}H$ **d.** $^{153}_{62}Sm$

23. $^{215}_{85}At \rightarrow {}^{4}_{2}He + {}^{211}_{83}Bi$

25. neutrons

31. gamma

37. internal: 10 mCi; external: 500 Ci

39. about 0.028% of lethal dose

41. gamma rays

43. ultrasonography and MRI

CHAPTER 4

PRACTICE EXERCISES

4.1 S^{2-}

4.2 a. $:\ddot{A}r:$ b. $\cdot Sr \cdot$ c. $:\ddot{F}:$ d. $:\ddot{N} \cdot$

     e. $K \cdot$ f. $:\ddot{S}:$

4.3 $Li \cdot + \cdot \ddot{F}: \rightarrow Li^+ + :\ddot{F}:^-$

4.4

        $\cdot \ddot{O}:$ $:\ddot{O}:^{2-}$

   $\cdot Al \cdot$ Al^{3+}

   $+ \cdot \ddot{O}: \rightarrow$ $+ :\ddot{O}:^{2-}$

   $\cdot Al \cdot$ Al^{3+}

        $\cdot \ddot{O}:$ $:\ddot{O}:^{2-}$

4.6 Ca_3N_2

4.9 copper(II) bromide

4.10 a. $:\ddot{B}r \cdot + \cdot \ddot{B}r: \rightarrow :\ddot{B}r:\ddot{B}r:$

   b. $H \cdot + \cdot \ddot{B}r: \rightarrow H:\ddot{B}r:$

   c. $:\ddot{I} \cdot + \cdot \ddot{C}l: \rightarrow :\ddot{I}:\ddot{C}l:$

4.12 N_2O_5

4.13 P_4Se_3

4.14 K_3PO_4

4.15 $Ca(CH_3CO_2)_2$

4.16 calcium carbonate

4.17 potassium dichromate

4.18

        H   H

   H—C—C—$\ddot{C}l:$

        H   H

4.19 $:\ddot{F}—\ddot{O}—\ddot{F}:$

4.20

       H

   H—P—H

       H

4.21 $:\ddot{O}::\ddot{N}:\ddot{O}:$

        $:\ddot{F}:$

4.22 linear

4.23 pyramidal

EXERCISES

1. noble gases

7. a. $\cdot \dot{C} \cdot$ **b.** $K \cdot$ c. $\cdot \dot{M}g \cdot$ **d.** $:\ddot{C}l \cdot$

   e. $\cdot \dot{N} \cdot$

9. a. 3+   **b.** 2−

13. $:\ddot{I} \cdot + \cdot \ddot{I}: \rightarrow :\ddot{I} \overset{\times}{\underset{\times}{}} \ddot{I}:$ ($\overset{\times}{\times}$ = bonding electrons)

15. a. ionic   **b.** polar covalent   **c.** ionic

17. a. nonpolar covalent   **b.** nonpolar covalent

   c. polar covalent

19. polar (all)

21. a. $\overset{\longrightarrow}{H—N}$ **b.** $\overset{\longrightarrow}{Be—F}$ **c.** $\overset{\longrightarrow}{P—Cl}$

23. polar $\overset{\times S \times}{\swarrow \quad \searrow}_{F \quad F}$ net dipole: $\overset{S}{\underset{F \downarrow F}{\swarrow \searrow}}$

25. C, N, and O

27. a. $Na^+ :\ddot{F}:^-$ **b.** $K^+ :\ddot{C}l:^-$ **c.** $K^+ :\ddot{F}:^-$

29. a. $3Na^+ :\ddot{N}:^{3-}$ **b.** $Al^{3+} 3:\ddot{C}l:^-$ **c.** $3Mg^{2+} 2:\ddot{N}:^{3-}$

       $:\ddot{F}:$

31. $:\ddot{F}:C:\ddot{F}:$

       $:\ddot{F}:$

33. a. sodium ion   **b.** magnesium ion   **c.** aluminum ion

   d. chloride ion   **e.** oxide ion   **f.** nitride ion

35. a. iron(III) ion (ferric)   **b.** copper(II) ion (cupric)

   c. silver ion

37. a. Br^- **b.** Ca^{2+} **c.** K^+ **d.** Fe^{2+}

39. a. sodium bromide   **b.** calcium chloride

   c. iron(III) chloride (ferric chloride)   **d.** lithium iodide

   e. potassium sulfide   **f.** copper(I) bromide (cuprous bromide)

41. a. N_2O **b.** P_4S_3 **c.** PCl_5 **d.** SF_6

43. a. carbon disulfide   **b.** dinitrogen tetrasulfide

   c. phosphorus pentafluoride   **d.** disulfur decafluoride

45. a. $^{\delta-}N—H^{\delta+}$ **b.** $^{\delta+}C—F^{\delta-}$ **c.** no partial charge

47. a.
$$H:\overset{\displaystyle H}{\underset{\displaystyle H}{C}}:\overset{..}{\underset{..}{O}}:H$$
b. $H:\overset{\displaystyle H}{N}:\overset{..}{\underset{..}{O}}:H$
c. $H:\overset{\displaystyle H}{C}:\overset{\displaystyle H}{N}:H$

d. $H:\overset{\displaystyle H}{N}:\overset{\displaystyle H}{N}:H$

49. a.
$$H-\overset{\displaystyle H}{\underset{\displaystyle H}{C}}-O-H$$
b. $H-\overset{\displaystyle H}{N}-O-H$

c. $H-\overset{\displaystyle H}{C}-\overset{\displaystyle H}{N}-H$
d. $H-\overset{\displaystyle H}{N}-\overset{\displaystyle H}{N}-H$

51. a. $:\overset{..}{\underset{..}{F}}:C:\overset{..}{\underset{..}{F}}:$ with $\overset{..}{\underset{..}{O}}:$ below
b. $:\overset{..}{\underset{..}{Cl}}:P:\overset{..}{\underset{..}{Cl}}:$ with $:\overset{..}{\underset{..}{Cl}}:$ below

c. $H:\overset{..}{\underset{..}{O}}:P:\overset{..}{\underset{..}{O}}:H$ with $:\overset{..}{\underset{..}{O}}:$ and H below
d. $H:C:::N:$

53. a. $[:\overset{..}{\underset{..}{Cl}}:\overset{..}{\underset{..}{O}}:]^-$
b. $[H:\overset{..}{\underset{..}{O}}:P:\overset{..}{\underset{..}{O}}:]^{2-}$ with $:\overset{..}{\underset{..}{O}}:$ below

c. $[:\overset{..}{\underset{..}{O}}:\overset{..}{\underset{..}{Cl}}:\overset{..}{\underset{..}{O}}:]^-$
d. $[:\overset{..}{\underset{..}{O}}:\overset{..}{\underset{..}{Br}}:\overset{..}{\underset{..}{O}}:]^-$ with $:\overset{..}{\underset{..}{O}}:$ below

55. a. $:N::\overset{..}{\underset{..}{O}}:$
b. $:\overset{..}{\underset{..}{Cl}}:B:\overset{..}{\underset{..}{Cl}}:$ with $:\overset{..}{\underset{..}{Cl}}:$ below

57. a. tetrahedral   **b.** bent
59. a. pyramidal   **b.** tetrahedral
61. a. pyramidal   **b.** bent
63. a. carbonate ion   **b.** monohydrogen phosphate ion
  c. permanganate ion   **d.** hydroxide ion
65. a. NH_4^+ **b.** HSO_4^- **c.** CN^- **d.** NO_2^-
67. a. $MgSO_4$ **b.** $NaHCO_3$ **c.** KNO_3 **d.** $CaHPO_4$
69. a. $Fe_3(PO_4)_2$ **b.** $K_2Cr_2O_7$ **c.** CuI **d.** NH_4NO_2
71. a. potassium nitrite   **b.** lithium cyanide
  c. ammonium iodide   **d.** sodium nitrate
  e. potassium permanganate   **f.** calcium sulfate
73. a. sodium monohydrogen phosphate   **b.** ammonium phosphate   **c.** aluminum nitrate   **d.** ammonium nitrate
77. ClO_2
79. AlP and Mg_3P_2

CHAPTER 5

PRACTICE EXERCISES

5.1 $P_4 + 6 H_2 \rightarrow 4 PH_3$

5.2 $4 V + 5 O_2 \rightarrow 2 V_2O_5$
5.3 $C_3H_8 + 5 O_2 \rightarrow 3 CO_2 + 4 H_2O$
5.4 $2 H_3PO_4 + 3 Ca(OH)_2 \rightarrow Ca_3(PO_4)_2 + 3 H_2O$
5.5 50.0 L
5.7 88.0 amu
5.8 121 g
5.9 a. 0.0640 mol   b. 0.776 mol   c. 1.04 mol
  d. 0.0103 mol
5.10 Molecular: 2 molecules of H_2S react with 3 molecules of O_2 to form 2 molecules of SO_2 and 2 molecules of H_2O
  Molar: 2 mol of H_2S react with 3 mol of O_2 to form 2 mol of SO_2 and 2 mol of H_2O
  Mass: 68 g of H_2S react with 96 g of O_2 to form 128 g of SO_2 and 36 g of H_2O
5.11 a. 1.59 mol   b. 305 mol   c. 0.6120 mol
5.12 0.334 g
5.13 136 g
5.14 8.02 g
5.15 21.4 g

EXERCISES

 5. 6.02×10^{23}; 12.04×10^{23}
 7. 44 g/mol
 9. 22.4 L (all)
11. a. 4   **b.** 4   **c.** 8   **d.** 6
13. a. 12   **b.** 8
15. 12 N, 6 P, 54 H, 24 O
17. balanced: a, d, e; not balanced: b, c
19. 1 molecule (mol) of C_3H_8 reacts with 5 molecules (mol) of O_2 to form 3 molecules (mol) of CO_2 and 4 molecules (mol) of H_2O; 44 g of C_3H_8 reacts with 160 g of O_2 to form 132 g of CO_2 and 72 g of H_2O.
21. a. $Cl_2O_5 + H_2O \rightarrow 2 HClO_3$
  b. $V_2O_5 + 2 H_2 \rightarrow V_2O_3 + 2 H_2O$
  c. $4 Al + 3 O_2 \rightarrow 2 Al_2O_3$
  d. $Sn + 2 NaOH \rightarrow Na_2SnO_2 + H_2$
  e. $PCl_5 + 4 H_2O \rightarrow H_3PO_4 + 5 HCl$
23. a. $Na_3P + 3 H_2O \rightarrow 3 NaOH + PH_3$
  b. $Cl_2O + H_2O \rightarrow 2 HClO$
  c. $2 CH_3OH + 3 O_2 \rightarrow 2 CO_2 + 4 H_2O$
  d. $3 Zn(OH)_2 + 2 H_3PO_4 \rightarrow Zn_3(PO_4)_2 + 6 H_2O$
  e. $C_3H_8 + 5 O_2 \rightarrow 3 CO_2 + 4 H_2O$
25. a. 16 g/mol   **b.** 84 g/mol   **c.** 352 g/mol
27. a. 156.9 g/mol   **b.** 178 g/mol   **c.** 294.2 g/mol
  d. 342.3 g/mol
29. a. 1.59 L   **b.** 80.5 L
31. 26.0 g $NaNO_2$
33. 0.0731 mol
35. a. 0.435 g   **b.** 47.0 g   **c.** 45.0 g   **d.** 615 g
37. a. 16.7 mol CO_2 **b.** 55.9 mol O_2
39. 2490 g NH_3
41. 11.3 g O_2
43. 2050 g

45. 3600 g
47. a. C   **b.** B   **c.** A
49. endothermic
51. rate increases
53. rate decreases
57. shift to the right
59. a, b, c: shift to the right; d: no change
63. 0.861 g
65. 2.20 g
67. 0.0521 g

CHAPTER 6

PRACTICE EXERCISES

6.1 $2 Zn + O_2 \rightarrow 2 ZnO$
6.2 $2 PbS + 3 O_2 \rightarrow 2 PbO + 2 SO_2$
6.3 a, b, c, d
6.4 b
6.5 b, c, d
6.6 none
6.7 a
6.8 a
6.9 **a.** $\underline{Se} + \boxed{O_2}$ **b.** $\boxed{CH_3C\equiv N} + \underline{H_2}$

   c. $\boxed{V_2O_5} + \underline{H_2}$ **d.** $\underline{K} + \boxed{Br_2}$

EXERCISES

3. elemental carbon, C; hydrogen, H_2; hydroquinone, $C_6H_4(OH)_2$
7. a. oxidized   **b.** reduced
9. neither
11. a. $4 \underline{Al} + 3 \boxed{O_2} \rightarrow 2 Al_2O_3$ **b.** $2 \underline{SO_2} + \boxed{O_2} \rightarrow 2 SO_3$
13. a. $\underline{Fe} + 2 HCl \rightarrow FeCl_2 + H_2$ **b.** $\underline{CS_2} + 3 O_2 \rightarrow CO_2 + 2 SO_2$
15. a. S oxidized, N reduced   **b.** I oxidized, Cr reduced
17. reduced
19. I^- oxidized, Cl_2 reduced
21. reduced
23. Zr oxidized by water
25. oxidized

CHAPTER 7

PRACTICE EXERCISES

7.2 1.33 atm
7.3 3.79 L
7.4 2.98 L
7.5 $-161\ °C$
7.7 1.25 atm
7.8 41 mL
7.9 0.67 L
7.10 22.2 L/mol
7.11 5.0 L
7.12 0.047 atm

EXERCISES

5. decreasing
7. at same temperature, same pressure
9. pressure increase (all)
11. a. 1500 torr   **b.** 380 mm Hg   **c.** 0.23 cm Hg
13. a. 150 lb/in.^{2} **b.** 14 cm H_2O **c.** 0.10 atm
15. 1000 mL
17. a. 9000 L   **b.** 19 hr
19. cavity expands, pressure decreases
21. 10 mL
23. 3.3 L
25. 2470 mm Hg
27. 0.48 m^3
29. 750 cc
31. 0.95 atm
33. 708 mm Hg
35. 95%
37. 22 L
39. 19 L
41. 1.2 atm
43. 390 K
45. 7.3×10^8 mol
47. 67%
49. O_2 to cells; CO_2 to lungs
51. 280 s

CHAPTER 8

PRACTICE EXERCISES

8.1 59.00 cal/g
8.2 23.6 kcal
8.3 48.0 kcal
8.5 180 cal
8.6 8500 cal

EXERCISES

5. N_2 (dispersion forces), HCl (dipole interactions), H_2O (hydrogen bonds)
7. a. dispersion forces   **b.** dipolar forces   **c.** hydrogen bonds
13. a. H_2S, H_2Se, H_2Te **b.** O_2, CO, H_2O **c.** CCl_4, CBr_4, CI_4
15. 7200 cal/mol
17. 96.8 cal
21. 44.5 cal
23. 7413 cal/mol vs. 7300 cal/mol (table)
25. ethanol
27. 72,250 cal
29. 16 kcal
31. lower density
33. ethane
35. Xe
37. 180 kcal

CHAPTER 9

PRACTICE EXERCISES

9.2 2.5 M

9.3 7.83 g

9.4 0.030 L

9.5 8%

9.6 12 mL of alcohol and 88 mL of H_2O

9.7 3%

9.9 Weigh out 3 g of H_2O_2 and dissolve it in 97 mL of H_2O.

EXERCISES

3. **a.** particle size less than 0.25 nm diameter   **b.** random distribution   **c.** solute not separable by filtration **d.** not observed   **e.** may be colored, must be clear

5. no

7. Ethanol has —OH group and can form hydrogen bonds.

9. saturated

11. 40 mL of acetic acid + water to make 2.0 L

13. 4% by mass

15. 5% by mass

17. 0.50 kg of NaCl + 4.5 kg of water

19. 1 ppb, 1 ppm, 1 mg/dL, 1%

21. **a.** 0.1 mL of drug diluted to 100 mL volume **b.** 0.0005 L (0.5 mL) diluted to 1 L volume

23. 2.0 M·

25. 3.0 M

27. 5.35 g of NH_4Cl + water to 100 mL of solution

31. **a.** 2   **b.** 1   **c.** 3   **d.** 2   **e.** 3   **f.** 4

33. **a.** 0.5 mol   **b.** 1.0 mol   **c.** 0.33 mol **d.** 0.50 mol   **e.** 0.33 mol   **f.** 0.25 mol

37. **a.** 0.1 M $NaHCO_3$ **b.** 1 M NaCl   **c.** 1 M $CaCl_2$ **d.** 3 M glucose

39. shrivel in 5.5% NaCl solution; unaffected by 0.92% solution

43. no

49. 1 g

CHAPTER 10

PRACTICE EXERCISES

10.1 nitrous acid; chloric acid

10.2 HNO_3

10.3 KOH

10.4 $Ca(OH)_2 + 2 HCl \rightarrow CaCl_2 + 2 H_2O$

EXERCISES

3. turns red litmus blue, feels slippery, tastes bitter, reacts with acid to form water and a salt

5. hydroxide ion (OH^-)

7. strong: NaOH (sodium hydroxide), KOH (potassium hydroxide), LiOH (lithium hydroxide); weak: NH_3 (ammonia)

9. a substance that releases hydrogen ions in water to yield hydronium ions
$HCl + H_2O \rightarrow H_3O^+ + Cl^-$

11. **a.** diprotic   **b.** monoprotic   **c.** diprotic **d.** triprotic   **e.** tetraprotic

13. hydroselenic acid

15. phosphorous acid

17. benzoic acid

19. **a.** $HAsO_3$ **b.** AsO_3^- **c.** $KAsO_3$

21. a substance that reacts with water to produce a base

23. $H_2CO_3 \rightarrow H_2O + CO_2$

25. $Na_2CO_3 + 2 H_3O^+ \rightarrow 3 H_2O + CO_2 + 2 Na^+$

27. **a.** base   **b.** acid   **c.** acid

29. **a.** HCl   **b.** H_2SO_4 **c.** H_2CO_3 **d.** HCN

31. **a.** LiOH   **b.** $Mg(OH)_2$ **c.** NaOH

33. $HCl + H_2O \rightarrow H^+ + Cl^-$, hydrochloric acid

35. H_2SO_4, acid

37. KOH, base

39. strong base

41. weak acid

43. weak acid

45. $NaOH + HCl \rightarrow NaCl + H_2O$

47. $Ca(OH)_2 + 2 HCl \rightarrow CaCl_2 + 2 H_2O$

49. $H_3PO_4 + 3 NaOH \rightarrow Na_3PO_4 + 3 H_2O$

51. taste becomes salty; litmus unaffected

55. milk of magnesia

57. H_2SO_4

59. chemical burns

CHAPTER 11

PRACTICE EXERCISES

11.1 153 mL of acid and 247 mL of water

11.2 3.8 mL of acid and 26.2 mL of water

11.4 0.028 M

11.8 11

11.10 8.57

11.11 neutral

11.12 acidic

11.13 basic

EXERCISES

3. **a.** 4.0 L   **b.** 2.8 mL

5. **a.** yes   **b.** yes   **c.** no   **d.** yes

7. $NH_4^+ + H_2O \rightleftharpoons NH_3 + H_3O^+$

9. basic

11. 0.1 M

13. 10.0 mL

15. 0.121 M

17. **a.** 2   **b.** 3

19. **a.** 1   **b.** 5

21. (17a) 12   (17b) 11   (18a) 10   (18b) 13

23. **a.** basic   **b.** acidic   **c.** neutral   **d.** acidic

27. bicarbonate/carbonic acid, phosphate, proteins
29. decrease
31. **a.** 2.5   **b.** 6.3   **c.** 9.1
33. 1.06
35. 7.34
37. 11.70

SPECIAL TOPIC B

EXERCISES

1. **a.** $HBO_2 \rightleftharpoons H^+ + BO_2^-$ **b.** $HClO_2 \rightleftharpoons H^+ + ClO_2^-$
   c. $HC_9H_7O_4 \rightleftharpoons H^+ + C_9H_7O_4^-$
   d. $H_2Se \rightleftharpoons H^+ + HSe^-$
3. 4.00×10^{-7} M
5. **a.** $K_a = \dfrac{[H^+][OCl^-]}{[HOCl]}$ **b.** $K_a = \dfrac{[H^+][C_6H_7O_6]}{[HC_6H_7O_6]}$
   c. $K_a = \dfrac{[H^+][HCO_2^-]}{[HCO_2H]}$
7. **a.** 1.3×10^{-4} M   **b.** 1.2×10^{-3} M
   c. 1.8×10^{-5} M
9. **a.** 7.7×10^{-11} M   **b.** 8.3×10^{-12} M
   c. 5.6×10^{-10} M
11. **a.** 6.7×10^{-4} M   **b.** 6.5×10^{-3} M
    c. 6.5×10^{-6} M
13. **a.** 1.5×10^{-11} M   **b.** 1.5×10^{-12} M
    c. 1.5×10^{-9} M
15. **a.** 6.2×10^{-10} M   **b.** 6.6×10^{-4} M
    c. 1.8×10^{-5} M
17. $[OH^-] = 1.8 \times 10^{-5}$ M; $[H^+] = 5.6 \times 10^{-10}$ M
19. 10
21. 9.21
23. 4.18
25. 4.34

CHAPTER 12

EXERCISES

5. **a.** K and Br_2 **b.** Li and Cl_2 **c.** Al and O_2
7. $HI + H_2O \rightarrow H_3O^+ + I^-$
13. **a.** base   **b.** acid   **c.** salt
15. **a.** $2 Na + 2 H_2O \rightarrow H_2 + 2 NaOH$
    b. $2 K + 2 H_2O \rightarrow H_2 + 2 KOH$
    c. $2 Li + 2 H_2O \rightarrow H_2 + 2 LiOH$
    d. $Ca + 2 H_2O \rightarrow H_2 + Ca(OH)_2$
21. yes
23. cathode
25. $CaCO_3 + 2 HCl \rightarrow H_2CO_3 + CaCl_2$
    $H_2CO_3 \rightarrow H_2O + CO_2$
29. high blood pressure
35. 7.9×10^{-7} M
37. no

SPECIAL TOPIC C

EXERCISES

3. unreactive; filled outer shell
9. Be; does not react with water or oxidize rapidly
11. Al
13. $\cdot Ga \cdot$ and Ga^{3+}
15. **a.** F^- **b.** I^- **c.** O^{2-} **d.** S^{2-} **e.** K^+
    f. Sr^{2+}
17. inhibits growth of bacteria on teeth, preventing plaque buildup
19. oxides of nonmetals that react with water, forming H_3O^+ ion
    $$SO_3 + H_2O \rightarrow H_2SO_4 + 2 H_2O \rightarrow 2 H_3O^+ + SO_4^{2-}$$

CHAPTER 13

PRACTICE EXERCISES

13.3 a. 3-methylhexane   b. 2,5-dimethylpentane
     c. 3-ethylhexane   d. 3-isopropylpentane

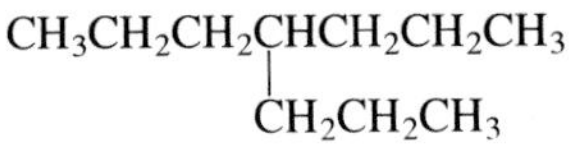

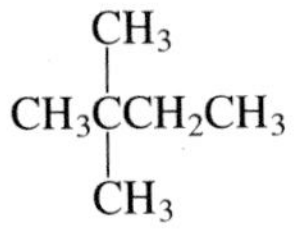

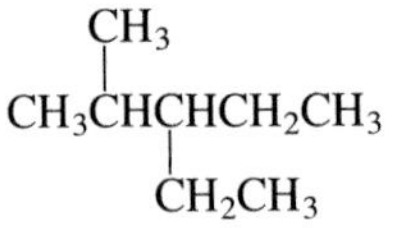

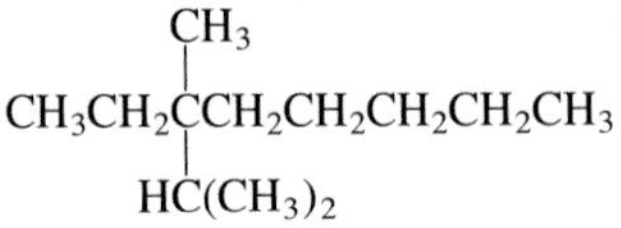

13.4

13.9

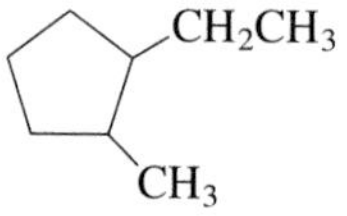

13.11 a. 4-ethyl-2-methyl-2-hexene
      b. 1-methylcyclohexene

13.12

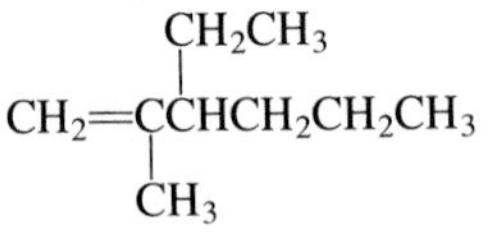

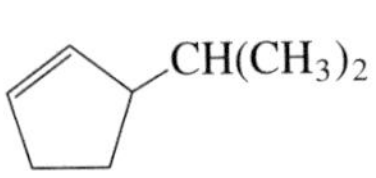

13.13

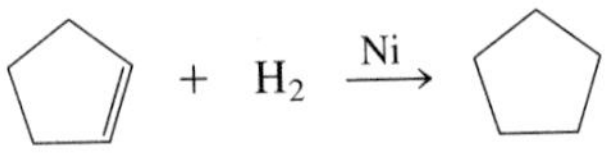

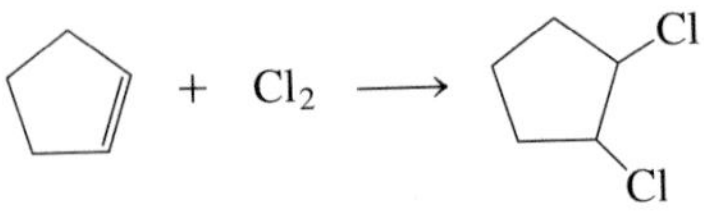

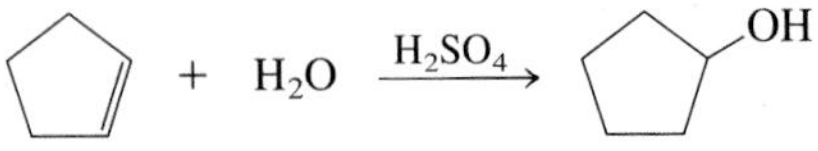

EXERCISES

3. organic: a, c, d; inorganic: b, e, f
5. organic
7. same; a, b, d; isomers: c, e
9. hexane, 2-methylpentane, 3-methylpentane, 2,3-dimethylbutane, 2,2-dimethylbutane
11. saturated: b, d; unsaturated: a, c
15. butyl bromide and 1-bromobutane, *sec*-butyl bromide and 2-bromobutane, *tert*-butyl bromide and 2-bromo-2-methylpropane, isobutyl bromide and 1-bromo-2-methylpropane
19. **a.** cycloheptane   **b.** cyclohexene
    c. methylcyclopentane   **d.** ethylbenzene
21. aromatic: a, d; aliphatic: b, c
25. correct names: **a.** 2,2-dimethylpropane
    b. 2,2,3-trimethylbutane   **c.** 3,5-dimethylheptane
    d. 3,4,5-trimethyloctane
29. **a.** 3-methylpentane   **b.** 2,4-dimethylbutane
    c. 3-methylpentane   **d.** 2,2,4,4,-tetramethylhexane
31. **a.** methylcyclopropane   **b.** 1,2-diethyl-4-methylcyclopentane   **c.** cyclobutene
    d. 3-ethylcyclohexene
37. **a.** pentane   **b.** $CH_3(CH_2)_4CH_3$ **c.** cyclohexane
    d. $CH_3(CH_2)_7CH_3$
49. **a.** $CH_4 + 2 O_2 \rightarrow CO_2 + 2 H_2O$
    b. $2 C_8H_{18} + 25 O_2 \rightarrow 16 CO_2 + 18 H_2O$
51. $2 C_6H_6 + 15 O_2 \rightarrow 12 CO_2 + 6 H_2O$; 132 g CO_2
53. plastic bags, bottles, toys, electrical insulation
55. double bond

CHAPTER 14

PRACTICE EXERCISES

14.1 1-methylcyclopentanol

14.2 **a.**

  b.

14.3 **a.** $CH_3CH_2CH_2CH_2CH_2OH \xrightarrow[H_2SO_4]{K_2Cr_2O_7}$

  b.

  c. $CH_3CH_2C(CH_3)_2 \xrightarrow[H_2SO_4]{K_2Cr_2O_7}$ No reaction

  (with OH on central C)

EXERCISES

3. **a.** 1-pentanol   **b.** 2-pentanol   **c.** 3-pentanol
   d. 3-methyl-1-butanol   **e.** 3-methyl-2-butanol

---

f. 2-methyl-2-butanol   **g.** 2-methyl-1-butanol
h. 2,2-dimethyl-1-propanol

5. butyl methyl ether, isobutyl methyl ether, *sec*-butyl methyl ether, *tert*-butyl methyl ether, ethyl propyl ether, ethyl isopropyl ether
7. **a.** 4,4-dichloro-2-butanol
   b. 3,3-dibromo-2-methyl-2-butanol
   c. 1-methylcyclohexanol   **d.** 3-ethyl-3-hexanol
17. **a.** about equally insoluble   **b.** *p*-cresol more soluble
23. $CH_3CH_2CH_2COOH$ from butyl alcohol; $CH_3CH_2C(\!=\!O)CH_3$ from isobutyl alcohol; $(CH_3)_2CHCOOH$ from *sec*-butyl alcohol; no reaction with *tert*-butyl alcohol
27. methanol, ethanol, 1-propanol
29. diethyl ether, 1-butanol, 1-octanol
31. pentane, diethyl ether, propylene glycol
37. $CH_3CH_2CH_2OH$, 1-propanol; $(CH_3)_2CHCH_2OH$, isobutyl alcohol; $(CH_3)_2CHCH_2CH_2OH$, 3-methyl-1-butanol; $CH_3CH_2CH(CH_3)CH_2OH$, 2-methyl-1-butanol
47. 23 kg

CHAPTER 15

PRACTICE EXERCISES

15.1 3,3-dimethylbutanal

15.2 $CH_3CH_2CHCH_2CHCH_2C\overset{\text{O}}{\underset{\text{H}}{\|}}$ (with Br and I substituents)

15.3 $CH_3CH_2CHCHCH_2CCH_2Br$ (with Br, O, and CH_2CH_3 substituents)

15.4 2-chlorocyclohexanone

15.6

EXERCISES

1. pentanal, 3-methylbutanal, 2-methylbutanal, dimethylpropanal
3. **a.** benzaldehyde   **b.** butanal   **c.** 4-methylpentanal
   d. 2-ethylbutanal   **e.** 3-hydroxypropanal
   f. 2,2,3-trichlorobutanal   **g.** 2-chlorobenzaldehyde
   h. 2,2-dimethyl-3-phenylpropanal
9. 2-propanol
11. acetaldehyde
15. yes: a, c, d; no: b, e
17. phenylhydrazine; orange precipitate (2-pentanone)
19. **a.** Cu(II) ion   **b.** Cu(II) ion   **c.** Ag(I) ion

21. benzaldehyde, cinnamaldehyde, biacetyl (see Figure 16.1)
25. alkene, alcohol, ketone
27. a, d
29. c, e

CHAPTER 16

PRACTICE EXERCISES

16.1 3-methylbutanoic acid

16.2 $CH_3CHCH CH_2CH_2C$
$\quad\quad\overset{\displaystyle Br}{|}\quad\quad\quad\quad\overset{\displaystyle O}{\|}$
$\quad\quad\underset{\displaystyle CH_3}{}\quad\quad\quad\quad OH$

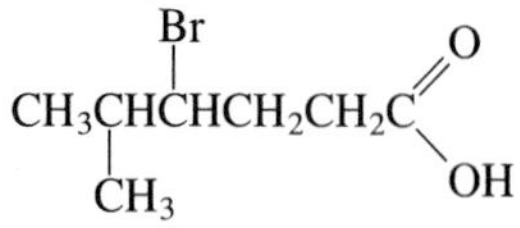

16.3 a.

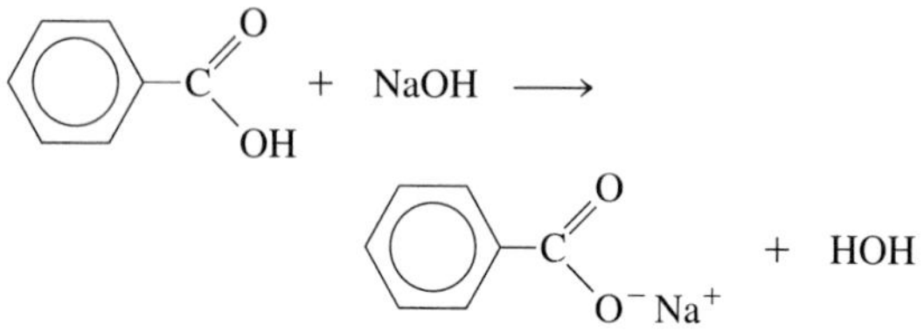

+ NaOH ⟶

+ HOH

b.

+ NaHCO₃ ⟶

+ HOH + CO₂

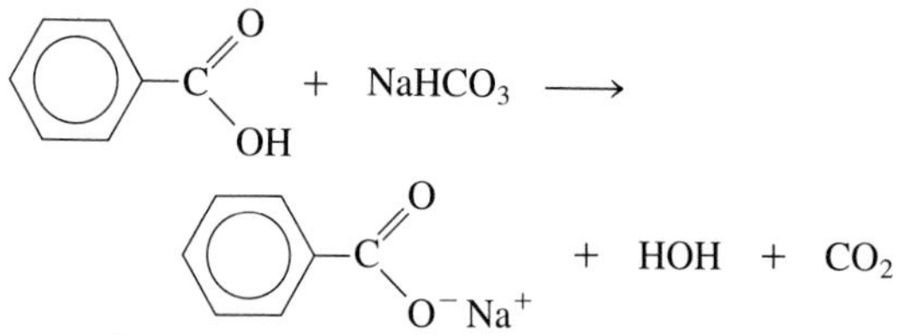

16.5 methyl benzoate

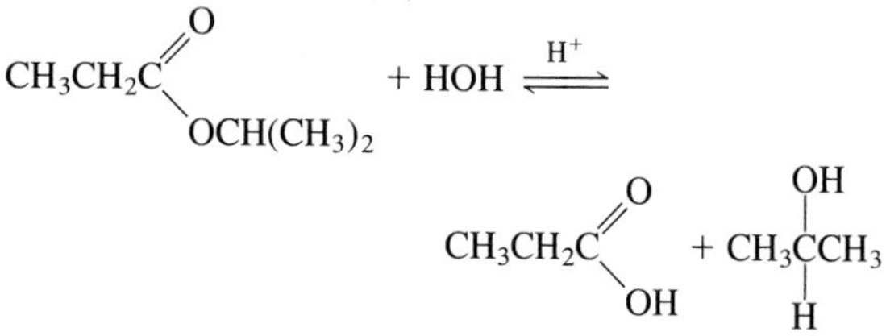

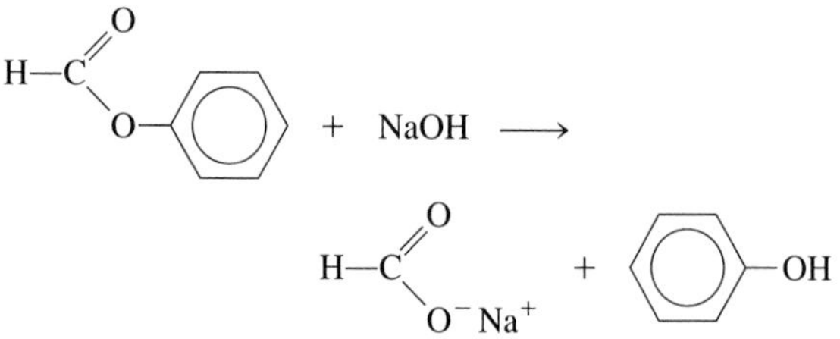

16.7 $CH_3CH_2CH_2C$ (structure with phenyl ester)

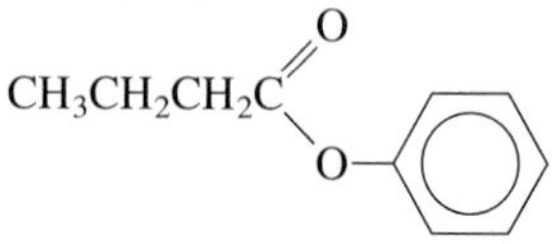

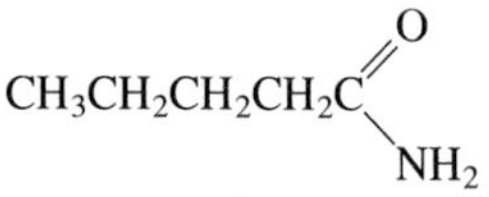

16.8 CH_3CH_2C + HOH $\overset{H^+}{\rightleftharpoons}$
$\quad\quad OCH(CH_3)_2$

CH_3CH_2C + CH_3CCH_3
$\quad\quad OH\quad\quad\quad\quad H$

16.9 $H-C$ + NaOH ⟶

$H-C$ + phenol
$\quad\quad O^-Na^+$

16.11 $CH_3CH_2CH_2CH_2C$
$\quad\quad\quad\quad\quad\quad NH_2$

16.12 CH_3CH_2C
$\quad\quad\quad\quad N(CH_3)_2$

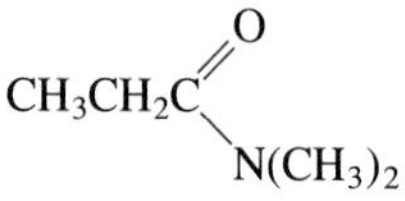

16.13

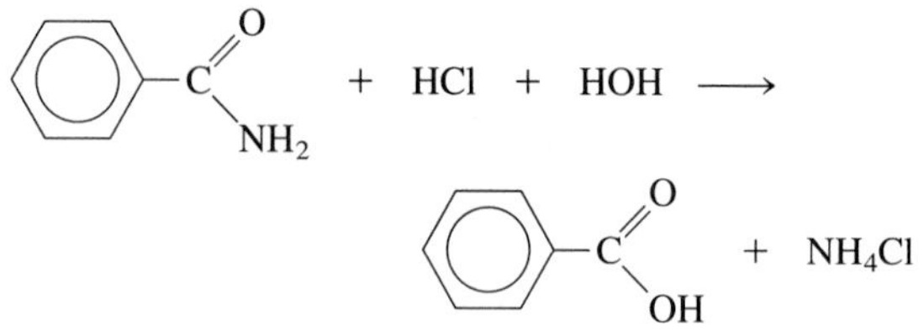

+ HCl + HOH ⟶

+ NH₄Cl

16.14

+ NaOH ⟶

+ NH₃

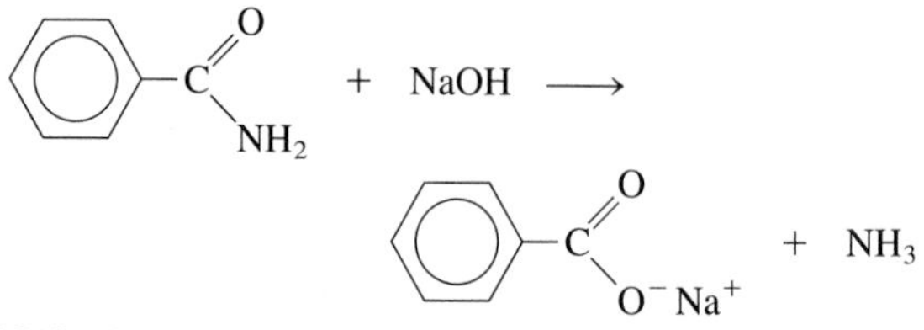

EXERCISES

7. a. decanoic acid   **b.** 4,4-dichloro-5-methylhexanoic acid   **c.** 3-ethyl-4-hydroxy-2-iodopentanoic acid   **d.** 4-bromobenzoic acid
9. a. lithium benzoate   **b.** ammonium butanoate
17. a. $CH_3CH_2CH_2OH$ **b.** $HOCH_2CH_2OH$ **c.** CH_3OH **d.** $(CH_3)_2CHCH_2CH_2OH$
21. a. benzamide   **b.** 2-methylbutanamide   **c.** acetamide   **d.** *p*-chlorobenzamide   **e.** *N,N*-dimethylpropionamide   **f.** *N*-phenylbutanamide
23. II
25. I
27. I
29. I
31. short-chain carboxylic acids; esters
45. a. $K_2Cr_2O_7$, H^+ **b.** $K_2Cr_2O_7$, H^+ **c.** NaOH
47. a. H_2O, HBr   **b.** KOH, H_2O
51. a. carboxylic acid   **b.** ketone, alcohol   **c.** ketone, alcohol   **d.** aldehyde, alcohol   **e.** ether, aldehyde   **f.** ether, aldehyde   **g.** ester   **h.** ether, ketone   **i.** ester   **j.** ester   **k.** ketone, alcohol   **l.** alcohol, aldehyde
53. a. pyrophosphoric acid   **b.** ethyl dihydrogen phosphate

SPECIAL TOPIC E

EXERCISES

7. isobutyl; ibu = isobutyl, pro = propanoic acid, fen = phenyl group
9. a. acetic anhydride   **b.** acetic anhydride

CHAPTER 17

PRACTICE EXERCISES

17.5 *p*-propylaniline

17.6

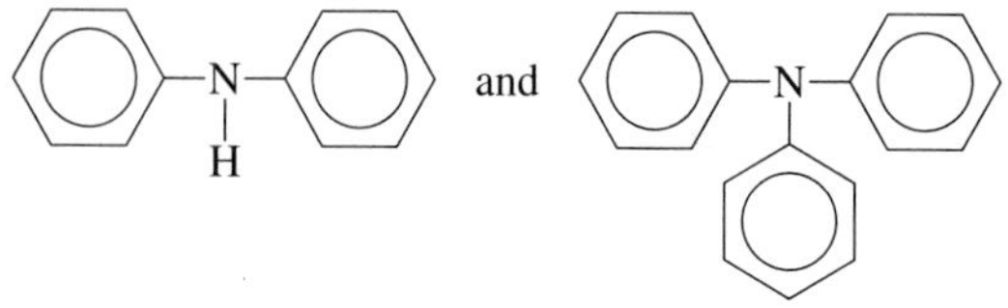

and

17.7

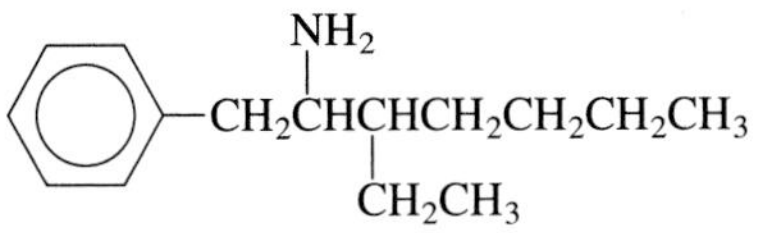

17.8 ethylammonium ion and tetraethylammonium ion

EXERCISES

1. $CH_3CH_2CH_2CH_2NH_2$, butylamine (1°); $CH_3CH_2CH_2NHCH_3$, methylpropylamine (2°); isobutylamine (1°), *sec*-butylamine (1°), methylisopropylamine (2°), *tert*-butylamine (1°), diethylamine (2°), dimethylethylamine (1°)
3. **a.** alcohol (1°) **b.** amine (1°) **c.** alcohol (2°) **d.** amine (2°) **e.** ether **f.** amine (2°) **g.** amine (2°) **h.** ether **i.** amine (3°) **j.** alcohol (3°) **k.** amine (1°) **l.** phenol
7. a cyclic compound having atom(s) other than carbon in the ring
11. **a.** *p*-nitroaniline **b.** ethylphenylamine
15. **a.** basic **b.** neutral **c.** acidic **d.** neutral
17. butylamine
19. propylamine
21. $CH_3CH_2CH_2NH_2$
31. no reaction
41. 23 mL

SPECIAL TOPIC F

3. **a.** tryptophan (precursor of serotonin) and tyrosine (precursor of norepinephrine and dopamine)
9. ketamine and phencyclidine
11. **a.** a six-member barbituric acid ring
15. phenol, amide, alcohol, amine
21. fluoride, ketone, amine, alcohol, chloride

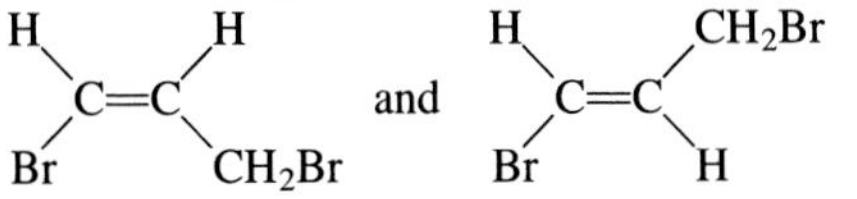

CHAPTER 18

PRACTICE EXERCISES

18.1

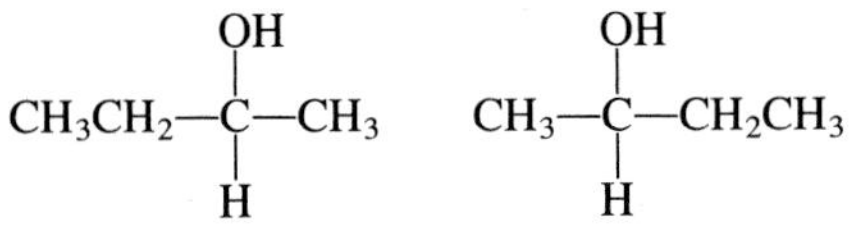

18.2

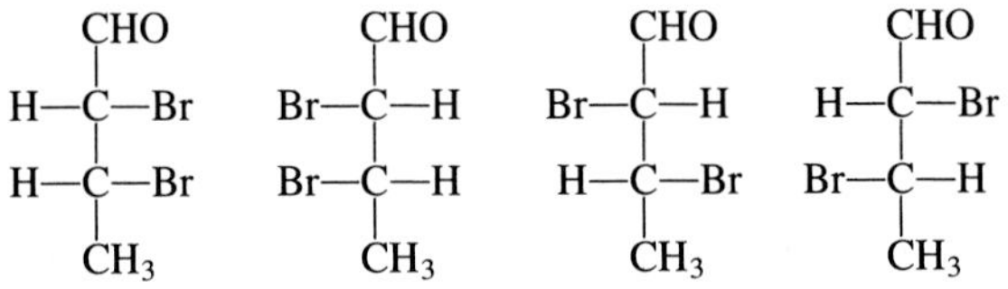

18.3

cis-1,2-dibromopropene *trans*-1,2-dibromopropene

cis-1,3-dibromopropene *trans*-1,3-dibromopropene

1,1-dibromopropene (no isomers) 2,3-dibromopropene (no isomers)

3,3-dibromopropene (no isomers)

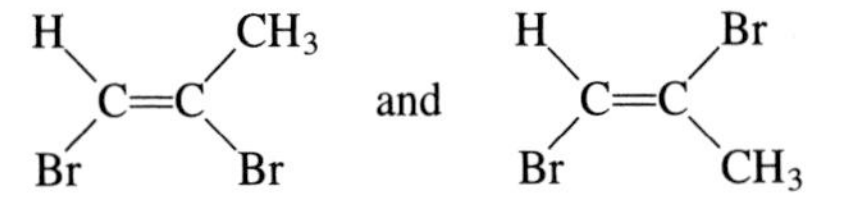

EXERCISES

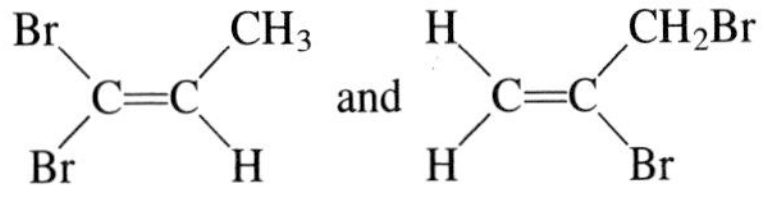

3. yes, yes
9. a, d
11. **a.** no **b.** no (meso) **c.** yes **d.** yes
15. 2-pentanol, 2-methyl-1-butanol, 3-methyl-2-butanol
19. eight
21. **a.** yes **b.** no **c.** no
31. two
33. $CH_3CH_2CH(CH_3)CHO$
35. **a.** $CH_3CH_2CH_2CH_2OH$, $CH_3CH_2CHOHCH_3$, $(CH_3)_2CHCH_2OH$, $(CH_3)_3COH$ **b.** $CH_3CH_2CHOHCH_3$

CHAPTER 19

3. **a.** D **b.** L **c.** L **d.** L
5. **a.** hexose **b.** pentose **c.** hexose **d.** triose
13. **a.** beta **b.** alpha **c.** beta **d.** alpha
15. d
17. **a.** glucose **b.** glucose **c.** glucose **d.** glucose and galactose **e.** glucose and fructose **g.** glucose
19. Ribose has —OH on C-2; deoxyribose has only H's.
21. a hemiacetal carbon not in a glycosidic link
23. galactose, glucose
25. yes
29. a, b, c, d, f, g, h, and i
31. amylose α-1,4 linked; cellulose, β-1,6; both polymers of glucose
33. **a.** $Ag(NH_3)_2^+$ **b.** CH_3OH, HCl **c.** H^+ or maltase
37. galactose and fructose
41. the L form of a ketopentose

CHAPTER 20

EXERCISES

7. saturated: a, c, d, f; unsaturated: b, e
9. four
11. double bond
13. corn oil
15. tristearin (both)
19. product: sodium oleates
21. yes

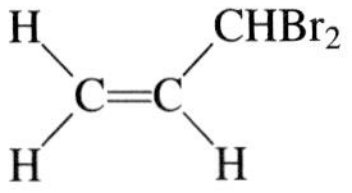

27. **a.** amide linkage   **b.** acetal linkage
29. polar ionic "head" and long nonpolar tails
33. glycerol: a, b, d; sphingosine, e; neither: c, f
35. a protein that extends partially or fully into lipid bilayer
37. transport across semipermeable membrane down concentration gradient aided by a protein
41. c, d
43. active transport
45. glycerol, stearic acid, oleic acid, phosphate group, choline

SPECIAL TOPIC H

3. **a.** (3)   **b.** (4)   **c.** (1)   **d.** (2)   **e.** (5)
5. pituitary
7. **a.** (6)   **b.** (4)   **c.** (5)   **d.** (1)   **e.** (2)
   f. (3)
9. steroids
11. ethynyl group
13. alcohol, aldehyde, alkene, ketone
15. alkene, ketone
19. regulation of blood pressure, inhibition of gastric secretions, relief of asthma and nasal congestion

CHAPTER 21

PRACTICE EXERCISE

page 550: a.

$$H_3N^+-CH-C(=O)-N(H)-CH-C(=O)-N(H)-CH-C(=O)O^-$$

with side chains: $HCCH_3$ / CH_3 ; CH_2 / CH_2 / $S-CH_3$; $HCOH$ / CH_3

b.

pyrrolidine ring $H-N^+(H)$ —C(=O)—N(H)—CH(—CH_2—CH_2—COO$^-$)... —C(=O)—N(H)—CH—CH_2—indole; with COO$^-$

c.

$$H_3N^+-CH-C(=O)-N(H)-CH-C(=O)-N(H)-CH-CH_2CH_2C(=O)NH_2$$

side chains: CH_2—(phenol ring with OH); $(CH_2)_4$ / $^+NH_3$; COO$^-$

EXERCISES

7. aspartic acid, glutamic acid
11. $H_3N^+-CH_2-COOH$
13. polypeptide of high (>10,000) molecular weight = protein

17. isoleucine, threonine, cystine, hydroxyproline
19. alpha helix secondary structure
21. salt linkage, hydrogen bond, disulfide linkage, hydrophobic interaction
23. globular: a, d, f; fibrous: b, c, e
25. globular
27. secondary, tertiary, quaternary
31. alanine stationary, histidine to cathode, aspartic acid to anode
35. Side chain acidic and basic groups function like buffer.
37. At isoelectric pH positive and negative charges are equal.
39. **a.** +   **b.** +   **c.** −
51. inside: a, e; outside: b, c, d, f

CHAPTER 22

EXERCISES

3. **a.** maltose   **b.** cellulose   **c.** proteins (peptides)
   d. lipids
7. intracellular: lactic acid dehydrogenase, fumarase, succinic acid dehydrogenase; extracellular: ptyalin, pepsin, trypsin
9. **a.** lyase   **b.** peptidase   **c.** transferase
   d. oxidoreductase   **e.** transferase   **f.** hydrolase
13. no
15. rate doubles
17. less active (both)
35. papain
37. creatine kinase, glutamic-oxaloacetic transaminase, lactate dehydrogenase
39. lactase; hydrolase
51. yes; a competitive inhibitor

SPECIAL TOPIC I

EXERCISES

3. ascorbic acid, vitamin C; ergocalciferol, vitamin D; cyanocobalamin, vitamin B_{12}; retinol, vitamin A; tocopherol, vitamin E
5. **a.** vitamin C   **b.** vitamin D   **c.** vitamin A
9. water-soluble: b, c; fat-soluble: a, c
11. fat-soluble
15. no
17. liver, red meat, eggs
19. polyunsaturated fatty acids in cell membranes; vitamin A
21. niacin
23. pantothenic acid
25. **a.** niacin, riboflavin   **b.** biotin   **c.** thiamine
    d. pyridoxine   **e.** folic acid   **f.** cyanocobalamin

CHAPTER 23

EXERCISES

3. **a.** ribonucleic acid, deoxyribonucleic acid   **b.** DNA
7. neither: a, d, f;   nucleoside: b, c;   nucleotide: e

9. a. adenine, guanine, thymine, cytosine, deoxyribose, phosphoric acid **b.** adenine, guanine, uracil, cytosine, ribose, phosphoric acid
19. hydrogen bonding
21. a. uracil **b.** cytosine **c.** adenine **d.** guanine
23. paired with a daughter strand
27. DNA, RNA
29. a. mRNA **b.** tRNA
31. TAAGC
33. AGGCTA
35. a. AAC **b.** CUU **c.** AGG **d.** GUG
37. a. Asn **b.** Leu **c.** Arg **d.** Val
39. Asn-Glu-Ser
45. Met-Tyr-His-Gly-Thr-Arg-Val-Leu-Leu-Ala-Asp-Gly
51. a. Thr **b.** Gln **d.** Ala

SPECIAL TOPIC J

EXERCISES

9. abnormal growth of new tissue
11. smoking
15. charcoal-grilled meat, cigarette smoke, automobile exhaust, coffee, burnt sugar
21. an agent that causes a mutation in DNA
23. antimetabolites, vinca alkaloids, sex hormones
25. adenine; by inhibiting the synthesis of adenine and guanine nucleotides

CHAPTER 24

EXERCISES

3. AMP is ribose linked to one phosphate group; ADP has a second phosphate, and ATP a third.
5. a, c, d (e somewhat)
7. hydrolysis
9. monosaccharides
11. starch breakdown to glucose by amylase in saliva
23. glycogen
31. a. both **b.** both in CO_2, neither in ethanol
37. a. acetaldehyde **b.** dihydroxyacetone phosphate and glyceraldehyde 3-phosphate **c.** enolase **d.** glucose 6-phosphate **e.** glucose 1-phosphate **f.** triose phosphate isomerase **g.** glucose 6-phosphate **h.** pyruvic acid **i.** phosphofructokinase **j.** glucose 6-phosphate **k.** 1,3-diphosphoglyceric acid **l.** phosphoglycerokinase **m.** 2-phosphoglyceric acid **n.** pyruvic acid decarboxylase **o.** phosphoenolpyruvic acid **p.** glucose 6-phosphate

CHAPTER 25

EXERCISES

3. CO_2
5. three

7. guanosine triphosphate
11. flavin, adenine, ribitol, ribose, phosphate; flavin and ribitol
15. a. a **b.** a, b, c, i **c.** a **d.** d, e, f, h, j **e.** d, e, h **f.** g
21. a. 87.6 kcal **b.** 43.8%
23. glycerol phosphate and malic acid–aspartic acid
25. 1,3-diphosphoglyceric acid $\rightarrow$ 3-phosphoglyceric acid phosphoenolpyruvic acid $\rightarrow$ pyruvic acid succinyl CoA $\rightarrow$ succinic acid
27. $C_6H_{12}O_6 + 6\ O_2 + 36\ ADP + 36\ P_i \rightarrow$
$$6\ CO_2 + 42\ H_2O + 36\ ATP + 423{,}000\ cal$$
29. O_2, H_2O
35. fats
37. aerobic
43. type IIB (pheasant), type I (heron)

CHAPTER 26

PRACTICE EXERCISES

26.1 47.5%
26.2 360 mol
26.3 35 miles
26.4 3.6 lb

EXERCISES

7. monoglycerides, fatty acids, free cholesterol
9. $C_{15}H_{31}COOH$ (palmitic acid), $C_{17}H_{35}COOH$ (stearic acid), $C_{17}H_{33}COOH$ (oleic acid), $C_{17}H_{31}COOH$ (linoleic acid)
19. dehydrogenation, hydration, thiolation
21. C-2 in acetyl CoA, C-4 and C-6 in butyryl CoA
29. 31
31. 407
35. fat reserves
41. energy intake exceeding energy output
45. 3500
51. 9.5 kg
53. 385 km

CHAPTER 27

EXERCISES

3. a. breaks down protein in stomach **b–d.** break down proteins in intestine **e.** breaks down di- and tripeptides in intestine **f.** converts trypsinogen to trypsin in intestine
7. a. at N-terminal end **b.** at C-terminal end
11. protein synthesis, hormone production, energy source (through breakdown)
13. becomes negative
15. 100 g
25. transfer of amine group from alanine to α-ketoglutarate
33. a. ammonia and hydrogen peroxide

35. a. leucine   **b.** α-ketoglutaric acid   **c.** α-ketoglutaric acid   **d.** 5-hydroxytryptophan   **e.** glutamic acid   **f.** glutamine
37. a. liver   **b.** kidney

CHAPTER 28

EXERCISES

3. blood, interstitial fluid, lymph
7. Na^+, K^+, Ca^{2+}, Mg^{2+}, HCO_3^-, Cl^-, HPO_4^{2-}, SO_4^{2-}

9. a. to combat infectious microorganisms, cancer cells, and chemical toxins   **b.** by B-lymphocytes   **c.** IgM, IgG, IgA, IgD, IgE   **d.** IgG
21. a. II   **b.** II   **c.** III   **d.** II
23. hemoglobin, cytochromes, catalase
27. 7.35–7.45
37. liquid uptake, perspiration, fever, diarrhea
41. mucus, lacrimal secretions, oily film
43. no; yes
45. diffusion
47. lower

Credits for Photographs

10.1 Richard Megna/Fundamental Photographs
10.5 Carey B. Van Loon
10.6 Adam Hart-Davis/SPL/Photo Researchers, Inc.
10.7 Robert Mathena/Fundamental Photographs

CHAPTER 11

Opener Richard Megna/Fundamental Photographs
11.3 Joel Florin/Courtesy of Microessential Laboratory
11.4 Richard Megna/Fundamental Photographs

CHAPTER 12

Opener Craig Hammel/The Stock Market
12.3 The Bettmann Archive
12.6 Carey B. Van Loon
12.12 Richard Megna/Kip Peticolas/Fundamental Photographs
12.15 Biophoto Associates/Photo Researchers, Inc.

SPECIAL TOPIC C

C.1 Courtesy of Goodyear
C.2 Roy Marsch/The Stock Market
C.4 Ellis Herwig/Stock Boston
C.5 Courtesy of U.S. Environmental Protection Agency

CHAPTER 13

Opener John Madre/The Stock Market
13.1 Richard Megna/Fundamental Photographs
13.10b Richard Megna/Fundamental Photographs
13.13 (left) Bates M.D./Custom Medical Stock Photo
13.13 (rt) Dan McCoy/Rainbow

CHAPTER 14

Opener Serallier/Rapho/Photo Researchers, Inc.
14.4 Richard Megna/Fundamental Photographs
14.5 Diane Schiumo/Fundamental Photographs
14.6 Prof. Marvin Lang and Mr. Gary Shulfer, University of Wisconsin–Stevens Point (C-115)

CHAPTER 15

Opener Ed Bock/The Stock Market

CHAPTER 16

Opener Tom Tracy/The Stock Market
16.1 Richard Megna/Fundamental Photographs

SPECIAL TOPIC E

E.3 Courtesy of U.S. Department of Justice Drug Enforcement Agency
E.4 Courtesy of Bureau of Narcotics and Dangerous Drugs, Washington DC
E.5 Courtesy of U.S. Department of Justice Drug Enforcement Agency
E.6 Courtesy of U.S. Department of Justice Drug Enforcement Agency
E.7 Courtesy of U.S. Department of Justice Drug Enforcement Agency
E.9 Courtesy of U.S. Food and Drug Administration

E.10 Courtesy of U.S. Department of Justice Drug Enforcement Agency
E.11 Courtesy of U.S. Department of Justice Drug Enforcement Agency
E.12 Courtesy of U.S. Department of Justice Drug Enforcement Agency

CHAPTER 17

Opener Manfred Kage/Peter Arnold, Inc.
17.6 Courtesy of The Metropolitan Museum of Art, Wolfe Fund, 1931. Catherine Lorillard Wolfe Collection (31.45)

SPECIAL TOPIC F

F.7 Courtesy of U.S. Department of Justice Drug Enforcement Agency

CHAPTER 18

Opener Richard Megna/Fundamental Photographs
18.13 Prof. Marvin Lang and Mr. Gary Shulfer, University of Wisconsin–Stevens Point (France B468)

CHAPTER 19

Opener Daniel Aubry/The Stock Market
19.1 Richard Megna/Fundamental Photographs
19.6 Courtesy of Brach's
19.9 Dr. Brian Eyden/SPL/Photo Researchers, Inc.
19.10 Fred Bavendam/Peter Arnold, Inc.

CHAPTER 20

Opener Jack Fields/Photo Researchers, Inc.
20.1 Richard Megna/Fundamental Photographs
20.3 Richard Megna/Fundamental Photographs
20.6 Richard Megna/Fundamental Photographs
20.7 Tom McCabe/Courtesy of U.S. Soil Conservation Service
20.18 SIU/Science Source/Photo Researchers, Inc.
20.19a&b Carolina Biological Supply Co.
20.19c Dan McCoy/Rainbow

CHAPTER 21

Opener Joyce Photographic/Photo Researchers, Inc.
21.1 Richard Megna/Fundamental Photographs
21.16c Richard Megna/Fundamental Photographs
21.17 CNRI/SPL/Photo Researchers, Inc.
21.21 Lab. of Molecular Biology/MRC/SPL/Photo Researchers, Inc.

CHAPTER 22

Opener Bucky Reeves/Photo Researchers, Inc.
22.19 Dr. Jeremy Burgess/SPL/Photo Researchers, Inc.
22.20 Prof. Marvin Lang and Mr. Gary Shulfer, University of Wisconsin–Stevens Point (Hungary 2699)

SPECIAL TOPIC I

I.3 Biophoto Associates/Science Source/Photo Researchers, Inc.

I.4   Courtesy of Marc Moldawer M.D., The Methodist Hospital, Texas Medical Center

I.5   Courtesy of Children's Hospital, Columbus OH

I.6   AP/Wide World Photos, Inc.

CHAPTER 23

Opener   Phil A. Harrington/Peter Arnold, Inc.

23.6   Richard Megna/Fundamental Photographs

23.9   The Bettmann Archive

23.10   CNRI/SPL/Photo Researchers, Inc.

23.21   Courtesy of Bayer AG, Germany

23.23   Courtesy of Eli Lilly and Company

SPECIAL TOPIC J

J.4   Hank Morgan/Rainbow

CHAPTER 24

Opener   Gerard Vandystadt/Photo Researchers, Inc.

24.14   AP/Wide World Photos, Inc.

CHAPTER 25

Opener   Spencer Grant/Photo Researchers, Inc.

25.6   David Madison/Duomo

25.7   David Madison/Duomo

CHAPTER 26

Opener   Gregory G. Dimijian/Photo Researchers, Inc.

26.7   Will & Deni McIntyre/Photo Researchers, Inc.

CHAPTER 27

Opener   Connie Hansen/The Stock Market

27.3   Nancy McKenna/Photo Researchers, Inc.

CHAPTER 28

Opener   Ted Horowitz/The Stock Market

28.1   Manfred Kage/Peter Arnold, Inc.

28.3   Courtesy of U.S. Public Health Service Centers for Disease Control

28.5   CNRI/SPL/Photo Researchers, Inc.

28.7a   Manfred Kage/Peter Arnold, Inc.

28.7b   Jackie Lewin, Royal Free Hospital/SPL/Photo Researchers, Inc.

28.11   Courtesy of Miles Laboratories

28.12   Dave Caulkin/AP/Wide World Photos, Inc.

Index

(f = figure, n = footnote or margin note, t = table)

TABLE OF ATOMIC WEIGHTS (BASED ON CARBON-12)

Element	Symbol	Atomic Number	Atomic Weight	Element	Symbol	Atomic Number	Atomic Weight
Actinium	Ac	89	(227)	Neon	Ne	10	20.2
Aluminum	Al	13	27.0	Neptunium	Np	93	237.0
Americium	Am	95	(243)	Nickel	Ni	28	58.7
Antimony	Sb	51	121.8	Niobium	Nb	41	92.9
Argon	Ar	18	39.9	Nitrogen	N	7	14.0
Arsenic	As	33	74.9	Nobelium	No	102	(254)
Astatine	At	85	(210)	Osmium	Os	76	190.2
Barium	Ba	56	137.3	Oxygen	O	8	16.0
Berkelium	Bk	97	(245)	Palladium	Pd	46	106.4
Beryllium	Be	4	9.01	Phosphorus	P	15	31.0
Bismuth	Bi	83	209.0	Platinum	Pt	78	195.1
Boron	B	5	10.8	Plutonium	Pu	94	(242)
Bromine	Br	35	79.9	Polonium	Po	84	(210)
Cadmium	Cd	48	112.4	Potassium	K	19	39.1
Calcium	Ca	20	40.1	Praseodymium	Pr	59	140.9
Californium	Cf	98	(251)	Promethium	Pm	61	(145)
Carbon	C	6	12.0	Protactinium	Pa	91	231.0
Cerium	Ce	58	140.1	Radium	Ra	88	226.0
Cesium	Cs	55	132.9	Radon	Rn	86	(222)
Chlorine	Cl	17	35.5	Rhenium	Re	75	186.2
Chromium	Cr	24	52.0	Rhodium	Rh	45	102.9
Cobalt	Co	27	58.9	Rubidium	Rb	37	85.5
Copper	Cu	29	63.5	Ruthenium	Ru	44	101.1
Curium	Cm	96	(245)	Samarium	Sm	62	150.4
Dysprosium	Dy	66	162.5	Scandium	Sc	21	45.0
Einsteinium	Es	99	(254)	Selenium	Se	34	79.0
Erbium	Er	68	167.3	Silicon	Si	14	28.1
Europium	Eu	63	152.0	Silver	Ag	47	107.9
Fermium	Fm	100	(254)	Sodium	Na	11	23.0
Fluorine	F	9	19.0	Strontium	Sr	38	87.6
Francium	Fr	87	(223)	Sulfur	S	16	32.1
Gadolinium	Gd	64	157.3	Tantalum	Ta	73	180.9
Gallium	Ga	31	69.7	Technetium	Tc	43	98.9
Germanium	Ge	32	72.6	Tellurium	Te	52	127.6
Gold	Au	79	197.0	Terbium	Tb	65	158.9
Hafnium	Hf	72	178.5	Thallium	Tl	81	204.4
Helium	He	2	4.00	Thorium	Th	90	232.0
Holmium	Ho	67	164.9	Thulium	Tm	69	168.9
Hydrogen	H	1	1.01	Tin	Sn	50	118.7
Indium	In	49	114.8	Titanium	Ti	22	47.9
Iodine	I	53	126.9	Tungsten	W	74	183.8
Iridium	Ir	77	192.2	Unnilennium	Une	109	(266)
Iron	Fe	26	55.8	Unnilhexium	Unh	106	(263)
Krypton	Kr	36	83.8	Unniloctium	Uno	108	(265)
Lanthanum	La	57	138.9	Unnilpentium	Unp	105	(262)
Lawrencium	Lr	103	(257)	Unnilquadium	Unq	104	(261)
Lead	Pb	82	207.2	Unnilseptium	Uns	107	(262)
Lithium	Li	3	6.94	Uranium	U	92	238.0
Lutetium	Lu	71	175.0	Vanadium	V	23	50.9
Magnesium	Mg	12	24.3	Xenon	Xe	54	131.3
Manganese	Mn	25	54.9	Ytterbium	Yb	70	173.0
Mendelevium	Md	101	(256)	Yttrium	Y	39	88.9
Mercury	Hg	80	200.6	Zinc	Zn	30	65.4
Molybdenum	Mo	42	95.9	Zirconium	Zr	40	91.2
Neodymium	Nd	60	144.2				

Parentheses around atomic weight indicate that weight given is that of the most stable known isotope.